THE HARLOW-SHAPLEY SYMPOSIUM ON GLOBULAR CLUSTER SYSTEMS IN GALAXIES

The cover photo is Schmidt camera view of Halley's comet, the galaxy Centaurus-A (with its globular cluster system) and the galactic globular Omega Centauri.
It was taken by William Liller, Instituto Isaac Newton, Ministerio de Education de Chile, on April 14.513, 1986, as part of NASA's International Halley Watch, Large Scale Phenomena Section. The telescope was a 20-centimeter f/1.5 Schmidt operated by Liller on Easter Island. The combined view of the most massive globular in the Galaxy with one of the most studied external galaxies and globular cluster systems is a fitting reminder of the themes of the IAU 126 Symposium. The simultaneous view of Halley's comet records the year of the meeting and the centennial year of Harlow Shapley's birth.

INTERNATIONAL ASTRONOMICAL UNION

UNION ASTRONOMIQUE INTERNATIONALE

THE HARLOW-SHAPLEY SYMPOSIUM ON
GLOBULAR CLUSTER SYSTEMS IN GALAXIES

PROCEEDINGS OF THE 126TH SYMPOSIUM OF THE
INTERNATIONAL ASTRONOMICAL UNION,
HELD IN CAMBRIDGE, MASSACHUSETTS, U.S.A.,
AUGUST 25-29, 1986

Edited by

JONATHAN E. GRINDLAY

*Harvard-Smithsonian Center for Astrophysics,
Cambridge, Massachusetts, U.S.A.*

and

A. G. DAVIS PHILIP

*Van Vleck Observatory and Union College,
Schenectady, New York, U.S.A.*

KLUWER ACADEMIC PUBLISHERS

DORDRECHT / BOSTON / LONDON

Library of Congress Cataloging in Publication Data
Harlow Shapley Symposium on Globular Cluster Systems in Galaxies (1986: Cambridge, Mass.)
   The Harlow Shapley Symposium on Globular Cluster Systems in Galaxies: proceedings of the 126th symposium of the International Astronomical Union, held in Cambridge, Massachusetts, August 25–29, 1986 / edited by Jonathan E. Grindlay and A. G. Davis Philip.
    p.   cm.
   At head of title: International Astronomical Union.
   Includes bibliographies and index.
   ISBN 90-277-2664-7.   ISBN 90-277-2665-5 (pbk.)
   1. Stars—Globular clusters—Congresses. 2. Shapley, Harlow, 1885–1972—Congresses. I. Grindlay, Jonathan E. II. Philip, A. G. Davis. III. Shapley, Harlow, 1885–1972. IV. International Astronomical Union. V. Title.
QB853.H37 1986
523.8′55—dc 19   87-35637
                                                                              CIP

CIP

Published on behalf of
the International Astronomical Union
by
Kluwer Academic Publishers, P.O. Box 17, 3300 AA Dordrecht, Holland.

Kluwer Academic Publishers incorporates
the publishing programmes of
D. Reidel, Martinus Nijhoff, Dr W. Junk and MTP Press.

Sold and distributed in the U.S.A. and Canada
by Kluwer Academic Publishers,
101 Philip Drive, Norwell, MA 02061, U.S.A.

In all other countries, sold and distributed
by Kluwer Academic Publishers Group,
P.O. Box 322, 3300 AH Dordrecht, Holland.

All Rights Reserved
© 1988 by the International Astronomical Union

No part of the material protected by this copyright notice may be reproduced or utilized in any form or by any means, electronic or mechanical, including photocopying, recording or by any information storage and retrieval system, without written permission from the publisher.

Printed in The Netherlands

TABLE OF CONTENTS

Preface   xvii

List of Participants   xxiii

Conference Photo   xxviii

List of Photos of Participants and Special Figures   xxx

    CHAPTER I   REVIEW PAPERS ON HARLOW SHAPLEY

    Session Chair: M. McCarthy   1

SHAPLEY'S DEBATE
    Michael Hoskin   3

SHAPLEY'S ERA
    Helen Sawyer Hogg   11

SHAPLEY'S IMPACT
    Owen Gingerich   23

    CHAPTER II   REVIEW PAPERS ON GLOBULAR CLUSTERS IN THE
    MILKY WAY

    Session Chairs: R. McClure, J. Cohen, and
                    J. Norris   35

AN OVERVIEW OF THE GLOBULAR CLUSTER SYSTEM OF THE GALAXY
    Robert Zinn   37

KINEMATICS OF THE GALACTIC GLOBULAR CLUSTER SYSTEM
    R. F. Webbink   49

GLOBULAR CLUSTER COLOR-MAGNITUDE DIAGRAMS
    James E. Hesser   61

THE OVERALL ABUNDANCES OF GLOBULAR CLUSTERS
    R. A. Bell   79

THE CHEMICAL INHOMOGENEITY WITHIN GLOBULAR CLUSTERS
    John Norris   93

AGES OF THE GALACTIC GLOBULAR CLUSTERS
    Don A. Vandenberg   107

GLOBULAR CLUSTER LUMINOSITY FUNCTIONS
    Pierre Demarque   121

GLOBULAR CLUSTERS AND FIELD HALO STARS
Bruce W. Carney   133

CHAPTER III: REVIEW PAPERS ON GLOBULAR CLUSTERS IN NEARBY GALAXIES

Session Chairs: G. Wallerstein, J. Watanabe
and V. Straizys   149

OLD GLOBULAR CLUSTERS IN THE MAGELLANIC CLOUDS
J. A. Graham   151

INTERMEDIATE-AGE MAGELLANIC CLOUD GLOBULAR CLUSTERS
Edward W. Olszewski   159

M 31 CLUSTER SYSTEM
F. Fusi Pecci   173

THE CLUSTERS OF M 33
C. A. Christian   187

THE NGC 5128 CLUSTER SYSTEM
Hugh C. Harris, Gretchen L. H. Harris
and James E. Hesser   205

DWARF SPHEROIDAL GALAXIES AND GLOBULAR CLUSTERS
G. S. Da Costa   217

AN OVERVIEW OF GLOBULAR SYSTEMS IN DISTANT GALAXIES
William E. Harris   237

THE M 87 GLOBULAR CLUSTER SYSTEM
John Huchra   255

CHAPTER IV REVIEW PAPERS ON EVOLUTION OF GLOBULAR CLUSTERS

Session Chairs: V. Trimble, H. Zinnecker,
M. Aurière and J. Goodman   269

THE EVOLUTION OF THE SYSTEM OF GLOBULAR CLUSTERS
Jeremiah P. Ostriker   271

TIDAL HEATING OF GLOBULAR CLUSTERS
David Chernoff and Stuart L. Shapiro   283

CLUSTER SWAPPING
Juan C. Muzzio   297

GALAXY FORMATION AND CLUSTER FORMATION
Richard B. Larson   311

THE ORIGIN OF GLOBULAR CLUSTERS
    S. Michael Fall and Martin J. Rees   323

SURFACE PHOTOMETRY OF GLOBULAR CLUSTERS
    S. Djorgovski   333

X-RAY BINARIES AND CLUSTER EVOLUTION
    Jonathan E. Grindlay   347

PRECOLLAPSE EVOLUTION OF GLOBULAR CLUSTERS
    Shogo Inagaki   367

AFTER CORE COLLAPSE, WHAT?
    Haldan Cohn   379

DISOLUTION OF STAR CLUSTERS IN GALAXIES
    Roland Wielen   393

    CHAPTER V   REVIEW PAPERS ON GLOBULAR CLUSTERS AS TRACERS AND HST

    Session Chairs: M. Rees and Y. Gnedin   409

INTERSTELLAR MATTER IN GLOBULAR CLUSTERS
    Morton S. Roberts   411

GLOBULAR CLUSTERS AS TRACERS OF THE GALAXY MASS DISTRIBUTION
    K. A. Innanen   423

GLOBULAR CLUSTERS AND PRIMORDIAL COMPOSITION
    Roger Cayrel   431

STELLAR EVOLUTION IN GLOBULAR CLUSTERS AND HST
    Alvio Renzini   443

SIMULATIONS OF HST OBSERVATIONS OF GLOBULAR CLUSTERS
    John N. Bahcall and Donald P. Schneider   455

    CHAPTER VI   REVIEW PAPERS SUMMARY

    Session Chair: J. Grindlay   465

GLOBULAR CLUSTER SYSTEMS IN GALAXIES: MAIN TRENDS AND FUTURE DIRECTIONS
    Sidney van den Bergh   467

    CHAPTER VII   POSTER PAPERS ON HARLOW SHAPLEY AND GLOBULAR CLUSTERS IN THE MILKY WAY

    Discussion Leaders: M. McCarthy and C. Pilachowski   475

HARLOW SHAPLEY: A VIEW FROM THE HARVARD ARCHIVES Barbara L. Welther	477
HARLOW SHAPLEY AND THE UNIVERSITY OF MISSOURI Charles J. Peterson	479
HARLOW SHAPLEY AND RED GIANT STARS Martin F. McCarthy S. J.	481
THE DEVELOPMENT OF A RED-GIANT BRANCH IN LOW TO INTERMEDIATE MASS STARS A. V. Sweigart, L. Greggio and A. Renzini	483
NEW MAIN-SEQUENCE LUMINOSITY FUNCTIONS FOR GLOBULAR CLUSTERS Robert D. McClure, Peter B. Stetson, James E. Hesser, Graham H. Smith, William E. Harris and Don A. VandenBerg	485
GLOBAL VERSUS LOCAL MASS FUNCTIONS Ivan R. King	487
A NEW SURVEY OF GLOBULAR CLUSTER STRUCTURAL AND LUMINOSITY PARAMETERS B. Cameron Reed and Charles J. Peterson	489
AXIAL RATIOS AND ORIENTATIONS FOR 100 GALACTIC GLOBULAR STAR CLUSTERS Raymond E. White and Stephen J. Shawl	491

ABUNDANCES IN STARS IN GLOBULAR CLUSTERS FROM PALOMAR CCD SPECTRA
    E. Myckki Leep, George Wallerstein and
    J. B. Oke
	493

THE METAL ABUNDANCE OF METAL-RICH GLOBULAR CLUSTERS
    Raffaele Gratton, Maria Lucia Quarta and
    Sergio Ortolani
	495

THE COMPOSITION OF WARM GIANTS IN M 71 AND M 5 Catherine A. Pilachowski and Christopher Sneden	497
THE INTEGRATED SPECTRA OF METAL-RICH GALACTIC GLOBULAR CLUSTERS: A TWO-PARAMETER FAMILY James A. Rose and Michael J. Tripicco	499
IUE INVESTIGATIONS AT THE CORE OF M 79 Bruce Altner	501
ONE-MICRON PHOTOMETRY OF OMEGA CENTAURI GIANTS Graeme H. Smith	503

ON THE BIMODAL DISTRUBUTIONS OF HORIZONTAL BRANCHES
      Young-Wook Lee, Pierre Demarque and Robert Zinn   505

BIMODAL DISTRIBUTIONS ON THE HORIZONTAL BRANCH
      Robert T. Rood and Deborah A. Crocker   507

HORIZONTAL-BRANCH STARS WITH STRONG HE LINES
      Deborah A. Crocker and Robert T. Rood   509

SPECTRA OF BHB STARS IN M 3, M 13 AND M 92
      A. G. Davis Philip and N. N. Samus   511

FOUR-COLOR MEASURES OF BHB STARS IN M 4, M 13 AND M 55
      A. G. Davis Philip   513

GLOBULAR CLUSTERS IN THE VILNIUS PHOTOMETRIC SYSTEM
      K. Zdanavičius   515

THE METALLICITY DISTRIBUTION FUNCTION OF HALO DWARFS
AND GLOBULAR CLUSTERS
      J. B. Laird, M. P. Rupin, B. W. Carney, D. W. Latham
      and R. L. Kurucz   517

THE SIMILARITY OF THE HALO FIELD K GIANT POPULATION
WITH THE GLOBULAR CLUSTER SYSTEM OF OUR GALAXY
      Kavan U. Ratnatunga   519

APPARENT ROTATION OF THE GALACTIC GLOBULAR CLUSTER SYSTEM
      J. Colin   521

ASTROMETRIC DISTANCES OF GLOBULAR CLUSTERS
      Kyle Cudworth and Ruth C. Peterson   523

ABSOLUTE PROPER MOTIONS AND SPACE MOTIONS OF GLOBULAR CLUSTERS
      H.-J. Tuchcolke, P. Brosche and M. Geffert   525

A SEARCH FOR OBSCURED GLOBULAR CLUSTERS
      S. Djorgovski   527

ON THE COLOR EXCESSES OF GLOBULAR CLUSTERS
      V. Straižys and R. Janulis   529

FIRST POSTER PAPER DISCUSSION   531

      CHAPTER VIII POSTER PAPERS ON CLUSTER SYSTEMS IN
      NEARBY GALAXIES

      Discussion Leader: G. Lyngå   539

A SEARCH FOR GLOBULAR CLUSTER CANDIDATES IN NGC 2403
      M. L. Malagnini, P. Santin, F. Bonoli, L. Frederici,
      F. Fusi Pecci and R. G. Kron   541

SEARCH FOR GLOBULAR CLUSTERS IN THE NEARBY GALAXIES II.
NGC 3109
  A. Blecha   543

A COMPLETE SAMPLE OF GLOBULAR CLUSTERS IN NGC 5128
  Ray Sharples   545

ASTRONOMICAL CATALOGUES IN THE M 31 REGION
  P. Battistini   547

THE BLUE STAR CLUSTERS OF M 31
  Paul Hodge   549

SPATIAL DISTRIBUTION OF GLOBULAR CLUSTERS IN M 31
  Jun-ichi Watanabe and Tomohiko Yamagata   551

FORMATION OF POPULOUS CLUSTERS FROM METAL-POOR GAS
IN THE MAGELLANIC CLOUDS
  T. Richtler and W. Seggewiss   553

THE DEVELOPMENT OF THE RED GIANT BRANCH IN MAGELLANIC
CLOUD CLUSTERS: PROGRESS REPORT
  R. Buonanno, C. E. Corsi, F. Fusi Pecci, L. Greggio,
  A. Renzini and A. V. Sweigart   555

THE AGE DISTRIBUTION AND AGE-METALLICITY RELATION OF STAR
CLUSTERS IN A NORTHERN REGION OF THE LMC
  Mario Mateo   557

PHOTOMETRIC MODELS FOR GLOBULAR CLUSTERS FROM POPULATION
SYNTHESIS
  R. Capuzzo Dolcetta   559

BVRI PHOTOMETRY OF STAR CLUSTERS IN THE BOK REGION OF THE
LARGE MAGELLANIC CLOUD
  William Liller and Gonzalo Alcaino   561

AGES AND METAL ABUNDANCES OF STAR CLUSTERS IN THE
MAGELLANIC CLOUDS
  Horace A. Smith, Leonard Searle and Armando Manduca   563

INTERNAL DYNAMICS OF MAGELLANIC CLOUD CLUSTERS
  Paul Papenhausen and R. A. Schommer   565

DO BINARY CLUSTERS EXIST IN THE LARGE MAGELLANIC CLOUD?
  D. Hatzidimitriou and R. K. Bhatia   567

ELLIPTICITIES OF GLOBULAR CLUSTERS IN THE ANDROMEDA
GALAXY
  Nedka M. Spassova, Anelia V. Staneva
  and Valery K. Golev   569

OBSERVED VARIATIONS IN THE DENSITY PROFILES OF STAR
CLUSTERS IN THE LMC
     M. Kontizas, D. Hatzidimitriou and M. Metaxa   571

RATIO OF EARLY TO LATE TYPE STARS IN SMC CLUSTERS
     E. Kontizas, M. Kontizas, A. Dapergolas
     and D. Hatzidimitriou   573

THE SMC CLUSTER LINDSAY 11
     J. Buttress, R. D. Cannon and W. K. Griffiths   575

ABUNDANCES OF YOUNG LMC CLUSTERS
     R. A. Schommer and Doug Geisler   577

THE ABUNDANCE OF THE LMC GLOBULAR CLUSTER NGC 2213
     Doug Geisler   579

DEEP PHOTOMETRY OF THE DRACO DWARF SPHEROIDAL GALAXY
     Bruce W. Carney and P. Seitzer   581

CCD PHOTOMETRY IN THE CORE OF THE FORNAX DWARF GALAXY
     Robert M. Light and P. Seitzer   583

A CANDIDATE FOR THE RECOVERED NOVA OF 1938 IN THE
GLOBULAR CLUSTER M 14
     Michael M. Shara, Michael Potter, Anthony F. J. Moffat,
     Helen Sawyer Hogg and Amelia Wehlau   585

THE ABSOLUTE LUMINOSITY OF RR LYRAE VARIABLES
     C. Cacciari, G. Clementini and L. Prevot   587

THE DISTANCES TO RR LYRAE VARIABLES
     Rodney V. Jones, Bruce W. Carney, David W. Latham
     and Robert L. Kurucz   589

DOUBLE-MODE RR LYRAE STARS IN IC 4499
     Christine M. Clement, James M. Nemec,
     Robert J. Dickens, Elizabeth A. Bingham   591

SHORT-PERIOD VARIABLES IN GLOBULAR CLUSTERS OF MODERATE
METALLICITY
     Martha L. Hazen   593

SECOND POSTER PAPER DISCUSSION   595

    CHAPTER IX: POSTER PAPERS ON CLUSTER SYSTEMS IN DISTANT
    GALAXIES, DEEP PHOTOMETRY AND CM DIAGRAMS

    Discussion Leader: D. Hanes   601

THE NUCLEI OF NUCLEATED DWARF ELLIPTICAL GALAXIES -
ARE THEY GLOBULAR CLUSTERS?

H. Zinnecker, C. J. Keable, J. S. Dunlop, R. D. Cannon and W. K. Griffiths	603
THE GLOBULAR CLUSTER SYSTEM OF M 87 Judith G. Cohen	605
THE CORE OF THE M 87 GLOBULAR CLUSTER SYSTEM Tod R. Lauer and John Kormendy	607
U PHOTOMETRY OF GLOBULAR CLUSTERS IN THE CENTRAL REGION OF M 87 E. V. Held and J.-L. Nieto	609
GLOBULAR CLUSTERS DETECTED IN THE COMA CLUSTER'S CENTRAL GIANT GALAXY NGC 4874 Laird A. Thompson and F. Valdes	611
GLOBULAR CLUSTERS IN DIFFERENT TYPES OF SPIRAL GALAXIES Hugh C. Harris, Gregory D. Bothun and James E. Hesser	613
GLOBULAR CLUSTERS IN LENTICULAR GALAXIES: NCG 3115 E. V. Held and M. Capaccioli	615
GLOBULAR CLUSTERS AS EXTRAGALACTIC DISTANCE INDICATORS: MAXIMUM LIKELIHOOD METHODS David A. Hanes and Donna G. Whittaker	617
PHOTOMETRY OF FAINT STARS IN GLOBULAR CLUSTERS USING THE SIX METER TELESCOPE N. N. Samus	619
HIGH PRECISION PHOTOMETRY OF 10,000 STARS IN M 3 R. Buonanno, A. Buzzoni, C. E. Corsi, F. Fusi Pecci and A. R. Sandage	621
DEEP CCD PHOTOMETRY IN M 5 Harvey B. Richer and Gregory G. Fahlman	623
PHOTOGRAPHIC PHOTOMETRY OF 4500 STARS IN M 30 G. Piotto, M. Capaccioli, S. Ortolani, L. Rosino, G. Alcaino and W. Liller	625
AN AUTOMATED HR DIAGRAM FOR NGC 6809 (M 55) Michael J. Irwin and Virginia Trimble	627
DEEP CCD PHOTOMETRY IN OMEGA CENTAURI AND NGC 3201 S. Ortolani	629
DEEP CCD PHOTOMETRY OF OMEGA CENTAURI R. G. Noble, J. Buttress, W. K. Griffiths, A. J. Penny, R. J. Dickens and R. D. Cannon	631

THE AGES OF GLOBULAR CLUSTERS DERIVED FROM BVRI CCD PHOTOMETRY
        Gonzalo Alcaino and William Liller   633

TURNOFFS AND AGES OF GLOBULAR CLUSTERS
        R. Buonanno, C. E. Corsi and F. Fusi Pecci   635

THE DYNAMICS OF GLOBULAR CLUSTERS IN HIGH ECCENTRICITY ORBITS
        R. K. Bhatia   637

MASS DISTRIBUTIONS OF GALAXIES WITH GLOBULAR CLUSTER SYSTEMS
        Kazutomo Takayanagi   639

THE DYNAMICS OF GLOBULAR CLUSTER SYSTEMS
        Natarajan Ramamani   641

THIRD POSTER PAPER DISCUSSION   643

    CHAPTER X   POSTER PAPERS ON FORMATION AND EVOLUTION OF GLOBULAR CLUSTERS

    Discussion Leader: I. King   653

A MULTICOLOR CCD SURVEY OF SOUTHERN GLOBULAR CLUSTERS
        Juan Forte and Mariano Méndez   655

THE STRUCTURE OF COLLAPSED CLUSTER CORES
        Phyllis M. Lugger, Haldan Cohn, Jonathan E. Grindlay,
        Charles D. Bailyn and Paul Hertz   657

RADIAL VELOCITY STUDY OF NGC 6712
        J. Grindlay, C. Bailyn, R. Mathieu and D. Latham   659

A SURVEY OF GLOBULAR CLUSTER VELOCITY DISPERSIONS
        Carlton Pryor, Robert D. McClure, J. M. Fletcher
        and James E. Hesser   661

ANISTROPY IN OMEGA CENTAURI AND 47 TUCANAE
        G. Meylan   663

EVOLUTION OF GLOBULAR CLUSTERS INCLUDING A DEGENERATE COMPONENT
        Hyung Mok Lee   665

EVOLUTION OF GLOBULAR CLUSTERS WITH TIDALLY-CAPTURED BINARIES THROUGH CORE COLLAPSE
        Thomas S. Statler, Jeremiah P. Ostriker
        and Haldan N. Cohn   667

BINARY INTERACTIONS IN STAR CLUSTERS
    Stephen L. W. McMillan   669

TIDAL EFFECTS ON STELLAR EVOLUTION IN CLOSE BINARIES
FORMED IN GLOBULAR CLUSTERS
    H. M. Antia, A. K. Kembhavi and A. Ray   671

THE EFFECTS OF STELLAR EVOLUTION AND GALACTIC TIDES ON
GLOBULAR CLUSTER EVOLUTION
    D. F. Chernoff, M. D. Weinberg and S. L. Shapiro   673

THE SPATIAL DISTRIBUTION OF SPECTROSCOPIC BINARIES AND
BLUE STRAGGLERS IN M 67
    Robert D. Mathieu and David W. Latham   675

EVIDENCE FOR MASS SEGREGATION IN NGC 5466
    James M. Nemec and Hugh C. Harris   677

ORIGIN AND RADIAL DISTRIBUTION OF FAINT BLUE HORIZONTAL-
BRANCH STARS
    Charles D. Bailyn, Jonathan E. Grindlay, Haldan Cohn
    and Phyllis M. Lugger   679

VARIABILITY OF OMEGA CENTAURI BLUE STRAGGLERS: CLUES TO
THEIR ORIGIN
    G. S. Da Costa and John Norris   681

A SEARCH FOR OPTICAL COUNTERPARTS OF GLOBULAR CLUSTER
X-RAY SOURCES
    Michel Aurière, Lydie Koch-Miramond,
    Claude Chevalier, Jean-Pierre Cordoni
    and Sergio Ilovaisky   683

LOW LUMINOSITY GLOBULAR CLUSTER X-RAY SOURCES
    Paul Hertz   685

EXOSAT OBSERVATIONS OF OMEGA CENTAURI
    Lydie Koch-Miramond and Michel Aurière   687

NEW METHODS FOR THE SEARCH FOR HOT GAS IN GLOBULAR CLUSTERS
    Yu N. Gnedin and T. M. Natsvlishvili   689

RADIAL VELOCITY PROFILES FOR ANISOTROPIC SPHERICALLY
SYMMETRIC CLUSTERS: AN EXAMPLE
    Herwig Dejonghe   691

LINEAR DENSITY WAVES IN GLOBULAR CLUSTERS
    Yousef Sobouti   693

ON GRAVOTHERMAL OSCILLATIONS
        Jeremy Goodman   695

COOLING AND FRAGMENTATION OF PROTO-GLOBULAR CLUSTER CLOUDS
        Francesco Palla and Hans Zinnecker   697

FORMATION OF GLOBULAR CLUSTERS AND THE FIRST STELLAR
GENERATION
        A. Di Fazio   699

FORMATION OF POPULATION III OBJECTS DUE TO COSMIC STRINGS
        Tetsuya Hara and Shigeru Miyoshi   701

FOURTH POSTER PAPER DISCUSSION   703

RANDOM QUOTES   711

NAME INDEX   715

OBJECT INDEX   723

SUBJECT INDEX   728

ADDRESSES OF PARTICIPANTS   741

Heretofore unpublished portrait of Harlow Shapley, probably from the late 1930's, found by Owen Gingerich in the Harvard University archives.

PREFACE

In the centennial year, 1985-86, of Harlow Shapley's birth, the study of globular clusters was no less important to the development of astronomy than in 1915, when Shapley first noted their concentration on the sky. By 1917 Shapley had used the properties of the system of globular clusters to complete the Copernican revolution and locate the solar system, and its Earth-bound observers, far from the center of the Galaxy and the globular cluster distribution. Seven decades later, in the year of these proceedings, globular cluster research and the study of the system of globular clusters in our own and distant galaxies is undergoing a renaissance of activity. The introduction of new observational tools, particularly CCD imagers and digital spectrographs, as well as powerful theoretical methods have transformed the study of globular clusters into one of the main line areas of modern astrophysics. Thus it seemed particularly appropriate to one of us, when considering how the Harvard College Observatory might mark the Shapley centennial, to propose and plan for an IAU Symposium on Globular Cluster Systems in Galaxies.

Planning for the Shapley Symposium, as it came to be called, was even more drawn out than the preparation of this volume. The Symposium was originally proposed to the IAU Secretariat in time for it to be held in August, 1985, so that it might occur in the centennial (calendar) year. The IAU turned this down, however, having banned all Symposia within a few months of the upcoming General Assembly meeting in New Delhi out of concern for finite travel budgets for the participants. Approval was promptly given for the Symposium to be held at the next available date proposed, August 1986, which still would allow the meeting to occur (so the Organizing Committee noted) within the 100th year of Shapley's birth in November 1885. Although Cambridge in August can be either stifling or cold and wet, the weather the week of the meeting (August 25-29, 1986) was as clear and crisp as the many excellent talks and discussions which took place in the meeting hall of the Harvard Science Center.

The intent of IAU Symposium 126 was to review the recent progress and future prospects in the studies of globular clusters in our own and external galaxies. Although there had been several recent meetings on globular clusters, including the 1984 IAU Symposium 113 on Dynamics of Star Clusters and the 1981 IAU Colloquium 68 on Astrophysical Parameters for Globular Clusters, no meeting had yet focused on the properties of the system of globular clusters as a whole. Given the remarkable progress in cluster observations with CCD detector systems on 4 m class telescopes, as well as the explosive growth of desk-top VAX class (e.g. microVAX) computers for both data analysis and theoretical modeling, the study of the nature and evolution of the cluster system in the Galaxy and beyond was now possible. Thus the Shapley Symposium was intended to highlight the large-scale properties of globular clusters and the formation and evolution of globular cluster systems in galaxies. Another motive in our original planning was to have a Symposium on globular clusters to highlight the prospects and problems for the new studies of globular clusters that would be carried out by the Hubble Space Telescope, originally scheduled for launch only a few months after the meeting. The Challenger tragedy delayed this, of course, but has not diminished the prospects that some of the most exciting science with HST will result from its observations of globular clusters.

Perhaps the best tribute to the success of the meeting is contained in the opening words of Sidney van den Bergh's summary paper at the end of this volume: "This was a really exciting conference !" Indeed, the meeting followed closely on and provided a forum for the discovery of the apparent correlation between cluster metallicity and the slope of the luminosity (and thus mass) function of its stars. Comprehensive studies of the globular clusters in the Magellanic Clouds, as well as M 31 and Local Group galaxies, were presented which would have done Shapley (who was called "Mr. Magellanic Clouds" by Bart Bok, as Helen Sawyer Hogg reminded us in her delightful historical reminiscences) proud. The meeting was intended to pick up on several hot topics which were just (barely) introduced at the IAU Symposium 113 on stellar dynamics: the evolution of globular clusters past their core collapse stage and the possible re-expansion, and eventual tidal disruption, of clusters, as well as the disruption of globular clusters (not necessarily post-core collapse) by tidal shocks in the disk of the galaxy and by encounters with giant molecular clouds. These processes have been "confirmed" by numerous contributions at this meeting, although it remains for the next IAU Symposium to include the hard observational evidence that cluster re-expansion and cluster disruption are indeed observed. Finally, the quantity and quality of new results on distant globular cluster systems was reassuring for the original objectives of the conference. It is likely that a comparable surge in new results will only be available well after the first images with HST or, perhaps, in the era of either new dedicated 4 m class or very large ground-based telescopes.

This meeting achieved its goals thanks to the hard work of a large number of people. Perhaps foremost among the many who planned for and worked hard during the meeting was the chairman of the Local Organizing Committee, Robert Davis, who handled the CfA VAX's, mailings, registration and billing jobs with aplomb. Martha Hazen arranged the housing and feeding of conference participants as well as provided numerous invaluable suggestions for the overall organization of the meeting; David Latham carried out the Herculean task of editing the poster-paper abstracts submitted and arranged for them to be reproduced in a most useful book and for the poster papers themselves to be displayed on the poster boards; John Huchra organized the coffee and doughnut breaks, the registration procedures, and other matters; Jacqueline Kloss arranged for the bus transport to the evening reception at the Gardner Museum; and Owen Gingerich arranged for our use of the Science Center facilities as well as, with Barbara Welther, the delightful reception and music at the Isabella Stuart Gardner

Museum in Boston. The following generous donors helped to make the Gardner reception possible: Owen and Miriam Gingerich, Martha Hazen, Jacqueline and Henry Kloss, Edward Lilley, George Mumford, Alan Shapley, Lloyd Shapley, Willis Shapley, Charles Whitney, and John Wolbach. The presence of Willis Shapley at the meeting and the reception forged a welcome link between the participants and Harlow Shapley.

The actual operations of the meeting went off smoothly thanks to a number of people. Marilyn Bibeau and David Plancon of the HCO business office are thanked for their long hours keeping the books and arranging for bills to be paid. Corbin Covault and John Flanagan helped with numerous tasks in running the Symposium, and Christie Karlin did much of the secretarial work in putting together the Proceedings. Mary Bongiovanni took care of the job of passing out the discussion sheets, typing them up and then passing them back to speakers for proofing. She and Kristina Philip typed many of the final manuscript pages.

We note that at this meeting the submission of some of the papers was done in a new way, namely by electronic means. Approximately $1/8^{th}$ of the papers were transferred to Schenectady or Cambridge either by BITNET or on a floppy disk. For these papers the editing process was made much more simple. Since all these papers could be printed on the same, laserjet, printer, the resulting book becomes more uniform in appearance. We expect that at future meetings the percentage of papers submitted in this manner will be much larger and at some point it will be possible to present the proceedings of a meeting with a completely uniform appearance. BITNET was also used for communications between the two editors and mitigated the effect of our being in two different cities.

Paul Hodge provided a most interesting, and entertaining, after-dinner talk at the conference banquet. Finally, Irwin Shapiro thoughtfully made available to the LOC resources of the CfA to help with the meeting and opened the conference with Derek Bok, who kindly took time out from his busy schedule to welcome the participants to Harvard in the year of its 350th anniversary.

The members of the Scientific Organizing Committee are listed below and are thanked for their assistance and patience in planning the Symposium:

> P. Demarque (USA)
> M. Hoskin (UK)
> K. Freeman (Australia)
> J. Grindlay (USA), Chair
> G. Lyngå (Sweden)
> R. Wielen (W. Germany)
> C. Pilachowski (USA)
> A. Renzini (Italy)
> V. Straižys (USSR)
> T. Van Albada (Netherlands)
> S. Van den Bergh (Canada)

The members of the Local Organizing Committee are thanked once again for their many efforts:

  R. Davis (Chair)   J. Huchra
  O. Gingerich   J. Kloss
  M. Hazen   D. Latham

The following conference participants chaired the scientific sessions, which are listed in the Table of Contents, and are thanked for leading the discussion in the sessions:

  J. McClure   V. Trimble
  J. Cohen   H. Zinnecker
  J. Norris   M. Aurière
  M. McCarthy   J. Goodman
  C. Pilachowski   D. Hanes
  G. Wallerstein   M. Rees
  J. Watanabe   Y. Gnedin
  V. Straižys   J. Grindlay
  G. Lyngå   I. King

The Symposium was co-sponsored by the following IAU Commissions:

  IAU Commission 33 - Structure and Dynamics of the Galactic System
  IAU Commission 37 - Star Clusters and Associations

The Symposium was supported financially by the IAU and by the Harvard College Observatory and Smithsonian Astrophysical Observatories, to whom we are grateful.

September, 1987

Cambridge, Mass.
Schenectady, N.Y.

            Jonathan E. Grindlay
            A. G. Davis Philip

               Editors

ASSORTED FIGURES FROM THE SYMPOSIUM

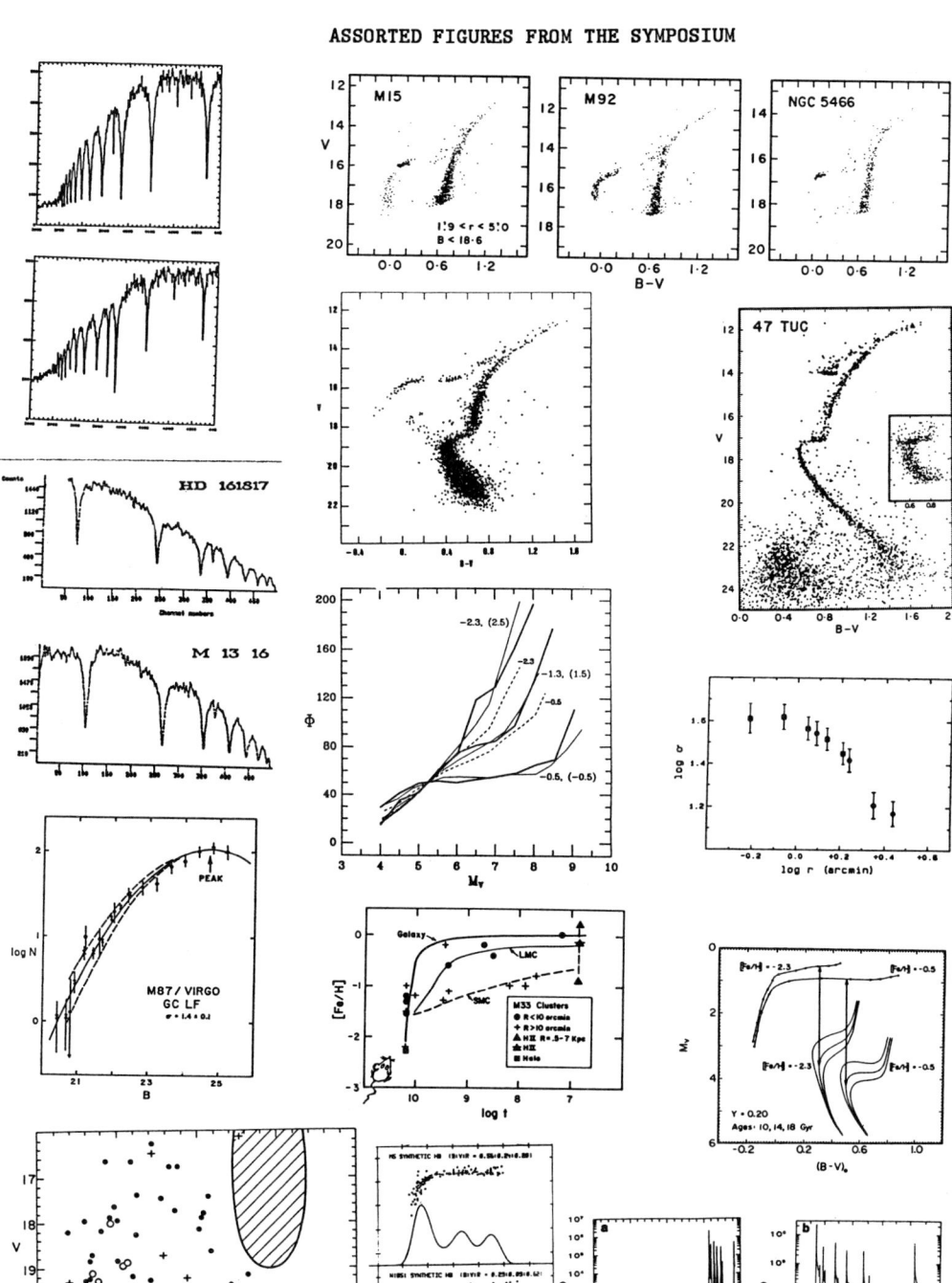

LIST OF PARTICIPANTS

Argentina

Mendez, M.   Astronomical Observatory of La Plata
Muzzio, J. C.   Astronomical Observatory of La Plata

Australia

Norris, J.   Australian National University
Sharples, R.   Anglo-Australian Observatory

Bulgaria

Spassova, N.M.   Bulgarian Acad. Sci.

Canada

Hanes, D. A.   Queen's University (Ontario)
Harris, G.   University of Waterloo
Harris, W.   McMaster Univ.
Hesser, J.   Dominion Astrophys. Obs.
Innanen, K.   York University
McClure, R. D.   Dominion Astrophysical Observatory
Mitalas, R.   University of Western Ontario
Pryor, C.   Univ. of Victoria
Racine, R.   Université de Montréal
Reed, B. C.   Saint Mary's University
Richer, H.   University of British Columbia
Sawyer-Hogg, H.   David Dunlap Observatory
Smith, G.   Dominion Astrophysical Observatory
VandenBergh, D.   University of Victoria
van den Bergh, S.   Dominion Astrophysical Observatory
Wehlau, A.   University of Western Ontario

Chile

Alcaino, G.   Instituto Isaac Newton
Geisler, D. P.   CTIO
Liller, W.   Vina del Mar

France

Cayrel, R.   Observatoire de Paris
Colin, J.   Observatoire de Besancon
Koch-Miramond, L.   CEN-Saclay
Aurière, M.   ESO and Obs. du Pic du Midi

India

Adur, B.   Nehru Center

Kilambi, G. C.   Osmania University
Ray, A.   Tata Inst. of Fundamental Research

Iran

Sobouti, Y.   Shiraz University

Italy

Gratton, R.   Astronomical Observatory of Rome
Battistini, P.   University of Bologna
Buonanno, R.   Astronomical Observatory of Monte Mario
Capuzzo Dolcetta, R.   Astronomical Observatory of Rome
Di Fazio, A.   Astronomical Observatory of Rome
Fusi Pecci, F.   University of Bologna
Greggio, L.   University of Bologna
Held, E. V.   Instituteo of Astronomy (Padua)
Lo Cascio, L.   University of Palermo
Malagnini, M. L.   University of Trieste
Ortolani, S.   Astronomical Observatory of Asiago
Renzini, A.   University of Bologna
Rossi, L.   Institute of Space Astrohysics

Japan

Hara, T.   Kyoto Sangyo University
Inagaki, S.   University of Kyoto
Takayanagi, K.   Ryukoku University
Watanabe, J.   University of Tokyo

Sweden

Lyngå, G.   Lund Observatory

Switzerland

Blecha, A.   Observatoire de Geneve

U. K.

Buttress, J.   University of Leeds
Griffiths, W.   University of Leeds
Noble, R. G.   University of Leeds
Rees, M.   Institute of Astronomy
Bhatia, R.   Royal Observatory (Edinburgh)
Hatzidimitriou, D.   Royal Observatory (Edinburgh)
Ramamani, N. R.   University of Edinburgh
Zinnecker, H.   Royal Observatory (Edinburgh)

U. S. A.

Ables, H.D.	U.S. Naval Obs.
Altner, .	Applied Research Corp.
Andersen, J.	Copenhagen Univ. Obs.
Armandroff, T.	Yale University
Bahcall, J. N.	Inst. for Advanced Study (Princeton)
Bailyn, C.	Harvard University
Baum, W. A.	Lowell Observatory
Bell, R. A.	Univ. of Maryland
Bond, H. E.	Space Telescope Science Institute
Brodie, J.	University of California
Cacciari, C.	Space Telescope Science Institute
Carney, B.	Univ. of North Carolina
Chernoff, D.	Cornell University
Christian, C. A.	Canada-France-Hawaii Telescope Corp.
Clementini, G.	Space Tel. Sci. Inst.
Cohen, J.	California Institute of Technology
Cohn, H. N.	Indiana University
Crocker, D. A.	Univ. of Virginia
Croswell, K.	Center for Astrophysics
Cudworth, K.	Yerkes Observatory
Da Costa, G.	Yale University
Davis, J.	Center for Astrophysics
Dejonghe, H.	Inst. for Advanced Study (Princeton)
Demarque, P.	Yale University Observatory
Djorgovski, S.	Center for Astrophysics
Dupree, A.	Center for Astrophysics
Field, C. A.	Univ. of Massachusetts
Friel, E.	Lick Observatory
Gingerich, O.	Center for Astrophysics
Goodman, J.	Inst. for Advanced Study (Princeton)
Graham, J.	Carnegie Inst. of Washington
Grindlay, J. E.	Center for Astrophysics
Harris, H.	U.S. Naval Observatory
Hartmann, L.	Center for Astrophysics
Hazen, M. L.	Center for Astrophysics
Hertz, P.	Naval Research Laboratory
Hodge, P. W.	University of Washington
Huchra, J.	Center for Astrophysics
Janes, K.	Boston University
Jones, K.	West Tisbury, Mass.
Jones, R. V.	Univ. of North Carolina
King, I.	Univ. of California
Kloss, J.	Cambridge, Mass.
Kron, G. E.	Pinecrest Observatory
Kurucz, R. L.	Center for Astrophysics
Laird, J. B.	Univ. of North Carolina
Larson, R.	Yale University
Latham, D. W.	Center for Astrophysics

Lauer, T. R.	Princeton University Observatory
Lee, H. M.	Princeton University Observatory
Lee, Y. -W.	Yale University
Light, R. M.	Yale University
Lilley, A. E.	Center for Astrophysics
Liu, P.	Boston University
Lugger, P.	Indiana University
Marschall, L. A.	Center for Astrophysics
Mateo, M.	University of Washington
Mathieu, R.	Center for Astrophysics
McMillan, S.	Northwestern University
Menzel, M. F.	Cambridge, Mass.
Meylan, G.	Univ. Calif. (Berkeley)
Mumford, G. S.	Tufts University
Murray, S.	Lick Observatory
Nemec, J. M.	Calif. Inst. of Technology
Olszewski, E.	Steward Observatory
Ostriker, J. P.	Princeton Univ. Obs.
Papaliolios, C.	Harvard University
Papenhausen, P.	Rutgers University
Penny, A.	Space Telescope Science Institute
Peterson, C. J.	University of Missouri
Peterson, R.	Visiting Scientist, Whipple Obs.
Philip, A. G. D.	Van Vleck Observatory
Pilachowski, C.	KPNO
Pound, M.	Boston University
Ratnatunga, K. U.	Inst. for Advanced Study (Princeton)
Roberts, M. S.	NRAO
Rood, R. T.	University of Virginia
Rose, J. A.	Univ. of North Carolina
Schild, R. E.	Center for Astrophysics
Schommer, R.	Rutgers University
Seitzer, P.	KPNO
Shapley, W. H.	Washington, DC
Smith, H. A.	Michigan State University
Statler, T. S.	Princeton University Observatory
Stecher, T. P.	NASA-Goddard Space Flight Ctr.
Sweigart, A. V.	NASA Goddard Space Flight Center
Thompson, L. A.	University of Hawaii
Trimble, V.	University of Maryland
Walker, M.	University of California
Wallerstein, G.	Univ. of Washington
Webbink, R.	Univ. of Illinois
Welther, B. L.	Center for Astrophysics
White, R. E.	University of Arizona
Whitney, C. A.	Center for Astrophysics
Wolbach, J. G.	Center for Astrophysics
Zinn, R. J.	Yale University

U. S. S. R.

Gnedin, Y. N.   Pulkovo Observatory
Samus, N. N.   Academy of Sciences of U. S. S. R.
Straižys, V. L.   Institute of Physics, Vilnius
Zdanavičius, K.   Institute of Physics, Vilnius

W. Germany

Seggewiss, D. W.   Observatorium Hoher List
Tucholke, H.   University of Munster
Wielen, R.   Astronomisches Rechen-Institut

Vatican City State

McCarthy, M. F.   Vatican Observatory

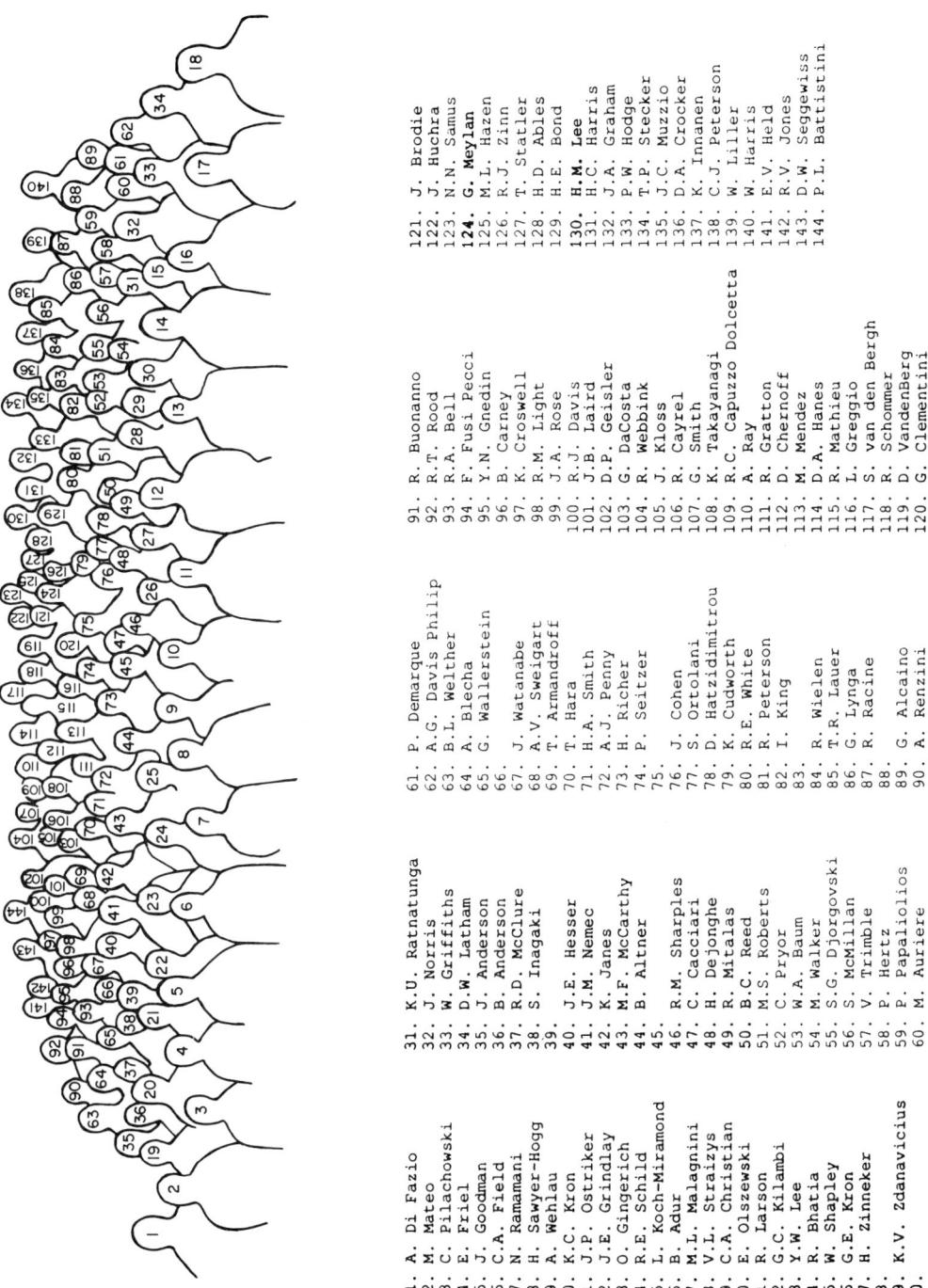

1. A. Di Fazio
2. M. Mateo
3. C. Pilachowski
4. E. Friel
5. J. Goodman
6. C.A. Field
7. N. Ramamani
8. H. Sawyer-Hogg
9. A. Wehlau
10. K.C. Kron
11. J.P. Ostriker
12. J.E. Grindlay
13. O. Gingerich
14. R.E. Schild
15. L. Koch-Miramond
16. B. Adur
17. M.L. Malagnini
18. V.L. Straizys
19. C.A. Christian
20. E. Olszewski
21. J. Larson
22. G.C. Kilambi
23. Y.W. Lee
24. R. Bhatia
25. W. Shapley
26. G.E. Kron
27. H. Zinneker
28. K.V. Zdanavicius
29.
30.
31. K.U. Ratnatunga
32. J. Norris
33. W. Griffiths
34. D.W. Latham
35. J. Anderson
36. B. Anderson
37. R.D. McClure
38. S. Inagaki
39.
40. J.E. Hesser
41. J.M. Nemec
42. K. Janes
43. M.F. McCarthy
44. B. Altner
45.
46. R.M. Sharples
47. C. Cacciari
48. H. Dejonghe
49. R. Mitalas
50. B.C. Reed
51. M.S. Roberts
52. C. Pryor
53. W.A. Baum
54. M. Walker
55. S.G. Djorgovski
56. S. McMillan
57. V. Trimble
58. P. Hertz
59. P. Papaliolios
60. M. Auriere
61. P. Demarque
62. A.G. Davis Philip
63. B.L. Welther
64. A. Blecha
65. G. Wallerstein
66.
67. J. Watanabe
68. A.V. Sweigart
69. T. Armandroff
70. T. Hara
71. H.A. Smith
72. A.J. Penny
73. H. Richer
74. P. Seitter
75.
76. J. Cohen
77. S. Ortolani
78. D. Hatzidimitrou
79. K. Cudworth
80. R.E. White
81. R. Peterson
82. I. King
83.
84. R. Wielen
85. T.R. Lauer
86. G. Lynga
87. R. Racine
88.
89. G. Alcaino
90. A. Renzini
91. R. Buonanno
92. R.T. Rood
93. R.A. Bell
94. F. Fusi Pecci
95. Y.N. Gnedin
96. B. Carney
97. K. Croswell
98. R.M. Light
99. J.A. Rose
100. R.J. Davis
101. J.B. Laird
102. D.P. Geisler
103. R. DaCosta
104. R. Webbink
105. J. Kloss
106. R. Cayrel
107. G. Smith
108. K. Takayanagi
109. R.C. Capuzzo Dolcetta
110. A. Ray
111. R. Gratton
112. D. Chernoff
113. M. Mendez
114. D.A. Hanes
115. R. Mathieu
116. L. Greggio
117. S. van den Bergh
118. R. Schommer
119. D. Vandenberg
120. G. Clementini
121. J. Brodie
122. J. Huchra
123. N.N. Samus
124. G. Meylan
125. M.L. Hazen
126. R.J. Zinn
127. T. Statler
128. H.D. Ables
129. H.E. Bond
130. H.M. Lee
131. H.C. Harris
132. J.A. Graham
133. P.W. Hodge
134. T.P. Stecker
135. J.C. Muzzio
136. D.A. Crocker
137. K. Innanen
138. C.J. Peterson
139. W. Liller
140. W. Harris
141. E.V. Held
142. R. Jones
143. D.W. Seggewiss
144. P.L. Battistini

LIST OF PHOTOS OF PARTICIPANTS AND SPECIAL FIGURES

Cover: Halley's Comet, Cen-A and Omega-Cen (photo by W. Liller)

Frontispiece: Harlow Shapley (Harvard University archives)   xvi

Assorted figures from the Symposium   xxi

Photograph of Symposium participants (photo by Steven Seron)   xxviii

Helen Sawyer Hogg and Willis Shapley remembering
Willis Shapley fielding questions on his father   2

Bob McClure running his session
Ron Webbink answering questions after his talk   36

Bruce Carney studies a particulary detailed poster after registering
Gary DaCosta and colleagues studying posters   150

Poster papers were also set up downstairs
Paul Hodge checking in   270

Martha Hazen and Morton Roberts in discussion
The crowd awaits the opening remarks   410

Josh Grindlay thanks attendees for not being in China
Another view of the Symposium participants   466

Drs. McCarthy, Gingerich, Hazen, and Alcaino examine historic plate
Bob McClure explains his pivotal poster to Rene Racine, Juan Forte
    and others   476

Charles Peterson and Martha Hazen pinning up
President Derek Bok officially welcomed the participants   540

Jim Hesser and a well-known globular cluster captivate the crowd
Roberto Buonanno with a cluster of globular luminaries   602

Judy Cohen fields the discussion after Roger Bell's talk
The poster papers received due attention and discussion   654

Irwin Shapiro opened the first IAU Symposium co-sponsored by the CfA
Core Collapse after the conference   712

The editors - J. Grindlay (R) and A. G. Davis Philip - after
    finishing their editing task in Cambridge
    (photo by Sandra Grindlay)   742

Back cover: Luminosity function vs. metallicity correlation
    from R. McClure et al

    (All photos by James Cornell unless otherwise credited)

Chapter I

Review Papers

Harlow Shapley

Helen Sawyer Hogg and Willis Shapley remembering

Willis Shapley fielding questions on his father

SHAPLEY'S DEBATE

Michael Hoskin
Cambridge University

The attempt to make three-dimensional sense of the Milky Way goes back to a most unlikely origin: the English antiquary of the early eighteenth century, William Stukeley, remembered today for associating the Druids with Stonehenge. Stukeley came from Lincolnshire and so was a fellow-countryman of Isaac Newton, and as a result he was privileged to talk with the great man from time to time. In his Memoirs of Newton Stukeley records one conversation they had in about 1720, in which Stukeley proposed that the Sun and the brightest stars of the night sky make up what we today would term a globular cluster, and this cluster is surrounded by a gap, outside of which lie the small stars of the Milky Way in the form of a flattened ring.

Stukeley's remarkable suggestion was recorded only in his manuscript memoirs, and had no effect on the subsequent history of astronomy. It chanced that the very same model was proposed a century later by John Herschel, who in 1833 in his A treatise on astronomy compared the appearance of "our own sidereal firmament and milky way" to Saturn and its ring. Herschel afterwards took his 20ft reflector to South Africa and between 1834 and 1838 subjected the southern skies to their first close scrutiny. He decided that the structure of the Galaxy is immeasurably more complex than his simple Saturn model had allowed; and that in fact the Sun and the brighter stars occupy a sparsely-populated region surrounded by a dense and complex ring of stars whose more distant windings extend out to the limits of telescopic vision and beyond. Herschel's unique status as the only astronomer in history systematically to examine the whole of the sky, both northern and southern hemispheres, with a major telescope, discouraged lesser mortals from theorising about the Milky Way for the next half-century. Theorising resumed when Jacobus Kapteyn realised that, as an astronomer without a telescope, he could nevertheless get access to a wealth of new information if he offered his services in measuring the positions of stars on David Gill's Cape Photographic Durchmusterung then in progress, and this he duly did with the help of teams of convicts from Groningen State Prison. Right up to his death in 1922 Kapteyn worked to determine the structure of the Galaxy. He was of course, like his predecessors, working outwards in all directions from the location of the observer in the solar system; and it could hardly be otherwise. Kapteyn was alert to the threat posed to his analysis by possible obscuration in the plane of the Galaxy, but the reddening of light that he thought would reveal such obscuration was not in fact observed. As a result, he believed that the whole of the Galaxy was accessible to his investigations when in fact he could see only our neighbourhood; and the more remote stars in the galactic plane seemed fainter and more thinly scattered than was in fact the case, so that the Galaxy seemed to be only a few thousand light years in size and the solar system fairly central.

A crucial clue lay hidden in the verbose prose of John Herschel's volume reporting his Cape observations. He speaks (p. 136) of "the extraordinary display of fine ... globular clusters" in the general direction of Sagittarius. "Here, in a circular space of $18°$ in radius, we find collected no less than thirty of these beautiful and exquisite objects.... are we to connect it with the very peculiar structure of the Milky Way in this particular part of its course, which is here unlike in its constitution to any other portion of that zone, and which passes diametrically across the circular area in question". The fact that the great majority of globular clusters lie in one half of the sky was remarked on in 1909 by the Swedish astronomer Karl Bohlin, and he proposed that the globular clusters are packed together close to the centre of the Galaxy and clustered symmetrically around it (so that the Sun is eccentric in position); and that this "cluster of clusters" is surrounded by the star clouds of the Milky Way together with the Sun and other isolated stars.

Bohlin went on to propose that the entire Galaxy is a former planetary nebula, the material at whose poles has dispersed to form spiral nebulae as the great planetary nebula has rotated. The material in the equatorial zone of the planetary nebula is still in position, though it has condensed to form the star clouds of the Milky Way.

Bohlin's bizarre paper, tucked away in the proceedings of a Swedish journal devoted to general science, and the work of an author as eccentric as the location he assigned to the Sun, carried little weight; and Bohlin's insight into the peculiar distribution of the globular clusters had little effect. It was however known some years later to the rising young star of American astronomy, Harlow Shapley. Shapley had been born in 1885, and after some experience as a teenage reporter on a newspaper had enrolled in 1907 at the University of Missouri, intending to study in the University's School of Journalism. Finding the school was not to open for another year, he looked round for something else to study, and hit on astronomy. When Shapley was in his third year as a student, Frederick H. Seares of the Laws Observatory offered him a teaching assistantship, and was sufficiently impressed by the young man later to recommend him to Princeton for a fellowship in astronomy. There Shapley became a pupil of Henry Norris Russell, a legendary orator with wide-ranging interests in astronomy and astrophysics. Russell put Shapley to work on eclipsing binaries, and in his doctoral dissertation Shapley discussed the orbits of ninety such binaries where only a handful had been studied previously. He also showed that if Delta Cephei were to be an eclipsing binary, as commonly believed, the two components would have to fall inside each other; instead he proposed that it was a single, pulsating star.

Meanwhile Seares had moved to Mount Wilson, where he told George Ellery Hale of Shapley's ability. Hale interviewed him and appointed him to the Mount Wilson staff. Solon I. Bailey of Harvard advised Shapley to use the 60-inch telescope to study variables in globular clusters. Bailey himself had already detected a number of Cepheids in globular clusters, and Henrietta Leavitt had earlier identified many Cepheids in photographs of the Small Magellanic Cloud. As early as 1908 Miss Leavitt had noted that the longer the period of the star, the brighter it appeared on her photographs, and therefore the more luminous it was. In a further paper in 1912 she set out her results graphically, and pointed out that if her photographic magnitude scale could be calibrated for absolute magnitude, the distance of the Cloud could be determined.

Shapley's discovery that Cepheids were not binary stars, whose appearance would depend upon the chance geometrical configuration relative to the observer, but pulsating stars displaying the results of intrinsic physical processes, must have brought home to him the plausibility of using Cepheids as distance indicators. Before long he was being allowed substantial periods of time on the 60-inch for his own research on variables in clusters, and he managed to calibrate Cepheids by means of an ingenious and not wholly convincing statistical argument. In 1915 he noted that globular clusters are largely concentrated into one hemisphere in galactic longitude, and he remarks on Bohlin's theory -- though we must remember that for Bohlin the globulars are densely concentrated at the centre of the Galaxy, and are themselves surrounded at a great distance by the ring of the Milky Way stars. But for Shapley in 1915 the accepted size of the Galaxy -- a few thousand light years in diameter -- was so small compared to the distances he was deriving for globular clusters that the jigsaw refused to fall into place. It seemed that globular clusters were "very distant systems, distinct from our Galaxy and perhaps not greatly inferior to it in size".

In the winter of 1917/18 came the breakthrough, as Shapley brought himself to reject the accepted size of the Galaxy. As he wrote to Eddington at the time, "The globular clusters outline the sidereal system." His conclusion was that we are some 60,000 light years away from the centre of the system of globulars, and therefore the same distance from the centre of the Galaxy, whose diameter he now estimated at 300,000 light years. This was a dramatic conclusion, based on an immense amount of detailed labour, but also on some questionable steps, as with the calibration of the Cepheids. Shapley was at this time acquiring an unfortunate reputation for irresponsible speculation. In 1919, for example, he downgraded the spiral nebulae to being insubstantial, gaseous bodies being driven off by radiation pressure from our Galaxy. Both Hale and Russell wrote to warn him of the risks he was taking with his standing as an astronomer. Russell wrote: "... I am sorry to see you join the company of those who advance theories that are 'startling if true'. There has been a great deal too much of this done in the last few years."

The enormous increase that Shapley proposed in the size of the Galaxy naturally aroused great controversy, for his methods challenged the traditional techniques for working steadily outwards from the solar system towards the more remote regions of the Galaxy. Shapley's big Galaxy also made it almost unthinkable that the innumerable spiral nebulae could be other such galaxies, and so Shapley had become caught up in the age-old problem of the status of the nebulae: were they enormous star systems, 'island universes', -- or were they nearby clouds of chaotic matter? Christopher Wren had taken the former position, Edmond Halley the latter. In the 1860s the spectroscope had shown that some nebulae are gaseous; but what of the tiny spirals, of which hundreds of thousands appeared on the Lick Observatory photographs?

Early in the twentieth century three compelling pieces of evidence argued against the spirals being island universes comparable to our Galaxy. First was the 'zone of avoidance': spiral nebulae were dense around the poles of the Galaxy, and avoided the plane of the Galaxy. Surely this must mean that the spirals were related to our Galaxy and not independent galaxies in their own right. Heber Doust Curtis of Lick had found the solution to this difficulty in the photographs he had taken of edge-on spirals, with dark lanes of dust clearly visible. Similar dust in the plane of our Galaxy would conceal from us any spirals that lay in low galactic latitudes.

Secondly there was the new star that had flared up in the Andromeda Nebula in 1885. At one stage the star, S Andromedae, was about one-tenth the brightness of the entire nebula, and so if the nebula was an island universe of millions of stars, this nova must have reached the brightness of hundreds of thousands of stars, and this in only a few hours. No processes known to physics could achieve such sensational results; much more likely that the nebula was a gas cloud that had encountered a passing star, and that the star had flared up as it passed through the cloud. Here again Curtis had found the answer. In 1917, on photographs of other spirals, he had come across further examples of novae; and these were fireworks on a much more modest scale. While Curtis was cautiously examining other photographs of these particular spirals, George W. Ritchey of Mount Wilson discovered a nova that was still visible. This sent a number of astronomers (including Shapley) hurrying to compare photographs of spirals, and several more were found. All were markedly less dramatic than S Andromedae and the Z Centauri of 1895, which began to be recognised as exceptional. As Curtis suggested in 1921, "a division into two magnitude classes is not impossible." If the novae recently discovered were comparable to galactic novae, then the spirals could well be distant enough and large enough to be independent island universes.

The third item of evidence was more recent: the huge recession velocities of spirals measured by Vesto M. Slipher. These were difficult to explain on any theory, but at least the velocities were much larger than those of any known components of our Galaxy.

As a result of all this, Curtis and many of his Lick colleagues were convinced that the spirals _were_ other island universes. But meanwhile, at Mount Wilson, a striking new piece of counter-evidence was at hand: Adriaan van Maanen, a meticulous worker, had used a stereocomparator to compare old and new photographs of M 101, and had concluded that the spiral was rotating in only a few thousands of years -- so rapidly that the outlying parts would need to travel impossibly fast _if_ the spiral was a huge island universe. Shapley was van Maanen's friend, and believed him. Curtis did not.

Although it was entirely possible that the spirals might be star systems, or island universes, that were orders of magnitude smaller than our Galaxy, the instinct of astronomers then as always was to pose the question in its simplest terms: was the Galaxy one of the spiral nebulae? If Shapley was correct, and the diameter of the Galaxy is ten times greater than had been thought, and therefore its volume a thousand times greater, the answer was much more likely to be No. And so it is not surprising that when in late 1919 Hale was looking for a topic for the annual lecture at the National Academy of Sciences in memory of his philanthropist father, William Ellery Hale, he should think of a debate, between Shapley and a suitable opponent. A few weeks later, in February 1920, Shapley received a telegram inviting him to debate with Heber Curtis of Lick Observatory, 45 minutes each on the scale of the universe.

The invitation could not have come at a more awkward time for Shapley, whose confidence in his ability as an astrophysicist was matched by his confidence in his ability as a future observatory administrator. A year earlier death had at last brought to an end Edward Pickering's 42-year reign as Director of the Harvard College Observatory. The moment he heard the news, Shapley decided he would try to be the next director, and wrote to Russell and Hale to enlist their support.

Both men were aghast at Shapley's naked ambition, and told him so in forthright terms. Russell remarked to Hale that Shapley "would not suffer if he pondered the old fairy tale about the man who got all sorts of good things from a magic fish whose life he had saved, -- until his wife wanted to be Pope!" Yet in fact both Hale and Kapteyn considered Shapley's scientific brilliance made him the best man for the job, although both realised that his youth and lack of experience in the management of staff would make it a risky appointment. Despite their misgivings, both these powerful advocates lobbied Harvard on Shapley's behalf. Hale's second choice was Frank Schlesinger of Allegheny Observatory, then in his late forties and a 'safe' appointment. His third was Russell himself, who would have been an obvious candidate save for one failing: he was a notoriously bad administrator. Harvard dithered, and when Hale's telegram reached Shapley the post had already been left unfilled for a full year. Shapley's own hopes had been encouraged a few weeks earlier when the President of Harvard telegraphed Mount Wilson to know if Shapley had plans to visit the East Coast. On learning that Shapley had no such plans, Harvard had sent one of their Regents to visit Shapley. The Regent knew nothing about science, but was much interested in a AAAS convention that Shapley had organised. The reason was that Harvard had no doubts about Shapley as a scientist, but wanted to reassure themselves as to his abilities as an administrator. Shapley, his ambition fuelled by the visit, wrote to Russell in great excitement, saying "I might say that I am very confident that Harvard is not too big for me and that the thing I could and would do there would be a credit to American astronomy".

Confident that he was under active consideration for the coveted directorship, Shapley read the telegram from Hale with dismay. He judged, correctly, that a delegation from Harvard would come to Washington to see how he performed under fire, and the Academy meeting would be make-or-break. He was then inexperienced as a public speaker, and in fact spoke poorly at the meeting -- Russell afterwards expressed the hope that if Shapley came to Harvard as his number two, he would offer a lecture course, for this "cultivates the gift of the gab, which he needs". Curtis was by contrast an excellent lecturer, and in defending the traditional position he had a much easier task. Shapley accordingly set about an exercise in damage limitation. He tried to defuse the encounter by turning it into a partnership rather than a confrontational debate; but here he reckoned without Curtis's Irish ancestry. Curtis relished "a good friendly scrap.... It might be far more interesting both for us and our jury, to shake hands, metaphorically speaking, at the beginning and conclusion of our talks, but use our shillelahs in the interim to the best of our ability." Shapley then sought to trivialise the occasion, by reducing the time allotted from 45 to a mere 35 minutes; but after much writing of letters -- happily for historians, the telephone was not then much used -- a compromise of 40 minutes was agreed. This still left him dangerously exposed, so he prepared a presentation so elementary that of the 16 typed pages that Shapley devoted to the topic of the evening, it took him seven to reach the definition of a light-year -- and this to the National Academy of Sciences! Curtis listened to Shapley with dismay, considered abandoning his own closely argued thesis, but decided at the last moment to go ahead.

Shapley's last defensive measure was to persuade Russell that Russell's own ideas would be equally under attack, and to arrange for the chairman to call Russell to speak first from the floor, to undo the harm that Curtis had done. In fact Russell made so substantial contribution that the question arose as to whether he should appear in the published version as a third contributor.

To analyse the detailed scientific arguments that eventually appeared in the published version would be to misrepresent the actual encounter, as many incautious historians have done in the past. Shapley took as the prime topic, the dimensions of the Galaxy, which he claimed was far larger than had been thought and unlike any other object known to observational astronomy; Curtis was more concerned with comparing the Galaxy to the spiral nebulae. To this extent it is true to say that both men were partly right and partly wrong. Curtis was convinced he had come off best, and this is certainly the case.

But what of Harvard? In writing to President Lowell in March to tell him of Shapley's forthcoming appearance in Washington, Hale began to have misgivings about his earlier support for Shapley, and took the opportunity to hedge his bet by arguing at length the case for yet another candidate, namely Seares, to whom, he said, he would confidently entrust his own observatory if Walter Adams were unavailable. Shapley, he said, was versatile, daring and industrious, and had shown brilliant and unusual qualities in his recent work; but he had not yet reached complete maturity or established his final place. His appointment at Harvard would involve a certain measure of risk. Nevertheless, Hale believed he would be a great success there.

The late introduction of Seares into the discussion was seen at Harvard as reflecting no credit on Hale, and only served to confirm the influential G.R. Agassiz in his support for Shapley. But then came the disaster that Shapley had foreseen: Agassiz attended the Shapley-Curtis debate, and wrote next day to Lowell to say that "Shapley lacks maturity and force, and does not give the impression of being a big enough personality for the position". Russell, who had impressed Agassiz when speaking from the floor at the debate and in private talks afterwards, "besides being more mature, has more balance more force and a broader mental range". And so, before long, Harvard decided to grasp the nettle and offer the directorship to Russell.

In a long and frank letter, Russell turned to Hale for advice. One problem was, "If I don't go to Harvard, who will?" Russell's solution was to have a team, to include a post tailor-made for Shapley: "A second astronomer, younger, and with modern ideas, to be called, to act as the Director's right hand man...." But in the end Russell turned the job down.

Harvard were now back to square one. Should they risk appointing Shapley, with all his limitations? A bizarre compromise was now proposed, whereby the distinguished geometer Julian Coolidge should take responsibility for the Observatory, but with Shapley invited to be leader in the scientific work. And so it was that Shapley found himself that November being invited to become, not Director, but "the technical man", with the title of Assistant Professor and Astronomer.

To be expressly denied the directorship, even when Harvard had proved incapable of filling the senior post, must have been a bitter disappointment. Shapley wrote for more information as to how the responsibilities would be divided between Coolidge and himself, but he had enough maturity quickly to realise that the arrangement would be an unworkable compromise, and he soon turned the offer down.

Once again Harvard were back to square one. But now Hale quickly stepped in, and with considerable statesmanship told Lowell that he was prepared to give Shapley a year's leave if he wished to go to Harvard on a trial basis. This gave all parties the opportunities they had been seeking. In April 1921 Shapley took up residence at Harvard; and he must have made a far better impression than expected, for that October he was at last appointed Director. The fisherman's wife had become Pope.

ACKNOWLEDGEMENTS AND BIBLIOGRAPHY

I am grateful to Harvard University Archives for access to documents relating to the appointment of a successor to Pickering and for permission to quote. My attention was kindly drawn to these documents by Owen Gingerich and the later paragraphs of this paper have greatly benefited as a result.

The text of Shapley's Washington paper, and of the typed slides summarising Curtis's argument, are reprinted with documentation and archival acknowledgements in M. Hoskin, 1976, J. Hist. Astron. **7**, 169. This also appears with other relevant material in M. Hoskin, 1982, Stellar Astronomy: Historical Studies, Science History Publications Ltd, Chalfont St Giles, Bucks, UK. On Stukeley, see M. Hoskin, 1985, J. Hist. Astron. **16**, 77. On John Herschel, see M. Hoskin, 1987, ibid. **18**, 1. On Kapteyn, see E.R. Paul., ibid. **17**, 155. On Shapley, see H. Shapley, 1969, Through Rugged Ways to the Stars, Scribner's Sons, New York, and the article by O. Gingerich in Dictionary of Scientific Biography. On the 'Debate' and its background, see R. Berendzen, R. Hart and D. Seeley, 1976, Man Discovers the Galaxies, Neale Watson Academic Publications, New York, and more especially R. Smith, 1982, The Expanding Universe: Astronomy's 'Great Debate' 1900-1931, Cambridge University Press.

SHAPLEY'S ERA

Helen Sawyer Hogg

David Dunlap Observatory
University of Toronto

I am greatly pleased to have a chance to speak at this Symposium about Harlow Shapley, the man who was probably responsible, more than any other, for the shape my life has taken.

In my talk this morning I am repeating a few of the remarks I made at the Centennial Celebration for the birthday of Harlow Shapley last October 17 at the Harvard College Observatory. Some of the people here today were present then, and I trust they will excuse the repetition. But my talk today has a different slant.

First I wish to sketch my personal background, and how I happened to arrive at the Harvard College Observatory in September, 1926. I was born in the nearby city of Lowell which at that time, 1905, had a mile of textile and leather mills stretching along the banks of the Merrimack river. I attended Mount Holyoke College in South Hadley, Massachusetts, in the Connecticut valley. In my junior year, under the influence of a remarkable teacher, Miss Anne Young, I switched a major in chemistry to one in astronomy. Professor Young was the niece of Prof. Charles Young of Princeton, author of the finest astronomy textbooks of the first quarter of this century, books still very useful for their factual information. For the total solar eclipse of January 24, 1925 Miss Young was able to get a special train to take all the college people to a golf links in Connecticut, inside the path of totality. There the glory of the spectacle seems to have tied me to astronomy for life, despite my horribly cold feet as we stood almost knee deep in snow.

A year later, when my graduation was only a few months away, the renowned Miss Annie J. Cannon of Harvard paid a visit to Mount Holyoke. (Figure 1). I still have the diary I kept in the 1920's, up to mid-1927 when apparently life became too complicated for me to keep up with a diary any longer. I read for 1926 Jan. 8 "Miss Young asked me to luncheon with Miss Cannon". For Jan. 11, "Miss Young asked me to introduce people to Miss Cannon Friday". For Jan. 16, "Lunch at Gateway with Miss Cannon. Good prospects for Harvard job". Jan. 22. "Miss

Young wrote Dr. Shapley a long letter". Feb. 2. "Letter from Dr. Shapley saying my chances pretty good". Feb. 24. "Letter from Dr. Shapley offering me Edward C. Pickering Fellowship at Harvard for next year ($600). I am so excited it's hard to put my mind to work!" Feb. 28 "Wrote Dr. Shapley accepting job." And on April 3 I went to Cambridge and saw Dr. Shapley "a nerve-wracking day".

Fig. 1 Miss Annie J. Cannon at Mount Holyoke College, South Hadley, Massachusetts, in January 1926. Photo by Helen Sawyer.

Fig. 2 Dr. Harlow Shapley dictating at his revolving desk in the spring of 1929. Photo by Frank Hogg.

On Wednesday, September 29 I appeared at the Observatory at 9 a.m. for work and had a long talk with Dr. Shapley. (Figure 2). "I am to help him on a book on star clusters." At Mount Holyoke I had developed a particular fondness for globular star clusters as my favorite celestial objects. It was almost too good to be true that now I was to work on them with the world's authority.

In the 1920's the Harvard Observatory was a veritable beehive of activity, night and day. Dr. Shapley had been Director for just under five years of his three decades in that post. Edward C. Pickering had died on February 3, 1919 after 42 years as Director, and Solon I. Bailey was in charge of the Observatory for almost three years following his death. In the yard eight or more telescopes were manned by technicians Frank Bowie and Henry Sawyer and some of the graduate students. (Figure 3). Along with those in the southern station they had made a collection of some half a million plates. The buildings had offices for eleven "Members of the Observatory" with positions senior enough to be listed by Solon I. Bailey (1932) in his excellent History and Work of the Harvard College Observatory. William H. Pickering worked at his own observatory in Jamaica, and John S. Paraskevopolous managed the southern station. There were a couple of dozen other workers who fell into two groups - a small group of elderly ladies some of whom, like Miss Florence Cushman, had started work there as early as 1888, and the larger group of newly starting graduate students and young assistants. (Figure 4). This staff was turning out an impressive number of publications which included the Harvard Annals, Circulars, Bulletins, Announcement Cards, Reprints and Reports, with the Monographs series just starting.

Fig. 3   Harvard College Observatory yard in the 1920's. Photo by Helen Sawyer.

Fig. 4   The young ladies of the Harvard College Observatory, about 1927. Cecilia Payne in rear, Emma T.R. Williams, second from right, Margaret Walton (Mayall) second from left.

So I became the first person that Dr. Shapley supervised for the doctorate in his own field of star clusters. He had then trained only one student for the doctorate in astronomy, Cecilia Payne, who had received her doctorate from Radcliffe College in 1925 with an outstanding thesis on "Stellar Atmospheres". I worked in an office adjacent to that of Miss Cannon, and day after day I heard her calling out the spectral classifications for the Henry Draper Extension, with the help of her invaluable assistant Margaret Walton Mayall. It was remarkable that Dr. Shapley encouraged me to have my name on so many papers right from my first year there, - a dozen publications, mostly in Harvard Observatory Bulletins, before my doctoral thesis.

In my very first weeks at the observatory, Dr. Shapley instructed me to get out various series of plates to estimate integrated diameters and magnitudes of globular clusters. I soon found that almost exactly one third, 34 of the 103 globular clusters then catalogued in our galaxy could be identified on a single small scale Harvard plate in the Sagittarius-Scorpius region and I marked them in white circles to indicate them, as shown in Figure 5, a slide kindly supplied by Martha Hazen and Barbara Welther from the Harvard collection. I have my first record book in which Dr. Shapley contributed observations as "HS" while i was "HBS". (Display of record book). When I left in 1931 Dr. Shapley kindly told me to keep the eyepiece I had been using for five years. I have now been using it for 60 years less one month. (Display of eyepiece). Most of the plate frames around the Observatory then were built for 8 x 10 inch plates, which were easy to use. The A series with the Bruce 24-inch refractor, 14 by 17 inches, are much more cumbersome to handle. The plates were considered practically sacred objects in those days. Anyone who had the misfortune to break one was supposed to sign on the plate cover for the misdeed. Miss Hodgdon, the plate keeper (Known as Hodgie) infuriated Cecilia Payne one day by leaving a broken plate on her plate table after writing in large writing on the cover "Miss Payne says she does not know who broke this plate".

Dr. Shapley's usual method of training was to send me instructions on yellow slips of paper, though frequently he worked directly with me, either in his office or mine. These instructions he had dictated with his dictaphone at all hours of the day and night, sitting at his fabulous, round, spinning desk with its many alcoves. His most faithful secretary, Miss Arville D. Walker typed and distributed the slips. They went to all workers at the observatory in a wide range of frequency. Not merely did they keep US on our toes, but they also showed how well up on things Dr. Shapley himself was, especially in his perusal of all the astronomical literature that came into the observatory. I had a fascinating example of this last fall when I was preparing a talk for the Shapley birthday celebration at the Observatory.

I found such a slip dated April 30, 1929 instructing me to make the corrections given in Astronomische Nachrichten 5622 for an earlier article in A.N. 5609. This proved to be a paper by A. Markov

Fig. 5 One third of the 103 globular clusters known in 1930 were marked with white circles by Helen Sawyer on a single Harvard small scale plate of the Sagittarius-Scorpius region.

Fig. 6 Henrietta Swope, Helen Sawyer, Frank Hogg and Adelaide Ames, in the spring of 1929, at Harvard College Observatory.

on "The Nature of Spiral and Gaseous Nebulae". I asked Barbara Welther here to explore the Harvard Observatory copy. At first she had a dismal report - the corrections had NOT been made, and I felt inadequate to be a scientist. But after pondering, I realized there were two sets of the Nachrichten here - one at the Observatory, one here in Harvard yard, and further delving by Barbara located the set originally at the Observatory - in which I had carried out Dr. Shapley's instructions.

It constantly surprises me that I was unaware then of what was actually happening around the Observatory, that this was the start of the Harvard-Radcliffe graduate school in astronomy. Master's degrees had been given for some time, mainly for work down at the Yard, I think. But the idea that this was the beginning of an epoch during which, by 1985, according to Owen Gingerich in the November issue of Harvard Magazine, 240 Ph.D's would be given in astronomy - that idea certainly never occurred to me. As I have mentioned, Cecilia Payne became the first, receiving the Ph.D. from Radcliffe in 1925, because Harvard did not then give doctorates in science to women, and her thesis "Stellar Atmospheres" became Harvard Monographs No. 1. Frank Scott Hogg, a Canadian with an honours degree from the University of Toronto became a graduate student at the same time as I did, and received the first doctorate in astronomy in 1929 given by Harvard University. (Figure 6). Cecilia Payne was his supervisor, and also trained Emma T.R. Williams (who married Alexander Vyssotsky) and who received her doctorate from Radcliffe in 1930. I received the fourth doctorate in 1931, again from Radcliffe. In retrospect I can see that it was the combination of the brilliant minds of Harlow Shapley and Cecilia Payne, with their originality, zeal and enthusiasm, that provided the stimulus needed to establish the graduate school here.

Another aspect of my association with Shapley that I realized belatedly, only after several decades, is that I worked with him at a time when he was very close to the peak of his scientific career. Shapley (1969) was less than two decades past his journalism career as he described it in his autobiographical memoirs "Through Rugged Ways To The Stars". His brilliant Mount Wilson researches with more than 80 papers on star clusters and cepheids, had had ample time to circulate around the world. Although, as you have just heard from Michael Hoskin, his decision on the nature of the spiral nebulae was in error - an error induced by the proper motions of Adriaan van Maanen - the picture he had drawn of the galaxy surrounded by globular clusters concentrated toward its center, was essentially correct, though improvements would be made to the scale. The controversial decision on the absorption of light in space had not yet been made. The depression had not yet hit, making the perennial seeking for funds even more difficult, as Willis Shapley described so vividly at the October Centennial meeting. And Shapley had not yet got into the controversial politics of the late 30's and 40's which would be a drain on his energy and capabilities, with his ever strong desire to help humanity, individually and collectively.

When he came to Harvard he gave up access to the largest telescopes in the world at Mount Wilson, but gained access to smaller scale photographs of the entire southern sky. These drew him increasingly into studies of the Magellanic Clouds and catalogues of galaxies. As Owen Gingerich (1985) has written in Harvard Magazine "at Harvard he never again made a first rate cosmological discovery, ... but few astronomers make even one breakthrough as brilliant as his." In one of his later flings of research he returned to his earlier field of binaries,

the subject of his doctoral thesis at Princeton under Henry Norris
Russell and presented in 1946 a paper on the W Ursae Majoris stars at
the Centennial Symposia of the Observatory. I asked my office mate at
the David Dunlap Observatory, Dr. Slavek Rucinski who is an expert on
these stars, how Shapley's paper then compares with present ideas. He
gave me a splendid technical analysis of various points in the paper,
which some of you might like to read. The general conclusion is that
the situation with various types of variables has changed so markedly
in the last forty years that a good comparison of Shapley's findings
then with opinions on these stars now is not really valid. For example,
18 W Ursae Majoris systems brighter than tenth magnitude were known
then; there are at least 43 now.

But Shapley was much more than a matter-of-fact scientist. I
found him to be one of the most fascinating individuals I have ever met.
He had an amazing breadth of knowledge and quickness of response.
Probably his earlier newspaper work had given him clever manipulation
of the English language, and his sense of humor was delightful.
Through all these years he had the help of another brilliant individual,
his wife, Martha Betz Shapley who helped in many ways, with his research
as well as heavy social duties, while she brought up her family of five
children. The Observatory parties at the Shapley home were greatly
enjoyed and widely known, perhaps culminating in the rehearsals and
performance of Pinafore in 1930, in which I myself was pretty much in
the part of producer because of my previous theatrical experience. Dr.
Shapley had shown me the old manuscript of the Observatory Pinafore
when it was found after being lost for many years, and he asked me if I
thought we could put it on - and we did!

Dr. Shapley put considerable effort into standing behind
various astronomical activities associated with the Observatory such as
the Bond Club, the AAVSO (American Association of Variable Star
Observers), and of course the weekly colloquia. In my first one of
these I heard Miss Payne speak as I recall, "On the Lifetime of an
Excited Hydrogen Atom". Atomic physics had been under way for only a
few years, and it embarrasses me still that I thought she was trying to
inject humor into her title for a technical subject of which I had
considerable ignorance.

Dr. Shapley was active in drawing distinguished visitors to
the Observatory. Willem Luyten had already joined the staff from the
Netherlands when I arrived, and Boris Gerasimovic came from Pulkova,
(and tragically disappeared after his return to Russia), and Ejnar
Hertzsprung from Leiden.

Shapley also strongly supported the American Astronomical
Society and it was at its winter meeting at Yale in 1927 that I became
a member of it and also was drawn into the branch of globular clusters
which I still pursue. At that meeting Dr. Jan Schilt of Columbia spoke
up with determination to say that he was tired of people talking about
the period-luminosity curve for Cepheids in globular clusters because

there was too little material to be meaningful. This was, of course, intended as a direct hit at Shapley. After I returned to Harvard from the meeting, I began looking up all the relevant references, and rather reluctantly concluded that I agreed with Schilt. So I started the pursuit to find more long period Cepheids in globular clusters, and to do this properly I needed to make a card catalogue of all available data to date. And 59 years later I am still keeping this catalogue up to date as I work toward the publication, with my colleague Christine Clement, of the <u>Fourth Catalogue of Variable Stars in Globular Clusters</u>.

In my final months at HCO in the spring of 1930 the epochal paper by Robert J. Trumpler proving the absorption of light in space from measures of the diameters of open clusters appeared in Lick Observatory Bulletin vol. 14, 1930. This was a sharp rebuff to Shapley's ideas, and he felt it deeply. For more than a decade he had refused to give in to the idea of such absorption. This was mainly because of the blue stars in the most distant globular cluster then known, NGC 7006. How could light be absorbed if it were still blue at the end of such a great distance? He asked me to review Trumpler's paper for a colloquium, knowing, I think that my empathy for the situation would lead me to deal with it as unabrasively as possible.

In September, 1930 Frank Hogg and I were married, and after an academic year I spent at Mount Holyoke, where, besides teaching full time I finished my doctoral thesis, and Frank spent at Amherst in research, we left Massachusetts in the summer of 1931 for the Dominion Astrophysical Observatory at Victoria, British Columbia. Dr. Shapley reminded me that the mails were still running, and we kept in touch for the rest of his life. In 1935 Frank and I moved to the University of Toronto where Frank was the Director of the David Dunlap Observatory and Chairman of the Department of Astronomy at the time of his death on January 1, 1951. Dr. Shapley made several visits there, speaking and capturing his hearers with his charm, honored by the highest circles of the city and university, and given an honorary degree from the U of T. When he came, my husband and I, junior astronomers at the time, traveled with the elite, such as Sir Robert and Lady Falconer (he was Past President of U of T) and Mrs. Dunlap who gave the Observatory to the University. In a letter I wrote to my family I give a long description of his visit to speak to the Royal Canadian Institute on November 16, 1935.

Later in 1939 I was on the special train with him for the opening of the McDonald Observatory in Texas. We visited the Shapley's in Dublin, N.H. a number of times, as my Massachusetts home was only 40 or so miles away.

And in 1952, Dr. Shapley invited me to teach elementary astronomy at Harvard Summer School, with David Heeschen (later the Director of National Radio Astronomy Observatory) as my first-rate helper. Once again I had many hours to enjoy Shapley's remarkable qualities.

Those of us who were there could never forget the Celebration of Dr. Shapley's 80th birthday November 9, 1965, at the American Academy of Arts and Sciences, then located in Brookline. Many people commented that Shapley often tried to do things in a flamboyant and spectacular way, and he certainly succeeded that night. For it was the night of the major electrical black out all over the eastern coast of the United States and into Canada. The dinner was for formal dress. I had just checked into the Commander Hotel and was about to get my formal attire out of my suitcase when presto - total darkness descended. It did not take me many minutes to realize this was serious, as there were no lights anywhere. Eventually some feeble ones came on in the hotel corridors. To make a long story short, the drive from Cambridge over to Brookline was horrendous, with no traffic lights. But for those who did arrive, the dinner was carried out beautifully because the caterer had a gas-fired truck, and hundreds of candles were already decorating the dining room tables. And Dr. Shapley was in excellent form.

In recognition of Shapley's 80th birthday, Peter van de Kamp arranged for a series of five papers to appear in the Publications of the Astronomical Society of the Pacific, vol. 77, 1965. Since this is now two decades in the past, these papers may not be familiar to all who are attending this symposium.

Dr. van de Kamp wrote on "The Galactocentric Evolution, a Reminiscent Narrative". His opening paragraph assessed Shapley's contribution in vivid terms.

"The galactocentric viewpoint developed by Harlow Shapley in 1917 represents a step forward in astronomical thinking and perception analogous to the introduction of the heliocentric viewpoint by Nicolaus Copernicus in 1543. The galactocentric viewpoint is now completely accepted and so taken for granted that it may be difficult for the younger astronomers to comprehend that it involved the overcoming of established systems and prejudices, and that this giant step forward was due not only to a substantial amount of observational work but above all to the insight, intuition, and courage of one individual, Harlow Shapley."

Helen Sawyer Hogg in "Harlow Shapley and Globular Clusters" outlined his most important papers on globular clusters and called attention especially to Shapley's work on Cepheid variables. "It was Shapley who named the curve 'Luminosity-period curve of Cepheid variables' in his paper 'On the Determination of the Distances of Globular Clusters'."

Carl Schalen wrote on "On Some Problems in Interstellar Absorption" and discussed the reasons Shapley obtained negative results in his studies of this.

"When Shapley started his pioneer investigations on the distributions in space of the globular clusters there were, according to results of J.C. Kapteyn, P.J. van Rhijn and others some indications of a possible small increase with distance of the color indices of stars, i.e. of a selective absorption in space. If this were real, stars in very distant objects like the globular clusters ought to be considerably reddened by the interstellar absorption. When studying the color indices of stars in Messier 13 (in 1915) Shapley unexpectedly discovered a large number of blue and white stars: out of 495 stars with well-determined color indices, 86 were found to be of color class b, and 63 of class a; and he did not find any exceptionally red stars. Shapley concluded that in the direction of Messier 13 the interstellar absorption must be extremely small. But in Shapley's words, 'That light scattering is absent in this direction is no proof that it does not occur elsewhere especially in low galactic latitudes where stars and diffuse nebulosity are concentrated' . . . In 1915-17, when Shapley made his famous studies of these clusters nothing was known about the distribution of interstellar matter in the Galaxy. R.J. Trumpler's investigation in 1930 and subsequent papers by P. van de Kamp and others showed that interstellar matter is concentrated in a relatively thin layer near the galactic plane. Therefore, even very distant objects in high galactic latitudes show a relatively small amount of absorption as the light path through the interstellar medium is short".

Bart J. Bok described "Shapley's Researches on the Magellanic Clouds". His article opened with the dramatic sentence "For thirty years, from 1922 to 1952, Harlow Shapley was 'Mr. Magellanic Clouds'." And declares that "Harlow Shapley was first and foremost the cosmographer of the Magellanic Clouds".

Bok also mentioned Shapley's importance in fostering Miss Cannon's classification work.

"Shapley was instrumental in promoting the work by Miss Cannon and Mrs. Mayall in the field of spectral classification. It was realized soon after the publication of the Henry Draper Catalogue that the overall spectral survey did not really reach as faint as it should in the region of the Magellanic Clouds. Three publications in the Henry Draper Extension Series contain very valuable basic material for spectral studies of the fainter stars in the Clouds".

Actually as an example of the great increase in material which the H.D. Extension provided, S.I. Bailey cited (op. cit. p.159) a field of about 80 square degrees in the star cloud in Cygnus where one plate had the spectra of 4490 stars, of which only 498 were in the Henry Draper Catalogue, and 3992 had been added by the Extension.

The final article by Hudson Hoagland, a biologist, "Harlow Shapley - Some Recollections" provided some personal glimpses, a summary of his life, and a fine description of his famous desk.

"Harlow Shapley has honorary degrees from seventeen institutions and he has been made a foreign honorary member of national academies of ten foreign countries. He has been awarded medals and prizes by the Vatican, India, Mexico, England, France, et al. The diversity of his interests is reflected by the fact that he is a trustee of the Massachusetts Institute of Technology, the Worcester Foundation for Experimental Biology, the Woods Hole Oceanographic Institution, and Science Service. He has been the president of eight national scientific organizations. These include the American Academy of Arts and Sciences, Science Clubs of America, the Society of the Sigma Xi, American Astronomical Society, and the American Association for the Advancement of Science.

"His famous desk at the Harvard Observatory was symbolic of his way of life. It was a desk in the form of a great wheel mounted on a vertical axle - a kind of rotating galaxy for ideas. Near the hub of the wheel were radially arranged compartments, cubby holes, and drawers. The disk of the wheel extended beyond the radius of these containers gave sample writing space for any position of the wheel. Thus sitting in one place, by turning the wheel, Shapley could bring before him any one of his divergent fields of interest. This marvelous desk thus allowed him, from his chair, to marshall the contents of half a dozen desk and files on as many topics by merely a twist of the wrist".

This morning I have tried to give you a view of Shapley's era from my personal association with him, which extended over four and a half decades. It is for the third speaker, Owen Gingerich, to sweep with a wider brush in describing "Shapley's Impact". But I would not like to end my own talk without some other opinion. The late Bart Bok, in obituaries of Harlow Shapley in the November, 1972 issue of Sky and Telescope and the Quarterly Journal of the Royal Astronomical Society, vol. 15, 1974 has given splendid summaries of Shapley's life and works. In the latter publication he writes "We can feel proud that our generation of astronomers produced a man who did for our Milky Way system what Copernicus achieved for our Solar System". And "Shapley was a great humanitarian, whose love for people and whose desire to assist them knew no bounds."

And now I will show you a few slides from my own old photo albums to give you a glimpse into Shapley's era as I knew it. Dr. Karl Kamper of the University of Toronto has expertly transformed some of the brownish photographs into fresh looking, but still quaint pictures.

The more significant of the slides appear as figures in this paper. Attention was drawn to Cecilia Payne, Emma T.R. Williams, and Margaret Walton (Mayall), Figure 4; to Frank Hogg, Helen Sawyer, (Hogg), Henrietta Swope and Adelaide Ames. (Figure 6.) Adelaide Ames whose name is well known for the Shapley-Ames catalogue of galaxies died tragically by drowning in 1932, but Henrietta Swope continued for half a century her researches on variable stars. After leaving Harvard she was on the staff of the Hale Observatories and in addition was a major benefactor of the Carnegie Southern Observatory at Las Campanas, Chile. Dr. Dorrit Hoffleit kindly provided material for one slide.

Some of my slides capture the gay spirit of the 1920's, before the great depression, before World War II, and before anyone had used the words "atomic bomb".

REFERENCES

Bailey, S. I. 1932 Harvard Observatory Monographs No. 4.
Gingerich, O. 1985 Harvard Magazine, Nov. Dec.
Shapley, H. 1969 Through Rugged Ways to the Stars, Charles
    Scribner's Son's, New York.
Trumpler, R.J. 1930 Lick Obs. Bull. 14, 154.

SHAPLEY'S IMPACT

Owen Gingerich

Harvard-Smithsonian Center for Astrophysics

ABSTRACT: Harlow Shapley's legacy can be divided into three aspects: his scientific contributions, the institutions he built, and his multi-faceted efforts to publicize astronomy. Today's public funding of science undoubtedly owes much to Shapley's enthusiasm for astronomy.

It was almost four decades ago that I came to work as a summer assistant for Harlow Shapley, an event destined to turn me toward professional astronomy. Thus perhaps even I can be considered a minor example of Shapley's impact. However, this doesn't make it much easier for me to summarize Shapley's influence in a few pages. On the other hand, the fact that I helped him prepare his memoirs, *Through Rugged Ways to the Stars*, does help. At the time I suspected that Shapley subscribed to the old adage that one must not let truth stand in the way of a good story, and so I took some trouble to check out those tales that loomed bigger than life in his account. To my surprise, I could verify one story after another. Yes, he really was the man who almost single-handedly added the S into UNESCO. Yes, he really did find academic places for numerous refugees from Nazi and Fascist tyranny in Europe. And yes, he did pickle in vodka an ant collected from Joseph Stalin's banquet table.

To probe into Harlow Shapley's legacy, I think we must briefly explore three aspects: his scientific contributions, the institutions he built, and his astonishingly wide reputation as a spokesman for astronomy. These areas blend seamlessly into each other, but let me try to separate them. I shall begin with a few remarks on his scientific achievements.

The period-luminosity relation was neither Shapley's discovery, nor was he the first to apply it. Yet, by the dramatic results he achieved with it, he effectively made the period-luminosity relation his own. Thus, when Edwin Hubble found the light curves for several M31 Cepheids, it was to Shapley at Harvard that he turned for the most up-to-date calibration of the relation. On February 27, 1924, Shapley responded to Hubble saying:

> *Your letter telling of the crop of novae and of the two variable stars in the direction of the Andromeda nebula is the most entertaining piece of literature I have seen for a long time.*

Note that Shapley didn't say "*in* the Andromeda nebula" but "in *the direction of* the Andromeda nebula." Shapley must have realized that his debating position of just a few years earlier was crumbling, and he made one final parry of resistance. Even though Shapley had in 1917 found one of the novae in M31 and had suggested a distance of a million light years, he had hastily back-pedaled after devising his new model of our galaxy. Without interstellar absorption our Milky Way seemed so enormous compared to the spirals that it appeared to be a cosmic unit of altogether different proportions. In the 1920s Shapley thought of the Kapteyn universe as one cloud of an enormous flattened assemblage, a supergalaxy as he called it. The familiar cross section with the globular clusters in their halo-like array was *not* the model Shapley had in mind then—the familiar old picture is from J. S. Plaskett's Halley Lecture of 1935. Even after Trumpler's discovery of interstellar absorption in 1930, the anomaly persisted; and I shall return to this question at the end of my paper.

Meanwhile, let me remark on Shapley's own scientific work at Harvard. Here he was hampered both by weather and by the lack of the big telescopes that he had used so effectively at Mount Wilson. Hence he followed in Pickering's footsteps as an astronomical administrator, organizing large scale surveys—of spectral classes, of variable stars, of stellar magnitudes; for himself, he chose galaxies. The Shapley-Ames catalog of bright galaxies was one result. Another was his forceful demonstration of the inhomogeneity of galaxy distributions, an area in which the CfA is once again in the forefront.

A result of these investigations, in 1937, was the discovery of the Sculptor and Fornax dwarf galaxies. One of the women who counted galaxies for him on these large A plates from the South African station noticed a peculiar peppering of fine stars, almost like a thumbprint, and called Shapley's attention to it. He soon recognized that it was a new class of objects, unlike any galaxies that had been seen before. In such a case we might well ask who really discovered the Sculptor system: Was it the person who master-minded the investigation, who raised the funds and set the survey into motion, or Sylvia Mussells, who went beyond her immediate instructions of marking galaxies and who noticed the smudge? Was it the person who found the smudge or the one who figured out its significance? These are questions that, in their many guises, historians of science must struggle with, but in general the historical interest lies not with the person who first saw an object or who first grasped a theoretical idea, but with the one who recognized and exploited its significance.

In our era of big science, such questions have increasing poignancy when one plants and another reaps. Harlow Shapley lived in a transitional age, already inherited from Pickering and even from George Ellery Hale, when an effective scientist-administrator could marshall forces beyond a single person's capacity, bringing not only to himself but to his colleagues or assistants resources that they might not otherwise have. Such was to a large extent Shapley's role at

Harvard, and if he himself made no discoveries as grand at Harvard as he had in those heady days at Mt. Wilson, we should hesitate to judge that it was a mistake for him to leave the clear skies of southern California. To very few is given the chance to change our conception of sidereal structure as much as Shapley once did, and to have asked him to do it again at Harvard is demanding rather much. And we must ask as well, did he not play a key role here as a catalyst for his colleague's discoveries? Surely the answer is yes! So this brings me to the second aspect of Shapley's legacy, his role as builder of institutions.

In four decades as director, Edward Pickering had turned the Harvard College Observatory into a world-class research establishment. But astronomy education at Harvard took place not on Observatory Hill, but with an entirely different staff in the astronomical laboratory, a frame building that stood just north of the Harvard Yard. Once in an undergraduate career there was an opportunity to see the real observatory, when seniors lined up two abreast to march up Garden Street to visit Pickering's domain. As for graduate students, there may have been some apprenticeships, but surely no formal degrees. Shapley's approach was obviously different, and not many years after his arrival he seized an opportunity to work toward an astronomy graduate program.

In May of 1922 Shapley had gone to the Rome IAU meeting, and later in the month he spoke briefly about the spectrographic work at HCO at the centenary meeting of the RAS in London. Among his fascinated listeners was Cecilia Payne, then a Cambridge undergraduate, who boldly told him after his lecture that "I should like to come and work under you." Shapley cheerfully agreed; but it undoubtedly surprised him when she eventually turned up in the American Cambridge.

After about a year Shapley suggested that Miss Payne turn in a thesis for a PhD at Radcliffe College. There was, however, no provision for advanced degrees in astronomy, and Theodore Lyman, chairman of the Harvard physics department, was adamantly opposed to a woman candidate in physics. How Shapley managed to have Payne's doctorate awarded in astronomy is not known, but this event in 1925 marked the birth of the graduate school in astronomy. Soon there were other graduate students, and doctorates including Helen Sawyer Hogg's. As the program expanded, Shapley recruited Harry H. Plaskett from Victoria to teach astrophysics, and on October 18, 1928 he included in a letter to Edwin Hubble the following typically light-hearted note:

> *Curiously enough I find that H H Plaskett has grave doubts about the large distances for extra-galactic nebulae. He is sufficiently serious about it that he and Miss Payne are to debate the subject at a colloquium two weeks from today. If he convinces me, I shall cable you.*

One of the educational innovations Shapley established, beginning in 1935, was the Harvard Summer School in Astronomy, which was deliberately designed to bring together physicists and astronomers. For this he used the famous and influential Michigan summer school in theoretical physics as his model. The Harvard program, the first of its kind in America, introduced graduate students such as Jesse Greenstein, Leo Goldberg, James Baker, and Lawrence Aller to scientists such as Jan Oort, Ira Bowen, Meghnad Saha, Bengt Edlen, H. P. Robertson, John Slater, Robert Marshak, and George Shortley.

Shapley was particularly proud that he has assembled such an international group at Harvard Observatory, and he delighted in showing a picture taken around 1940 with students or staff from 13 foreign countries. This sort of enthusiastic internationalism, which is so essential to the International Astronomical Union, was eventually to get him in trouble with the House Un-American Activities Committee and with Senator McCarthy, episodes that are also part of Shapley's impact.

I could mention also his role in reforming the American Academy of Arts and Sciences, or the long hours he spent reviewing fund proposals for the American Philosophical Society and for Sigma Xi. These and many more topics you can find in his autobiography, *Through Rugged Ways to the Stars*. If you knew Shapley, you can hear him talking in the book, a raconteur with a splendid combination of vanity and a sly facetiousness that enabled him to poke fun at himself. I always thought he carried a continual air of amazement that a farm boy from Missouri could become a celebrity and meet presidents and dictators and even the Pope.

Something of that quality carried over to the public platform, where Shapley eventually became an engaging speaker and also enough of a journalist to know what made good headlines. He frequently spoke of his own most brilliant achievement, from heliocentric to galactocentric, and I often heard him say that in the old days at Mount Wilson he had no idea of the *philosophical* implications of his research. In fact, that was patently false, as the following letter to George Ellery Hale, written on January 19, 1918, shows:

So the center has shifted: egocentric, lococentric, geocentric, heliocentric.

Now, getting nearer those serious things that the heartless corporation pays me to investigate, we know that for the last few decades some observing and unheeded astronomers have noted that the Milky Way is not a great circle and that it is brighter in some places than in others. But that is not necessarily our fault. We are not responsible for the imperfections of this Universe. We have indeed held on to that heliocentric center doggedly, in spite of increasing doubts when we troubled to think on the subject analytically. If the center got away from us, we feared that Man, the Ultimate Purpose of Creation, would loose [sic] his hold on all things. Instead of MAN and the universe it might become man and The Universe.

Harlow Shapley achieved widespread public fame by speaking articulately and imaginatively about the vastness of space, the peripherality of man, and the industry of ants. The Harvard Archives contains an astonishing number of press clippings, beginning in great quantities with his appointment to Harvard in 1921. Let me select a handful to illustrate the extent of his fame.

Here is a telegram signed "Clarence" from Dayton, Tennessee on July 10, 1925:

Distinguished colleagues of yours have suggested you might be willing to come to testify for defense at Dayton Tennessee next week in the case of State of Tennessee versus Professor Scopes STOP We of the defense would be delighted to add you[r] authority to our position STOP Your expenses will be paid STOP Will you wire me directly at Dayton and I will let you know what day you will be needed.

Shapley did not go to Dayton for the famous Scopes trial, but the nature of the invitation is an indicator of his public fame in the 1920s.

In the next decade, we find his picture of the cover of *Time* magazine, on July 29, 1935, in connection with an article on the Paris Congress of the IAU. (Shapley was the second astronomer to gain this distinction, the first having been Arthur Eddington on April 16, 1934.)

"Harvard's Harlow Shapley: From ants on his desk to the Great Nebula in Andromeda."

RANKIN IN ROW WITH SHAPLEY

Harvard Professor, Accused of Contempt, Charges 'Gestapo Tactics' by House Committee

DR. HARLOW SHAPLEY (right) studies with his attorney, Thomas H. Eliot, the report the Harvard astronomer was reading before House committee today when Chairman Rankin became aroused, allegedly snatching the paper from the witness' hands.

WASHINGTON, Nov. 14 (UP) — Representative John E. Rankin, Dem., of Missis-

Headline in a Boston tabloid, 1946.

In the mid-1940s Shapley's name loomed large in the press on account of his tift with Congressman Rankin and the House Un-American Activities Committee. Shapley's firmly held internationalism and his stubborn independence had attracted the attention of congressional witch-hunters, and so he had been called to Washington to testify. He was not allowed to have a lawyer present in the "Star Chamber" hearings before the congressman, but he recorded the session in the shorthand he had picked up as a fledgling reporter in Kansas decades earlier; Rankin left the hearings saying he had never had a witness treat the committee with such contempt. Banner headlines four centimeters high proclaimed "RANKIN IN ROW WITH SHAPLEY"—little other identification being needed for the local Boston public.

In the 1950s he not only tangled with Senator McCarthy, but also with Immanuel Velikovsky, the Princeton doctor and author of the best selling *Worlds in Collision*. Had Macmillan listed the book as science fiction rather than science, they might have avoided trouble from some of the astronomers. Shapley sent a few letters to the publisher about it, complaining that if *Worlds in Collision* were a science book, Macmillan should have had it refereed. Perhaps his harshest but most amusing comments were written to them on January 25, 1950:

> *If I remember correctly, several years ago Dr. Velikosky met me in a New York hotel. He sought my endorsement of his theory. I was astonished. I looked around to see if he had a keeper with him... I tried, rather futilely, to explain that if the earth could be stopped in such a short period of time, ... it would have made impossible that he and I could meet together in a building in New York City less than four thousand years after this tremendous planetary event.*
>
> *Dr. V seemed very sad. But somehow I felt he was feeling sorry for me and the thousands of other American physical scientists and geologists and historians who have been so, so wrong.*

Velikovsky's fans claimed that Shapley was instrumental in organizing a boycott against Macmillan, but if so, the Archives show no trace of it.

It was only a few years earlier, when the world was at war, that it came time to celebrate the 400th anniversary of the publication of Copernicus' *De Revolutionibus*. Poland was in bondage, but the American Polish community was determined not to let the anniversary pass without notice, so a grand affair was scheduled in New York. It was symbolic of Shapley's public impact that he was the man who read President Roosevelt's greeting to the distinguished guests, and it was to his observatory that a remarkable painting of Copernicus was given. It still hangs in the entrance to Building A, a magnificent gift to the man who took the step from heliocentric to galactocentric.

Today the scientific principle of mediocrity is often called the Copernican principle: we should not expect to be in the center of things, or in the most splendid galaxy. Even Shapley's galactocentric move might today be called a Copernican maneuver. Shapley was clearly troubled by the fact that the Milky Way seemed to be the biggest of them all, and I remember how, in a graduate cosmogony course that he helped to teach in the spring of 1952, he assigned to Frank Orrall the topic of why the globular clusters in M31 were only half the size of those in our own galaxy. Neither Shapley nor any of the students could come up with an answer. I mention this to show what a viselike grip a wrong assumption can have when it seems to fit so coherently into the rest of the picture.

As you all know, only a few months later at the Rome IAU meeting Walter Baade announced that two different types of Cepheids had been mixed together in Shapley's Mount Wilson work, and that when these were sorted out, M31 was twice as far as had been supposed, the anomalous globular clusters fell into place, and the Milky Way was no longer the unqualified king of the galaxies. Only in retrospect was it noticed by the overwhelming majority of astronomers that the Copernican principle of mediocrity might have pointed the way.

Baade had the advantage of the world's largest telescopes at his disposal, and he was also one of the pre-eminent observers of our century. Would Shapley have made the discovery if he had stayed on at Mount Wilson? Trying to rewrite history as it might have been is a rather fruitless exercise. Did Shapley lose out in his science by abandoning the giant telescopes of the West? Probably not—as I have noted, few have the chance to make even one discovery as grand—and few have as fine a chance to build astronomy through education, activism, and public appeal. I suspect we all owe much in the public funding of science—foreign as that was to Shapley's instincts—because of this one man's multi-faceted efforts through the press, through the pioneering Harvard radio talks on astronomy (beginning in the 1920s), through the Harvard books on astronomy, and through his own ubiquitious appearances on the lecture circuit to arouse in an interested public a curiosity and fascination with topics astronomical. Thus it is entirely appropriate that the American Astronomical Society has instituted the Shapley lectureships to help in the public understanding of astronomy. I think it was inappropriate for the International Astronomical Union to refuse to recognize the Shapley centennial when it fell directly on the dates of the Congress in India last November, but fortunately we are now having the opportunity to commemorate this remarkable man under IAU as well as Harvard auspices with a topic that was always dear to his heart, globular clusters.

ACKNOWLEDGMENTS

I should like to thank the Harvard University Archives for permission to quote from letters in the Shapley collection, and the Shapley family for making many of the materials available to me and to the Archives. Michael Hoskin and Richard Berendzen originally showed me important pieces of the astronomical correspondence related to Shapley's career, which I deeply appreciate.

DISCUSSION

TRIMBLE: Could we ask for a show of hands of those persons who knew, or met, Shapley personally.

EDITOR'S NOTE: About a third of those present, heavily concentrated towards the front of the room, raised their hands.

HANES: You commented on Shapley's bemusement in early 1952 over the discrepant sizes of globular clusters in M 31 and the Galaxy, subsequently resolved when Baade announced the recognition of two separate kinds of Cepheids. However, de Vaucouleurs has pointed out that there was some work in the years leading up to 1952 that independently suggested the need for revision in the extragalactic distance scale. Did Shapley have any intimations of this?

GINGERICH: I am glad that you asked this question, because I specifically researched part of this point in preparing my paper. After Baade's announcement at the Rome IAU, Henri Mineur in Paris claimed that nearly a decade earlier (in 1944) he had shown the difference in zero points for the RR Lyrae stars and the Cepheids. However, he apparently had no inkling that two different P-L relations were involved, and he did not understand or believe this result enough to make a simple prediction such as: you ought to be looking for RR Lyrae stars about a magnitude and a half fainter in the Magellanic Clouds, which is precisely what Shapley did after he heard Baade's announcement.

Meanwhile Knut Lundmark had in 1948 argued that by using the novae and globular clusters as distance indicators, the distance to M 31 came out to 1,700,000 light years. He admitted that a new proper motion analysis of 100 Cepheids gave the same zero point calibration that Shapley had got, so this must have been the dilemma that Shapley was worrying about when he tried to figure out why the M 31 globulars had the "wrong" luminosity.

It seems to me that in both these cases there is a parallel with the situation so astutely analyzed by Otto Struve concerning whether Bessel or his great-grandfather deserved the credit for finding the first stellar parallax. Struve concluded that astronomers could know that his illustrious ancestor had the right answer first (in 1837) only <u>after</u> Bessel had done his work in 1838.

I think that neither Shapley, who followed the literature fastidiously, nor anyone else could really appreciate what Mineur's paper had hidden in it until the dramatic and convincing demonstration by Baade that there were two different variable star populations. Similarly, it was hard to appreciate that Lundmark was on the right track until the riddle of the Cepheids had been cracked.

OSTRIKER: We learned from Helen Sawyer Hogg that Cecilia Payne was the first Ph.D in Astronomy at Harvard. What other U. S. institutions were there awarding Ph.D degrees at that time or earlier in Astronomy?

GINGERICH: Before World War II only six graduate institutions in America granted doctorates in Astronomy: Yale, Michigan, Berkeley, Chicago, Princeton and Harvard. Quite possibly Harvard was the last to join this list.

DEMARQUE: The first Ph.D in Astronomy was awarded at Yale in 1877.

LILLER: You mentioned that you listened to a course from Shapley in 1952. Isn't it true that Shapley never taught a course while he was director?

GINGERICH: Shapley rarely gave a formal course, and this one was actually a team-taught course on stellar evolution with Bok, Payne-Gaposchkin, and Whipple sharing the lectures with Shapley. He did, however, offer an "Introduction to Cosmogony" course in the 1949 summer school, and one of the things I did that summer was to make lantern slides for the course. One evening he came by my office and told me that he had once tried to make his own lantern slides, for the first AAS meeting he had attended, at Pittsburgh in 1912. He said that after he got to the meeting and realized how much more elegant everyone else's slides were, he spent all his lunch money to have his slides remade professionally!

ZINNECKER: Was Shapley the first to discover dwarf spheroidal galaxies, and did he compare their loose structure to the much more compact structure of globular clusters?

GINGERICH: Certainly yes as far as our local family is concerned. At the time there wasn't anything else like the Sculptor and Fornax systems.

RICHER: Could you detail the reasons why Shapley was brought in front of the McCarthy hearings?

GINGERICH: Dr. Shapley's oldest son, Willis, is here today and can answer that better than I.

SHAPLEY, W.: Rankin was a Mississippi Congressman who had become Chairman of the House Un-American Activities Committee in the years immediately after World War II. This Committee was notorious in its indiscriminate labelling of academic and other liberals as "communists" and in its unfairness to the rights of those accused. Harlow Shapley had been outspoken in support of liberal causes of the period, including civil rights and opposition to the "Cold War" policies of the Government, which he viewed as provocative of a military confrontation between the US and the USSR. Like many others he was publicly branded a "communist" by the Committee and then subjected to questioning in secret session. His resistance to this process was also highly publicized at the time and was instrumental in bringing about rules changes giving the witnesses the right to have counsel present at all hearings of the Committee.

LILLER: Was it true that Shapley was asked to run for Governor of Massachusetts in 1948 by the leftist Henry Wallace?

SHAPLEY, W.: In 1948 Harlow Shapley had joined former Vice President Henry Wallace and others to form the liberal-oriented Progressive Party to support Wallace for President. The Massachusetts organizers of the Progressive Party asked him to be their candidate for governor, but he declined. He broke with the Progressive Party after their convention at which the Communist element within the party demonstrated their control of it by voting down a platform plank that would have condemned Communist as well as Fascist dictatorships.

McCARTHY: Shapley was writing at the end of his active research work in the years just before and just after his retirement at Harvard (1952). He saw how important red giants and supergiants were for the study of structure and evolution. It was a time of great changes in astronomy (The P-L relation zero point was revised by Baade; new telescopes were coming into use in Southern Hemisphere). Limitations for Shapley were the lack of spectral types for faint(!) 14th-18th mag stars and the lack of radial velocities to discriminate members from field stars. He used "what he had" and looked forward to larger telescopes and better detectors to study evolution effects and population differences in the clouds.

HANES: What happened to the famous rotating desk of Harlow Shapley? It seems like a marvelous idea for efficient work?

GINGERICH: Shapely's rotating desk was originally acquired by Pickering from the Gift of 1905. Now Dr. Shapely's son Alan has it in Boulder, but it will eventually be returned to the Observatory and set up in a museum room for that period.

Chapter II

Review Papers

Globular Clusters in the Milky Way

Bob McClure running his session

Ron Webbink answering questions after his talk

AN OVERVIEW OF THE GLOBULAR CLUSTER SYSTEM OF THE GALAXY

Robert Zinn

Yale University Observatory

1. INTRODUCTION

Harlow Shapley (1918) used the positions of globular clusters in space to determine the dimensions of our Galaxy. His conclusion that the Sun does not lie near the center of the Galaxy is widely recognized as one of the most important astronomical discoveries of this century. Nearly as important, but much less publicized, was his realization that, unlike stars, open clusters, HII regions and planetary nebulae, globular clusters are not concentrated near the plane of the Milky Way. His data showed that the globular clusters are distributed over very large distances from the galactic plane and the galactic center. Ever since this discovery that the Galaxy has a vast halo containing globular clusters, it has been clear that these clusters are key objects for probing the evolution of the Galaxy. Later work, which showed that globular clusters are very old and, on average, very metal poor, underscored their importance. In the spirit of this research, which started with Shapley's, this review discusses the characteristics of the globular cluster system that have the most bearing on the evolution of the Galaxy.

2. SUBSYSTEMS

Roughly 150 globular clusters have been identified in the Galaxy, and the majority of them are distributed in a roughly spherical volume around the galactic center. It is very important to ascertain whether all globular clusters belong to this halo population or whether the cluster system contains subsystems with distinct properties that reflect different origins.

Starting with the suggestion by Baade (1958) that the metal-rich globular clusters belong to a disk system, there have been a number of investigations of the correlations between metal abundance, spatial distribution, and kinematics to see whether the clusters divide into distinct groups. It was clear from the first spectroscopic surveys of the cluster system (Mayall 1946; Morgan 1956, 1959; Kinman 1959) that the metal-rich clusters are more concentrated towards the galactic center than the metal-poor ones, but until recently the data on the metallicities, distances, and radial velocities of the clusters near the galactic center were inadequate to tell whether the metal-rich clusters are a separate population or simply the metal-rich tail of the halo population. Now there is substantial evidence that the metal-rich clusters constitute a distinct disk system (Zinn 1985), and some of this evidence is reviewed here (see also Zinn 1986).

The distribution of the distances of the globular clusters from the galactic plane shows a sharp discontinuity near [Fe/H] = -1 (on the metallicity scale of Zinn and West 1984 which is used throughout this paper). The clusters more metal poor than [Fe/H] = -1 are spread throughout the halo to distances approaching 20 kpc from the galactic plane (the much more distant clusters, e.g., Pal 3 and 4, are discussed later), whereas the more metal rich ones are confined to within 3.2 kpc of the plane. This concentration of the metal-rich clusters near the galactic plane is not simply a consequence of their concentration toward the galactic center, for there are several metal-rich clusters, e.g., M 71 and NGC 5927, that lie at substantial distances from the galactic center and yet close to the galactic plane. Using Frenk and White's (1982) method of analyzing the positions of the clusters on the sky, Zinn (1985) has shown that the metal-rich and metal-poor clusters near the galactic center as well as those at angular distances greater than 20 degrees from the center have different flattenings toward the galactic plane and that these differences are statistically significant (>95% confidence). This analysis suggests that the dividing line between the flattened and essentially unflattened groups occurs at [Fe/H] = -0.8, and hence, that roughly 25% of the cluster sample belongs to a disk system, while the remaining 75% belongs to the halo.

The kinematical properties of the clusters also change near [Fe/H] = -0.8 (Zinn 1985), as expected if the flattening of the metal-rich group is real. The clusters more metal poor than -0.8 are rotating slowly as a group around the galactic center (V_{rot} = 50 ±22 km/s) and have a large line-of-sight velocity dispersion (σ_{los} = 113 ±9 km/s). In contrast, the metal-rich clusters have the large rotation (V_{rot} = 153 ±29 km/s) and small dispersion (σ_{los} = 71 ±11 km/s) of a disk system. The more recent analysis by Hesser, Shawl, and Meyer (1986) of slightly different samples of clusters also shows that the metal-rich and metal-poor clusters have distinctly different kinematics. Cudworth (1984) has measured the proper motion of the metal-rich cluster M 71, and as expected from these analyses, its space velocity is typical of the old disk stellar population.

Additional evidence for two populations of clusters is provided by the distribution of the clusters over [Fe/H]. This distribution is bimodal (Harris and Canterna 1979), and the valley between its two peaks occurs near the dividing line in metallicity between the halo and disk subsystems (Freeman and Norris 1981; Zinn 1985).

There may be two populations of metal-poor clusters. Harris (1976) and Zinn (1985) have noted that the decline in the number density of globular clusters (ϕ) with increasing galactocentric distance (R) is erratic beyond R = 20 kpc. From 3 to 20 kpc, ϕ decreases as $R^{-3.5}$. It has a steeper decrease between 20 and 35 kpc, and between 35 and 60 kpc, there are no known clusters. There are, however, seven clusters in the range 60 to 110 kpc (Pal 3, Pal 4, Pal 14, Pal 15, AM-1, Eridanus, NGC 2419). This number is much larger than expected if ϕ declines smoothly beyond 20 kpc. With the exception of NGC 2419, this group consists of very low luminosity clusters that are hard to see on sky survey plates. It is conceivable, therefore, that additional clusters of this kind remain to be discovered, which would make the behavior of even more peculiar. One interpretation of these data is that the galactic halo ends at R $\approx$ 40 kpc and the more distant clusters constitute a separate population of objects (Harris 1976). The recent work of Saha (1985) on the density of RR Lyrae variables in the halo appears to be consistent with this hypothesis, for he has found that their number density also falls off more rapidly between 20 and 33 kpc (the limit of his survey) than at smaller distances. Obviously, it is important to extend as far as possible this survey and others of different samples of halo stars.

3. METALLICITY GRADIENTS

If the whole sample of globular clusters in the Galaxy is considered, then there is a steep decline in the mean value of [Fe/H] with increasing R because the metal-rich globular clusters are concentrated near the galactic center. But, as was pointed out above, there exist distinct halo and disk subsystems of clusters that undoubtedly evolved in different ways. If these systems are considered separately, then it is much less clear that gradients in metal abundance exist in either one.

3.1 The Halo Gradient

The question of the size of the gradient in the halo population of globular clusters has had a controversial history. Searle and Zinn (1978) concluded that there is no sign of a metallicity gradient in the halo beyond R = 8 kpc. Harris and Canterna (1979) and more recently Pilachowski (1984) have contested this conclusion, for they found evidence for a gradient running from the galactic center out to 100 kpc. The large gradient found by these authors is a consequence of their samples containing the metal-rich disk clusters and the poor quality of the [Fe/H] measurements then available for the clusters more distant than R = 60 kpc. This gradient nearly disappears once the disk clusters are removed from the sample of halo clusters and the most recent measurements of [Fe/H] are adopted for the very distant clusters (see Zinn 1985 and 1986).

There is nonetheless substantial evidence for a small gradient in the halo. The clusters in the zones $0 < R < 7$ kpc and $7 < R < 40$ kpc have significantly different distributions over [Fe/H] (Zinn 1985), even though in these zones the mean values of [Fe/H] are not very different (-1.50 ±0.05 and -1.69 ±0.08 respectively). In the $7 < R < 40$ kpc zone, there is no evidence for a gradient, although with a sample of only 36 clusters, which includes nearly all the known clusters in this zone, it is impossible to rule out a very small gradient. The clusters more distant than 60 kpc have a distribution over [Fe/H] that is essentially the same as the clusters in the 7<R<40 kpc zone, which may be evidence that they do not constitute a separate population after all. At every value of R in the halo population, there is a wide range of [Fe/H] which has a standard deviation of approximately 0.3. This large dispersion and the small size of the gradient suggest that the collapse of the halo was chaotic and did not proceed with a smooth buildup of metals as its radius shrank.

The metallicity gradient exhibited by the RR Lyrae variables in galactic fields is in good agreement with the globular cluster observations. The observations of RR Lyrae variables in the $7 < R < 40$ kpc zone (Butler, Kinman, and Kraft 1979; Butler, Kemper, Kraft and Suntzeff 1982; Saha 1985) have not revealed any evidence for a gradient. There is a slightly steeper gradient between the variables in the $0 < R < 7$ and $7 < R < 40$ kpc zones (see Kinman, Kraft, Friel and Suntzeff 1985) than between the globular clusters. Zinn (1986) has argued that this may not be evidence for a difference between the cluster and field populations, but a consequence of the influence that the second parameter effect has on the [Fe/H] distribution of the RR Lyrae variables.

3.2 The Disk Gradient

In his description of the disk system, Zinn (1985) presented some meager evidence for metallicity gradients with distance from the galactic plane and with R. More recently, Armandroff and Zinn (1986) have measured the infrared Ca II triplet in the spectra of many of the most metal-rich clusters. While the agreement with previous estimates of metallicity by other means is generally very good, the agreement with the values that Malkan (1982) obtained from infrared photometry is poor, and several of the clusters that he suggested were very metal rich turn out not to be so extreme. When these revisions are made to the data considered by Zinn (1985), the evidence for metallicity gradients disappears. This does not mean that gradients do not exist. The distance moduli of many of the metal-rich clusters have not been measured or are uncertain, and until these deficiencies are corrected, one cannot be certain that distance errors have not smeared out the gradients.

4. HORIZONTAL-BRANCH MORPHOLOGY

Theoretical calculations have shown that the morphology of the horizontal branch (HB) is sensitive to a large number of parameters, including cluster age, helium abundance, metallicity (primarily the abundances of C, N, and O), and the rate of core rotation (affects the size of the He burning core and the amount of mass lost on the giant branch). If all of these parameters save metallicity were fixed or were monotonic functions of metallicity, then the HB's of globular clusters would be well behaved. When going from metal-poor to metal-rich clusters, one would see the bulk of the HB stars shift from the blue side of the instability strip to the red side. Instead one sees a large scatter in HB morphology at each metallicity, which indicates that something besides metallicity is varying from cluster to cluster. This additional parameter is called the second parameter, and it is a very real possibility that this effect is actually caused by more than one of the candidate second parameters mentioned above.

What links the second parameter effect to the evolution of the Galaxy is the fact that the size of the effect varies with galactocentric distance (see Zinn 1986 for a review). The effect is largest among the very distant clusters (60 < R < 110 kpc). Of the seven clusters in this zone, which are all metal poor ([Fe/H] < -1), the HB's of 5 are very red (i.e., they resemble that of 47 Tuc, a much more metal rich cluster) while the remaining 2 have the blue HB's expected of metal-poor clusters. The variation in HB morphology among the clusters in the 7 < R < 40 kpc zone is smaller, but still appreciable at every metallicity. In the 0 < R < 7 kpc zone, there is little evidence that the second parameter effect exists at all, for HB morphology varies smoothly with [Fe/H].

The origin of this variation with R has not been identified. Although Peterson's (1985) discovery of a correlation between the frequency of rotation among the HB stars in a cluster and its HB morphology suggests that core rotation is a second parameter, it is not clear why the amount of core rotation should vary from cluster to cluster in the outer halo and at the same time should be nearly uniform among the clusters of the same metal abundance in the inner halo. Core rotation is nonetheless a leading candidate for the origin of other peculiarities of HB morphology (see, for example, Buonanno, Corsi and Fusi Pecci 1985). Searle and Zinn (1978) have argued that differences in age can explain in a straight forward way the dependence of the second parameter effect on R. In their picture, the inner halo evolved rapidly, while the outer halo formed over a ≈ 4 billion year period (see also Zinn 1980).

Unfortunately, the differences in age predicted on the basis of HB morphology are smaller than the precisions with which clusters can be dated. To illustrate this, let us consider the clusters NGC 288 and 362, which have very nearly the same metallicities ([Fe/H] = -1.40 and -1.27, respectively, Zinn and West 1984) and yet have very different HB's (B/(B+R) = 1.00 and 0.05, respectively, Zinn 1980). Buonanno's (1986) compilation of measurements of the difference in bolometric magnitude between the HB and the main-sequence turn off, ΔM_{bol}(TO-RR), lists values of 3.50 and 3.31, respectively, for 288 and 362. If one or another of these clusters has an age near 15 billion years, as commonly found for globular clusters (e.g., Buonanno 1986), then the difference in ΔM_{bol}(TO-RR) between them suggests that they differ in a age by about 2.4 billion years, in the sense that 288 is the older cluster (assuming they have the same abundance of He). Theoretical calculations of synthetic HB's indicate that this difference in age can account for the observed difference in HB morphology (see fig. 1 in Rood and Seitzer 1982). However, this cannot be considered proof that the second parameter is age, for the uncertainty in each of the values of ΔM_{bol}(TO-RR) is ± 0.15 (Buonanno 1986); hence, there is no reason to believe that their turn offs actually differ in luminosity. If measurements of the same precision could be obtained for the red HB clusters in the 60 < R < 110 kpc zone, then there is some hope that the age hypothesis could be tested. On the basis of their HB morphologies, these clusters are expected to be ≥ 4 billion years younger than blue HB clusters of the same metallicity. Their turn offs are, however, too faint to be measured at present with the required precision.

5. AGES AND THE AGE-METALLICITY RELATION

Despite the progress produced by the advent of CCD detectors and the calculations of new theoretical isochrones, the dating of globular clusters has reached a plateau and very desirable quantities, such as the age-metallicity relation, remain beyond our grasp. To describe the current impasse, it is convenient to discuss the dating procedure that compares the luminosity of the main-sequence turn off with values given by theoretical isochrones. Because this method is insensitive to errors in the reddenings of the clusters and in the colors of the isochrones, it may be superior to the method of fitting the isochrones directly to the color-magnitude diagram (CMD). In any case, the following uncertainties afflict both methods.

Observationally, the luminosity of the main-sequence turn off is estimated in two steps. From a cluster's CMD, the quantity ΔM_V(TO-RR) is measured. Then a value for the absolute magnitude of the RR Lyrae variables (M_V(RR)), appropriate for the cluster's metal abundance, is adopted. The addition of these quantities yields, of course, the luminosity of the turn off, and this in turn yields an age from the theoretical calculations once a value of the abundance of He (Y) is adopted. The major uncertainty in the age that one obtains stems from the current debate regarding the dependence of M_V(RR) on [Fe/H]. For example, with the assumption that M_V(RR) is the same for all metallicities, one obtains a very large range in age (22.6 to 13.6 billion yrs, if M_V(RR) = +0.6) over the metallicity range of the halo, while with the assumption that ΔM_V(RR)/Δ[Fe/H] = 0.35 (Sandage 1982), one obtains no variation in age at all (see Zinn 1986). Both of these M_V(RR)-[Fe/H] relations are supported by observations, and it is impossible at present to rule out either one. In addition, there is the possibility that ΔM_{bol}(TO-RR) varies with [Fe/H]. While the present observations suggest that it does not (Buonanno 1986), a variation that yields a 3 billion year difference in age between [Fe/H] = -2.5 and -0.8 cannot be excluded. Furthermore, the adopted value of Y may be incorrect, and Y may vary with [Fe/H]. The isochrones may contain errors and may be for the wrong mixture of elements (e.g., wrong CNO/Fe). At present, therefore, the ages of globular clusters do not provide a precise chronology for the collapse of the Galaxy.

In spite of these problems, their ages do yield precious information. As many authors have remarked (e.g., Sandage 1982), the ages of globular clusters appear to be inconsistent with the age of the universe unless the value of the Hubble constant is no larger than about 50 km/sec/Mpc. In addition, the ages that are obtained for the two clusters of the metal-rich disk system that can be dated at present, M71 and 47 Tuc, are identical to the ages obtained for some halo clusters, which suggests that there was not a large gap in time between the formation of the halo and the galactic disk, or at least the thick disk.

6. ARE THE GLOBULAR CLUSTERS GOOD TRACERS OF THE GALAXY'S STELLAR POPULATIONS?

This is not an idle question, because substantial evidence exists for differences between the globular cluster and stellar populations of giant elliptical galaxies (see the review by W. Harris 1987). If the properties of the globular clusters in the Galaxy are not shared by the field stars, then it would appear that under certain conditions the formation of globular clusters was favored over stars, as suggested, for example, by Fall and Rees (1985).

Recently, there have been a number of comparisons of the spatial distributions, kinematics, and metal abundance distributions of globular clusters and samples of stars, with the result that, at present, there appears to be no firm reason to reject the hypothesis that they are members of the same populations (see Hartwick 1983; Zinn 1985, 1986; Norris 1986; Freeman 1986 and references therein). The globular clusters that are more metal poor than [Fe/H] = -0.8 appear to be similar in these properties to halo stars. The more metal-rich clusters have the metallicities and kinematics of the Gilmore and Reid (1983) thick disk population (see Norris 1986; Freeman 1986). The scale height of the metal-rich clusters appears to be smaller than that inferred for the thick disk stars ($\approx$500 pc as opposed to $\approx$1500 pc, see Zinn 1985), but its measurement is uncertain.

Norris (1986) has recently analyzed the kinematics of a large sample of metal-poor stars (i.e., [Fe/H] $\geq$ -0.6), and has found that the dividing line between the metal-poor halo and the more metal-rich thick disk occurs at [Fe/H] = -1.2. This value is significantly lower than the value indicated by the globular clusters (-0.8), and it is important to see if this difference is real or is merely a consequence of the errors in the measurements of [Fe/H], which are presumably larger in Norris's heterogeneous sample. To do this, I have compiled a sample of 148 RR Lyrae variables near the Sun that have well determined radial velocities and spectroscopically measured values of Preston's (1959) ΔS parameter (the literature is ripe with ΔS values inferred from photometry, and many of these disagree with more recent spectroscopic ones). The variables have been put into the following groups: ΔS ≤ 2, 2 < ΔS ≤ 4, 4 < ΔS ≤ 6, 6 < ΔS ≤ 08, and ΔS < 8, and the rotational velocity and line-of-sight velocity dispersion of these groups have been calculated by the Frenk and White (1980) technique, assuming the Local Standard of Rest rotates at 220 km/s. To transform the ΔS measurements to the [Fe/H] scale of Zinn and West (1984), the equation: [Fe/H] = -0.16 ΔS - 0.41 (Zinn 1986) was used, which differs slightly from the one used by Norris for the same purpose. According to this equation, the typical standard deviation of a ΔS measurement translates into a deviation of ≈ 0.2 dex in [Fe/H], which is comparable to the precision of the measurements for the globular clusters.

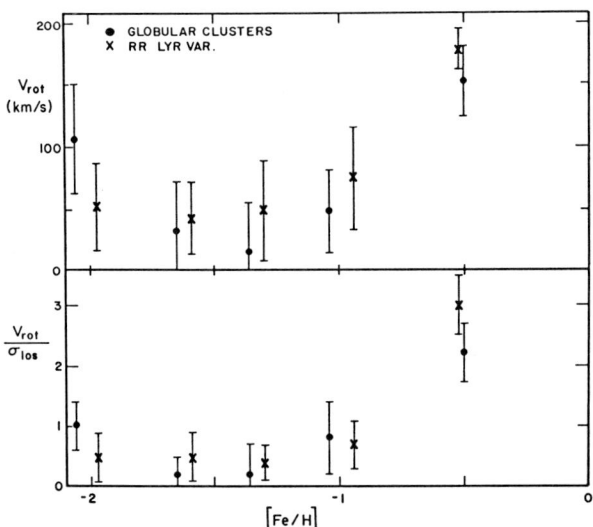

Fig. 1. For groups of globular clusters and RR Lyrae variables, the mean rotational velocity (V_{rot}) and the ratio of V_{rot} to the line-of-sight velocity dispersion (σ_{los}) are plotted against the mean metallicity of the group.

Fig. 1 compares the calculations for the groups of RR Lyrae variables with ones for groups of globular clusters that have approximately the same mean metallicities (102 clusters in total, metallicities and radial velocities from Zinn 1985). The quantity V_{rot}/σ_{los} is a diagnostic of rotational flattening, which is expected to be less than one in slowly rotating systems and greater than one in rotating disk systems. The data in Figure 1 indicate that the rotational properties of the clusters and the variables are very similar. In both samples, the change from halo to disk-like rotation appears to occur at [Fe/H] < -1, and [Fe/H] = -0.8 appears to be a reasonable choice for the dividing line.

The data in Fig. 1 show that there is not a significant correlation between rotation and [Fe/H] below [Fe/H] = -0.8. Norris (1986) found the same effect in his much larger sample of objects below [Fe/H] = -1.2. The lack of a correlation is what one expects on the basis of the observed absence of a steep metallicity gradient among the halo globular clusters and RR Lyrae variables, and it is consistent with Searle and Zinn's (1978) hypothesis that the halo formed out of the merger of several small systems. Sandage (1986), on the other hand, in his analysis of the metallicities and kinematics of a large sample of subdwarfs, has found a very significant correlation between rotation and [Fe/H], which he interprets as strong evidence in favor of Eggen, Lynden-Bell, and Sandage's (1962) picture of progressive metal enrichment as the collapse of the halo proceeded. Until the origin of the discrepancy between these analyses is identified (see Freeman 1986 for one possibility), it is not clear which of these pictures of the formation of the halo best fits the data.

This research was supported by NSF grant AST-8304034.

REFERENCES

Armandroff, T. E. and Zinn, R. 1986 in preparation.
Baade, W. 1958 in Stellar Populations, D. J. K. O'Connell, ed., North Holland, Amsterdam, p. 303.
Buonanno, R. 1986 Mem. Soc. Astron. Italiana, in press.
Buonanno, R., Corsi, C. E. and Fusi Pecci, F. 1985 Astron. Astrophys. 145, 97.
Butler, D., Kemper, E., Kraft, R. P. and Suntzeff, N. B. 1982 Astron. J. 87, 353.
Butler, D., Kinman, T. D. and Kraft, R. P. 1979 Astron. J. 84, 993.
Cudworth, K. M. 1984 Astron. J. 90, 65.
Eggen, O. J., Lynden-Bell, D. and Sandage, A. 1962 Astrophys. J. 136, 748.
Fall, M., and Rees, M. 1985 Astrophys. J. 298, 18.
Freeman, K. C. 1986 in Stellar Populations, C. A. Norman, A. Renzini, and M. Tosi, eds., Cambridge University Press, in press.
Freeman, K. C. and Norris, J. 1981 Ann. Rev. Astron. Astrophys. 19, 319.

Frenk, C. S. and White, S. D. M. 1980 *Monthly Notices Roy. Astron. Soc.* **193**, 295.
Frenk, C. S. and White, S. D. M. 1982 *Monthly Notices Roy. Astron. Soc.* **198**, 173.
Gilmore, G. and Reid, N. 1983 *Monthly Notices Roy. Astron. Soc.* **202**, 1025.
Harris, W. E. 1976 *Astron. J.*, **81**, 1095.
Harris, W. E. 1987 in *IAU Symposium No. 126, Globular Cluster Systems in Galaxies*, J. E. Grindlay and A. G. D. Philip, eds., Reidel, Dordrecht, p. 237.
Harris, W. E. and Canterna, R. 1979 *Astrophys. J.*, **231**, L19.
Hartwick, F. D. A. 1983 *Mem. Soc. Astron. Italiana*, **54**, 51.
Hesser, J. E., Shawl, S. J. and Meyer, J. E. 1986 *Publ. Astron. Soc. Pacific* **98**, 403.
Kinman, T. D. 1959 *Monthly Notices Roy. Astron. Soc.* **119**, 538.
Kinman, T. D., Kraft, R. P., Friel, E. and Suntzeff, N. B. 1985 *Astron. J.*, **90**, 95.
Malkan, M. A. 1982 in *IAU Colloq. No. 68, Astrophysical Parameters for Globular Clusters*, A. G. D. Philip and D. S. Hayes, eds., L. Davis Press, Schenectady, p. 533.
Mayall, N. U. 1946 *Astrophys. J.* **104**, 290.
Morgan, W. W. 1956 *Publ. Astron. Soc. Pacific* **68**, 509.
Morgan, W. W. 1959 *Astron. J.* **64**, 432.
Norris, J. 1986 *Astrophys. J. Suppl.* **61**, 667.
Peterson, R. C. 1985 *Astrophys. J. Letters* **294**, L35.
Pilachowski, C. A. 1984 *Astrophys. J.*, **281**, 614.
Preston, G. W. 1959 *Astrophys. J.*, **130**, 507.
Rood, R. T. and Seitzer, P. O. 1982 in *IAU Colloq. No. 68, Astrophysical Parameters for Globular Clusters*, A. G. D. Philip and D. S. Hayes, eds., L. Davis Press, Schenectady, p. 369.
Saha, A. 1985 *Astrophys. J.* **289**, 310.
Sandage, A. 1982 *Astrophys. J.* **252**, 553.
Sandage, A. 1986 in *Stellar Populations*, C. A. Norman, A. Renzini and M. Tosi, eds., Cambridge University Press, in press.
Searle, L. and Zinn, R. 1978 *Astrophys. J.*, **225**, 357.
Shapley, H. 1918 *Publ. Astron. Soc. Pacific* **30**, 42.
Zinn, R. 1980 *Astrophys. J.*, **241**, 602.
Zinn, R. 1985 *Astrophys. J.*, **293**, 424.
Zinn, R. 1986 in *Stellar Populations*, C. A. Norman, A. Renzini and M. Tosi, eds., Cambridge University Press, in press.
Zinn, R. and West, M. J. 1984 *Astrophys. J. Suppl.* **55**, 45.

DISCUSSION

COHEN: Although there are uncertainties of 2×10^9 years in globular cluster ages, if one assumes they all have same He abundance, the relative ages between clusters should be better. (If one is not willing to assume this, all cluster distances are suspect, anyway.) You count a mean age difference of at least 2×10^9 years between cluster at 7 kpc and at 40 kpc to explain the second parameter problem. Increasing numbers of CM diagrams for globular clusters in outer halo now exist and show the same age for globular clusters. Given better statistics, can one eliminate age as the second parameter?

ZINN: I will stick by my comment that the present measurements of the ages of globular clusters lack sufficient precision to test the hypothesis that cluster age is the dominant second parameter. The age range is predicted to be $\sim 2 \times 10^9$ for the clusters in the $7 < R < 40$ kpc zone, which is comparable in size to the most optimistic claims made on the errors with which clusters can be dated. The ages of the very distant clusters with red horizontal branches are predicted to be as much as 4×10^9 yrs younger than blue horizontal-branch clusters. While, in principle, this difference should be measurable, it is not possible at present to construct CM diagrams of the required precision for the very distant clusters.

OSTRIKER: Could you say something further about the similarity or difference between the halo stars and the halo globular clusters? Specifically, is it possible, from chemical considerations, that most of the halo stars originated in dissolved globular clusters?

ZINN: The existing data suggest that the field stars and the globular clusters have similar metallicity distributions, kinematics, and spatial distributions, therefore, I believe that it is reasonable to hypothesize that the field stars are debris of disrupted clusters or that the field stars and the globular clusters originated in larger systems, the proto-dwarf galaxies that I mentioned, which dissolved into the galactic halo.

ZINNECKER: On your picture of accretion of proto-dwarf galaxies to form the galactic halo: isn't there additional support for this picture from the finding that there seem to be subsystems of clusters having distinctly difference space motions (prograde and retrograde)?

ZINN: I am not convinced that there is firm evidence for prograde and retrograde groups of clusters (See Zinn, R. 1985, *Astrophys. J.*, 293, 424.). The problem with this, as I see it, is that for most clusters it is very difficult to tell the direction of rotation from radial velocity measurements alone. Obviously, if we had in addition proper motion data, which may be provided some day by observations from spacecraft, we could tell the direction of rotation, and then it would be very interesting to look for groups of clusters that have similar space motions.

KING: For the distant galactic globulars, we need not wait for HST, which will provide resolving power but will do relatively little for limiting magnitude. They should be done with ground-based observations, concentrating on good seeing. Our existing telescopes will collect enough photons; reduction time is more important than telescope time.

ZINN: To date the very distant clusters require photometry of ~2% precision for V ~ 25-26. This is very difficult, in my opinion, from the ground because it requires truly exceptional seeing. Christian and Heasley (1986, Astrophys. J.) have produced, from data obtained in good seeing, a very pretty CM diagram for the distant globular cluster Pal 4. However, this diagram does not place strong constraints on the age of Pal 4. I'm not sure one can do much better than this at the present time.

CARNEY: To partially answer Jerry's question and previewing my talk later, the field and cluster stars seem to show about the same overall metallicity distribution and detailed elemental abundance patterns. At least, there is no strong evidence they differ significantly.

KINEMATICS OF THE GALACTIC GLOBULAR CLUSTER SYSTEM

R. F. Webbink

University of Illinois

ABSTRACT: Constraints on cluster kinematics proper motions, radial velocities and tidal radii are reviewed. Analysis of the cluster radial velocity distribution suggests a rotation law for the system in which the specific angular momentum is nearby independent of galactocentric distance, and the residual velocity dispersion is isotropic. However, the absence of severely tidally truncated clusters indicates that nearly radial orbits are absent from this distribution. The kinematic properties of the remote halo clusters remain largely indeterminate. Absolute proper motions measured directly with respect to background galaxies and quasars are needed to determine the kinematics of these objects, and also to elucidate the process of tidal stripping.

1. INTRODUCTION

The kinematics of the galactic globular cluster system have long been a subject of great interest for two important reasons: (i) they serve as a dynamical probe of the galactic mass distribution; and (ii) they provide a fossil record of the early dynamical and chemical evolution of the Galaxy. Historically, globular clusters have enjoyed a favored status over single stars for these purposes, because they are easily identifiable entities over the entire Galaxy, and because they are usually bright enough to be studied spectroscopically by photographic means. Recent advances in detector technology promise to make kinematic studies of distant halo giants feasible, mitigating some of these advantages, but globular clusters remain the probes of choice, despite their limited numbers, because their distance scale is much more secure than that of single stars.

Most of our present knowledge of cluster kinematics has been derived from studies of their radial velocities. These have the advantage of being relatively quick and easy to obtain. Very high accuracy is now possible using modern cross-correlation techniques (whether digital or analog), which are very well-suited to the late-type giants which dominate cluster light. This advance has made possible detailed studies of stellar kinematics within clusters, as described elsewhere in this Symposium. Nevertheless, radial velocity data alone carry some significant drawbacks for studies of kinematics of the cluster system: First, information regarding the tangential velocities of clusters (with respect to the galactic center) is only accessible for clusters near the solar circle or inside it. These velocity components make no significant contribution to the observed radial velocities of clusters far outside the solar circle. Second, with only one-dimensional velocity data for individual clusters, one can obtain only a statistical description of cluster orbits within the Galaxy. Furthermore, the number of clusters known is so small that only the first and second moments of the velocity distribution are statistically significant: kinematic details are washed out.

In principle, cluster proper motions could provide us with much more information regarding their kinematics. In the first place, they are two-dimensional data. Combined with radial velocities, which are now available for the vast majority of clusters, they constitute a complete kinematic description of the cluster system. Moreover, they provide the only observational constraint on rotation of the cluster system about the $\ell = 0$, $b = 0$ axis. (It should be recalled that the Magellanic Stream is nearly perpendicular to this axis [Wannier and Wrixon 1972]. Any direct kinematic evidence of its dynamical interaction with the galactic globular cluster system will be manifested observationally in proper motions of the affected clusters.) The difficulties in obtaining absolute proper motions of clusters are of course well-known: A long time base is both desirable and necessary in most cases. (This is itself a distinct handicap in modern astronomy.) The potentials for systematic errors are legion, making reductions difficult and time-consuming. The accuracy of most studies of cluster proper motions to date is poor, owing to the large and unavoidable dispersion in proper motions of the (foreground) reference stars. This statistical uncertainty is further compounded by the inadequacy of current models for the kinematics of field stars, which renders the correction from relative to absolute proper motions highly uncertain.

An indirect constraint on cluster orbits comes from their observed limiting radii. These limiting radii are generally attributed to the tidal limit imposed by the galactic gravitational field (von Koerner 1957, King 1962), and thus, in principle, provide information about gradients in that field (and hence local mass densities) of a different kind from that obtained through dynamical modeling. Indeed, it should be possible to constrain possible cluster orbits rather strongly through their use. In practice, limiting radii can be readily estimated from star counts (e.g. King, et al. 1968; Peterson and King 1975), although the cluster contribution to the total star density is rarely distinguishable above background beyond one-half the limiting radius. Unfortunately, the accuracy of this method (or indeed of any other method so far devised) is badly degraded by a high or variable background of field stars and by differential reddening. As a result, large uncertainties remain in the determinations for individual clusters, particularly those at low galactic latitude. Moreover, the precise limiting radius extrapolated from a given set of observations depends upon the internal kinematics of the cluster (see, e.g., Gunn and Griffin 1979), and the tidal field implied by that limit depends upon the total cluster mass-to-light ratio. The most serious interpretational problems in exploiting limiting radii, however, come from (i) severe inadequacies in current theories of tidal stripping, which have yet to deal satisfactorily with the problems of eccentric cluster orbits, internal dynamical evolution of clusters, or the fact that marginally unbound stars may not be lost from clusters for many dynamical timescales; and (ii) the fact that the kinematical implications of tidal radii depend on the galactic mass model — the kinematical and dynamical problems are not separable.

Let us turn now to the question of what has been learned of cluster kinematics by these methods.

2. PROPER MOTIONS

Historically, the first attempts to detect the motions of globular clusters (and other "nebulae" as well) were astrometric ones. During the latter half of the 19th Century, efforts were made at nearly every major observatory to secure accurate positions for these objects visually, positions which might have served as a basis for proper motion studies. Unfortunately, these efforts were devoted almost exclusively to the determination of centroid positions, and not those of individual stars, and so are too ill-defined to be of use for kinematic purposes, except to show that the globular clusters were not nearby objects.

The modern era of globular cluster astrometry began with the work of van Maanen (1925, 1927) at Mt. Wilson, and Balanowsky (1928) at Pulkovo, who used photographic plates to study the proper motions of stars in and around several bright Northern Hemisphere clusters. Since this pioneering work, numerous studies of individual clusters have been published (albeit almost exclusively for Northern Hemisphere clusters). For the most part, these have been relative proper motion studies, and so are affected by serious uncertainties in the reduction to absolute proper motions. Hallermann (1965) attempted to measure the proper motions of 11 clusters directly in the NFK coordinate system, Meurers and Prochazka (1969) tied that of NGC 6838 (MT1) directly to the FK4 system, and Brosche and coworkers have recently measured the proper motions of NGC 4147 (Brosche, et al. 1986) and NGC 5466 (Brosche and Geffert 1983) with respect to field stars with known proper motions in the Lick extragalactic reference system. However, the only attempts to determine cluster proper motions directly with respect to extragalactic objects have been a series of studies at Pulkovo of NGC 6205 (M 13) referred to the nucleus of NGC 6207 (Gamalej 1948; Fatchikin 1952; Kadla 1963). These studies have reached inconsistent results, apparently because of difficulties in defining the nucleus of NGC 6207.

In Figure 1 are illustrated the proper motions of the 15 globular clusters and 2 dwarf spheroidal satellites of the Galaxy for which published data are available. Relative proper motions have been reduced to an absolute (inertial) frame assuming the standard open for solar motion ($A = 270°$, $D = +30°$); this reduction may differ considerably from that following from use of the Lick apex (Vasilevskis and Klemola 1971), especially for clusters in the vicinity of these apexes (e.g., Cudworth 1976a,b, 1979a,b; Cudworth and Monet 1979). Proper motions in fundamental coordinate systems were corrected to FK4 (Nowacki 1935; Fricke, et al. 1963), as appropriate, and the equinox correction to FK5 (Fricke 1982) was then applied. The results from different published sources have then combined into weighted means.

The reflex proper motion of the cluster system due to the net rotation of the cluster system with respect to the local standard of rest is clearly discernable in Figure 1. However, the residual proper motions of individual clusters, once this effect is removed, are mostly of doubtful significance.

3. RADIAL VELOCITIES

The relative ease and precision with which cluster radial velocities can be determined over galactic distances has made them the preferred avenue for kinematic studies. Since the pioneering work of Strömberg (1925), numerous statistical analyses of cluster motions based on their radial velocities have been published, including important papers by Edmondson (1935), Mayall (1946), Perek (1954), von Hoerner (1955), Kinman (1959), Matsunami (1964), Woltjer (1975), House and Wiegandt (1977), Hartwick and Sargent (1978), Frenk and White (1980), Pier (1984), Rodgers and Paltoglou (1984), Zinn (1985), Hesser, Shawl, and Meyer (1986), and Norris (1986). Radial velocities have, at this writing, been published for a total of 115 galactic globular clusters, as well as for all 7 dwarf spheroidal satellites of the Galaxy, making this the most extensive (as well as the most reliable) body of kinematic data presently available.

Recent studies based on radial velocities, while emphasizing different details, have arrived at a fairly consistent picture of the global kinematics of the cluster system. In the following discussion, the numerical results quoted are those derived by the author from the analysis of the 85 clusters (plus NGC 6569) and 4 dwarf spheroidals contained in his radial velocity catalogue (Webbink 1981). To the extent that they are comparable, however, these

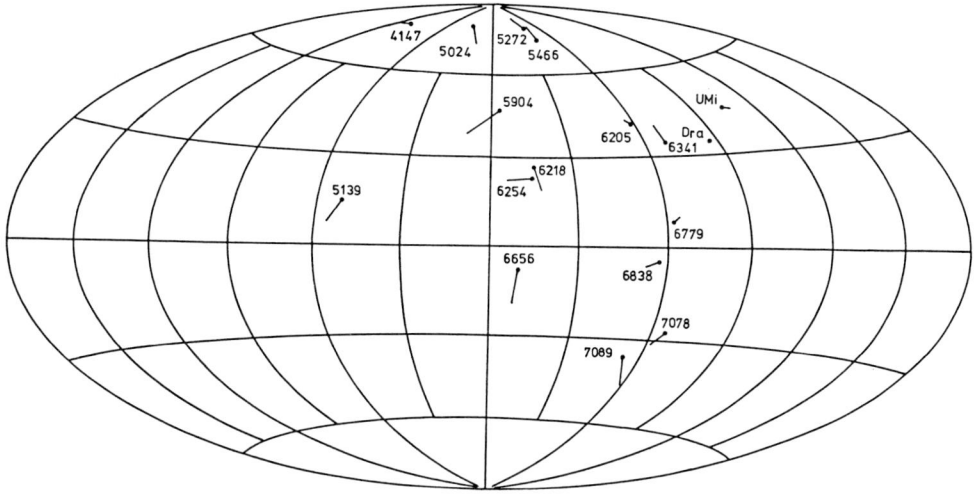

Fig. 1. Absolute proper motions of galactic globular clusters and dwarf spheroidal satellites. The present positions of the clusters are plotted as points in galactic coordinates ($\ell = 0°$, $b = 0°$ at the center of the figure), with lines indicating the proper motion for 3.6 Myr (i.e., $1° = 1$ mas/yr).

results are entirely consistent with the findings of all of the recent studies cited above.

3.1 Rotation of the cluster system

In order to explore the rotation of the cluster system, we divide the 90 clusters and dwarf spheroidals into six spherical shells, each containing 15 objects. For this purpose, the distances to individual clusters are based on the assumption that $M_V(HB) = +0.60$, and the distance to the galactic center is taken to be $R_\odot = 8.8$ kpc (Harris 1976). The clusters within each shell are assumed to rotate about an axis perpendicular to the galactic plane (the $b = +90°$ axis) with uniform linear velocity, with the circular velocity at the Sun taken to be $\Theta_0 = 220$ km s^{-1} (Gunn, Knapp, and Tremaine 1979). The results of this exercise are listed in Table I.

TABLE I.
ROTATION OF THE CLUSTER SYSTEM

$\bar{R}/R_\odot$	Θ_{cl} (km s^{-1})	σ_{los} (km s^{-1})
0.24±0.01	120±64	100±19
0.44±0.02	68±51	102±17
0.74±0.02	67±37	90±8
1.11±0.04	28±58	145±23
1.87±0.08	31±140	125±10
7.94±1.27	231±303	139±17

As was to be expected, the rotation curve of the cluster system (Θ_{cl}) is not well-defined much beyond the solar circle, but there clearly exists differential rotation within the cluster system as a whole. For the inner four bins, a trend in which the linear rotation velocity increases towards the galactic center is apparent, even though the data for individual bins have large uncertainties. This trend was suspected by Frenk and White (1980), and elaborated considerably by Zinn (1985). Also apparent in the third column of Table I is the abrupt increase in the line-of-sight velocity dispersion beyond the solar circle first noted by Frenk and White. This phenomenon can be explained either by an increase in the total velocity dispersion, or by a slightly elongated velocity ellipsoid in which the major axis points radially toward the galactic center.

On the strength of the above evidence, we may attempt a second solution to the cluster rotation law, treating the entire system as a single entity. We assume a rotation law of the form

$$\Theta_{c1}(R) = \Theta_{c1}(R_\odot) (R/R_\odot)^\alpha$$

and permit as well a rate of expansion of the same form:

$$\Pi(R) = \Pi(R_\odot) (R/R_\odot)^\alpha.$$

The axis about which the cluster system rotates is left as a free parameter, with components Θ_z about the ($b = +90°$)-axis and Θ_y about the ($\ell = 90°$, $b = 0°$)-axis. (As noted in the Introduction, rotation Θ_x about the ($\ell = 0°$, $b = 0°$)-axis is indeterminate from radial velocities alone.) In addition, we may treat Θ_0, the local circular velocity, as a free parameter in the solution: it is indeterminate only if $\alpha = 1$. The best-fit parameters thus obtained are:

$\alpha = -0.98 \pm 0.51$

$\Theta_0 = 203 \pm 27$ km s^{-1}

$\Theta_z(R_\odot) = 34 \pm 14$ km s^{-1}

$\Theta_y(R_\odot) = -1.4 \pm 9.2$ km s^{-1}

$\Pi(R_\odot) = 8.0 \pm 8.8$ km s^{-1}

Clearly, there is no significant rotation of the cluster system about the y-axis, nor is there evidence of any net expansion or contraction of the cluster system, contrary to the conclusions of Clube and Watson (1979). If we therefore set $\Theta_y(R_\odot) = \Pi(R_\odot) = 0$, we obtain:

$\alpha = -1.08 \pm 0.97$

$\Theta_0 = 196 \pm 27$ km s^{-1}

$\Theta_z(R_\odot) = 25 \pm 13$ km s^{-1}

It appears that the cluster system rotation law is approximately one in which the mean specific angular momentum is independent of galactocentric distance. Note also that the deduced local lag in rotation of the cluster system, $\Theta_z(R_\odot) - \Theta_0$, is practically independent of the assumed rotation law.

3.2 Velocity ellipsoid of the cluster system

We may now remove the systematic effects which the rotation of the cluster system contributes to the observed radial velocities, according to the final set of parameters deduced above, and examine the properties of the cluster velocity dispersion tensor (cf., e.g., Ogorodnikov 1965). We adopt a spherical polar coordinate system in which the Π-axis points from the galactic center to the cluster position, the Θ-axis points in the direction of galactic rotation, parallel to the galactic disk, and the Φ-axis is orthogonal to the other two. (Note that the Θ-Φ notation is reversed from that of Norris [1986], but the same as that of Pier [1984].) The deduced dispersion tensor is listed in Table II.

TABLE II.
THE GLOBULAR CLUSTER VELOCITY DISPERSION TENSOR

	σ_{ij}^2 (in 10^4 km s^{-1})		
j\i	Π	Θ	Φ
Π	+1.42±0.27	-0.09±0.33	-0.14±0.37
Θ	-0.09±0.33	+1.37±0.43	+0.08±0.51
Φ	-0.14±0.37	+0.08±0.51	+0.90±0.65

The off-diagonal components of this tensor are all consistent with zero, implying that the principal axes of the velocity ellipsoid (resolved in this way) coincide with those of the adopted coordinate system. The values for the diagonal components agree well with those found by Pier (1984), and Norris (1986) for the cluster system, except for Pier's inexplicably large value for $\sigma_{\theta\theta}^2$. They differ from those found by Woolley (1978; see also Pier 1984) for halo RR Lyrae stars ($\sigma_{\pi\pi}^2$ = 2.11±0.54; $\sigma_{\theta\theta}^2$ = 1.55±0.54; $\sigma_{\phi\phi}^2$ = 0.50±37) only in having a somewhat smaller radial component. Indeed, within the errors, the cluster velocity ellipsoid is isotropic, as deduced by Frenk and White (1980).

4. TIDAL LIMITS

The identification of the limiting radius of globular clusters with the galactic tidal cutoff (von Hoerner 1957; King 1962) opened the possibility that this information could be turned to advantage in exploring cluster kinematics. The physical argument is a simple one: the cluster achieves tidal equilibrium only when it has been stripped down to those members whose orbits remain bound to the cluster in the presence of the strongest tidal field experienced by the cluster, namely, that at perigalacticon. Given an estimate of the total cluster mass (from its integrated luminosity and mass-to-light ratio), the maximum tidal stress, and hence perigalactic distance, can be calculated. This method has been applied to numerous individual clusters, and in recent years to the cluster system as a whole (Peterson 1974; Rastorguev and Surdin 1980; Seitzer and Freeman 1981; Innanen, Harris, and Webbink 1983). As noted in the Introduction, however, this method is plagued not only by large observational uncertainties, but also by unsolved theoretical problems and the dependence of answers on an assumed galactic potential. In recent years, a controversy has arisen over whether clusters ever actually achieve tidal equilibrium, and on what timescale (see, e.g., Seitzer 1985; Angeletti and Giannone 1984; Angeletti, Capuzzo-Dolcetta, and Giannone 1984; and references therein), and over the relationship between the observed spatial cutoff and the tidal energy cutoff (Innanen, Harris, and Webbink 1983).

Notwithstanding the unsolved problems which complicate a naive interpretation of tidal radii, there exist some properties of the distribution of tidal radii which appear to carry significant kinematical implications which are independent of the resolution of these controversies. Innanen, Harris, and Webbink (1983) point out that any truly isotropic velocity dispersion implies the existence of an asymmetric tail to the distribution of cluster tidal mass densities at any given galactocentric distance. The tail extends toward high densities, and represents very compact clusters on nearly radial orbits. The prominence of this tail depends on the nature of the galactic potential, but even for a very "soft" potential (one giving a flat galactic rotation curve) the complete absence of such a tail in the observed distribution of cluster tidal mass densities implies that the population of extant clusters must have a severe deficiency of low angular momentum (i.e., highly eccentric) orbits. Evidently, the portrayal of the cluster distribution in velocity-space as an ellipsoid is very misleading, but the existence of this hole along the Π-velocity axis is not (and cannot be) revealed from a moment analysis of the observed radial velocities. It is possible that this deficiency is related to the larger radial component of the RR Lyrae velocity ellipsoid (see above).

5. PROSPECTUS

We stand at a crossroads in kinematic studies of the galactic globular cluster system. Accurate radial velocities are now known for the great majority of clusters — there is little room for expansion of this data base. Nevertheless, serious questions remain regarding the detailed form of the cluster velocity distribution, particularly outside the solar circle (with its attendant dynamical implications for the mass of the Galaxy), and regarding the physical significance of cluster tidal radii.

The key to further progress is clearly to obtain <u>absolute proper motions</u> of globular clusters directly with respect to background galaxies and quasars. At high galactic latitude, these objects outnumber field stars at magnitudes ($V \gtrsim 20$; Kron 1980) now within reach observationally. The time baselines needed for useful results, even for distant halo clusters, are not unreasonable, provided that background objects can be identified in sufficient numbers, and their relative positions established with sufficient accuracy:

$$\Delta t \approx 4.8 \text{ yr } \left(\frac{D}{10 \text{ kpc}}\right)\left(\frac{\epsilon_{xy}}{0\rlap{.}''10}\right)\left(\frac{\epsilon_v}{10 \text{ km s}^{-1}}\right)^{-1}\left[\left(\frac{n_f}{100}\right)^{-1} + \left(\frac{n_{cl}}{100}\right)^{-1}\right]^{-1/2},$$

where D is the cluster distance, ϵ_{xy} the single-coordinate uncertainty in relative position, ϵ_v the single-coordinate uncertainty in tangential velocity, n_f the number of background reference objects, and n_{cl} the number of cluster stars. The technology is within reach and the need is clear.

This research was supported in part by National Science Foundation grant AST 83-17916. The author also gratefully acknowledges the hospitality of the Institute of Astronomy, University of Cambridge, where the kinematic analysis of cluster radial velocities described above were first undertaken.

REFERENCES

Angeletti, L., Capuzzo-Dolcetta, R., and Giannone, P. 1984, Astron. Astrophys., 138, 404.
Angeletti, L., and Giannone, P. 1984, Astron. Astrophys., 138, 396.
Balanowski, J. 1928, Izv. Gl. Astron. Obs. Pulkovo, 11, 167.
Brosche, P., and Geffert, M. 1983, Astron. Astrophys., 127, 415.
Brosche, P., Geffert, M., Klemola, A.R., and Ninkovic, S. 1986, preprint.
Clube, S.V.M., and Watson, F.G. 1979, Monthly Notices Roy. Astron. Soc., 187, 863.
Cudworth, K.M. 1976a, Astron. J., 81, 519.
_____. 1976b, Astron. J., 81, 975.
_____. 1979a, Astron. J., 84, 1312.
_____. 1979b, Astron. J., 84, 1866.
Cudworth, K.M., and Monet, D.G. 1979, Astron. J., 84, 774.
Edmondson, F.K. 1935, Lowell Obs. Bull., 3, 143.
Fatchikin, N.V. 1952, Izv. Gl. Astron. Obs. Pulkovo, 19, pt. 1, 150.
Fricke, W. 1982, Astron. Astrophys., 107, L13.
Fricke, W., Kopff, A., Gliese, W., Gondolatsch, F., Lederle, T., Nowacki, H., Strobel, W., and Stumpff, P. 1963, Veröff. Astron. Rechen-Inst. Heidelberg, No. 10.
Frenk, C.S., and White, N.E. 1980, Monthly Notices Roy. Astron. Soc., 193, 295.
Gamalej, N.V. 1948, Izv. Gl. Astron. Obs. Pulkovo, 17, pt. 6, 27.
Gunn, J.E., and Griffin, R.F. 1979, Astron. J., 84, 752.
Gunn, J.E., Knapp, G.R., and Tremaine, S. 1979, Astron. J., 84, 1181.
Hallermann, L. 1965, Dissertation, Univ. Bonn.
Harris, W.E. 1976, Astron. J., 81, 1095.
Hartwick, F.D.A., and Sargent, W.L.W. 1978, Astrophys. J., 221, 512.
Hesser, J.E., Shawl, S.J., and Meyer, J.E. 1986, Publ. Astron. Soc. Pacific, 98, 403.

House, F., and Wiegandt, R. 1977, Astrophys. Space Sci., **48**, 191.
Innanen, K.A., Harris, W.E., and Webbink, R.F. 1983, Astron. J., **88**, 338.
Kadla, Z.I. 1963, Astron. Zh., **40**, 691 (English transl: 1964, Soviet Astron.-A.J., **7**, 528).
King, I.R. 1962, Astron. J., **67**, 471.
King, I.R., Hedemann, E., Hodge, S.M., and White, R.E. 1968, Astron. J., **73**, 456.
Kinman, T.D. 1959, Monthly Notices Roy. Astron. Soc., **119**, 157.
Klemola, A.R., and Vasilevskis, S. 1971, Publ. Lick Obs., **22**, pt. 2.
Kron, R.G. 1980, Astrophys. J. Suppl., **43**, 305.
Matsunami, N. 1964, Publ. Astron. Soc. Japan, **48**, 191.
Mayall, N.U. 1946, Astrophys. J., **104**, 290.
Meurers, J., and Prochazka, F. 1969, Ann. Univ.-Sternw. Wien, **28**, 211.
Norris, J. 1986, Astrophys. J. Suppl., **61**, 667.
Nowacki, H. 1935, Astron. Nachr., **255**, 301.
Ogorodnikov, K.F. 1965, Dynamics of Stellar Systems (Oxford: Pergamon), p. 45.
Perek, L. 1954, Contr. Astron. Inst. Brno., **1**, No. 12.
Peterson, C.J. 1974, Astrophys. J. Letters, **190**, L17.
Peterson, C.J., and King, I.R. 1975, Astron. J., **80**, 427.
Pier, J.R. 1984, Astrophys. J., **281**, 260.
Rastorguev, A.S., and Surdin, V.G. 1980, Astron. Tsirk., No. 1102.
Rodgers, A.W., and Paltoglou, G. 1984, Astrophys. J. Letters, **283**, L5.
Seitzer, P. 1985, in IAU Symposium No. 113, Dynamics of Star Clusters, J. Goodman and P. Hut, eds., Reidel, Dordrecht, p. 343.
Seitzer, P., and Freeman, K.C. 1981, in IAU Colloquium No. 69, Astrophysical Parameters for Globular Clusters, A.G.D. Philip and D.S. Hayes, eds., L. Davis Press, Schenectady, p. 185.
Strömberg, G. 1925, Astrophys. J., **61**, 353.
van Maanen, A. 1925, Astrophys. J., **61**, 130.
_____. 1927, Astrophys. J., **66**, 89.
von Hoerner, S. 1955, Zeit. Astrophys., **35**, 255.
_____. 1957, Astrophys. J., **125**, 451.
Wannier, P.G., and Wrixon, G.T. 1972, Astrophys. J. Letters, **173**, L119.
Webbink, R.F. 1981, Astrophys. J. Suppl., **45**, 259.
Woltjer, L. 1975, Astron. Astrophys., **42**, 109.
Woolley, R. 1978, Monthly Notices Roy. Astron. Soc., **184**, 311.
Zinn, R. 1985, Astrophys. J., **293**, 424.

DISCUSSION

KING: The existing tidal radii are only a first try, from star counts. One should now do a second generation study, using color-magnitude arrays to eliminate most of the field stars.

WEBBINK: Anything which distinguishes cluster stars from field stars will no doubt improve estimates of tidal radii.

KING: For proper motion studies, QSO's are essential; one QSO is worth twenty galaxies.

WEBBINK: QSO's are certainly far superior as reference objects, but I think one will still require large numbers of background objects to ensure good proper motion solutions.

COHEN: The new Palomar Sky Survey will include a very short exposure in an effort to extend the system of stars with absolutely known positions to the faintest stars measurable on PSSII. This should enable absolute proper motions to be determined from plates from large reflectors.

NORRIS: Your analysis of globular cluster kinematics makes no distinction between the disk and halo groups of Zinn. Have you repeated the analysis for the halo group alone?

WEBBINK: I have not attempted to do so. The number of free parameters in the solution is so large that I doubt one can obtain meaningful solutions from a much smaller sample, and the tangential components of the velocity ellipsoid become indeterminate unless the sample includes a large number of clusters inside the solar circle, where the Sun appears nearby at right angles from the galactic center, as seen from the cluster. For distant halo clusters, we see only the galacticentric radial component.

CAYREL: My question is about the best observational accuracy attainable in proper motion measurement for globular clusters. Is there a significant statistical gain due to the fact that each globular cluster is made of many point sources and not only one as for a normal single star proper measurement?

WEBBINK: In principal, I suppose what you say must be correct, but I would imagine that variations in seeing would make it very difficult to take advantage of this structure, and most background galaxies will undoubtedly have low signal-to-noise detections.

GLOBULAR CLUSTER COLOR-MAGNITUDE DIAGRAMS

James E. Hesser

Dominion Astrophysical Observatory
Hertzberg Institute of Astrophysics

ABSTRACT. This review concentrates on new generalities emerging from this, the second "golden era" of B,V color-magnitude diagrams (CMDs), and remarks about future work in other photometric systems.

1. OVERVIEW

Within the context of this *Symposium*, the interpretation of CMDs of the Galactic globular clusters is central to the development of population synthesis models that will facilitate interpretation of the properties of more distant globulars (observable only by their composite light). As well they *should* serve as key reference points for evaluating the reliability of synthesis techniques in unravelling the integrated light of galaxies. Simultaneous consideration of the spatially resolvable Magellanic Cloud clusters and Dwarf Spheroidal galaxies (reviewed in this volume by Graham, Olszewski and Da Costa) is, of course, also of the utmost importance. As attested to by numerous contributed papers at this *Symposium*, we are embarked upon a new voyage of discovery about the basic constituents of Galactic globular clusters. This review will attempt to illustrate how the CMD fundamentally contributes to understanding star formation and evolution in metal-deficient material by examining: (i) basic morphological properties with emphasis on little-explored phases of stellar evolution; (ii) improving globular cluster ages and, hence, the lower limit to the age of the Universe; and (iii) the nature of globular cluster main-sequence luminosity functions.

The growth in observational material borders on being explosive (Fig. 1).

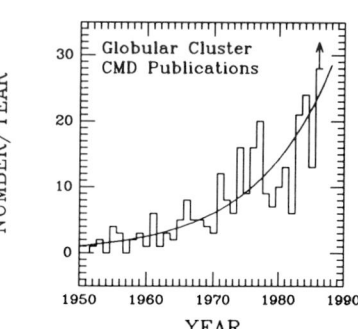

Fig. 1. The annual rate of publication of CMD papers, ignoring unpublished studies, abstracts, etc. (Peterson 1986a). The curved line represents a doubling of the rate on an eight year time scale.

J. E. Grindlay and A. G. Davis Philip (eds.),
The Harlow-Shapley Symposium on Globular Cluster Systems in Galaxies, 61–78.
© *1988 by the IAU.*

Although the wealth of material forces me to limit myself primarily to newer, CCD-based CMDs, even then I will be able to mention only a small fraction of the work being done worldwide. More complete references are in Peterson's (1986a) comprehensive CMD bibliography, as well as in the recent reviews of Caputo (1985), Castellani (1985), Sandage (1986), and the proceedings of this, the 1986 Rome and Baltimore conferences.

2. TECHNIQUES AND LIMITATIONS

Despite small sizes, relatively large pixels, occasional random and systematic surface imperfections, charge-transfer problems, etc., CCDs have, since Richer and Fahlman's (1984) first CMD, virtually replaced the use of photoelectrically calibrated photographic plates in determination of globular cluster CMDs. Their linearity, dynamic range, and sensitivity have dramatically reduced the *internal* photometric scatter (Fig. 2). Studies with the modern detectors have also revealed the presence of apparent *systematic* errors in much of the older work, in spite of Herculean efforts in the original studies to avoid them (see, e.g., Harris et al. 1983a; Hesser 1983; Heasley and Christian 1986, as well as Buonanno et al. 1983).

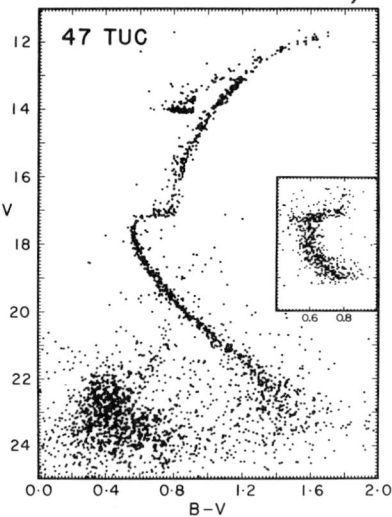

Fig. 2. Comparison of the 47 Tuc main-sequence from CCD photometry (Harris and Hesser 1986) with that (inset) from traditional techniques (Hesser and Hartwick 1977) illustrates a dramatic improvement in internal errors. The giant branch is from the photographic photometry of Lee (1977) and Hesser and Hartwick. Note the turnoff and giant branch of the SMC halo in the lower left-hand portion.

Software techniques share at least equal responsibility for the vastly improved CMDs. Indeed, the fact that CCDs allow us to work to much fainter limits in more crowded fields than before, presents us with fundamentally new, unique problems of data reduction. The key ingredient in the new software is replacement of fixed-aperture techniques with a point spread function empirically defined from uncrowded images in each frame or picture. By so tailoring the measurement aperture, the signal-to-background ratio is maximized and deconvolution of overlapping images becomes feasible. Another significant advance arises from the ease with which "artificial" stars having known magnitudes and positions can be inserted into the original data frames to evaluate errors and completeness factors. Since the appearance of Tody's (1981) "user-friendly" RICHFLD program at Kitt Peak in the late 1970's, many different codes have become available; to mention but a few, Buonanno et al. 1983, DAOPHOT (Stetson 1984, 1987), Walker 1984, Penny and

Dickens 1986, Lupton and Gunn 1986, and ESO's MIDAS/INVENTORY programs. Relatively few direct comparisons of results from different codes based on similar data have yet appeared, but the ones that have are encouraging (e.g., Gratton and Ortolani 1986a); it will be particularly valuable to have results from Green's (1985) comparison of various codes on *identical* data.

Walker (1986) has given an excellent review of the combined power of the new techniques. To get the best results on fundamental parameters such as ages, my colleagues and I continue to be concerned with the need to pay excruciatingly close attention to possible photometric zero-point errors in broad-band CCD photometry. Although the small internal scatter and high quantum efficiency in CCD photometry makes them seem almost magical, they present the user with all the challenges inherent in first-class photoelectric photometry, with which they share a common enemy: the Earth's atmosphere. Moreover, each chip, camera and filter combination has its own personality. Accordingly it is essential to get many standards each night covering both a wide color and air-mass range, to observe some Pop II tie-in fields, and to make observations of program fields on several nights. Even with these precautions, zero points based on transfers from standard stars outside the cluster CCD frame *may* be randomly in error by amounts up to 0.04 mag (see Table 2 of McClure et al. 1987). The need to reduce many independent frames taken on different nights seems unavoidable if we are to obtain the highest possible *external* precision from the ground with the new techniques.

One of the most exciting aspects of the new techniques is their potential for determining luminosity functions in crowded globular cluster fields. It may not, however, be widely recognized, how enormous is the ratio of reduction to observing time required to perform the deconvolution photometry (100:1 or more on VAX-class machines). For best results, multiple reductions are required with different test stars generated from the empirical point spread function in order to determine completeness factors. In addition, when the goal is to reach either extremely faint magnitudes or to study clusters at lower Galactic latitudes, observations and analysis of comparison star fields are necessary to estimate field star contributions independently of Galactic halo models. Both of these requirements impact substantially upon observing and reduction time.

Finally, the strong effect of distance on the quality of results (Fig. 3) must be born in mind. With 4-m telescopes we are able to get superb data capable, in principle, of age discrimination to ± 1 Gyr at the turnoff to distances of $\lesssim 20$ kpc, beyond which our discriminatory powers decrease rapidly. A similar degradation sets in at lower Galactic latitudes, where crowding and field star contamination dominate our observations. Proper motion studies can, for some clusters, provide a remarkable transformation in the observed CMD (see Fig. 4). For the opportunities and challenges presented by many distant or nuclear bulge objects, however, HST will be essential.

3. ASPECTS, OLD AND NEW, OF CMD MORPHOLOGY

In spite of occasionally valiant efforts, most older CMDs (particularly photographic ones) suffer from often unquantifiable incompleteness, photometric uncertainties, and biases towards larger radii. The potential of a thorough survey, with modern detectors and software, of globular cluster luminosity functions at magnitudes *brighter* than the turnoff deserves emphasis: an order-of-magnitude improvement is now possible for brighter portions of cluster luminosity functions.

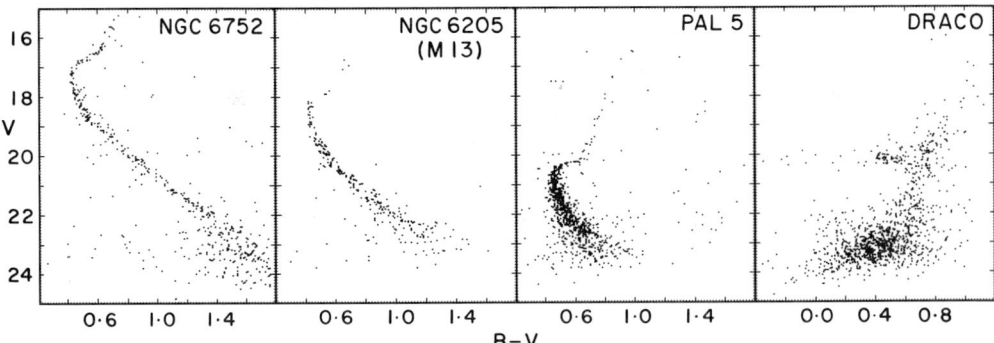

Fig. 3. CMDs from deep CCD exposures illustrate the effect on photometric quality at the turnoffs imposed (primarily) by increasing distance. From left to right: NGC 6752 (Penny and Dickens 1986); M13 (Richer and Fahlman 1986; see also Lupton and Gunn 1986); Pal 5 (Smith et al. 1986), and Draco (Carney and Seitzer 1986; see also Stetson et al. 1985).

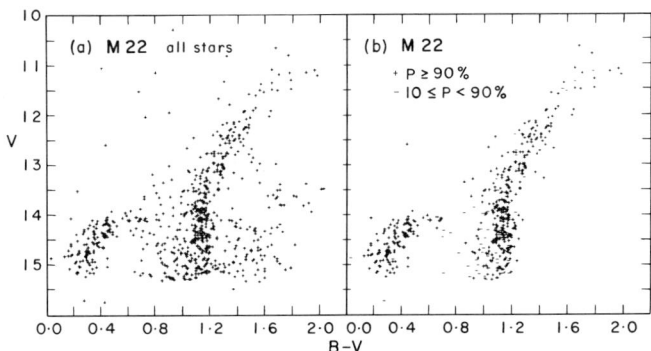

Fig. 4. The beneficial effects of proper motions for clarifying populations within lower-latitude globulars is well illustrated by the case of M22 (Cudworth 1986), at $l = 10°, b = -8°$.

Since brighter stars contribute significantly to the integrated light, a knowledge of their relative numbers and properties is essential for many problems. Some hints of what may result from such work are now described.

3.1 Correlations With Metallicity ([M/H])

The well-known correlation between CMD morphology and metallicity is evident upon comparing Fig. 2 (a cluster with $[M/H] \sim -0.8$) with the much more metal-deficient objects in Figs. 3—5. As metallicity increases, the slopes of the giant and asymptotic branches decrease, the giant-branch and turnoff colors become redder, and there is a tendency for the horizontal branch to become redder. Marked changes arising from [M/H] differences in the shape and position of subgiant and turnoff sequences are predicted by the isochrones and found in the observations (VandenBerg, this conference).

3.2 The As-Yet Unpredictable Horizontal Branch (HB)

Twenty years after its recognition, a convincing explanation for the "second parameter problem" (Sandage and Wildey 1967, van den Bergh 1965,1967), in which clusters with similar metallicities exhibit dramatically different color distributions of HB stars, remains elusive. One aspect of the observational situation is illustrated in Fig. 5 for three clusters for which the data indicate the same ages, Y and [M/H] values. M92 is thought to be typical of very metal-deficient clusters, while M15 and NGC 5466 show some red HB stars and more variables. The blue gap in the M15 HB (near B−V∼0.05), which sets in where the NGC 5466 HB ends and is not seen in M92, and the long, blue tail are striking features requiring explanation. Buonanno et al. (1985) conclude from their exhaustive analysis of (presently undetectable) differences in age, Y, [O/H], core and envelope masses, rotation, etc., that the solution to the "second parameter problem" may well lie in determining the appropriate weight to be applied to each of many competing parameters within each cluster, possibly as a function of position within the Galaxy. The survey called for in §3 would provide invaluable new constraints on the second parameter problem, to which I'll return in §3.7.

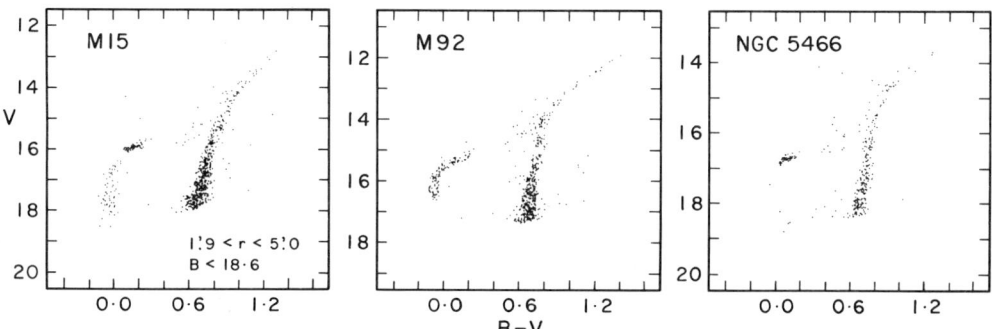

Fig. 5. Three clusters, each having [M/H]= −2.1 ± 0.2, show remarkably different HBs in these *photographic* CMDs from Buonanno et al. (1985).

Clusters having M15-like discontinuous HBs with an extremely blue tail are much more common than once recognized, with attendant importance for integrated light studies of globulars and old stellar populations, ionization of halo gas, etc. (Welch and Code 1980, van Albada et al. 1981, de Boer 1985, Castellani and Cassatella 1986). Such clusters now include (by NGC) 288 (Buonanno et al. 1984), 1851 (Stetson 1981), 1904 (Harris et al. 1983b; see also Heasley et al. 1986, Gratton and Ortolani 1986b), 2419 (Nemec et al. 1986b), 2808 (Harris 1974), 4833 (Nemec et al. 1986a), ω Cen (Da Costa et al. 1986), 6121 (Lee 1977), 6284, 6293 and 6333 (Harris and Harris 1986), and 6752 (Cannon 1981).

As Demarque describes elsewhere in this volume, gaps in luminosity functions are not restricted to globular cluster HBs. Gaps are common on the lower giant branches of metal-poor, but apparently not metal-rich, clusters (King et al. 1985). Such discontinuities provide extremely important constraints on key mixing events at earlier stages of stellar evolution.

3.3 Blue Straggler Stars (BSS)

Until recently the clearest evidence for a BSS sequence lying above and blueward of the main-sequence turnoff in a Galactic globular was M3 (Sandage 1953), with some BSS being suspected in the low-latitude cluster M71 (Arp and Hartwick 1971). Some clusters (47 Tuc—Hartwick and Hesser 1974, NGC 6752—Cannon 1981) definitely seem *not* to have them. Radial velocity membership of BSS is now established in M3 (Chafee and Ables 1983) and ω Cen (Da Costa et al. 1986); while strong sequences are visible in CMDs of NGC 5053 (Nemec and Cohen 1986) and NGC 5466 (Nemec and Harris 1986), and puzzling cooler candidates are identifed in E3 (van den Bergh et al. 1980, McClure et al. 1985). BSS are also known in three dwarf spheroidal galaxies, Sculptor (Da Costa 1984), Ursa Minor (Olszewski and Aaronson 1985) and Draco (Carney and Seitzer 1986). Explanations advanced for BSS include youth, binary mass transfer, main-sequence lifetime extension, etc. ω Cen and NGC 5466 provide much needed new data for this controversial phase of stellar evolution. In the latter cluster, the BSS are more centrally concentrated than the subgiant stars of the same magnitude (Fig. 6), with the inference that their masses are $\sim 1.3 M_\odot$. A recent epoch in which single, relatively massive stars were formed in this low density cluster seems unlikely, leading Nemec and Harris to favor scenarios involving mass-transfer binaries or coalesced stars. Clearly, if BSS are generally more centrally concentrated, available CMDs are very incomplete for them.

Another interesting aspect emphasized in the Da Costa et al. and Nemec and Harris work concerns the possible relationship between unusual variable stars, the BSS and extended blue HBs (§3.2). NGC 5466 is the only Galactic globular known to contain an anomalous Cepheid of the type found in dwarf spheroidal galaxies, while the ω Cen BSS, E39, is a dwarf Cepheid (AI Vel) variable. In their study of clusters containing short-period Cepheids, Harris and Harris (1986) have discovered very blue HBs, which they suggest supports the idea that anomalous Cepheids and similar UV bright stars come from the extreme blue end of the HB (whatever produces that!).

3.4 White Dwarf (WD) Stars

Penny and Dicken's (1986) Fig. 13 illustrates the current difficulties in locating this predicted constituent of globulars with ground-based data. Bahcall (1985) and Renzini (1985) have shown the potential of HST for identifying cluster WDs, an opportunity with important ramifications for understanding cluster dynamics and determining distances (Fusi Pecci and Renzini 1979).

3.5 Cluster Cores

Central X-ray sources, possible color gradients and interest in dynamical processes all drive the study of cluster centers reviewed elsewhere herein by Djorgovski and by Grindlay. Spatial resolution limitations from the ground frequently hamper exploration of populations at the centers of globular clusters, but some recent progress has been made and new surveys are being undertaken. The series of photometric studies by Aurière, Cordoni and their colleagues (e.g., Aurière et al. 1986 and Cordoni and Aurière 1984) illustrate the difficulties and potential rewards, one of the latter being the discovery of a strongly variable UV-bright star which seems

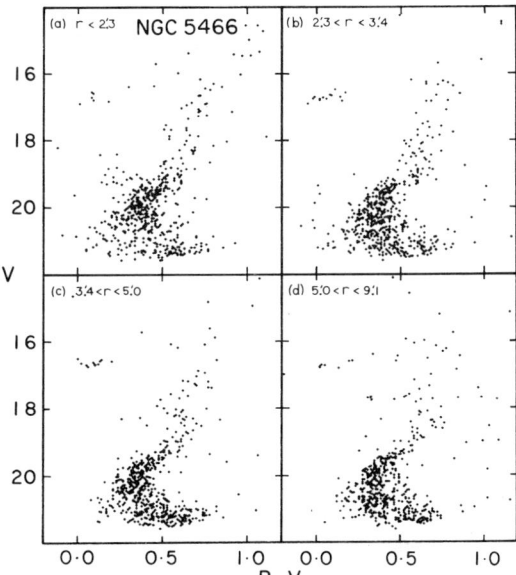

Fig. 6. *Photographic* CMDs as a function of radius for NGC 5466 (Nemec and Harris 1986). Note the large BSS population in the central region, whose area is $\sim 1/15$th that of the other three regions, and a nearly complete absence of faint blue HB stars.

to be the optical counterpart to the M15 X-ray source (Charles et al. 1986). Preliminary results from a related program of CCD observations from Mauna Kea of metal-poor clusters by Rose and Stetson (1986) may indicate a number of new supra-HB and UV-bright stars in their central regions, which will necesitate a revision of the already-impressive statistics of those stars (last compiled by Harris et al. 1983), and further clarify their, possibly underappreciated, role in the integrated light of galaxies.

3.6 Metal-Rich Clusters

This review has concentrated on metal-poor clusters, largely in reflection of how many fewer metal-rich ones have (or can be!) thoroughly studied from the ground, because of their preferential location in the (crowded) bulge region of the Galaxy. From spectroscopy of the composite light of the central regions of nine globular clusters having $[M/H] > -0.8$, Rose and Tripicco (1986) find large cluster-to-cluster differences in CN strengths. They also find that clusters having stronger CN strengths invariably show weak Sr II/Fe I ratios normally indicative of a dwarf-star population. To account for this dichotomy among the integrated spectral properties of the metal-rich globulars they speculate on the possibility of a 10 Gyr age spread among the clusters (the dwarf-dominated being younger), or on the presence of Campbell's (1986) coalesced binaries. While I suspect a more conventional explanation will ultimately prove tenable, their surprising results emphasize anew the importance of photometric (and, where appropriate, proper motion—see Fig. 4) studies of the highest possible spatial resolution for Galactic bulge clusters of *all* metallicities.

3.7 Sparse Clusters and Binary Sequences

Most of the Galaxy's lower luminosity clusters lie at substantial distances from the center. An example of their CMDs has already been given (Fig. 3), and in Fig. 7 six more are displayed as a function of distance. At ~120 kpc, AM-1 (Aaronson et al. 1984) is the most distant globular thought to be associated with the Galaxy, while at ~10 kpc the sparse E3 cluster is relatively nearby. The predominance of red HBs among these low [M/H] clusters, i.e., the classical second-parameter effect—§3.2, is evident in Figs. 3 and 7, and led Searle and Zinn (1978) to suggest that the second parameter might be age differences correlated with R_{gc}. With the exception of the much younger stars seen in some (but not all) dwarf spheroidal systems, the best available data for outer halo systems favor the same ~15 Gyr age for them as found for nearer globulars. We probably must await age discrimination at the sub-Gyr (!) level to evaluate age as the second parameter in the outer-halo clusters. However, the discussion under §3.2 has already indicated the complexities of the second-parameter situation. Indeed, as H. Smith and Perkins (1982) have stressed, clusters such as NGC 6171 and 6723, with R_{gc}'s ~3 kpc, provide evidence for the second-parameter effect among *inner* halo clusters (see also G. Smith and Hesser 1986). (Clusters such as AM-4 and E3 avoid the second-parameter problem altogether by having no HBs at all!)

New vistas on physical processes, especially dynamical ones, are opened as a consequence of the lower densities in sparse clusters. For instance, CMD studies reveal that the main sequences of Pal 5 (Fig. 3) and E3 (Fig. 7) are truncated, presumably in part by tidal stripping. The search for binary stars in globulars has been long with modest returns (cf. Trimble 1980, Grindlay–this volume). Evidence for binaries composed of similar-mass stars forming a parallel sequence ~0.75 mag above the main sequence has been seen in E3 (McClure et al. 1986, Gratton and Ortolani 1986a), which raises the suspicion that such pairs might more easily form or, once formed, survive in very sparse clusters. However, no such sequence is seen in Pal 5 (Smith et al. 1986), possibly because of increased observational scatter at the faint magnitude levels involved. Contrary to the suggested association between sparse clusters and binaries, a much more centrally concentrated and massive cluster, M68, shows some indications of main-sequence binaries (McClure et al. 1987). Buonanno (1986) has remarked that he and his colleagues find in their careful CCD photometry main-sequence widths that exceed those explained by photometric errors, which they attribute to binaries, so maybe one doesn't have to be from Victoria to see the effect! Clearly, very high S/N data for large samples extending to faint magnitudes will go far to explore the reality and frequency of main-sequence binaries in Galactic globulars.

4. AGES

Because ages are covered throughly by VandenBerg elsewhere in this *Symposium*, my remarks will concisely reflect a few concerns from my perspective as an observer. It should come as no surprise that the daunting task of setting the lower limit to the age of the Universe from globular cluster CMDs is not amenable to "quick and dirty" solutions: only the most thorough analyses based upon a homogeneous set of the highest S/N-ratio data stand any chance of success. As we newcomers embark on this ambitious quest, we can do no better than to bear in mind the scope and care with which Sandage has approached the necessary observations on this topic throughout his career. With CCD data *relative* age discrimination at the Gyr level is within reach for nearer globulars. However, present estimates are not achieving *absolute* precisions better than a few Gyr for reasons

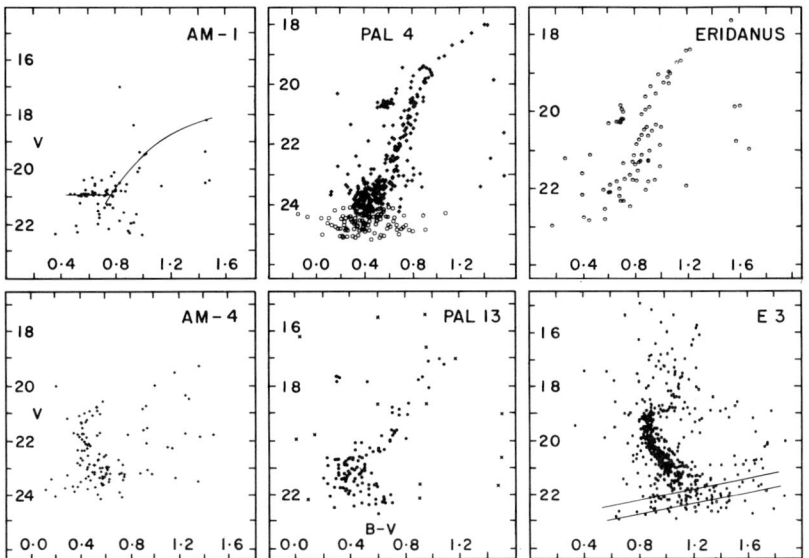

Fig. 7. CMDs for sparse halo clusters, based, except as noted, on CCD data. R_{gc}'s (in kpc) from Webbink (1985) are given following the name. Top row, from left: AM-1, 118 (Aaronson et al. 1984); Pal 4, 96 (Christian and Heasley 1986, see also Reed and Harris 1986); Eridanus, 90 (Da Costa 1985). Bottom row, from left: AM-4, 26, the poorest globular known (Inman and Carney 1986); Pal 13, 26 (*photographic* from Ortolani et al. 1985); E3, 10 (McClure et al. 1985; see also Gratton and Ortolani 1986a). Note the ordinate/abscissa ratios in the Pal 4 and Eridanus diagrams differ from the usual 1:5 value.

largely—but beware the caveats of §2—not having to do with the CCD photometry itself (e.g., Flannery and Johnson 1982, Heasley and Christian 1986, VandenBerg 1986 and this volume). As a consequence, conflicting results on the systematics of globular cluster ages surface in the latest analyses (Gratton 1985, Sandage 1986, Alcaino and Liller 1986, Peterson 1986b). The latter analysis, based upon the most complete sample consisting of 37 clusters with turnoff-region photometry (much of it photographic), yields 15.1 ± 0.5 Gyr (s.e.m.) with no evidence for differences in age between inner and outer halo clusters or with [M/H].

While there is an enormous amount of work still ahead, I think that there are many reasons for optimism. Once free of the Earth's atmosphere, the quality of the photometry will, with a well-calibrated HST, improve substantially in several regards. As well, the fundamental physical parameters required for interpreting the CMDs are showing steady improvement. Recent observations on the primordial Y value give strong reasons for adopting 0.24 ± 0.01 (Boesgaard and Steigman 1985, Caputo 1985, Kunth 1986, Pagel et al. 1986). Uncertainties in [M/H] and [CNO/H] values are still too large, but can be reduced further, and more distant clusters can come under scrutiny with the next generation of telescopes. New techniques for distance determinations involving the turn-up of the luminosity function at faint magnitudes (VandenBerg 1986) or white dwarfs (Fusi Pecci and Renzini 1985) offer promise for clarification of long-standing debates over HB absolute magnitudes, as will future studies of RR Lyraes and HBs with CCDs. Something that is long overdue, however, is a major attack on reddening determinations *independent* of cluster stars or integrated properties, particularly for those clusters with superbly defined turnoffs. (For instance, application of $uvby-\beta$ photometry to large samples of foreground stars might substantially improve E(B−V)s for many clusters.)

5. MAIN–SEQUENCE LUMINOSITY FUNCTIONS (LF)

As alluded to in §2, the determination of LFs is a very time consuming task much more suited to CCDs than to photographic plates. The initial results are very intriguing and suggest we may be learning something quite new about star formation processes in the early Galaxy. The basic observational results are summarized in Fig. 8, which is based upon work by Lupton and Gunn (1986), Penny and Dickens (1986), Richer and Fahlman (1984, 1986), as well as our own, and is taken from McClure et al. (1986). The figure compares (heavy solid lines) the

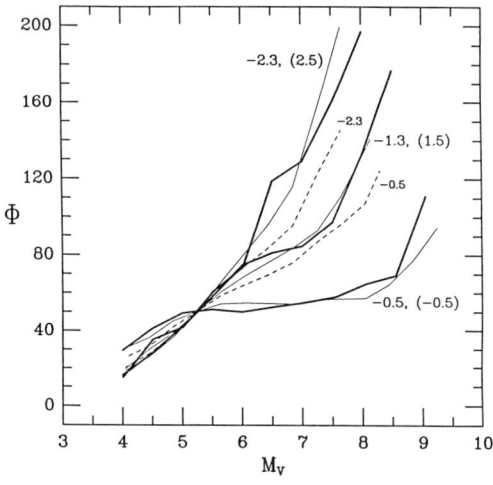

Fig. 8. Mean LFs for globulars of differing [M/H]s are compared with theoretical ones (see text).

average LF from deep CMDs of clusters with [M/H] ~ -2.3 (M15), ~ -1.3 (M5, M13, NGC 6752) and ~ -0.5 (M4, 47 Tuc) with theoretical LFs for the appropriate [M/H] and the usual, power-law, mass-function slope, x (in parentheses). The dashed curves represent additional LFs for $x = 1.5$ and the values of [M/H] shown. Within the small sample, there appears to be a strong metallicity dependence to the present-day mass function for these globulars suggesting that a universal (Salpeter) value is unsuitable for globular clusters. The result is independent of the [M/H] scale adopted. No correlations are evident between the inferred x and concentration class, central relaxation time, or the radius at which the observations were secured. Multi-component King models (Pryor et al. 1987) imply that dynamical corrections, while important, do not significantly alter the slope of the relation between x and [M/H]. Consequently, it appears that relation reflects, at least in part, the properties of the initial mass function in the proto-Galaxy. Observations of M92, M12 and NGC 6362 analyzed recently show a similar correlation of LF slope with [M/H], as described in a contributed paper herein.

In a thought-provoking paper, Smith and McClure (1986) have investigated one side of the "chicken and egg" questions, "does low metallicity favor production of low mass stars (which would run contrary to standard ideas)?" or "does the IMF determine the chemical properties of the halo clusters?". In the latter case, metal-poor clusters might have been formed in proto-clouds with steep mass functions and achieved self-enrichment with little mass loss. On the other hand, proto-clouds with flat mass functions would have produced more than enough heavy elements to enrich themselves to the [M/H]s seen in metal-rich clusters; indeed, a truncation of the mass function for more massive stars would be required to limit the enrichment to the observed levels. McClure (1986) has also noted than a correlation with x might be found in Hartwick's (1986) new description of the Galactic halo, in which case the flat LF clusters would represent the *disk* of the Galaxy, while those with steeper LFs would represent the inner and outer halo components Hartwick identifies from halo RR Lyrae star distributions.

6. OTHER PHOTOMETRIC SYSTEMS AND FUTURE OUTLOOK

The scarcity of CMDs on systems other than UBVRI arises, in large measure, because the necessary network of standards suitable for CCD work is simply *not* available for any system other than UBVRI. Nevertheless, I wish in closing to stress the formidable powers of other systems to provide answers to many of the astrophysical problems we study. An obvious example would be IR photometry, which has been extensively and effectively applied in single-channel mode (Frogel et al. 1983, Lloyd-Evans 1983, and references therein).

Recent observations on the $uvby - \beta$ system by Anthony-Twarog (1986) of NGC 6397 provide a CMD and color-color diagrams (Fig. 9). In the former the turnoff-region is well delineated in the complete sample she has analyzed. When she selects only those stars deemed to have the highest quality (from DAOPHOT's χ statistic), she is also able to identify the turnoff in the the $m_1, b - y$ diagram. If the error of $< m_1 >$ can be reduced to ± 0.01mag, [M/H] can, in principle, be

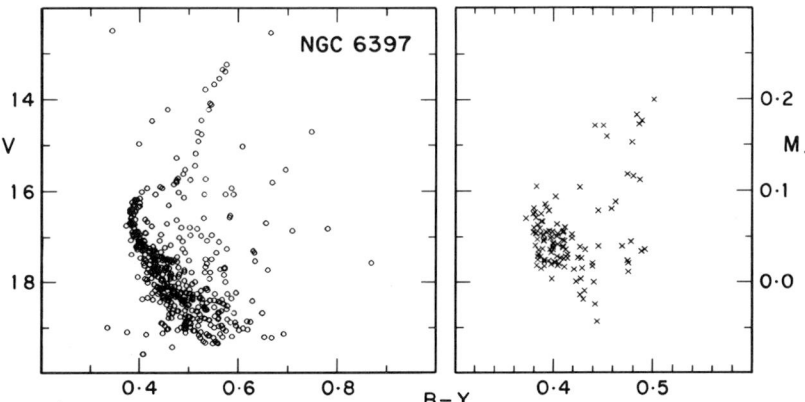

Fig. 9. CMD and color-color plot on the $uvby - \beta$ system from Anthony-Twarog (1986); for the latter, a subset of stars believed to have the most reliable photometry is used to show detection of the turnoff. Note that the zero points are based on only four stars from the photometry of Ardeberg et al. (1983).

determined to ±0.1dex. However, achieving this goal will require much better zero points than presently possible with the available standards. Encouraging concordance with theoretical isochrones is also obtained in the $c_1, b - y$ diagram.

Cohen (1985) has presented CMDs on the Thuan-Gunn (1976) system for NGCs 5466, 6229 and 7006. While the scatter arising from her short exposures on these distant clusters is large, her data demonstrate the potential of the system. Bell and VandenBerg (1986) have computed synthetic spectra and colors for the T-G system (Fig. 10), demonstrating how the system's CMDs offer the possibility of simultaneously constraining age, [M/H], and T_{eff}.

Another under-explored system that appears to offer considerable opportunities for the study of globulars, particularly in the integrated light from clusters around distant galaxies, is the Washington system (Canterna 1976, Harris and Canterna 1977). Its attractiveness stems, in part, from its broad, yet filter-defined, passbands which yield sensitive abundance information, as Geisler (1986) and Canterna et al. (1986) have convincingly demonstrated. A theoretical color study for the Washington system similar to Bell and VandenBerg's (*op. cit.*) would strongly influence the decision on the part of observers regarding where to put their energies in the laborious development of suitable standards for study of Pop II objects using non-UBV systems. To reiterate: there is no doubt that the lack of networks of faint equatorial standards suitable for calibrating CCD observations on these systems is delaying their wider application to globular cluster photometry.

The time also seems ripe to develop synthetic integrated globular cluster models incorporating the features being defined by the CCD CMDs: such models will be essential for analysis of the most distant globulars. All told, the future of CMD work on Galactic globulars is an exciting one whose impact will be felt on many areas of astrophysical research. HST, the Keck telescope, Tektronix 2048 CCDs, etc. all promise advances not only on the topics mentioned above (and the even more numerous ones not mentioned!), but also on those we do not yet imagine.

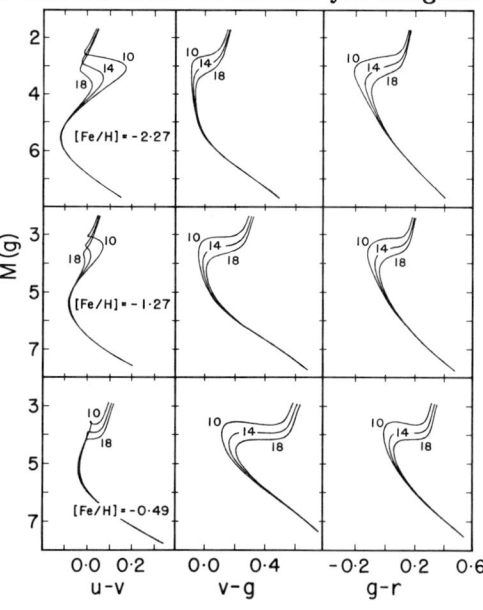

Fig. 10. An adaptation of Bell and VandenBerg's (1986) Fig. 6 depicting the behavior of isochrones for Thuan–Gunn photometry for Y=0.2, ages of 10, 14 and 18 Gyr and the [M/H]'s indicated. Similar calculations for the Washington system would be extremely valuable.

Were Shapley and his contemporaries Hertzsprung and Russell with us today, I'm certain they would be as excited about present and future prospects for research based on CMDs as those of us exploiting the new technology are!

Acknowledgements. My deepest thanks go to: the many colleagues around the world who answered my requests for information; my collaborators whose contributions and support make it all possible and so enjoyable; CTIO, KPNO and CFHT for access to their superb facilities; the IAU for travel support; Dave Duncan for artwork; and to Betty, and to Rebecca and Gillian, for accepting with grace my traveling to Boston on the eve of our 23rd wedding anniversary and their 15th birthday, respectively.

REFERENCES

Aaronson, M., Schommer, R. A. and Olszewski, E. W. 1984 Astrophys. J. 276, 221.
Alcaino, G. and Liller, W. 1986 Mem. Soc. Astron. Ital. in press.
Anthony-Twarog, B. 1986 Private communication.
Ardeberg, A., Lindgren, H. and Nissen, P. E. 1983 Astron. Astrophys. 128, 194.
Arp, H. C. and Hartwick, F. D. A. 1971 Astrophys. J. 167, 499.
Aurière, M. Maucherat, A., Cordoni, J. -P., Fort, B. and Picat, J. P. 1986 Astron. Astrophys. 158, 158.
Bahcall, J. N. 1985 in IAU Symposium No. 113, Dynamics of Star Clusters, J. Goodman and P. Hut, eds., Reidel, Dordrecht, p. 481.
Bell, R. A. and VandenBerg, D. A. 1986 Astrophys. J., in press.
Boesgaard, A. M. and Steigman, G. 1985 Ann. Rev. Astron. Astrophys. 23, 319.
Buonanno, R. 1986 Mem. Soc. Astron. Ital., in press.
Buonanno, R., Buscema, G., Corsi, C. E., Iannicola, G. and Fusi Pecci, F. 1983 Astron. Astrophys. Suppl. 53, 1.
Buononno, R., Corsi, C. E. and Fusi Pecci, F. 1985 Astron. Astrophys. 145, 97.
Buononno, R. Corsi, C. E. and Fusi Pecci, F., Alcaino, G. and Liller, W. 1984 Astrophys. J. 277, 220.
Campbell, B. 1986 Astrophys. J. 307, in press.
Cannon, R. D. 1981 in IAU Colloquium No. 68, Astrophysical Parameters for Globular Clusters, A. G. D. Philip and D. S. Hayes, eds., L. Davis Press, Schenectady, p. 501.
Canterna, R. 1976 Astron. J. 81, 228.
Canterna, R., Geisler, D., Harris, H. C., Olszewski, E. and Schommer, R. 1986 Astron. J., in press.
Caputo, F. 1985 Rep. Prog. Phys. 48, 1235.
Carney, B. W. and Seitzer, P. 1986 Astron. J. 92, 23.
Castellani, V. 1985 Fund. Cosmic Phys. 9, 317.
Castellani, V. and Cassatella, A. 1986 in Scientific Accomplishments of the IUE, Y. Kondo, ed., Reidel, Dordrecht, in press.
Chafee, Jr., F. H. and Ables, H. D. 1983 Publ. Astron. Soc. Pacific

95, 835.
Charles, P. A., Aurière, A., Ilovaisky, S. A. and Koch-Miramon, L. 1985 IAUC No. 4146.
Christian, C. A. and Heasley, J. N. 1986 Astrophys. J. 303, 216.
Cohen, J. G. 1985 Astron. J. 90, 2254.
Cordoni, J. -P. and Aurère, M. 1984 Astron. Astrophys. Suppl. 58, 559.
Cudworth, K. M. 1986 Astron. J. 92, 348.
Da Costa, G. S. 1984 Astrophys. J. 285, 483.
Da Costa, G. S. 1985 Astrophys. J. 291, 230.
Da Costa, G. S., Norris, J. and Villumsen, J. V. 1986 Astrophys. J., in press.
de Boer, K. S. 1985 Astron. Astrophys. 142, 321.
Flannery, B. P. and Johnson, B. C. 1982 Astrophys. J. 263, 166.
Frogel, J. A., Cohen, J. G. and Persson, S. E. 1983 Astrophys. J. 275, 773.
Fusi Pecci, F. and Renzini, A. 1985 in Astronomical Uses of the Space Telescope, F. Machetto and F. Tarenghi, eds., ESO Munich. p. 181.
Geisler, D. 1986 Publ. Astron. Soc. Pacific 98, 762.
Gratton, R. G. 1985 Astron. Astrophys. 147, 169.
Gratton, R. G. and Ortolani, S. 1986a Astron. Astrophys. Suppl. in press.
Gratton, R. G. and Ortolani, S. 1986b Astron. Astrophys. Suppl. in press.
Green, E. M. 1985 Private communication.
Harris, H. C. and Harris, W. E. 1986 Private communication.
Harris, H. C. and Canterna, R. 1977 Astron. J. 82, 798.
Harris, W. E. 1974 Astrophys. J. Letters 192, L161.
Harris, W. E. and Hesser, J. E. 1986, in preparation.
Harris, W. E., Hesser, J. E. and Atwood, B. 1983a Astrophys. J. Letters 268, L111.
Harris, W. E., Hesser, J. E. and Atwood, B. 1983b Publ. Astron. Soc. Pacific 95, 951.
Harris, H. C., Nemec, J. M. and Hesser, J. E. 1983 Publ. Astron. Soc. Pacific 95, 256.
Hartwick, F. D. A. 1986 in Cambridge Symposium, in press.
Hartwick, F. D. A. and Hesser, J. E. 1974 Astrophys. J. Letters 194, L129.
Heasley, J. N. and Christian, C. A. 1986 Astrophys. J. 307 in press.
Heasley, J. N., Janes, K. A. and Christian, C. A. 1986 Astron. J. 91, 1108.
Hesser, J. E. 1983 Mem. Soc. Astron. Ital. 54, 361.
Hesser, J. E. and Hartwick, F. D. A. 1977 Astrophys. J. Suppl. 33, 361.
Inman, R. T. and Carney, B. W. 1986 Astron. J., in press.
King, C. R., Da Costa, G. S. and Demarque, P. 1985 Astrophys. J. 299, 674.
Kunth, D. 1986 Publ. Astron. Soc. Pacific 98, in press.
Lee, S. -W. 1977 Astron. Astrophys. Suppl. 27, 381.

Lloyd Evans, T. 1983 Monthly Notices Roy. Astron. Soc. 204, 945.
Lupton, R. H. and Gunn, J. E. 1986 Astron. J. 91, 317.
McClure, R. D. 1986 Private communication.
McClure, R. D., Hesser, J. E., Stetson, P. B. and Stryker, L. L. 1985 Publ. Astron. Soc. Pacific 97, 665.
McClure, R. D., VandenBerg, D. A., Smith, G. H., Fahlman, G. G., Richer, H. B., Hesser, J. E., Harris, W. E., Stetson, P. B. and Bell, R. A. 1986 Astrophys. J. Letters, in press.
McClure, R. D., VandenBerg, D. A., Hesser, J. E., Stetson, P. B. and Bell, R. A. 1987 Astron. J., submitted.
Nemec, J. M. and Cohen, J. G. 1986 Private communication.
Nemec, J. M. and Harris, H. C. 1986 Astron. J., submitted.
Nemec, J. M., Richer, H. B. and Fahlman, G. G. 1986a, Private communication.
Nemec, J. M., Richer, H. B. and Stetson, P. B. 1986b, Private communication.
Olszewski, E. W. and Aaronson, M. 1985 Astron. J. 90, 2211.
Ortolani, S., Rosino, L. and Sandage, A. 1985 Astron. J. 90, 473.
Pagel, B. E. J., Terlevich, R. J. and Melnick, J. 1986 Publ. Astron. Soc. Pacific 98, in press.
Penny, A. J. and Dickens, R. J. 1986 Monthly Notices Roy. Astron. Soc. 220, 845.
Peterson, C. J. 1986a Publ. Astron. Soc. Pacific, submitted.
Peterson, C. J. 1986b Publ. Astron. Soc. Pacific, submitted.
Pryor, C. P., Smith, G. H. and McClure, R. D. 1987 Astron. J., in press.
Reed, B. C. and Harris, W. E. 1986 Astron. J., 91, 81.
Renzini, A. 1985 Astron. Expresss 1, 127.
Richer, H. B. and Fahlman, G. G. 1984 Astrophys. J. 277, 227.
Richer, H. B. and Fahlman, G. G. 1986 Astrophys. J. 304, 273.
Rose, J. A. and Stetson, P. B. 1986, Private communication
Rose, J. A. and Tripicco, M. J. 1986 Astron. J., 92, 610.
Sandage, A. 1986 Ann. Rev. Astron. Astrophys., in press.
Sandage, A. and Wldey, R. 1967 Astrophys. J. 150, 469.
Searle, L. and Zinn, R. J. 1978 Astrophys. J. 225, 357.
Smith, H. A. and Perkins, G. J. 1982 Astrophys. J. 261, 576.
Smith, G. H. and McClure, R. D. 1986 Astrophys. J., in press.
Smith, G. H. and Hesser, J. E. 1986 Publ. Astron. Soc. Pacific, in press.
Smith, G. H., McClure, R. D., Stetson, P. B., Hesser, J. E. and Bell, R. A. 1986 Astron. J. 91, 842.
Stetson, P. B. 1981 Astron. J. 86, 687.
Stetson, P. B. 1984 DAOPHOT User's Manual, DAO, Victoria.
Stetson, P. B. 1987 Publ. Astron. Soc. Pacific, in preparation.
Stetson, P. B., VandenBerg, D. A. and McClure, R. D. 1985 Publ. Astron. Soc. Pacific 97, 908.
Thuan, T. X. and Gunn, J. E. 1976 Publ. Astron. Soc. Pacific 88, 543.
Tody, D. 1981 RICHFIELD Photometry Program User's Guide, Kitt Peak National Observatory, Tucson.
Trimble, V. 1980 in IAU Symposium No. 85, Star Clusters, J. E.

Hesser, ed., Reidel, Dordrecht, p. 259.
van Albada, T. S., de Boer, K. S. and Dickens, R. J. 1981 Monthly Notices Roy. Astron. Soc. 195, 591.
VandenBerg, D. A. 1986 Mem. Soc. Astron. Ital., in press.
van den Berg, S. 1965 J. Roy. Astron. Soc. Canada 59, 153.
van den Berg, S. 1967 Astron. J. 72, 70.
van den Bergh, S., Demers, S. and Kunkel, W. E. 1980 Astrophys. J. 239, 112.
Walker, A. R. 1984 Monthly Notices Roy. Astron. Soc. 209, 83.
Walker, A. R. 1986 in IAU Symposium No. 118, Instrumentation and Research Programmes for Small Telescopes, P. L. Cottrell and J. B Hearnshaw, eds., Reidel, Dordrecht.
Webbink, R. F. 1985 in IAU Symposium No. 113, Dynamics of Star Clusters, J. Goodman and P. Hut, eds., Reidel, Dordrecht, p. 541.
Welch, G. A. and Code, A. D. 1980 Astrophys. J. 236, 798.

DISCUSSION

WALLERSTEIN: There is a great similarity between the Thuan-Gunn System and the Washington System. By just adding an I color, the Thuan-Gunn system is almost the same as MT_1T_2.

OSTRIKER: What is the location of blue stragglers?

HESSER: The prominent blue straggler sequence I showed from Nemec and Harris in NGC 5466 consists of stars almost entirely located in the central regions. Nemec and Cohen find a similar result for NGC 5053, another cluster of low central density.

GRINDLAY: I call your attention to the poster paper by Bailyn, et al. which shows that the faint blue horizontal-branch stars appear to be more centrally condensed than the BHB in ω Cen. We have interpreted this as evidence for binary mergers of white dwarfs and main sequence stars in the cluster core. These systems may be easier to recognize than blue stragglers in crowded cluster cores. My question is - when you state faint blue horizontal-branch stars are not observed in 47 Tuc, have you searched into the cluster core?

HESSER: A CCD frame pair (totaling 30 sec of integration and requiring more than 100 hours of VAX 11/780 analysis time) located at 2.5 arcmin. from the center does not show any (or at most only 1 or 2) blue straggler candidates.

ZINN: You said that there is a connection between very blue horizontal-branch stars, blue stragglers, and anomalous Cepheids. There are, however, systems such as the Draco and Sculptor dwarf spheroidal galaxies that contain blue stragglers and anomalous Cepheids and do not contain very blue horizontal-branch stars. This suggests to me that there may be no connection between the very blue horizontal-branch stars and these other types of stars.

HESSER: My statement reflected my understanding of recent results from the galactic globular clusters, particularly the work of DaCosta, Norris and Villmusen, Nemec and H. Harris, and H. Harris and W. Harris, many of whom are here and can better comment than I.

FUSI PECCI: In a poster paper on M 3 we find the blue stragglers to be less centrally concentrated.

HESSER: I warned you there was controversy on this topic!

LILLER: As a footnote to Dr. Hesser's plea for accurate calibration of CCD magnitude determinations, I should like to urge observers to set up photoelectric standards in the same CCD field. This work can be done numerous times - on 6 or 8 or more nights - using relatively small telescopes, thereby reducing errors of transfer.

HESSER: You have stated clearly the point I was trying to make and, as I remarked, stands at the heart of Sandage's approach to the fundamental problem of calibration. Each part of the system, camera, chip and filters, has its own "personality" which must be very carefully calibrated.

THE OVERALL ABUNDANCES OF GLOBULAR CLUSTERS

R. A. Bell

University of Maryland

1. INTRODUCTION

At a meeting on globular clusters held five years ago, Zinn(1981) made the following comments:

"In summary, there are clearly large differences in metal line strength between 47 Tuc giants and the giants in the other clusters. These differences suggest that there are substantial differences in metal abundance, which conflicts with the measurements obtained from echelle spectrograms, and there appears to be no simple way of reconciling these results. Until this is done, the metallicity scale for globular clusters hangs in limbo."

Is this still true? Can we understand the reasons why different methods for getting the abundances of globular cluster stars give different results? It now appears that a plausible reason for some of the discrepancy between the high dispersion results and the results given by other methods has been found. I will give this explanation subsequently in the talk. However, prior to this, I will briefly discuss the various methods used to obtain abundances for individual cluster stars and for clusters as a whole. There are a whole variety of methods ranging from observations of individual stars to studies of the geometry of color magnitude diagrams to measurements of the integrated light of the cluster. Zinn and West (1984) have intercompared the results from many of these methods, and their conclusions are very similar to mine.

2. SPECTROSCOPIC OBSERVATIONS OF INDIVIDUAL STARS

2.1 High Dispersion Spectroscopy.

The queen of the methods is clearly that of high dispersion spectroscopy. This is virtually the only method which allows us to find the abundances of individual elements. The first high dispersion analyses of globular cluster stars used photographic spectra obtained with the coude' spectrograph on the 200 inch telescope. Helfer, Wallerstein and Greenstein (1959) found [Fe/H] < -2.0 for M 92 and < -1.3 for M 13. Owing to the faintness of the stars, exposure times of over 12 hours were needed to obtain the spectra. Probably because of this, stars in other globular clusters were analyzed only after about twenty years had passed. Cohen (1978, 1979) found [Fe/H] = -1.6 for M 13, -1.8 for M 3, -2.35 for M 92 and -2.20 for M 15, in essential agreement with the previous results. Her subsequent work for M 67 (Cohen 1980a, 1980b), where she obtained [Fe/H] = -0.3, did not raise any eyebrows, but the result for M 71 was much more surprising and controversial, since she obtained [Fe/H] = -1.3. While many people probably felt that the M 67 result was a bit low, it was not alarmingly so and did give a lot of weight to the M 71 value. Cohen's M 71 result was also in agreement with a study of 47 Tuc, where Pilachowski, Canterna and Wallerstein (1980) found [Fe/H] = -1.2.

The high dispersion work continued with analyses of other globular clusters, including Omega Cen and M 22, and the very extensive paper by Pilachowski, Sneden and Wallerstein (1983) gave results for stars in seven clusters as well as re-analyzing other data. Pilachowski (1984) gives the results of high dispersion analyses for stars in 24 clusters. Stars in additional clusters have been studied by Geisler (1984) but few details of the analysis are given.

2.2 Low Dispersion Spectroscopy

This can be subdivided into two sets of observations, the ΔS measurements and the rest. The ΔS system, invented by Preston and used extensively by Butler and, more recently, by Smith (see Smith 1984a for a review), defines the difference between the spectral types of an RR Lyrae variable, judged from the K line and judged from the hydrogen lines, to be the quantity ΔS. For an individual star, ΔS does depend on pulsation phase and so, for abundance work, observed ΔS values have to be corrected to a constant phase. ΔS ranges from about 14 for metal poor stars to 0 for metal rich ones. The K line is chosen as the metal abundance indicator because there are few other metallic lines easily visible in low dispersion spectra of metal poor RR Lyraes. In practice, a vexing problem is presented by the interstellar K line, whose velocity and strength both affect ΔS.

In most applications, the ΔS measurements have been calibrated in terms of [Fe/H] from studies of field RR Lyraes (Butler 1975), although Manduca (1981) has published a spectrum synthesis calibration. The two calibrations are in reasonable agreement except at the metal-rich end, where Manduca finds that a given value of ΔS corresponds to a lower metal abundance.

The ΔS method has been applied to numerous clusters, Smith (1984b) giving a list of abundances. The most metal rich clusters in the list are NGC 6712 and NGC 6723, at [Fe/H] = -0.57 and -0.68, respectively. Unfortunately the method cannot be used for M 71, which does not contain RR Lyraes, and the results for 47 Tuc are uncertain. While there are three RR Lyraes near 47 Tuc, the question of membership is uncertain for at least two of them and the abundance is not settled for the third star, V9 (Smith 1984b). Smith finds that V9 has an abundance of [Fe/H] = -0.8 ±0.2 but Keith and Butler (1980) found it to have -1.1 ±0.14.

The current tendency with low dispersion digital spectra is to compute spectral indices (e.g. Suntzeff 1980, Canterna, Harris and Ferrall 1982, Burstein, Faber and Gonzalez 1986) although graphical comparisons of observation and calculations are also made. The indices are plotted versus a reddening corrected color. These index, color diagrams are then used in either of two ways. In the first way, the calibration is provided by other observations e.g. high dispersion spectroscopy or ΔS. In this method, the low dispersion data are essentially being used to interpolate between the standard clusters. Alternatively, the low dispersion spectra provide a "ranking" of the cluster abundances. The second means of calibration is again a spectrum synthesis one. The synthetic spectra of various models are analyzed in the same way that the observations are and the resulting indices and colors of the models are directly compared with the observations. This calibration method has advantages and disadvantages. It is clearly a very flexible method, since the synthetic spectra can be computed for any desired abundance, temperature and gravity. It can also utilize the results of stellar interior calculations. On the other hand, without carrying out detailed checks of the synthetic spectra, the accuracy of the results is not known. This is particularly true for the analysis of observations in the ultra-violet of metal-rich stars.

Some of these data illustrate the abundance differences found using high and low dispersion methods. Bell (1984) gives synthetic spectra for the wavelength interval 3600-4600 Å, intended to represent the spectra of stars on the giant branches of clusters with metal abundances [A/H] = -0.5, -1.0, -2.0 and -3.0. These spectra have been used to compute indices which can be compared with the observations of Canterna, Harris and Ferrall. It is quite clear from plots of M(V) versus indices such as mHK, mCa, mG and mCN (which are measures of the strength of H and K, Ca I 4226, the G band and the 3883 Å CN) that the 47 Tuc stars always have stronger spectral features than do NGC 288 stars. The calibration, made on the basis of synthetic spectra, suggests that 47 Tuc has a metal abundance of [M/H] = -0.8 while NGC 288 has an abundance of perhaps -1.1 and M 15 has [M/H] = -2.0. The results from high dispersion work quoted by Pilachowski (1984) for 47 Tuc are -1.09 ±0.2, -1.0 ±0.2 and -0.95 ±0.25 and, for NGC 288 and M 15, the results quoted are -0.95 ±0.15 and -1.76 ±0.25, respectively. If anything, this suggests that 47 Tuc is more metal poor than NGC 288, contrary to the low dispersion spectroscopic result.

Synthetic spectra, as shown by Bell (1984), were the main reason why it was very hard for me to accept a very low abundance for 47 Tuc. When we (Dickens, Bell and Gustafsson 1979) were studying 47 Tuc we compared our synthetic spectra with our low dispersion observations. It was apparent that the Ca I 4226 line in the cooler stellar spectra was always too weak in the spectra computed with [A/H] - 0.5 and always too strong when compared with spectra computed for [A/H] = -1.0.

3. PHOTOMETRIC OBSRVATIONS OF INDIVIDUAL STARS

3.1 The UBV system.

It is certainly possible to obtain some idea of the metal abundances of a star from its ultra-violet excess and this result has been employed for clusters (Sandage 1970, Carney 1979). However, I think that this approach is probably now more useful in exploratory work on field stars than it is for cluster stars.

3.2 The Washington system.

This system uses four filters, C, M, T_1 and T_2. The filters are centered at approximately 4000, 5000, 6000 and 8000 Å and have FWHM of 1000 Å. This width makes the system a potentially attractive one for observing faint stars. The index $(T_1 - T_2)$, corrected for reddening, is used to deduce temperatures while the index $(M-T_1)$ is intended to serve as an abundance indicator. It has been stated that (C-M) is an indicator of molecular band strength as well as a general abundance indicator. I do not believe that the last conclusion is supported by any calculations and more recent observational work (Geisler 1986a) seems inconclusive on this point.

A plot of $(M-T_1)$ versus $(T_1 - T_2)$, derived from synthetic spectrum calculations, using the photoelectric sensitivity functions of Geisler and Kapranidis (1983), shows that the abundance sensitivity is relatively weak, particularly for metal poor stars ([M/H] <-1.0). Much greater abundance sensitivity is seen in the (C-M), $(T_1 - T_2)$ diagram.

3.3 The DDO system.

Hesser, Hartwick and McClure (1977) found cluster abundances from the positions of stars in C(38-42), C(45-48) and C(42-45), C(45-48) diagrams, using a calibration based upon M 92, M 13 and M 71 as standard clusters. Janes(1979) provided a calibration based upon the difference in C(45-48) between a cluster star and a Population I star of the same C(42-45).

Many of the DDO system filters were chosen to measure the strengths of molecular bands e.g. the 3883 and the 4215 CN bands (the C38 and C41 filters) and the G band (the C42 filter). This is a photometric system which, like high dispersion spectroscopy, can be used to find the abundances of particular species such as CH and CN.

In particular, any use of the C(42-45), C(45-48) diagram which is based upon an empirical calibration does assume that the depletion of carbon is the same in all clusters of that overall metal abundances. Bell, Dickens and Gustafsson (1979) used this diagram to assess the carbon abundances of stars in NGC 6397 and M 92.

3.4 The Searle and Zinn (1978) observations.

This method uses spectral scans of stars in 18 pass bands. After correction for reddening, a least squares fit is made to the fluxes between 5000 and 7620 Å. This fit is used to define a fiducial continuum and the magnitude differences between this continuum and the observed magnitude are summed over the interval between 3800 and 4840 Å, to yield the quantity S. The mean value of S at M(V) = -1.0 is <S> and abundances are found from [Fe/H] = 3.73 <S> - 3.05.

This system has been analyzed by Bell and Gustafsson (1983) using synthetic spectra. Their results agree with those of Searle and Zinn to within +0.35 in [M/H] for all clusters in the sample. The Bell and Gustafsson result for M71 is [M/H] = -0.5.

This system is a very interesting one, giving at least the possibility of searching for anomalies in the abundances of different elements. It is unfortunate that one or two galactic clusters were not observed in order to tie down the metal-rich end.

3.5 TiO Observations.

TiO bands are seen in Population I stars with T_{eff} < 4000 K. The strengths of the TiO lines, do, of course, depend strongly on the temperature as well as the abundances of the star. A model with (T_{eff}/log g/[A/H]) 3800/0.5/-0.5 is found to have a stronger TiO line spectrum than 4000/0.75/0.0 has (the models referred to are either those published by Bell et al. 1976 or have been computed using the programs of Gustafsson et al. 1975). The TiO lines in the spectrum of the model 4250/1.5/0.0 are over ten times weaker than those of 4000/0.75/0.0.

Mould and his collaborators (see Mould and Bessell 1982) have used TiO line strengths to obtain cluster abundances. They use two filters, at 7120 Å and 7540 Å, to measure TiO band strength and another filter at 10175 Å, which is used with the 7540 filter to give a continuum gradient. The observations of 47 Tuc give [M/H] = -0.5 ± 0.1 whereas measurements of M 71 give [M/H] = -0.3 ±0.2, using the calibration of Johnson, Mould and Bernat (1982) in both cases. The important feature of this work is that it gives an abundance ranking for the most metal rich clusters. The derivation of abundances does depend upon the assumption that [Ti/H] = [O/H] = [A/H], unless the abundances of the individual elements are known from other work. In such cases a correction to the uniform depletion abundances can be found provided the depletion of oxygen by CO and SiO formation can be allowed for. The calibration by Johnson, Mould and Bernat uses gravities which are about 1.0 dex higher than those appropriate for Population I giants and globular cluster giants. In view of this and the dependence of TiO band strength on surface gravity at a given T_{eff} and abundance, and the great sensitivity of TiO band strength to details of the models, it is possible that the calibration of the models could be slightly improved.

3.6 The Strömgren system.

It may appear strange to discuss a system which has been used only sparingly in globular cluster research apart from work on blue horizontal-branch stars (see Philip 1987). The observations of Gustafsson and Ardeberg (1981) were used to find an abundance for 47 Tuc while the photometry of turn off region stars in NGC 6397 by Ardeberg, Lindgren and Nissen (1983) does confirm the very low abundances of this cluster. If more precise photometry of cluster stars in the turn off region can be obtained on the Strömgren system (and the Thuan-Gunn system), it would be very valuable in studies of the ages of the clusters and in studies of the properties of very metal-poor faint field stars.

4. THE PERIODS OF RR LYRAE STARS

Sandage (1982) has extended earlier work by Arp (1955) to show that the mean periods of the RR Lyrae variables in a cluster are strongly correlated with the cluster abundance. In addition to providing another ranking parameter, this result also forms a challenge to those calculating stellar interior models.

5. COLOR MAGNITUDE DIAGRAMS

A number of properties of a cluster color-magnitude (CM) diagram can be used as abundance indicators. These are: S, the slope of the line from the horizontal-branch - subgiant branch intersection to a point 2.5 mag. in V up the giant branch; $(B-V)_{o,g}$, the intrinsic color of the subgiant branch at the level of the horizontal branch; ΔV, the V magnitude difference between the HB and the GB measured at $(B-V)_o = 1.4$.

In practice, the calibration of these quantities is carried out using clusters of "known" abundance although it could be done using a combination of stellar evolution and stellar interior techniques (some authors have compared their calibration with the stellar evolution results).

There are some difficulties associated with the use of these indicators. Any observed color, such as $(B-V)_g$, must be corrected for reddening before use in this way. This requires knowledge of both

E(B-V) and the color excess ratios for other colors e.g. E(J-K)/E(B-V) for (J-K) and so on. (Alternatively, if one is confident enough of the cluster abundance, $(B-V)_{o,g}$ could be used to obtain the reddening.) A scarcity of stars on the GB may affect the determination of ΔV.

Rather than give a resume of current results for these indices, I simply refer to Bell and Gustafsson (1983) which gives plots of the abundance indicators versus abundances obtained from an analysis of Searle and Zinn's (1978) photometry, and go on to discuss a variant of this approach.

Frogel, Cohen and Persson (1983 and references cited therein, hereafter FCP) and their collaborators have published JHK and CO and H_2O measurements for giant stars in about 60 clusters. The CO measurements are a rich source of astrophysical data. The JHK colors have been used to determine two quantities: $(J-K)_o$ is the color of a cluster GB read at $M(K_o) = -5.5$; $\log T_{eff}(GB)$ is obtained at $M(Bol) = -3.0$. The color $(J-K)_o$ is thus analogous to $(B-V)_{o,g}$ except that cluster distances must be available. It refers to more luminous stars. FCP have calibrated their work using [Fe/H] values of Cohen (1983) and the integrated light measurements of Zinn (1980).

6. INTEGRATED LIGHT MEASUREMENTS

Zinn (1980) has made integrated light observations of 79 clusters, using filters very similar to those of the uvgr system (Thuan and Gunn 1976) and two filters which measure the absorption in the region of the H and K lines. These filters are centered at 3910 and 3955 Å, respectively and have FWHM of 180 and 90 Å. A measurement of the H and K lines is needed to study clusters which may be more metal deficient than M 92. The data are used to construct an index, Q39, which is reddening free and a measure of metal line strength. The Q39 indices for M 92 and 47 Tuc are -0.047 and 0.304, respectively. In a later paper, Zinn and West (1984) have measured other spectral features in integrated light and have correlated these features with Q39. Zinn and West also give a convenient table of abundances found by various means and discuss the anomalous abundances which some clusters have when measured in some systems.

The calibration of this quantity is again done using a set of calibrating clusters and there will again be the problem with interstellar H and K lines. Any measurement of the integrated light from a globular cluster is obviously affected by the relative numbers of stars in the different regions of the HR diagram. If we make measurements in the IR, for example, the giant stars will supply the most light, whereas UV measurements are affected by the hotter HB stars. Unusually large or small numbers of these stars will affect the measurements of clusters in our galaxy and will give results which are discordant with those given by other methods. Smith (1984b) has derived a correction to Q39 abundances based upon the relative numbers of blue and red horizontal-branch stars. While such a correction does improve the fit between Q39 and ΔS abundances, it is not clear if such a correction is valid for clusters of all abundances. Manduca (1983) has studied this problem using synthetic spectra. While the effect of the HB is undeniably a disadvantage, this does give us an idea of the problems likely to be encountered in analyzing observations of extragalactic globular clusters.

7. WHY WERE DIFFERENT RESULTS OBTAINED FOR METAL RICH CLUSTERS?

During the period 1980-1985, there were some criticisms of the high dispersion results (c.f. the discussion of several papers at IAU Colloquium 68 in addition to the comments by Zinn quoted above) but the only papers by high dispersion spectroscopists which, to my knowledge, criticized other high dispersion results were those of Cohen (1983) and Bessell (1983). Cohen's criticism was not based on high dispersion spectroscopic data but used some moderate dispersion spectral scans. She found that her clusters, which included M 71, 47 Tuc, NGC 3201, NGC 6171 and NGC 6352, had abundances which differed by 1 dex while Pilachowski, Sneden and Green (1981) found all these clusters to have an abundance of -1.1 ± 0.2 dex. Cohen also argued that the abundance of 47 Tuc was -0.6 ±0.1, although this result was again not based on high dispersion spectroscopy.

Why are the high dispersion results different from those given by other methods? At least part of the answer appears to be given in a recent paper by Geisler (1986b), who has obtained new spectra of four of the six stars observed by Pilachowski, Canterna and Wallerstein (1980) (hereafter PCW) and Cottrell and Da Costa (1981). While the data were obtained with an echelle spectrograph, Geisler has combined the data from a number of different orders to obtain a single spectrum covering 1200 Å. The presence of TiO bands in the spectrum of both the PCW stars is quite evident. While PCW noted the presence of TiO bands at wavelengths longer than the limit of their tracings, they did not recognize it as an absorber in the wavelength region which they analyzed. Spectrum synthesis calculations reveal that TiO lines are present over the wavelength interval 5000-7000 Å that is used for much of the high dispersion spectroscopy.

While a curve of growth analysis of a star is a fairly direct process to undertake, its use of the equivalent widths of weak lines does place very heavy demands upon the quality of the observational data. The dispersion employed must be high enough that the instrumental profile does not significantly add to the broadening of the spectral lines and, in addition, the signal/noise must be high enough that the continuum is well defined. In a situation where TiO lines are present in the spectra but are not allowed for in the measurements, it is in general likely that the continuum will be wrongly drawn and that the equivalent widths measured for the lines being analyzed will be too small. This is not true in all cases. If an iron line coincides with one of the stronger TiO lines, such as a Q branch line of $Ti^{48}O$, then the blending will cause the measured equivalent width to be too large. The possibility of this occurring for the [OI] lines at 6300 and 6363 Å must be examined. The effect of this blending will, of course, depend upon the Ti and O abundances of the star and will also depend strongly upon its temperature. In fact, the bluest (B-V) color at which TiO bands are seen in the spectra of globular cluster stars has been used as an abundance indicator.

It also appears to be quite possible that the unsuspected presence of TiO bands has affected Cohen's original observations of the M 71 stars, since the objects which she observed all have $T_{eff} < 4100$ K and the wavelength interval covered (5200-6800 Å) contains numerous TiO bands. The M 67 stars which Cohen observed have T_{eff} greater than 4300 K and model calculations show that TiO lines would not be expected in their spectra. The models do, however, show a delicate balance between abundance and temperature. The TiO lines in the spectrum of 3800/0.5/-1.0 are quite weak and would be unlikely to affect equivalent width measurements of iron and other metal lines. It remains to be seen if models can predict the correct strength of TiO lines in Population I giants and confirm that TiO is present in sufficient strength in the spectra of 47 Tuc giants with an abundance of [Ti/H] = [O/H] = -0.8. Most of the critics of the high dispersion work did focus their criticism on the possibility of errors in continuum location, noting that the use of echelle spectrographs did make this problem a much more difficult one than it was with conventional spectra. Bell and Gustafsson (1982) remarked that there were systematic differences in equivalent width in lines in different echelle orders, in terms of how well they fitted the model atmosphere curve of growth. This problem may still be plaguing the analyses of the more metal rich stars.

8. ABUNDANCES - A RECOMMENDATION

At this point it seems appropriate to tabulate the abundances of individual clusters. In order to do this, the following sets of results are intercompared: the high dispersion results, taken from Pilachowski (1984); the ΔS results, from Smith (1984b); the infra-red CM diagrams results; the Q39 results from Zinn (1980); results from $(J-K)_o$, calibrated using the results of Bell and Gustafsson (1983). Some results from the TiO work of Mould are also quoted. The reddening (from Zinn 1980) is given, since the abundances which are derived from colors are reddening dependent, while both Q39 and ΔS must be affected by interstellar lines. The results are given in Table I. Clusters where only Q39 abundances are available are not tabulated. Comments on some individual clusters follow, when the results from the different methods are sufficiently different.

NGC 288: the low abundance from Q39 is probably caused by the very large number of blue horizontal branch stars;
   NGC 362: the high dispersion abundance is slightly high;
   NGC 4833: the high dispersion abundance is somewhat high;
   NGC 5272 (M 3): the result from $(J-K)_o$ is slightly high;

TABLE I

Metal Abundances for Globular Clusters

Cluster	HD	ΔS	IR	Q39	(J-K)	E(B-V)
104(47 Tuc)	-1.09		-0.59	-0.64	-0.60	0.06
	-1.0					
	-0.95					
0288	-0.95		-1.13	-1.60	-0.90	0.
0362	-0.87	-1.15	-1.39	-1.18	-1.20	0.04
1261			-1.03	-1.24	-1.30	-1.00	0.
1851			-1.07	-1.26	-1.34	-1.00	0.06
1904 (M 79)			-1.43	-1.65	-1.73	-1.60	0.
2298			-1.50	-1.76	-1.84	-1.40	0.15
2808	-1.06			-1.48	-1.25	-0.90	0.22
3201	-1.00	-1.33	-1.67	-1.41	-1.30	0.27
	-1.19					
4590 (M 68)		-1.94	-1.96	-2.10	-2.00	0.02
4833	-1.37	-1.62	-1.94	-1.86	-1.80	0.34
5024 (M 53)	-1.90	-1.85	-1.98	-1.94	-2.20	0.02
5272 (M 3)	-1.57	-1.57	-1.47	-1.67	-1.20	0.02
5286			-1.85	-1.71	-1.60	0.24
5466	-1.60					
5634		-1.61		-1.87		0.04
IC4499		-1.33				
5897		-1.49	-1.74			
5904 (M 5)	-1.13	-1.08	-1.49	-1.59	-1.30	0.03
	-1.2					
5927			-0.18	-0.23	-0.50	0.42
6121 (M 4)		-1.24	-0.72	-1.40	-0.60	0.37
6171 (M 107)	-1.05	-0.83	-0.92	-0.97	-0.70	0.38
6205 (M 13)	-1.44	-1.03	-1.47	-1.70	-1.20	0.01
6218 (M 12)				-1.55		0.16
6229	-1.30			-1.48		0.01
6254 (M 10)	-1.32		-1.54	-1.68	-1.10	0.26
6266 (M 62)		-1.16		-1.26		0.48
6284		-0.91		-1.40		0.29
6304				-0.40	-0.10	0.49
6341 (M 92)	-2.06	-2.18	-2.01	-2.12	-2.00	0.03
6342				-0.47		0.46
6352	-0.80		-0.38	-0.66		
6362	-1.00	-0.95	-1.05	-1.06	-0.70	0.10
6397	-2.21		-1.84	-2.17	-1.90	0.16
6402 (M 14)		-1.11		-1.49		0.59
6637 (M 69)			-1.20	-0.49	-0.70	0.10
6656 (M 22)	-1.78	-1.70	-1.81	-1.72	-1.50	0.37
	-1.94					
	-1.67					
6681 (M 70)		-1.23		-1.62		0.08
6712		-0.57		-1.14	-0.70	0.39
6715 (M 54)		-1.27		-1.41		0.15
6723	-1.10	-0.68		-1.26		0.05
6752	-1.32		-1.35	-1.55	-1.20	0.00
6838 (M 71)	-1.30		-0.60	-0.40	-0.40	0.26
	-0.81					
	-1.07					
	-0.7					
	-0.57					
6981 (M 72)		-1.27		-1.56		0.07
7006			-1.55	-1.53	-1.30	0.05
7078 (M 15)	-1.76	-2.04	-2.21	-2.07	-2.40	0.08
7089 (M 2)		-1.43		-1.75		0.03
7099 (M 30)		-1.96		-2.15		0.01

The second, third, fourth, fifth and sixth columns of the table are metal abundances ([Fe/H] or [M/H]) deduced from high dispersion spectroscopy (Pilachowski 1984), ΔS work (Smith 1984b), IR color magnitude diagrams (Frogel, Cohen and Persson 1983), Zinn and West's (1984) or Zinn's (1980) values from Q39 and the present results, found from $(J-K)_o$ using the Bell and Gustafsson (1983) calibration.

NGC 5904 (M 5): the Q39 value is low and the $(J-K)_o$ value is high;

NGC 5927: the $(J-K)_o$ abundance is quite high but hinges upon the adopted value of the reddening;

NGC 6121 (M 4): the higher reddening of 0.47, which is implied by the RR Lyrae period shift versus (J-K) diagram, gives an [M/H] from (J-K) of - 0.9;

NGC 6205 (M 13): Q39 gives a low abundance;

NGC 6362: the RR Lyrae period shift versus abundance relation suggests a higher reddening which yields [M/H] = -1.0 from (J-K);

NGC 6553: the high reddening casts doubt on the $(J-K)_o$ value. TiO data would be of interest;

NGC 6637 (M 69): The adopted reddening is from Mould, Stutman and McElroy (1979). The TiO lines are as strong as in Pop I giants of the same (7540 - 10175) color so the abundance must be high. Geisler's

(1984) result of [Fe/H] = -1.42 seems astonishingly low;
  NGC 6712: Frogel (1985) emphasizes the fact that the CO bands are as strong as in field giants and stronger than in 47 Tuc stars. This could be explained by a high C12/C13 ratio in the 6712 stars;
  NGC 6838 (M 71): the Q39 and (J-K) values are in good agreement with the TiO result. The high dispersion results range from -1.35 (Pilachowski et al. 1983) to -0.57 (Bessell 1983);
  NGC 7078 (M 15): the relatively low abundance from (J-K) supports an increase in the reddening to E(B-V) = 0.12;

ACKNOWLEDGMENTS

  This work has been supported by the National Science Foundation under Grant AST85-13872. Some of the calculations were carried out using the SDSC Cray, under NSF Grant AST85-09915, while others were made using a Ridge 32C computer, also purchased with NSF funds. I am grateful for this support.

REFERENCES

Ardeberg, A., Lindgren, H. and Nissen, P. E. 1983 Astron. Astrophys. 128, 194.
Arp, H. C. 1955 Astron. J. 60, 317.
Bell, R. A. 1984 Publ. Astron. Soc. Pacific, 96, 518.
Bell, R. A., Dickens, R. J. and Gustafsson, B. 1979 Astrophys. J. 229, 604.
Bell, R. A., Gustafsson, B., Eriksson, K. and Nordlund, A. 1976 Astron. Astrophys. Suppl., 23, 37.
Bell, R. A. and Gustafsson, B. 1982 Astrophys. J., 255, 122.
Bell, R. A. and Gustafsson, B. 1983 Mon. Not. Roy. Astron. Soc., 204, 249.
Bessell, M. S. 1983 Publ. Astron. Soc. Pacific, 95, 94.
Burstein, D., Faber, S. M. and Gonzalez, J. J. 1986 Astron. J., 91, 1130.
Butler, D. S. 1975 Astrophys. J., 200, 68.
Canterna, R., Harris, W. E. and Ferrall, T. 1982 Astrophys. J., 258, 612.
Carney, B. W. 1979 Astrophys. J., 233, 211.
Cohen, J. C. 1978 Astrophys. J., 223, 487.
Cohen, J. C. 1979 Astrophys. J., 231, 751.
Cohen, J. C. 1980a, in IAU Symposium 85, Star Clusters, J.E.Hesser, ed., Reidel, Dordrecht, p. 385.
Cohen, J. C. 1980b Astrophys. J., 241, 981.
Cohen, J. C. 1983 Astrophys. J., 270, 654.
Cottrell, P. L. and Da Costa, G. S. 1981 Astrophys. J. Lett., 245, L79.
Dickens, R. J., Bell, R. A. and Gustafsson, B. 1979 Astrophys. J., 232, 428.
Frogel, J. A., Cohen, J. C. and Persson, S. E. 1983 Astrophys. J.,

275, 773.
Frogel, J. A., 1985 Astrophys. J., 291, 581.
Geisler, D. 1984 Astrophys. J. Lett., 287, L85.
Geisler, D. 1986a Publ. Astron. Soc. Pacific, 98, 762.
Geisler, D. 1986b Astrophys. J. Lett., 304, L41.
Geisler, D. and Kapranidis, S. 1983 Astron. J., 88, 461.
Gustafsson, B. and Ardeberg, A. 1978 in Astronomical Papers dedicated to Bengt Strömgren, A. Reiz and T. Andersen, eds, Copenhagen University Observatory, p 145.
Gustafsson, B., Bell, R. A., Eriksson, K. and Nordlund, A. 1975, Astron. Astrophys., 42, 407.
Helfer, H. L., Wallerstein, G. and Greenstein, J. L. 1959 Astrophys. J., 129, 700.
Hesser, J. E., Hartwick, F. D. A. and McClure, R. D. 1977 Astrophys. J.Suppl., 33, 471.
Jones, K. A. 1979 Problems of Calibration of Multicolor Photometric Systems, A. G. Davis Philip, ed., Dudley Obs. Report No. 14, p. 103
Johnson, H. R., Mould, J. R. and Bernat, A. 1982 Astrophys. J., 258, 161.
Keith, D. and Butler, D. 1980 Astron. J., 85, 36.
Manduca, A. 1981 Astrophys. J., 245, 258.
Manduca, A. 1983 Bull. Am. Astron. Soc., 15, 647.
Mould, J. R. and Bessell, M. S. 1982 Astrophys. J., 262, 142.
Mould, J. R., Stutman, D. and McElroy, D. 1979 Astrophys. J., 228, 423.
Philip, A. G. D. 1987 in IAU Symposium No. 126, Globular Cluster Systems in Galaxies, J. E. Grindlay and A. G. D. Philip, eds., Reidel, Dordrecht, p. 513.
Pilachowski, C. 1984 Astrophys. J., 281, 614.
Pilachowski, C., Canterna, R. and Wallerstein, G. 1980 Astrophys. J. Lett., 235, L21.
Pilachowski, C., Sneden, C., and Green, E. M. 1981 in IAU Colloquium 68, Astrophysical Parameters for Globular Clusters, A. G. Davis Philip and D. S. Hayes, eds., L. Davis Press, Schenectady, p 97.
Pilachowski, C., Sneden, C., and Wallerstein, G. 1983 Astrophys. J. Suppl., 52, 241.
Sandage, A. R. 1970 Astrophys. J., 162, 841.
Sandage, A. R. 1982 Astrophys. J., 252, 553.
Searle, L. S. and Zinn, R. 1978 Astrophys. J., 225, 357.
Smith, H. A. 1984a Publ. Astron. Soc. Pacific, 96, 505.
Smith, H. A. 1984b Astrophys. J. 281, 148.
Suntzeff, N. B. 1980 Astron. J. 85, 408.
Thuan, T. X. and Gunn, J. E. 1976 Publ. Astron. Soc. Pacific, 88, 543.
Zinn, R. 1980 Astrophys. J. Suppl. 42, 19.
Zinn, R. 1981 in IAU Colloquium 68, Astrophysical Parameters for Globular Clusters, A. G. Davis Philip and D. S. Hayes, eds., L. Davis Press, Schenectady, p. 45.
Zinn, R. and West, M. J. 1984 Astrophys. J. Suppl. 55, 45.

DISCUSSION

NEMEC: The observed main sequences of many globular clusters, derived using well calibrated CCD photometry, are often redder by as much as 0.10 mag than the main sequences calculated using theoretical interior plus atmosphere models. Do you see a resolution to this problem?

BELL: There are some comments on this problem in a paper on the Thuan Gunn system by VandenBerg and myself. There does appear to be a very good agreement between observed and predicted colors and temperatures for field F dwarfs so I would not expect a problem in transforming T_{eff} to color for the metal-poor cluster stars. There are possible stellar evolution reasons for color shifts as a function of a high abundance of oxygen. It is not clear that the color transformations used to convert CCD (B-V) values to (B-V) values on the Johnson system are valid for metal-poor stars. We also have to be more fussy about reddening than we have been in the past.

CHRISTIAN: There is one field in which a (B-V) shift is not reconciled with the models. This is the M 92 field that has been calibrated with sequences observed photoelectrically. These sequences have been observed with a number of CCD's with the same filter set, so they have been calibrated "properly". The cluster also has low reddening and obviously low metallicity. So although calibration of CCD photometry can be a problem I do not think it can explain away the effect in this case.

BELL: There is always the question of the oxygen abundance which will affect the tracks. I look forward to seeing the paper giving the M 92 results.

SCHOMMER: Two responses to your comments on the Washington System. I agree that $M-T_1$ has relatively low sensitivity to [Fe/H], and requires very precise photometry. In a recent Astron. J. article by Canterna, et al., a new calibration of the system is presented, and the CM diagram is proposed as a very sensitive abundance indicator, especially useful between $0 < [A/H] < -1$. Secondly, in defense of the carbon sensitivity, I merely mention that the first CH star in a dwarf galaxy was found by this system, and subsequently C stars were found in all dwarf spheroidals.

BELL: The CM diagram is clearly a better abundance indicator than $M-T_1$, but I am skeptical about the supposed sensitivity to CH and CN. I look forward to seeing the new work.

ZINN: If the period shifts of Sandage are plotted against Q_{39}, a very tight relationship is obtained. Both of these quantities are reddening independent, and their good agreement suggests that they are measures of the same thing, which appears to be metallicity. The poorer agreement between Q_{39} and $(J-K)_0$ which is evident in your diagram may not be, therefore, a consequence of the sensitivity of Q_{39} to variations in horizontal-branch morphology, but a consequence of the sensitivity of $(J-K)_0$ to errors in the reddenings of the clusters.

BELL: I think you misunderstood my remark. I quite agree that the period shift gives a very tight correlation with abundance. The point of showing the period shift, (J-K) diagram is to argue that it would be used as a determination of cluster reddening, by arguing that there should be a tight correction between period shift and (J-K). The same is true for $(B-V)_g$ versus abundance.

WALLERSTEIN: There is a difference between Fe/H and the total metal abundance. In particular Ti and O repeatedly appear to be less deficient than iron. Hence TiO yields small deficiencies of Ti and O, in agreement with high dispersion spectroscopy.

BELL: I think one must be particularly cautious about the overabundances of individual elements. The strengths of H and K and Ca I 4226 low dispersion spectra of in NGC 362 and 47 Tuc stars contradict the high dispersion results. The [O I] lines at 6300 and 6363 may be blended with TiO lines and the equivalent widths may be measured as being too strong. Calculations of CO band strengths do not support large overabundances of oxygen. The similarity of the TiO band strengths at a given color in stars in some clusters and Pop I stars supports Zinn's (1980) abundance scale and the one proposed in this talk.

THE CHEMICAL INHOMOGENEITY WITHIN GLOBULAR CLUSTERS

John Norris

Mount Stromlo and Siding Spring Observatories
Australian National University

1. INTRODUCTION

Twenty years ago it was believed by most astronomers that globular clusters were chemically homogeneous - where by homogeneous one means that the outer layers of all stars within a given cluster are the same to within a few tens of percent. Today it is possible to defend the case that no Galactic globular cluster has this characteristic. The reason that this phenomenon has exercised the minds of so many groups in the past 15 years is exciting and obvious: if one can ascertain which are the relevant physical processes in operation, one stands to gain significant insight into both the way in which globular clusters formed and/or the way in which individual low mass stars evolve and mix the products of their nucleosynthesis into their outer layers. A second important driver at the back of the minds of workers in this field is the possible ramifications of an understanding of the phenomenon: for example, if one concludes that the abundance anomalies are being driven today by some particular effect (angular momentum, magnetic fields, interactions within binary systems, stellar collisions - or whatever) this may lead to insight into other important globular cluster phenomena (eg bimodal horizontal branches, gaps at the base of the giant branch, horizontal branch rotation, etc.)

The purpose of the present review is twofold. First, an effort is made to summarize the systematic trends which have been deduced from the wealth of observational material. Second, the constraints which this imposes on the physical processes at work are discussed at some length.

2. CANONICAL WISDOM

The basic ideas which shaped the early expectations of chemical homogeneity are best illustrated by consideration of globular cluster color-magnitude diagrams. The first important point is that the small color spread seen on the giant branch of most systems indicates that there can be little range in the abundance of elements such as Si, Mg, and Fe which play an important role in determining the opacity in the outer layers of red giants. For most objects this sets the limit $\Delta[Fe/H] < 0.15$. The second point is that standard stellar evolution theory (see Iben 1967, 1974) explains the principle sequences in color-magnitude diagrams very well, suggesting that only minor modifications in surface abundances are to be expected during evolution.

The basic concepts are as follows. First one has the long-lived (15 Gyr) main sequence phase of hydrogen burning. After core hydrogen exhaustion a star leaves the main sequence (burning hydrogen in a shell) and evolves relatively quickly to the base of the giant branch. At this point, according to standard theory, the convective envelope reaches its greatest inward extent and actually mixes up material which has undergone CNO processing into the outer layers of the star. The basic difference between prediction and observation is one of considerable degree - for while the models predict changes in C and N of order a few tens of percent (Faulkner and Iben 1967; Da Costa and Demarque 1982) observation requires modifications by factors of up to 10-50. More of this later. The star now proceeds up the giant branch until the electron degeneracy of its helium core is lifted by the ignition of helium at the tip of the giant branch. Most theoreticians report no upheaval of the surface layers at this point, though the recent work of Deupree and Cole (1983a,b), which includes hydrodynamical effects, suggests the potential for considerable rearrangement. After some 100 Myr of core helium burning on the horizontal branch, the star retraces its steps up the asymptotic giant branch (AGB), where it undergoes helium shell flashes until it finally exhausts its supply of hydrogen and evolves to higher temperature on a short timescale on its way to becoming a white dwarf. The important fact about all standard stellar evolution calculations of the phases after the helium flash is that no prediction of major surface abundance rearrangement has been made.

3. THE REAL WORLD

Life is not that simple. Figure 1 shows spectra in the wavelength range $\lambda\lambda 3800$ - 4500A for 4 stars of comparable brightness in the remarkable cluster ω Centauri. This object possesses giants with enormous carbon enhancements (CH stars), with carbon deficiences (weak-G-band stars), and large CN enhancements (CN-strong stars) as shown here. ω Cen also exhibits an anomalously wide upper giant branch (Woolley et al. 1966; Cannon and Stobie 1973) indicative of a spread in heavy element abundance, together with a large range in Ca II H and K line strengths found by Freeman and Rodgers (1975) among the RR Lyrae variables.

Fortunately ω Cen (along with M22 to a lesser extent) is unique. Most clusters are much simpler, exhibiting little variation in heavy element abundance, and to some extent insight into the basic problem has come from investigations of the more homogeneous systems. We shall confine our attention initially to these more homogeneous clusters (Section 3.1) and then return to ω Cen in Section 3.2.

The present review makes no attempt to cover the early observational material on the problem. The reader is referred to the Annual Reviews articles by Kraft (1979) and Freeman and Norris (1981), and that of McClure (1979) for a comprehensive set of basic references on the topic. (During this conference the author was presented with a preprint by G.H. Smith entitled 'The Chemical Inhomogeneity of Globular Clusters' which will appear in P.A.S.P. This is the most detailed and

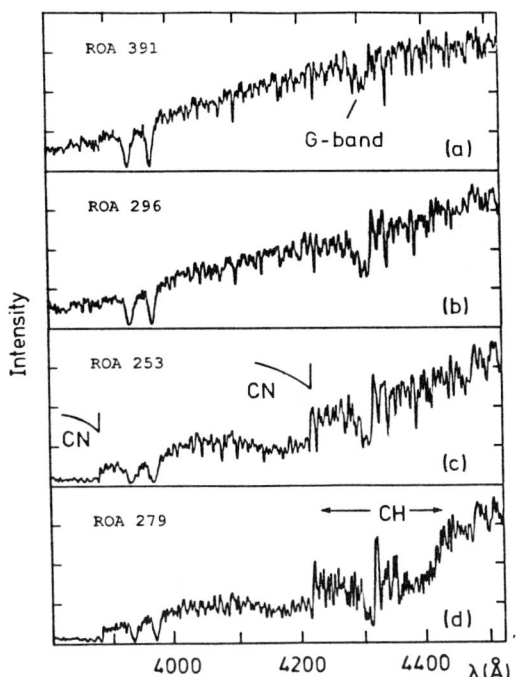

Fig 1. Spectra of 4 giants of similar luminosity in ω Cen. (a) ROA 391 is an example of a weak-G-band star, (b) ROA 296 is a 'normal' star having no carbon or nitrogen anomaly, (c) ROA 253 is a classic CN-strong star, and (d) ROA 279 is an example of a relatively hot CH star.

comprehensive discussion of the topic yet seen by the author and is recommended to any researcher interested in the subject.) Suffice it here to say that it was Harding (1962), Osborn (1971), and Zinn (1973), respectively, who first drew attention to the fact that CH and CN enhancements, together with CH depletions, exist in globular clusters. Subsequent investigations have provided a massive data bank on clusters of all abundances, and for stars at all phases of evolution subsequent to the main sequence. The only phase of evolution where systematic observational material does not exist is that of the main sequence. The aim of the present work is to describe the systematic abundance patterns which have emerged from these efforts with a view to constraining the possible origin of the observed anomalies.

Space precludes detailed description of techniques. Model atmosphere spectrum synthesis analysis (see eg Bell, Dickens, and Gustafsson 1979) of the features of molecules containing C, N, and O have played the major role, with the relevant species determined in large degree by the overall abundance of the cluster. Thus, for [Fe/H] < -1.8, the G band (CH) and the NH features at λλ3350-3360A are the main features analyzed. In the range -1.8 < [Fe/H] < -1.0 the violet CN bands degrading blueward of λ3883Å supplement CH and NH, while at higher abundances the blue CN bands near λ4216Å are adopted in preference to the (by this time) very strong violet bands. For oxygen one is faced with the basic problem that the available features (OI lines near λ6300Å) are extremely weak (~ 30 mÅ), while the CO bands near 2.4 μm are observed only with difficulty.

3.1 Clusters Other Than ω Cen and M22

For convenience we split the discussion into three parts – first comes C and N, then O, and finally the behavior of Na and Al, which seems inextricably connected with that of the CNO group.

3.1.1 Carbon and Nitrogen

3.1.1.1 The Dependence on Cluster [Fe/H]

Bell and Dickens (1980) first pointed out that the level of carbon depletion on the upper giant branch is greatest in clusters of lowest abundance. This remarkable effect is shown in Figure 2. One is

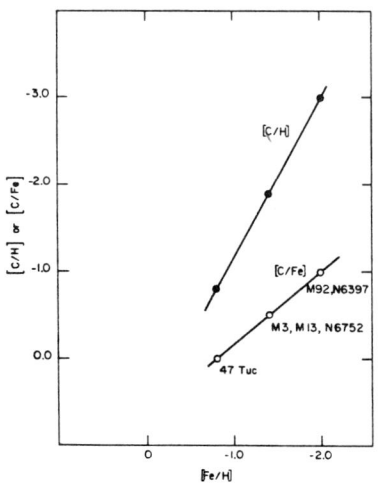

Fig 2. The dependence of [C/H] and [C/Fe] on cluster [Fe/H] (for upper giant branch stars) from Bell and Dickens (1980), showing that carbon depletion is most marked in clusters of lowest heavy element abundance. (Reproduced from Astrophysical Journal.)

thus seeking a process which operates most efficiently in a low abundance environment. As these authors point out, this phenomenon is just what might be expected from the meridional mixing model of Sweigart and Mengel (1979, hereafter SM). For giant branch stars SM examined the mass extent of the regions just outside the hydrogen burning shell in which CN and ON nuclear processing has occurred, and noted that for stars of lower heavy element abundance these regions become larger and further removed from that in which hydrogen has been substantially converted into helium. They argue that mixing of CNO-cycle processed material to the surface will thus proceed more readily in low abundance stars than in those of higher abundance. It is very instructive to look at their Figures 1-3. Note that this property of the extent of the CN and ON processed zones will assist any mixing process in low abundance stars.

3.1.1.2 Dependence on Evolutionary Status

It was evident from the very first that carbon depletion in the most metal weak systems ([Fe/H] ~ -2.0) is in some way connected with evolutionary status. The AGB of M92 was the site where Zinn (1973) identified the weak-G-band phenomenon; Bell, Dickens, and Gustafsson (1979), using the power of spectrum synthesis, next showed that the effect was marked at the tip of the giant branch in the most metal weak clusters; and finally Carbon et al. (1982) and Langer et al. (1986) have shown a pattern of increasing carbon depletion with increasing brightness along the red giant branch of M92, as reproduced in Figure 3.

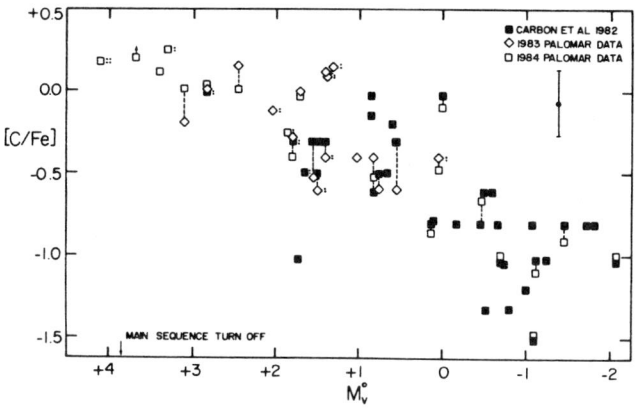

Fig 3. The dependence of [C/Fe] on M_v on the giant branch of M92, from Langer et al. (1986). (Reproduced from Astronomical Journal.)

Clearly this result receives a most natural explanation in terms of CN processing and mixing on the giant branch. It should be noted for completeness that some of the nitrogen overabundances found higher on the giant branch by Carbon et al.(1982) are so large (of order 20-30) as to require (i) additional processing in the ON cycle (which requires more extreme conditions) followed by mixing (SM), or (ii) primordial variations in C followed by CN processing and mixing (Carbon et al. 1982), or (iii) primordial variations in N (Norris and Pilachowski 1985). Unfortunately we know nothing of the behavior of C and N at the base of the giant branch of M92, because of the difficulty in making the necessary observations.

For the more metal rich clusters the situation appears somewhat different. In 47 Tucanae, with [Fe/H] = -0.7, Bell, Hesser, and Cannon (1983) have demonstrated the existence of CN anomalies right down to the main sequence turnoff, consistent with nitrogen abundance variations of order 5. They also find no evidence for a difference in the abundance ranges of C and N from the turnoff to the tip of the giant branch. As demonstrated by Da Costa and Demarque (1982), who considered several ad hoc main sequence mixing possibilities, the existence of a range in nitrogen abundance at the main sequence turnoff is more suggestive of a primordial origin than of a mixing process.

It should also be noted that the observations of C and N variations low on the giant branches of M92 and 47 Tuc are in disagreement with the meridional mixing model of SM. These authors predict that no mixing should occur until much higher on the giant branch when the hydrogen burning shell has moved out through the composition discontinuity left in the stellar envelope by the convective envelope at its deepest penetration on the lower giant branch. This disagreement between observation and the meridional mixing model is found in all clusters where the test has been made (Norris and Smith 1984, and references therein).

3.1.1.3 Systematics at a Given Point on the Giant Branch

Another approach to the problem has been to investigate a relatively large number of stars at roughly the same phase of evolution in a number of clusters. The results for the systematics of violet CN on the giant branch for 12 clusters with −1.9 < [Fe/H] < −1.3 are shown in Figure 4, which presents generalized histograms of the cyanogen excess of individual red giants, δCN, from Norris (1987). Note the

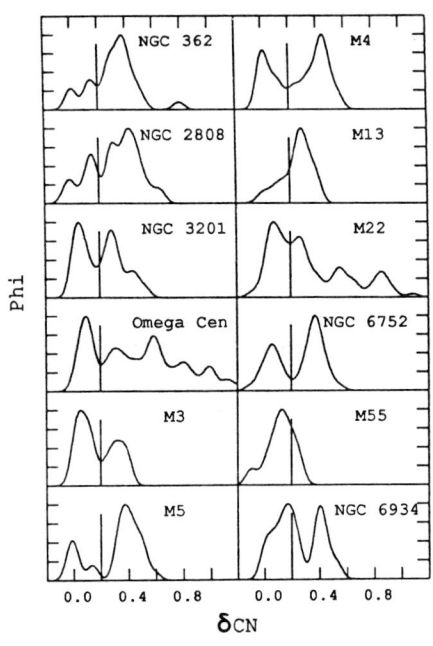

Fig 4. The generalized histograms of CN-excess, δCN, for 12 clusters. Note the large range in morphology of the distributions and the high incidence of bimodality. The vertical lines at δCN = 0.20 are used to separate CN-strong and CN-weak stars to define the parameter r = number of CN-strong stars/ number of CN-weak stars for use in Figure 7.

large range of morphologies exhibited and the propensity towards bimodality. The latter effect is an interesting phenomenon, being suggestive of two populations. This is, however, not necessarily so. Suntzeff (1981), Langer (1985), and Smith and Bell (1986) have suggested that this results from saturation effects in the violet CN bands as N is enhanced by the mixing to the surface of the products of the CN cycle. In principle this works, though it remains somewhat of a challenge to explain the degree on CN enhancement. If, for example, one compares the results for M4 ([Fe/H] = −1.3) with the spectrum synthesis calculations of Smith and Bell (1986) for [Fe/H] = −1.0 one finds that in M4 the difference in the cyanogen index between the CN-strong and CN-weak stars is ∼ 0.4 (see Figure 4), while that predicted from CN processing is only ∼ 0.2 (Smith and Bell 1986). A couple of points should be made here. First, similar spectrum synthesis calculations by the author suggest that the difference of $\Delta[Fe/H] \sim 0.3$ between the observations and calculations will have little effect on the comparison. Second, it might be fruitful to consider the role of ON processing in this problem.

A further important result that has come from this type of investigation is that in many clusters there is an anticorrelation between the behavior of the CN bands and that of the CH features (Norris, Freeman and Da Costa 1984, and references therein). The simplest explanation of this phenomenon clearly lies in the operation of the CN cycle – with the most favored site being the interior of the star itself. It is extremely difficult to explain within a primordial framework (Smith and Norris 1981b), and by any postulated accretion mechanism such as that of D'Antona, Gratton, Chieffi (1983). (See Norris, Freeman, and Da Costa 1984). The anticorrelation seems to hold in clusters with $[Fe/H] > -1.8$, with ω Cen and M22 being the exceptions.

3.1.2 Oxygen

It seems reasonable to suggest that the systematics of oxygen are not yet understood. This results from the weakness of the available spectroscopic features at optical wavelengths. The severity of the problem of the weakness of the OI lines is nicely illustrated by the work of Leep, Wallerstein and Oke (1986, Figure 1). The CO bands in the infrared have not yet been utilized in a definitive way to attack this problem. See Bell and Dickens (1980) for an example of the problems encountered.

Several comments may be made on the available data. First, such is the observational difficulty of the problem that most investigations are aimed at determining the mean oxygen abundance of clusters (see ,eg, Pilachowski, Wallerstein, and Leep 1980). Second, in an effort to see if oxygen was correlated with C and N, Cottrell and Da Costa (1981) reported that to within 0.1 dex there is no difference in the oxygen abundance of the CN-weak and CN-strong groups in 47 Tuc. Finally, Leep, Wallerstein, and Oke (1986) report that one out of five giants studied in M13 has an oxygen underabandance of a factor of at least 5 relative to the other four. (We note for future reference that the oxygen deficient star in question is II-67, the very object for which Peterson (1980) reported an overabundance of sodium of 0.7 dex. It is also CN-strong (Norris and Pilachowski 1985), while of the other four objects one (II-76) is known to be CN-weak.) Clearly much effort needs to be expended before a clear picture on this problem will emerge.

3.1.3 The Heavier Elements (In Particular Na and Al)

It has become evident that in almost all cases where one finds nitrogen enhancements the features of Al and Na are also enhanced. (See Norris and Smith 1983; Norris and Pilachowski 1985; and references therein.) The effect for Al is shown in Figure 5, which presents spectra of a CN-strong, CN-weak pair of stars in NGC 6752. It is immediately obvious that the mixing of CN-processed material cannot directly produce overabundances of this type. Cottrell and Da Costa

(1981) sought an explanation in terms of primordial production in the first generation 5-10 $M_\odot$ AGB stars, based on the stellar evolution calculations of Iben (1975, 1976).

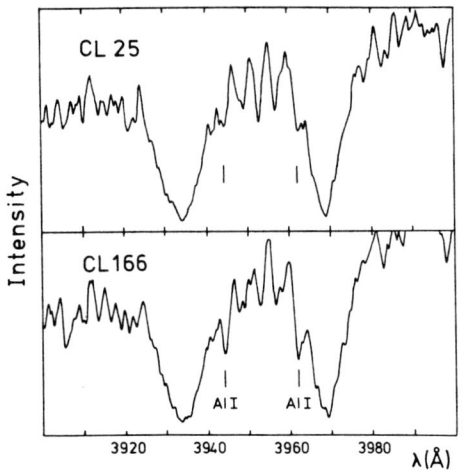

Fig 5. Comparison of the spectra of two red giants in NGC 6752 having similar effective temperature and gravity in the region of the Al I lines at $\lambda\lambda 3944$ and 3961Å. CL25 is CN-weak while CL166 is CN-strong. See Norris et al. (1981) for details. (Reproduced form Astrophysical Journal.)

3.2 ω Centauri

ω Cen is the most massive of the Galactic globular clusters, and shows the most extreme chemical inhomogeneity. Its color-magnitude diagram clearly indicates that it differs basically from most other Galactic globular clusters. On the giant branch one sees a spread in color of $\Delta(B-V) \sim 0.4$ (Woolley et al. 1966; Cannon and Stobie 1973) in comparison with much smaller ranges (< 0.03) seen in other systems. As noted earlier, the origin of this spread lies in a range of heavy element abundance. The red giants exhibit a large range in the strengths of the molecules formed from C and N (Harding 1962; Norris and Bessell 1975,1977; Dickens and Bell 1976; and Bessell and Norris 1976). Dickens and Bell (1976) also recognized the existence of stars with anomalously strong features of the s-process elements Sr and Ba, while Lloyd Evans (1977a,b) catalogued the coolest stars which exhibit TiO and enhanced Ba. Cohen (1981) and Gratton (1982) reported a large and complex range of heavy element abundances from model atmosphere analysis of high dispersion spectra. The light elements and the rare earths (plus Ba) appear enhanced relative to the iron peak elements in the more metal-rich ω Cen stars.

Two schools of thought developed to explain these observations. On the one hand the variations of Ca are difficult to understand except as the result of primordial effects, since this element is not synthesized in low mass stars. On the other hand anomalies involving C, N, Sr, and Ba are usually associated with mixing phenomema. This was complicated by the observation that at a given luminosity there is a direct correlation between the behavior of CN and Ca (Norris and Bessell 1977; Norris and Freeman 1983, Figure 7). It is important to note, however, that the observations still permit considerable spread in CN at a given Ca line strength.

Another important constraint on the problem is given by the work of Persson et al. (1980) who demonstrated a dichotomous behavior of the infrared bands of CO, suggestive of two distinct phenomena at work. This was followed up by Cohen and Bell (1986, hereafter CB) who analyzed complementary low resolution optical spectra of most of the Persson et al. sample. From spectrum synthesis analysis of the available material, CB conclude that there are three groups of stars in the cluster. First there are the well known carbon stars. Second there is a group of objects in which nitrogen is enhanced by factors of 5-15 and in which carbon and oxygen may be depleted by factors of as much as 10. The third group comprises objects in which C and N are normal to within a factor 2. There is a wide range in heavy element abundance in groups 2 and 3.

The author suggests that this classification is a considerable oversimplification brought about by the large selection effects inherent in the Persson et al. sample. It is his belief that while the nitrogen rich group may have the properties suggested by CB, it is at the extreme of a (probably) continuous distribution. This is best illustrated by the statement that of the 72 stars analyzed by CB only two of the N rich group were chosen in an unbiased manner.

This criticism having been made, it should also be stated that work by Paltoglou and Norris (1987) supports the result of CB that the nitrogen rich stars are oxygen deficient. Figure 6 shows spectra near λ6300 of the normal stars ROA 43 and 53 and the nitrogen rich objects ROA 100 and 150. The four objects have similar T_{eff} and log g, while [Fe/H] varies by ~ 0.2 within the group. Despite this, the OI line is clearly seen in ROA 43 and 53 but not in ROA 100 and 150. These data support the suggestion of CB that not only CN but also ON cycling has acted in the stars having extreme nitrogen enhancements in ω Cen. The preliminary analysis of Paltoglou and Norris finds a strong anticorrelation between oxygen and nitrogen for a sample of 12 stars.

We complete the discussion with one further interesting fact. The nitrogen rich stars exhibit large enhancements of Na and Al relative to iron. Cohen (1981) reports [Na/Fe] = 0.8 and [Al/Fe] = 1.4 for ROA 253, a classic nitrogen enriched object. This is exactly the same result noted in the more normal clusters. In ω Cen there is one further important piece of information. As noted above, oxygen appears anticorrelated with nitrogen. The preliminary results of Paltoglou and Norris (1987) also show the positive correlation between Na and N, and by implication an anticorrelation of oxygen and Na. It should also be recalled that a similar effect exists for II-67 in M13 (see Section 3.1.2.). The obvious conclusion is that, if these results stand, Na (and presumably Al) are in some way enhanced in the mixing process which leads to oxygen depletion, or at the very least the spectral features of Na and Al are modified by phenomena associated with the results of that mixing process.

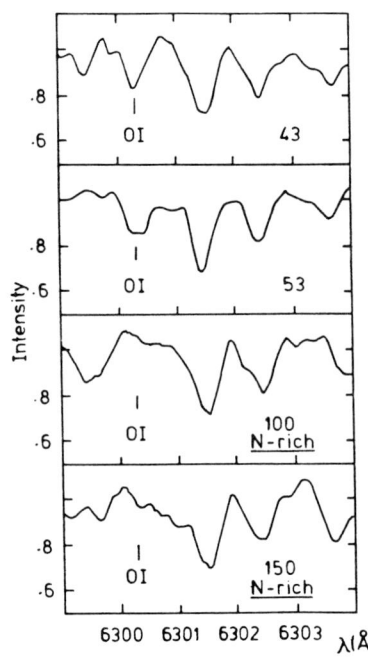

Fig 6. The behavior of OI line strength in four giants of different nitrogen abundance in ω Cen, from the work of Paltoglou and Norris (1987). (The number to the right is the ROA designation.) Note that the OI line is not detected in the N-rich stars, in support of the suggestion of Cohen and Bell (1986) that ON-cycle processing and subsequent mixing has occurred in these objects.

4 DEPENDENCE ON GLOBAL CLUSTER PROPERTIES

The author knows of no cluster in which the necessary observations have been made that does not exhibit C and/or N variations at least (roughly) the factor of 2 level. Investigations aimed at correlating cluster abundance 'anomalies' with global cluster properties such as mass, central density, etc have not been particularly successful. Smith and Norris (1981a, Figure 13) show that a range in CN strengths occurs for all present-day cluster masses. Anomalies exist at all cluster central densities. Even the ghostly Palomar 5 has roughly equal numbers of CN-strong and CN-weak stars (Smith 1985).

While searching for such correlations the author (Norris 1987) found a possible connection between the degree of CN enhancement and apparent cluster flattening, for clusters in which sufficient violet CN data were available. Figure 7 shows the result, where r is defined as the ratio of CN-strong to CN-weak stars in a given cluster, and ε is the apparent flattening of the system. (The adopted dividing line between CN-weak and CN-strong stars is shown in Figure 4.) It was suggested that angular momentum is the driver producing both the flattening of the clusters and, via internal rotation in individual stars, the greater degree of CN enhancement in the more flattened systems.

5 ORIGIN OF THE ANOMALIES

It should be evident from the above that at least two

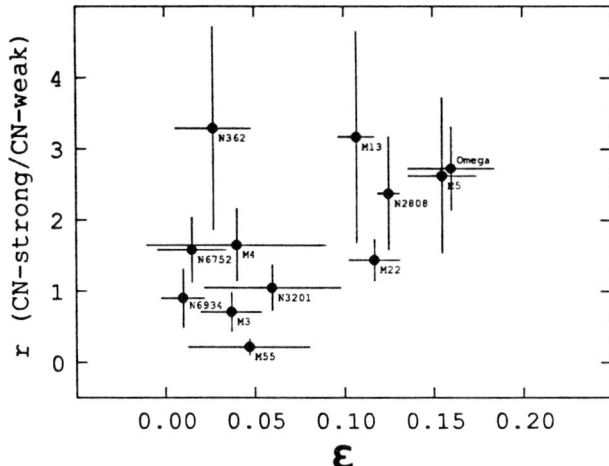

Fig 7. The dependence of the degree of cyanogen enrichment, r, on cluster flattening, ε. This diagram leads to the suggestion that angular momentum may play a role in driving abundance anomalies. (From Norris 1987.)

fundamentally different processes have operated to produce the phenomena described above. In ω Cen there can be little doubt that primordial abundance variations existed 10 Gyr ago. It is the author's opinion that no amount of invention can explain the calcium variations seen in the RR Lyrae stars in that system in any other way. At the other extreme there can be no explanation other than mixing processes for the steady downward march of carbon abundance as one proceeds up the giant branch of M92. It is thus not a question of either/or, but of how much of each.

The simplest picture which suggests itself at the present stage is the following. Some agency causes the stars in globular clusters to supermix (a word borrowed from Iben to denote mixing more extreme than that predicted by standard stellar evolution theory). That agency is not determined, but the number one suspect is angular momentum in stellar interiors. Here is a problem crying for a theoretical solution. The observations of M92 suggest that the process becomes important in clusters of lowest abundance by the time the stars reach $M_v = 2$ on the giant branch. What remains unclear is whether such a process can operate with the required efficiency to produce N anomalies at the main sequence turnoff as is indicated by the observations of 47 Tuc. This remains one of the unanswered questions concerning the evolution of low mass stars at this time.

On the other side of the coin there are some clusters, the clearest cases of which are ω Cen and M22, in which primordial variations also exist. In these objects the situation is complicated by the fact that one observes, for the red giants, not only the primordial variations but also the superposition upon them of mixing effects.

Several problematic observations remain. Foremost is the positive correlation of nitrogen on the one hand with Na and Al on the other. This phenomenon appears to exist in all clusters. Four possible explanations may be suggested. First, complicated primordial variations involving some or all of C, N, Na, and Al exist in most clusters. In principle at least, this could perhaps be solved by future observation of main sequence stars. The second possibility is that non standard evolution at the helium core flash might explain the result. Deupree and Cole (1983a,b) have suggested that during this event Na and Al might be produced within certain bubbles deep within the envelope of the star. Difficulties with this explanation have been described by Norris and Pilachowski (1985) and will not be repeated here. Third, it is not impossible that atmospheric effects related to the mixing status of a star (eg non-LTE effects, boundary-temperature differences induced by different molecular densities, different turbulence, magnetic fields, different rotation, etc.) may lead to spurious strengthening of Na and Al in stars which appear to have enhanced nitrogen. Finally, the preliminary result, noted above, of an oxygen, Na anticorrelation by Paltoglou and Norris (1987) for the giants in ω Cen is certainly suggestive of a causal relationship between the behavior of Na and phenomena associated with mixing. Since s-process element enhancement is also seen in the nitrogen enriched stars one must now address the possibility that neutron addition processes, similar to those described by Iben (1975, 1976) in the context of intermediate mass AGB stars, may be responsible for some of the anomalies.

REFERENCES

Bell, R. A. and Dickens, R. J. 1980 Astrophys. J. 242, 657.
Bell, R. A., Dickens, R. J. and Gustafsson, B. 1979 Astrophys. J. 229, 604.
Bell, R. A., Hesser, J. E. and Cannon, R. D. 1983 Astrophys. J. 269, 580.
Bessell, M. S. and Norris, J. 1976 Astrophys. J. 208, 369.
Cannon, R. D. and Stobie, R. S. 1973 Monthly Notices Roy. Astron. Soc. 162, 207.
Carbon, D. F., Langer, G. E., Butler, D., Kraft, R. P., Suntzeff, N. B., Kemper, E., Trefzger, C. F., Romanshin, W. 1982 Astrophys. J. Suppl. 49, 207.
Cohen, J. G. 1981 Astrophys. J. 247, 869.
Cohen, J. G. and Bell, R. A. 1986 Astrophys. J. 305, 698.
Cottrell, P. L. and Da Costa, G. S. 1981 Astrophys. J. Letters 245, L79.
Da Costa, G. S. and Demarque, P. 1982 Astrophys. J. 259, 193.
D'Antona, F., Gratton, R. and Chieffi, A. 1983 Mem. Soc. Astron. Italiana 54, 173.
Deupree, R. G. and Cole, P. W. 1983a Astrophys. J. 269, 676.
Deupree, R. G. and Cole, P. W. 1983b Private Communication.
Dickens, R. J. and Bell, R. A. 1976 Astrophys. J. 207, 506.
Dickens, R. J., Bell, R. A. and Gustafsson, B. 1979 Astrophys. J. 232, 428.
Faulkner, J. and Iben, I. Jr. 1967 Nature 215, 44.
Freeman, K. C. and Norris, J. 1981 Ann. Rev. Astron. Astrophys.

19, 319.
Freeman, K. C. and Rodgers, A. W. 1975 Astrophys. J. Letters 201, L71.
Gratton, R. G. 1982 Astron. Astrophys. 115, 336.
Harding, G. A. 1962 Observatory 82, 205.
Iben, I. Jr. 1967 Ann. Rev. Astron. Astrophys. 5, 571.
Iben, I. Jr. 1974 Ann. Rev. Astron. Astrophys. 12, 215.
Iben, I. Jr. 1975 Astrophys. J. 196, 525.
Iben, I. Jr. 1976 Astrophys. J. 208, 165.
Kraft, R. P. 1979 Ann. Rev. Astron. Astrophys. 17, 309.
Langer, G. E. 1985 Publ. Astron. Soc. Pacific 97, 382.
Langer, G. E., Kraft, R. P., Carbon, D. F., Friel, E. and Oke, J. B. 1986 Publ. Astron. Soc. Pacific 98, 473.
Leep, E. M., Wallerstein G. and Oke, J. B. 1986 Astron. J. 91, 1117.
Lloyd Evans, T. 1977a Monthly Notices Roy. Astron. Soc. 178, 345.
Lloyd Evans, T. 1977b Monthly Notices Roy. Astron. Soc. 181, 591.
McClure, R. D. 1979 Mem. Soc. Astron. Italiana 50, 15.
Norris, J. 1987 Astrophys. J. Letters in press.
Norris, J. and Bessell, M. S. 1975 Astrophys. J. Letters 201, L75.
Norris, J. and Bessell, M. S. 1977 Astrophys. J. Letters 211, L91.
Norris, J., Cottrell, P. L., Freeman, K. C. and Da Costa, G. S. 1981 Astrophys. J. 244, 205.
Norris, J. and Freeman, K. C. 1983 Astrophys. J. 266, 130.
Norris, J., Freeman, K. C. and Da Costa, G. S. 1981 Astrophys. J. 244, 205.
Norris, J. and Pilachowski, C. A. 1985 Astrophys. J. 299, 295.
Norris, J. and Smith, G. H. 1983 Astrophys. J. 272, 635.
Norris, J. and Smith, G. H. 1984 Astrophys. J. 287, 255.
Osborn, W. 1971 Observatory 91, 223.
Paltoglou, G. and Norris, J. 1987 in preparation.
Persson, S. E., Frogel, J. A., Cohen, J. G., Aaronson, M. and Matthews, K. 1980 Astrophys J. 235, 452.
Peterson, R. C. 1980 Astrophys J. Letters 237, L87.
Pilachowski, C. A., Wallerstein, G. and Leep, E. M. 1980 Astrophys. J. 236, 508.
Smith, G. H. 1985 Astrophys. J. 298, 249.
Smith, G. H. and Bell, R. A. 1986 Astron. J. 91, 1121.
Smith, G. H. and Norris, J. 1981a Astrophys. J. 254, 149.
Smith, G. H. and Norris, J. 1981b Astrophys. J. 254, 594.
Suntzeff, N. B. 1981 Astrophys. J. Suppl. 47, 1.
Sweigart, A. V. and Mengel, J. G. 1979 Astrophys. J. 229, 624.
Woolley, R. v. d. R. 1966 Roy. Obs. Ann., No. 2.
Zinn, R. 1973 Astrophys. J. 182, 183.

DISCUSSION

NEMEC: The recent finding by Nemec, Nemec and Norris (1986 <u>Astron. J.</u> 92), and others, that double-mode RR Lyrae stars are not found in the highly flattened globular cluster ω Cen, despite the apparent similarity of the primary periods of these stars to the primary (first overtone) periods of the double-mode RR Lyrae stars in M 15, seems to us might also be caused by star-to-star angular momentum differences.

BELL: Have you considered the possibility of CH lines blending with the Al lines in the H and K wavelength region?

NORRIS: There are two points. First, since one sees an anticorrelation of CN and CH, the effect of CH blending would be to decrease the Al line strengths in CN strong objects - just the opposite of the positive CN, Al correlation which is observed. Second, in her high dispersion analysis of the CN strong star ROA 253 in ω Cen, Cohen showed a positive CN, Al correlation based on Al lines in the red region of the spectrum.

DEMARQUE: I have two comments. First I would not give up too soon on the Sweigart Mengel mechanism to explain the progressive enrichment of N along the giant branch. Rotation could be responsible for internal mixing in many ways other than meridional circulation, which should occur at a lower luminosity. Second, in reference to N variations found in stars on the turnoff in 47 Tuc, there is, I believe, some recent work by Spite and Spite which shows that N-rich halo stars near the main sequence have the same Li abundance as N-poor stars. In my mind, these observations show unequivocally that mixing must be responsible for the N variations near the main sequence of 47 Tuc.

NORRIS: 1. I did not mean to imply that I think that the SM mechanism should be discarded completely. Rather, I would hope that the dependence of the degree of cyanogen enrichment on cluster flattening might spur the theoreticians, such as yourself, to renew their efforts in tackling the difficult problem of stellar rotation. 2. I think you are comparing two different phenomena. The stars studied by the Spites have [Fe/H] $\simeq$ -2.0, and represent only a few percent of a complete sample (see Bessell and Norris 1982, <u>Astrophys. J. Letters</u> 263, L29.). In 47 Tuc roughly half of the stars are CN strong. I think we are talking about apples and oranges.

AGES OF THE GALACTIC GLOBULAR CLUSTERS

Don A. Vandenberg

University of Victoria

ABSTRACT: Using stellar evolutionary models which have been computed for the preferred set of element abundances, ages of 13-14 Gyr are derived for the Galactic globular clusters. No significant (> 1.5 Gyr) spread in age is found as a function of either metal content or galactocentric distance. The lack of an age-metallicity relation is shown to be contingent upon the assumption that [O/Fe] is anticorrelated with [Fe/H]: the use of scaled solar abundance calculations would predict that the most metal-poor systems are about 3×10^9 yr older than the most metal-rich clusters. Some evidence in support of high oxygen-to-iron ratios is discussed.

1. INTRODUCTION

The goal of much of the globular cluster (GC) research that is presently being conducted is to answer two basic questions: (1) what is the highest cluster age?, and (2) are all globulars coeval? The answer to the first of these sets a lower limit to the age of the universe - thereby constraining cosmological models - while that for the second tells us whether the collapse of the Galactic halo was rapid or slow - thus providing vital input into our understanding of the evolution of galaxies. Unfortunately, in spite of a lot of effort, precise answers to these two questions have yet to be forthcoming. However, considerable progress is being made and it seems clear that the uncertainties in the derived ages are steadily being reduced. This review presents a status report of our current understanding.

2. GLOBULAR CLUSTER AGES

In principle, the preferred way to determine the age of a given globular cluster is to perform a main-sequence fit of appropriate isochrones onto photometric data. In practice, however, this cannot yet be regarded as the most reliable method. Too many things have to be known too accurately - not only the temperature scale of the models and the transformation relations between T(eff) and color, but also the reddening of the cluster and its metallicity. On the theoretical side, predicted effective temperatures depend sensitively on the assumed structure of the outer atmosphere and on the treatment of convection in the stellar interior - neither of which is well understood. In addition, it is well known that synthetic colors, particularly for cool/metal-rich stars, err in the sense of being too blue by a few to perhaps several hundredths of a magnitude in B-V (e.g., Magain 1983). Since the ratio of magnitude to color along a zero-age main sequence (ZAMS) is $\simeq 5$, every 0.02 mag error in either the predicted colors or in the observed cluster reddening would translate to an error of about 0.1 mag in the derived distance modulus and hence approximately 1.5 Gyr in age.

The tremendous range in "observed" metallicities for many clusters is also a major problem. For example, according to the compilation by Hesser and Shawl (1985), support may be found in the literature for a metal content for a well-studied system like NGC 6752 anywhere between [m/H] = -1.09 and -1.66. Certainly, the interpretation of its very tight main-sequence locus on the (V,B-V)-plane, which has recently been obtained by Penny and Dickens (1986) and by Buonanno et al. (1986), is going to depend on the composition which is assumed in the models. And if the computations for a particular [m/H] don't fit, is it because there are problems with the theory or simply that the wrong metallicity was initially assumed?

In order to avoid virtually all of these problems, Iben and Renzini (1984) have argued that the best way at the present time to determine ages is from a calibration of the magnitude difference between the horizontal branch and the main-sequence turnoff, at the color of the turnoff. Figure 1 illustrates what this method involves. Since the

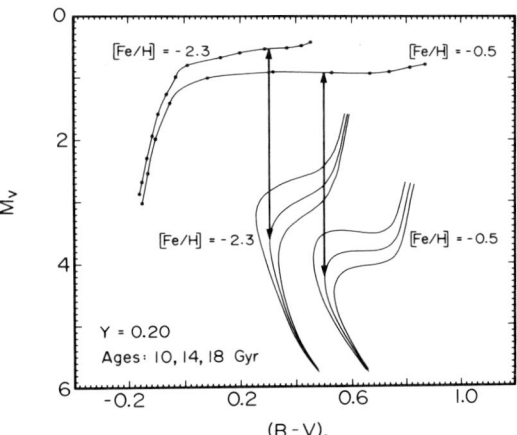

Fig. 1. - Plot of VandenBerg and Bell (1985) isochrones for the noted ages and metallicities, along with fully consistent zero age horizontal branches (VandenBerg 1986). Arrows illustrate the definition of the age-dependent quantity ΔV_{TO}^{ZAHB}.

luminosity (though not the color) of a computed zero-age horizontal branch (ZAHB) star is largely independent of the mass of the red-giant precursor, it is essentially independent of age. (Recall that the canonical explanation for the observed wide range in color of HB stars is that, while all such stars have nearly the same core mass, their envelope and hence total masses differ because of prior mass loss – either during evolution up the red-giant branch or as a result of the helium flash event itself.) Given then that the luminosity of the ZAHB is not a function of age, while that of the turnoff obviously is, the magnitude difference between these two features is clearly an excellent indicator of age. A particularly desirable attribute of this quantity is that, for a given age, it is not very sensitive to uncertainties in composition (note the example given in Figure 1).

In practice, there are some important limitations to this method. Many globulars show only a very blue or a very red horizontal branch and it is by no means a straightforward task, especially in the former instance, to estimate the location of the ZAHB at the color of the turnoff. Also, the distribution of stars at the turnoff is often vertical for several tenths of a magnitude with the result that the precise magnitude of the turnoff is difficult to define. Primarily because of these two uncertainties, the observed magnitude difference in most clusters cannot be estimated to within ± 0.15 – 0.2 mag, which means that precise age estimates are not possible.

The lower panel of Figure 2 illustrates a calibration, from theory, of the magnitude difference between the ZAHB and the turnoff as a function

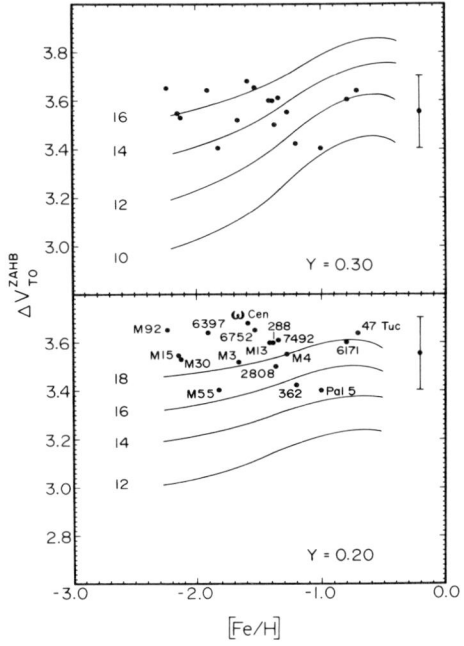

Fig. 2. – Theoretical calibration, for two Y values, of the magnitude difference between the ZAHB and the main-sequence turnoff, at the color of the turnoff, as a function of [Fe/H] and age. The latter is indicated in units of 10^9 yr to the left of each solid curve. Filled circles give the observed data for a number of GCs as reported by Buonanno (1986).

of [Fe/H] and age. Superimposed on this diagram are the observed values of ΔV_{TO}^{ZAHB} recently reported by Buonanno (1986) for 17 GCs for which he and his colleagues have obtained CCD photometry. The metallicity scale which they adopt and which is assumed here is largely that given by Zinn (1985). It is clear that if the helium content is close to Y = 0.20 and the heavy elements have solar number abundance ratios (which are assumed in the VandenBerg and Bell (1985) isochrones), then the globular clusters are predicted to have a very high age indeed. It is tempting to conclude that a number like 18 Gyr or slightly older is the preferred value, but one must keep in mind the large error bar associated with each datum and be wary of possible biases regarding how one actually chooses to extrapolate to the ZAHB location at the color of the turnoff - when none of the observed HB stars have such colors - and how the turnoff magnitude is selected when the distribution of stars is vertical. Such subjective sources of error may well affect the mean ΔV_{TO}^{ZAHB} value which is obtained. The main point of this plot is that there is no evidence, from this particular sample of GCs, for ages younger than about 14 Gyr (if Y = 0.2) or for large cluster-to-cluster variations in age.

It is worth pointing out that a fairly wide range in galactocentric radius (Rg) is encompassed by these clusters, from 3.9 kpc for NGC 6171 to 18.7 kpc for NGC 7492 (Webbink 1985). Clearly, on the basis of their measured ΔV_{TO}^{ZAHB} values, the ages of these two clusters must be nearly the same. In the latest review of globular cluster ages, Gratton (1985) suggested that there was a significant age-Rg relation among the globulars, primarily on the basis of available observations of NGC 7006 and the three Palomar clusters, numbered 5, 12, and 13. However, new CCD photometry of NGC 7006 (Rg = 36.3 kpc) by Cohen (1985) has demonstrated that its turnoff is fully consistent with usual estimates of GC ages (though the data is still sufficiently imprecise not to exclude the possibility that it is younger by 3-4 Gyr). Regarding Pal 5 (Rg = 16.5 kpc), which has the lowest ΔV_{TO}^{ZAHB} value of those plotted in Figure 2, it is apparent from Figure 3 that its C-M diagram is not distinctly different from that of M5,

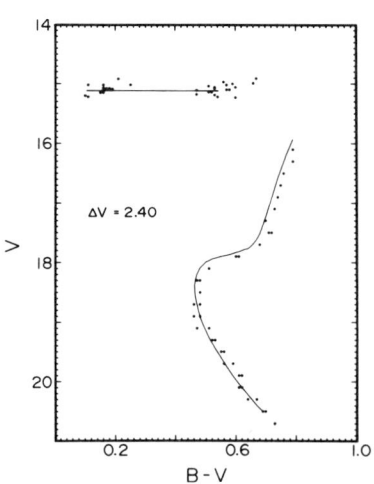

Fig. 3. - Superposition of the fiducial sequence for Pal 5 (solid curve) by Smith et al. (1986) onto that for M5 (dots) by Richer and Fahlman (1986). The M5 HB observations are due to H. Arp (Ap.J., 135, 31, 1962): their accuracy has been confirmed by Richer and Fahlman. The distance moduli of the two clusters, which have similar metallicities and reddenings, are assumed to differ by 2.4 mag.

which has a similar metal abundance and reddening, but is much nearer at Rg = 6.6 kpc. Considering the uncertainties in fitting the sparse (but tightly-defined) horizontal branch of Pal 5 to that of M5, the ages of the two clusters are probably the same to within about 10^9 yr.

In the case of Pal 12 (Rg = 16 kpc), the photographic photometry by Harris and Canterna (1980) certainly indicates a small ΔV_{TO}^{ZAHB}; but closer inspection of the data suggests that it may suffer from some calibration problems. In the left-hand panel of Figure 4, the published

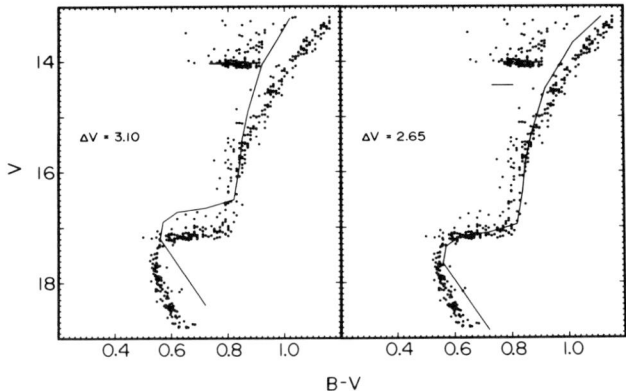

Fig. 4. - Comparison of the fiducial sequence for Palomar 12 (solid curve) with the composite C-M diagram of 47 Tuc. The Pal 12 locus was shifted horizontally by 0.02 mag, to allow for reddening differences, and by the two indicated values of ΔV in the vertical direction.

mean locus of Pal 12 is superimposed on the composite C-M diagram of 47 Tucanae (Harris and Hesser 1986) such that the mean horizontal-branch luminosities agree. But this produces a large discrepancy near the main sequence - one which is impossible to reconcile. Regardless of whether or not two clusters have different ages, their unevolved main sequences must be nearly coincident on the C-M plane if allowances are made for differences in distance and reddening and if, as is believed to be the case here, the two systems have similar metallicities. The right-hand panel shows that acceptable agreement of the lower main sequences can be obtained if the cluster moduli are assumed to differ by 2.65 mag - but then the horizontal branch of Pal 12 is much fainter than that of 47 Tuc. This would be just as hard to understand. Certainly the preferred explanation is that there are problems with the main-sequence photometry; which could well be possible since it relied on the calibration of secondary images produced by the Racine wedge. A follow-up study using the CCD is obviously in order.

Since the Pal 13 photometry used by Gratton (1985) in his analysis has not yet been published, no comment can be made of the reliability of the derived age for this system. But on the basis of the available evidence, the dependence of age on galactocentric distance appears to be rather small. What about a relation between the age of a cluster and its metallicity? The lower panel of Figure 2 does give the impression that the most metal-poor clusters are a little older than the most metal-rich systems. Such a trend is very much more pronounced if the observed ΔV_{TO}^{ZAHB} values are plotted on the theoretical calibration for Y = 0.30, as given in the upper panel of Figure 2. In this case, the size of the effect would seem to be about 4-5 Gyr over a 1.5 dex range in [Fe/H] - though again it should be emphasized that part of the trend may be associated with how one estimates the luminosity of the ZAHB and the main-sequence turnoff. An interesting way to view these results is that, if all GCs have the same age and if the present models are basically correct, then the cluster helium content and the metallicity must be anticorrelated - i.e., higher Y in more metal-poor systems - just as Sandage (1982) has found necessary to explain the correlation which he discovered between the period of an RR Lyrae star and its metallicity.

Thusfar, we have made use of only one small aspect of observed C-M diagrams and surely such things as the morphology of stellar distributions in different evolutionary phases will provide important constraints on our interpretation of the data. A good example is given in Figure 5, where theoretical ZAHBs for [Fe/H] $\simeq$ -2.25, Y = 0.24, and three different

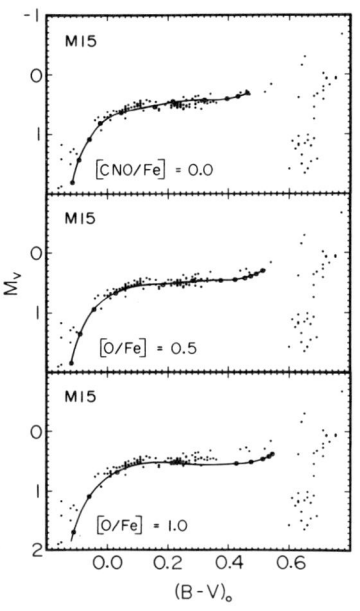

Fig. 5. - Fits of theoretical ZAHBs for [Fe/H] $\simeq$ -2.25 and three different values of [O/Fe] to published photometry of M15. Differences in oxygen abundance affect the shape and extent of the ZAHB locus redward of B-V $\simeq$ 0.2.

assumptions regarding the oxygen abundance are compared with published photometry for M15 by Sandage (1970) and by Bingham et al. (1984). If one remembers that a ZAHB should provide a good fit to the lower bound of the observations, then one must conclude that the best fits are found when it is assumed that oxygen is significantly overabundant with respect to iron. In fact, if the stars in M15 are similar to the metal-poor field dwarfs and giants which have been subjected to detailed abundance analyses (see the review by Sneden 1985), then there is reason to believe that the cluster oxygen-to-iron ratio should be up by about a factor of 5 compared to the solar value. Therefore, another important variable, [O/Fe], has to be taken into account in the stellar models.

Clearly further refinements of cluster age estimates must involve the entire C-M diagram. The magnitude difference between the HB and the MS turnoff is arguably still the best constraint on age, but improved determinations of this quantity can be obtained by fitting the model loci to the observed stellar distributions. The preferred procedure for doing so is best explained with the aid of the example given in Figure 6. This

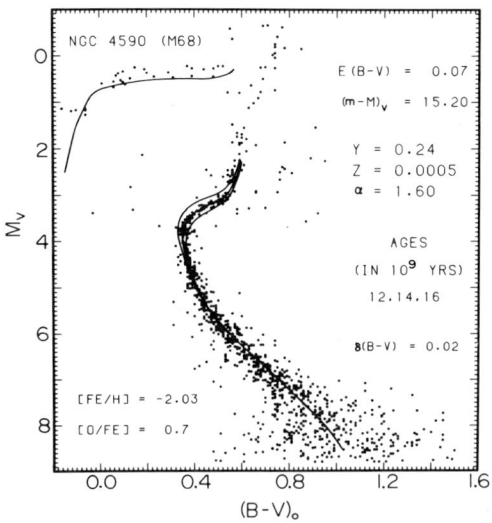

Fig. 6. - Fit of appropriate models (solid curves) to recent observations of NGC 4590 by McClure et al. (1986).

illustrates one of the finest CCD C-M diagrams yet obtained for a globular cluster: unlike several other examples of such results, enough bright stars were observed to be able to very accurately define the shape of the turnoff region and the subgiant branch. This is, of course, critically important for the determination of precise ages. In order to ensure that color uncertainties have a negligible effect on the ages derived, the absolute luminosity (and hence distance) scale should first be established by matching the observed and predicted ZAHBs. Granted, in this particular instance, the observational ZAHB is not well defined. However, it was noticed in the analysis of the M68 data that its appearance on the (V,B-V)-plane is morphologically identical to that of M15. The much more populous HB in the latter cluster was therefore used as an additional constraint on the location of the ZAHB of M68. Once a satisfactory match of the horizontal branches is found, then the isochrones can be shifted in color, if necessary and by whatever amount is necessary, until they provide a best-fit of the main-sequence turnoff observations.

Here, the isochrones which are used have been calculated for the primordial helium abundance, Y = 0.24 (e.g. Boesgaard and Steigman 1985), and what we believe to be the appropriate [Fe/H] and [O/Fe] values for M68. It is clear that an exceedingly fine match of the theory to the observations is obtained for an age close to 14 Gyr. What is particularly encouraging is that the shape of the model loci reproduces that of the data over the entire range of the fit. Even the steepening slope of the faintest stars seems to be consistent with the model predictions - which is further evidence that the adopted distance is close to being correct. There does appear to be a zero-point error in the color scales in the sense that the colors of the isochrones are too blue by 0.02 mag (if the adopted E(B-V) value is realistic). In fact, such a color error is suggested by the subdwarfs.

Figure 7 illustrates a comparison of representative isochrones for three metallicities, from the VandenBerg and Bell (1985) compilation,

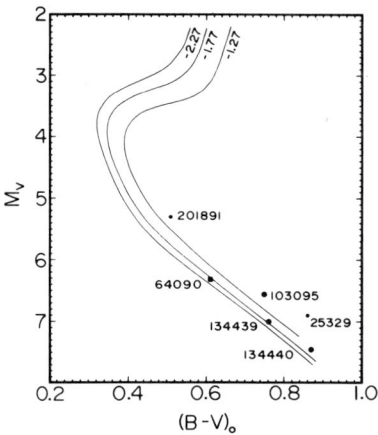

Fig. 7. - Comparison of 16 Gyr isochrones for the metallicities noted at the top of each curve with the Carney (1979) subdwarfs which have the best-determined properties. Revised parameters for some of these were provided by G.S. Da Costa (private communication). HD 64090 has [Fe/H] ≃ -1.75; the [Fe/H] values for the five other stars scatter about -1.34. Large symbols indicate that the relative uncertainty of the parallax determination is less than 10 percent.

with the Carney (1979) subdwarfs which have the most accurate parameters. Lutz-Kelker corrections have not been applied to the data - but this is arguably the proper procedure (see Richer and Fahlman 1986). For the present fit, one finds that, in the mean, the isochrones are too blue by about 0.02 mag. Alternatively, if cluster distances are found by the main-sequence fitting of the VandenBerg-Bell isochrones to the data, they will be about 0.1 mag smaller than those based on the subdwarf standards. But the subdwarf sequences are obviously not well-defined - better data is urgently needed - and it may well be that the models actually have larger errors than the present analysis suggests.

Returning to the question of whether or not there is an age-metallicity relation, the lower right-hand panel of Figure 8 illustrates a best-fit of isochrones for Y = 0.24 and [Fe/H] = -0.65, assuming scaled-solar abundances of the heavy elements, to the 47 Tuc data. Although not shown, the distance was derived by matching predicted and observed hori-

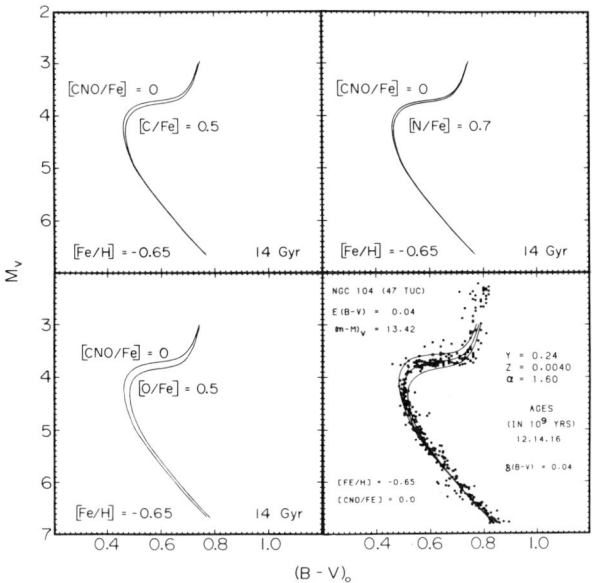

Fig. 8. — Comparisons of computed isochrones for the same Y, [Fe/H], and age (0.24, -0.65, and 14 Gyr, respectively) but for different assumed C, N, and O abundances. The lower right-hand panel illustrates the adopted fit of the [CNO/Fe] = 0.0 models for the indicated ages to the Harris and Hesser (1986) C-M diagram for 47 Tucanae.

zontal branches, which can be accomplished with little uncertainty. The fit at the turnoff indicates an age of 14 Gyr. The other three panels show that higher C, N, or O abundances tend to make the turnoff for a given age somewhat fainter. That is, the effect of enhanced [CNO/Fe] is to make the age corresponding to a given turnoff luminosity younger. The effect is most pronounced if oxygen is enhanced because it is the dominant constituent of the CNO group while the variation of nitrogen has little effect because it is the least abundant of the three elements. Thus, there is no inconsistency, for example, between the observation, on the one hand, that 47 Tuc has a very tight C-M diagram, and the abundance work by Bell et al (1983), on the other, which found star-to-star nitrogen abundance variations by about a factor of five. According to our present understanding (e.g., see Sneden 1985), carbon and nitrogen scale with iron - except perhaps at extremely low metallicities. Oxygen, on the other hand, is believed to become steadily overabundant with respect to iron as the [Fe/H] value drops though the trend may flatten below [Fe/H] = -1. If the adopted metal abundance for 47 Tuc is correct, then on the basis of field star studies, one expects an oxygen enhancement [O/Fe] ≃ 0.3 - in which case, the derived cluster age would be closer to 13 Gyr. This is about 1 Gyr smaller than our best estimate for M68 (and M15, which seems to be morphologically identical to M68 on the C-M plane) and one could argue that there is some evidence for an age-metallicity relation among the Galactic globular clusters. However, such a small age difference is obviously within the uncertainty. One therefore concludes that if there is a variation of age with metallicity, it must be very slight.

A final point about Figure 8: note that a redward color correction of 0.04 mag was applied to the models in order to reproduce the observed colors of stars in 47 Tuc. This could well be the expected error of the VandenBerg and Bell (1985) isochrones for metal-rich compositions. If a metallicity of [Fe/H] = -0.8 were assumed instead of [Fe/H]= -0.65, then the corresponding theoretical loci would have to be adjusted by 0.08 mag to yield a similar fit. Such a large color correction seems rather unlikely. This suggests that the metal content of 47 Tuc must be greater than [m/H] = -0.8, at least in the mean. If iron is underabundant, then there must be enhancements in other elements to partially take the place of iron as a main source of opacity and blanketing.

Oxygen has played a critical role in the present analysis. Indeed, had it not been assumed that [O/Fe] is higher in metal-poor systems than in metal-rich ones, a significant age-metallicity relation would have been found: M68 would have been predicted to be about 3×10^9 yr older than 47 Tuc. Some further support (besides Fig. 5) for the possibility that [O/Fe] is high in metal-poor clusters is given in Figure 9. This illustrates the superposition of computed ZAHBs for [Fe/H] = -2.25 and

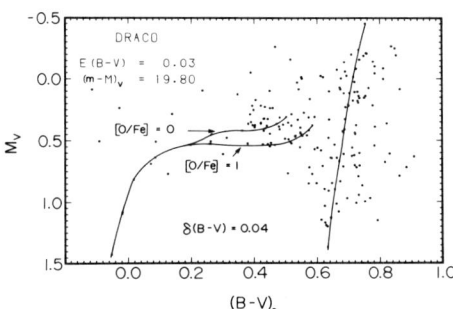

Fig. 9. - Overlay of theoretical loci for [Fe/H] = -2.25 and the noted [O/Fe] values on the recent C-M diagram for Draco by Carney and Seitzer (1986). The computed RGBs have identical locations in the two cases.

two different [O/Fe] values on new photometry for Draco, a dwarf spheroidal galaxy that has an anomalously red HB for its (M92-like) metal abundance. Note that a considerably-expanded luminosity scale has been adopted to exaggerate the differences between the two ZAHBs: their separation in luminosity at $(B-V)_0 = 0.4$ is actually only 0.12 mag. But this may prove to be another point in favor of enhanced oxygen models since the latter will predict a reduced magnitude difference between metal-poor and metal-rich RR Lyrae stars compared to usual findings. The important point of this figure, however, is that scaled-solar-abundance calculations cannot explain the reddest horizontal-branch stars (cf. Rood and Seitzer 1982) - regardless of whether or not the Draco stars span a wide range in age. The ZAHBs which have been plotted have nearly reached their maximum possible colors - the assumption of even higher masses (younger turnoff ages) will cause the HB loci to bend back toward the blue. No such problem exists if [O/Fe] $\simeq$ 1: a red HB is expected.

3. SUMMARY

Recent studies (Sandage 1982, Flannery and Johnson 1982, Janes and Demarque 1983, VandenBerg 1983, Gratton 1985) have found ages > 15 Gyr for the oldest Galactic GCs. On the basis of new calculations which assume $Y = 0.24$ and enhanced [O/Fe] ratios, this investigation suggests that the maximum age of the globulars is close to 13-14 Gyr. Moreover, any variation of age with galactocentric radius or with metallicity appears to be within the limits of detectability. That is, the age spread seems to be less than ~ 1.5 Gyr.

REFERENCES

Bingham, E. A., Cacciari, C., Dickens, R. J. and Fusi Pecci, F. 1984 Monthly Notices Roy. Astron. Soc. **209**, 765.
Boesgaard, A. and Steigman, G. 1985 Ann. Rev. Astron. Astrophys. **23**, 319.
Buonanno, R. 1986 Mem. Soc. Astron. Italiana in press.
Buonanno, R., Corsi, C. E., Iannicola, G. and Fusi Pecci, F. 1986 Astron. Astrophys. **159**, 189.
Carney, B. 1979 Astrophys. J. **233**, 877.
Carney, B. and Seitzer, P. O. 1986 Astron. J. **92**, 23.
Cohen, J. G. 1985 Astron. J. **90**, 2254.
Flannery, B. P. and Johnson, B. C. 1982 Astrophys. J. **263**, 166.
Gratton, R. G. 1985 Astron. Astrophys. **147**, 169.
Harris, W. E. and Canterna, R. 1980 Astrophys. J. **239**, 815.
Harris, W. E. and Hesser, J. E. 1986 in preparation.
Hesser, J. E. and Shawl, S. J. 1985 Publ. Astron. Soc. Pacific **97**, 465.
Iben, I. Jr. and Renzini, A. 1984 Phys. Rep. **105**, 329.
Janes, K. A. and Demarque, P. 1983 Astrophys. J. **264**, 206.
Magain, P. 1983 Astron. Astrophys. **122**, 225.
McClure, R. D., VandenBerg, D. A., Hesser, J. E., Stetson, P. B. and Bell, R. A. 1986 preprint.
Penny, A. J. and Dickens, R. J. 1986 Monthly Notices Roy. Astron. Soc. **220**, 845.
Richer, H. B. and Fahlman, G. G. 1986 preprint.
Rood, R. T. and Seitzer, P. O. 1982 in IAU Colloquium 68, Astrophysical Parameters for Globular Clusters, A. G. D. Philip and D. S. Hayes, eds., L. Davis Press, Schenectady, p. 369.
Sandage, A. 1970 Astrophys. J. **162**, 841.
Sandage, A. 1982 Astrophys. J. **252**, 553.
Smith, G. H., McClure, R. D., Stetson, P. B., Hesser, J. E. and Bell, R. A. 1986 Astron. J. **91**, 842.
Sneden, C. 1985 in Production and Distribution of the C, N, and O Elements, I. J. Danziger, F. Matteucci, and K. Kjar, eds., European Southern Observatory, Garching bei Munchen, p. 1.
VandenBerg, D. A. 1983 Astrophys. J. Suppl. **51**, 29.
VandenBerg, D. A. 1986 in preparation.
VandenBerg, D. A. and Bell, R. A. 1985 Astrophys. J. Suppl. **58**, 561.
Webbink, R. F. 1985 in IAU Symposium 113, Dynamics of Star Clusters, J. Goodman and P. Hut, eds., Reidel, Dordrecht, p. 541.
Zinn, R. 1985 Astrophys. J. **293**, 424.

DISCUSSION

ALCAINO: The isochrones of VandenBerg and Bell (1985) are given for Y = 0.2 and 0.3. You have mentioned that the best current estimate of the primordial helium abundance is Y = 0.24. Could you explain the basis on which this value is preferred?

VANDENBERG: A value of Y ~ 0.24 had been obtained from the analysis of extragalactic H II regions by Sargent and colleagues. Such a value has also been found to be consistent with lithium abundance determinations of Pop. II stars by F. Spite. Boesgaard and Steigman provide a recent summary of our present understanding in the 1985 Annual Review of Astronomy and Astrophysics.

TRIMBLE: By the fit to M 68, all the dots for the horizontal branch are above the line.

VANDENBERG: Yes, the ZAHB must necessarily be fitted to the lower bound of the observed HB distribution because subsequent evolution will tend to make the stars move above the ZAHB.

BELL: Can you estimate how much of the changes due to changing [O/Fe] are due to changes in opacity and how much is due to changes in energy generations?

ROOD: I have done such calculations. I was surprised to find that opacity was the most important factor in determining the turnoff age. Even though the CNO cycle dominates just after core hydrogen exhaustion I think the gradient set up in hydrogen during the earlier stages is the determining factor.

NORRIS: Given that blue horizontal-branch stars have been found by Ruth Peterson to have anomalously high rotational velocities, and that these effects are generally neglected in standard stellar evolution calculations, would you comment on how internal rotation in low mass stars might affect the results you have shown us?

VANDENBERG: Rotation has not been studied properly. From simplistic considerations, it is commonly believed that rotation will delay the onset of the helium flash, leading to higher core masses and smaller envelope masses than the canonical models, thereby favoring the formation of brighter and bluer horizontal-branch stars. In addition one expects rotation to increase turnoff ages since rotation will tend to support the star, thereby leading to lowered nuclear burning rates. But the tightness of observed CM diagrams must constrain the amount of rotation which is allowed. Clearly more work is needed.

SCHOMMER: Would you or Roger care to comment on the possible origin of the 0.02 - 0.04 (B-V) shifts you apply? Does this depend on the color or effective temperature, or is this result the best value for the turnoff region?

VANDENBERG: Roger has shown that his color - T_{eff} relations look pretty good (Though, these may still be the source of the small problem with the colors.). Problems with the T_{eff} scale of the models may well be the most likely explanation of the discrepancy. While infrared photometry has tended to confirm the predicted T_{eff}'s of red giant stars in globular clusters, confirmation of turnoff temperatures has yet to be obtained. Until this happens, one must be wary of the possibility that predicted T_{eff}'s (and hence synthetic colors) are somewhat in error, and one should continue to rely on luminosity criteria to derive distances and ages. I suspect that the error will become systematically larger as [Fe/H] increases. However, the error seems to be primarily in the zero-point. As shown in Fig. 6, the systematic variation of computed colors for a given [Fe/H] compares very well that observed.

CAYREL: The ages for M 68 and 47 Tuc you have were based on the ΔV^{ZAHB}_{TO} to mag. differences. Is it not true to say that, from the same diagrams you have shown, the age derived from main-sequence fitting would be the same?

VANDENBERG: Main-sequence fitting would lead to somewhat older ages since the models appear to be bluer than the observations. If theoretical loci are too blue, then derived distances are too small and ages too high. However, the uncertainties in color seem to be small enough that the estimated ages will not differ by more than 3 Gyr (or smaller for low-metallicity clusters) if the main-sequence method is employed.

ZINN: I have two questions. First, if I understand what you have done correctly, you have derived the distance moduli of the clusters by matching the observed horizontal branches with ones generated by theoretical calculations. What will happen to your conclusions about the age-metallicity relationship if instead you adopt a shallower or a steeper dependence of HB luminosity on [Fe/H]. For example, the one that Sandage found from his analysis of the Oosterhoff effect. Second, several independent workers, including myself, have concluded that there is a significant range in metallicity in the Draco dwarf spheroidal galaxy. If you accept these results, is it still necessary to conclude from the redness of the HB in Draco that it has an overabundance of oxygen?

VANDENBERG: 1. If all HB stars have $M_v \simeq 0.6$, then the metal-rich clusters must be younger than metal-poor clusters. The reverse would be true if Sandage's relation between $M_v(RR)$ and [Fe/H] is assumed, in the context of the present models since his relation is steeper than my predictions, from which a constant age scenario has been deduced.

2. Higher metallicities would reduce the color discrepancy but probably not remove it. To first order, a ZAHB with [Fe/H] = -2.25 and [O/Fe] = 1.0 would be similar to one with [Fe/H] ≃ -1.6 and [O/Fe] = 0.0. The enhanced oxygen mix has an effective $Z = 5 \times 10^{-4}$. The [Fe/H] value corresponding to the latter is higher than the upper limit of the observed range in Draco metallicities.

GRINDLAY: Suppose in three years time we knew that the disk globulars really were formed at a different time than the halo clusters, as their apparently distinct population might suggest. What would be the major effects and implications for your models?

VANDENBERG: There are so many possible variables; for example, perhaps low [Fe/H] clusters mix more of their envelopes than metal-rich clusters (for some unknown reason) during red giant branch evolution. Such a possibility is suggested by the work of the Lick group on M 92. In this case, the helium abundances would be larger in low-metallicity clusters than in those of high metal content (recall Sandage's result). This would affect derived distances and ages.

GLOBULAR CLUSTER LUMINOSITY FUNCTIONS

Pierre Demarque

Yale University Observatory

ABSTRACT: The use of luminosity functions in the following areas is reviewed: (a) the determination of the helium content and ages of the globular clusters; (b) the testing of stellar structure theory; and (c) the determination of the initial mass function of globular clusters.

I. INTRODUCTION

   Together with color-magnitude (CM) diagrams, the luminosity functions (LF) of globular clusters (GC) have historically been the primary interface between observation and theory in the study of stellar evolution. It was realized early that LF's provide us with the means to test empirically the rate of stellar evolution (Sandage 1957), and to derive the stellar birthrate function (Salpeter 1955; Schwarzschild 1958).
   During the next quarter century, however, there were few attempts to compare stellar evolution theory directly with observed LF's, or to use LF's to derive the properties of GC's (see Paczynski's 1984 discussion). The traditional techniques of photographic photometry were too laborious, insufficiently accurate, and subject to too many unavoidable systematic errors. At the same time, the development of photoelectric photometry made more precise calibration of the main sequence turnoff possible, albeit for stars observed one-at-a-time. The main research emphasis then naturally shifted to the construction and interpretation of CM-diagrams, and remarkable progress was made in our understanding of stellar evolution and in the sophistication of our determinations of GC ages (Sandage 1969; Demarque and McClure 1977; Demarque 1980; Sandage 1982; Janes and Demarque 1983; VandenBerg 1983; Green et al. 1987).
   There were exceptions, however. Some attempts were made during that period to evaluate the GC initial helium content using the subgiant LF. I shall return to this topic in the next section. In a detailed study, Green (1981) constructed LF's for the brighter end of the giant branch of a number of GC's, based on the available data in

the literature. Green's LF's, which were by necessity coarsely binned, agreed fairly well with the predictions of theory, although a small number of clusters appeared to be deficient in bright giants.

In the last few years, thanks to new computerized techniques which exploit the full power of PDS microdensitometers and of CCD panoramic detectors (Stetson 1979a, 1979b), it has become possible to construct high-precision LF's. As a result, we are now witnessing a veritable revolution in GC LF research, which promises to transform our understanding of a number of key areas in stellar evolution, stellar dynamics, and the evolution of stellar systems.

Most of the interest is currently concentrated in three directions:

(1) Researchers look to LF's to help set more stringent constraints not only on the helium abundance, but also on the metallicities and ages of individual globular clusters. The aim here is (a) to answer the cosmologically significant question of the age of the oldest stars, and (b) to understand better the formation and chemical evolution of the galactic halo.

(2) Accurate GC LF's can teach us about the internal structural changes taking place in the interiors of low mass stars as they evolve. We are now finding fine features on the giant branch LF which are unexplained. There are also unexplained CNO abundance variations, which have been known for some time among GC giants (Kraft 1979; Freeman and Norris 1981), some of which are probably due to rotationally induced internal mixing (Sweigart and Mengel 1979). The resolution of these problems would greatly strengthen our confidence in current models and in the cluster chronology which we derive from them.

(3) The LF for the unevolved main sequence yields the slope of the initial mass function (IMF) in each GC. One hopes thus to gain information on the processes of star formation and on the parameters which control the form of the IMF. An understanding of these parameters, together with reliable evolutionary lifetimes for stars of different chemical compositions and ages, is essential to unravel the history of the Galaxy and nearby systems (Larson 1986; Scalo 1986). The same data are also necessary to interpret the integrated light of distant stellar systems through the construction of population models for these systems (Tinsley 1980; Gunn et al. 1981).

2. THE MAIN SEQUENCE TURNOFF AND SUBGIANT PHASES

2.1 Helium.

The initial aim, in the mid-and late-sixties, was to distinguish between "low" Y, i.e. $Y < 0.1$, as seemed appropriate if all the helium in the Universe had been synthesized in stars, or "high" Y, i.e. $Y < 0.25$, if helium production had taken place in a primordial fireball (Hoyle and Tayler 1964). When it became clear that the spectroscopy of blue horizontal-branch (HB) stars could not resolve this problem [(i.e. the low surface helium abundances observed in the atmospheres of these stars are not representative of the initial Y in GC's (Sargent and Searle 1967)], several attempts were made to derive helium abundances from the slope of the subgiant LF in GC's (Simoda and Kimura 1968; Simoda and Tanikawa 1970; Hartwick 1970; Faulkner 1972).

Figures 1 and 2, taken from the work of Green et al.(1987), illustrate the effect of a change in Y on the GC LF, for a given metallicity. Provided a sufficiently large sample of stars can be measured (about two thousand stars in a three-magnitude interval above the turnoff), an accuracy of 0.05 in Y can be achieved (Cole et al. 1983). Of course, comparison of observation with theory is not only limited by the size of the observed sample. At a higher level of precision, some uncertainties in the theory are also likely to become a factor, such as the effect of molecular opacities, which affect calculated radii, and the rotational state of the interior, which may cause sufficient mixing and structural changes to modify evolutionary rates [e.g. see the work of Endal and Sofia (1981) on the evolution of the rotating Sun].

2.2 Metallicities.

The subgiant LF also depends on metallicity, as illustrated in Figures 2 and 3, where two sets of theoretical LF's with different metallicities, but the same helium content, are shown. Although not a strong test of metallicity by itself, the LF provides a useful consistency check when used in conjunction with a fit in the CM-diagram, both of the turnoff and the HB, and together with spectroscopic data and other population indices such as variable stars [see e.g. the work of VandenBerg (1987) described in these Proceedings].

2.3 Ages.

In principle, main sequence and subgiant LF's present major advantages for GC age determinations:
  (a) the LF is only weakly affected by the uncertainties in the calculated radii of models for late-type stars (the mixing-length problem in the convective envelope).
  (b) the nuclear and opacity data which control the rate of evolution near the main sequence are relatively well known at this point, and the effects of their uncertainties are minimized in LF's. This point is well illustrated by a comparison of LF's constructed over a period of years by a number of independent workers, using different opacity data and physical assumptions. The agreement is close and better than for the turnoff position in the HR-diagram [e.g. the work of Simoda and Iben (1970), Ciardullo and Demarque (1977), VandenBerg (1986), Ratcliff (1986) and Green et al.(1987)].

Figures 1 - 3 show the sensitivity of the LF on age, for a given set of composition parameters. We see that the LF depends equally sensitively on age and on chemical composition. Thus together with an independent spectroscopic or photometric metallicity estimate, the LF

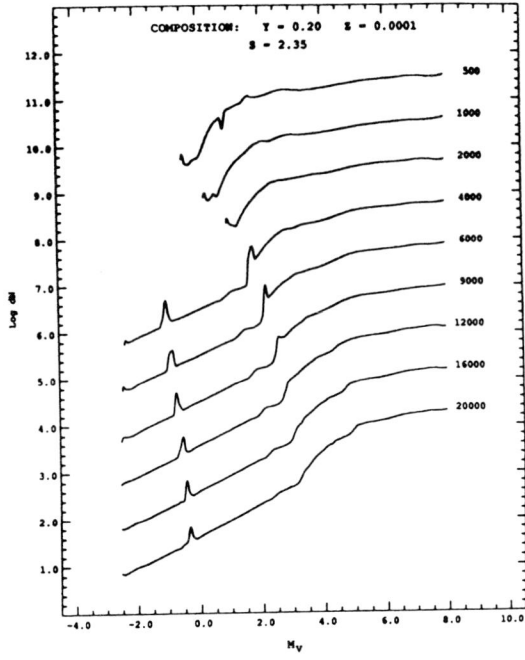

Fig. 1. Theoretical LF's for Y = 0.20 and Z = 0.0001, from Green et al (1987). The mass function is the Salpeter function, with S = 2.35. Ages in Myrs are shown for each LF.

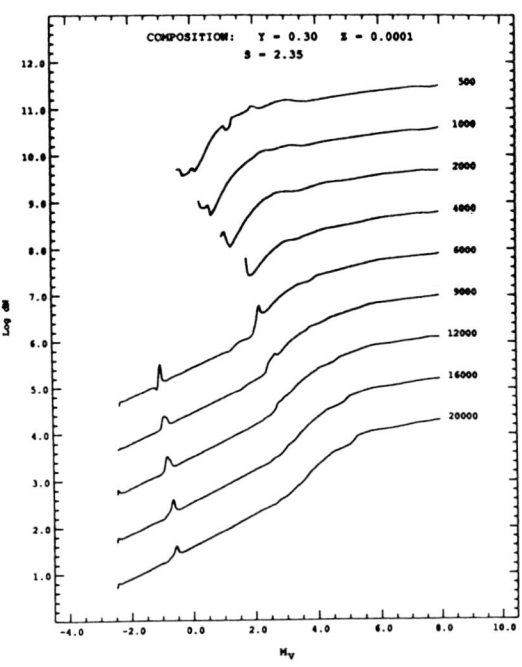

Fig. 2. Same as Fig. 1, but for Y = 0.30 and Z = 0.0001.

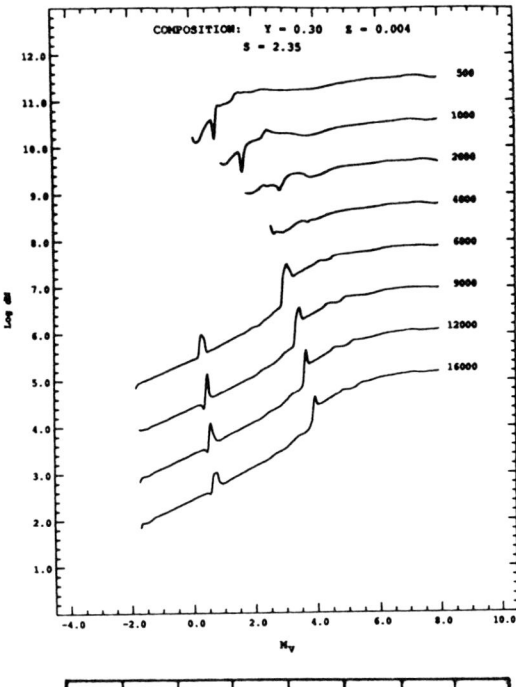

Fig. 3. Same as Fig. 1, but for Y = 0.30 and Z = 0.004.

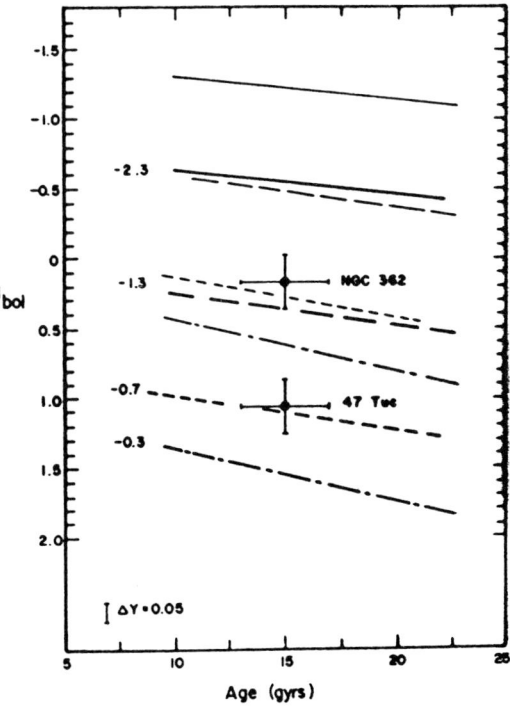

Fig. 4. Bolometric magnitude of the LF peak of a function of cluster age and [Fe/H] for models with convective overshoot (thick lines) and without convective overshoot (thin lines), after King et al.(1985). In each case Y = 0.25. The observed position of the LF peak for 47 Tuc and NGC 362 are marked with error bar estimates.

provides an important consistency check on a GC's age. How accurate these LF ages are is unknown at this point, since they are also of course subject to the limitations of present theory and the possible effects of rotation and internal mixing.

So far, for GC age determinations in the Galaxy, CM-diagrams have been used nearly exclusively, simply because it has not been practical until recently to construct LF's down to faint magnitudes. In the context of extra-galactic research, where CM-diagrams are much more difficult to construct, LF's are likely to become a more powerful probe of age than CM-diagrams. We note that LF's have already been used very effectively in studies of the stellar population within the Large Magellanic Cloud (Butcher 1977; Stryker 1984).

3. GAPS AND PEAKS IN THE LF AS PROBES OF STELLAR STRUCTURE

Observers have for a long time searched for gaps and peaks in the CM-diagrams of star clusters. Except near the main sequence turnoff where large samples of stars are usually available, this pursuit has been fraught with dangers because of the errors due to traditional photographic photometry and to small number statistics. Bahcall and Yahil (1972) have emphasized this last problem in their reanalysis of the giant branch data for M15 in which Sandage, Katem and Kristian (1968) had suggested the existence of gaps on the giant branch. But as mentioned earlier, it is now possible to do much better than a decade ago for at least some clusters using current techniques, which permit more accurate measurements of a statistically significant number of stars, down to fainter magnitudes (Da Costa and Villumsen 1981; King et al 1985).

We will consider here in turn two fine structure features which have been detected on the giant branch LF of some GC's: first a peak, discussed by Thomas (1967) and Iben (1968), which is predicted by standard stellar evolution theory; and then a gap (or dip), the interpretation of which is still unclear.

3.1 The Thomas peak.

This peak is due to a temporary slowdown in the rate of luminosity increase (for some chemical compositions, there is even a temporary decrease in luminosity), along the giant branch. This change in evolutionary rate is the result of the passage of the hydrogen-burning shell through a step in the run of the helium abundance created earlier deep in the envelope of the star. The step itself is due to the deepening of the convective envelope into layers in which some hydrogen has previously been converted into helium, during the main sequence phase. It coincides in the star with the point of maximum mass penetration of the convective envelope, which occurs near the base of the giant branch in the HR-diagram.

Figures 1 - 4 show that the Thomas peak luminosity is a function of helium content, metallicity and age. Figure 4, borrowed from the work of King et al (1985), illustrates the interdependence predicted by theory between these three variables and how, given any two of them, the third one can be derived from the position of the peak.

A careful comparison of the observed peak position in 47 Tuc with standard theoretical models for the correct composition parameters, shows that the convection zone depth of standard red giant models is too shallow. To reproduce the observations, one must introduce some overmixing at the bottom of the convection zone, of the order of one pressure scale height (King et al.1985). The same conclusion is reached for the NGC362 data of Harris (1982).

3.2 The gap near the base of the giant branch.

A gap has been found near the base of the giant branch of all GC's of low metallicity for which we have statistically significant data, i.e. by Cannon and Lee (1981) in NGC6752, by Da Costa and Villumsen (1981) in Omega Centauri, and by Buonanno et al,(1984) in NGC288. On the other hand, 47 Tuc, the only metal-rich cluster which has been similarly searched shows no such feature in its LF.

This gap must correspond to a temporarily rapid phase of evolution. No such rapid phase is encountered in standard evolutionary tracks. Yahil and van den Horn (1985) have noted that at this point there is a discontinuous transition between two different classes of solutions of the hydrostatic equations of stellar structure. It is not impossible that this transition could give rise in some stars to a phase of rapid internal readjustment which is not adequately described by the numerical models.

By analogy with the Thomas peak, a discontinuity in chemical composition through which the hydrogen burning shell might pass could also cause the observed gap. Armandroff and Demarque (1984) have proposed a plausible, although adhoc, model based on the creation of a discontinuity in He3 abundance due to rapid internal mixing. Unfortunately, this mechanism, like the previous one, does not explain in an obvious way the apparent dependence of the gap's magnitude on metallicity.

At the writing of this paper, it appears that the solution of the gap problem may be contained in still preliminary calculations of helium diffusion in the envelopes of extreme halo population stars, which predict a rapid dredge-up phase of the diffused helium back to the stellar surface as the convective envelope deepens at the base of giant branch (Deliyannis and Demarque 1987). One of the attractive features of this model is that the metallicity dependence of the gap magnitude is a natural consequence of the different rates of internal helium diffusion in metal-rich and metal-poor dwarfs in this temperature range. In this connection, I should mention that several researchers in our group are now concurrently studying the effects of convective, diffusive and rotationally induced mixing on the evolution of low mass stars. We hope that this work will clarify the factors which determine the luminosity of the Thomas peak, the nature of the gap, as well as the origin of the variable CNO abundances observed among GC stars (Freeman and Norris 1981).

4. THE LOWER MAIN SEQUENCE LF AND THE IMF

Very recently, a veritable information explosion in the area of faint star photometry in crowded fields has occurred, thanks to the advent of sophisticated software such as DAOPHOT. Deep main sequence LF's have been published by Fahlman et al.(1985), McClure et al.(1985, 1986), Buonanno et al.(1986), Lupton and Gunn (1986), Richer and Fahlman (1986), Penny and Dickens (1986), and Smith et al.(1986). In most cases, the exponent of the IMF has been derived, assuming a simple power law. McClure et al.(1986), on the basis of fitted power laws for seven clusters, concluded that the exponents decrease with increasing metallicity. However, the very metal-poor cluster NGC6397, studied earlier by Da Costa (1982), violates this trend. A meaningful comparison between clusters is complicated by dynamical effects, most particularly mass segregation which affects the star counts. Dynamical models, such as the multi-component King (1966) models studied by Da Costa and Freeman (1976) and Gunn and Griffin (1979), already have shed some light on this problem, but it may be that increasingly sophisticated models will have to be constructed to answer the IMF question conclusively. In particular, we need more detailed estimates of the effects of mass segregation in the outer layers of the cluster, where most of the star counts for LF's are made. Work in this area will not only improve our understanding of cluster dynamics, but will also certainly have far-reaching implications for our understanding of the history of stellar populations in the Galaxy.

Our research on stellar LF's has been supported in part by grants AST83-06143 from the National Science Foundation, and NAGW-778 from the National Aeronautics and Space Administration.

REFERENCES

Armandroff, T. E. and Demarque, P. 1984 Astron. Astrophys. 139, 305.
Bahcall, J. B. and Yahil, A. 1972 Astrophys. J. 177, 647.
Buonanno, R., Corsi, C. E., Fusi Pecci, F., Alcaino, G. and Liller, W. 1984 Astrophys. J. 277, 220.
Buonanno, R., Corsi, C. E., Iannicola, G. and Fusi Pecci, F. 1986 Astron. Astrophys. 159, 189.
Butcher, H. R. 1977 Astrophys. J. 216, 372.
Cannon, R. D. and Lee, S. W. 1981 in IAU Colloquium 68, Astrophysical Parameters for Globular Clusters A. G. D. Philip and D. S. Hayes, eds., L. Davis Press, Schenectady, p. 501.
Ciardullo, R. B. and Demarque, P. 1977 Trans. Yale Univ. Obs. 33.
Cole, P. W., Demarque, P. and Green, E. M. 1983 in ESO Workshop on Primordial Helium P. A. Shaver, E. Kunth and K. Kjar, eds., p. 235.
Da Costa, G. S. 1982 Astron. J. 87, 990.
Da Costa, G. S. and Freeman, K. C. 1976 Astrophys. J. 84, 752.
Da Costa, G. S. and Villumsen, J. V. 1981 in IAU Colloquium 68, Astrophysical Parameters for Globular Clusters A. G. D. Philip and

D. S. Hayes, eds, L. Davis Press, Schenectady, p. 527.
Deliyannis, C. and Demarque, P. 1987, to be published.
Demarque, P. 1980 in IAU Symposium 85, Star Clusters J. E. Hesser, ed., Reidel, Dordrecht, p. 385.
Demarque, P. and McClure, R. D. 1977 in The Evolution of Galaxies and Stellar Populations B. M. Tinsley and R. B. Larson, eds., Yale University Observatory, p. 199.
Endal, A. S. and Sofia, S. 1981 Astrophys. J. 243, 625.
Fahlman, G. G., Richer, H. B. and van den Bergh, D. A. 1985 Astrophys. J. Suppl. 58, 225.
Faulkner, J. 1972 Nature Phys. Sci. 235, 27.
Freeman, K. C. and Norris, J. 1981 Ann. Rev. Astron. Astrophys. 19, 319.
Green, E. M. 1981 Ph.D. dissertation, Univ. of Texas, Austin.
Green, E. M., Demarque, P. and King, C. R. 1987 Revised Yale Isochrones and Luminosity Functions in preparation.
Gunn, J. E. and Griffin, R. F. 1979 Astron. J. 84, 752.
Gunn, J. E., Stryker, L. L. and Tinsley, B. M. 1981 Astrophys. J. 249, 48.
Harris, W. E. 1982 Astrophys. J. Suppl. 50, 573.
Hartwick, F. D. A. 1970 Astrophys. J. 161, 845.
Iben, I. J. 1968 Nature 220, 143.
Janes, K. A. and Demarque, P. 1983 Astrophys. J. 264, 206.
King, C. R., Da Costa, G. S. and Demarque, P. 1985 Astrophys. J. 299, 674.
King, I. R. 1966 Astron. J. 71, 64.
Kraft, R. P. 1979 Ann. Rev. Astron. Astrophys. 17, 309.
Larson, R. B. 1986 Monthly Notices Roy Astron. Soc. 218, 409.
Lupton, R. H. and Gunn, J. E. 1986 Astron. J. 91, 317.
McClure, R. D., Hesser, J. E., Stetson, P. B. and Stryker, L. L. 1985 Publ. Astron. Soc. Pacific 97, 665.
McClure, R. D., van den Bergh, D. A., Smith, G. H., Fahlman, G. G., Richer, H. B., Hesser, J. E., Harris, W. E., Stetson, P. B. and Bell, R. A. 1986 Astrophys. J. 307, L49.
Paczynski, B. 1984 Astrophys. J. 284, 670.
Penny, A. J. and Dickens, R. J. 1986 Monthly Notices Roy. Astron. Soc. 220, 845.
Ratcliff, S. 1986, preprint.
Richer, H. B. and Fahlman, G. G. 1986 Astrophys. J. 304, 273.
Salpeter, E. E. 1955 Astrophys. J. 121, 161.
Sandage, A. R. 1957 Astrophys. J. 125, 422.
Sandage, A. R. 1969 Astrophys. J. 157, 515.
Sandage, A. R. 1982 Astrophys. J. 252, 553.
Sandage, A. R., Katem, B. and Kristian, J. 1968 Astrophys. J. 153, L129.
Sargent, W. L. W. and Searle, L. 1967 Astrophys. J. 150, L33.
Scalo, J. M. 1986 Fund. Cosmic Phys. 11, 1.
Schwarzschild, M. 1958 Structure and Evolution of the Stars Princeton Univ. Press, Princeton.
Simoda, M. and Iben, I. J. 1970 Astrophys. J. Suppl. 22, 81.
Simoda, M. and Kimura, H. 1968 Astrophys. J. 151, 133.

Simoda, M. and Tanikawa, K. 1970 Publ. Astron. Soc. Japan 22, 143.
Stetson, P. B. 1979a Astron. J. 84, 1056.
Stetson, P. B. 1979b Astron. J. 84, 1149.
Stryker, L. L. 1984 Astrophys. J. Suppl. 55, 127.
Sweigart, A. V. and Mengel, J. G. 1979 Astrophys. J. 229, 624.
Thomas, H.-C. 1967 Zeit. Astrophys. 67, 420.
Tinsley, B. M. 1980 Fund. Cosmic Phys. 5, 287.
Vandenberg, D. A. 1983 Astrophys. J. Suppl. 51, 29.
Vandenberg, D. A. 1987 in IAU Symposium No. 126, Globular Cluster Systems in Galaxies, J. E. Grindlay and A. G. D. Philip, eds., Reidel, Dordrecht, p. 107.
Yahil, A. and van den Horn, L. 1985 Astrophys. J. 296, 554.

DISCUSSION

CAYREL: Do the luminosity functions that you have shown include the horizontal-branch peaks of the evolution?

DEMARQUE: No, the peaks that you saw on the luminosity functions are the Thomas peak.

COHEN: If the observers tell you that Na and Al appear to be mixed to the surface from the stellar interior, somewhere between the main sequence and the AGB phase, could you accommodate this within the framework of your models somehow?

DEMARQUE: No, of course not! These Na and Al variations must be primordial.

LILLER: The gaps in the sub-giant branch occur close to the point where the onset of chromospheric activity takes place. Have you and Armandroff looked into atmospheric effects for an explanation for the gaps?

DEMARQUE: This is an interesting point. All Armandroff and I did, regarding the surface conditions, was to experiment with different mixing lengths in the convection zone. Indeed, if these is angular momentum hidden in the interior of these stars, this is a phase of evolution where enhanced magnetic activity might be expected since as the convection zone deepens, dynamo activity should be increased.

ROOD: Monte Carlo simulations done by Debe, Zworkin and myself show that reliable determination of the bump location using the differential luminosity function require samples of 1000-2000 stars in the upper 4-1/2 mags. of the RGB. The bump is best identified in the log of the integrated luminosity function.

DaCOSTA: We have 1200 stars in that range.

ROOD: That's probably enough. Besides the higher metallicity of 47 Tuc makes the bump bigger and at lower luminosity than in most of our simulations. That helps.

GLOBULAR CLUSTERS AND FIELD HALO STARS

Bruce W. Carney

University of North Carolina

ABSTRACT. Recent work on the chemistry and kinematics of the field halo population stars is reviewed, including the metallicity distribution function, elemental abundance patterns, primordial abundances, and their relations with stellar kinematics. The important role played by these stars in determining the ages of the globular clusters is discussed. A comparison is made between the kinematic and chemical properties of the field and cluster stars to ascertain if they share a common history.

1. INTRODUCTION

Zinn and West (1984) list 34 globular clusters beyond the solar galactocentric distance but inside 25 kpc. Summing their integrated absolute magnitudes and upon adopting $M/L = 1.7\ M_\odot/L_\odot$, appropriate for both high- and low-concentration clusters (Illingworth 1976; Peterson and Latham 1986), we find the "local" plus "outer" halo clusters total about $10^7\ M_\odot$. If we integrate the Bahcall and Soneira (1984) model for the spheroid number density from 7 to 25 kpc, and assume a median stellar mass of $0.3\ M_\odot$, we find $1.7 \times 10^9\ M_\odot$ of field halo stars. They thus outnumber the cluster population by over 100 to 1, and (probably) have a much larger ratio in the number of independent origins sampled. The consequent proximity of the field stars has made them invaluable in the study of the more distant globular clusters and the stars they contain, and for the questions we hope the halo population can answer.

2. PROSPECTING FOR FIELD HALO STARS

Clusters are (usually) easy to recognize, but the more numerous field stars are hard to identify amidst the sea of field disk stars. They are distinguishable generally by low metallicities, high velocities, or both (see Sandage 1986a for the fascinating history). Norris (1986) has compiled from the literature a list of about 1200 objects with [Fe/H] < -0.6, and which have been identified without any kinematical biases. Complementing his work are the two recent studies of Lowell Proper Motion Catalogue stars, the first by Sandage (Sandage and Kowal 1986; Fouts and Sandage 1986; Sandage 1986b), and the second by Carney and Latham (1986c; hereafter CL). These surveys have strong kinematic biases but no metallicity biases. Together, they contain almost 1500 stars with $\mu \geq 0\rlap{.}''26$ yr^{-1}. Stock (1984, 1985) has identified high velocity stars independently of metallicity using objective prism radial velocities, but there has been little follow-up work.

3. AGES

Theory and observations of clusters yield age estimates for the halo population. Despite major successes (c.f., VandenBerg, this meeting), more definitive results for the ages of the oldest stars and the relations between age, metallicity, and galactocentric distance will rely on improvements in the cluster distance scale. With accurate distances, we can rely only on the main sequence turn-off luminosity, not the convection-sensitive temperatures used in matching isochrones to color-magnitude diagrams. The otherwise unobservable helium abundance could then be inferred from the horizontal branch luminosity (§4.3).

Distances to clusters are currently estimated using isochrones (although the use of the surface temperatures as an extra parameter is undesirable), or "standard candles", which are calibrated using the nearer field stars. Assuming for the moment that field and cluster stars share common histories (see §6), there are three classes of standard candles. First, main sequence fitting has been utilized, with the calibration accomplished using the seven field stars with accurate trigonometric parallaxes ($\sigma_\pi/\pi < 0.2$), following Sandage (1970), Carney (1979, 1980), and Richer and Fahlman (1984), or by using larger samples of field stars and a statistical parallax (Carney 1979). HIPPARCOS and the Hubble Space Telescope will greatly improve the π_{trig} calibration, while the CL survey will soon provide a much-improved π_{stat} re-determination.

Second, RR Lyrae variables are valuable, for they are relatively bright and do not differ greatly in absolute magnitude. The goal, however, is to determine M_v to within $\pm 0\rlap{.}^m 1$, and test whether M_v varies with metallicity. Currently, statistical parallax results (Strugnell, Reid, and Murray 1986; Hawley et al. 1986; Barnes and Hawley 1986) show $M_v \sim +0\rlap{.}^m 80 \pm 0\rlap{.}^m 15$, with no metallicity dependence, when reddening is treated in a consistent fashion. The lack of a relation between M_v and [Fe/H] does not agree with the cluster work of Sandage (1982) and his interpretation of the period-luminosity-amplitude relation he discovered. Baade-Wesselink analyses of field stars (c.f., Burki and Meylan 1986; Jones et al. 1986, 1987) similarly do not show any strong M_v-[Fe/H] dependence, and, based on the work of Jones et al. (1986, 1987), $<M_v(RR)>$ is probably close to $+0\rlap{.}^m 90$. Such a faint absolute magnitude suggests even greater ages for the clusters than derived heretofore, and the lack of a metallicity dependence also implies a relation between cluster metallicity and age, with the more metal-rich clusters being much younger. These results, if correct, pose interesting tests for stellar evolution theory.

Third, a cluster white dwarf locus, if it can be well delineated by HST, will give us a composition-insensitive means to derive a distance. Once again, however, halo field white dwarfs must be used to calibrate the M_v vs. color index relations, lest we repeat Shapley's mistake in applying disk Cepheid absolute magnitudes to the Cepheids in globular clusters. Such work is underway.

In the future, cluster distances may yet be obtained directly, using the Baade-Wesselink method for the cluster variables, by comparing the internal cluster radial velocity and proper motion dispersions (e.g., Cudworth and Peterson, this meeting), or even by direct trigonometric parallaxes using proposed space-based interferometers, which could in principle yield accurate results down to parallaxes of 5 to 10 microarcseconds.

4. CHEMICAL ABUNDANCES

4.1 The Metallicity Distribution Function

The metallicity distribution function, $\phi(Z)$, measures the history of the halo's chemical enrichment and the processing of gas through stars, and which may be compared with the predictions of models of galactic evolution, such as those of Hartwick (1976) and Searle (1979). Agreement was considered to generally be good, but Bond (1981) argued that while the observed $\phi(Z)$ agrees with the models at intermediate and higher metallicities, there is a significant lack of very low metallicity stars and clusters. He suggested that the halo population (or the low-mass part of it we can still sample at this epoch) began with a "basement" metallicity of [Fe/H] ~ -2.6. This is not consistent with Big Bang predictions unless the lower mass stars took longer to form than the stars of "Population III". Hartwick (1983) reanalyzed the cluster and field $\phi(Z)$ and concluded the data agreed with the models. Recently, Beers, Preston, and Shectman (1985; 1986) reported some results of their searches for extremely low metallicity stars. They have found several stars with [Fe/H] apparently between -3.0 and -4.0, and so have argued that the results agree well with simple models that predict dN/dZ ~ constant. The survey was biased against more metal-rich stars (they claim completeness below about

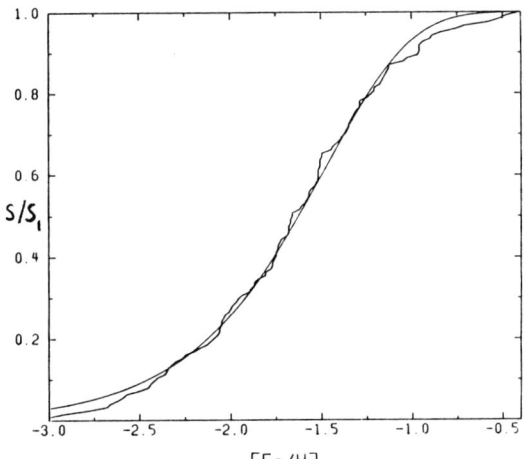

Fig. 1. The cumulative metallicity distribution function S/S_1 vs. [Fe/H] for a simple one-zone model with mean [Fe/H] = -1.5 (smooth curve), and CL stars V < -220 km sec^{-1}.

[Fe/H] = -1.5), but as Howard Bond has pointed out, comparisons of their results with model predictions depend upon an accurate placement of this upper metallicity limit, beyond which their results are biased.

Basically, large samples are required to test the models, for if dN/dZ is constant, we expect stars with [Fe/H] = -4 to be 100 times less abundant than stars with [Fe/H] = -2. The CL survey is one such step, for it is large (almost 400 stars with [M/H] < -1), has no metallicity bias, and the metallicities are well determined since they are obtained from high-resolution (although low S/N) spectra. Figure 1 shows a comparison of the cumulative metallicity distribution function, S/S_1 (Hartwick 1976), for a simple model with [M/H] = -1.5, and a subset of CL stars (those with V < -220 km sec^{-1} = retrograde orbits) which are kinematically selected to be halo stars, done in collaboration with John Laird and Michael Rupen. Agreement is very good, which implies that simple one-zone models explain satisfactorily the chemical evolution of the bulk of the halo population.

4.2 Abundance Patterns

The proximity of the halo's field stars allows us to conduct high-S/N, high-resolution, spectroscopic studies of elemental abundance patterns. The larger number of available halo stars also means we may study lower metallicity objects than found in clusters. For example, clusters have [Fe/H] $\gtrsim$ -2.4, whereas the field stars extend to [Fe/H] ~ -4.0 (Carney and Peterson 1981; Bessell and Norris 1984; Peterson, Kurucz, and Carney 1986). Results for a sample of three dozen field red giants have been published by Luck and Bond (1985), and for field dwarfs by Peterson (1981), Magain (1985), and Francois (1986). Of course, there are many other studies involving smaller samples. Lambert (1986) has excellently reviewed the behavior of the light elements Na through Ca. Sneden (1986) has reviewed CNO abundances, while Spite and Spite (1985) have reviewed all known abundance results for halo stars, with the results divided into families of elements and production processes. The α-nuclei (Mg, Ca, Si) rise relative to iron as [Fe/H] declines from -0.2 to about -1.2, then level off down to -2.0. At extremely low metallicities, [α/Fe] may again rise as [Fe/H] declines. C, N, and O do not show equivalent behavior, which suggests they have different production sites and a time-variable rate of the importance of these sites. Briefly, [O/Fe] behaves like the α nuclei, while [N/Fe] is roughly constant (although the data are sparse at the lowest metallicities). [N/Fe] is abnormally high, however, in at least four cases (~ 5% of the sample), probably due to mass transfer or patchiness in the galactic nucleosynthesis processes (Bessell and Norris 1982; Laird 1985; Spite and Spite 1986). [C/Fe] ~ 0.0 down to [Fe/H] ~ -1.8, below which it appears to rise (Tomkin, Sneden, and Lambert 1985), perhaps due to an increasing contribution of explosive nucelosynthesis at the earliest epochs. Other signatures of explosive nucleosynthesis have been claimed in the "odd-even" [Al/Mg] vs. [Mg/H] results of Peterson (1981; but see also Arpigny and Magain 1983) and the very careful magnesium isotope abundance work of Tomkin and Lambert (1980) on HD 103095, and Lambert and McWilliam (1986) on ν Indi. The "odd-even' effect may also show breaks at [Fe/H] ~ -1.2 and -2.0 (Lambert 1986). A related but rather remarkable result for [Ni/Fe] has been claimed by Luck and Bond (1983, 1985). They found it, like [C/Fe], to be roughly solar until [Fe/H] ~ -2.0, below which it rises. The result remains disputed, however (Sneden and Parthsarathy 1983; Leep and Wallerstein 1981; Barbuy, Spite, and Spite 1985; Magain 1985; Peterson, Kurucz, and Carney 1986).

Based on very high S/N spectra of two very metal-poor field giants, Sneden and Parthsarathy (1983) and Sneden and Pilachowski (1985) found s-process elements deficient with respect to iron, as expected, but that the r-process elements-to-iron ratio is variable, perhaps suggesting the importance of "local" supernovae in the enrichment processes. This is reminiscent of the implications of the [N/Fe] variations. It would be very interesting to extend such work to more stars. With lower S/N, but a much larger sample, Luck and Bond (1985) have confirmed that [s-process/Fe] declines with [Fe/H], but only for [Fe/H] $\leq$ -1.5.

In summary, there appears to have been a change in the nucleosynthesis patterns (including C, s-process, and possibly α-nuclei, oxygen, and Ni) that sets in at the epoch when [Fe/H] had risen to about -2.0, and then again (α-nuclei, oxygen, "odd-even" balance) at about -1.2, due to, perhaps, a change in the initial mass function (which governs the frequency of Type II supernovae) or the first appearance of Type I supernovae. The continued declining behavior of [O/Fe] and [α/Fe] as [Fe/H] rises until old disk metallicities are reached at [Fe/H] ~ -0.2 also suggests a change in the supernovae type, or, perhaps, relative rates, when the disk formed. More work should be undertaken, directed especially at the lowest metallicity stars to probe the earliest phases of nucleosynthesis and at the halo/disk transition, and at stars with extreme kinematical properties as a means of studying the history of nucleosynthesis throughout the Galaxy.

4.3 Primordial Abundances

Roger Cayrel will review later in this meeting the primordial abundances of the halo population, as determined almost entirely from spectroscopic studies of low metallicity field stars.

Perhaps the most impressive results have been those of the Spites (c.f., Spite and Spite 1986 and references therein), who have measured lithium abundances in halo stars. Standard hot Big Bang models predict such low lithium abundances only if $\Omega \ll 1$.

Helium is an extremely important element, both for cosmology and age-dating of clusters, but it is very difficult to measure. The only direct measurements are for the planetary nebula K648 in the cluster M15 (Hawley and Miller 1978; Adams et al. 1984), and the three field halo planetaries 49+88°1 and 108-76°1 (Hawley and Miller 1978) and 61+41°1 (Barker and Cudworth 1984), all of which show normal helium abundances (i.e., disk-like Y ~ 0.3) for these heavily-evolved objects. Indirect methods must be used for less-evolved stars. These rely on accurate luminosity determinations and stellar evolution theory. Zero-age horizontal-branch luminosities are functions primarily of the core mass and Y. Although M_{core} may be estimated via turn-off masses and mass-loss, ZAHB luminosities are not yet well enough determined (§3). On the other hand, main sequence luminosities are also affected by M, Y, Z, and age. Thus spectroscopy to determine Z and distance measures of lower-mass (i.e., unevolved) stars can be utilized to estimate Y, if the masses are known. Although μ Cas has received much attention (Lippincott 1981; McCarthy 1984; Russell and Gatewood 1984; Pierce and Lavery 1985), it is an old disk star. CM Dra (Lacy 1977; Paczyński and Sienkiewicz 1984) is a much better candidate. Still, the derived helium abundances remain imprecise. By introducing another variable, T_{eff}, Y may be estimated by comparing field star L-T_{eff} data with model isochrones. While results are plausible (Y = 0.23 $\pm$ 0.04 for [O/Fe] = +0.6: Carney 1979), use of T_{eff} introduces uncertainties that may be difficult to resolve. Halo star mass determinations thus remain an important goal, and field stars are the only possible sources. The CL survey has discovered a dozen metal-poor double-lined spectroscopic binaries, and perhaps one of them may yet prove useful for such mass determinations.

5. KINEMATICS

5.1 Θ_o

To determine space velocities in the Galaxy's non-rotating frame, we must remove the contributions to the perceived motions due to the solar peculiar velocity and especially that of the Local Standard of Rest's circular velocity, Θ_o. Gunn, Knapp, and Tremaine (1979) used 21 cm data to estimate Θ_o = 220 km sec^{-1}. If this is approximately correct, stars near the Sun but with velocities of this order directed perpendicular to the LSR apex define a kinematically-selected non-rotating ensemble, which may also be used to estimate Θ_o. The CL survey includes 149 such stars, all of which are metal-poor, with a mean V velocity of -222 $\pm$ 8 km sec^{-1}. We will therefore adopt Θ_o = 220 km sec^{-1}.

5.2 Kinematics vs. Metallicity

Eggen, Lynden-Bell, and Sandage (1962; hereafter ELS) pioneered such studies, finding clear trends between metallicity (inferred from the normalized ultraviolet excess, $\delta(U-B)_{0.6}$) and orbital eccentricity (projected onto the plane), as well as orbital angular momentum and the W velocity (i.e., perpendicular to the disk). They concluded that the Galaxy's collapse and early metal-enrichment were very rapid, comparable to a free-fall or orbital timescale (few x 10^8 years). With the recent completion of major new surveys (Norris 1986; Sandage 1986; CL), the relationships between kinematics and metallicity may be re-addressed.

The fundamental ELS result, that projected orbital eccentricity, e, correlates with metallicity, has been challenged by Norris, Bessell, and Pickels (1984) and Norris (1986), who utilized kinematically unbiased surveys and found 15% to 20% of the metal-poor stars had e < 0.4. ELS found 0%. Sandage's (1986) new results, like those of ELS, rely on a kinematically-selected sample, again do not show a substantial number of low-eccentricity, low-metallicity stars, although the result is not zero, either. On the other hand, the similarly defined CL sample confirms Norris's result. Figure 2 shows their results as a plot of spectrocopically-derived metallicity vs. a three-dimensional orbital eccentricity, which was computed by Luis Aguilar by numerically integrating the U, V, and W velocities in a Bahcall and Soneira (1984) model Galactic potential. Two basic populations are seen, but the large range in eccentricities shown by the metal-poor dwarfs argues for a halo

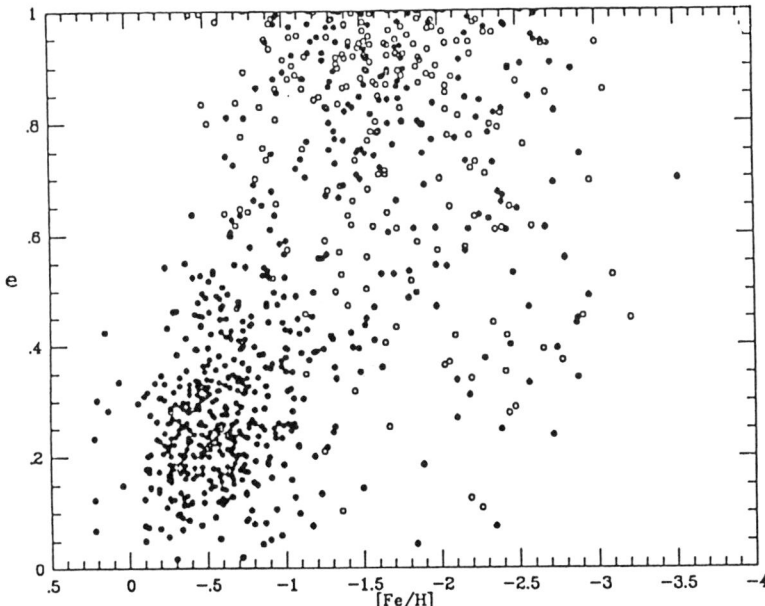

Fig. 2. The three-dimensional orbital eccentricity vs. spectroscopic [Fe/H] for stars in the CL survey.

formation and evolution timescale much longer than that espoused by ELS.

One may also compare the behavior of the mean U, V, and W velocities with mean metallicity. For brevity, we focus here upon only the second, as manifested by the mean rotational velocity, $<v_{rot}> = \Theta_0 - <V>$, and compare it to metallicity. In Figure 3 we show the results of Norris (1986), Carney and Latham (1986a: 174 kinematically unbiased metal-poor red giants), Sandage (1986), and the CL survey. Although Sandage (1986) has argued that the metal-poor stars show a monotonic change of v_{rot} vs. [Fe/H], the figure suggests instead that a major change occurred when [Fe/H] had risen to about -1.4. It is provocative that this is also the time when the α nuclei, oxygen, the odd-even effect, and perhaps even the s-process element abundances relative to iron underwent a major change.

5.3 Kinematics vs. Distance

Field stars and clusters are excellent test particles with which to probe the Galaxy's mass distribution, using either velocity dispersions (i.e., the velocity ellipsoid) or the total space velocity in the non-rotating rest-frame, $v_{RF} = [U^2 + (V + \Theta_0)^2 + W^2]^{\frac{1}{2}}$.

Velocity dispersions must be computed using kinematically unbiased samples. The local velocity ellipsoid for metal-poor stars has been derived by Norris (1986) using many data sources. He finds $\sigma_U = 131 \pm 6$,

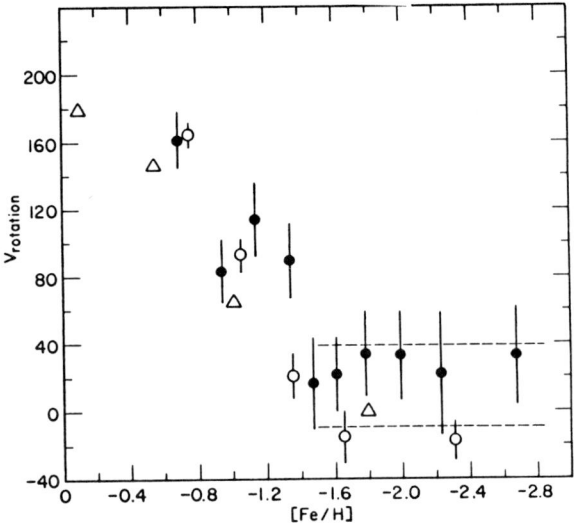

Fig. 3. The average rotational velocity binned in metallicity, with data taken from CL (o); Norris 1986 (o); Sandage 1986 (Δ); and Carney and Latham 1986a (dashed lines border the result for a sample of metal-poor red giants).

σ_V = 106 ± 6, and σ_W = 85 ± 4 km sec^{-1} (not including the red giant results of Carney and Latham 1986a, who found σ_R = 154 ± 18, σ_θ = 102 ± 27, and σ_ϕ = 107 ± 15 km sec^{-1}). The local halo velocity ellipsoid is thus anisotropic.

There is a dispute as to how the halo velocity dispersion varies with distance. Ratnatunga and Freeman (1985) studied distant metal-poor giants and found a low velocity dispersion, 60 ± 8 km sec^{-1} toward the South Galactic Pole at a mean distance of 14 kpc. Norris (1986) discussed the problem, and his results showed that the velocity ellipsoid changes at large distances, but that the changes may be related to Galactocentric distance rather than height above/below the plane. Thus spherical polar rather than cylindrical polar coordinates (as suggested by Ratnatunga and Freeman) are the natural frame. Further, the radial velocity studies of nearby and distant blue horizontal branch (*i.e.*, metal-poor) stars disagreed with the Ratnatunga and Freeman results (Pier 1984; Sommer-Larsen and Christensen 1985), although they supported the decline in the velocity dispersion with increasing Galactocentric distance.

Radial velocities of distant clusters (Hartwick and Sargent 1978; Lynden-Bell, Cannon, and Godwin 1983; Olszewski, Peterson, and Aaronson 1986) may be used to estimate the Galaxy's mass distribution, M(r), although assumptions must be made about the clusters' orbital shapes and the sample is small. By applying the local velocity ellipsoid to the outer halo clusters, Norris (1986) concluded M(r) ~ 3 x 10^{11} M$_\odot$ at r =

35 kpc, or about four times that contained within the LSR orbit.

Individual objects' rest frame velocities may also be used to set lower limits to the Galaxy's total mass, assuming they are bound to the Galaxy. If R15 (Hawkins 1983) proves to be an RR Lyrae at $R_{GC} = 59$ kpc with $v_{rad} = -465$ km sec^{-1}, the Galaxy's total mass must exceed $1.4 \pm 0.2 \times 10^{12} M_\odot$. Carney and Latham (1986b) estimate the local value of the Galactic escape velocity to be ≥ 500 (and may exceed 550) km sec^{-1}, in which case (for $\theta_0 = 220$ km sec^{-1}) the Galaxy's total mass exceeds that within the LSR orbit by a factor of 5 (or 8).

6. DO CLUSTERS AND FIELD STARS SHARE A COMMON HISTORY?

One expects or hopes for an affirmative answer to this question, since much of our understanding of the Galactic halo and its history depends on the study of clusters, which we have seen comprise only a very small fraction of the total halo contents.

The age(s) of the field stars cannot be determined accurately, but differences of the order of 30% can be ruled out. Figure 1 of Sandage (1983) and Figure 5 of Sandage and Kowal (1986) shows the blue limit of local high ultraviolet excess (i.e., metal-poor) proper motion stars resembles that of the turn-offs of comparably low-metallicity globular clusters ($B-V \leq 0^m.4$, $U-B \sim -0^m.2$). Figure 3 of Sandage (1983) shows a V vs. B-V diagram for SA 45, and that the blue limit again occurs at about $B-V \sim 0^m.35 - 0^m.40$.

The metallicity distributions of the field and the clusters differ somewhat. In Figure 4 we show S/S_1 vs. [Fe/H] for the field halo dwarfs of CL (with $V < -220$ km sec^{-1}) and the globulars beyond solar orbit, with clusters' metallicities taken from Zinn (1985). The field contains more metal-poor stars as well as more metal-rich stars than the clusters, which implies that $\phi(Z)$ is broader for the former. A Kolmogorov-Smirnov test indicates an 81% chance that there are two different populations, which is suggestive, although not completely convincing. The metallicity scales have not been derived by exactly the same means. Peterson, Kurucz, and Carney (1986) have also suggested a possible downward revision in [Fe/H] for the lowest metallicity clusters.

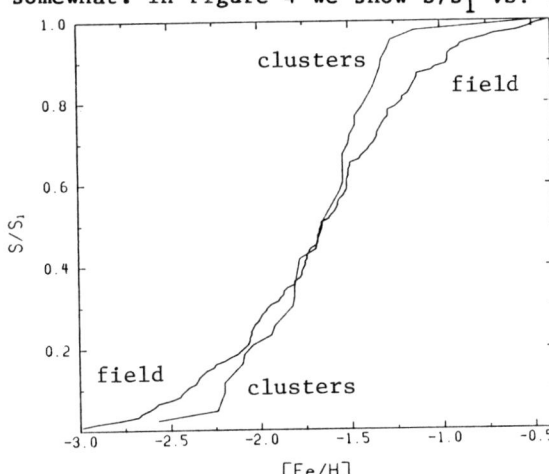

Fig. 4.   Comparison of globular cluster and field star metallicity distributions.

The elemental abundance patterns agree fairly well. Kraft (1986) has discussed the field vs. cluster CNO abundances finding the same general enhancement of [O/Fe] in both groups. CN abundances are vulnerable to mixing-induced variations and will not be considered. The clusters appear to divide into oxygen-rich and oxygen-poor samples, however, whereas the field red giants show a continuous range. Again, this may be an artifact of the analyses. For the heavier elements, Table I below is a compilation for a few key elements: the light α-rich Mg and Ca, the odd-Z Na, the light and heavy s-process species Sr, Y, Zr; and Ba, and the r-process element Eu. Abundances derived from spectroscopic analyses of cluster giants were taken uncritically from several sources (Cohen 1978, 1979, 1981; Gratton 1982; Pilachowski, Sneden, and Wallerstein 1983; Pilachowski, Wallerstein, and Leep 1980; Pilachowski et al. 1982), and were divided into three groups on the basis of mean metallicity: [Fe/H] = -1.0 to -1.5 (NGC 2808, NGC 3201, NGC 4833, NGC 6752, M5, and M10; [Fe/H] = -1.5 to -2.0 (M3, M13, and M22); and [Fe/H] < -2.0 (NGC 6397, M92, and M15). I have restricted the field star comparisons to include only red giants, with data collected from the large survey of Luck and Bond (1985), supplemented by Leep and Wallerstein (1981), Gratton (1983), Sneden and Parthsarathy (1983), and Sneden and Pilachowski (1985). All together, 45 field stars are included. Agreement is, in general, quite good. The Eu results are very difficult to compare, since only one or two lines are usually available,

TABLE I.

Comparative Abundances of Clusters and Field Stars

	Clusters/Field Stars*		
<[Fe/H]> =	-1.20/-1.26	-1.63/-1.78	-2.25/-2.40
[Na/Fe] =	+0.1/[+0.2]	[+0.2]/-0.1	+0.25/[+0.1]
[Mg/Fe] =	+0.3/+0.4	+0.1/+0.4	[+0.1]/+0.4
[Ca/Fe] =	+0.3/+0.4	[+0.35]/+0.4	+0.45/+0.5
[Sr,Y,Zr/Fe] =	-0.2/-0.1	+0.3/-0.1	-0.2/-0.45
[Ba/Fe] =	-0.3/+0.1	-0.1/-0.2	-0.1/[-0.6]
[Eu/Fe] =	-0.3/[+0.8]**	+0.35‡/+0.34‡	.../-0.3

* Brackets indicate scatter larger than observational errors allow.
** Based on two stars, with [Eu/Fe] = +0.4 and +1.5.
‡ Based on only one star.

only a few stars have been studied, and, as noted in §4.2, there may be local variations.

Finally, we compare cluster and field star kinematics, restricting the latter to include only those stars selected by means independent of kinematics. Table II repeats Norris's (1986) analysis of the motions of metal-poor ([Fe/H] < -1.2) clusters and field stars, and include Webbink's (this meeting) similar analysis of cluster kinematics. The

TABLE II.
The Field Star and Cluster Velocity Ellipsoids

	v_{rot}	σ_r	σ_θ	σ_ϕ	\<e\>	Reference
Clusters	40 ± 25	129 ± 19	131 ± 26	124 ± 19	0.48 ± 0.03	Norris
		119 ± 12	117 ± 19	95 ± 37		Webbink
Field	37 ± 11	129 ± 8	108 ± 13	96 ± 8	0.47 ± 0.05	Norris

eccentricity in the table is a three-dimensional one, not the projected value used by ELS. The two populations do not obviously different.

In summary, the field and cluster stars do not differ greatly in age, appear to share the same dynamical properties, but may have experienced somewhat different chemical enrichment histories.

ACKNOWLEDGEMENTS

Some of the work reported herein has been supported by grants from the National Science Foundation (AST-8111938 and AST-8312849) to the University of North Carolina.

REFERENCES

Adams, S., Seaton, M. J., Howarth, I. D., Auriere, M. and Walsh, J. R. 1984 Montly Notices Roy. Astron. Soc. 207, 471.
Arpigny, C. and Magain, P. 1983 Astron. Astrophys. 127, L7.
Bahcall, J. N. and Soneira, R. M. 1984 Astrophys. J. Suppl. 55, 67.
Barbuy, B., Spite, F. and Spite, M. 1985 Astron. Astrophys. 144, 343.
Barker, T. and Cudworth, K. M. 1984 Astrophys. J. 278, 610.
Barnes, T. G., III and Hawley, S. L. 1986 Astrophys. J. Letters 307, L9.
Beers, T. C., Preston, G. W. and Shectman, S. 1985 Astron. J. 90, 2089.
Beers, T. C., Preston, G. W. and Shectman, S. 1986 Bull. Amer. Astron. Soc. 17, 803.
Bessell, M. S. and Noris, J. 1982 Astrophys. J. Letters 263, L29.
Bessell, M. S. and Noris, J. 1984 Astrophys. J. 285, 622.
Burki, G. and Meylan, G. L. 1986 Astron. Astrophys. 156, 131.
Carney, B. W. 1979 Astrophys. J. 233, 877.

Carney, B. W. 1980 Astrophys. J. Suppl. 42, 481.
Carney, B. W. and Latham, D. W. 1986a Astron. J. 92, 60.
Carney, B. W. and Latham, D. W. 1986b in IAU Symposium No. 117, Dark Matter in the Universe, G. R. Knapp and J. Kormendy, eds., Reidel, Dordrecht.
Carney, B. W. and Latham, D. W. 1986c Astron. J., in press.
Carney, B. W. and Peterson, R. C. 1981 Astrophys. J. 245, 238.
Cohen, J. G. 1978 Astrophys. J. 223, 487.
Cohen, J. G. 1979 Astrophys. J. 231, 751.
Cohen, J. G. 1981 Astrophys. J. 247, 869.
Eggen, O. J., Lynden-Bell, D. and Sandage, A. 1962 Astrophys. J. 136, 748 (ELS).
Fouts, G. and Sandage, A. 1986 Astron. J. 91, 1189.
Francois, P. 1986 Astron. Astrophys. 160, 264.
Gratton, R. G. 1982 Astron. Astrophys. 115, 171.
Gratton, R. G. 1983 Astron. Astrophys. 123, 289.
Gunn, J. E., Knapp, G. R. and Tremaine, S. D. 1979 Astron. J. 84, 1181.
Hartwick, F. D. A. 1976 Astrophys. J. 209, 418.
Hartwick, F. D. A. 1983 Mem. Soc. Astron. Italiana 54, 51.
Hartwick, F. D. A. and Sargent, W. L. W. 1978 Astrophys. J. 221, 512.
Hawkins, M. R. S. 1983 Nature 303, 406.
Hawley, S. A. and Miller, J. S. 1978 Astrophys. J. 220, 609.
Hawley, S. L., Jeffreys, W. H., Barnes, T. G., III and Lai, W. 1986 Astrophys. J. 302, 626.
Illingworth, G. 1976 Astrophys. J. 204, 73.
Jones, R. V., Carney, B. W., Latham, D. W. and Kurucz, R. L. 1986 Astrophys. J., in press.
Jones, R. V., Carney, B. W., Latham, D. W. and Kurucz, R. L. 1987 Astrophys. J., in press.
Kraft, R. L. 1986 in Proc. ESO Workshop on Production and Distribution of C, N, O Elements, p. 21.
Lacy, C. M. 1977 Astrophys. J. 218, 444.
Laird, J. B. 1985 Astrophys. J. 289, 556.
Lambert, D. L. 1986 preprint.
Lambert, D. L. and McWilliam, A. 1986 Astrophys. J. 304, 436.
Leep, E. M. and Wallerstein, G. 1981 Monthly Notices Roy. Astron. Soc. 196, 543.
Lippincott, S. L. 1981 Astrophys. J. 248, 1053.
Luck, R. E. and Bond, H. E. 1983 Astrophys. J. Letters 271, L75.
Luck, R. E. and Bond, H. E. 1985 Astrophys. J. 292, 559.
Lynden-Bell, D., Cannon, R. D. and Godwin, P. J. 1983 Monthly Notices Roy. Astron. Soc. 204, 87P.
Magain, P. 1985 Astron. Astrophys. 146, 95.
McCarthy, D. W. 1984 Astron. J. 89, 433.
Norris, J. E. 1986 Astrophys. J. Suppl. 61, 667.
Norris, J. E., Bessell, M. S. and Pickles, A. G. 1984 Astrophys. J. Suppl. 58, 463.
Olszewski, E. W., Peterson, R. C. and Aaronson, M. 1986 Astrophys. J. Letters 302, L45.

Paczynski, B. and Sienkiewicz, R. 1984 Astrophys. J., 286, 332.
Peterson, R. C. 1981 Astrophys. J. 244, 989.
Peterson, R. C., Kurucz, R. L. and Carney, B. W. 1986 Astrophys. J. submitted.
Peterson, R. C. and Latham, D. W. 1986 Astrophys. J. 305, 645.
Pier, J. R. 1984 Astrophys. J. 281, 260.
Pierce, M. J. and Lavery, R. J. 1985 Astron. J. 90, 647.
Pilachowski, C. A., Sneden, C. and Wallerstein, G. 1983 Astrophys. J. Suppl. 52, 241.
Pilachowski, C. A., Wallerstein, G., Leep, E. M. and Peterson, R. C. 1982 Astrophys. J. 263, 187.
Ratnatunga, K. U. and Freeman, K. C. 1985 Astrophys. J. 291, 260.
Richer, H. B. and Fahlman, G. G. 1984 Astrophys. J. 277, 227.
Russell, J. L. and Gatewood, G. D. 1984 Publ. Astron. Soc. Pacific 96, 429.
Sandage, A. 1970 Astrophys. J. 162, 841.
Sandage, A. 1982 Astrophys. J. 252, 553.
Sandage, A. 1983 in Kinematics, Dynamics, and Structure of the Milky Way, W. L. H. Shuter, ed., Reidel, Dordrecht, p. 315.
Sandage, A. 1986a Ann. Rev. Astron. Astrophys., in press.
Sandage, A. 1986b Astron. J., submitted.
Sandage, A. and Kowal, C. 1986 Astron. J. 91, 1140.
Searle, L. 1979 in Liege Colloquium No. 22, Les Elements et leurs Isotopes dans L'Univers, Univ. Liege Inst. D'Astrophys., p. 437.
Sneden, C. 1986 in Proc. ESO Workshop on Production and Distribution of C, N, O Elements, p. 1.
Sneden, C. and Parthasarathy, M. 1983 Astrophys. J. 267, 757.
Sneden, C. and Pilachowski, C. A. 1985 Astrophys. J. Letters 288, L55.
Sommer-Larsen, J. and Christensen, P. R. 1985 Monthly Notices Roy. Astron. Soc. 219, 537.
Spite, M. and Spite, F. 1985 Ann. Rev. Astron. Astrophys. 23, 225.
Spite, F. and Spite, M. 1986 Astron. Astrophys. 163, 140.
Stock, J. 1984 Rev. Mex. Astron. Astrof. 9, 77.
Stock, J. 1985 Rev. Mex. Astron. Astrof. 11, 49.
Strugnell, P., Reid, N. and Murray, C. A. 1986 Monthly Notices Roy. Astron. Soc. 220, 413.
Tomkin, J. and Lambert, D. L. 1980 Astrophys. J. 235, 925.
Tomkin, J., Lambert, D. L. and Balachandran, S. 1985 Astrophys. J. 290, 289.
Tomkin, J., Sneden, C. and Lambert, D. A. 1986 Astrophys. J. 302, 415.
Zinn, R. J. 1985 Astrophys. J. 293, 424.
Zinn, R. J. and West, M. J. 1984 Astrophys. J. Suppl. 55, 45.

DISCUSSION

LATHAM: If the halo field stars have all originated from globular clusters, what do you expect for the frequency of binaries?

OSTRIKER: If the halo were produced primarily by dissolving globular clusters (an extreme possibility I do not believe) the close binary function would be higher in the field, since binaries are preferentially from clusters.

LAIRD: For the comparison of field dwarf and cluster metallicity distributions, there are 43 clusters and 36 stars used (since an additional kinematic criterion was applied to bias the sample toward outer halo stars. This mimics the selection of clusters with galactocentric radius > 7 kpc). There are about 3 stars of the 36 with [Fe/H] < -2.7.

STATLER: I'd like to clarify Jerry's answer concerning the fraction of binaries lost from clusters. This depends on how they are lost: if it is by evaporation through the tidal limit, the numbers of binaries will be very small. If tidal truncation is unimportant, then stars are lost by ejection from the center, and the fraction of those that are binaries can be anything up to 50%, this number being larger for harder binaries. A question then is: what are the orbital velocities of the binaries that you have observed?

SCHOMMER: Some 10 years ago, Kraft et al. found that $\Delta S = 2$ field RR Lyrae had a significantly different mean period than the $\Delta S = 2$ cluster variables. I believe the field $<P> \sim 0^d45$, white $<P> \sim 0^d55$ in the clusters. Is this still true?

CARNEY: Yes so far as I'm aware.

WALLERSTEIN: Surveys by Kraft and Saha find very few RR Lyraes with periods greater than 0.6 days in the globular clusters, especially the Oosterhoff II clusters. Also type II Cepheids are common in globular clusters but nearly absent from the halo. Since type II Cepheids are seen only in globulars with blue-horizontal branches, this indicates that the halo is a red horizontal-branch population.

CARNEY: There are quite a few local and distant field blue HB stars known (e.g., Pier 1984, Sommer-Larsen and Christensen 1985). It would be interesting to perform relative $1/V_{max}$ tests to test your idea.

HARRIS: At high galactic latitudes, the ratio of the numbers of Mira variables to numbers of RR Lyraes is higher than it is in clusters, and many of the Miras have periods longer than 250 days. Both facts are indicative of a more prominent metal-rich population in the halo field than in cluster, and are consistent with the few Population II Cepheids found in the halo field. Possibly a kinematically-selected sample is not fully sensitive to this population.

CARNEY: I don't see why a kinematically-selected sample would miss these stars. Our metallicity distribution does show more metal-rich stars as well as more metal-poor stars than in clusters. The metal-rich excess would help explain the Miras. I do not know, but would like to know, the kinematic properties of those Miras, however.

NORRIS: I was pleased to see that our results for V_{rot} vs. [Fe/H] agreed so well, especially at the low abundance end, in contrast with the recent result of Sandage. Would you care to comment on the difference between your and his results which are both based on kinematic samples?

CARNEY: Since our survey is a little deeper, we have a larger sample of higher velocity stars. We also have spectroscopic metallicities (and correspondingly somewhat better photometric parallaxes and hence kinematics); whereas Allan has relied only upon the ultraviolet excess, $\delta(U-B)_{0.6}$. We have also included reddening whereas he has not, although I believe its influence to be small. In any event, if there is a real population discontinuity, observationally-introduced scatter will transform it into a smooth trend. We, like you, see signs of a discontinuity.

PHILIP: I was interested in your statement that 22% of the halo stars were binaries. In my work on the early-type halo stars I have not run into any case of an A-type FHB star in a binary system What are the spectral types of your binary halo stars? Are there any A types among them?

CARNEY: Our stars are predominantly F, G, and K dwarfs, not post main-sequence stars, so our results don't apply directly to your objects. However, Dave Latham and I have commented (Astron. J. **91**, 60, 1986) that metal-poor field red giants also show a non-zero (> 10%) binary fraction, so I'd expect the field HB stars to contain some binaries, although the periods will probably exceed 100 days. Let me remind you the searches among the HB stars are not very complete, too. The field halo dwarf binary fraction was thought to be near zero for years until detailed studies such as ours were done.

PHILIP: Then it does seem to be true that there is no case yet known of a binary A-type Halo star.

COHEN: Since you mentioned planetary nebulae, I should say that Neugebauer, Soifer, Gillettt and I have found a new PN near the center of M 22 while checking up on an IRAS scan. It is probably impossible to get an He abundance from this object.

CARNEY: That is too bad.

ZINNECKER: When you discuss the metallicity distribution of halo field stars, there is the possibility that mixing in the interstellar gas of the protogalaxy is not complete. In other words: there may be a metallicity dispersion in the halo gas at any given time. Would that help to interpret the data? And one more question. There has been the suggestion in the literature that the halo gas would be polluted by supernova ejecta from massive star formation in the Virgo Cluster that preceded Pop II in our galaxy. Would anyone like to comment on this suggestion?

CARNEY: That may be part of it, but our sample is drawn from stars that originated throughout the galaxy and are only now nearby. I'd expect the clusters and field stars to then share the same spatially-averaged metallicity distribution function. Your second question was answered in a short note by George Wallerstein. Perhaps he will respond.

WALLERSTEIN: The suggestion that the first metals in our galaxy came as dust particles from Virgo does not fly, because the metals in extremely metal-poor stars do not correlate with those that are locked up in the interstellar dust.

Chapter III

Review Papers

Globular Clusters in Nearby Galaxies

Bruce Carney studies a particulary detailed poster after registering

Gary DaCosta and colleagues studying posters

OLD GLOBULAR CLUSTERS IN THE MAGELLANIC CLOUDS

J. A. Graham

Dept. of Terrestrial Magnetism
Carnegie Institution of Washington

ABSTRACT: The old globular clusters in the Magellanic Clouds are important links between our understanding of globular clusters in our own galaxy and similar unresolved objects in more distant galaxies. The Cloud clusters spread over a large range in age. Several contain RR Lyrae variable stars. High weight abundance data are needed for individual cluster members as well as deeper color-magnitude diagrams.

1. INTRODUCTION

Harlow Shapley's contribution to Magellanic Cloud research was immense. Under his direction, the Harvard College Observatory dominated this field at a time when few other observatories realized its importance. Shapley's vision was clear and far-seeing. "In more ways than one," he wrote, "the Magellanic Clouds serve as a gateway to the Metagalaxy - to the outer and overall aggregate of galaxies" (Shapley 1956). In his book, Star Clusters (Shapley 1930), he reported how Miss Cannon had noted that many of the Cloud globular clusters had surprisingly early spectral types; a characteristic which will be much discussed at this meeting. Shapley's interest in the Magellanic Clouds and his drive towards their exploration continued through his years as Director at Harvard. However the road was not always easy and progress was limited by the instrumentation available at the time. For example, without photoelectric standards of brightness and color, it was difficult to do much towards constructing the color-magnitude diagrams of the quality we have come to take for granted.

In the context of this meeting, the Cloud clusters are indeed links in our understanding of the Galaxy and of the Universe as a whole. We can easily resolve the component stars in a Cloud cluster for individual studies. In a parallel way, we can examine the integrated brightness, spectrum and velocity of the cluster as a unit; something that can be quite hard to do for nearby objects. But for the clusters in more distant galaxies, this latter type of measurement is often all we can do and prior understanding of nearby examples is essential.

2. IDENTIFICATION OF OLD CLUSTERS

There are not very many old globular clusters in the Magellanic Clouds in the sense that we understand the term in the Galaxy. One can easily count them off on the fingers of both hands. This makes my job today a fairly easy one but the next speaker, Ed Olszewski, has a much harder task. Some months ago, Ed and I decided that the borderline age should be 10 Gyr with Ed concentrating on the intermediate age clusters which are of a type unknown in our galaxy and, as well, dominate the statistics. Because of the relative rarity of old clusters, it is important to review how they can be identified. In this group, every addition counts. We shall group the discovery process under 3 headings.

2.1 Color-Magnitude Diagrams

Historically, the successful assembly of color-magnitude diagrams for Magellanic Cloud clusters has been the spur to further investigation of their characteristics. The technique here was pioneered by Arp (1961) and by Gascoigne (1962,1966). It was shown that while some clusters looked very similar to old globular clusters in our galaxy, others resembled more the intermediate age open clusters while a third class appeared to have no counterparts in the Galaxy at all. Until the arrival of fast linear detecting arrays like the CCD, this work was limited by the inability to push below the limits of the main-sequence turn-off. Now that this can easily be done, the whole history of the cluster system in the Magellanic Clouds is coming into sharper focus. A good example is the cluster Hodge 11. Earlier work (Gascoigne 1966, Walker 1979) was not able to distinguish a blue horizontal branch from a blue main sequence. As we know now, the presence of a background component of field stars with different age contributed to the confusion. Now it is very clear and agreed upon by everybody that Hodge 11 is one of the oldest and most metal-poor of all globulars in the Clouds. (Andersen, Blecha and Walker 1984, Stryker, Nemec, Hesser and McClure 1984). The most recent color-magnitude diagrams for old clusters support the "short" distance scale for the Magellanic Clouds (Andersen, Blecha and Walker 1985) in agreement with the result of Schommer, Olszewski and Aaronson (1984) for two intermediate-age clusters.

2.2 Integrated Photometry and Spectra

From early on, it was evident that some age discrimination could be made from UBV photometry alone (Gascoigne 1965). However a major procedural advance came from the work of Searle, Wilson and Bagnuolo (1980) who showed how it was possible to use intermediate band photometry to order the Magellanic Cloud clusters in a one-dimensional sequence. The character of the spectral variations along the sequence suggested that both chemical abundance and cluster age vary as the sequence is traversed so that clusters of the extreme type VII are the most metal poor and are old enough to come into the discussion in this paper. The method is not tied specifically to any particular photometric system. Frenk and Fall (1982) showed that UBV photometry suffices if variable interstellar reddening is not an important factor. The SWB classification system certainly seems to work. Clusters such as NGC 2210 and NGC 1786 classed as type VII were subsequently shown to have characteristics of old globular clusters (Graham 1981,1985; Hesser, McClure and Harris 1984). The SWB classification for Hodge 11 was confirmed when good color-magnitude diagrams became available.

Rabin (1982) has shown what can be accomplished from the direct analysis of integrated spectra themselves. From measurements of Balmer-line and K-line strengths, he concluded that, at a given metallic line strength, the Cloud clusters have consistently stronger Balmer lines. There is a strong age dependence. Flower (1984) pointed out that the contribution of a small, stochastically variable number of red giant stars might distort this age dependence.

2.3 RR Lyrae Variable Stars

The presence of RR Lyrae stars in a cluster does not in itself guarantee advanced age. However, their appearance in many old clusters whose ages have been determined by other methods, underlines that they are, to say the least, useful flags which can easily be spotted with small telescopes (Graham and Nemec 1984). Nearly all SWB class VII clusters have RR Lyrae stars. A notable exception is Hodge 11 which has none simply because the horizontal branch is so blue that no stars are to be found in the instability strip. Walker (1984) has discussed the use of CCD observations of the RR Lyrae stars in NGC 2210 as indicators of the LMC distance.

3. AGE SPREAD IN THE CLUSTER SYSTEM

The age spread of the old globular clusters in the Magellanic Clouds seems to be greater than that in the Galaxy. This is not surprising in view of the existence of so many intermediate age populous clusters. It does mean that we have a good opportunity in the Magellanic Clouds to study the effects of age as well as of metallicity on the properties of a cluster a a whole. In the Galaxy the age differences are small and are very difficult to decouple from the abundance spread. In the Clouds this may not be the case. Both effects need to be understood if we are to make sense of the integrated properties of unresolved clusters in more distant systems.

It is appropriate to review here the characteristics of two clusters which have been studied in considerable detail in recent years. They are NGC 2257 in the Large Cloud and NGC 121 in the Small Cloud. Color-magnitude diagrams which extend below the main sequence turnoff have been published, surveys have been made for RR Lyrae stars and good abundance determinations are available for individual cluster members. Some properties are summarised in Table I.

Table I

Comparison of Two Old Clusters

	NGC 121 (SMC)	NGC 2257 (LMC)	Sources
Age(Gyr)	11 ± 1	15 ± 2	1, 2
[Fe/H]	−1.4 ± 0.1	−2.0 ± .2	1, 2
P_{ab}(RR Lyr)	0.55 d	0.58 d	3, 4
H. Branch	red	blue	5, 6
Pec. Stars	1 C, 1 SRd	Double mode RR Lyraes	7, 4

References to Table I

1. Stryker, L.L., Da Costa, G.S. and Mould, J.R. 1985, Astrophys J. 298, p.544.

2. Stryker, L.L. 1983, Astrophys. J. 266, p. 82.

3. Graham, J.A. 1975, Pub. A.S.P., 87, p.641.

4. Nemec, J.M., Hesser, J.E. and Ugarte, P. 1985, Astrophys J. Suppl. 57, p.287.

5. Tifft, W.G. 1963, Monthly Notices Roy. Astron. Soc. 125, p.199.

6. Gascoigne, S.C.B. 1966, Monthly Notices Roy. Astron. Soc. 134, p. 59.

7. Feast, M.W. and Lloyd Evans, T. 1973, Monthly Notices Roy. Astron. Soc. 164, p. 15P.

4. DYNAMICS OF THE OLD CLUSTER SYSTEM

No review of the cluster system of the Magellanic Clouds would be complete without reference to the famous paper by Freeman, Illingworth and Oemler (1973) on the kinematics of the globular cluster system of the Large Magellanic Cloud. They were able to show that the youngest clusters formed a flattened system with low line-of-sight velocity dispersion and that this system shared the rotation solutions previously found for the young stars and the associated HI and HII. In addition they found that the older clusters with ages >1 Gyr also appeared to rotate as a flattened disk system but with a different line of nodes. This peculiarity persists to the oldest clusters discussed in this paper, but with less certainty owing to the small size of the sample. The question is raised as to whether, in view of this result, the Large Cloud has a halo population in the dynamical sense at all. Evidence to date is still fragmentary but it does suggest that the old stars may also be confined to a disk like structure (Graham 1975a, 1975b, 1977). We are conscious of the need to get velocity information for many more old objects but we should beware of trying to force-fit too much order and regularity into a galaxy which is certainly distorted and warped (Alvarez, Aparici and May 1987) especially in regions far from what little mass concentration there is in the central bar.

From dynamical arguments, Chun (1978) and Elson and Freeman (1985) suggest that the old clusters, NGC 1835, NGC 2257 and NGC 2210, may have masses ($\approx 5 \times 10^4 M_\odot$) which are low compared to those of globular clusters in our galaxy.

5. SUMMARY REMARKS

Looking towards the future both in this symposium and, in the longer term, towards the Hubble Space Telescope and beyond, I want to reiterate the importance of studying in detail the characteristics of globular clusters near at hand if we are to understand the integrated properties of distant globulars in remote galaxies. We need to know more precisely how abundance and age correlate with other properties such as the height of the horizontal branch above the main sequence turn-off and about the morphology of color magnitude diagrams as a whole. All these are important factors which determine the integrated characteristics. A good start has been made by Gascoigne, Bessell and Freeman (1981) and by Cohen (1982). It is not as simple a job as it looks. Gascoigne (quoted by Stryker 1983) has pointed out that a large fraction of the giants near a cluster can be field stars. In the Magellanic Clouds there are several cases where the cluster stars are distinctly older or younger than the surrounding field stars (Stryker 1983; Mould, Da Costa and Crawford 1984; Rich, Da Costa and Mould 1984) so considerable care has to be exercised in choosing the program stars and it is important to observe more than one or two stars per cluster to avoid the occasional interloper from the field. High weight data for well studied clusters are needed to make the foundation a solid one.

ACKNOWLEDGEMENT

I would like to thank Johannes Andersen for bringing me up-to-date on some references that I had earlier missed.

REFERENCES

Alvarez, H., Aparici, J. and May, J. 1987 Astron. Astrophys., in press.
Andersen, J., Blecha, A. and Walker, M. F. 1984 Monthly Notices Roy. Astron. Soc., 211, 695.
Andersen, J., Blecha, A. and Walker, M. F. 1985 Astron. Astrophys., 150, L12.
Arp, H. C. 1961 Science, 134, 810.
Chun, M. S. 1972 Astron. J., 82, 1062.
Cohen, J. G. 1982 Astrophs. J., 258, 143.
Elson, R. A. W. and Freeman, K. C. 1985 Astrophys. J., 288, 521.
Flower, P. J. 1984 in IAU Symposium No. 108, Structure and Evolution of the Magellanic Clouds, S. van den Bergh and K. S. de Boer, eds., Reidel, Dordrecht, p. 31.
Freeman, K. C., Illingworth, G. and Oemler, A. 1983 Astrophys. J., 272, 488.
Gascoigne, S. C. B. 1962 in Problems of Extra-Galactic Research, G. C. McVittie, ed., Macmillan, New York, p. 49.
Gascoigne, S. C. B. 1965 in Symposium on the Magellanic Clouds, J. V. Hindman and B. E. Westerlund, eds., Mount Stromlo Observatory, p. 66.
Gascoigne, S. C. B. 1966 Monthly Notices Roy. Astron. Soc., 134, 59.
Gascoigne, S. C. B., BEssell, M. S. and Norris, J. E. 1981 in IAU Colloquium No. 68, Astrophysical Parameters for Globular Clusters, A. G. D. Philip and D. S. Hayes, eds., L. Davis Press, Schenectady, p. 223.
Graham, J. A. 19875a Publ. Astron. Soc. Pacific, 87, 641.
Graham, J. A. 1975b Irish Astron. J., 12, 45.
Graham, J. A. 1977 Publ. Astron. Sco. Pacific, 89, 425.
Graham, J. A. and Nemec, J. M. 1984 in IAU Symposium No. 108, Structure and Evolution of the Magellanic Clouds, S. van den Bergh and K. S. de Boer, eds., Reidel, Dordrecht, p. 37.
Hesser, J. E., McClure, R. D. and Harris, W. E. 1984 in IAU Symposium No. 108, Structure and Evolution of the Magellanic Clouds, S. van den Bergh and K. S. de Boer, eds., Reidel, Dordrecht, p. 47.
Mould, J. and Aaronson, M. 1980 Astrophys. J., 240, 464.
Nemec, J. M., Hazen-Liller, M. L. and Hesser, J. E. 1985 Astrophys. J. Suppl., 57, 329.
Nemec, J. M., Hesser, J. E. and Ugarte, P. 1985 Astrophys. J. Suppl., 57, 287.
Rabin, D. 1982 Astrophys. J., 261, 85.
Rich, R. M., Da Costa, G. and Mould, J. 1984 Astrophys. J., 286, 517.

Schommer, R. A., Olszewski, E. W. and Aaronson, M. A. 1984
    Astrophys. J. (Letters), **285**, L53.
Searle, L., Wilkenson, A. and Bagnuolo, W. G. 1980 Astrophys. J.,
    239, 803.
Shapley, H. 1930 in Star Clusters, McGraw-Hill, New York, p. 183.
Shapley, H. 1956 American Scientist, **44**, 73.
Stryker, L. L. 1983 Astrophys. J., **266**, 82. Stryker, L. L.,
Nemec, J. M., Hesser, J. E. and McClure, R. D. 1984
    in IAU Symposium No. 108, Structure and Evolution of the
    Magellanic Clouds, S. van den Bergh and K. S. de Boer, eds.,
    Reidel, Dordrecht, p. 43.
Suntzeff, N. B., Friel, E., Klemola, A., Kraft, R. P. and Graham, J. A.
    1986 Astron. J., **91**, 275.
Walker, A. R. 1985 Monthly Notices Roy. Astron. Soc., **212**,
    343.
Walker, M. F. 1979 Monthly Notices Roy. Astron. Soc., **186**,
    767.

DISCUSSION

COHEN: Private communication from Freeman indicates that he no longer believes the model put forward by Freeman, Illingworth and Oemler where the older clusters form à disk inclined with respect to the younger clusters and the HI gas.

ZINN: Has Ken Freeman or anyone else evidence for a halo population in the LMC from kinematics?

COHEN: The existing radial velocities for the old LMC clusters are probably not accurate enough for a definite answer. I am planning to determine the velocities of a higher accuracy.

RICHER: A comment: We have observed the velocities of a number of hydrogen deficient carbon stars in the LMC which appear to be old when we compare them with those known in the Galaxy. Their dynamics can not be fit to that of a disk whether that shown by HI or any other. This is in disagreement with the Freeman, Oemler and Illingworth result.

SEGGEWISS; There is a recent paper by Elson and Freeman (1985) dealing with mass determination. They found a range of $2 \times 10^4 - 7 \times 10^4$ $M_\odot$ for old globular LMC clusters. These values are close to Galactic globular clusters with a mean total mass of 7×10^4 $M_\odot$ for the more star-rich and compact ones.

INTERMEDIATE-AGE MAGELLANIC CLOUD GLOBULAR CLUSTERS

Edward W. Olszewski

Steward Observatory, Univ. of Arizona

ABSTRACT. In this paper, I discuss some of the new facts that have been learned about Magellanic Cloud clusters, mostly thanks to new detectors and associated reduction code. I first show the extent of the LMC cluster system, in order to note that studies of age, abundance, and kinematics of the cluster system have been missing clusters to the north and south of the Hodge and Wright atlas, and to point out that star formation has gone on in places far from present day neutral hydrogen. I will concentrate on the intermediate age clusters (10^8–10^{10} y) in the discussion concerning new stellar evolution results, neglecting the 10^7 y clusters and 30 Doradus. I further restrict my choice of topics to 1) the luminosity of clump giants, 2) the youngest possible RR Lyrae stars, and 3) the patterns and history of cluster formation. The discussion of abundances of Cloud clusters leads readers to the excellent poster papers presented at this meeting.

1. THE EXTENT OF THE MAGELLANIC CLOUD CLUSTER SYSTEMS

I will start this discussion by reminding readers what the Magellanic Clouds look like, and how far out the cluster systems extend. Our typical conception of the LMC, for instance, is shown in Figure 1 of Alcaino and Liller (1984), which shows the LMC bar, the 30 Doradus region, and goes north to slightly beyond Shapley's Constellation III. Another example, the color picture in Sky and Telescope (April, 1984, p304) shows an even more restricted region. Casual inspection of the SRC J plates will also lead to a similar conclusion, for the ionized gas is most evident near the bar, and the density of stars drops off rapidly south, east, and west of the bar, and a couple of degrees north of Constellation III.

This picture of the LMC is incomplete, as the following three pictures will show. The first is in Schommer, Olszewski, and Aaronson (1986), and is a reproduction of two Canterbury Atlas (Doughty, Shane, and Wood 1972) prints on which we have labelled some clusters. What I want to mention here, for I will return to the cluster E2 below, is

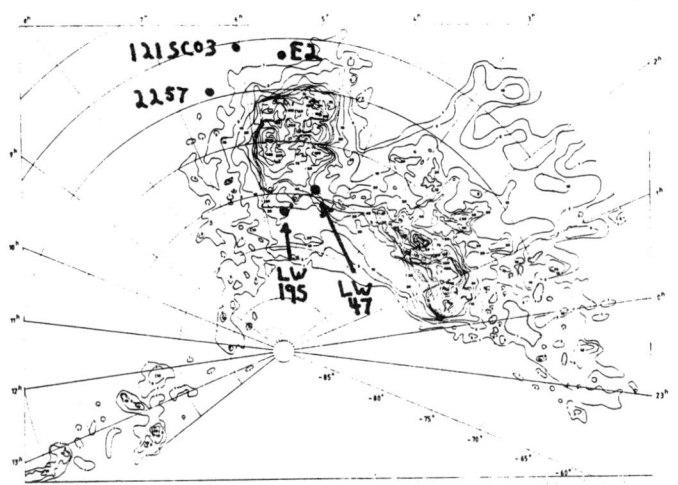

Fig. 1. This is the Magellanic Cloud H I distribution shown by Mathewson and Ford in IAU Symposium 108. I have added the positions of the clusters E2, 121SC03, NGC 2257, LW 47, and LW 207, which are mentioned in the text.

---

that there are LMC clusters many degrees from the bar, far beyond the picture of the LMC described above. Lynga and Westerlund (1963, their Figure 1) showed this more than 20 years ago. They also noted that a southern 'gap' in clusters followed by a more southerly increase in cluster density corresponded in position with features noted in de Vaucouleurs' (1955) low resolution deep photographs. The second picture, which I reproduce here as Figure 1, is the LMC and SMC neutral hydrogen distribution given in Mathewson and Ford (1984), with some of the clusters from this review marked. Note that the northernmost clusters extend beyond the detected neutral hydrogen (the limit is 10^{19} atoms cm^{-2}). The third picture, Figure 2, is reproduced from Freeman, Illingworth, and Oemler (1983), which was a compendium and analysis of the kinematics of the LMC cluster system. I have added to their figure the outline of the Hodge and Wright (1967) atlas, NGC and IC clusters outside the atlas, the clusters in the Olszewski, Harris, and Schommer (1987) catalog of outer LMC clusters, the region studied in Mateo's paper in this symposium, and the Reticulum system.

I think that there are two important points about LMC clusters to be made here. The first is that while the more populous clusters inside the boundaries of the Hodge and Wright atlas are reasonably well sampled, though not complete, there are 18 NGC and IC clusters outside the atlas. The boundaries of the atlas were chosen by the available plate material, which covered 'most of the recognized area of the LMC.' The publications resulting from Hodge's thesis (1960, 1961) identified the red and blue clusters within the atlas as well as a few others (e.g., NGC 1868). Aside from NGC 2257, 1466, and 1841, which were

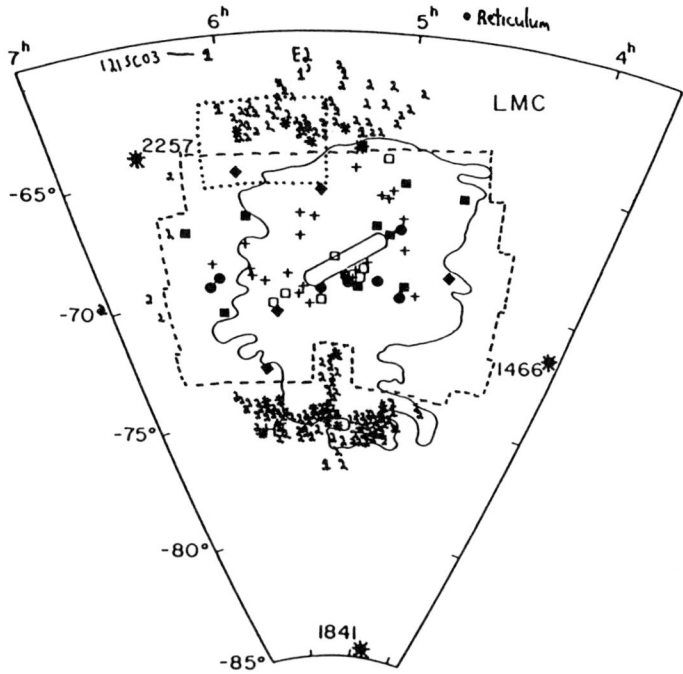

Fig. 2. A map of the LMC, taken from Freeman, Illingworth, and Oemler (1983). The dashed line is the boundary of the Hodge and Wright (1967) atlas; stars are NGC and IC clusters outside the Hodge and Wright atlas; ones are outer clusters mentioned in the text; twos are outer clusters from Olszewski, Harris, and Schommer (1987); the dotted line is the approximate boundary of the Mateo (this volume) study.

---

already well known clusters, essentially all of the clusters in the Freeman, Illingworth, and Oemler (1983) and Searle, Wilkinson, and Bagnuolo (1980) studies are within the boundaries of the atlas or from Hodge's thesis. What of these other NGC and IC clusters? Many are as bright as representative inner clusters, and are in less crowded fields. It seems to me that the problem was a lack of suitable chart material (the SRC J plates have only recently become available; without marked charts finding clusters is tedious), and I think the lack of recognition of the value of the Lyngå and Westerlund survey. I hope that this 'Hodge Atlas bias' can be corrected in future work.

The second point is that with the detector revolution of the past few years, we can relatively easily measure CMDs and velocities of any clusters, not just the most populous ones. It is not necessary to restrict studies to the clusters which have colors compiled in van den Bergh (1981), which in general represent some linear combination of the least crowded and most luminous subset of the subset of clusters within

the Hodge and Wright atlas. Mateo (this volume) has surveyed all
clusters in a small northern region of the LMC, determining accurate
ages from isochrone fitting, and deriving an age-metallicity relation
for this spatially defined sample. In this way, one can avoid sampling
the tail of the cluster distribution.

I have generally ignored the SMC in the above discussion. The
Hodge and Wright (1977) SMC atlas does seem to cover the entire SMC, as
represented by the clusters. In order to have a more complete sample
of the clusters, however, higher resolution plates must be examined
(SRC J) or obtained (Hodge 1987). In the SMC, almost all of the
clusters which are reasonably populous and reasonably uncrowded now
have CCD color-magnitude diagrams, mostly thanks to Da Costa, Mould,
and collaborators. Some of the results will be mentioned below. Most
of the SMC clusters are in or near the bar, which makes CMD work very
difficult, and cluster membership hard to deduce.

My last point about the clusters in the context of the two
galaxies as a whole is illustrated by de Vaucouleurs and Freeman's
(1972) review. Their Figure 9a, if we did not know was of the LMC/SMC,
is of some galaxies which have obviously undergone some strong
interaction (among other things, remember that the SMC has a 'wing' and
that the Magellanic Stream exists). I don't mean to make this case too
strongly, but I think that we should remember that the LMC and SMC do
not exist as island universes in majestic isolation. If we truly wish
to understand the cluster system in the Clouds, we need to look for
evidence that interactions (see the cover of IAU 108) have affected
the clusters, perhaps helping to make the outermost clusters, and
almost certainly influencing the outer cluster dynamics. We should
also remember that projection effects can distort our picture and that
dispersion along the line of sight may be significant, especially for
the SMC.

2. STELLAR EVOLUTION

In this section, I will discuss some results which have come from
a number of good CMDs of Cloud clusters. We now can actually look at
specific stages of stellar evolution in clusters of different age,
which finally vindicates the optimism which Magellanic Cloud workers
have had for the potential of the confrontation of observation with
stellar evolution theory. Perhaps surprisingly, most of the data to be
discussed come from SMC clusters.

2.1 Clump Giants

The clump giants are the more massive, hence younger, analog of
the globular cluster horizontal branch, and are burning helium in their
cores. As we do for the horizontal branch stars, in order to
understand the clump, we need to ask questions like: 1) how do clump
giant magnitudes vary with metallicity; 2) how do magnitudes and masses
(or mass loss from main sequence star to clump giant) vary with age; 3)

how does the morphology of the clump vary with metallicity or mass; 4) how well do models predict the observed features and can the model results be reconciled with our prejudices; and 5) can the study of clumps help solve other fundamental questions about the Clouds? I don't think that we can answer all these questions, but a good start has been made.

The questions about metallicity really must await more clusters, and of course are entwined with the question of age. Mateo and Hodge (1985) have argued that M_V(clump) is a constant, if the cluster in question has an age $> 3 \times 10^8$ y. Olszewski, Schommer, and Aaronson (1986) have tested this hypothesis by plotting the apparent mean R magnitude of the clump of six SMC clusters versus age. The clusters span an age of 1.5 - 12 Gyr; the results are shown in Figure 3, both for the short SMC modulus of 18.7, and for this modulus corrected for the presumed tilt of the SMC at the position of each cluster. Using the

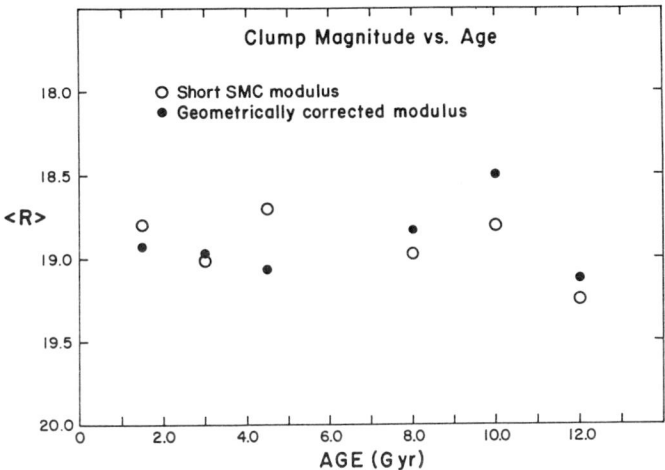

Fig. 3.   A plot of the apparent R magnitude of the giant branch clump versus derived cluster age for several SMC clusters. Open circles are measured from CMDs, while filled circles are magnitudes corrected for the presumed geometry of the SMC.

data in Seidel, Da Costa, and Demarque (1986) gives similar results. Formally, m_R(clump) = ~ 18.8 + ~ 0.02* (age in Gyr), with a correlation coefficient of ~ 0.4. As Olszewski, Schommer, and Aaronson, and Seidel, Da Costa, and Demarque point out, the severe tilt of the SMC can cause interpretative problems for individual clusters. At present there is little evidence for a variation of M_V(clump) with age; more CMDs need to be made, and much more work needs to be done on the distances to and the tilts of the Magellanic Clouds.

Seidel, Da Costa, and Demarque (1987) have gathered CMDs made by Da Costa and collaborators to investigate clump giants, determining their masses, and the amount of mass loss as a star evolves from the main sequence to the clump. The mass loss is calculated by determining the turnoff mass from conventional isochrone fits to CMDs, while the clump mass comes from a new grid of models of core helium burning stars.

If the short modulus to the SMC is adopted, the older clusters seem to lose $\sim 0.2\ M_\odot$ of material, while the ~ 1 Gyr clusters lose $\sim 0.6\ M_\odot$. This latter large mass loss is somewhat unexpected, but is consistent with the conclusion in Aaronson and Mould (1985) that quantitative agreement between the observed luminosity of the AGB with theory is possible if mass loss on the AGB increases with increased luminosity for initial masses greater than $1.5\ M_\odot$. I point the reader to the lengthy discussion in Seidel, Da Costa, and Demarque.

Not only did the Seidel, Da Costa, and Demarque study teach us about clump giants and mass loss, but it was able to make a consistency argument about the correctness of the long and short distance moduli to the Clouds. If the long moduli are adopted, the derived clump giant masses are in general larger than their progenitor masses. Either the theory is wrong, or more likely, another reason exists for believing the short scale. (I point out that the poster papers at this conference discussing Galactic RR Lyraes also point to the short scale.)

2.2 The Youngest Known RR Lyrae Stars

This discussion is adapted from that given in Olszewski, Schommer, and Aaronson (1987). The problem can be stated simply: Lindsay 1 has an age of ~ 10 Gyr and has no RR Lyraes while NGC 121 is ~ 12 Gyr old and contains them. Have we discovered the approximate age of the onset of the RR Lyrae phenomenon, and is there any evidence for 'young' RR Lyraes in other galaxies? Again, remarkably, we are dealing with SMC clusters.

The CMDs of L1 (Olszewski, Schommer, and Aaronson 1987) and NGC 121 (Stryker, Da Costa, and Mould 1985), both have well defined main sequence turnoffs; their ages are as well determined as for any other Magellanic Cloud clusters. That either cluster contains RR Lyraes is somewhat surprising, since the giant branch clumps are quite red. Both clusters have approximately the same metal abundance.

The absence of RR Lyrae stars in L1 is based on the unsuccessful blinking of plates by Gascoigne (1966); no modern study such as Graham and Nemec (1984) has been made for this cluster. If we accept this pair of clusters as representative, we conclude that a cluster can make RR Lyrae stars at an age of 10 Gyr $< t_{RR} < 12$ Gyr, for [Fe/H] ~ -1.3.

Evidence for young RR Lyraes in other galaxies includes the metal rich disk RR Lyraes in the Milky Way (Taam, Kraft, and Suntzeff 1976; Strugnell, Reid, and Murray 1986). These form a kinematically distinct sample of RR Lyraes, which can best be called 'old disk.' Current wisdom maintains that the old disk is several billion years younger than the Galactic globular clusters.

The Carina dwarf galaxy provides evidence which the present set of observations cannot unambiguously interpret. Mould and Aaronson (1983) deduced that the bulk of the stellar population of Carina was ~ 7 Gyr old, with only a small old population. Saha, Monet, and Seitzer (1986) have now found ~ 50 RR Lyraes in Carina, which can either argue for a (small) old population or for younger RR Lyrae stars.

Clearly, the Magellanic Cloud clusters provide the best limits on the youngest possible age of RR Lyrae stars. Comments about increasing the sample of L1-, N121-aged clusters will be made below.

2.3 Post-AGB Stars

In the oral version of this paper, I presented a crude argument which suggested that for every 10 AGB stars we should expect one post-AGB star, and asked the question, 'where are the post-AGB stars?' Renzini made a comment which can be read in the questions following this paper which said that my numbers were far too optimistic. This section is therefore revised.

Renzini and Voli (1981) estimate the AGB lifetime of ~ 1 $M_\odot$ stars to be ~ 2×10^6 y. Paczynski (1970) calculated lifetimes of post-AGB stars (nuclei of planetaries in his case) to be approximately a few x 10^4 y. Given the total number of ~ $1-2 \times 10^2$ AGB stars found in the extensive surveys of Mould and Aaronson (see Aaronson and Mould 1985), and the total luminosity surveyed of ~ 3×10^6 $L_\odot$, we expect to have ~ 1 post-AGB star in these ~ 60 clusters. What post-AGB stars will look like is not clear to me. Will they be hot, blue cores of stars which have lost their envelopes (planetary nebulae nuclei) or stars highly obscured by their own dusty ejected atmospheres (RAFGL objects)? Do the latter evolve into the former?

3. PATTERNS AND HISTORY OF STAR FORMATION

3.1 The Strange Age Distributions of the Oldest Clusters

Figure 4 shows the spatial distribution of the oldest clusters in the SMC, those with B-V > 0.5, U-B > 0.0, from van den Bergh's (1981) compilation of integrated colors. To this figure, I have added the cluster names and ages, the latter determined in most cases from isochrone fits. Note two important properties of these clusters: 1) NONE are as old as Galactic globular clusters; and 2) most have ages between 3-12 Gyr.

This age distribution of the oldest populous clusters is very different from that in the LMC. If we examine Hodge's (1984) list of 'genuine' globulars, we find that NGC 1466 may not be an LMC member; NGC 1841 is very distant in projection, with no CMD to the level of the

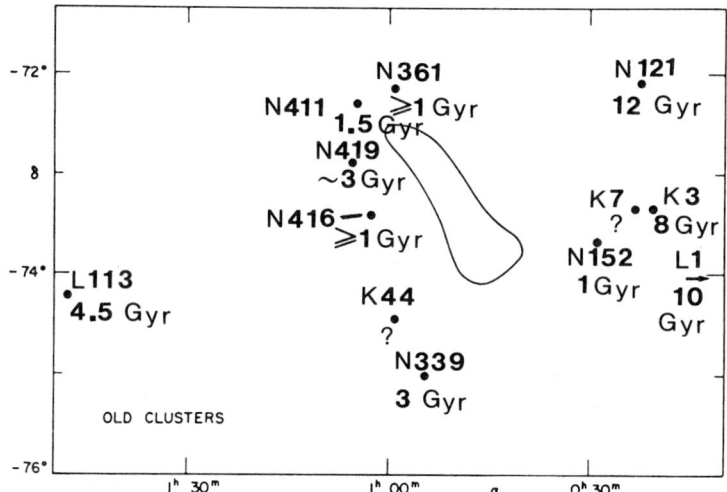

Fig. 4. The distribution of the oldest SMC clusters, from van den Bergh (1981). I have added the cluster identifications and derived ages.

main sequence yet published; NGC 2257 and 2210 have not yet been published with isochrone fits; NGC 1786 and 1835 are not being attempted; and NGC 121 (SMC) is significantly younger than Galactic globulars. Given the discussion of RR Lyraes above, there is currently no compelling evidence for globular-cluster-aged clusters in either Cloud, although NGC 2257 is probably that old.

Of all the other populous LMC clusters studied, none are older than 3-4 Gyr; I know of no good candidates for populous clusters in the 5-10 Gyr age range. Why are there no such clusters in the LMC, yet several in the SMC? Perhaps cluster destruction is more important in the LMC for even these most populous clusters. It would be nice to have good models of the LMC tidal field and of cluster destruction (but note Schommer's comments in this conference on observed vs. computed tidal radii of Cloud clusters).

To slightly complicate matters, Mateo, Hodge, and Schommer (1986) have shown that the sparse cluster ESO 121SC03, the northernmost LMC cluster, has an age of ~ 10 Gyr. This cluster is not in any list of integrated photometry. Uncovering more of these (seemingly) rare clusters will not be easy, but is necessary if we are to understand both cluster destruction and possible bursts in cluster formation.

3.2 The E2, 121SC03, N2257 Region

As Figure 2 shows, the clusters E2 (Schommer, Olszewski, and Aaronson 1986), ESO 121SC03 (Mateo, Hodge, and Schommer 1986), and NGC 2257 (see Hesser, McClure, and Harris 1984) all exist far to the north of the LMC bar, beyond measured neutral hydrogen. E2 is ~ 2 Gyr old, 121SC03 ~ 10 Gyr, and NGC 2257 ~ 15 Gyr. How this region of the LMC has made clusters at such different epochs is hard for me to understand. Could some tidal mechanism be at work?

3.3 A Catalog of Distant LMC Clusters; CMDs of Distant LMC Clusters

Olszewski, Harris, and Schommer (1987) have compiled a catalog of clusters outside the Hodge and Wright (1967) atlas, from examining SRC J plates. This catalog contains ~ 150 clusters, most rather sparse, but 18 of which have NGC or IC numbers. This catalog is virtually identical to the subset of Lyngå and Westerlund clusters which are outside the Hodge and Wright atlas, with a few important additions to the north.

We present here (Figure 5) CMDs of LW 177, 195, 399, 47, and 207, the first three to the north, and the last two to the south of the Hodge and Wright atlas. Along with E2, we now have six clusters all with ages 2-3 Gyr. These clusters were picked to be reasonably populous, very distant, and reasonably uncrowded, excepting E2, which was observed because E1 and E3 were amusing clusters.

Mateo (this volume) has derived ages for ~ 30 clusters in his northern LMC region; he finds a peak at 2-3 Gyr as well. This may be the most fundamental and least ambiguous way to see if there were bursts of cluster formation. Certainly if we continue to find many 2-3 Gyr clusters it will be hard to accept the results of Elson and Fall (1986), which are based on very poorly determined ages.

3.4 Metallicities and the Age-Metallicity Relation

This section would best be written in a year or two, for the poster papers at this meeting show that many different techniques are being successfully used to derive metallicities of Magellanic Cloud clusters. I will state a prejudice, advertise some of the work presented at this meeting, and point out a result that astonishes me.

At the Schenectady meeting in 1981, we saw three attempts to derive ages and/or abundances of Magellanic Cloud clusters: Searle and Smith (1981), who derived their results from integrated spectrosopy; Hodge (1981), who attempted to compile and evaluate ages and abundances from all available techniques; and Cohen (1981), who measured spectral indices of individual stars. The clusters in these studies are scattered all about the face of the LMC; they are essentially the clusters measured by Freeman, Illingworth, and Oemler (1983) which are displayed in Figure 2. I've always been baffled by what is meant by an

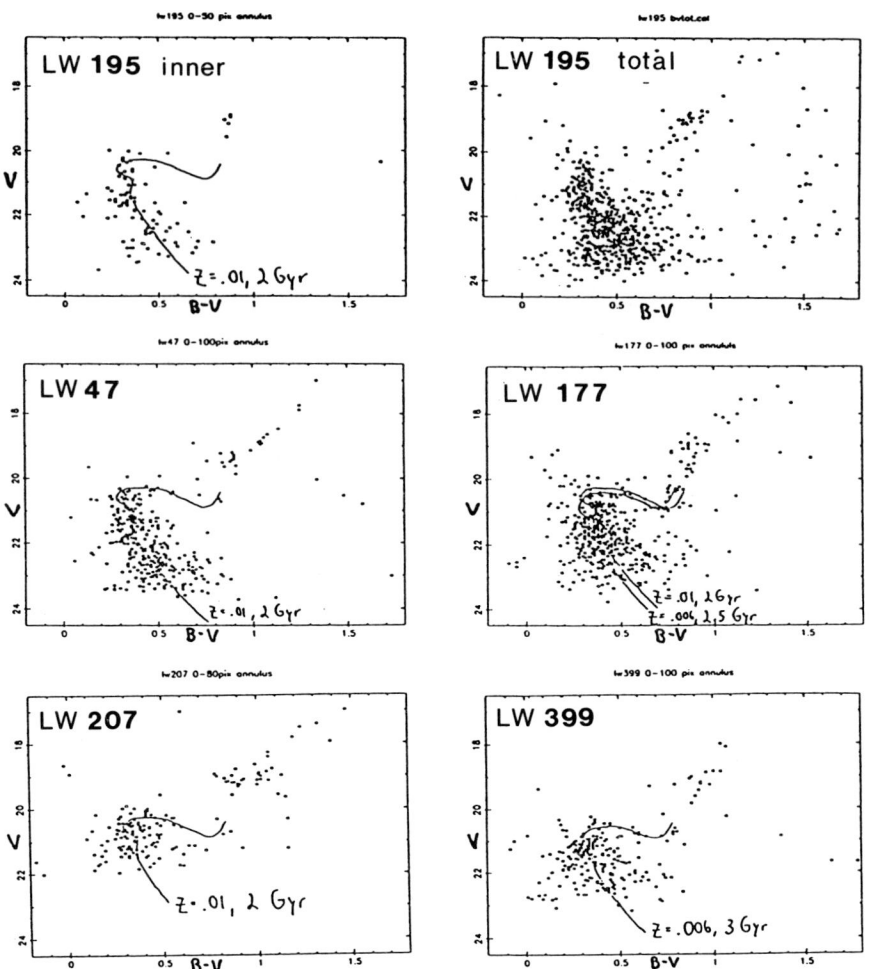

Fig. 5. CMDs of several distant LMC clusters, whose positions are given in Figure 2. These plots are from work being done by Olszewski, H. Harris, and Schommer.

age-metallicity relation for the entire LMC, given that star formation seems to appear in a place for a while then disappear (Hodge 1973). Are we assuming something very simplistic about the LMC, or is it well enough mixed, or is the time-averaged rate of star formation constant enough that there is one age-metallicity relation?

Good ages are getting easy to come by, with many good CCD CMDs, the understanding of the limitations of SWB class, and better analysis of integrated spectra (Smith, Searle, and Manduca, this volume). Abundances are also now getting more certain, partly because of new instrumentation, which allows new spectral regions to be profitably used (see Armandroff, this volume); partly because of better calibrations and wider understanding of special photometric systems (Geisler, this volume; Schommer and Geisler, this volume); and partly because of the number of complementary techniques being used. I note that all the studies are giving similar abundances for clusters of similar age, excepting the work of Richtler and Seggewiss (this volume).

Mateo (this volume) reports on a study of all the star clusters, populous or otherwise, in a small northern section of the LMC (see Figure 2). He derives ages for ~ 30 clusters and metallicities for ~ 16, from CCD CMDs, isochrone fitting, and integrated colors. The resultant age-metallicity diagram can be seen in his poster paper in this volume; this diagram has a variety of other authors' relations sketched in. I am astonished at the similarities between the two data sets. Mateo's region is far from the LMC bar, with part of the region beyond detectable neutral hydrogen. There had to be a lot of star formation in this remote piece of the LMC or a lot of mixing. When we understand how this particular age-metallicity relation came to be, we'll be a lot closer to an understanding of the LMC.

Bob Schommer and I have worked closely together on many of the topics discussed here. I'd like to specially thank both him and Paul Hodge for help and influence for the past twelve years. This paper came into being by assimilating the work of many people, especially Mario Mateo and Gary Da Costa. Hugh Harris, Jim Hesser, and Marc Aaronson have also contributed in many ways. I appreciate a travel grant from the IAU which helped defray some of my costs of attending IAU 126, and acknowledge NSF grant AST 83-16629 for support of all my research.

REFERENCES

Aaronson, M. and Mould, J. 1985 Astrophys. J. 288, 551.
Alcaino, G. and Liller, W. 1984 in IAU Symposium 108, Structure and Evolution of the Magellanic Clouds S. van den Bergh and K. S. De Boer, eds., Reidel, Dordrecht, p. 49.
Cohen, J. G. 1981 in IAU Colloquium 68, Astrophysical Parameters for Globular Clusters, A. G. D. Philip and D. S. Hayes, eds., L. Davis Press, Schenectady, p. 229.
Cohen, J. G. 1982 Astrophys. J. 258, 143.
de Vaucouleurs, G. 1955 Astron. J. 60, 126.
de Vaucouleurs, G. and Freeman, K. C. 1972 Vistas in Astronomy 14, 163.
Doughty, N. A., Shane, C. D. and Wood, F. B. 1972 The Mount Johns Observatory Sky Survey and the Canterbury Atlas Mount Johns Observatory, University of Canterbury, Christchurch, New Zealand.
Elson, R. A. W. and Fall, S. M. 1985 Astrophys. J. 299, 211.
Freeman, K. C., Illingworth, G. and Oemler, A., Jr. 1983 Astrophys.

J. 272, 488.
Gascoigne, S. C. B. 1966 Monthly Notices Roy. Astron. Soc. 134, 59.
Graham, J. A. and Nemec, J. M. 1984 in IAU Symposium 108, Structure and Evolution of the Magellanic Clouds S. van den Bergh and K. S. de Boer, eds., Reidel, Dordrecht, p. 37.
Hesser, J. E., McClure, R. D. and Harris, W. E. 1984 in IAU Symposium 108, Structure and Evolution of the Magellanic Clouds S. van den Bergh and K. S. de Boer, eds., Reidel, Dordrecht, p. 47.
Hodge, P. W. 1960 Astrophys. J. 131, 351.
Hodge, P. W. 1961 Astrophys. J. 133, 413.
Hodge, P. W. 1973 Astron. J. 78, 807.
Hodge, P. W. 1981 in IAU Colloquium 68, Astrophysical Parameters for Globular Clusters A. G. D. Philip and D. S. Hayes, eds., L. Davis Press, Schenectady, p. 205.
Hodge, P. W. 1984 in IAU Symposium 108, Structure and Evolution of the Magellanic Clouds S. van den Bergh and K. S. de Boer, eds., Reidel, Dordrecht, p. 7.
Hodge, P. W. 1987 preprint.
Hodge, P. W. and Wright, F. W. 1967 The Large Magellanic Cloud Smithsonian Press, Washington.
Hodge, P. W. and Wright, F. W. 1977 The Small Magellanic Cloud Univ. of Washington Press, Seattle.
Lynga, G. and Westerlund, B. E. 1963 Monthly Notices Roy. Astron. Soc. 127, 31.
Mateo, M. and Hodge, P. W. 1985 Publ. Astron. Soc. Pacific 97, 505.
Mateo, M., Hodge, P. W. and Schommer, R. A. 1986 preprint.
Mathewson, D. S. and Ford, V. L. 1984 in IAU Symposium 108, Structure and Evolution of the Magellanic Clouds S. van den Bergh and K. S. de Boer, eds., Reidel, Dordrecht, p. 125.
Mould, J. R. and Aaronson, M. 1983 Astrophys. J. 273, 530.
Olszewski, E. W., Harris, H. C. and Schommer, R. A. 1987 in preparation.
Olszewski, E. W., Schommer, R. A. and Aaronson, M. 1986 preprint. Pac
1970 Acta. Astron. 20, 47.
Renzini, A. and Voli, M. 1981 Astron. Astrophys. 94, 175.
Saha, A., Monet, D. G. and Seitzer, P. 1986 Astron. J. 92, 32.
Schommer, R. A., Olszewski, E. W. and Aaronson, M. 1986 Astron. J. in press.
Searle, L. and Smith, H. A. 1981 in IAU Colloquium 68, Astrophysical Parameters for Globular Clusters A. G. D. Philip and D. S. Hayes, eds., L. Davis Press, Schenectady, p. 201.
Searle, L., Wilkinson, A. and Bagnuolo, W. G. 1980 Astrophys. J. 239, 803.
Seidel, E., Da Costa, G. S. and Demarque, P. 1986 preprint.
Strugnell, P., Reid, N. and Murray, C. A. 1986 Monthly Notices Roy. Astron. Soc. 220, 413.
Stryker, L. L., Da Costa, G. S. and Mould, J. R. 1985 Astrophys. J. 298, 544.
Taam, R. E., Kraft, R. P. and Suntzeff, N. B. 1976 Astrophys. J. 207, 201.
van den Bergh, S. 1981 Astron. Astrophys. Suppl. 46, 79.

DISCUSSION

MCCARTHY: First, tell us what is the globular cluster distribution to the east and west of the Hodge Wright Atlas of the LMC. Why are there so few to the east and west and so many to the north and south? Second, How do your RV measures of stars in ESO 121-03 compare with the RV for the Magellanic Clouds.

OLSZEWSKI: The LMC cluster system extends outside the Hodge and Wright atlas in the north and south, but not in the east and west. We searched outside the Atlas in all directions for several degrees beyond the last discovered cluster. Our discovery of clusters to the north and south is mostly an artifact of the plates Hodge and Wright chose to put in their atlas, and partly because of the better resolution of the SRC plates. In answer to your second question no velocities are known; we have time in Jan. 1987.

SCHOMMER: I would like to mention that many of these outer clusters had been catalogued by LW, that is Lynga and Westerlund in 1963 (MNRAS). We found some additional, fainter, objects, but I'm afraid some of us had failed to read this paper before, or at least appreciate its significance.

LYNGA: I can mention that the LW survey was made over a field of 15 x 15° where we tried to find all clusters with a plate scale of 2 min of arc/mm.

OLSZEWSKI: Most of the clusters we found were in your list; we found ~ 25 new ones, mostly again due to improved resolution.

RENZINI: To answer your question about the absence of Post-AGB stars in the Magellanic Cloud clusters; one expects to find one of such star for every five million solar luminosities of cluster light. One should then expect to inspect about 100 clusters before finding just one Post-AGB star.

SMITH: It is worth noting that the most metal poor SMC stars yet identified are field red giants and RR Lyraes. Perhaps the absence of SMC clusters older than 12 Gyr is telling us that star formation took place outside of large clusters before that time.

OLSZEWSKI: Most of the old SMC clusters seem to have [Fe/H] ~ -1.3!

RICHER: The nearby galactic globulars show almost dispersionless main sequences down to a few magnitudes below the turnoff. The CM diagrams you showed have very wide main sequences. An honest error for the ages of the galactic globulars is at least ±3 Gyr. What error would you attach to the ages of the LMC clusters you discussed?

OLSZEWSKI; I can't quantify this, but it's my feeling that anyone in the audience who works in Magellanic Cloud cluster CM diagrams would get the same age to ± 1 or 2 Gyr, given the same data and isochrones. It's my impression that systematic errors are what's discussed most in Galactic globular cluster age determinations.

DiFAZIO: If we inspect more closely the graph where you compared J. Cohen's metallicity-age relation to the data, since you only have one point in the lower-right end of the graph, and given the shape of the distribution of the other points, I think Cohen's relation cannot be said to fit the data very well; which I think can be fitted pretty well by a straight line too. Do you agree?

OLSZEWSKI; All I wanted to state was my amazement that this distant place in the LMC had an age-metallicity relation similar to that given by Cohen for the inner clusters. There had to be a lot of star formation in the last 10 Gyr, which qualitatively contradicts our Astronomy 101 notion of where the important star formation in the LMC occurs.

M 31 CLUSTER SYSTEM

F. Fusi Pecci

Dept. of Astronomy, Bologna

ABSTRACT: The present status of the search for globular clusters in M31 is reviewed and some outstanding properties of the cluster system as a whole are briefly discussed.

1. INTRODUCTION

The production of a complete and uncontaminated sample of clusters in M31 still represents an extremely difficult observational task since cluster candidates in M31 have sizes comparable to the seeing disk (10pc ~ 3.3 arcsec). Any revision of each existing sample has thus shown the presence of some level of both incompleteness and contamination.

Given the uniqueness of M31 for studying a very populous cluster system, many different attempts have been made by various authors to improve the search technique. It is thus important to understand whether the various approaches applied present the same pros and cons or whether they can be considered at least in part complementary. In fact, if they were to cover the selection of the whole spectrum of potential candidates, a coordinated effort might lead to eventually obtaining a complete sample (down to a given magnitude) necessary to study in detail the properties of the whole M31 cluster system and to compare it with those found in other galaxies.

Most of the searches in M31 are essentially based on visual inspection of images to distinguish clusters from stars and other types of non-stellar objects on a morphological basis. Pure eye-selections have been applied up to the search made with Kitt Peak plates by Sargent et al. (1977). In order to improve the selection, several complementary search techniques have been added by various groups (see list of references in Table I).

2. THE SEARCHES

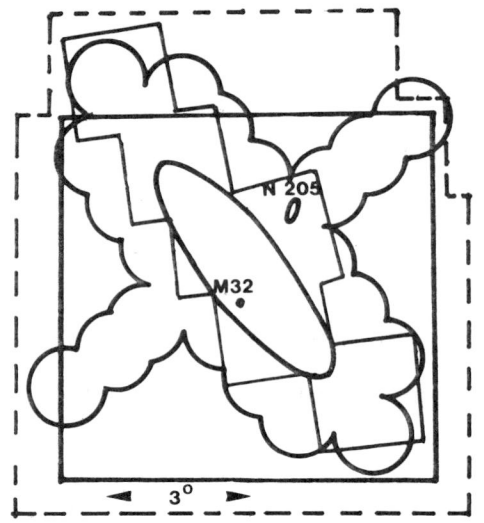

Fig. 1. Map of the areas covered by the 3 latest major searches.
a) composite circles: Sargent et al.(1977);
b) composite squares: Crampton et al.(1985);
c) large square and large dotted area: Battistini et al.(1986).

Table 1.
List of searches for globular clusters in M31

Reference	cluster names
Hubble 1932	H
Seyfert and Nassau 1945	
Mayall and Eggen 1953	M
Hiltner 1958	
Kron and Mayall 1960	
Johnson 1961	J
Vetesnik 1962	V
Baade and Arp 1964	B
Sandage 1971	S
Sharov 1973	Sh
Alloin et al. 1976	
Karimova and Sharov 1977	Sh
Sargent et al. 1977	KP,M31C,G
van den Bergh 1977	vdB
Hodge 1979	
Battistini et al. 1980	Bo
Huchra et al. 1982	CfA
Crampton et al. 1985	DAO, G
Wirth et al. 1985	WSB
Battistini et al. 1986	Bo

Table II

Results of the 'Bo-Survey' (Battistini et al. 1986)

Homogeneity: Excellent
Contamination: < 10 - 15 %
Completeness: ~ 100% allowed by 'morphological method'   > 80% absolute (down to V = 18)

Class A	(very high confidence candidates)	254
Class B	(high condidence candidates)	99
	Total (high confidence candidates)	353
Class C	(plausible candidates, probability < 50%)	152

'Out of field' (candidates in other lists)	31
'Rejected in field' (inserted in previous lists)	
miscellaneous non-stellar objs = Bo D	51
too faint for classification in Bo-plates	18
rejected	75

Table III

Comparison with the previous lists

Bo-Class	V	KP	Bo80	G	DAO	Bo86
A	182	232	190	248	0	254
B	30	55	43	82	2	99
C	8	25	49	57	27	152
D	11	10	0	43	34	218
'rejected'	22	15	6	52	36	0
'out of field'	3	18	0	27	10	0
Total	256	355	288	509	109	(723)

V = Vetesnik 1962   KP = Sargent et al. 1977
Bo80 = Battistini et al. 1980   G = Crampton et al. 1985
DAO = Crampton et al. 1985 (108 new candidates)
Bo86 = Battistini et al. 1986

Fig. 1 shows a map of the areas systematically surveyed for globular clusters by the three latest searches. They have been carried out independently of any previous identification. Revisions "a posteriori" of all the candidates previously known in the considered areas have always been made. Table II presents the results of our latest survey (Battistini et al. 1986). Table 3 gives the comparison with other main samples by showing our classification of all the objects included in the various lists. A close inspection of the tables shows that updated efforts to find globular clusters in M31 using "morphological" criteria have led to a critical revision of the list rather than to a significant increase of the number of candidates. In fact, the total number of candidates has been increased mainly by increasing the limiting magnitude of the survey and by extending the studied area rather than by finding a conspicuous set of new candidates. In order to properly derive the total number of globular clusters in M31 one must correct for: a) contamination by spurious objects, b) incompleteness for the area of the sky not surveyed yet, c) incompleteness at faint magnitudes (V > 18) and in the highly reddened regions, d) incompleteness due to the possible losses of highly compact clusters (see sect. 3), e) asymmetries or peculiarities in the shape of the cluster luminosity function (here assumed to be a Gaussian with σ =1.2 mag, see van den Bergh 1985 for a discussion). Many uncertainties still affect the various quoted steps, nevertheless we estimate N(tot) = 500 ± 50. This figure for the total number of clusters leads to values of S, the specific globular cluster frequency (Harris and van den Bergh 1981), ranging from 0.4 to 1.5 according to the slightly different asssumptions made on N(tot) and the total absolute luminosity of M31. As is well known, this confirms that S is higher in ellipticals than it is in spirals. The parameter S for the spheroid, probably more meaningful, turns out to be of about 5 ± 3.

3. THE "MISSING" CLUSTERS IN THE BULGE OF M31

The frequency distribution of clusters in M31 as a function of projected galactocentric distance is presented in Fig. 2 (taken from Wirth et al. 1985). Several relations have been derived up to now to describe the projected density profile of the cluster system. They differ from one another mainly due to differences in the sample used, but the main conclusion is susbtantially unaffected: the distibution follows an $R^{1/4}$ law rather well apart from the central region (r < 3kpc) where a flattening is evident using all the available lists. However, the actual existence of this flattening is still an open question. Harris and Racine (1979) noticed that if this were due to incompleteness in the survey of Sargent et al.

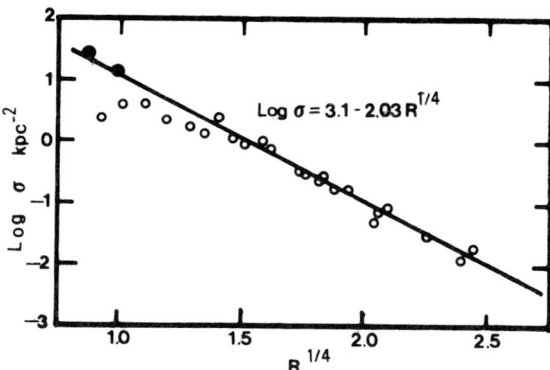

Fig. 2. The logarithm of the surface density of globular clusters in M31 as determined by Harris and Racine (circles) fitted to a de Vaucouleurs $R^{1/4}$ law. The inclusion of the "missing globulars" found by Wirth et al. would imply the values represented by the filled circles.

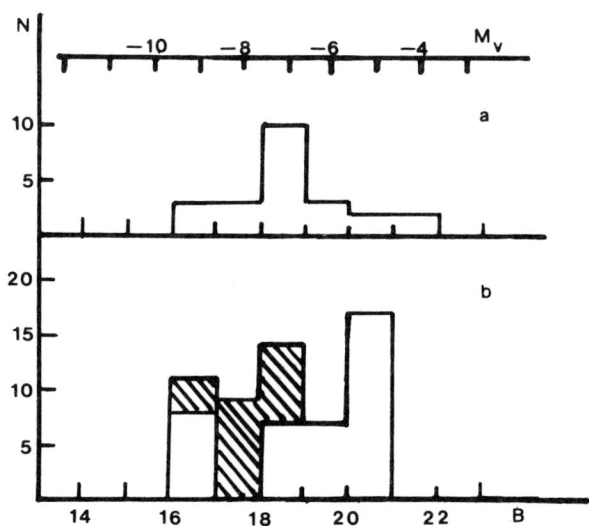

Fig. 3. a) histogram of the distribution in magnitude of the clusters in our own Galaxy having a core radius less than 0.4 pc. b) histogram of the distribution of the "excess-images" which should be the "missing globulars" according to Wirth et al. (1985); the known globular clusters are shown by shading.

(1977), then some 140 objects including 20 brighter than V = 16 would have been missed in the inner 30 arcmin. The survey of Battistini et al. (1980) has slightly increased the number of candidates in the central region. However, even using that list, the flattening will remain. By comparing the Bahcall-Soneira model with the luminosity distribution of all the images detected in M31's bulge down to a B limiting magnitude of 21, Wirth et al. (1985) have found an excess of bright images in the luminosity range of globular clusters at M31's distance. They conclude thus that, if the optical candidates prove to be clusters, the derived flattening may simply be an observational effect due to the loss of very compact clusters not detectable with morphological techniques in the central regions. The two filled circles in their plot presented in Fig. 2 would show the claimed distribution. Battistini et al. (1986) have cast some doubts on the possibility that 20 new cluster candidates for the "missing" clusters can actually be detected. In particular, there is no doubt that some very compact clusters (like M80 in the Milky Way) have been lost in all the morphological surveys, however, the excess-count method used by Wirth et al. gives quite uncertain results when applied to M31 due to:(i) the statistical fluctuations in the counts; (ii) the uncertainties in the model used for the counts of foreground stars, and (iii) the possible contamination due to the brightest resolved objects in M31. Moreover, as shown in Fig. 3, the histogram of the distribution in magnitude of the "excess-images" in M31 and that of the very compact clusters in our own Galaxy looks highly different.

In conclusion, since the globular cluster candidates have been found mainly by morphological criteria, it is likely that all the available lists still suffer from similar biases which might affect some of the indications presently drawn from the study of the whole cluster system in M31. However, we do not believe that the discussed flattening might be totally due to incompleteness.

4. GENERAL CONSIDERATIONS AND DEDUCTIONS

1. The globular cluster candidates in M31 have been selected on the basis of their appearance without regard to their ages. This implies that one can use the word "globular" only in the "morphological" sense. As a consequence, these clusters should be compared with both the globular and the bright open clusters in the Milky Way (MW) and in the outer galaxies. In particular, it is important to estimate what fraction of the MW clusters we could classify as "globular" if we were using the same criteria used to select M31 candidates. Taking into account most of the factors possibly involved in this "simulation" we believe that "MW globular cluster candidates" would closely

resemble the sample actually observed in M31.

2. The fraction of the clusters bluer than $(B-V)_o = 0.5$ (corresponding to the bluest globular in our own Galaxy) is decreasing along the sequence LMC - M33 - MW - M31. This suggests that there is a systematic variation which correlates to the galaxy-type sequence, like the bulge-to-disk ratio does.

3. The mean $(B-V)_o$ of clusters in M31 is slightly larger than in the MW. Even if strong uncertainties in the reddening of individual clusters are present, most of the available lists and methods converge toward this evidence. In particular, Harris and Racine (1979) found $\langle(B-V)_o\rangle$ = 0.72 and $\langle(B-V)\rangle_o$ = 0.74 for MW and M31 globulars respectively. As well known, this may imply a slightly higher mean metallicity for the M31 clusters.

4. The cluster luminosity functions do not differ significantly in M31 and in the Galaxy. A more or less symmetric Gaussian distribution peaked at $M = -7.2 \pm 0.2$ and $\sigma = 1.2$ mag fits the data. However, as shown and discussed by van den Bergh (1985), the cluster luminosity function appears to be correlated with the galactocentric distance. This may induce some caution on the use of globular clusters as distance indicators.

5. METALLICITY AND KINEMATICS OF M31 GLOBULAR CLUSTERS

It still remains difficult to determine whether there is a clear-cut metallicity gradient in the M31 cluster system. Most of the recent studies seem to suggest that only a mild radial gradient may exist and that, at the same time, if the gradient does exist, it is small compared to the observed metallicity dispersion at all galactocentric distances. Fig. 4 obtained from new IR data added to the whole set of previous IR measures (Bonoli et al. 1986) confirms this deduction. Moreover, one also has to notice that it is hard to deconvolve the possible metallicity gradient from the reddening law within M31.

The studies of the kinematical properties of M31 globular clusters lead to highly different conclusions if the still unpublished data presented by Searle (1984) will be confirmed. In fact, as can be seen in Table IV, all the previous data converge toward a scenario where the metal rich clusters lie in a rapidly rotating disk (within about 10 kpc of the center), and the metal poor clusters are in a slowly rotating halo, as in our Galaxy (Zinn 1985). Searle (1984), by increasing the sample, has found that: i) the M31 cluster system as a whole rotates with a low mean rotational velocity in the same sense as the disk, ii) the velocity dispersion of the clusters is large, and roughly one third of the clusters are in retrograde rotation with respect to the disk, iii) these rotational properties are independent of metallicity. These results cannot support the picture that globular clusters form early in the collapse of a single gaseous mass. It is thus clear that any significant difference in the [Fe/H]-rotational velocity relation between M31 and the Galaxy may give basic hints on the study of the early evolutionary phases of the galaxies.

Table IV
Kinematics of M31 globular clusters

Ref.	v(rot)	N(obj)	σ	Conclusions
				M31
vdB69			high	Objs near the nucleus with high metallicity: members of disk population ?
HS74		3	179±49	Q⩽-0.46 Correlation between:
		5	152±32	Q⩽-0.44 velocity dispersion
		13	133±18	Q⩽-0.40 and
		27	116±11	Q⩽-0.25 metallicity
		39	118± 9	Q⩽-0.06
HSvS82	160±40	32		14 objs with x<-15'
				18 objs with x>+15'
	negleg.	29	130	|x| ⩽ 15' (peak-to-peak)
F83	200	26	90	[Fe/H]>-0.6, Rapidly rot disk
	little	30	90	[Fe/H]<-0.6, Slowly rot system (data from HSvS82)
S84	60	100	160	Rotational properties independent of metallicity. Roughly 1/3 of all clusters are in retrograde rotation respect to the disk.
				GALAXY
FW80	60±26		116±10	
Z85	152±29		71	[Fe/H]>-0.8 Disk system
	50±23		114	[Fe/H]<-0.8 Halo population

vdB69 = van den Bergh 1969   HS74 = Hartwick and Sargent 1974
HSvS82 = Huchra et al. 1982   F83 = Freeman 1983
S84 = Searle 1984   FW80 = Frenk and White 1980
Z85 = Zinn 1985

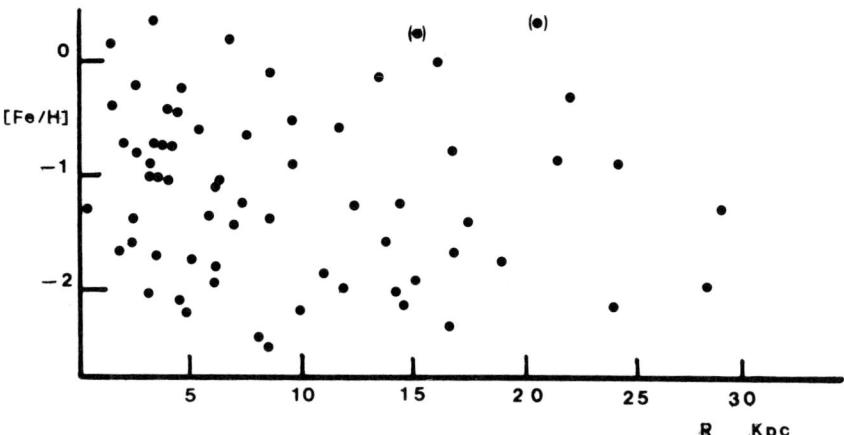

Fig. 4 The metal abundance of individual globular clusters in M31 as a function of projected galactocentric distance. [Fe/H]-values have been obtained through a (V-K)-[Fe/H] calibration based on the Zinn (1985) scale for galactic globular clusters.

6. THE STELLAR POPULATIONS OF M31 GLOBULAR CLUSTERS

One of the main reasons why the stellar populations in globular clusters are the subject of continuous study is the hope that we can describe galaxies of composite metallicity (Z) and age (t) by constructing models based on the studies of the integrated light of individual globular clusters of known Z and t. Searle first noticed that, at first approximation, from U to K the overall spectral distributions of the integrated light of individual globular clusters in M31 form a one-parameter family, and that Z is the parameter needed to rank the spectral flux distributions. This means that the integrated spectra of old populations of the same t and Z should be essentially the same. However, intrinsic differences in the integrated spectra of globular clusters, galactic nuclei, and elliptical galaxies have been found (Burstein 1985, and references therein) which have led to an extremely complex framework whose interpretation seems far from being reached. On the other hand, the present uncertainties in stellar population syntheses (Renzini 1986) do not permit the definition of stringent constraints to the models used for the comparison and interpretation of the observed indices and colors. In particular, the globular clusters of M31 present a wide set of spectroscopic indices systematically different both from the galactic globulars and from the elliptical galaxies (see Burstein 1985, for a review). **Table V** reports a quick summary of this observational evidence. Many alternative interpretations have been suggested (Burstein 1985, O'Connell 1986, Renzini 1986, and references therein). It is clear that a simultaneous fit of all the properties displayed by each individual spectral feature is impossible with the available models. However, there has been a growing claim that a substantial difference in age must be considered between the M31 and the MW globulars, in the sense that M31 globular clusters might be younger by 3 - 10 billion years. We still believe that no significant difference in age is present and that other mechanisms and phenomena (i.e. different "chemical trajectories", see Renzini 1986) may explain the quoted peculiarities. A complete understanding of these aspects is crucial for any use of globular clusters as basic tool for cosmological studies.

TABLE V
Summary of observed peculiarities in the integrated light of 'Old Stellar Populatons'

Plot (1) .vs. (2)	Regime 1	Regime 2	For fixed value of index (2), Regime 2 is
H_β .vs. Mg_2	M31 GC M32 Nuc M31 Nuc Ellipt.	MW GC	▽
CN .vs. Mg_2	MW GC M32 Nuc M31 Nuc Ellipt.	M31 GC	△
⟨Fe⟩.vs. (J-K)o			--
CN4170.vs.(J-K)o	MW GC M32 Nuc	M31 GC M31 Nuc	△
CO,H_2O.vs.(J-K)o	M31 GC MW GC	Ellipt. M31 Nuc M32 Nuc	△
CaII H,K	MW GC	M31 GC	△
UV 1500A	MW GC	Ellipt. M31 (Bo158 ??)	△
Sr II	M32	MW GC ([Fe/H]↑)	△

REFERENCES

Alloin, D., Pelat, D. and Bijaoui, A. 1976 Astron. Astrophys. **50**, 127.
Baade, W. and Arp, H. C. 1964 Astrophys. J. **139**, 1027.
Battistini, P., Bonoli, F., Braccesi, A., Fusi Pecci, F., Malagnini, M. L. and Marano, B. 1980 Astron. Astrophys. Suppl. **42**, 357.
Battistini, P., Bonoli, F., Braccesi, A. F., Federici, L., Fusi Pecci, F., Marano, B. and Borngen, F. 1986 Astron. Astrophys., in press.
Bonoli, F., Delpino, F., Federici, L. and Fusi Pecci, F. 1986 Astron. Astrophys., submitted.
Burstein, D. 1985 Publ. Astron. Soc. Pacific **97**, 89.
Crampton, D., Schade, D. J., Chayer, P. and Cowley, A. P. 1985 Astrophys. J. **228**, 494.
Freeman, K. C. 1983 in Internal Kinematics and Dynamics of Galaxies E. Athanassoula, ed., Reidel, Dordrecht, p. 359.
Frenk, C. S. and White, S. D. M. 1980 Monthly Notices Roy. Astron. Soc. **193**, 295.
Harris, W. E. and Racine, R. 1979 Ann. Rev. Astron. Astrophys. **17**, 241.
Harris, W. E. and van den Bergh, S. 1981 Astron. J. **86**, 1627.
Hartwick, F. D. A. and Sargent, W. L. W. 1974 Astrophys. J. **190**, 283.
Hiltner, W. A. 1958 Astrophys. J. **128**, 9.
Hodge, P. W. 1979 Astron. J. **84**, 744.
Hubble, E. 1932 Astrophys. J. **76**, 44.
Huchra, J., Stauffer, J. and Van Speybroeck, L. 1982 Astrophys. J. Letters **259**, L57.
Johnson, H. L. 1961 Astrophys. J. **133**, 109.
Karimova, D. K. and Sharov, A. S. 1977 Sov. Astron. Letters **3**, 207.
Kron, G. E. and Mayall, N. U. 1960 Astron. J. **65**, 581.
Mayall, N. U. and Eggen, O. J. 1953 Publ. Astron. Soc. Pacific **65**, 24.
O'Connell, R. W. 1986 Publ. Astron. Soc. Pacific **98**, 163.
Renzini, A. 1986 in Stellar Populations C. Norman, A. Renzini and M. Tosi, eds., in press.
Sandage, A. R. 1971 in Nuclei of Galaxies J. K. O'Connell, ed., North Holland Publishing Company, Amsterdam, p. 222.
Sargent, W. L. W., Kowal, S. T., Hartwick, F. D. A. and van den Bergh, S. 1977 Astron. J. **82**, 947.
Searle, L. 1984 in Annual Report of the Director, Mt. Wilson and Las Campanas Obs. 1983-84, p. 32.
Seyfert, C. K. and Nassau, J. J. 1945 Astrophys. J. **102**, 377.
Sharov, A. S. 1973 Sov. Astron. A. J. **17**, 174.
van den Bergh, S. 1969 Astrophys. J. Suppl. **19**, 145.
van den Bergh, S. 1985 Astrophys. J. **297**, 361.
Vetesnik, M. 1962 Bull. Astron. Inst. Czech. **13**, 180.
Wirth, A., Smarr, L. L. and Bruno, T. L. 1985 Astrophys. J. **290**, 140.
Zinn, R. J. 1985 Astrophys. J. **293**, 424.

DISCUSSION

DiFAZIO: You showed the comparison between the luminosity functions of clusters further and closer than 10 kpc; can you tell us if the comparison was made by normalizing the areas, taking a dimensionless abscissa (such as L/L_{max}) and then using a statistical test, or just comparing the two absolute magnitude histograms? The conclusions can differ substantially and only the first method gives a quantitatively reliable answer.

FUSI PECCI: I have taken the plot from the paper by Crampton et al. 1985 and, as far as I know, no normalization has been made.

HANES: Sidney van den Bergh (some years ago) deduced, from the distributions of color and magnitudes for globular clusters in M 31, that the reddening laws were not the same in M 31 and the Milky Way. Did you (a), find evidence for this or (b), consider this in doing reddening corrections?

FUSI PECCI: At the present stage of the project we have not yet obtained any independent estimate of the reddening of individual clusters in M 31. When necessary, we have used the reddenings obtained by Searle (See Frogel et al. 1980) if available, or the Harris and Racine (1979) approach which assumes (as a first approximation) the same reddening law in M 31 and the Milky Way.

LAUER: The distribution of clusters around M 31 does appear to be flat in the center. The Wirtanen and Searle objects pull a $R_{1/4}$ law out of a hat only by comparing excess stellar images near the nucleus with a relationship defined elsewhere by clusters selected on morphological grounds. This is mixing apples and oranges.

van den BERGH: In the Galaxy, globular cluster radii increase with galactocentric distance. If the M 31 clusters behave in the same way then it should be more difficult to distinguish clusters from stars near the M 31 nucleus than it is farther out.

GRINDLAY: If the total number of M 31 globular clusters is relatively constant, but the individual clusters have "changed" in the most recent surveys you discussed, then is it not possible that the complex (or lack of) correlations in M 31 cluster properties (as opposed to galactic globular clusters) might be due to the inhomogeneity still in the sample?

FUSI PECCI: The bulk of the cluster population (200-250 "bright" objects) forms a sample substantially unaffected by any revision. Since almost all the photometric and spectroscopic data on individual clusters come from this sample, the "changes" brought to the list have not affected the "growing" scenario. The degree of inhomogeneity in the search has been highly reduced by the latest survey and its influence on the reduction of any correlation (or lack of) should thus be low.

COHEN: Based on unpublished high spatial resolution images, many of the outermost M 31 globular clusters are spurious, consisting of either small galaxies or small random groupings of galactic stars.

FUSI PECCI: Searle (see Harris and Racine, 1979) has found that about 20% of the Sargent et al. (1977) list are spurious (particularly in the outer regions). In our survey, we have rejected some of their candidates because they seemed to be galaxies (usually the brightest and most rounded object in a very distant cluster of galaxies); but we have not found any indication of contamination due to groupings of galactic stars. There is no doubt however, that only very high spatial resolution images and/or spectroscopic observations will reduce the degree of contamination to a negligible level.

SCHOMMER: Several years ago, while taking spectra of M 33 clusters, Carol Christian and I observed a few M 31 clusters. We found that they had definitely different indices than the oldest clusters in M 33 or the Milky Way. This got us in some trouble with certain authors, but we naively interpreted this to be an age effect.

THE CLUSTERS OF M 33

C. A. Christian

University of Hawaii and
Canada-France-Hawaii Telescope Corporation

1. INTRODUCTION

Star Clusters are essential tools in studies of stellar and galactic evolution and in cosmology. In reference to investigations of stellar populations, star clusters are used to probe the chemical enrichment, kinematical and dynamical history of a galaxy. The relationship between the characteristics of a cluster population and the morphology of the parent galaxy is enigmatic, however.

In our Galaxy the two groups of clusters, the globulars and the open clusters, are quite distinct in spatial distribution, age, chemistry mass, etc. The sample of open clusters is fairly restricted due to severe selection effects inherent in studies of the Galactic disk (c.f. Janes and Adler 1982). In M31, it appears that globular clusters share some similarities with their galactic counterparts but there are important differences. Little is known about the M31 population equivalent to the galactic open clusters due to the difficulty in observing the highly inclined M31 disk. A first look at the global characteristics of cluster populations in galaxies comes from studies of the Magellanic Clouds (MC) where it was found that few globulars exist but that a large population of massive "globular like" young to intermediate aged clusters exists.

This information uncovers many puzzles concerning galactic structure and evolution as well as problems in stellar evolution. One is lead to speculate about the parameters that lead to cluster formation, for example for galactic globular clusters there appear to be relationships between metallicity, mass functions, tidal truncation and possibly orbital parameters (see Pryor, Smith, and McClure

1986 and references therein) that are clues to cluster formation processes.

For the LMC it has been suggested that the populous star clusters are confined to disks possibly inclined as a function of age (Freeman, Illingworth and Oemler 1983; hereafter FIO). In M31 and the Galaxy it appears that except for globulars these clusters may be rare although selection effects may exacerbate the discovery of such objects. One naturally turns to the Sc galaxy M33 for clues to these riddles. The modest inclination of the galaxy (~60°) allows the global properties of a cluster population to be examined which is particularly of interest because it is the only Local Group spiral that can be studied so extensively.

2. PREVIOUS WORK ON M33 CLUSTERS

2.1 Catalogues and photometry

The first hint that M33 clusters were not completely analogous to M31 and galactic globulars came from the photoelectric photometry of Hiltner (1960) and Kron and Mayall (1960) who discovered that several of the brightest clusters are bluer than even the bluest globular clusters in the Galaxy. The cluster population was also surveyed extensively by A. Sandage and P. Osmer by examination of Palomar 5m plates taken in three colors (A. Sandage, unpublished). These plates were also used by Humphreys and Sandage (1980) for a study of bright blue and red stars in M33 and for the studies of M33 Cepheids (Sandage and Carlson 1983 and references therein) where printed reproductions of several of the fields can be seen. Over 500 clusters were found covering wide range of ages.

More recently, a catalog of clusters was compiled by R. Schommer and myself (Christian and Schommer 1982, hereafter CSI) that included over 250 nonstellar objects. The objects were selected from Kitt Peak National Observatory (KPNO) 4m Richey-Chrétien and 4m prime focus plates. We specifically looked for nonstellar objects with uniform symmetric profiles with the goal of finding MC type star clusters. BVR photometry was obtained for a sample of 60 of the objects, which when combined with the previous photometry suggested that the clusters are fairly uniformly distributed in (B-V) from 0.0 mag to 0.8 mag. The apparent magnitudes of the surveyed objects suggest that there is a substantial population of clusters brighter than $M_V = -5$ at <u>all ages</u>. Compared to the LMC it appears that the formation of these clusters is less episodic than in the LMC. The intermediate aged M33 clusters are much brighter (and hence more massive) than the most populous clusters known in the galaxy such as M67 and NGC2158. Therefore it appears that the M33 cluster population is quite distinctive compared with the other Local Group galaxies. We were therefore tempted to suggest that the distinctive characteristics of M33 cluster population are causally related to the galaxy's intermediate mass and morphology which must determine the star formation history in the galaxy.

2.2 Spectroscopy

While the photometric studies were indicative of the M33 cluster population further investigation as to the identities and the age-metallicity kinematic relationships is important. Intermediate resolution spectroscopy (~10Å) was obtained at KPNO with the Intensified Image Dissector Scanner (IIDS) on the 4m (Christian and Schommer 1983, hereafter CSII). Twenty M33 clusters were observed covering the range of (B-V) sampled in the photometric surveys. Spectrophotometric indices measuring $H\beta$, Mg b, the Ca H and K lines and the 4200Å CN features indicated that the photometric results were correct: the M33 massive clusters occur at the full range in ages with metallicities from [Fe/H] = -2.0 to nearly solar, with metallicity roughly correlated with age. These results agree with an independent studies by Cohen, Persson and Searle (1984) and Sharov and Lyutyj (1984).

The kinematics of the cluster system appears to be more typical of a spiral (disk) galaxy than in the LMC in that the oldest clusters appear to have "random" velocities when compared to the HI disk rotational velocities, but clusters with ages $<10^{10}$ years appear to follow the disk rotation. These results suggest that kinematically the clusters "know" they are in a spiral galaxy but that the environment for formation (and disruption) of clusters does not inhibit continuous production of massive systems.

3. RECENT INVESTIGATIONS

3.1 Catalog

The catalog of M33 clusters in CSI, while extensive, cannot be considered complete due to selection effects involving the plate material used and our cataloging procedure which avoided regions of dust and star formation. In fact as noted in CSII some of the brightest clusters were found close to the galaxy's nucleus by R. Racine on short 3.6m plates taken at the Canada-France-Hawaii Telescope (CFHT) in very good seeing.

Also, it was known that an extensive catalog of clusters had been compiled at Palomar. Dr. A. Sandage kindly lent the catalog material including prints and plates to me in order that a new catalog may be published. With the assistance of D. Ward (University of Hawaii) the two catalogs (Palomar and CSI) were cross referenced and all objects identified were reclassified into two bins solely on the basis of morphology. The first group represents clusters as specified in the CSI catalog; namely objects with non-stellar symmetric profiles. Nearly all of the CSI objects identified as "clusters" are in this group. In addition a large percentage of objects designated as "UFO" in CSI, a handful of objects identified as "H" in CSI (see CSII), and roughly 40% of the Palomar clusters are designated cluster candidates.

The second group which have asymmetric or extremely diffuse morphology or which are clearly imbedded in nebulosity were classified as "younger" objects. The majority of objects classified as "H" in CSI, a substantial number of "UFO" objects and the remainder of the Palomar clusters fall in the second group.

The spatial distributions of the two groups (Figure 1) are revealing. The younger objects do preferentially lie within regions that correspond to the most active star formation regions. The mamority of the objects appear to be truly imbedded in nebulosity although it is possible that some of the objects are "clusters" superimposed on star formation regions. The uniform morphology clusters are in clearer regions of the disk including the outlying regions. The catalog is relatively complete to V = 19.0 but there are significant numbers of objects that can be identified to V ~ 20.

The spatial distribution for the most part is fairly uniform except it can be seen in Figure 1 that there is a lack of clusters in the SE quadrant of M33. This is an area notably deficient in star formation regions and disk population possibly due to obscuration. There are a few outlier clusters, one notable example is C39 (Figure 1 in CSI) an intermediate aged cluster (CSII) situated in this region of M33 that appears devoid of a significant stellar population.

The new catalog is being prepared for publication and will include cross references to the previous catalogs.

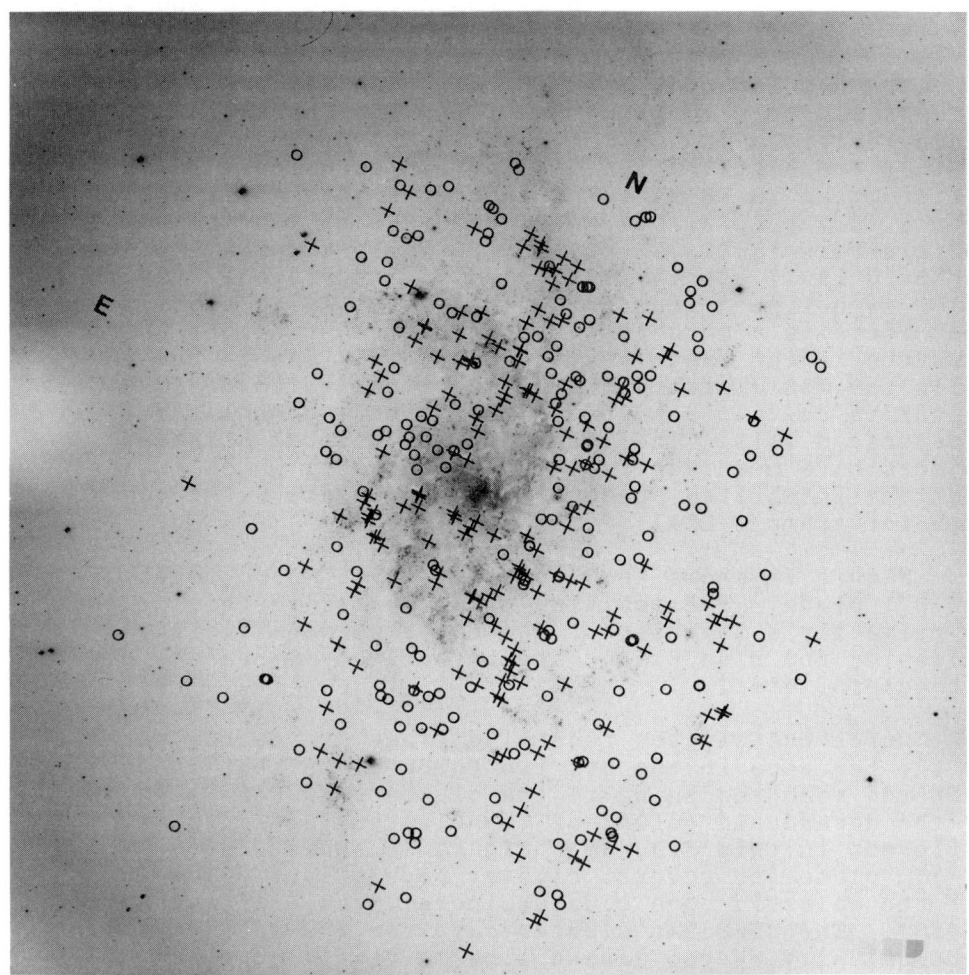

Fig. 1. Spatial distribution of objects classified as clusters (circles) and younger objects (crosses).

3.2 Optical Photometry

It is now of interest to study the cluster population of M33 in even greater depth. To this end CCD frames in B, V and I were obtained at the University of Hawaii 2.2m using the Galileo/IFA 500^2 TI 3 phase CCD. Thirteen frames with 128 objects from the new catalog have been measured with an aperture photometry algorithm. The average seeing on the two photometric nights was ~0.8 arcsec so that the extended profiles of the many of the clusters are quite obvious. The CCD B-V photometry agrees well (0.02 mag) with the previous photoelectric and video camera data for the clusters in common.

The new integrated color-magnitude diagram (CMD) for all M33 clusters is shown in Figure 2. The basic characteristics of the CMD seen in CSI are reinforced here in that the distribution of clusters is uniform in (B-V). A historgram in (B-V) color indicates that for the bins 0 to 0.2, 0.2 to 0.4, 0.4 to 0.6 and 0.6 to 1.0 the number of M33 clusters is 21, 26, 25 and 23 while for the LMC (van den Berg 1981) are 44, 43, 7 and 29. Admittedly the (B-V) color is not on a linear age scale, but it is significant that there is a much smaller percentage of bright LMC clusters in the range (B-V) = 0.2 to 0.4 mag in the LMC while the M33 distribution is much more uniform. M33 does appear to contain a number (~20) of true globular clusters as well. The identities of several of these objects have been verified by spectroscopy which indicates that the integrated spectral types are late F through K (see for example CSII).

Figure 3a shows the distribution of M33 clusters in the BVI plane. The position of a cluster in this diagram is primarily a function of age but also metallicity, reddening and blue star content (horizontal branch, blue stragglers, etc.).

Unfortunately the reddening lines are nearly parallel to the sequence in the BVI two color plane, but it appears that the mean reddening to the clusters is E(B-V) ~0.09 so the dereddened colors of the clusters are only slightly different in this diagram. Figure 3b shows the $(B-V)_0$ vs $(V-I)_0$ plane for several cluster populations. The data for the galactic open clusters comes from Kron and Mayall (1960). The globular cluster data for the Galaxy, shown schematically by the dashed line in this figure are Hanes and Brodie (1985). The galactic open clusters and globulars blend with the M33 sequence. The bulk of the M33 tend to be bluer in (B-V) at a given (V-I) than galactic clusters but the scatter is large.

A rough age calibration could be derived by considering the ages of the galactic clusters derived from Janes' and Adler's (1982) compendium. The ages assigned to M33 clusters agree for the most part with the ages for the M33 clusters in CSII and with the ages assigned by Sharov and Lyutyj (1984) from UBV photometry. By assigning ages according to the line shown in Figure 3b the age distribution of the clusters can be examined. After assigning ages to the clusters, an S index for each object can be derived using Elson and Fall's relation (1985b;

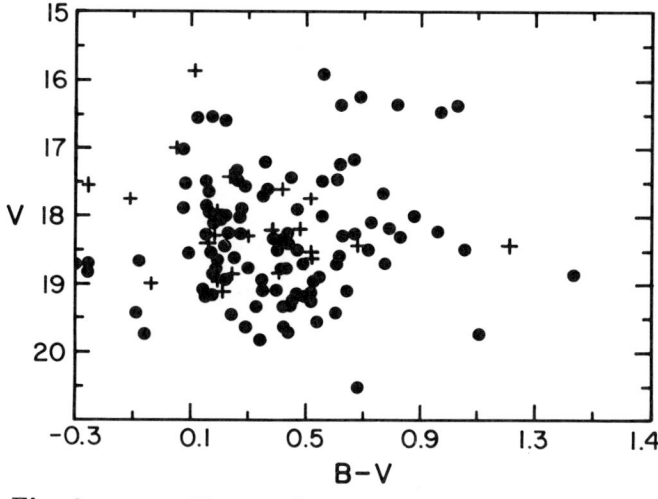

Fig. 2. The new integrated color-magnitude diagram of M33 clusters. The distribution of clusters appears to be reasonably uniform in (B-V) when compared to a similar diagram for LMC clusters (CSI).

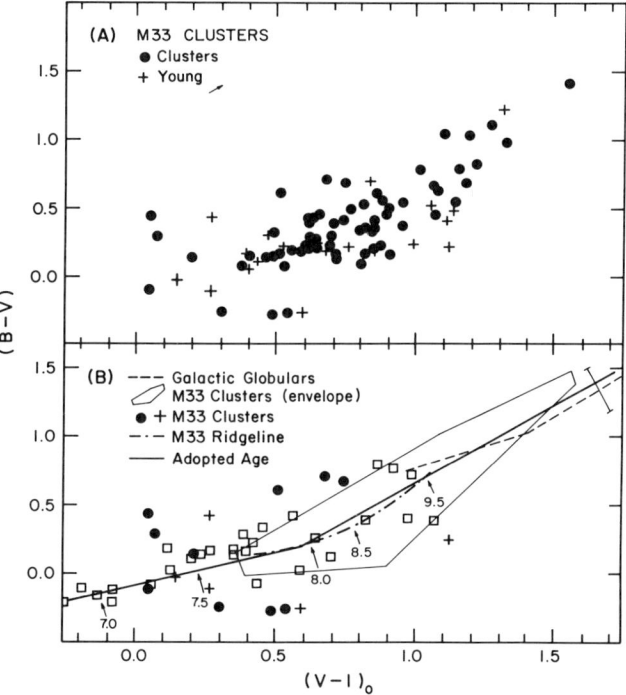

Fig. 3. Two-color BVI diagram from Christian and Schommer (1987). In (a) M33 clusters are shown (dots) and younger objects (crosses). In (b) most of the M33 data is shown schematically as the area enclosed by the solid line. The ridge line of M33 clusters is shown by the dot-dashed line, while the younger objects are shown explicitly because they scatter widely in this plane. The data for galactic open clusters from Kron and Mayall (1960) are shown as boxes, and the locus of galactic globulars from Hanes and Brodie (1985) is shown as a dashed line where the error bar indicates the scatter in the data. Finally the bold line shows the line chosen for the age calibration. Ages are indicated.

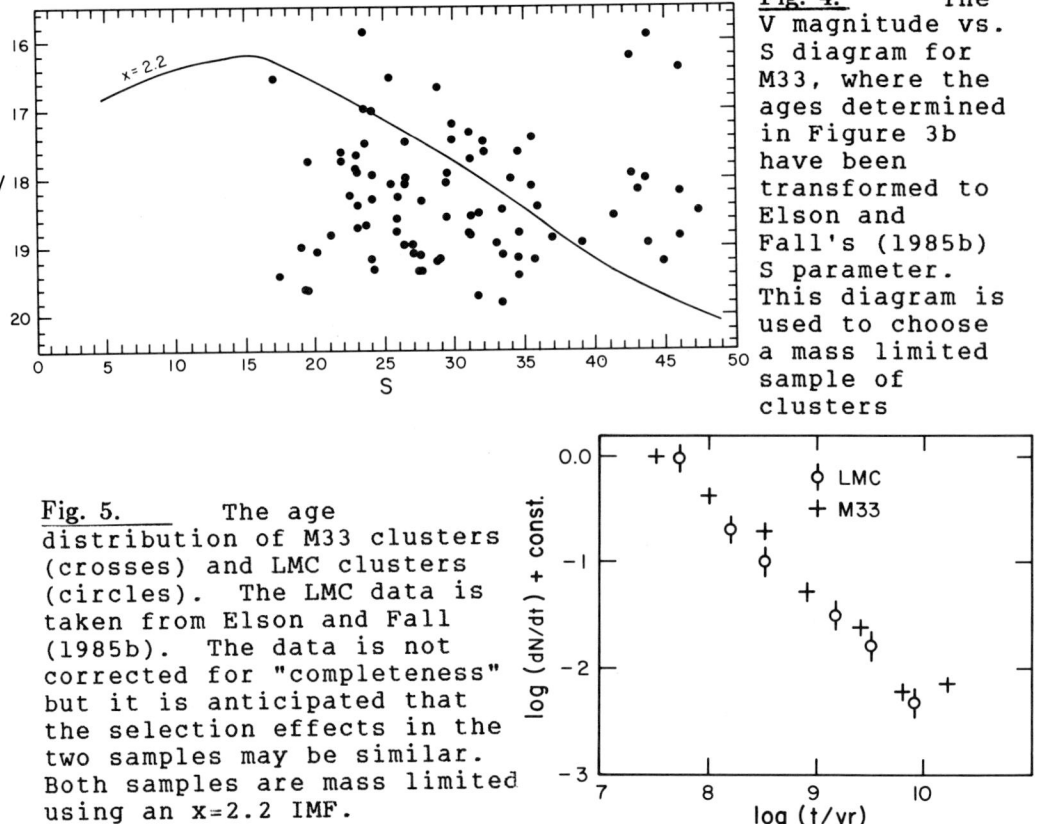

Fig. 4. The V magnitude vs. S diagram for M33, where the ages determined in Figure 3b have been transformed to Elson and Fall's (1985b) S parameter. This diagram is used to choose a mass limited sample of clusters

Fig. 5. The age distribution of M33 clusters (crosses) and LMC clusters (circles). The LMC data is taken from Elson and Fall (1985b). The data is not corrected for "completeness" but it is anticipated that the selection effects in the two samples may be similar. Both samples are mass limited using an x=2.2 IMF.

their Figure 2). The magnitude vs. S figure (Figure 4) analogous to Elson and Fall's Figure 3 is used to select the mass limited sample of M33 clusters. The sample is selected with the x = 2.2 IMF similar for the LMC sample. The uncorrected age distributions for the LMC and M33 clusters are shown in Figure 5 where it is seen that the two distributions are nearly identical. Considering that the selection effects may be similar this figure implies that the age distributions in the two galaxies are much flatter than an IMF with x = 2.2. As Elson and Fall (1985b) discuss, only the richest clusters in the Galaxy follow this relation. One wonders whether the fainter open clusters in either M33 or the LMC follow the IMF with x = 2.2 as the galactic open clusters do, i.e., that the flatness in the age distribution is a function of mass.

Elson and Fall also argue that the smoothly declining age distribution suggests that the number of true globulars in the LMC is indeed small. However in M33 the age

distribution seems to flatten out as shown by the last
point in Figure 6, that is, the number of globular aged
clusters does not follow smoothly from the other data.
However selection effects (i.e. the globulars are
preferentially observed?) could enhance this last data
point. As Elson and Fall mention this diagram does not
indicate that star formation in the LMC has been "bursty"
yet the histogram of clusters in (B-V) suggests otherwise.
It is clear that more complete samples may aid in
determining whether the cluster formation histories in
Local Group galaxies has been episodic or not.

3.3 Infrared Photometry

Persson et al, have shown that infrared photometry can
be a tracer of carbon star content in integrated systems as
well as being used to establish a longer wavelength base
for stellar population studies. Cohen, Persson and Searle
(1984) have used a number of photometric indices including
infrared colors to examine a few M33 clusters. Aaronson,
Riecke and Christian (unpublished) have also obtained JHK
photometry for some of the brighter clusters. The sample
is still small but the photometry suggests that some M33
clusters may indeed contain carbon stars. In collaboration
with H. Zinneker and R. Cannon a more extensive survey will
be accomplished for population synthesis to probe the
existence of carbon stars in M33 clusters as well as to
examine the metallicities of clusters as measured by the
(J-K) color (Frogel et al. 1980). The more uniform
distribution of ages in M33 should allow a reasonable test
of carbon star production rates as well as examination of
metallicity and population effects in cluster systems.

4. OTHER RESULTS

4.1 Luminosity Function

Elson and Fall (1985a) have discussed the properties
of the luminosity function of LMC clusters in relation to
galactic open clusters. They find that the differential
luminosity function of the LMC clusters, which is dominated
by clusters younger than 10^{10} years resembles the
luminosity function for galactic open clusters (Figure 6).
They further argue that luminosity functions of young to
intermediate aged clusters are biased at the faint ends due
to observational limits and that there is no evidence that
the LMC "populous" clusters follow the universal luminosity
function of galactic and M31 <u>globular</u> clusters. In fact
Elson and Fall (1985a) maintain that the luminosity
function of the LMC clusters follow a power law $\phi(B)$ ~
$L-\alpha$ where α~1.5. The differential luminosity

function of the M33 clusters is also shown in Figure 6 as well where an apparent blue distance modulus $(m-M)_B$ = 24.6 has been adopted (Christian and Schommer 1986). It is seen that the M33, LMC and galactic open clusters all have similar differential luminosity functions, rising in the similar way.

The luminosity histograms for the (B-V) color bins <0.3 mag, 0.3 to 0.6 mag and <0.6 mag are shown in Figure 7. The luminosity functions for clusters with ages <10^{10} years are clearly rising at V=19.0 mag. There is evidence for "fading" in the intermediate aged group in that the brightest clusters with 0.3 < (B-V) < 0.6 are fainter in V than the brightest young clusters.

The luminosity histogram of the measured globulars, (B-V) > 0.6 mag (Table I) is shown in Figure 7c. The luminosity function does not resemble the gaussian galactic globular or M31 globular luminosity functions although the number of objects (25) is very small for statistical purposes. In particular the four clusters brighter than V = 16.4 mag (three of which have been verified with spectroscopy) stand out. These four objects are located at projected distances <9 arcmin (~1.8 kpc) from the center of the galaxy. The mean V magnitude of all the red clusters is ~18.0 mag with σ=1.1 mag. It is possible that some of the bluer clusters in the sample are in fact intermediate aged clusters, so assuming that the mean reddening is E(B-V)=0.09 mag and taking the red clusters with $(B-V)_o$ > 0.6 mag one obtains <V> ~ 17.8 mag with σ = 1.0. The dispersion here is similar to that for galactic and M31 globulars yet the implied mean absolute usual magnitude $<M_V>$ is ~-6.7 mag assuming an apparent B distance modulus of 24.6 mag to M33. [This assumes $(m-M)_o$ = 18.5 mag for the LMC, a relative distance modulus of 5.7 from the LMC to M33 (Christian and Schommer 1986) and a mean reddening of E(B-V) = 0.09 mag for the red clusters]. It is interesting to note that the mean <V> magnitude for clusters at projected galactocentric distances <10 arcmin (~2kpc) is 17.4 while the more distant objects have <V>=18.4. While it is true that there are three primary selection effects influencing the data; namely (1) the small number of objects surveyed; (2) the selection of fields for the CCD survey and (3) the identification of objects in the central parts of M33 is biased towards brighter objects due to severe background contamination, it is unlikely that many bright objects at large radii have been missed. In fact most of the bright distant red objects in the M33 field identified by ourselves and Melnick and D'Odorico (1978) have turned out to be galaxies as determined from their

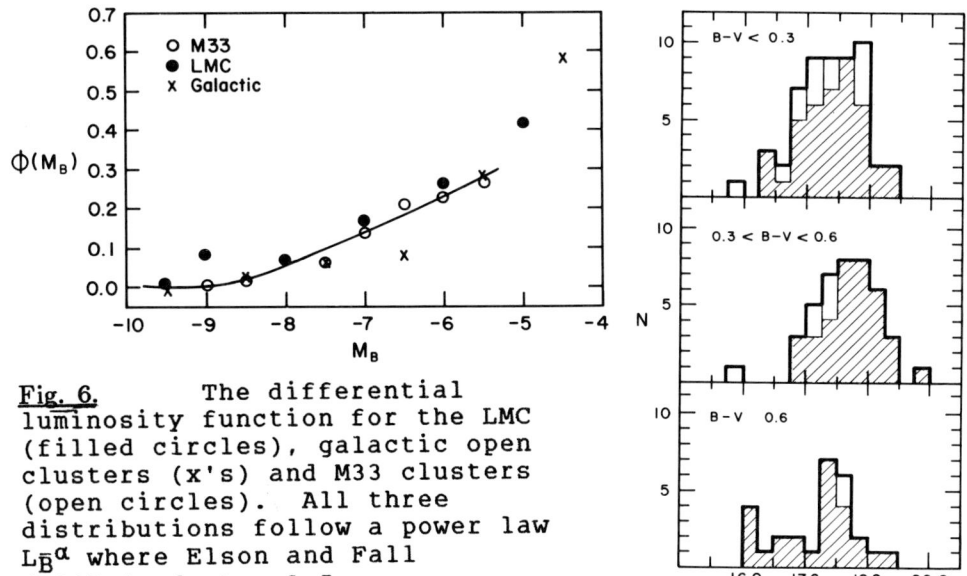

Fig. 6. The differential luminosity function for the LMC (filled circles), galactic open clusters (x's) and M33 clusters (open circles). All three distributions follow a power law L_B^α where Elson and Fall (1985a) adopt $\alpha=1.5$.

Fig. 7. V luminosity histograms for three color groups: (a) (B-V) <0.3 mag, (b) 0.3 < (B-V) <0.6 mag and (c) (B-V) > 0.6 mag (globulars).

radial velocities (Table II) and most distant positively identified bright cluster is C39 which is an intermediate aged cluster. It should be noted that this trend for the brightest clusters to be more centrally concentrated has been reported for the M31 globulars by Crampton et al (1985).

4.2 Kinematics

In the LMC it appears that the clusters are distributed in disk like structures where the inclination of the disk varies with age. Considering that M33 is a bona fide spiral galaxy, it is of interest to study the kinematic of the cluster population as a function of age. Velocities of several M33 clusters have been obtained (Cohen, Persson and Searle 1984; CSII). In addition several more velocities have been obtained with the MMT Schectograph and reduced with a cross correlation technique. A summary of all velocities with rms error σ < 50 km/sec appear in Table 3 where the clusters have been divided into the same color bins as for Figure 7. It is obvious from column 4 in this table that the young and intermediate aged clusters follow the HI disk reasonably well. The red clusters are much more randomized suggesting a halo structure as in other spiral galaxies. The red clusters do not exhibit any rotation at levels <50 km/sec. The youngest clusters tend to follow more closely the HI rotation, but more spectra are required to study in detail the kinematics of each age group.

4.3 Chemical Enrichment

Aaronson (1986) has discussed the chemical enrichment histories of the Galaxy, the LMC and the SMC as traced from the stellar populations. From the current data it appears that the Galaxy underwent rapid chemical enrichment while the enhancement in the MC was much more "leisurely" as can be seen in Figure 8, adapted from Aaronson's Figure 1. A look at the chemical enrichment history of M33 can be derived from spectroscopy of the clusters with ages assigned from the BVI data as described above. Metallicities are estimated from the M_{HK} vs Hβ data in CSII. M33 has a radial abundance gradient as evidenced from HII region surveys described by Pagel and Edmunds (1981). These data are also shown in Figure 8. In anticipation of metallicity differences between inner and outer regions in M33 the clusters are plotted so that clusters at projected radii <10 arcmin are distinguished from the inner clusters. Also shown is the data point from Mould and Kristian's (1986) study of the halo population in M33.

It is clear that at each age there is a dispersion in the abundances. It also appears that the inner population follows an enrichment history similar to the LMC while the outer objects have a history more similar to the SMC. That is, the chemical enrichment history is a function of radius in M33. In the mean M33 is more metal poor than the solar neighborhood at a given age although at the current epoch solar metallicities appear to have been achieved in the inner most regions. The oldest stellar population, namely the globulars and the halo population cover a range in metallicity from 2.2 < Fe/H < -1.0 with no known clusters as enriched as the galactic globular 47 Tuc or the red M31 globular clusters.

For some of the intermediate aged clusters the assigned metallicities suggest some clusters in M33 are more metal rich than LMC clusters. In fact comparison of several clusters (eg. U83, C27 and H21: CSII) do appear to have stronger metal lines than LMC clusters (eg. NGC2209: Rabin 1982) at a given Hβ strength.

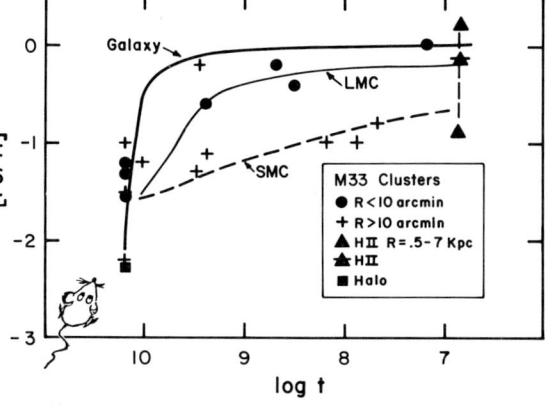

Fig. 8. The chemical enrichment history of the Galaxy, LMC and SMC (adapted from Aaronson 1986). The data for M33 suggest that the inner regions of the disk were enriched more rapidly than the outer regions.

5. SUMMARY

The investigation of star clusters in M33 has been very useful as a probe of the stellar population in that galaxy. The main results to date have been:

(1) M33 contains a substantial population of massive clusters that are more or less evenly distributed in color. The color distribution appears more uniform than for LMC clusters, but the mass limited (uncorrected) age distributions of M33, the LMC and the high richness class galactic open clusters are similar.

(2) JHK photometry suggests that several of the clusters may have carbon stars although further investigation is required to fully map the carbon star production rate in M33 clusters as a function of age.

(3) The optical luminosity functions of three main color groups, (B-V) < 0.3 mag, 0.3 < (B-V) < 0.6 mag and (B-V) > 0.6 mag (globulars) show that the luminosity functions are still rising at V = 19.5 and there is some evidence for luminosity fading in the intermediate color group. The luminosity function for the globulars does not resemble the "universal" gaussion function found in M31 and the galaxy. The mean V magnitude is much fainter than expected although the selection effects may dominate the small sample. There also is evidence that the mean <V> for the inner clusters is higher than that for the outer (R>10arcmin) clusters. It is true that fainter clusters may be missed in the inner regions but it is very unlikely that bright clusters have not been found in the relatively clean outer regions of the M33 disk.

The differential luminosity function of all M33 clusters resembles strongly the function for LMC and galactic open clusters following an exponential of the form

$\phi(B) \propto L^{-\alpha}$ where $\alpha = 1.5$.

(4) The chemical enrichment history of M33 appears to be less dramatic than that for the galaxy. The outer part of the disk shows a gradual enrichment similar to the SMC while the inner regions are more similar to the LMC. Some clusters at a given age and Hβ strength appear to be stronger lined than corresponding LMC clusters. The chemical enrichment history of M33 appears to be a function of position in the disk.

(5) The kinematics of the clusters, based on a small sample of objects suggests that the globular clusters have randomized velocities but no appreciable rotation in excess of 50 km/sec. The velocities of young and intermediate aged clusters are disk like, so at some point M33 experienced a disk collapse as for other spiral galaxies. The chemical enrichment history is to be studied in greater detail to detect any evidence of the collapse on the enrichment. It may be that the inner disk has collapsed rapidly causing rapid enrichment but he outer disk has collapsed more slowly.

The work related to the M33 cluster photometry and spectroscopy is being done in collaboration with R. Schommer. We wish to acknowledge support of NSF grant AST 8412515.

REFERENCES

Aaronson, M. 1986 in Stellar Populations, Local Group Dwarf Galaxies: The Red Stellar Content (STSI).
Christian, C. A. and Schommer, R. A. 1982 Astrophys. J. Suppl. 49, 405 (CSI).
Christian, C. A. and Schommer, R. A. 1983 Astrophys. J. 275, 92 (CSII).
Christian, C. A. and Schommer, R. A. 1986 Astron. J., submitted.
Christian, C. A. and Schommer, R. A. 1987 in preparation.
Cohen, J., Persson, S. E. and Searle, L. 1984 Astrophys. J. 281, 141.
Crampton, D., Cowley, A., Schade, D. and Chayer, P. 1985 Astrophys. J. 288, 494.
Elson, R. and Fall, M. 1985a Publ. Astron. Soc. Pacific 97, 692.
Elson, R. and Fall, M. 1985b Astrophys. J. 299, 211.
Freeman, K., Illingworth, G. and Oemler, G. 1983 Astrophys. J. 272, 288 (FIO).
Frogel, J., Persson, S. E. and Cohen, J. 1980 Astrophys. J. 240, 785.
Hanes, D. and Brodie, J. 1985 Monthly Notices Roy. Astron. Soc. 214,

491.
Hiltner, W. 1960 Astrophys. J. 131, 163.
Humphreys, R. and Sandage, A. 1980 Astrophys. J. Suppl. 44, 319.
Janes, K. A. and Adler, D. 1982 Astrophys. J. Suppl. 49, 425.
Kron, R. G. and Mayall, N. 1960 Astron. J. 65, 581.
Melnick, J. and D'Odorico, S. 1978 Astron. Astrophys. Suppl. 34, 349.
Mould, J. and Kristian, J. 1986 Astrophys. J. 305, 591.
Pagel, B. and Edmunds, M. 1981 Ann. Rev. Astron. Astrophys. 19, 77.
Persson, S. E., Aaronson, M., Cohen, J., Frogel, J. and Matthews, K. Astrophys. J. 266, 105.
Pryor, C., Smith, G. and McClure, R. 1986 preprint.
Sandage, A. and Carlson, G. 1983 Astrophys. J. Letters 267, L25.
Sharov. A. S. and Lyutyj, V. M. 1984 Pis'ma Astron. Zh. 10, 654.
van den Bergh, S. 1981 Astron. Astrophys. Suppl. 46, 79.
VandenBerg, D. 1985 Astrophys. J. Suppl. 58, 711.

TABLE I.
Red Clusters

Cluster	V	radius(')
U 49	16.25	8.3
R 13	16.36	2.9
R 12	16.38	4.5
R 15	16.38	4.9
R 14	16.48	3.2
M 9	17.12	10.0
C 20	17.16	18.4
U 77	17.19	5.7
H 38	17.25	10.6
H 21	17.46	9.0
C 36	18.00	13.9
C 38	18.10	17.9
C 18	18.17	15.9
H 10	18.23	10.1
C 3	18.26	19.1
C 32	18.29	6.6
U 137	18.30	14.3
C 21	18.60	11.3
U 23	18.70	11.0
C 9	18.71	15.7
S 24	18.49	18.5
S 72	18.49	7.0
S 247	18.87	19.9
U 67	19.11	6.8
S 161	19.43	9.7
S 160	19.73	10.0
U 7	20.51	16.2

TABLE II.
Compendium of Velocities

I.D.	B-V	Vel +/-	V-V(HI)	ref
THE GOOD		SIG <+/- 50 KM/SEC		
Young				
M 4	0.21	-255	10	CPS
M 6	0.21	-245	15	CPS
U 62	0.12	-220,-255	15,20	CSII, CPS
U 138	0.18	-116 +/- 41	<5	MMT
Intermediate				
C 27	0.37	-210	24	CSII
C 39	0.56	-145	0	CSII
Old				
C 20	0.77	-160,-62	85	CSII,MMT
H 21	0.61	-165	40	CSII
H 38	0.83	-200	80	CSII
U 49	0.68	-185,-125,-91	50,100	CSII, CPS, MMT
M 9	0.72	-300	115	CSII
R 12	0.77	-190	12	CSII
R 14	0.68	-125	62	CSII
U 137	0.83	-7 +/- 29	100	MMT
H 10	0.96	-284 +/-39	65	MMT
THE BAD		(Galaxies)		
MD18		20,500		MMT
MD57		10,260		MMT
MD58		37,066		MMT
MD53		>20,000		MMT
U 90		30,500		MMT
THE UGLY				
Many		(To be remeasured)		

Notes: MMT = MMT spectra obtained by Huchra, Bothun, Schommer, and Christian
CPS = Cohen, Persson, and Searle (1984)
MD = Melnick and D'Odorico (1978)

DISCUSSION

ALCAINO: It is interesting to notice that the color-magnitude diagram of the integrated light of the clusters in M 33 does not show the clear dichotomy among blue and red objects seen in both Magellanic Clouds, in spite of the similarities suggested in your talk about the clusters in these galaxies. Are the four bright (V~16) red (B-V>0.6) clusters shown in your luminosity function of red clusters true globular clusters?

CHRISTIAN: The histogram in (B-V) indicates that the LMC clusters are not as evenly distributed as the M 33 clusters so that one may think that M 33 has had a much more continuous cluster formation rate than the LMC. The four bright red clusters have been measured spectroscopically. Their integrated spectral types are G0, G4 and late F.

RICHER: Your comment about the M 33 clusters being a good place to look for the onset of the carbon star phenomenon, is this based on the fact that the age distribution is more continuous in M 33 than that for the LMC?

CHRISTIAN: Yes, it seems that the distribution of (B-V) colors for the clusters suggests that there are plenty of clusters at each age. The CCD data sample about 30% of the galaxy so one anticipates that there are many clusters at a given age. It would be interesting to sample the clusters with a fine age resolution and look for the onset of carbon star production.

KING: Will these clusters be good HST targets for the study of horizontal branches? And could you remind us of the distance modulus of M 33? I believe Sandage has recently revised it on the basis of a correction to Hubble's photometry.

CHRISTIAN: Some of the M 33 clusters would be excellent targets for color-magnitude diagram studies. I know of a few groups writing proposals to do this. The distance modulus we used was 24.6 in V-apparent. the revised modulus proposed by Sandage based on Hubble's data is a complicated issue and is a subject of a new preprint discussing CCD data relative to Hubble's data. The full description of this work is beyond the scope of the M 33 cluster work presented here.

ZINN: If I remember the recent work of Elson & Fall correctly, they found that the age distribution of the clusters in the LMC was similar to that of the open clusters in the Milky Way, except that there is a tail of the LMC distribution indicating that: it has relatively more clusters with ages ~ 5×10^9 yrs. (I don't remember the precise age range). Can you clarify whether or not you find a similar effect among the M 33 clusters?

CHRISTIAN: The age distribution for the LMC and galactic open clusters apparently matches for younger clusters, i.e., those younger than ~10^9 yr. The data is normalized at the young end. In the vicinity of 10^9 yr. the applied completeness corrections become significant. The flatness of the corrected counts could largely be due to the completeness corrections. The uncorrected counts indicate that the number of globular clusters in the LMC is expected to be small. Whether the M 33 data really turns up at the oldest data point is not clear because the CCD data was selected perhaps preferentially near globular clusters. The M 33 data turns up at ages ~ 5×10^9 yrs only if the LMC completeness corrections are valid and applicable to M 33 data. The match between uncorrected counts in M 33 and the LMC at all ages could be taken as evidence that the age distributions are the same if one believes that the selection and completeness effects are the same in the two samples. However, it is disturbing that the uncorrected counts match as well as they do because the M 33 sample was assembled in a relatively "cavalier" way with rather crude ages so one wonders how valid these age distributions are. Secondly, some LMC fields show that the history of star formation may be episodic yet this is not seen in the age distributions. We can see a significant difference in the (B-V) histograms of the M 33 and LMC clusters so that one would expect the age distributions to be <u>different</u>. It will be interesting, eventually, to study the detailed age distributions with finer age resolution as a function of position in both galaxies. As well, we want to know the age distribution of the massive (high richness class) clusters in the galaxy which are apparently similar to the LMC clusters and by reference the M 33 clusters. It would also be interesting to study the age distributions of "classical" open clusters in all three galaxies.

NEMEC: Do you see any evidence for rotation in the M 33 star cluster system? And, would your survey have identified outlying globular clusters such as NGC 2419 in the Galaxy?

CHRISTIAN: At the level of ~50 km/sec we have not detected rotations in the globular clusters. For young clusters we find V/σ ~ 5, that is they rotate with the disk. The mean velocity difference for clusters younger than 10^{10} years is 12 km/sec from the projected line-of sight HI velocity. The red clusters have $V\sigma$ ~ 0.5 but the sample is small. We intend to obtain more velocities this fall. We would have difficulty finding objects that are very compact in any part of the Galaxy. If all the "missing" red clusters are located in the outer regions, they would need to be exceedingly compact yet very bright, 16.5 - 17.5 in V. I still suspect that the missing clusters, if they exist, are in the inner regions that are difficult to sample.

MCCARTHY: Do you detect any traces of TiO in any of your spectral data from M33?

CHRISTIAN: No. The spectra were mostly in the blue region and the features were H and K, G Band, Mg etc.

ZINNECKER: I'd like to reemphasize that the clusters in M 33 constitute perhaps the best template system for observational population synthesis (optical and infrared), given their wide spread in age and metallicity and given the favorable face-on orientation of this disk galaxy.

CHRISTIAN: I agree and I anticipate that very exciting work will be done in the next few years. With higher resolution we will be in a good position to understand the evolutionary history of an entire disk galaxy.

RAMAMANI: How does the extent of the globular cluster system in M 33 compare with the extent of the Galaxy?

CHRISTIAN: The red clusters are centrally concentrated in the Galaxy. The intermediate age clusters are found near the outer regions. Even if the missing red clusters are found, they are likely to be near the central region, where the selection effects are more severe.

HATZIDIMITRIOU: What kind of errors do you have for the integrated V magnitude and (B-V) color for the clusters of your sample?

CHRISTIAN: Approximately 0.04 mag in V - the error is, of course, worse for fainter magnitudes - and 0.05 to 0.07 mag in (B-V).

GRAHAM: Any data on ellipticity of the globular clusters?

CHRISTIAN: No. The seeing for the CCD data was very good (0.8 arcsec) so that the clusters are clearly seen to have extended profiles. We have not seen any obvious ellipticity (except for the objects which turned out to be galaxies) but marginal ellipticity has not been really tested for.

THE NGC 5128 CLUSTER SYSTEM

Hugh C. Harris[†]

U. S. Naval Observatory

Gretchen L. H. Harris[†]

University of Waterloo

James E. Hesser[†]

Dominion Astrophysical Observatory
Hertzberg Institute of Physics

ABSTRACT: We review what has been learned about the globular clusters in NGC 5128 (Cen A). Two independent programs (our work at CTIO and Sharples' work at the AAT) have confirmed 87 objects as clusters in NGC 5128 from their velocities. We discuss the colors of the clusters, which are indicative of a wide range of metallicities, their velocities, and their origin as genuine globular clusters in this elliptical galaxy.

1. INTRODUCTION

The globular cluster systems in our Galaxy and M31 are well studied, and provide much of the data for the discussions at this conference. The information acquired about cluster systems in other spiral galaxies is more fragmentary because of their much larger distances and the greater observational difficulties for such systems. Because NGC 5128 is the closest large elliptical galaxy, it can, in some respects, play the same role for studying the clusters in elliptical galaxies that our Galaxy and M31 play for the clusters in spirals. We are fortunate that NGC 5128 is at a distance where three techniques for finding clusters all have some effectiveness: (1) the largest and brightest clusters have barely nonstellar images on plates or CCD frames taken in good seeing, (2) the galaxy's radial velocity ($\sim$550 km s^{-1}) allows velocities to be used to confirm membership of clusters in NGC 5128, and (3) star counts give statistical information about the total population of clusters and their distribution in the galaxy. No *single* technique is as effective as might be desired: at closer distances the third technique breaks down and the second only discriminates against background galaxies, while at larger distances the first technique is hopeless for ground-based observers and the second becomes very difficult. The combination, however, is effective.

[†]Visiting astronomer, Cerro Tololo Inter-American Observatory, National Optical Astronomy Observatories, operated by AURA, Inc., under contract with the National Science Foundation.

This paper will concentrate on what has been learned about the NGC 5128 cluster system based primarily on two research programs. We have been pursuing cluster identification from Cerro Tololo Inter-American Observatory for six years using several approaches. Recently, Sharples has achieved excellent results using the multi-aperture fiber-fed spectrograph on the AAT to obtain spectra and velocities of more than 40 objects at a time. Half of the clusters have been confirmed during the 1986 observing season, so most of the present analysis is preliminary. Previous work has been reviewed at IAU Colloquium 68 (Hesser et al. 1981) and in the Astrophysical Journal (Hesser et al. 1984; G.Harris et al. 1984; H.Harris et al. 1984; Hesser et al. 1986).

2. IDENTIFICATION OF CLUSTERS

There are now 87 confirmed clusters in NGC 5128. Many were flagged as candidates by their nonstellar appearance on CTIO prime-focus plates during exhaustive visual searches of many plates by us, following the lead of Graham and Phillips (1980) and van den Bergh et al. (1981). Some, although missed in the visual searches, were flagged as nonstellar during an analysis of scans of the same plates with the COSMOS microdensitometer at Edinburgh (H.Harris et al. 1984). Radial velocities obtained at CTIO have confirmed membership in NGC 5128 for 47 clusters identified by their nonstellar character. Three more clusters were confirmed through velocities measured at CTIO of a random sample of objects projected near NGC 5128. Finally, 44 clusters (7 in common with the CTIO studies) were confirmed from their velocities by Sharples (1987) at the AAT as part of observations of a sample of 227 objects projected near NGC 5128. The nonstellar character of most of these clusters is *quite* subtle. However, a combination of careful visual inspection and COSMOS or PDS image analysis has proven very successful at finding clusters: six of the seven candidates published by Hesser et al. (1984) have now proven to be clusters, and fifteen of the twenty top candidates observed at CTIO this year were found to be clusters.

A histogram of the velocitites of all objects observed in NGC 5128 shows a large peak near 0 km s^{-1} (with a range from -192 to $+221$ km s^{-1}) and a broad distribution from 282 to 860 km s^{-1}. The bimodal nature of the distribution has allowed the decision about membership in NGC 5128 to be made without ambiguity for most objects. It is possible for foreground stars in our Galaxy's halo to have high velocities overlapping with the distribution of NGC 5128 velocities. However, such stars are sufficiently rare that probably none have been observed and misclassified. No background galaxies have been found in the random samples (although Sharples' sample excludes objects with very extended images). A few galaxies have been observed at CTIO among the cluster candidates with extended images, but their high velocities (of order 10,000 km s^{-1}) cause no confusion. Therefore, velocity discrimination is indeed very effective for assigning membership in NGC 5128. Another possible worry is that high luminosity stars in the spheroid of NGC 5128 have been observed and misclassified as clusters. This cannot be the case for the objects which were found by their extended images (47 objects out of 87). However, for objects found in the random samples, this sort of misclassification may be occurring occasionally. It is probably not happening often because the colors of clusters found in the random samples are all intermediate, while a magnitude-limited sample of supergiants might be expected to have a wider range of colors.

3. THE CLUSTER LUMINOSITY FUNCTION

The numbers of *fainter* clusters have been estimated from star counts (both visual and automated) and are described by G.Harris et al. (1984). Combining star-count results with the individual confirmed clusters at the bright end gives the total cluster luminosity function shown in Figure 1. The figure includes all 77 confirmed clusters with projected distances from NGC 5128 of $1.6 \leq R \leq 16.1$ arcmin, in order to compare with the star counts in the same region. Also shown is a gaussian function with a peak $M_V = -7.1$ and $\sigma_V = 1.35$, similar to the luminosity functions found for cluster systems in the Galaxy and M31 by van den Bergh (1985). Assuming that the clusters in NGC 5128 obey the same function, then the best fit is obtained with a distance of 3.2 Mpc (with $E(B-V) = 0.1$) and a total of 900 clusters within the radial limits. (The total at *all* radii is then 1500, assuming that the clusters follow the halo light.) The distance could range from 2.5 Mpc to 4.0 Mpc and still remain marginally consistent with the data and these assumptions. The agreement between the data and the fitted curve is very satisfactory, the confirmed clusters being noticeably (but not surprisingly) incomplete for $V \geq 18$. Whether the clusters in NGC 5128 do, in fact, obey the adopted gaussian function can best be decided after measuring the distance to NGC 5128 independently, although this measurement is difficult to make with the required accuracy of $\leq 20\%$. A distance near 3 Mpc has recently been suggested by the relatively low velocity dispersion measured for the bulge light (Wilkinson et al. 1986), and a small distance is implied by the magnitude of Supernova 1986G currently being analyzed by several groups.

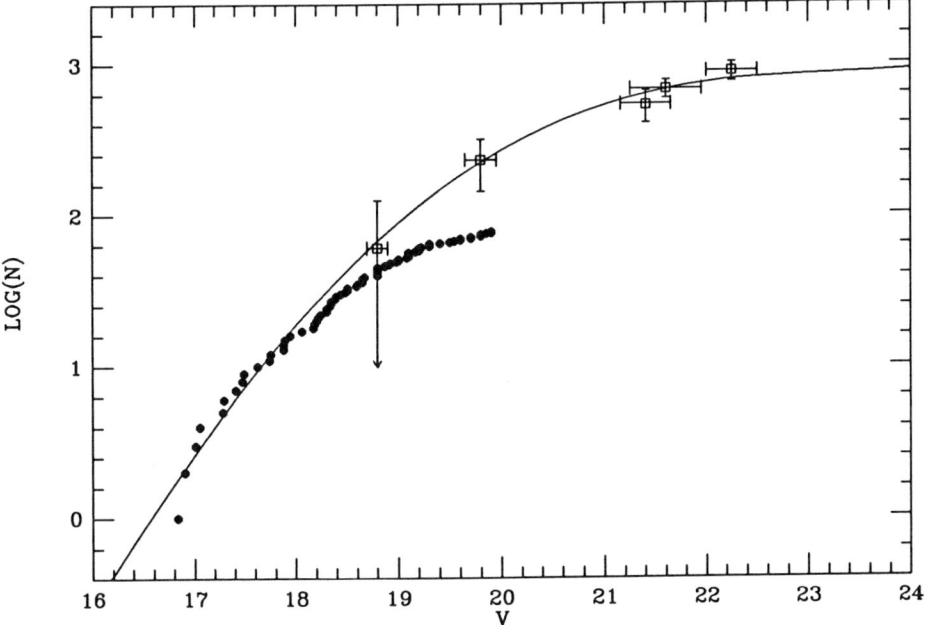

Fig. 1. Cumulative luminosity function of clusters with $1.6 \leq R \leq 16.1$ arcmin. Confirmed clusters are shown by filled circles. Results from image counts are shown by open squares. The line is a gaussian fit to the data (see text).

Alternatively, if the distance is as large as 5 Mpc (a value often used in the literature), then the clusters must have a luminosity function with some combination of a brighter peak, a larger dispersion in the luminosity function, and/or a non-gaussian distribution.

Are many bright, compact clusters being missed in our searches, particularly in view of the tendency for the inner clusters to be more compact (Hesser et al. 1984), or are the star counts in error, giving a misleading luminosity function in Figure 1? Four arguments suggest not. (Some of this evidence is discussed more fully by H.Harris et al. 1984.) (1) The repeated recovery of extended objects during several independent searches is good enough that we think we have found most of the bright, extended objects. (2) Most of the candidates that we have classified as only very marginally extended have turned out to have low velocities indicative of their being foreground stars. (3) The size distribution of objects with $V \leq 18$ measured in the COSMOS analysis did not show the clump of very marginally extended objects near the search limit that we would expect to see if many compact clusters were being missed. (4) Few clusters have been found from velocities of random objects near NGC 5128. (The selection of objects observed at CTIO has excluded extremely blue and red objects, so is actually not totally random. See Hesser et al. 1984.) Only 3 clusters were found among 32 random objects observed at CTIO having $16.0 \leq V \leq 18.5$ and projected in two fields just northeast and southwest of the dust lane, and one resolved cluster lies in these fields. In Figure 2, the left panel shows the color-magnitude diagram of these fields. At a radius of 5 arcmin from NGC 5128, the star counts show 2.3 objects arcmin^{-2} with $16.0 \leq V \leq 18.5$, and for the distance and cluster population found in the preceding paragraph, we expect 2.7 clusters arcmin^{-2} at this radius, of which 4.5% should be within the magnitude range. Therefore, we expect 4 clusters within this magnitude range in the two fields with random objects observed at CTIO, and 4 clusters are found. The star counts have now been strengthened by Sharples' data.

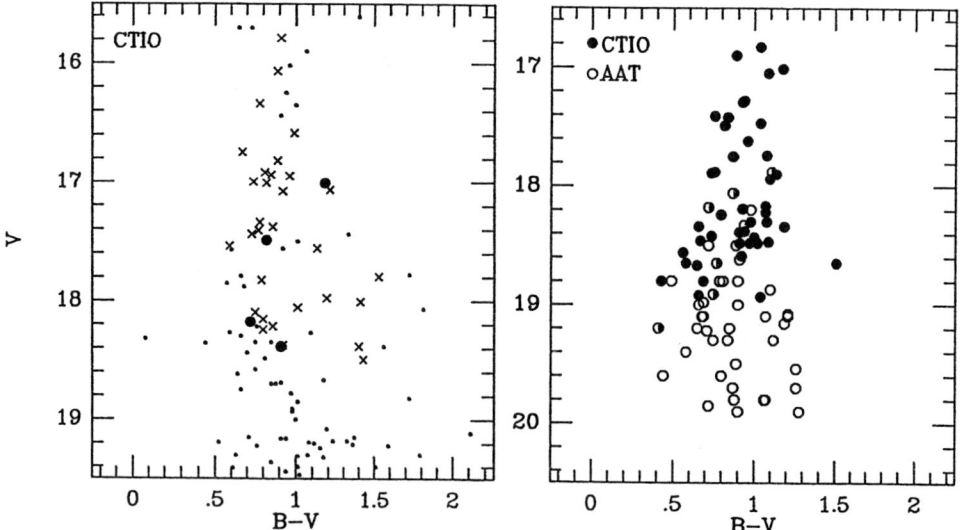

Fig. 2.   Left panel: objects in two fields observed at CTIO (filled circles are clusters, crosses are stars). Right panel: all confirmed clusters.

Only 29 clusters were found among 194 objects with $18.0 \leq V \leq 19.5$ observed at the AAT, and only one cluster brighter than 18.5 was found that had not already been identified and confirmed at CTIO from its nonstellar appearance. The star counts show 2.7 objects arcmin^{-2} with $18.0 \leq V \leq 19.5$, and we expect 15% of all clusters to have these magnitudes. Therefore, we expect 33 of the 194 AAT objects to be clusters. The consistency between the numbers of clusters expected and found adds confidence that the observed luminosity function in Figure 1 is not too far in error. Therefore, the luminosity distribution of the 87 confirmed clusters provides not only a lower limit, but also probably a very good approximation, to the true luminosity function at the bright end (brighter than $V \sim 18$).

4. CLUSTER COLORS

The right panel of Figure 2 shows the color-magnitude diagram of all confirmed clusters in NGC 5128. They have measured B−V colors between 0.41 and 1.28. The one exception is Cluster 42, lying 6 arcminutes west of the center of NGC 5128, with B−V = 1.51. It may lie behind part of the dust lane. A few other confirmed clusters may also be reddened by dust in NGC 5128, but most of the clusters lie well away from most detectable dust (Pennington 1984), so the distribution of cluster colors is probably not affected much by internal reddening. All clusters have a foreground reddening of E(B−V) $\approx$ 0.10 (van den Bergh 1976), and this amount has been subtracted from the observed colors in the following discussion. The NGC 5128 clusters appear to cover a larger range of colors than do the globular clusters in our Galaxy. The two distributions are shown in Figure 3, where the observed colors for Galactic globulars have been taken from Peterson (1986) and reddenings have been taken from Webbink (1985). The same result is found using colors for Galactic globulars from Reed (1985).

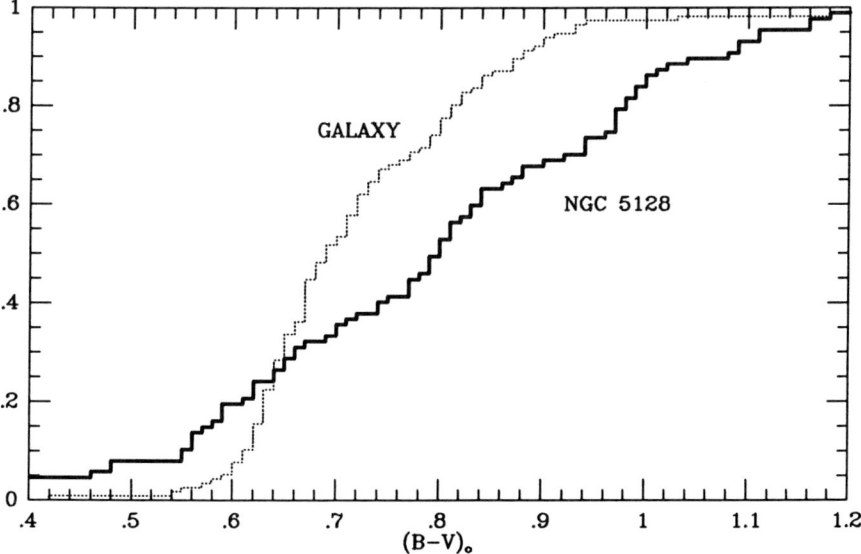

Fig. 3. Cumulative distribution of colors, corrected for reddening, for clusters in NGC 5128 compared with our Galaxy.

The blue end of the color distribution in NGC 5128 is similar to that in our Galaxy with the exception of four clusters in NGC 5128 with $0.40 \leq B-V \leq 0.50$. Cluster 46 = G331 has been identified as a relatively blue cluster independently by Sharples and by us, and it appears nonstellar, so it, at least, is a real blue cluster. They *may* be younger than true globulars. They all lie at projected radii ≤ 10 arcmin, so they could have an origin associated with the dust lane. However, none lie very close to the dust lane – they are not the blue objects noted by van den Bergh (1976) along the dust lane – nor are they the blue knots identified along the optical jet by Blanco et al. (1975), although some do lie in the region of the optical jet. They are not as blue as many clusters in the LMC and M33 (van den Bergh 1981; Christian 1987). Clusters in the LMC with similar B−V colors have ages of 3×10^8 to 3×10^9 years (Hodge 1983; Mateo 1987). However, a comparison with LMC clusters should take into account possible metallicity differences; the present abundances of HII regions in NGC 5128 are metal-rich but uncertain (Dufour et al. 1979; Phillips 1981), suggesting ages younger than LMC clusters with the same $(B-V)_o$. Therefore, intermediate ages near 3×10^8 years are plausible for these relatively blue clusters.

The clusters in NGC 5128 extend considerably redder than those in the Galaxy: more than 30% of the clusters are redder than $(B-V)_o = 0.90$, while in the Galaxy only 10% are so red. There are two reasons for believing that these clusters (with possibly a couple of exceptions) do not have unusually high reddening: their infrared colors (Frogel 1984) indicate a cool, metal-rich population, rather than a highly reddened population, and several of these red clusters lie well out in the halo where it is unlikely that they are affected by internal dust. We are attempting to measure the line strengths of our CTIO spectra to investigate this question further. Many of the blue SIT spectra are noisy enough that they do not give meaningful results, and the red CCD spectra of the Ca triplet lines, while having better signal-to-noise, do not show enough sensitivity in the Ca lines to be useful. However, the better spectra show a correlation between the line strengths and B−V colors, with the redder clusters at least as strong-lined as metal-rich Galactic globulars.

Comparing the color distribution with other galaxies would be desirable. Two possibilities are M31 and M87, but comparisons do not appear to be meaningful at present. The colors of Class A and B clusters in the southeast half of M31 (where reddening within M31 is a minimum) taken from Battistini et al. (1986) show a distribution even more extended than in NGC 5128. It is likely that some younger disk clusters, reddened clusters, and background galaxies are being included in the M31 sample. Similarly, the B−R colors of clusters in M87 in Zones 1, 2, and 3 from Strom et al. (1981), transformed to B−V, also show a distribution more extended than in NGC 5128. Spectroscopy by Mould et al. (1986) shows that the bluest and reddest objects are generally not clusters, but more data are needed to define the true color distribution of M87 clusters.

The JHK colors measured and discussed by Frogel (1984) provide strong evidence that the red clusters are indeed quite metal rich. The red (B−V) and (V−K) colors imply high metal abundance, while the lack of very red (J−K) colors indicates a lack of carbon stars in the clusters and hence an old age ($\geq 10^{10}$ yr). It appears that the distribution of metallicities of clusters in NGC 5128 is similar at the metal-poor end to clusters in the Galaxy, but includes a higher fraction of metal-rich clusters extending to higher metallicities than in the Galaxy.

Earlier work gave the tentative suggestion that a gradient in the cluster colors might exist (Hesser et al. 1984), implying a metallicity gradient in the cluster system. Figure 4 shows that no gradient is visible with the more complete

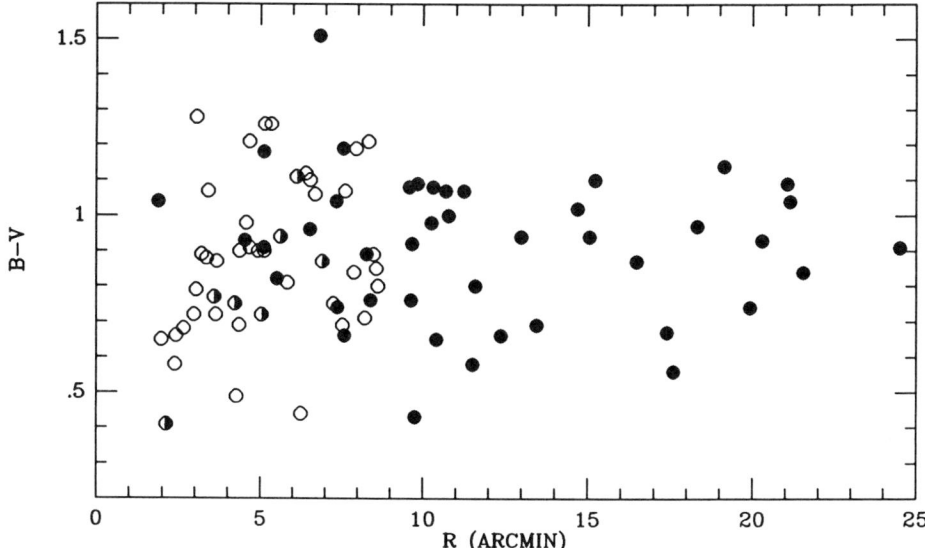

Fig. 4. Colors of confirmed clusters vs. projected distance from the center of NGC 5128. Filled circles are clusters observed at CTIO, and open circles at AAT.

data now available. Possibly a few points in Figure 4 are biasing this conclusion because of either the presence of younger, bluer clusters at small radii or the higher likelihood of internal reddening at small radii. However, with the identification of clusters with $(B-V)_o \sim 1.0$ at projected radii of 15 to 20 arcmin, it appears unlikely that any gradient other than a very weak one can exist in the halo of NGC 5128.

5. CLUSTER VELOCITIES

Globular cluster velocities provide an excellent probe of the dynamics of the halo of galaxies. Hesser et al. (1986) suggested that the outer clusters in NGC 5128 (projected more than about 12 arcminutes from the center) reflected the potential of the unperturbed elliptical galaxy, but that the inner clusters showed effects from the interaction or merger which is thought to have produced the dust lane. In the year since that paper was prepared, independent data have become available from our CCD red spectra taken at CTIO and from Sharples' spectra taken at the AAT. These data sets provide necessary checks against errors in velocities (systematic, or random due to low signal-to-noise) from the earlier SIT spectra, as well as more than doubling the number of clusters with measured velocities.

The present data are shown in Figure 5. They have not yet been analyzed for significant trends, but some preliminary results can be indicated. The surprisingly large mean velocity of the inner clusters reported by Hesser et al. (1986) is not supported by the much larger sample of (primarily fainter) inner clusters. However, the possible rotation of the cluster system, both about the minor axis (NE approaching) and about the major axis (SE approaching) is still marginally evident. The mean velocity of the clusters on the southeast side of the major axis is smaller than the mean on the northwest side by 60 km s^{-1}. This difference

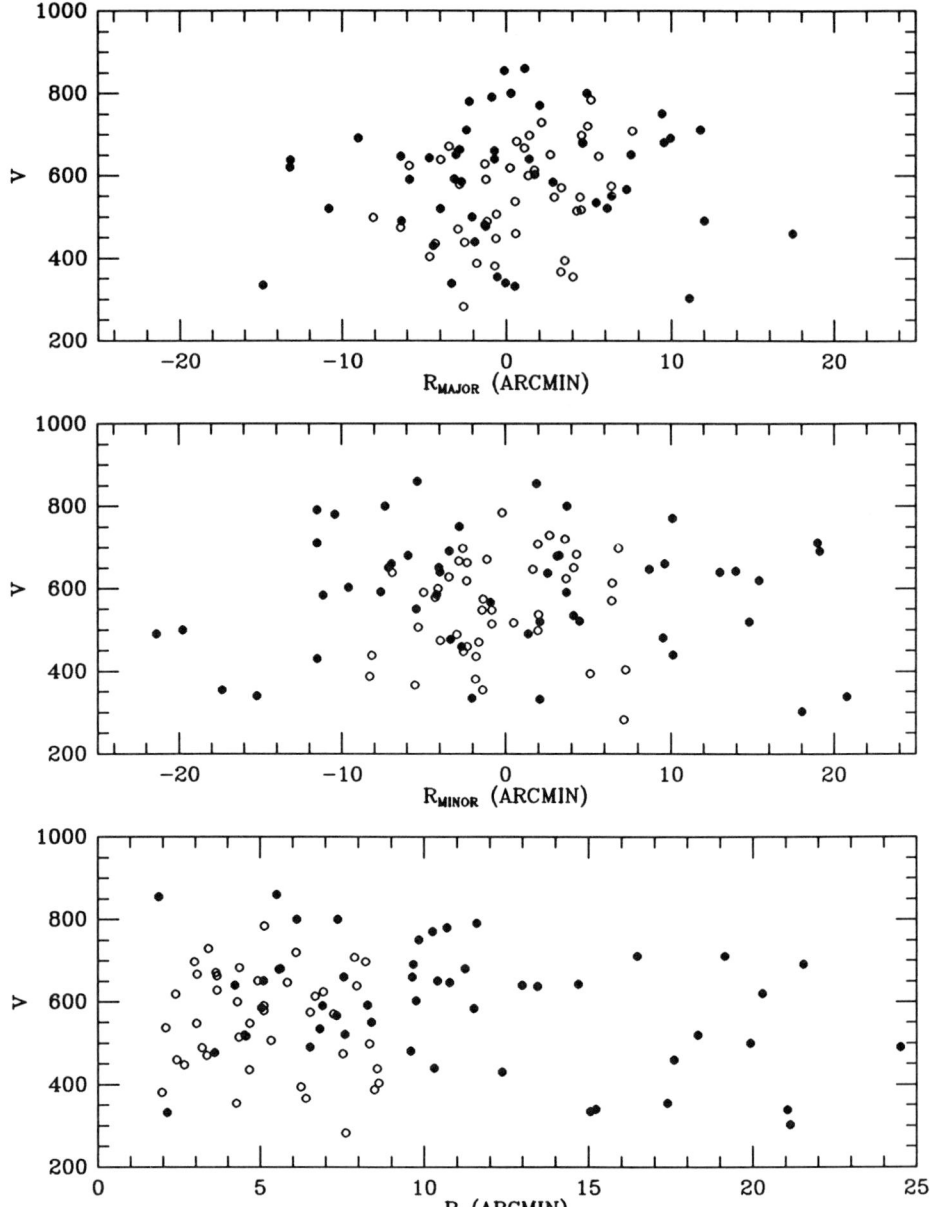

Fig. 5. Radial velocities of all confirmed clusters from the CTIO sample (filled circles) and from the AAT sample (open circles). The top panel shows velocities vs. projected distance from the major axis of NGC 5128 (with SE toward the left and NW toward the right), the middle panel is vs. distance from the minor axis (NE to the left, SW to the right), and the bottom panel is vs. radial distance.

is probably significant, but is certainly much less than the amplitude of the disk rotation curve (approximately 600 km s^{-1}). A solution for systematic rotation is needed to access its reality, and will be done.

The velocity dispersion of the clusters is nearly independent of radius (Figure 5): for radii less than 5 arcmin and radii greater than 10 arcmin, the observed dispersions are 130 ± 20 and 154 ± 20 km s^{-1}, respectively. Using the projected mass estimator (Heisler et al. 1985) with $f_{PM}=48/\pi$, the mass is 7×10^{11} M$_\odot$ for a distance of 3.2 Mpc (scaling as D/3.2), the magnitude is V=5.9 (van den Bergh 1976), and the mass/light ratio is 17 M$_\odot$/L$_\odot$ (scaling as 3.2/D) within a radius of ~20 kpc.

6. ORIGIN OF THE CLUSTERS

In comparing the clusters in NGC 5128 with globular clusters in other galaxies, or in using the clusters to estimate the distance or mass of NGC 5128, for example, we are implicitly assuming that the clusters are, in fact, genuine globular clusters representative of a normal, large elliptical galaxy. Arguments for this assumption are of two types: first, consistency arguments that the properties (colors, magnitudes, spectral types, sizes, spatial distribution, etc.) of the clusters are generally what are expected for a normal globular cluster system; and second, difficulties in understanding the origin of the cluster system in ways other than formation early in the history of a large elliptical. Arguments of the first type have been presented as part of several papers (Hesser et al. 1984; H.Harris 1984; this paper).

Within the framework of the interaction picture for NGC 5128, with a relatively low mass, gas-rich galaxy falling into NGC 5128 some 3×10^8 years ago (Tubbs 1980), not more than a small fraction of the clusters we observe can have come from the intruding galaxy, because of the low halo mass expected for such an intruder and the lack of disk-like rotation seen either in the bulge light of NGC 5128 or in the cluster system. Nor can more than a small fraction have been formed during the interaction, because of their red colors and their spatial distribution away from the dust lane. The strongest arguments that some of the clusters are *not* normal globulars may be that more clusters tend to be found in the northeast and southwest quadrants of the inner bulge (Sharples 1987), that the cluster system may share some small amount of disk-like rotation, and that some clusters are bluer than Galactic globulars with ages consistent with age estimates for the dust lane. However, even these arguments would suggest that not more than a small fraction of the confirmed clusters are other than normal globulars. The multiple shells seen at large radii in NGC 5128 suggest that one or more interactions may have occurred longer ago than 3×10^8 years (Malin et al. 1983), events presumably unrelated to the present dust lane. It is more difficult to argue for or against a connection between such early interactions and the properties of the clusters. However, the overall evidence indicates to us that the large majority of clusters presently being studied probably do, indeed, represent globular clusters in a normal elliptical galaxy.

ACKNOWLEDGEMENTS

We most grateful to Dr. R. Sharples for sharing his results in advance of publication.

REFERENCES

Battistini, P., Bonoli, F., Braccesi, A., Federici, L., Fusi Pecci, F. and Marano, B. 1986 preprint.
Blanco, V. M., Graham, J. A., Lasker, B. M. and Osmer, P. 1975 Astrophys. J. Letters 198, L63.
Christian, C. A. 1987 in Globular Cluster Systems in Galaxies J. E. Grindlay and A. G. D. Philip, eds., Reidel, Dordrecht, p. 187
Dufour, R. J., van den Bergh, S., Harvel, C. A., Martins, D. H., Schiffer, F. H., Talbot, R. J., Talent, D. L. and Wells, D. C. 1979 Astron. J. 84, 284.
Frogel, J. A. 1984 Astrophys. J. 278, 119.
Graham, J. A. and Phillips, M. M. 1980 Astrophys. J. Letters 239, L97.
Harris, G. L. H., Hesser, J. E., Harris, H. C. and Curry, P. J. 1984 Astrophys. J. 287, 175.
Harris, H. C., Harris, G. L. H., Hesser, J. E. and MacGillivray, H. T. 1984 Astrophys. J. 287, 185.
Heisler, J., Tremaine, S. and Bahcall, J. N. 1985 Astrophys. J. 298, 8.
Hesser, J. E., Harris, H. C., van den Bergh, S. and Harris, G. L. H. 1981 in Astrophysical Parameters for Globular Clusters A. G. D. Philip and D. S. Hayes, eds., L. Davis Press, Schenectady, p. 467.
Hesser, J. E., Harris, H. C., van den Bergh, S. and Harris, G. L. H. 1984 Astrophys. J. 276, 491.
Hesser, J. E., Harris, H. C. and Harris, G. L. H. 1986 Astrophys. J. Letters 303, L51.
Hodge, P. W. 1983 Astrophys. J. 264, 470.
Malin, D. F., Quinn, P. J. and Graham, J. A. 1983 Astrophys. J. Letters 272, L5.
Mateo, M. 1987 in Globular Cluster Systems in Galaxies J. E. Grindlay and A. G. D. Philip, eds., Reidel, Dordrecht, p. 557.
Mould, J. R., Oke, J. B. and Nemec, J. M. 1986, preprint.
Pennington, R. L. 1984 A Surface Photometric Study of NGC 5128, Rice University (Ph.D. Thesis).
Peterson, C. J. 1986 Publ. Astron. Soc. Pacific 98, 192.
Phillips, M. M. 1981 Monthly Notices Roy. Astron. Soc. 197, 659.
Reed, B. C. 1985 Publ. Astron. Soc. Pacific 97, 120.
Sharples, R. M. 1987 in Globular Cluster Systems in Galaxies J. E. Grindlay and A. G. D. Philip, eds., Reidel, Dordrecht, p. 545.
Strom, S. E., Forte, J. C., Harris, W. E., Strom, K. M., Wells, D. C. and Smith, M. G. 1981 Astrophys. J. 245, 416.
van den Bergh, S. 1976 Astrophys. J. 208, 673.
van den Bergh, S. 1981 Astron. Astrophys. Suppl. 46, 79.
van den Bergh, S. 1985 Astrophys. J. 297, 361.
van den Bergh, S., Hesser, J. E. and Harris, G. L. H. 1981 Astron. J. 86, 24.
Webbink, R. F. 1985 in IAU Symposium No. 113, Dynamics of Star Clusters, J. Goodman and P. Hut, eds., Reidel, Dordrecht, p. 541.
Wilkinson, A., Sharples, R. M., Fosbury, R. A. E. and Wallace, P. T. 1986 Monthly Notices Roy. Astron. Soc. 218, 297.

DISCUSSION

GRAHAM: Did the new CCD velocities based on the λ 8500 Ca II lines agree with the earlier SIT velocities?

HARRIS, H: For most objects the agreement was good (within the estimated errors). For a couple of objects the CCD velocities differed by more than expected, apparently due to a tail on the error distribution of SIT velocities.

HESSER: A high percentage of the most discrepant velocities appear to be due to velocities from single, relatively short exposures from the SIT vidicon. Velocities from multiple SIT exposures appear to agree much better with our CCD or Sharples' IPCS velocities.

RACINE: How many candidates were identified on the basis of their non stellar appearance and for what fraction of those do you have spectroscopic confirmations?

HARRIS, H: Approximately 75% of the non stellar cluster candidates that we have observed spectroscopically have proven to be clusters. Forty-seven clusters have been confirmed this way.

GNEDIN: Your last slide shows "Cluster mag. being formed by the jet". What do you mean by this? Is there a real connection between clusters and jets?

HARRIS, H.: I included this possibility as a (quite speculative) mechanism to explain Sharples' finding more clusters in the NE and SW quadrants of the inner halo.

GRAHAM: Some blue clusters are being formed out in the NE radio lobe. One is on the S end of knot A (cf Blanco et al. 1975). Another is knot D identified in that paper. Both are slightly non stellar, blue continuum objects without strong emission lines in their spectrum.

HUCHRA: Why did you pick such a large value of the projection factor ($48/\pi$) in the projected mass formula of Bahcall and Tremaine? In the outskirts of the galaxy, the cluster orbits are more likely to be circular. Bahcall and Tremaine in that, suggest using a factor of $24/\pi$.

HARRIS, H.: A small factor may be appropriate. We have no information on the orbits of the clusters in NGC 5128.

HANES: Some decades ago Shklovsky commented that he saw more globular clusters in the quadrant of the jet in M 87 than in the other three quadrants. Later (deeper) analysis seems to have ruled that out, but it is provocative that you seem to be finding similar effects here.

DI FAZIO: Did you notice any correlation between the shape of the projected distribution of Cen A clusters and the axis of the jet or of the radio lobes? If yes, can this be due to the relatively small number of objects that you considered?

HARRIS, H.: Our CTIO sample of clusters has a slightly elliptical distribution oriented with the outer isophotes of NGC 5128. The combined CTIO and AAT samples may be elliptical, but we have not yet estimated the statistical significance.

DWARF SPHEROIDAL GALAXIES AND GLOBULAR CLUSTERS

G. S. Da Costa

Yale University Observatory

ABSTRACT: Recent observational results for dwarf spheroidal galaxies are reviewed and discussed. In particular, the differences in stellar populations between dwarf spheroidal galaxies and globular clusters are highlighted. It seems most probable that the origin and evolution of dwarf spheroidal galaxies was very different from that of globular clusters.

1. INTRODUCTION

In 1938 Shapley (1938a) announced the discovery of "A Stellar System of a New Type" in the constellation of Sculptor. The collection of faint images just visible at the limit of a 3 hour Bruce telescope plate was at first thought to be an extended cluster of galaxies, but subsequent 60-inch plates revealed the individual member objects to be faint stars, not galaxies. This system was novel in that it had the smooth density profile of a low central concentration globular cluster, yet if its brightest stars were assumed to have the same absolute magnitude as those in globular clusters, the Sculptor system had to be much larger and much more distant than any galactic globular cluster known at that time. Of course, this system is now known as the Sculptor dwarf spheroidal (dSph) galaxy.

Some months later Shapley (1938b) reported the existence of a second such system in the constellation of Fornax. However, an examination of over 150 small scale plates taken in South Africa, which covered more than 15,000 sq. degrees of sky at galactic latitudes above 20°, revealed no additional systems (Shapley 1939). In fact, discovery of additional systems had to await the Palomar survey of the northern sky; two systems in Leo were announced by Harrington and Wilson (1950) and a further two, Draco and Ursa Minor, were listed by Wilson (1955). A single additional southern object, Carina, was added by the SRC southern sky survey (Cannon, Hawarden and Tritton 1977) and except for the obscured regions at low galactic latitudes, the census of dSph galaxies associated with the Galaxy is considered complete. Three dSph galaxies associated with M31 are also known (van den Bergh 1972a,b; 1974).

In his original article Shapley (1938a) argued that the most appropriate description for the Sculptor system was as "a super-cluster of the globular type" and this view, that dSph galaxies are simply very low density analogues of globular clusters, has prevailed for many years. However, in recent years evidence has accumulated which indicates that this view is too simple, and that in fact, dSph galaxies and globular clusters have as many differences as they do similarities. In this review then, it is these differences that will be the focus.

2. VARIABLE STARS

The variable star content of dSph galaxies has been reviewed recently by Zinn (1980, 1985) and so only new results will be discussed here.

2.1 RR Lyrae Variables

Large samples of RR Lyrae variables have been discovered in the Draco, Sculptor, Ursa Minor and Leo II galaxies (Baade and Swope 1961; van Agt 1967, 1973, 1978; Swope 1967) which until recently was all the dSph galaxies in which adequate searches have been made. To this list we can now add Carina since Saha, Monet and Seitzer (1986) have just published detailed photometry for 53 variables, mostly RR Lyrae stars, that are believed to be members of this dSph galaxy. As has been found for Draco and Leo II (see for example Zinn 1980), the mean period of the ab-type variables in Carina (0.62 days) is intermediate between Oosterhoff type I ($<P_{ab}> = 0.55d$) and type II ($<P_{ab}> = 0.65d$), though as noted by Sandage (1982), the Oosterhoff types for galactic globular clusters in reality form a "bimodal continuum" rather than a strict dichotomy.

In this context it is worth mentioning the work of Nemec (1985) who, by analyzing the photographic photometry of Baade and Swope (1961), identified and studied 10 double mode RR Lyrae variables in Draco. From their period ratios, he found that nine of these stars have masses near 0.65 M_o and were indistinguishable from the double mode variables in the Oosterhoff type II cluster M15. The tenth star however, has a derived mass of 0.55 M_o and is indistinguishable from the double mode variables in the type I cluster M3. In the period-amplitude diagram, this star also shows a shift from the fiducial M3 relation that is smaller than that of the other double mode stars by an amount equal to, within the uncertainties, the period shift between M3 and M15. Thus, in addition to being 0.1 M_o less massive, this star probably has an abundance larger than the mean of the others by about 0.5 dex, the difference in [Fe/H] between M3 and M15. Indeed both the period-amplitude diagram and the period-rise time diagram for the entire sample of variables (Nemec 1985) offer strong support for the existence of an abundance range in Draco, a result that will be discussed further below.

2.2 Anomalous Cepheids

The first indication of a difference in stellar populations between globular clusters in dSph galaxies came with the recognition of a new class of variable stars in Draco (Baade and Swope 1961). These stars are known as anomalous Cepheids because they fail to obey the period-luminosity relations for either Type I or Type II Cepheids. Anomalous Cepheids are relatively much more common in dSphs than in globular clusters; at least 25 such stars are known in the Draco, Ursa Minor, Sculptor, Leo I and Leo II systems (Zinn and Searle 1976, van Agt 1967, Swope 1968, Kunkel and Demers 1977, Hodge and Wright 1978), whereas the globular clusters contain but a single example, V19 in NGC 5466 (Zinn and Dahn 1976). Anomalous Cepheids are also found in the Small Magellanic Cloud but not in the LMC.

Both theoretical (Demarque and Hirshfeld 1975, Hirshfeld 1980) and observational analyses (Norris and Zinn 1975, Zinn and Searle 1976, Zinn and King 1982, Smith and Stryker 1986) indicate that these variables have masses in the range of 1 - 2 $M_\odot$. Further, in order that they evolve into the instability strip, they are also required to be quite metal-poor ([Fe/H] < -1.3 approximately). Explanations for the larger masses are that either the stars result from a population of 1-3 billion year old stars, or they are the result of mass transfer in a binary system of old stars (Renzini, Mengel and Sweigart 1977). At present there seems no obvious way to select between these competing hypotheses for the dSph anomalous Cepheids. However, the discovery of a large number of blue stragglers in NGC 5466 (Nemec and Harris 1986) suggests the binary mass transfer hypotheses is appropriate for NGC 5466 V19 while, following the discussion of Smith and Stryker (1986), the "young metal-poor stars" explanation seems reasonable for the SMC variables.

An anomalous Cepheid has now been discovered in Fornax also by Light, Armandroff and Zinn (1986). Though this star lies near one of the Fornax globular clusters, it is most likely a member of the field population of this galaxy. Undoubtedly more extensive searches will reveal additional such stars. This leaves Carina as the one remaining dSph in which no anomalous cepheids have yet been identified. However, given that every other dSph contains at least one, it seems highly unlikely that Carina would lack them. Indeed it is quite possible that such stars have already been identified. The variable star survey of Saha et al. (1986) contains 8 short period variables that are 0.5 to 1.5 mag brighter than the Carina RR Lyraes. If these stars are assumed to be halo field variables, then their density is 2 orders of magnitude greater than that expected (Saha et al. 1986). However, despite this discrepancy, Saha et al. did not explore in any detail the possibility that these stars are Carina anomalous Cepheids. In Fig. 1a, the data for these stars, taken from Saha et al. (1986), are plotted in the (log P, $<M_B>$) plane on the assumption that they are members of Carina, along with equivalent data for the anomalous Cepheids in other dSph galaxies taken from the sources cited above. Fig. 1b shows the same stars in the (log P, A_B) plane. At least 4 of the Carina variables (nos. 1, 29, 33 and 129)

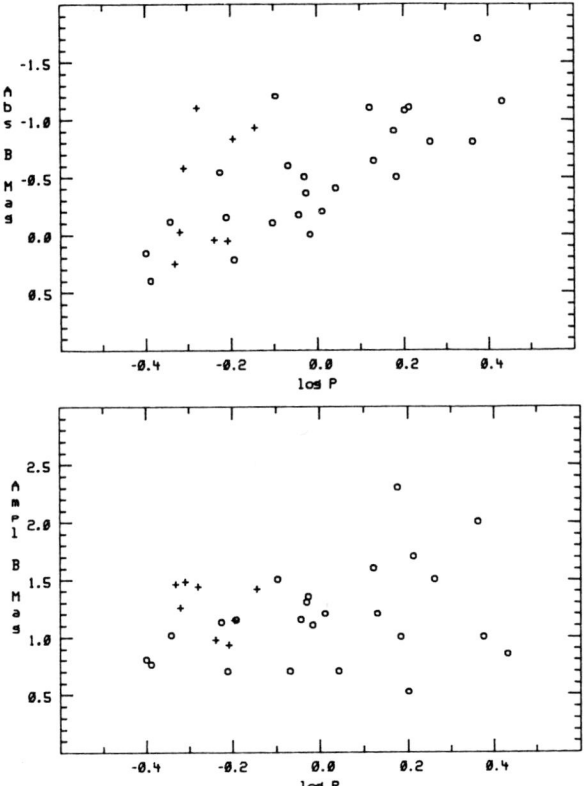

Fig. 1. (a) Upper panel: Period-Luminosity relation for all known anomalous cepheids in dSph galaxies (open symbols) and for the "bright" RR Lyraes in the Carina field (plus signs) plotted as if they are members of the Carina system. (b) lower panel: Period-Amplitude diagram. The symbols are the same as for Fig. 1a.

fall within the scatter outlined by the known anomalous cepheids in both diagrams and thus deserve to be considered as likely candidates for anomalous Cepheids in this dSph galaxy. In this context it is also worth noting that since the plates used by Saha et al. (1986) were taken in a single 4 night run, their probability of defining light curves and periods for variables with periods longer than about 0.9d is rather low. Hence it is quite possible that additional longer period anomalous cepheids are also present.

2.3 Type II cepheids

The paper of Light et al. (1986) also reports the discovery of the first Type II Cepheids in a dSph galaxy. Such stars are found only in galactic globular clusters with blue horizontal branches (Harris 1985) so it is reassuring that the variable discussed by Light et al. appears to be a member of Fornax globular cluster 2, which Buonanno et al. (1985) have shown to have a blue horizontal branch. The lack of such stars in the other dSph systems is consistent with their generally red horizontal branches except of course for Ursa Minor. In this situation it is possible to argue, as is done for the globulars with blue horizontal branches that also lack Type II Cepheids, that the observed absence of such stars is due to statistical fluctuations caused by the short evolutionary lifetime of such stars.

3. ABUNDANCE VARIATIONS

With the well established exception of ω Centauri and the possible exception of M22, all galactic globular clusters are chemically homogeneous with respect to heavy element abundances. The available results for dSph galaxies however, suggest that in general they are not chemically homogeneous, possessing instead substantial abundance ranges.

Evidence to suggest abundance ranges is usually found from one of two distinct methods which, given the controversy that has at times enveloped this subject, are worth describing here. The simplest method seeks to test whether the observed color widths of principal features in c-m diagrams are greater than that expected from the photometric errors alone. In principal this method has the advantage of allowing use of a large sample of stars, but in practice its application is complicated by: (a) uncertain knowledge of the true photometric errors, (b) the probable inclusion of AGB stars if the sample being investigated is at magnitudes above that of the horizontal branch, and (c) by the relatively low sensitivity of broad band colors to abundance variations at low abundances. While (b) can be corrected by using stars fainter than the horizontal branch, effects (a) and (c) remain.

The second method is spectraphotometry of individual red giant stars in the dSph galaxy. Here the principal difficulty, aside from those inherent in measuring features in low S/N spectra of cool luminous stars, is simply gathering a sufficiently large sample of stars for meaningful analysis. This method should yield more precise abundance measures, especially for low abundances, than photometry. A variant of this method is the application of the ΔS method to the variables in dSph galaxies. While this technique has seen little use because of the faintness of the stars involved, it is known to give reliable results for galactic globular clusters (Smith 1984) and the coming new generation of large aperture telescopes may make it practical for dSph stars. At present, this kind of analysis has been employed only for anomalous Cepheids in Sculptor (Smith and Stryker 1986).

3.1 Draco

This dSph galaxy was the first for which a claim of a substantial abundance range was made (Zinn 1978). Since Zinn's work there has been a succession of papers (e.g. Kinman, Kraft and Suntzeff 1981; Smith 1984; Stetson 1984; Aaronson and Mould 1985) all of which have argued for a real abundance spread in this dSph galaxy, with estimates of the total abundance range present reaching as high as 1 dex ($-2.7 <$ [Fe/H] < -1.7, approximately). Bell (1985) however, has contested this result, arguing principally from Stetson's data that the apparent abundance spread in Draco is the result of underestimated observational errors.

This issue is further complicated by the recent photometry of Carney and Seitzer (1986). These authors claim that in the c-m diagram, the giant branch in the magnitude interval $20.5 < V < 21.5$ (equivalent to $+1.0 < M_V < +2.0$; i.e. below the horizontal branch) shows an intrinsic total range of 0.19 mag in (B-V). This corresponds to an abundance range of approximately 0.8 dex if the Zinn and West (1985) calibration of $(B-V)_{o,g}$ is assumed to apply, and in particular remain linear, at abundances less than that of M92, to which the mean Draco lower giant branch color corresponds. However, it appears these authors have underestimated their photometric errors. Using the 20 stars in common between their fields 1 and 2 in the magnitude interval $20.5 < V < 22.0$, the (B-V) single observation 1σ error is calculated as 0.057 mag. The standard deviation of the (B-V) colors for the 54 giants in the interval $20.5 < V < 21.5$, selected as described in Carney and Seitzer (1986), is 0.058 mag; thus contrary to the claims of Carney and Seitzer (1986), their photometry analyzed in this fashion provides no evidence to support the existence of an abundance spread in Draco. However, as reported elsewhere in this volume, a reanalysis using the error estimate calculated above but with a larger sample of stars ($20.5 < V < 22.0$ instead of $20.5 < V < 21.5$) and with a ridge line defined by the giant branch itself rather assuming a fixed (B-V) color, indicates that the lower giant branch in Draco does have an intrinsic width.

Further support for the existence of an abundance spread in Draco is the correlation found by Aaronson and Mould (1985) between the $(J-K)_o$ and $(V-K)_o$ color residuals from the M92 giant branch at constant luminosity, and the spectroscopically determined residuals at constant color from the M92 ridge line in the (m_{HK}, $(B-V)_o$) diagram of Kinman, Kraft and Suntzeff (1981).

3.2 Ursa Minor

The majority of the authors cited above have also observed stars in Ursa Minor, but the largest sample of spectroscopically observed stars is that of Suntzeff et al. (1986). Aside from one very metal-poor ([Fe/H] = -3.5 dex) star, the spread in abundance sensitive parameters in this sample is consistent with the observational errors alone; certainly there is no compelling evidence for any abundance spread similar in size to that claimed for Draco. This result is

supported by the photometry: neither the lower giant branch data of
Olszewski and Aaronson (1985) nor the new upper giant branch
photometry of Cudworth, Olszewski and Schommer (1986) indicate any
substantial color width in excess of the errors.

3.3 Fornax

The Fornax dSph is unique among the galactic dSph galaxies in
that it has 5, possibly 6, globular clusters associated with it (Hodge
1961, 1969). These clusters appear in all respects to be comparable
to galactic globular clusters (see Buonanno et al. 1985 and the
references therein). In particular, these clusters cover a
considerable range in abundance, from approximately [Fe/H] = -2.2
(cluster 3) to [Fe/H] = -1.2 dex (cluster 4; Zinn and Persson 1981).

Buonanno et al. (1985) have also produced c-m diagrams for two
field regions in this dSph galaxy. In these diagrams the giant branch
is quite broad, as first noted by Demers et al. (1979). After
consideration of their photometric errors, Buonanno et al. (1985)
conclude first, that the mean abundance of the stars in these fields
is [Fe/H] = -1.4 dex, and second, that there is an intrinsic abundance
spread. The abundance spread is characterized by a standard deviation
$\sigma([Fe/H]) = 0.3$ dex and an apparent total abundance spread of
approximately $-2.0 < [Fe/H] < -0.9$, though given that these values are
drawn from a c-m diagram, it is possible that more metal poor stars
exist in this dSph galaxy. Interestingly, the mean abundance of the
field population is approximately 0.4 dex more metal rich than the
mean abundance of the globular clusters. See also the paper of Light
and Seitzer in this volume.

3.4 Sculptor

The suggestion of an abundance range in this dSph galaxy was
first put forward by Norris and Bessell (1978). These authors, based
on their own analysis of the giant branch photometry of Kunkel and
Demers (1977) and on spectra of 2 giants, concluded that there was an
abundance range of perhaps 0.6 dex in this dSph galaxy. A range of Ca
II H and K lines strengths among Sculptor giants was confirmed by the
narrow band imaging of Smith and Dopita (1983) but unfortunately these
authors gave no actual estimate of the abundance spread implied by
their results. The CCD photometry of Da Costa (1984) also supported
the results of Norris and Bessell (1978) but the number of stars in
the appropriate part of the c-m diagram was not large. Yet more
support is provided by the ΔS results of Smith and Stryker (1986); the
3 anomalous cepheids studied show an abundance range of 0.4 dex.

In Fig. 2 preliminary results from a new study of the abundance
range in Sculptor are presented (Armandroff, Da Costa and Zinn 1986).
The upper panel shows the location of 16 radial velocity members of
Sculptor in the M_I, $(V-I)_o$ plane along with equivalent photometry for
a number of giants in the well-studied galactic clusters M15, M2, NGC
1851 and 47 Tuc. The photometry of all stars was obtained at CTIO
with the CCD detector. The middle panel shows a plot of the sum of

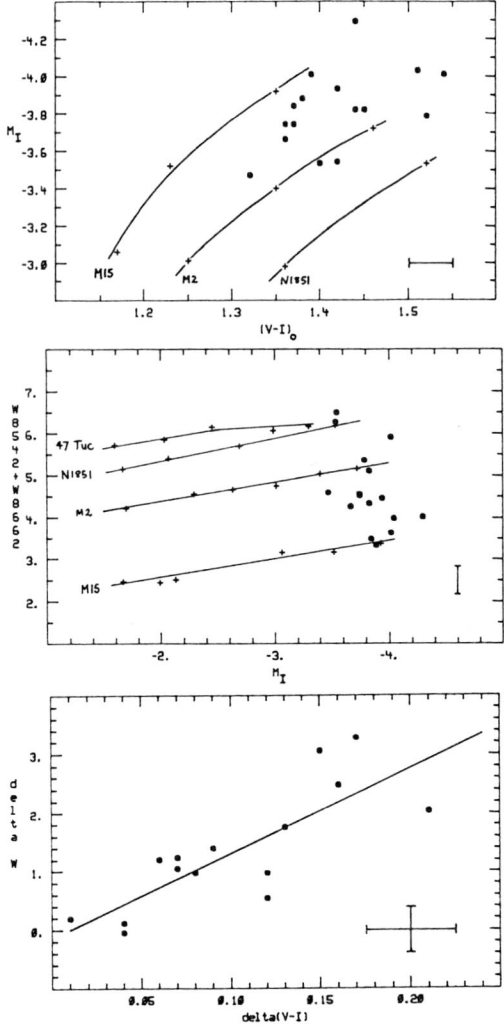

Fig. 2. Upper panel: $M_I,(V-I)_o$ color-magnitude diagram for 16 radial velocity members of the Sculptor dSph galaxy. Also shown are the giant branches of three galactic globular clusters. Middle panel: the total equivalent width (Å) of the CaII IR-triplet lines at $\lambda 8542$ and $\lambda 8662$ plotted against M_I for the Sculptor stars. Constant abundance lines are defined by the galactic globular cluster stars. Lower panel: the deviation in equivalent width from the M15 ridge line in the middle panel, at constant M_I, plotted against the deviation in $(V-I)_o$ color from the M15 giant branch in the upper panel, again at constant M_I. The correlation coefficient is 0.80.

the equivalent widths of the Ca II infrared triplet lines at λ8542 and λ8662 against M_I for both the globular cluster and Sculptor stars. Again the data were obtained at CTIO. Finally the bottom panel shows a plot of residual equivalent width, measured at constant M_I from the M15 fiducial line, against residual $(V-I)_o$ color, from the M15 giant branch, for the Sculptor stars.

A number of points can be made from these diagrams. These include: (a) the Sculptor star (H264) with $M_I = -4.3$ has, using the $(M_{bol}, (V-I)_o)$ relation of Mould, Kristian and Da Costa (1984), $M_{bol} = -3.75$ and thus is an upper-AGB star. (b) Both the photometry and the spectroscopy indicate that the mean abundance of the Sculptor stars is [Fe/H] = -1.8 ± 0.05 dex if the Zinn and West (1985) abundance values are adopted for the calibrating clusters. (c) The good correlation between the equivalent width and color residuals indicates that the scatter shown by the Sculptor stars in both the upper and middle panels is not due to observational errors alone. Although the effect of different gravities, temperatures and abundances combine to produce ambiguity in the interpretation of the abundances of the most metal rich Sculptor stars in the middle panel, the combined data are consistent with a total abundance range of approximately 0.6 dex, from [Fe/H] = -2.1 to [Fe/H] = -1.5 dex. The dispersion σ([Fe/H]) is approximately 0.2 dex.

3.5 Comment

It has been established for sometime now that the mean metal abundances of the dSph galaxies are tightly correlated with their luminosities. The dwarf elliptical companions of M31 studied by Mould, Kristian and Da Costa (1983, 1984), namely NGC 147 and NGC 205, also fall on this relation which is shown in Fig. 3. The galactic globular clusters however, show no such relation and this difference is yet another clue that the origins of globular clusters and dSph galaxies were different. Interestingly, it also appears likely that the abundance dispersions in these galaxies are correlated with their mean abundance (or luminosity). The data are presented in Fig. 4. For NGC 205 and NGC 147, the values are taken from Mould et al. (1983, 1984); note that the value for NGC 205 is a lower limit since this galaxy contains stars more metal rich than 47 Tuc, the most metal rich cluster is used in their calibration. For Fornax, the results of Buonanno et al. (1985) are used, while the results described above are used for Sculptor. For Draco the value is the mean of the dispersions calculated from Zinn (1978), 0.16 dex, which is probably an underestimate if the abundances of the most metal poor stars have been overestimated, and that, 0.30 dex, from Kinman, Kraft and Suntzeff (1981), calculated without any attempt to correct for abundance errors. The Ursa Minor value is an attempt to allow for one very metal poor star in a sample with an otherwise small dispersion (Suntzeff et al. 1986). The mean metal abundances are typically determined to 0.2 dex or better; the errors in the dispersions are poorly known but probably do not exceed ±0.1 dex.

Naively such a correlation is perhaps expected; more generations

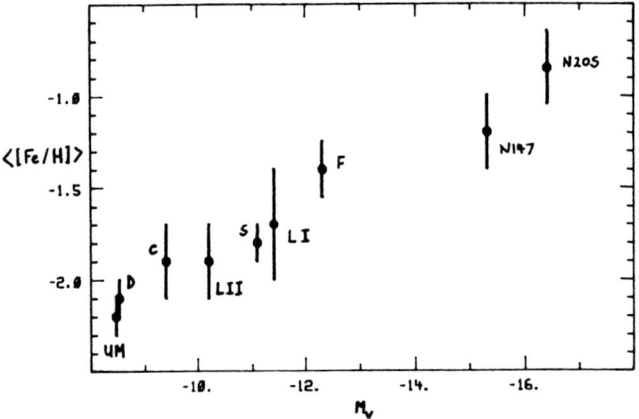

Fig. 3. The mean metal abundances of the 7 galactic dSph galaxies and of the M31 dE companions NGC147 and NGC205 plotted against the absolute visual magnitudes of these systems.

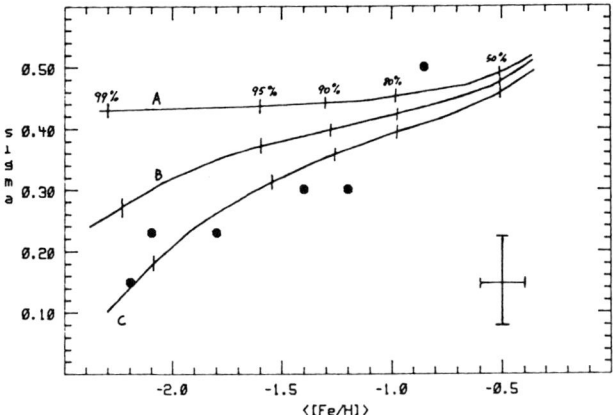

Fig. 4. The abundance dispersion σ([Fe/H] plotted against mean abundance <[Fe/H]>. In order of increasing abundance, the filled symbols are for Ursa Minor, Draco, Sculptor, Fornax, NGC147 and NGC205. Typical errors for both quantities are indicated. The solid curves are predictions of the simple model of chemical evolution assuming a yield equal to the solar abundance. Curve A: initial abundance $z_o = 0.0$; Curve B: $z_o = 2 \times 10^{-5}$; Curve C: $z_o = 6 \times 10^{-5}$. The vertical bars indicate the amount of initial mass lost from the system.

of stellar processing meaning not only a higher mean abundance but
also a larger abundance dispersion. In an attempt to quantify this
statement, Fig. 4 also shows the results of some crude calculations.
The simple "closed box" model of chemical evolution was employed with
the addition that evolution was assumed to halt when the stellar
abundance reached a predetermined Z_{max}. As noted by Zinn (1978), in
such models, assuming a yield equal to the solar abundance (as done
here) requires that the evolution cease when only a small fraction of
the original mass of gas has been turned into stars if the metal
abundance is to be low. The curves, labeled A, B, and C were
calculated assuming "primordial" abundances of 0.0, 0.001 Z_{sun} and
0.003 Z_{sun}, respectively. The fractions of original mass lost, which
are generally large, are also indicated. With the possible exception
of curve C, none of these models give a good representation of the
observations. This is undoubtedly the result of the unrealistic
assumptions (e.g. instantaneous recycling) that underlie the simple
model of chemical evolution. However, like the (M_v, <[Fe/H]>)
relation, the data of Fig. 4, and future more precise versions, can be
used to provide constraints on more detailed models of the chemical
evolution of dSph galaxies.

4. AGE AND AGE VARIATIONS

It is clear from the many very accurate main sequence c-m
diagrams for globular clusters now available, that the range of age
present among the stars in any one cluster must be very small. That
does not seem to be the case for dSph galaxies. Following the
discovery of upper-AGB, and therefore intermediate age, carbon stars
in the Fornax dSph (Aaronson and Mould 1980), there have been a
plethora of papers concerning the ages, age ranges and carbon star
content of dSph galaxies. The situation regarding carbon stars has
been thoroughly discussed recently by Aaronson and Mould (1985) and
little can be added to their conclusions: five of the seven galactic
dSph galaxies have extended giant branches and therefore contain stars
younger than the globular clusters. This appears also to be the case
for at least one of the M31 dSph companions (Aaronson et al. 1985).
Only Draco and Ursa Minor lack stars at luminosities above that of the
first giant branch tip. Since these are also the lowest luminosity
systems, and since the evolutionary lifetime of upper-AGB stars is
short, it is by no means inconceivable that this apparent lack of
upper-AGB stars is simply a statistical effect. This point will be
discussed further below.

4.1 Main Sequence Color-Magnitude Diagrams

Color-Magnitude diagrams that reach the main sequence are
currently available for 5 dSph galaxies and the results from each of
these data will now be examined. The least ambiguous result is that
for Ursa Minor (Olszewski and Aaronson 1985) which, it will be
recalled, is the only dSph known to have a blue horizontal branch.

The c-m diagram is well fit by the ridge lines of the old metal poor globular cluster M92 and there is little, if any, indication of the presence of a substantial younger population. An age of 16 ± 2 billion years is derived from isochrone fits. The c-m diagram however, does contain a small number of blue stragglers which will be discussed further below.

The results for Draco, a dSph galaxy with a red horizontal branch, are however, less clear cut. Stetson, VandenBerg and McClure (1985) conclude from a rather small sample of stars that Draco is not measurably younger than the galactic globular clusters or Ursa Minor. The precision of their data however, is rather low. However, photometry of a much larger sample of Draco stars, with somewhat better accuracy, is presented by Carney and Seitzer (1986). These authors conclude from both a c-m diagram and a luminosity function comparison with M92, as well as from isochrone fits, that Draco may in fact contain a sizeable population of stars that is perhaps a few billion years younger than the globular clusters of similar abundance (and Ursa Minor). The existence of this population will of course go some way towards explaining the horizontal branch difference with Ursa Minor, but not the whole way since there is evidence that a significant old population is also present. Once again there is evidence for a sizeable blue straggler population.

For Sculptor, Da Costa (1984) has argued from both a comparison with the M92 ridge lines in the (V,B-V) plane and from isochrone fits in the (R,B-R) plane, which reduces the sensitivity to photometric color errors, that the bulk of the stars in this dSph galaxy are 2 - 3 billion years younger than the globular clusters. An age difference of this size is quite capable of explaining the predominantly red horizontal branch of this system. Further, although an older population is certainly not ruled out by these data, it is not likely to be dominant. As for Draco and Ursa Minor, there is also a notable blue straggler population in this dSph galaxy.

The IR-photometry of carbon stars in Carina (Mould et al. 1982) suggested that this dSph galaxy had a substantial intermediate age population. This was confirmed by the CCD photometry of Mould and Aaronson (1983) who derived a turnoff age of only 7 billion years for the majority of the stars in their c-m diagram. Further, comparison with theoretical luminosity functions apparently left little room for any older population. However, the discovery of RR Lyrae variables in this dSph galaxy (Saha et al. 1986) shows that a population of stars with ages in excess of 10 billion years must be present. The relative importance of this older population however, is apparently not large. Depending on the number of RR Lyraes produced per old giant progenitor, the fraction of old stars could be as low as 2%; a direct comparison with Draco suggests a 15% contribution in line with estimates from the carbon star numbers of approximately 70% intermediate age population.

As noted above, the Fornax globular clusters appear to be indistinguishable from their galactic counterparts, even to the existence of blue horizontal branches in their c-m diagrams (Buonanno et al. 1985). This would appear to be *prima facie* evidence that

Fornax contains very old stars. Yet this dSph galaxy also contains a large number of carbon stars whose luminosities are the brightest of any known in the dSph galaxies, indicating ages as young as 3 billion years or less (e.g. Aaronson 1986). The main sequence progenitors of these stars are seen in the deepest c-m diagrams of Buonanno et al. (1985); hence it seems inescapable that star formation has proceeded in Fornax over most of its life. Based again on the numbers of carbon stars, the fraction of this younger (i.e. age < 10 billion years) population is estimated as 15 - 25 percent of the total (Aaronson 1986). Similar fractions apparently also apply for the Leo I and Leo II systems (Aaronson 1986).

In summary then, it appears that the degree of on-going star formation in dSph galaxies, by which is meant the formation of stars younger than 10 billion years, is highly variable, ranging from negligible or non-existent in Ursa Minor, Draco and Sculptor, to significant for Leo I, Leo II and Fornax, to dominant for Carina. In a similar sense, as typified by the sequence Ursa Minor, Sculptor, Carina, it is also clear that the epoch of formation for the bulk of the stars varies from dSph to dSph.

4.2 Blue Stragglers: Binaries or Young Stars?

As noted above, the main sequence c-m diagrams of the Ursa Minor, Draco and Sculptor dSph galaxies all show a population of blue stragglers and it is of interest to inquire into the origin of these stars. Are they, for example, intermediate age main sequence stars, which would indicate on-going star formation at very low levels in these dSph galaxies, or are they the result of mass transfer in old binary systems? Since blue stragglers are known to occur in at least some globular clusters, the existence of similar stars in the dSph galaxies does not of itself require them to be of intermediate age. However, by investigating the consequences, the validity of the assumption that they are intermediate age stars may be investigated.

The stars in question, if interpreted as normal main sequence stars, would appear to have ages from 1 to 3 billion years upwards. The subsequent evolution of such stars should produce (a) upper AGB carbon stars and (b) at least for the more massive progenitors, anomalous cepheids since the abundances in these dSph galaxies are sufficiently metal poor. Turning first to the upper AGB stars and noting that (a) the total population of possible main sequence progenitors, assumed to be all stars with indicated ages less than 10 billion years, is approximately 300, 700 and 1400 stars in Ursa Minor, Draco and Sculptor (Olszewski and Aaronson 1985, Carney and Seitzer 1986, Da Costa 1984); and (b) that the number of upper AGB stars is somewhere between 1/500 and 1/2000 times the number of main sequence progenitors (Da Costa 1984; Olszewski and Aaronson 1985). The predicted numbers of upper AGB carbon stars are then approximately 0.4, 0.6 and 2 stars, respectively. The observed numbers are 0, 0, and at least 1 (Aaronson and Mould 1985). Hence there is no obvious conflict here.

Second, if following Carney and Seitzer (1986), blue stragglers brighter than about $M_V = +2.5$ are assumed to have sufficient mass to become anomalous Cepheids, and if anomalous Cepheids are assumed to have an evolutionary lifetime of approximately 50 million years (Hirshfeld 1980), then the number of anomalous cepheids should be approximately 1/40 the number of main sequence progenitors. The total numbers of bright blue stragglers in these dSph galaxies is approximately 45, 70 and 200 leading to predicted numbers of anomalous cepheids of 1, 2 and 5 stars. The observed numbers are 3, 5 and 3 stars. Hence once again, despite the obvious large uncertainties in these estimates, there is no conflict with the hypothesis that the blue stragglers observed in the Ursa Minor, Draco and Sculptor dSph galaxies are young single stars.

In summary then, while there is no evidence against the old binary mass transfer scenario, it is also true that there is no evidence that rules out the competing hypothesis, namely that the blue stragglers are single intermediate age main sequence stars. Certainly for Sculptor, where the existence of upper AGB (intermediate age) carbon stars is confirmed, it seems likely that this second hypothesis is correct, and that therefore, this dSph galaxy continued to form stars at a very low rate long after the major epoch of star formation.

5. DARK MATTER

The globular clusters of our Galaxy have typical visual mass-to-light ratios of the order of 2 to 3 in solar units. These M/L values can be adequately explained without the need to postulate the existence of any additional "dark matter" other than the remnants of evolved stars (e.g. Illingworth 1975, Gunn and Griffin 1979; Lupton, Gunn and Griffin 1985). On the other hand, perhaps the most intriguing new result for dSph galaxies is the suggestion that the M/L values, for at least some dSph galaxies, maybe an order of magnitude or more higher. This would suggest that dSph galaxies contain substantial amounts of dark matter in an analogous fashion to the halos of spiral galaxies. Such a result has a number of important astrophysical applications aside from the obvious ones for the formation and evolution of dSph galaxies. For example, the dark matter almost certainly could not be "hot", i.e. in the form of massive neutrinos, if it can be confined on scales the size of dSph galaxies (Tremaine and Gunn 1979; Lin and Faber 1983).

Since Aaronson's (1983) original measurement of an unexpectedly large velocity dispersion for Draco, a number of further studies have been published. These are summarized in Table I, which gives velocity dispersion measurements for 5 dSph galaxies; the samples on which the entries are based are described in the notes. Also listed in the Table are the corresponding central mass-to-light ratios which, with the exception of the recent studies of Armandroff and Da Costa (1986) and Aaronson and Olszewski (1986b), have been taken directly from Kormendy (1986). As discussed by Kormendy (1986), central M/L values are to be preferred to "global" values in order to minimize the impact of assumptions regarding the internal dynamics of dSph galaxies.

Further, use of central M/L values also minimizes the effect of errors in the limiting radii of these systems, which are often poorly determined.

Table I
Dwarf Spheroidal Velocity Dispersions

dSph	Reference	$\langle v^2 \rangle^{1/2}$ (km/s)	$(M/L)_v$	Notes
Draco	A086a	9 ± 2	37 ± 11	1
Ursa Minor	A086a	11 ± 3	100 ± 40	2
Carina	SF85	6 ± 2	10 ± 6	3
Sculptor	SF85	6 ± 3	6 ± 3	4
	AD86	6.2 ± 1.2	6.0 ± 2	5
Fornax	SF85	6 ± 2	1 ± 0.5	6
	A086b	8 ± 3	2 ± 1	7

References: A086a, A086b = Aaronson and Olszewski (1986a, 1986b); SF85 = Seitzer and Frogel (1985); AD86 = Armandroff and Da Costa (1986).
Notes:
1. Multiple epoch observations; 3 Carbon stars and 8 K-giants; 2 velocity variables (1 Carbon star, 1 K-giant).
2. Multiple epoch observations; 2 Carbon stars and 8 K-giants; 3 velocity variables (2 carbon stars, 1 K-giant)
3. Single epoch; 6 Carbon stars
4. Single epoch; 3 Carbon stars
5. Single epoch; 16 K-giants
6. Single epoch; 5 Carbon stars
7. 3 Fornax globular clusters.

Table I also lists the uncertainty in the velocity dispersion measures. In most instances this uncertainty stems entirely from the small numbers of stars observed. Uncertainties are also given for the central M/L values. In this case, again following Kormendy (1986), the listed uncertainties reflect only the velocity dispersion errors and do not include the effects of uncertainties in the central surface brightnesses or in the core radii of these systems. These latter quantities however, especially the central surface brightnesses, are difficult to establish observationally and they are therefore also subject to large uncertainties. Consequently, the M/L errors listed in Table I are very much lower limits; for example, Armandroff and Da Costa (1986) have shown that the uncertainty in the central M/L value for Sculptor, a relatively well-studied system, increases by almost 50 percent when uncertainties in all quantities, not just in the velocity dispersion, are considered.

It is also worth emphasizing that the central M/L values in Table I have been calculated using the observed velocity dispersions; no corrections for the expected decrease in velocity dispersion with distance from the galaxy center or for projection effects have been applied, since to do so would require the adoption of a (possibly incorrect) model for the dynamics. Fortunately, since the stars observed are generally within 1 or 2 core radii of the center, these corrections are probably not large. Nevertheless it must be kept in mind that the M/L values listed in Table I are actually lower limits to the true central values.

Given all these caveats, can any reliable conclusions be drawn from these results? The answer is undoubtedly yes and the implications have been extensively discussed by Kormendy (1986). Briefly, despite the large uncertainties, the conclusion that both Draco and Ursa Minor contain large amounts of dark matter seems difficult to avoid. This is less obviously the case for Sculptor and Carina, while for Fornax, there is no requirement to invoke the presence of dark matter at this time. The result for Carina however, deserves some further attention since this dSph galaxy is dominated by a population that is less than half the age of the bulk of the stars in the other systems. Simple calculations show that Carina's M/L value should be increased by a factor of 1.5 to 2 to compensate for this age difference when comparing it with those for the other dSph galaxies. Thus if its measured velocity dispersion is confirmed by further observations, Carina will join Draco and Ursa Minor as a dSph galaxy apparently containing a substantial amount of dark matter.

Regarding Sculptor and in particular Fornax, it is important to remember that the tabulated values are central M/L values. If the core radius of the dark matter exceeds the core radius of the visible stars, or equivalently, if the central density of dark matter is significantly less than that of the stars, as seems likely in Fornax (Kormendy 1986), then it is difficult to learn much about the presence of dark matter from this type of observation. What is required in these cases is a measurement of the velocity dispersion at large r/r(core), a difficult but not impossible observational task.

6. SUMMARY

It should be clear from the preceeding discussion that in fact dSph galaxies do not have much in common with globular clusters. They share only the common properties of similar density profiles, at least for the lowest central concentration globular clusters, a lack of gas and current star formation, and of course, the general characteristic of a basically "old" stellar population. Instead it appears more appropriate to consider dSph galaxies in the same context as dwarf ellipticals and dwarf irregulars, for as shown by Kormendy (1985) for example, all these systems follow the same correlations between absolute magnitude, central surface brightness and core radius (see also Dekel and Silk 1986). The observations in dSph galaxies of ongoing star formation over a substantial fraction of a Hubble time,

albeit at variable rates, of large internal abundance ranges, the
clear correlation between mean metal abundance and absolute magnitude,
and the possibility of dark matter halos all add further support to
this idea. In this respect, it is perhaps appropriate to end by once
again quoting from Shapley's 1938a paper announcing the discovery of
Sculptor. In it Shapley offers as an alternative to his preferred
description of Sculptor as a "super cluster of the globular type" the
following: "or...a Magellanic Cloud [i.e. a dwarf irregular galaxy]
devoid of its characteristic bright stars, clusters and luminous
diffuse nebulosity". These latter remarks may well be close to the
truth.

REFERENCES

Aaronson, M. 1983 Astrophys. J. Letters 266, L 11.
Aaronson, M. 1986 in Stellar Populations, A. Renzini and
 M. Tosi, eds., Cambridge University Press, p. 45.
Aaronson, M. and Mould, J. 1980 Astrophys. J. 240, 804.
Aaronson, M. and Mould, J. 1985 Astrophys. J. 290, 191.
Aaronson, M. and Olszewski, E. W. 1986a in IAU Symposium
 Symposium No. 117, Dark Matter in the Universe, J. Kormendy
 and G. R. Knapp, eds., Reidel, Dordrecht.
Aaronson, M. and Olszewski, E. W. 1986b Astron. J. 92, 580.
Aaronson, M., Gordon, G., Mould, J., Olszewski, E. W. and Suntzeff,
 N. B. 1985 Astrophys. J.Letters 296, L7.
Armandroff, T. E. and Da Costa, G. S. 1986 Astron. J. 92, in
 press.
Armandroff, T. E., Da Costa, G. S. and Zinn, R. 1986, in preparation.
Baade, W. and Swope, H. H. 1961 Astron. J. 66, 300.
Bell, R. A. 1985 Publ. Astron. Soc. Pacific 97, 219.
Buonanno, R., Corsi, C. E., Fusi Pecci, F., Hardy, E. and Zinn, R.
 1985 Astron. Astrophys. 152, 65.
Cannon, R. D., Hawarden, T. G. and Tritton, S. B. 1977 Monthly
 Notices Roy. Astron. Soc. 180, 31p.
Carney, B. and Seitzer, P. 1986 Astron. J. 92, 23.
Cudworth, K. M., Olszewski, E. W. and Schommer, R. A. 1986 Astron.
 J., in press.
Da Costa, G. S. 1984 Astrophys. J. 285, 483.
Dekel, A. and Silk, J. 1986 Astrophys. J. 303, 39.
Demarque, P. and Hirshfeld, A. W. 1975 202, 346.
Demers, S., Kunkel, W. and Hardy, E. 1979 Astrophys. J.
 232, 84.
Gunn, J. E. and Griffin, R. F. 1979 Astron. J. 84, 752.
Harrington, R. G. and Wilson, A. G. 1959 Publ. Astron. Soc.
 Pacific 62, 118.
Harris, H. C. 1985 in IAU Colloquium No. 82, Cepheids: Theory
 and Observation, B. F. Madore, ed., Cambridge Univ. Press,
 Cambridge, p. 232.
Hirshfeld, A. W. 1980 Astrophys. J. 241, 111.

Hodge, P. W. 1961 Astron. J. 66, 83.
Hodge, P. W. 1969 Publ. Astron. Soc. Pacific 81, 875.
Hodge, P. W. and Wright, F. W. 1978 Astron. J. 83, 228.
Illingworth, G. 1975 in IAU Symposium No. 69, Dynamics of Stellar Systems, A Hayli, ed., Reidel, Dordrecht, p 151.
Kinman, T. D., Kraft, R. P. and Suntzeff, N. B. 1981 in Physical Processes in Red Giants, I. Iben Jr. and A. Renzini, eds., Reidel, Dordrecht, p. 71.
Kormendy, J. 1985 Astrophys. J. 295, 73.
Kormendy, J. 1986 in IAU Symposium No. 117, Dark Matter in the Universe, J. Kormendy and G. R. Knapp, eds., Reidel, Dordrecht.
Kunkel, W. and Demers, S. 1977 Astrophys. J. 214, 21.
Lupton, R., Gunn, J. E. and Griffin, R. F. 1985 in IAU Symposium No. 113, Dynamics of Star Clusters, J. Goodman and P. Hut, eds., Reidel, Dordrecht, p. 19.
Light, R. M., Armandroff, T. E. and Zinn, R. 1986 Astron. J. 92, 43.
Lin, D. N. C. and Faber, S. M. 1983 Astrophys. J. Letters 266, L21.
Mould, J. and Aaronson, M. 1983 Astrophys. J. 273, 530.
Mould, J., Cannon, R. D., Aaronson, M. and Frogel, J. A. 1982 Astrophys. J. 254, 500.
Mould, J., Kristian, J. and Da Costa, G. S. 1983 Astrophys. J. 270, 471.
Mould, J., Kristian, J. and Da Costa, G. S. 1984 Astrophys. J. 278, 575.
Nemec, J. M. 1985 Astron. J. 90, 204.
Nemec, J. M. and Harris, H. C. 1986, preprint.
Norris, J. and Bessell, M. S. 1978 Astrophys. J. Letters 225, L49.
Norris, J. and Zinn, R. 1975 Astrophys. J. 202, 335.
Olszewski, E. W. and Aaronson, M. 1985 Astron. J. 90, 2221.
Renzini, A., Mengel, J. G. and Sweigart, A. V. 1977 Astron. Astrophys. 56, 369.
Saha, A., Monet, D. G. and Seitzer, P. 1986 Astron. J. 92, 302.
Sandage, A. 1982 Astrophys. J. 252, 553.
Seitzer, P. and Frogel, J. A. 1985 Astron. J. 90, 1796.
Shapley, H. 1938a Harvard College Obs. Bull., No. 908, 1.
Shapley, H. 1938b Nature 142, 715.
Shapley, H. 1939 Proc. Nat. Acad. Sci. 25, 565.
Smith, G. H. 1984 Astron. J. 89, 801.
Smith, G. H. and Dopita, M. A. 1983 Astrophys. J. 271, 113.
Smith, H. A. 1984 Publ. Astron. Soc. Pacific 92, 505.
Smith, H. A. and Stryker, L. L. 1986 Astron. J. 92, 328.
Stetson, P. B. 1984 Publ. Astron. Soc. Pacific 96, 128.
Stetson, P. B., VandenBerg, D. A. and McClure, R. D. 1985 Publ. Astron. Soc. Pacific 97, 908.
Suntzeff, N. B., Olszewski, E. W., Kraft, R. P., Friel, E., Aaronson, M. and Cook, M. 1986, in preparation

Swope, H. H.   1967   Publ. Astron. Soc. Pacific 79, 439.
Swope, H. H.   1968   Astron. J. 73, S204.
Tremaine, S. and Gunn. J. E.   1979   Phys. Rev. Letters 42, 467.
van Agt, S. L. T. J.   1967   Bull. Astron. Inst. Neth. 19, 275.
van Agt, S. L. T. J.   1973   in IAU Colloquium No. 21, Variable Stars in Globular Clusters, J. D. Fernie, ed., Reidel, Dordrecht, p. 35.
van Agt, S. L. T. J.   1978   Publ. David Dunlap Obs. 3, 205.
van den Bergh, S.   1972a   Astrophys. J. Letters 171, L31.
van den Bergh, S.   1972b   Astrophys. J. Letters 178, L99.
van den Bergh, S.   1974   Astrophys. J. 191, 271.
Wilson, A. G.   1955   Publ. Astron. Soc. Pacific 67, 27.
Zinn, R.   1978   Astrophys. J. 225, 790.
Zinn, R.   1980   in Globular Clusters, D. A. Hanes and B. F. Madore eds., Cambridge Univ. Press, Cambridge, p. 191.
Zinn, R.   1985   Mem. Soc. Astron. Ital. 56, 223.
Zinn, R. and Dahn, C. C.   1976   Astron. J. 81, 527.
Zinn, R. and King, C. R.   1982   Astrophys. J. 262, 700.
Zinn, R. and Persson, S. E.   1981   Astrophys. J. 247, 849.
Zinn, R. and Searle, L.   1976   Astrophys. J. 209, 734.
Zinn, R. and West, M. J.   1985   Astrophys. J. Suppl. 55, 45.

DISCUSSION

NEMEC: You failed to mention the important structural differences in the central characteristics as summarized by Kormendy (1985).

DA COSTA: The (M/L) values were central (M/L) values calculated following Kormendy's precepts. For Draco, Ursa Minor and Carina, the values were taken directly from his paper.

COHEN: How do you know Draco and Ursa Minor, the least luminous dwarf spheroidal galaxies, are not currently being torn apart?

DA COSTA: Perhaps Ed Olszewski would like to comment.

OLSZEWSKI: If Draco and Ursa are being ripped apart, the stars with high velocity move very far in 10^9 yr. We would have to catch both Draco and Ursa in a unique part of their life, a necessarily small part of their life (since at 10km/s the crossing time is 10^8 years), which seems unlikely.

CAYREL: Could you say something about the globular clusters associated to the dwarf spherical galaxies?

DA COSTA: The only galactic dwarf spheroidal galaxy containing globular clusters is Fornax. The clusters show an abundance range of about 1 dex and have CM diagrams that show blue horizontal-branch stars. Essentially they appear identical to galactic globular clusters.

CARNEY: The proper way to compare the spread in Draco's (B-V) giant colors due to observational errors versus due to a metallicity spread is to derive σ's in the same magnitude range, unlike what we did and you showed. When it is done rigorously, as in our poster paper, the metallicity spread seems real.

DA COSTA: If the giant branch width is calculated exactly as outlined in your published paper, then there is no evidence for any intrinsic width. If however, you increase the sample, by going 0.5 mag fainter, calculate relative to a ridge line, etc., I agree you will find that an intrinsic width probably is present. The calculation is however rather sensitive to which outlying stars are included.

CARNEY: Comparing anomalous Cepheids in low density dwarf spheroidal galaxies vs. high density globular clusters is not quite fair. Perhaps one should instead compare to the low density halo field. There is at least one known local field anomalous Cepheid. Does Hugh Harris know of any more?

DA COSTA: I think it must be rather difficult to identify anomalous Cepheids in the galactic halo since you generally lack distance information. Especially for shorter period stars, anomalous Cepheids are not easily distinguished from RR Lyrae stars.

HARRIS, H.: XZ Ceti may be a field anomalous Cepheid, but its distance and luminosity have not yet been determined. Many other field Type II Cepheids could be anomalous, but we have distance estimates for only a few.

INAGAKI: What do you think about Saito's (1980? PASJ) formation picture of dwarf spheroidal galaxies? He considered that dwarf spheroidal galaxies were more massive before (10 to 100 times) and lost mass. So dwarf spheroidal galaxies were nearer to dwarf Es than to globular clusters.

DA COSTA: I think its generally agreed that dwarf spheroidal galaxies have lost a lot mass during their formation; this seems a reasonable way to account for their low density.

WEBBINK: It seems to me that all of the evidence cited in favor of an age spread within individual dwarf spheroidal galaxies might as well be explained by supposing they have some initial population of binaries. Can you give an argument to rebut this hypothesis?

DA COSTA: The strongest evidence in support of an age range is the existence of luminous carbon stars in 5 of 7 galactic dwarf spheroidal galaxies (and 1 M 31 companion). These stars appear indistinguishable from similar stars in intermediate-age Magellanic Cloud clusters.

VAN DEN BERGH: In thinking about dwarf spheroidal galaxies it is, perhaps, a good idea to remark that NGC 147, NGC 185 and NGC 205 probably belong to the same "form family".

AN OVERVIEW OF GLOBULAR CLUSTER SYSTEMS IN DISTANT GALAXIES

William E. Harris

McMaster University

ABSTRACT: Information is currently available for globular cluster systems in almost 50 galaxies from the Milky Way to the Coma Cluster. The observed features of these systems are reviewed, with emphasis on (a) their total populations (specific frequencies), (b) spatial structures compared with the underlying halo light, (c) photometric characteristics (luminosity function and metallicity). The combined evidence suggests strongly that globular clusters are likely to have formed in a rather sharply defined epoch clearly before the bulk of the halo field stars, and that this epoch was more active in E galaxies, especially those in rich environments. Finally, the special role of the supergiant E's at the centers of rich galaxy clusters is reviewed.

1. INTRODUCTION

In 1918, a paper by Harlow Shapley appeared in the Astrophysical Journal containing a discussion of the overall space distribution of the globular clusters in the Milky Way. For Shapley, this summary culminated years of patient accumulation of data on the Galactic globular clusters, one object after another, until a global picture of the entire halo began to emerge. But this paper can also now be seen as the first stride into a completely new realm: Shapley was the first to study a globular cluster <u>system</u> as a complete entity. Shapley's essential vision that the cluster system formed the skeletal outline of the entire Galaxy remains remarkably valid seventy years later.

Discovery and study of globular cluster systems (GCS's) in distant, major galaxies - M 31 and beyond - began with Hubble's work (1932). However, the basic nature of the problem from an observer's point of view is that globular clusters are very faint relative to their parent galaxies, and the galaxies concerned are very distant; so substantial progress in the observations was slow until the current decade, with the notable exception of the Virgo System survey by Hanes (1977a,b). During previous conferences in the past decade, the only reviews relevant to globular cluster systems in galaxies beyond the Local Group are those given by Hanes (1980) and by Racine (1980). It is an encouraging sign of the rapid growth of this field that a large

TABLE I

GLOBULAR CLUSTER SYSTEMS IN GALAXIES

NGC	Type	MVT	Specific Frequency		Environment*
147	E5	-14.9	4.4	± 2.0	Local Group
185	E3	-15.2	5.8	2.2	"
205	E5	-16.4	2.2	0.8	"
221	E2	-16.4	(< 1.5)		"
524	E1	-21.8	5.6	2.0	CfA 13
1052	E4	-20.2	3.0	0.6	HG 44
1374	E0	-19.4	6.4	3.0	HG 17 (Fornax)
1379	E0	-19.7	4.4	1.2	"
1387	SB0	-19.9	3.7	1.0	"
1399	E1/cD	-20.9	16	4	"
1404	E1	-20.6	1.2	0.3	"
3115B	dE,N	-16.7:	6.6	3.0	N3115 companion
3226	E2	-19.6	6.1	2.1	CfA 58
3311	E0/cD	-22.4	(20:)		Hydra I
3377	E5	-19.8	2.8	0.6	CfA 68 (Leo)
3379	E1	-20.8	1.7	0.8	"
3607	E1	-20.9	3.4	2.4	CfA 77
4278	E1	-20.3	8.3	1.0	CfA 94
4340	SB0	-20.1	7.1	2.4	CfA 106 (Virgo)
4374	E1	-21.8	5.7	0.7	"
4406	E3	-22.0	5.4	0.7	"
4472	E4	-22.8	5.5	0.7	"
4486	E1	-22.5	15	4	"
4526	E7	-21.5	6.6	0.9	"
4564	E6	-20.2	9.0	2.5	"
4596	SB0	-20.6	5.5	0.9	"
4621	E5	-21.3	5.8	1.0	"
4636	E1	-21.6	11	1	"
4649	E2	-22.2	6.1	0.6	"
4697	E6	-21.9	4.1	0.5	"
4874	E0	-22.9	(15:)		CfA 113 (Coma)
4889	E4	-23.3	(5:)		"
5128	Ep	-22.1	3.0	0.5	HG 19 (Centaurus)
5813	E1	-21.3	7.6	1.7	CfA 150
5846	E0	-21.7	4.5	2.6	"
Galaxy	Sbc:	-19.8:	2.2	± 0.5	Local Group
224	Sb	-19.6:	4.5	1.5	"
2683	Sb	-18.9:	(8	4)	(isolated)
3115	S0	-21.2	2.0	0.5	(isolated)
4565	Sb	-21.4:	1.0	0.5	CfA 94
4594	Sa	-22.1	3.0	0.3	HG 41

* CfA = Geller and Huchra 1983; HG = Huchra and Geller 1982.

part of this Symposium contains discussions of the GCS's in numerous galaxies from the Milky Way all the way to the Coma Cluster. Most badly needed of all, strong contributions are beginning to come from the theorists on the topics of cluster formation and dynamical evolution of galactic halos. Numerous fundamental questions (How old are the clusters relative to the rest of the halo? Were the clusters responsible for "seeding" the enrichment of the entire halo? How do these factors depend on galaxy environment? What were the conditions that produced the extraordinary GCS's surrounding the supergiant ellipticals at the centers of rich clusters?) will be answerable in detail only with well constructed models that do not yet exist.

2. TOTAL POPULATIONS AND SPECIFIC FREQUENCIES

This review will highlight what we have learned about the overall properties of GCS's observationally since 1980 (see Harris and Racine 1979 for a literature review up to that point). In 1980, a survey program designed to investigate the GCS's in a widely chosen sample of large galaxies was begun, primarily with the wide-field prime focus camera of the CFHT (Harris and van den Bergh 1981; and later papers in that series). By counting the projected number density of starlike images around each galaxy as a function of radius (ideally until the counts drop to a constant "background" level), we derive both the total population of clusters belonging to the galaxy and their radial distribution in the halo (see, e.g., Harris 1986 or Harris and van den Bergh 1981 for a more complete discussion of the methodology). A summary list of currently available data for more than 40 GCS's is given in the adjacent table (excluding only those in Sc/Irr galaxies), and additional new ones are reported in other papers at this meeting.

The first totally new result of the survey program was a comparison of the total populations in the cluster systems, and a test of the earlier belief (Hanes 1977, Harris and Racine 1979) that the number of clusters N found in a galaxy was directly proportional to the total spheroidal luminosity L. The single clear exception known at that time was the enormously populous GCS in M 87, which was thought to be related somehow to its central location in Virgo (Harris and Smith 1976; van den Bergh 1977). A new quantity called the specific frequency, S, (equal to the number of clusters per unit galaxy luminosity; cf. Harris and van den Bergh 1981) was defined to help the process of comparison. The average S-value is plainly not uniform from one galaxy to another; that is, the strict N(L) proportionality rule is followed only in a rough sense. The most obvious variations appear to be correlated with galaxy environment (Fig. 1), such that systems in rich surroundings like the Virgo and Fornax groups have a mean S almost twice as large as those in the smaller, sparser groups which can be found in and around the Local Supercluster. At least part of this difference might be due to the varying importance of galaxy mergers in different environments (Harris 1981), but this approach is not likely to explain the low <S> observed for the disk galaxies in these groups too. Close but less catastrophic encounters between galaxies are capable of rearranging the numbers of halo stars and clusters held by big vs. small galaxies (see Muzzio et al. 1984; Muzzio 1986), but here again it is doubtful whether the full range of the observed S-values can be produced this way, especially the more extreme M 87-type anomaly. The most likely explanation (see the more extensive discussion of alternatives by Harris 1986) may instead require that E galaxies in different locations went through early cluster formation epochs of different intensities, or of different durations; and that the halos of disk galaxies were not as prolific at halo cluster formation as were ellipticals. Other clues to the GCS formation period will be added in the sections following.

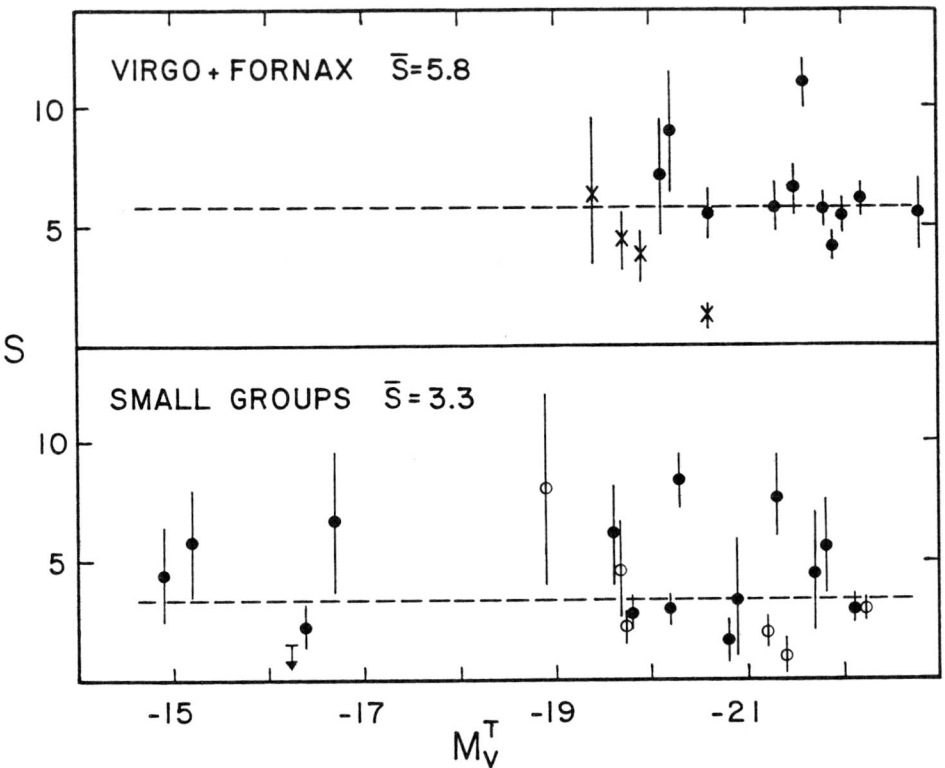

Fig. 1. Specific frequency S plotted vs. galaxy luminosity M_v^T for 36 galaxies. Top panel: dots represent Virgo E/S0 systems, crosses show the Fornax ellipticals. Bottom panel: dots represent E/S0 systems found in small galaxy groups, while open circles represent disk galaxies (in this case M_v^T refers to the total spheroid luminosity excluding the disk). Dashed lines show the mean (weighted inversely as the internal errors) for each group; the systems in the richer environments (at top) have a higher <S> by almost a factor of two.

3. SPACE DISTRIBUTIONS AND RADIAL STRUCTURES

The classical assumption (e.g. Blaauw 1965) is that globular clusters and halo field stars are equally useful tracers of the same oldest Population II. A different view was first expressed by Harris and Racine (1979) that "... the globular cluster systems have significantly larger characteristic sizes than do the parent galaxies ... this may indicate that the clusters were formed at an earlier epoch than were the stars of the spheroidal component, i.e. at a time when the protogalactic material was less centrally condensed."

The most dramatic evidence for this distinction is probably to be found in M 87 itself, where we now have a projected radial profile all the way from the center out to its remotest outskirts (Fig. 2). The difference in slope between the cluster distribution and the halo surface intensity persists everywhere through the galaxy, and provides compelling evidence that, at least in M 87, we are looking at two

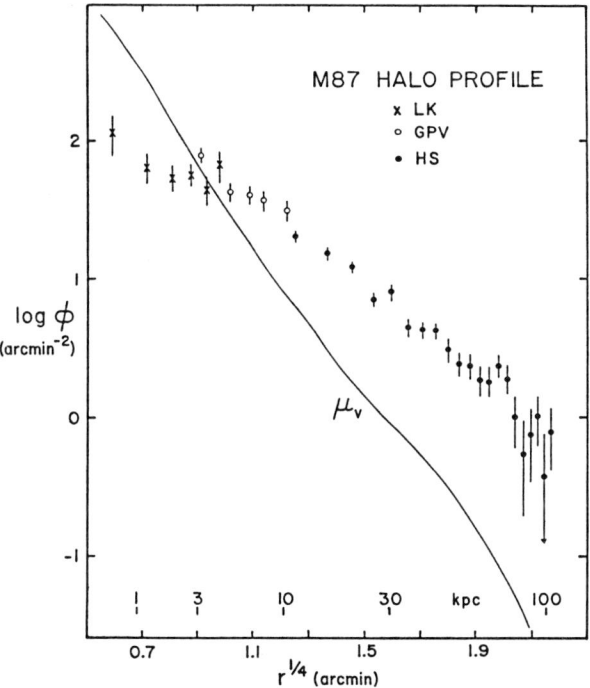

Fig. 2. Radial distribution of globular clusters in M 87. Here ϕ is the number of clusters per square arcmin brighter than B = 23.6, versus galactocentric radius r (the scale in kpc assumes d = 16 Mpc). Crosses are from Lauer and Kormendy (1986); open circles from Grillmair et al. (1986); and filled circles from Harris (1986). The solid line is the V surface intensity of the M 87 halo, from Oemler (1976).

separate populations. Several recent discussions (Lauer and Kormendy 1986; Grillmair et al. 1986; Harris 1986) make it extremely doubtful that such an extended GCS profile could be generated artificially by an ongoing process of dynamical evolution. Preferential destruction of the inner clusters by dynamical friction operates effectively only within 1-2 kpc of the center; and adding large numbers of clusters in the outer parts due to galaxy-galaxy interactions would add similar proportions of halo field stars as well, so that both types of populations would maintain the same radial profiles to first order.

We should, of course, keep in mind that the M 87 GCS is an anomalous case. We need to search for a general process (as opposed to something designed especially for one object) only if we find the same pattern showing up elsewhere. However, the same radial distribution is found in M 49, a Virgo galaxy of the same size and structure as M 87 but well away from the Virgo center and with a thoroughly "normal" specific frequency (Harris 1986). It should also be emphasized that, because the M 87 and M 49 radial distributions are nearly identical in shape (Harris 1986), the basic M 87 "anomaly" of a specific frequency 3 times higher than average exists _everywhere_ through its halo. Thus we cannot solve this anomaly by, for example, adding in large numbers of clusters only in the outer parts of the M 87 halo (either by formation or by tidal accretion) while leaving the inner halo alone. Something to do with the formation of the parent galaxy must be involved in making the GCS more spatially extended than the main halo.

Interestingly, there are two known cases of GCS's in central cD-type galaxies: NGC 1399 in Fornax (Hanes and Harris 1986b) and NGC 3311 in Hydra I (Harris et al. 1983). In both of these, there is good agreement between the halo light profile and the GCS profile - not because the GCS is any different from the M 87 and M 49 cases, but because the cD envelope is more extended than a normal elliptical. For the more normal E's, a few less certain cases are known (cf. Harris and van den Bergh 1981, and Hanes 1977a) in which the GCS appears to have a broader space distribution than the galaxy light; significantly, no E galaxy is known where the reverse is true.

The data for GCS's in two large disk galaxies, NGC 4594 (Harris et al. 1984) and NGC 3115 (Hanes and Harris 1986a) are informative in a different way. Here, the isophotes for each galaxy are quite plainly flattened along the major axis over the spatial region where globular clusters are observed in large numbers (r < 30 kpc). Yet in both cases, the clusters lie in a distribution that is (within the count statistics) indistinguishable everywhere from spherical; that is, the cluster system is rounder than the underlying halo. Acceptable agreement between the radial profiles of the GCS and the halo light is obtained only if the surface intensity along the minor axis is used. Interestingly, H. Harris et al. (1987) report that in another edge-on Sa disk system, NGC 7814, the GCS is flattened much like the spheroid light (by rotation?), yet is more spatially extended than the spheroid!

One last, less certain result has emerged recently in which halo cluster systems seem to mimic a trend with luminosity that the galaxies themselves follow. This is that the GCS appears to become less centrally concentrated with increasing galaxy luminosity (Harris 1987). In terms of a power-law index $a = \Delta \log \sigma / \Delta \log r$ where σ is the projected number of clusters per unit area around the galaxy, the rate of decline of a is about 0.3 (toward lower absolute value) for every 1-magnitude increase in galaxy luminosity. Both the slope of this trend, and the scatter around it, strongly resemble plots of profile slope vs. luminosity for the galaxies themselves (e.g. Strom and Strom 1978), but just displaced toward lower central concentration by $\Delta a = 0.4$.

4. LUMINOSITY FUNCTIONS AND PHOTOMETRIC COLORS

Globular clusters have long been potentially attractive as extragalactic distance indicators, since: (a) they are as luminous as other conventionally used standard candles such as Cepheids, HII regions, supergiants, or novae; (b) they do not suffer from the internal reddening and crowding that affect all Population I indicators; (c) they do not require the long time series of repeated measurements that all variable stars do; and (d) they afford the chance to construct a pure Population II distance scale related ultimately to the assumed luminosity of the RR Lyrae stars. Nevertheless, they will become truly useful as standards only if we can show that they follow a 'universal' luminosity distribution in different galaxies (or, if their intrinsic LF does differ from one type of galaxy to another, we at least need to understand exactly how these differences behave systematically).

The assumption is often made (e.g. Hanes 1977a; Harris and Racine 1979; van den Bergh 1985) that the LF for globular clusters is Gaussian in $n(M_v)$. This use of a normal distribution is (so far) strictly an empirical approximation with no rigorous theoretical basis, but there are more reasons for using it than just numerical convenience. Fig. 3 shows the observed LF for M 87 (van den Bergh et al. 1985) combined with the sum of 6 large Virgo ellipticals (Hanes 1977a) to produce a combined LF spanning almost 5 magnitudes and apparently reaching just past the peak or "turnover" point. It is clear that a straight power law - as has sometimes been proposed (Tremaine 1976; Racine 1980) - will not match this distribution. The normal curve superimposed on the M 87/Virgo data has a dispersion of 1.4 ± 0.1 mag, and the peak near B = 24.7 corresponds to a Virgo distance of 16 Mpc (or H_o = 80 km/s/Mpc). This fit, or others close to it (e.g. Hanes 1987) clearly provides a satisfactory description of the Virgo data.

It is not yet clear just how similar to the M 87/Virgo distribution the LF's of globular cluster systems in other galaxies actually are. The reasons for this are that a small sample size obscures the intrinsic properties of the LF (in the case of our Local Group members, including the Milky Way), or that insufficiently deep or complete photometry is available to reach close to the turnover point (in the case of more distant major ellipticals or spirals). However, at least the intrinsic dispersion of the LF is known to be similar enough in other galaxies to keep them competitive as an interesting distance indicator: for the Milky Way and M 31, van den Bergh (1985) finds σ = 1.2 to 1.3 magnitudes, and for NGC 5128, G. Harris et al. (1984) estimate σ = 1.4 mag. Deep CCD photometry now in progress for the GCS's around 3 more Virgo giant ellipticals and several other small ones in other groups (Harris and Allwright 1987; Hanes and Harris 1987) should soon help answer the question whether the dispersion or peak luminosity of the LF vary significantly between E galaxies. If both σ

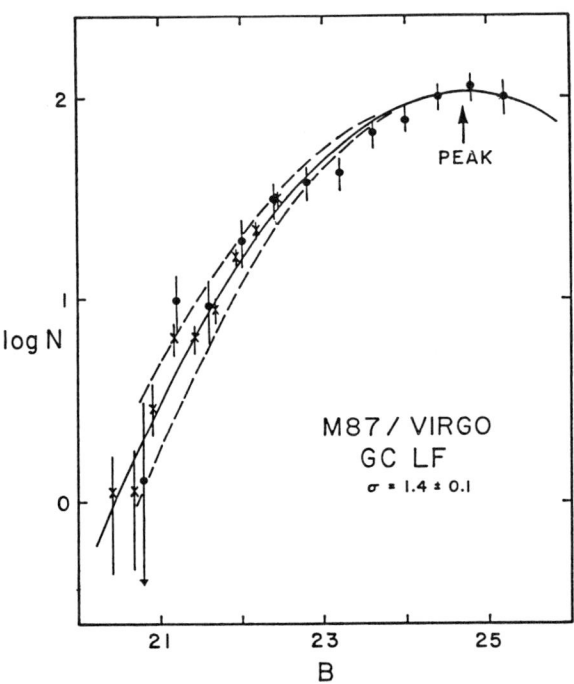

Fig. 3. Luminosity distribution of the globular clusters in the Virgo ellipticals. Crosses represent the sum of globular cluster populations in 6 large Virgo E's, from Hanes (1977a); dots are the faint M 87 sample measured by van den Bergh et al. (1985). A Gaussian curve with peak ("turnover") at B = 24.7 and dispersion 1.4 mag (solid line) is shown fitted to the data, along with similar curves for dispersions of 1.3 and 1.5 mag (dashed lines).

and M_B (peak) turn out to be sensibly constant over the range of parent galaxies that have already been observed (more than a factor of 1000 in total luminosity, and over Hubble types from E to Sb), then the GCS's will also provide important constraints on modelling the protogalactic epoch in which they were formed. It may be that the clusters belonged to such an early stage of the protogalactic collapse that their mass spectrum of formation was relatively uninfluenced by the type or size of galaxy that would eventually emerge (but see Fall and Rees 1985).

In terms of photometric colors and metallicity characteristics, it has been known for some time (Hanes 1977b, Harris and Racine 1979) that clusters in many galaxies are roughly similar, with a slight trend for those in bigger galaxies to be somewhat more metal-rich (redder). An important new step was taken with the work of Strom et al. (1981) and Forte et al. (1981), who concluded from their multicolor photometry of the Virgo ellipticals that the globular clusters at a given galactocentric radius were bluer (and by hypothesis more metal-poor) than the halo light at the same location. When combined with the evidence on their extended spatial distribution (sec. 3 above), this was the second major piece of evidence that the clusters might belong to a somewhat older population than the bulk of the halo stars. The clusters might even be responsible for the first phase of metal enrichment for the entire halo (De Young et al. 1983). Shortly after that, it was pointed out (Harris 1983) that just the same color difference between the clusters and the halo light could also be found in the dwarf ellipticals of the Local Group. Multicolor BVI and Gunn-Thuan CCD photometry currently in progress (Cohen 1986; Harris and Allwright 1987) for the GCS's in several Virgo ellipticals should be able to investigate the reality of this effect further.

As the evidence concerning the differences between galactic spheroids (halos) and their embedded cluster systems has accumulated, support has grown for the single explanation that the clusters are the older of the two, even though they make up only a tiny fraction (less than 1%) of the total halo light. A summary is as follows:

- Globular cluster systems tend to be more spatially extended than normal E galaxies (sec. 3). For the two known GCS's in cD galaxies, it is interesting that good agreement with the halo structure is seen, but this feature may well be a result of whatever merger or capture events have produced the extended cD structure, as opposed to something arising from the initial formation process.

- Globular cluster systems in some disk galaxies are noticeably rounder in projection than the galaxy halo, and match the radial structure of the halo reasonably well only along the minor axis (sec. 3). - Globular clusters in a wide range of E galaxies (Virgo gE's, Local Group dE's) are bluer (and by hypothesis of lower metallicity) than the halo stars at the same projected galactocentric radius. - The globular cluster system in M 87 appears to form a kinematically hotter subsystem (larger velocity dispersion) than the halo light at the same projected radius (Mould et al. 1986; Huchra and Brodie 1986).

If there is a particular, somehow unique, early epoch which is especially suitable to globular cluster formation, then we might envisage that the ellipticals in general got an earlier start in this epoch than the disk galaxies, and that the E's that found themselves in rich environments benefitted even more (Harris 1986). In this way, the observed mean differences in specific frequency (sec. 2), as well as the more extended space distribution, lower metallicity, and kinematics, could all be a natural outcome of the protogalactic collapse as long as this special epoch is earlier than the bulk of the halo star formation by about a free-fall time. The only specific model which gives some theoretical underpinning to such a picture is the recent discussion of Fall and Rees (1985), which gives a plausible reason for the cluster-sized units condensing out significantly earlier than the majority of the material: the gas in smaller clumps or lower-density zones would be prevented by heating from its surroundings from entering gravitational instability and collapse until later on.

5. THE CENTRAL SUPERGIANTS AND THE CLUSTER FORMATION EPOCH

Even if we have arrived at a coherent overall description of the evolutionary role of GCS's in normal galaxies, there is more to the story. The keenest recent interest of all has been attracted to the extraordinary properties of the systems that are found in the supergiant ellipticals located at the centers of rich galaxy groups. There are now four such systems known: M 87 in Virgo, NGC 3311 in Hydra I, NGC 1399 in Fornax (Harris 1986, Hanes and Harris 1986), and NGC 4874 in the Coma Cluster (Thompson and Valdes 1987; Harris 1987). These stand apart in only one respect from the "normal" GCS's described above: they have a specific frequency 3 or more times higher than average. Forming the sheer numbers of clusters that are seen in these dominant central galaxies (M 87, for example, probably holds more than 20,000 clusters) will require special conditions of formation over and above anything that has yet been suggested, in the Fall-Rees scenario or elsewhere. Despite some interesting recent attempts to explain such systems by dynamical simulations involving tidally disrupted neighbors (see Muzzio 1986), there does not seem any clear way to arrive at such a system by taking an existing normal galaxy and then capturing swarms of additional clusters (at all radial distances!) to the exclusion of other types of halo stars. It seems necessary to build them in from the start.

The Coma Cluster galaxies deserve special mention here since, at a mean redshift near 7000 km/s, they are the closest members of an outstandingly rich and dense (Abell class I) cluster. Detecting globular clusters in the two biggest Coma ellipticals, NGC 4874 and 4889, has long been a challenging goal for photometrists in this field. [Comparing the mean redshifts of Coma and Virgo (see table below) shows that Coma should be $\simeq$ 5.6 times (or 3.7 mag) more distant than Virgo.

If the brightest M 87 clusters appear in large numbers for B > 21, then in Coma they should appear at B ≈ 24.7 with their peak near B = 28.4.]

Some modest success in this direction now seems to have been achieved. Thompson and Valdes (1987) have obtained CCD frames with the ISIS camera at the CFHT on Mauna Kea which show a clear excess of starlike images in a small (1') field just off the center of NGC 4874. Harris (1987) has used the RCA1 CCD camera at the CFHT prime focus to obtain deep images of both NGC 4874 and the other Coma supergiant, NGC 4889. The GCS's around both these galaxies do seem to be resolved; preliminary results for their LF's are displayed here in Fig. 4. (It should be emphasized that the fit shown to the M 87/Virgo LF is not a direct distance measurement for Coma, but simply a consistency test that the Coma systems resolve at the magnitude level expected if they are intrinsically similar to the Virgo globular clusters.) Notably, NGC 4889 does not have as large a GCS population as NGC 4874. Comparing the observed numbers in similar areas around each galaxy indicates that their ratio of specific frequencies is S(4874)/S(4889) = 3 ±1. The S-values themselves are highly uncertain because only the top magnitude of the cluster LF is seen, and a large extrapolation is needed to predict the total population; nevertheless, N4874 clearly has a specific frequency similar to that of M 87 or higher.

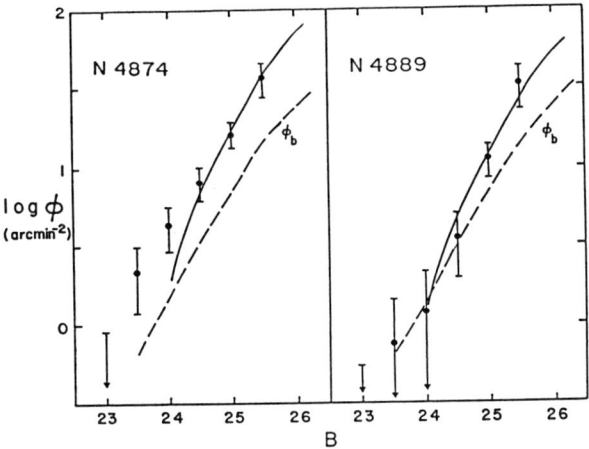

Fig. 4. Preliminary luminosity functions for the globular cluster systems in the supergiant Coma ellipticals, NGC 4874 and 4889 (Harris 1987). Here ϕ is the projected number density of measured images near each galaxy (excluding clearly nonstellar images, and corrected statistically for background objects) per quarter-magnitude interval B. The solid line represents the M 87/Virgo globular cluster LF, shifted fainter by 3.7 magnitudes and fitted to the data; the dashed line ϕ_b is the comparison-field LF which was subtracted from the raw data.

Results for all four central supergiants are summarized in Table II below. Note that in every case, the radial velocity (V_g) of the galaxy is within 1/4 of a standard deviation of the mean velocity (V_{cl}) of its cluster (in the table, $\Delta V = V_g - V_{cl}$, expressed in units of the central velocity dispersion of the cluster). In other words, it appears that all these giants are sitting rather quietly enthroned at the dynamical centers of their surroundings. It is important to note that the anomaly of an outstandingly populous GCS appears to be very strongly connected with position: no large galaxies other than the four listed here are known to have it; and conversely, all the giant E galaxies that can clearly be identified as central objects in rich surroundings (and that have been observed so far) do have it.

TABLE II

Globular Cluster Systems in Central Giant Ellipticals*

Object	M_V^T	V_g	V_{cl}	$|\Delta V|$	S
M 87(N4486)	-22.5	1190	1000	0.23 σ	15
NGC 1399	-20.9	1430	1460	0.06 σ	16
NGC 3311	-22.4	3575	3420	0.17 σ	20:
NGC 4874	-22.9	7120	6950	0.14 σ	15:

* Here M_V^T is the galaxy luminosity assuming HO = 75; S is the specific frequency of the globular cluster system.

Clearly, the story of their formation must run a little differently than for other galaxies. One possibility is perhaps to be found in the comment by Larson (1987) that globular clusters are likely to form within the dense cores of larger, bound units that might have been of dwarf-galaxy size (e.g. $10^8 M_\odot$ or so). Thus the protogalaxies at the centers of big groups would need to have a process which especially encouraged the formation of these dense cores all through the gas-rich proto-halo. It is not unlikely that these same protogalaxies underwent especially early and violent nuclear activity (the quasar phenomenon?), either by being the first objects in their environments to collapse or by suffering a uniquely high interaction rate with the other protogalaxies around it (cf. Stocke and Perrenod 1981; Yee and Green 1984; De Robertis 1985). The resulting shocks moving through the halo gas might then have started proto-globular cores forming in unusual numbers (Gunn 1980; McCrea 1982). At least on a qualitative level, this approach would not conflict with the Fall-Rees picture that heating from the surrounding medium simultaneously prevented the less massive clouds from condensing until their metal abundance was higher or the surroundings were cooler.

An alternative idea mentioned, e.g., by Fall and Rees, is that heavy gas infall and the development of a cooling flow in these big galaxies allowed their epoch of cluster formation to continue long past the normal galaxy formation stage. This is unlikely, for the reason that the period of cluster formation must have finished clearly before the bulk of the halo started condensing - otherwise, both types of objects would now have the same radial distribution and metallicity, which is contrary to the observations. The existing measurements on radial distribution, specific frequencies, colors and metallicities, and kinematics (see sec. 4 above) for GCS's in large galaxies already place difficult constraints on appropriate models for their formation. It seems necessary to construct a model in which the clusters formed in a rather narrowly defined epoch, clearly before the rest of the halo but after the even more extended "dark" matter, almost independent of galaxy size or type but with some noticeable influence from environment. If we could understand why these remarkable systems exist, we are likely to understand much more about the entire galaxy formation process.

For stimulating ideas related to these topics, I am indebted to D. Hanes, R. Larson, R. Pudritz, M. Rees, and numerous other colleagues. M. Hazen and H. Liller provided invaluable hospitality during this conference. This work was supported by an NSERC (Canada) operating grant to the author.

REFERENCES

Blaauw, A. 1965 in Galactic Structure, A.Blaauw and M.Schmidt, eds., U. Chicago Press, Chicago, p.435.
Cohen, J. 1986 preprint.
De Robertis, M. 1985 Astron.J. 90, 998.
De Young, D. S., Lind, K., and Strom. S. E. 1983 Publ. Astron. Soc. Pacific 95, 401.
Fall, S. M. and Rees, M. J. 1985 Astrophys. J. 298, 18.
Forte, J. C., Strom, S. E. and Strom, K. M. 1981 Astrophys. J. Letters 245, L9.
Geller, M. J., and Huchra, J. P. 1983 Astrophys. J. Suppl.52, 61.
Grillmair, C. J., Pritchet, C. J. and van den Bergh, S. 1986, Astron. J. 91, 1328.
Gunn, J. E. 1980 in Globular Clusters, D. Hanes and B. Madore, eds., Cambridge Univ. Press, Cambridge, p.301. Hanes, D. A. 1977a Mem. Roy. Astron. Soc. 84, 45; and 180, 309.
Hanes, D. A. 1977b Mon. Not. Roy. Astron. Soc. 179, 331.
Hanes, D. A. 1980 in Globular Clusters, D.Hanes and B.Madore, eds., Cambridge Univ. Press, Cambridge, p.213.
Hanes, D. A. 1987 in IAU Symposium No. 126, Globular Cluster Systems in Galaxies, J. Grindlay and A. G. D. Philip, eds., Reidel, Dordrecht, p. 617.
Hanes, D. A. and Harris, W. E. 1986a Astrophys. J. 304, 599.
Hanes, D. A. and Harris, W. E. 1986b Astrophys. J. 309 (in press).
Hanes, D. A. and Harris, W. E. 1987 in preparation.

Harris, G. L. H., Hesser, J. E., Harris, H. C. and Curry, P.J. 1984
    Astrophys. J. 287, 175.
Harris, H. G., Bothun, G. D. and Hesser, J. E. 1987 in IAU
    Symposium No. 126, Globular Cluster Systems in Galaxies, J. E.
    Grindlay and A. G. D. Philip, eds., Reidel, Dordrecht, p. 613.
Harris, W. E. 1981 Astrophys. J. 251, 497.
Harris, W. E. 1983 Publ. Astron. Soc. Pacific 95, 21.
Harris, W. E. 1986 Astron. J. 91, 822.
Harris, W. E. 1987, preprint.
Harris, W. E. and Allwright, J. W. B. 1987 in preparation.
Harris, W. E., Harris, H. C. and Harris, G. L. H. 1984 Astron. J.
    89, 216.
Harris, W. E. and Racine, R. 1979 Ann. Rev. Astron. Astrophys. 17,
    241.
Harris, W. E. and Smith, M. G. 1976 Astrophys. J. 207, 1036.
Harris, W. E., Smith, M. G. and Myra, E. S. 1983 Astrophys. J. 272,
    456.
Harris, W. E. and van den Bergh, S. 1981 Astron. J. 86, 1627.
Hubble, E. 1932 Astrophys. J. 76, 44.
Huchra, J. P. and Brodie, J. 1986 preprint.
Huchra, J. P. and Geller, M. J. 1982 Astrophys. J. 257, 423.
Larson, R. 1987 in IAU Symposium No. 126, Globular Cluster Systems in
    Galaxies, J. Grindlay and A. G. D. Philip, eds., Reidel, Dordrecht,
    p. 311.
Lauer, T. R., and Kormendy, J. 1986 Astrophys. J. Letters 303, L1.
McCrea, W. H. 1982 in Progress in Cosmology, A. Wolfendale, ed.,
    Reidel, Dordrecht, p.239.
Mould, J. R., Oke, J. B. and Nemec, J. M. 1986 Astron. J. (in press).
Muzzio, J. C. 1986 Astrophys. J. 301, 23.
Muzzio, J. C., Martinez, R. E. and Rabolli, M. 1984 Astrophys. J.
    285, 7.
Oemler, A. 1976 Astrophys. J. 209, 693.
Racine, R. 1980 in Star Clusters, IAU Symposium No. 85, J.E.Hesser,
    ed., Reidel, Dordrecht, p.369.
Shapley, H. 1918 Astrophys. J. 48, 154.
Stocke, J. T. and Perrenod, S. C. 1981 Astrophys. J. 245, 375.
Strom, S. E., Forte, J. C., Harris, W. E., Strom. K. M., Wells, D. C.
    and Smith, M. G. 1981 Astrophys. J. 245, 416.
Strom, S. E. and Strom, K. M. 1978 Astron. J. 83, 732.
Thompson, L. A. and Valdes, F. 1987 preprint.
Tremaine, S. 1976 Astrophys. J. 203, 345.
van den Bergh, S. 1977 Vistas Astron. 21, 71.
van den Bergh, S. 1985 Astrophys. J. 297, 361.
van den Bergh, S., Pritchet, C. J. and Grillmair, C. J. 1985,
    Astron. J. 90, 595.
Yee, H. K. C. and Green, R. F. 1984 Astrophys. J. 280, 79.

DISCUSSION

ZINN: It appears to me that most, if not all, of the evidence that the globular clusters constitute a more distended and bluer population than the halo stars comes from observations of elliptical galaxies. Most of the evidence that globular clusters and halo stars are very similar in metallicity, spatial distribution, and kinematics comes from observations of one disk galaxy, the Milky Way. Is there an example of a disk galaxy in which the globular cluster population is definitely more distended and bluer than the population of halo stars?

HARRIS, W.: As far as the metallicity of the halo population is concerned, there are no other disk galaxies aside from the M 31 and the Milky Way for which metallicities or photometric colors are available yet. For the space distributions, H. Harris and J. Hesser show that the clusters around NGC 7814, an edge-on Sa disk galaxy, form a more extended space distribution than the halo light; and as far as NGC 4594 and 3115 are concerned, the cluster distribution agrees with the light distribution reasonably well only along the minor axis, which really suggests that the cluster system is intrinsically more extended.

SCHOMMER: I am rather worried about a similar problem. The brightest outer cluster in M 33 has $(B-V) = 0.51$ $((B-V)_0 <0.45)$. I think this is a massive intermediate age cluster and not a true globular cluster, color measures are clearly important; the differences Strom et al. noted could easily be age and not metallicity. There is some, admittedly controversial, evidence that E galaxies have an intermediate age stellar population.

HARRIS, W.: You've noticed that in my summary I neglected the Sc and irregular galaxy systems; in the later-type galaxies with bigger halos that problem is probably less severe. However, it does point up the need for more color information in all these systems.

OSTRIKER: Many of the differences that you find between the elliptical light distribution and the globular cluster system may be due to evolution of the cluster system. In centrally concentrated galaxies the inner clusters are very effectively destroyed.

HARRIS, W.: Most of the statement that the radial distribution of the clusters vs. the halo stars are different relies on the observed regions more than 10 kpc from the galaxy center and not the nuclear region. It will be interesting to see if the destruction mechanisms are effective that far out, or farther.

KING: This question is meant with a spirit of caution rather than challenge. How can you be sure of a Gaussian fit when there is only one point on the other side of the peak?

HARRIS, W.: My comments about the "Gaussian" nature of the distribution were pragmatic ones rather than any statement about the true shape of the whole distribution. For example, vandenBergh (1985 Astrophys. J.) has suggested that the luminosity function for the galactic globular clusters is somewhat asymmetric when the faint half is included. However, the clusters fainter than the peak point are somewhat irrelevant to the distance fitting problem since they are never seen in the more distant galaxies. That means that we're primarily interested in the shape of the brighter half alone, which is why I showed the fit to the Virgo luminosity function the way I did.

BAUM: You called attention to the unusually high specific frequencies of globular clusters associated with central galaxies in four clusters (including NGC 4874 in Coma) and you mentioned the "stationarity" of those galaxies (judged from radial velocities) relative to the rest of the cluster. Yet in the Coma Cluster there are two central galaxies, so neither may be a unique stationary infall locus. The other central galaxy in Coma is NGC 4889. Do you have globular cluster data for NGC 4889? Is it different from NGC 4874?

HARRIS, W.: Yes, I have new CCD data (from the CFHT this March) for both NGC 4874 and 4889. The key result is that 4874 seems to have about 3 ± 1 times as many clusters (excess images) than 4889, which would make it analogous to the M 87/M 49 situation in Virgo. The velocity of NGC 4889 is a few hundred km/sec farther away from the Coma mean than NGC 4874, and also NGC 4874 is the center of a bright radio source, so on balance it looks as if NGC 4874 is more likely to be the central object.

CAYREL: You have used a specific number per unit of visual absolute magnitude. Is it true that S would be more stable versus the morphological type if defined with respect to something more related to the mass of the galaxy?

HARRIS, W.: A specific frequency defined relative to mass instead of total light might well be more interesting, but would be harder to define, since the mass distribution is so much more extended than the galaxy light and it's not clear how you would cut it off at some radius to get a total mass that was meaningful. Getting a total luminosity for just the halo (e.g in a disk galaxy) is difficult too, but we haven't been able to think of anything better to use as yet.

HANES: I refer Ivan King to my poster paper, where I show that one can fit Gaussians to truncated data sets with remarkable confidence via maximum likelihood methods. It is time that some people interested in H_o have alternatively supported and then criticized the use of globular clusters as distance indicators, but let me record here that de Vaucouleurs has always been very positive about their use ever since he first used them in his 1977 Nature paper.

ROBERTS: You've described globular clusters as having a Gaussian like luminosity function. It is worth noting that the luminosity functions for stars in our galaxy, galaxies themselves, and clusters of galaxies do not have such a peaked shape. If there is a universal luminosity function we can surely learn much from these differences.

HARRIS, W.: As you suggest, the luminosity function of the clusters as we see them now is a combination of how they formed plus what has happened since then to destroy them selectively. So we should be able to learn about both processes. If we really want to use globular clusters as distance indicators, we need to do a lot more work in understanding the detailed similarities and/or differences from one galaxy to another. The main difference here compared with the other indicators like Cepheids and supergiants is that the people working on those other indicators have spent the time to find out what the first and second order characteristics of their standard candles actually are like from one galaxy to another. Because of the lack of data for enough galaxies to enough depth, we have not got that far with the globular clusters so far.

THE M 87 GLOBULAR CLUSTER SYSTEM*

John Huchra

Harvard-Smithsonian Center for Astrophysics

ABSTRACT: We review early and recent photometric studies of the M87 globular cluster system. We also describe recent high quality spectroscopic studies that have been used both to measure the metallicities of M87 globular clusters and to determine the mass of M87's halo.

1. HISTORY

Studies of globular cluster systems around galaxies are useful for many purposes: determining extragalactic distances, probing of galaxy halos and testing theories of galaxy and galaxy halo formation. The M87 system has been particularly heavily studied despite the difficulty of observing faint objects in an extended galaxy halo at the distance of Virgo. M87 is a unique object in our neighborhood. It is the nearest brightest cluster galaxy. It sits at the center of an extended halo of X-ray gas (eg. Fabricant and Gorenstein 1983), although it is *not* at the center of the *galaxy* distribution (Huchra 1985). It has a veritable plethora of globular clusters — the current estimate of clusters around M87 is $\sim 20,000$, giving M87 one of the highest specific frequencies known (eg. Harris 1986). Virgo is also a key cluster in the study of the extragalactic distance scale.

Although nearby extraglactic globular cluster systems were discovered as early as 1930 (Hubble 1932), discovery of the M87 remained for the construction of the great, 5-meter Hale telescope (Baum 1955, Sandage 1961). Most of the early work on the cluster system concentrated on its photometric properties, ie. colors, magnitudes, luminosity function, and the comparison of these properties with those of the much more easily studied cluster systems of our own galaxy and M31. Following Baum's (1955) suggestion, much effort has also been devoted to the use of the M87 cluster system to derive the distance to the Virgo cluster and thus determine the local value of the Hubble constant.

* This work based on observations taken with the MMT, a joint facility of the Smithsonian Institution and the University of Arizona.

Early photographic efforts to catalog the globular clusters around M87 were able only to distinguish clusters outside the inner few arc minutes of the galaxy; more recent observations with CCD's have allowed the study of the distribution into the galaxy core and thus determination of the system strucural parameters. Most recently, spectroscopic studies are being used to determine both the chemical and kinematical properties of the cluster system. Results of these studies can be compared to the predictions of several different theories of the origin and evolution of globular cluster systems around galaxies and around M87 in particular.

1.1 Photometric Properties

The earliest detailed study of the M87 system was by R. Racine (1968a,b). Racine studied the magnitudes colors of 1000 objects within 7.5' of the galaxy center and brighter than $B = 23.5^m$. Most were also outside 3.5'. Racine concluded that the colors of these clusters were consistent with those of galactic globulars and that the bright end of their luminosity function was similar to those of the galaxy and M31. Racine derived a value of the Virgo cluster distance modulus, $(m - M)_B = 30.7 \pm 0.2$. Racine also noted that the spatial distribution of the clusters in the outer parts of the halo was similar to the galaxy photometric profile.

Further photometric studies have been made by Hanes (1971, 1977, etc.), Ables, Newell and O'Neil (1974), Strom et al. (1981), and Cohen (1986). Hanes derived luminosity functions for the cluster systems of M87 and several other Virgo galaxies in an attempt to study the form of the cluster luminosity function and derive the Virgo distance modulus. Hanes work suggests that the number of globular clusters is proportional to the *mass* of the parent galaxy. Ables et al. (1974) presented electronographic BV photometry for $\sim$ 100 clusters to accurately calibrate exisitng photographic photometry.

Strom et al. (1981) used photographic UBR photometry of $\sim$ 1700 clusters between 1.'5 and 9.'0, and brighter than $B = 23.5^m$ to investigate their color distribution. They confirmed Hanes' luminosity function, discovered a radial color gradient in the clusters (clusters become bluer with increasing radius), discovered a *difference* of $\Delta(U - R) \sim 0.5^m$ between the mean cluster color and the galaxy halo color at the same radius, and found that the cluster luminosities were *not* correlated with color. Strom et al. (1981) interpret the blue colors of the globular clusters to show both a radial metallicity gradient and a difference between the mean cluster metallicity and the halo metallicity. This color difference has recently been confirmed by Cohen (1986, this conference) with CCD photometry.

Membership of the brighter candidate clusters has also been derived by Prociuk (1976), who used proper motions to remove foreground stars.

1.2 Globular Clusters and the Virgo Distance

Many authors starting with Baum's (1955) original suggestion have used the luminosity function of M87's globular clusters to derive a distance to M87 and thus to the Virgo Cluster. Their use has fallen in and out of favor as the debate about the Hubble constant raged. Table I summarizes some of the values of the Virgo distance moduli that have been derived over the years:

TABLE I
Virgo Distance Moduli
Derived by Globular Clusters

Baum (1955)	30.2
Racine (1968b)	30.7
Sandage (1968)	31.1
deVaucouleurs (1977)	30.2
Hanes (1977)	30.4
Hanes (1979)	30.7
van Den Bergh (1985)	31.2

The use of globular cluster systems as distance indicators is the subject of many other reviews.

1.3. Spatial Distribution

In addition to the early work of Baum (1955) and Racine (1968a,b), the spatial distribution of M87's globular clusters was also studied photographically by Harris and Smith (1976) and Strom et al. (1981). All the photographic work seemed to confirm the similarity of the spatial distribution of the globular cluster system, $N(r)$, with the distribution of light in the halo, $\Sigma(r)$. Recent CCD studies of clusters in the inner few arc minutes of M87 by van den Bergh et al. (1985; Grillmair, Pritchet and van den Bergh 1986) and by Lauer and Kormendy (1986) now have shown that this similarity is an artifact due to the poor sampling of the clusters at small radii — because photographic plates lack dynamic range — and the similarity of $R^{1/4}$ power laws at large radii. The cluster system is much more extended than the galaxy halo, with an effective radius, r_e, somewhere between 10' and 20' (Grillmair et al. 1986) – 5 to 10 times that of the galaxy halo (eg. de Vaucouleurs and Nieto 1979). Similarly the core radius, r_c, of $\sim$ 90" derived by Lauer and Kormendy (1986) is also much larger than that of M87, $r_c \sim$ 7".

1.4 Spectroscopy

The first crude spectra of M87 globular cluster candidates were taken by Racine, Oke and Searle (1978). These were low quality, low resolution 200" Multichannel scanner observations of three objects. The brightest M87 globular clusters are barely above 20th magnitude in B. From these spectra, Racine, Oke and Searle were able to estimate that the mean metallicity, $[Fe/H]$, of these three clusters was $\sim$ -0.7, or one-fifth solar. Since then several groups have tried to obtain higher quality spectra of larger numbers of objects. Huchra and Brodie (1984) found that most of the bright ($B < 19.5^m$) candidates in the list of Strom et al. (1981) were either foreground stars or members of a backgound cluster of galaxies at a redshift of $z \sim 0.09$. Hanes and Brodie (1986) observed five of the brighter globular clusters from Hanes' list and derived a mean $[Fe/H] = -0.5 \pm 0.4$. Huchra and Brodie (1986) and Mould, Oke and Nemec (1986) have observed larger numbers of objects to both measure $[Fe/H]$ and study the dynamics of the globular cluster system. The former study will be reported on below.

2. THE MMT STUDY

Huchra and Brodie (1984; 1986; Brodies and Huchra 1987) have recently obtained spectra of ~ 30 globular cluster candidates surrounding M87. The spectra were obtained with the MMT, and cover the range 3200Åto 7000Åat $\sim$ 7Åresolution. The S/N obtained was sufficient to measure velocities accurate to $\sim 75\ km/s$ and to measure line indices for metallicity determinations.

Cluster candidates were taken from the lists of Hanes(1971) and Strom et al. (1981). Globular clusters were identified via their velocities and spectroscopic properties. Almost all of the brightest objects, $B < 19.5^m$, are foreground galactic stars or background galaxies. The fraction of candidates that were *bona fide* globulars rose dramatically below $B \sim 20.5^m$. Figure 1 shows the velocity histogram for the objects with velocities less than 3000 km/s. There is an almost clear separation between clusters and galactic stars. We have observations of 10 clusters.

2.1 Dynamics

The velocity data can be used to derive two different estimates of the mass of the M87 halo – the Virial theorem estimator and the projected mass estimator (Bahcall and Tremaine 1981). The Virial theorem mass for a cluster of N objects is:

$$M_{VT} = \frac{3\pi N}{2G} \frac{\sum_i^N V_i^2}{\sum_{i<j} 1/r_{ij}} ,$$

r_{ij} is the separation between the ith and jth galaxy, V_i is the velocity difference between the ith galaxy and the mean cluster velocity. The projected mass for N objects surrounding a central mass is

$$M_P = \frac{f_p}{GN} (\sum_i^N V_i^2 r_i) ,$$

where r_i is the separation of the ith galaxy from the centroid. The quantity f_p depends on the distribution of orbital eccentricities for the galaxies and is equal to $32/\pi$ for radial orbits and $16/\pi$ for isotropic orbits. Bahcall and Tremaine (1981) favor the value $24/\pi$ for the projection factor. The parameters we derive are given in Table II.

Zinn, R. 1985, *Astrophys. J.* **293**, 424.

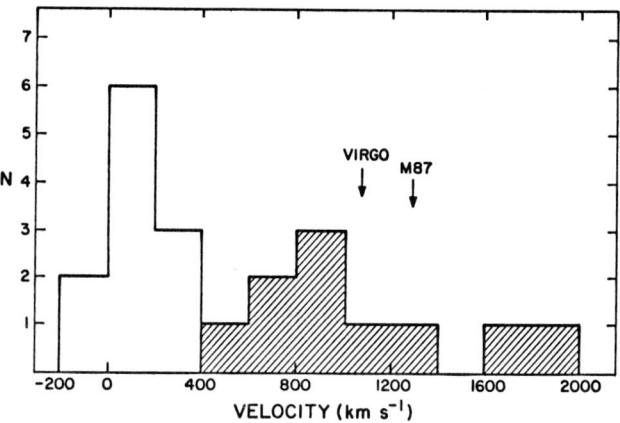

Fig. 1. Velocity histogram for objects around M87. The objects identified as globular clusters are crosshatched.

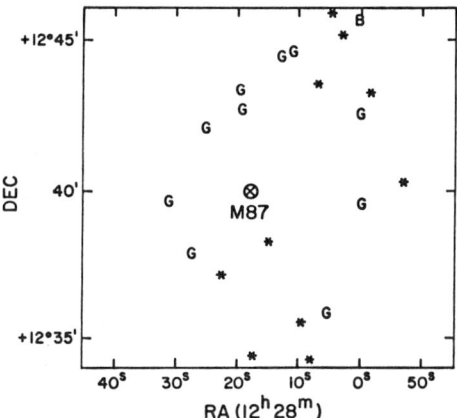

Fig. 2. Spatial distribution of objects observed around M87. Globular clusters are marked 'G', galactic stars are '*', and the compact companion galaxy, NGC 4486B, is marked 'B'.

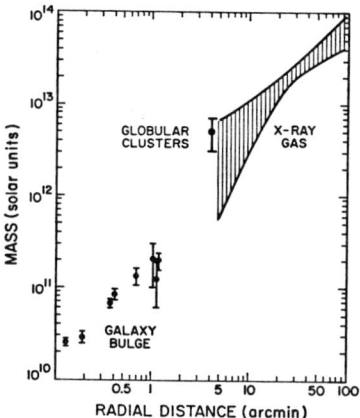

Fig. 3. Mass estimate from the globular cluster system compared to the mass versus radius derived from the X-ray halo and from inner stellar velocity dispersion measurements (Sargent et al. 1978).

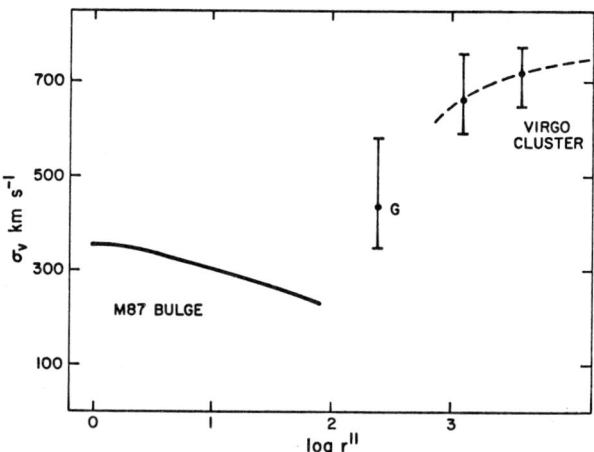

Fig. 4. Velocity dispersion as a function of radius for the M87-Virgo Cluster system.

TABLE II
DYNAMICAL PARAMETERS

	Virial Theorem	Projected Mass
Mass	$5.31 \times 10^{12}\ M_\odot$	$6.08 \times 10^{12}\ M_\odot$
error	$\pm 1.97 \times 10^{+12}\ M_\odot$	$\pm 2.24 \times 10^{+12}\ M_\odot$
R_H	$14.7\ kpc$	
R_P		$16.6\ kpc$
CT	9.0×10^{-4} Hubbles	
$<v>$	$1090 \pm 262\ km\ s^{-1}$	
$<\sigma_v>$	$436^{+143}_{-72}\ km\ s^{-1}$	

R_H and R_P are the mean harmonic radius and the projected radius, respectively. CT is the cluster system crossing time in units of the Hubble time (we assume a *distance* to M87 of $15.7\ Mpc$, which, combined with a Virgo infall velocity of $250 km/sec$ implies a Hubble constant of $82\ km/s/Mpc$). The photometry of Aaronson and Mould (1981) implies that $L \sim 4.1 \times 10^{10}\ L_\odot$ inside $18\ kpc$, which means that the enclosed, integrated mass-to-light ratio, $(M/L)_B \sim 150$. In an annulus between 2' and 5', $M_P \sim 5 \times 10^{12}$, and $L \sim 1.4 \times 10^{10}$, so $(M/L)_{B,ann} \sim 360$. The mass we derive is in good agreement with the mass derived from studies of the X-ray halo (Fabricant and Gorenstein 1983, see Figure 3), although the cluster velocity dispersion is almost a factor of two higher than the predicted stellar velocity dispersion (Stewart et al. 1984). The large velocity dispersion of the outer cluster system combined with the extened distribution of the clusters relative to the galaxy bulge probably indicates that the cluster system is dynamically evolved and that clusters at large radii are predominantly on circular orbits. We believe that it is also important to remember that M87 lies at the center of the Virgo cluster — the velocity dispersion of the globular clusters is higher than that of the stars in the halo but half that of the galaxy cluster (Figure 4).

3.2 Metallicity

Cluster metallicities can also be derived from narrow band indicies and colors (Burstein et al. 1984; Hanes and Brodie 1986; Huchra et al. 1986). The best of these are the Mgb and MgH indices, the G band, the blue and ultraviolet CN features and the H+K index, although H+K tends to sature above $[Fe/H] \sim -1$. The mean metallicity we derive is $[Fe/H] \sim -0.7 \pm 0.3$, which is in good agreement with the values of -0.5 ± 0.4 of Hanes and Brodie (1986) and of -1.2 ± 0.2 of Mould et al. (1986), although those two determinations do not agree. The mean metallicites of several globular cluster systems (Huchra et al. 1986) as well as that of 11 Virgo dwarf galaxies are compared in Table III.

TABLE III
GLOBULAR CLUSTER SYSTEM METALLICITIES
[Fe/H] Zinn 1985 Scale

Fornax Globulars	-1.8
Virgo Dwarf Galaxies	-1.6
Mean of Galactic Globulars	-1.4
M31 Globulars	-1.3
M87 Globulars	-0.7

The data for M87 appear to confirm the trend of increasing mean metallicity with galaxy luminosity/mass. The metallicities of individual clusters are spread over a large range, as in the galaxy and M31. We do not find any evidence for a radial gradient in metallicity (Figure 5), but the data are too sparse to perform a strong test.

3. SUMMARY

3.1 Distance to Virgo and H_o

Almost all of the early work gave H_o between 80 and 100 $km/s/Mpc$ uncorrected for Virgo infall. The most recent attempt by van den Bergh (1985), which includes the effect of Virgo flow, gave ~ 75 $km/s/Mpc$. This value is not far from that currently advocated by the Infrared Tully-Fisher consortium (Aaronson et al. 1986).

In recent reviews of the distance scale problem, however, Sandage and Tammann give the globular cluster luminosity function a grade of $D+$ as a distance indicator. The major difficulty is that most of the well studied systems outside the local group are in elliptical galaxies, while the calibrating systems are all in Sb spirals — the galaxy and M31. This author also wishes to urge caution regarding use of the M87 system until a better understanding of the formation and evolution of cluster systems around giant, X-ray gas rich, central cluster elliptical galaxies is achieved.

3.2 Chemical Probe of Galaxy Halo

If the colors for the M87 clusters are related to metallicity in the same way as in the galaxy and M31 (a pretty safe bet), then the work of Racine (1968b), Strom et al. (1981) and Cohen (1986) tells us that:

A. The M87 clusters are similar in properties and range of properties to galactic and M31 globular clusters.
B. There is a metallicity gradient in both the cluster system and halo.
C. The globular clusters are more metal poor than the halo at the same projected radius.

In addition, spectroscopic studies indicate that the M87 clusters are probably slightly more metal rich on average than galactic or M31 clusters. The values of $[Fe/H]$ derived in the recent studies range between -1.2 and -0.5. This agrees with

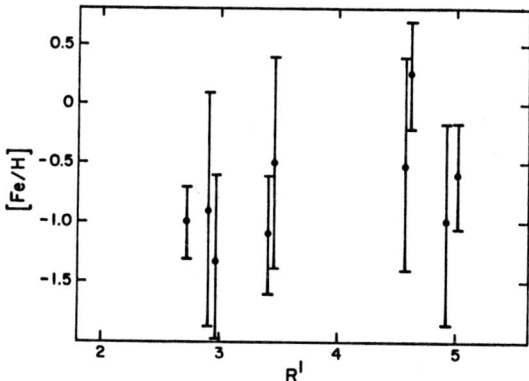

Fig. 5. Metallicity estimates for individual M87 globular clusters plotted as a function of radial distance from the galaxy center.

the general trend of mean metallicity versus luminosity seen in local group globular cluster systems and in elliptical galaxies.

3.3 Dynamical Probe of Galaxy Halo

Globular clusters can be used as 'test particles' to study halos of galaxies. The structure of the cluster system can describe the shape of the gravitational potential well and the kinematics can be used to measure the mass of the galaxy's halo. Globular clusters themselves probably contribute only slightly to the overall mass of the halo.

Although early work on the structure of the M87 cluster system indicated that at large radii, $\mu(r)_{Halo} = N(r)_{Glob}$, the best recent work shows that $R_c^G \gg R_c^H$ as well as $R_e^G \gg R_e^H$. The gloubular cluster system is much more extended than the galaxy halo.

The recent kinematic studies of small numbers of clusters find that the velocity dispersion of the cluster system is $\sim 400\ km/s$. This is higher than that of the galaxy bulge but still lower than that of the surrounding cluster of galaxies. The galaxy mass estimated from these studies is slightly less than $10^{13}\ M_\odot$ inside a radius of 20 kpc. This is in good agreement with the mass of the halo estimated by assuming that the X-ray halo is in hydrostatic equilibrium. The integrated mass-to-light ratio out to this radius is then $(M/L)_B \sim 150$, in solar units.

3.4 Formation of Globulars and the Galaxy Halo

Several theories for the formation of globular cluster systems have been proposed over the years, and all have their attractive features and problems. M87 is a key galaxy in these theories because of its very large specific frequency of clusters (it has about 5 times as many as its sister galaxy M49, which is, in fact, the brightest galaxy in the Virgo cluster (Aaronson and Mould 1981). Some theories have been proposed specifically *for* M87, because of its special place at the galaxy cluster core. The theories can be roughly categorized as:

- A. Formation in Place
    1. Halo/Globulars Coeval (Peebles and Dicke 1968)
    2. Clusters before Halo (Fall and Rees 1985)
- B. Stripping and Accretion - (eg. Forte et al. 1982)

The metallicity difference between clusters and halo is an argument in favor of either accretion or formation of clusters before the halo. The acrretion process, however, is probably *not* efficient enough to create *all* of M87's system (eg. Muzzio et al. this conference). The metallicity versus galaxy luminosity relation also argues against accretion of stripped globulars from dwarf galaxies.

In any case, the cluster system distribution and kinematics indicate that it is very likely that the M87 cluster system has evolved (eg. Chernoff et al. 1986, and this conference). Evolution will make it even more difficult to disentangle the origins of the cluster system.

REFERENCES

Aaronson, M., Bothun, G., Mould, J., Huchra, J., Schommer, R. and Cornell, M. 1986 Astrophys. J. 302, 536.
Aaronson, M. and Mould, J. 1981 Publ. Astron. Soc. Pacific 93, 20.
Ables, H., Newell, E. and O'Neill, E. 1974 Publ. Astron. Soc. Pacific 86, 311.
Bahcall, J. and Tremaine, S. 1981 Astrophys. J. 244, 805.
Baum, W. 1955 Publ. Astron. Soc. Pacific 67, 328.
Brodie, J. and Huchra, J. 1987 in preparation.
Burstein, D., Faber, S., Gaskell, M. and Krumm, N. 1984 Astrophys. J. 287, 586.
Chernoff, D., Kochanek, C. and Shapiro, S. 1986 Astrophys. J., in press.
de Vaucouleurs, G. 1977 Nature 266, 126.
de Vaucouleurs, G. and Nieto, J.-L. 1979 Astrophys. J. 230, 697.
Fabricant, D. and Gorenstein, P. 1983 Astrophys. J. 267, 535.
Fall, M. and Rees, M. 1985 Astrophys. J. 298, 18.
Forte, J. C., Martinez, R. E. and Muzzio, J. C. 1982 Astron. J. 87, 1465.
Grillmair, C., Pritchet, C. and van den Bergh, S. 1986 Astron. J. 91, 1328.
Hanes, D. A. 1971 M. Sc. Thesis, U. of Toronto.

Hanes, D. A. 1977 Mem. Roy. Astron. Soc. 84, 45.
Hanes, D. A. and Brodie, J. 1986 Astrophys. J. 300, 279.
Harris, W. E. and Smith, M. G. 1976 Astrophys. J. 207, 1036.
Harris, W. E. 1986 Astron. J. 91, 822.
Hubble, E. P. 1932 Astrophys. J. 76, 44.
Huchra, J. 1985 in The Virgo Cluster, O.-G. Richter and B. Binggelli, eds., ESO, Garching, p. 181.
Huchra, J. and Brodie, J. 1984 Astrophys. J. 280, 547.
Huchra, J. and Brodie, J. 1986 Astron. J., in press.
Huchra, J., Stauffer, J. and Brodie, J. 1986, in preparation.
Lauer, T. and Kormendy, J. 1986 Astrophys. J. Letters 303, L1.
Mould, J. R., Oke, J. B. and Nemec, J. M. 1986 Astron. J., in press.
Peebles, P. J. E. and Dicke, R. 1968 Astrophys. J. 154, 891.
Prociuk, I. 1976 M. Sc. Thesis, Univ. of Toronto.
Racine, R. 1968a Publ. Astron. Soc. Pacific 80, 326.
Racine, R. 1968b Journal Roy. Astron. Soc. Canada, 62, 367.
Racine, R., Oke, J. B. and Searle, L. 1978 Astrophys. J. 223, 82.
Sandage, A. 1961 in The Hubble Atlas of Galaxies, Carnegie Inst. of Washington, Washington, DC.
Sandage, A. 1968 Astrophys. J. Letters 152, L149.
Sargent, W. L. W., Young, P. J., Boksenberg, A, Shortridge, K., Lynds, C. R. and Hartwick, F. D. A. 1978 Astrophys. J. 221, 831.
Stewart, G. C., Canizares, C. R., Fabian, A. C. and Nulsen, P. E. J. 1984 Astrophys. J. 278, 536.
Strom, S. E., Forte, J. C., Harris, W. E., Strom, K. M., Wells, D. C. and Smith, M. G. 1981 Astrophys. J. 245, 416.
van den Bergh, S., Pritchet, C. and Grillmair, C. 1985 Astron. J. 90, 585.

DISCUSSION

MUZZIO: Our more recent results suggest that massive central galaxies may increase their globular cluster population indeed, but by not more than about 40%; thus, I do not think now that it is possible for M 87 to get three times the "normal" population by taking its globular clusters from other galaxies. Besides, I have just found that our original idea that the central giant galaxies capture globular clusters from dwarf galaxies was wrong: the mean magnitude difference between the donor galaxy and the capturing galaxy is about 1.4 mag and one cannot have a large metallicity difference from so similar galaxies.

HUCHRA: I agree. It is not likely that the very large frequency of clusters around M 87 is due entirely to capture, but the process probably does operate at some level and it will be the larger galaxies that contribute the most.

GNEDIN: Yesterday H. Harris told about possible connection between globulars and jets. What is the situation in this case? Is there a possible correlation between jets and globular clusters?

HUCHRA: I don't know. Most of the globular clusters we've studied are, in fact, not in the quadrant of the jet.

VANDENBERGH: Could you say anything about your brightest object for color (B-V) = 1.35?

HUCHRA: It is possible that it is a star (galactic). It is also possible that the photometry is wrong.

HESSER: How low is the velocity of your brightest (and reddest) cluster?

HUCHRA: It is one of the lower velocity objects, $V_{Heliocentric} \sim 700$ km/s.

GOODMAN: You did not tell us how you selected your objects for spectroscopic study, except to say that they were bright and red, but in view of the fact that only a third of the objects turned out to be globular clusters, can you say something about the completeness of the photometric surveys of cluster systems?

HUCHRA: This is a toughie since our "selection" criteria changed with time and observing conditions. Initially we observed in magnitude order (brightest to faintest) and almost all of the very bright objects were stars or background galaxies. As we got smarter (i.e. removing stars by eliminating high proper motion objects, strange colors) and went fainter - the fraction of globular clusters went way up. Almost all objects fainter than 21st are indeed globular clusters. The photometric surveys aren't bad when you get fainter.

HANES: I'm very excited to see these reliable velocities coming out, especially as one who ha tried and failed to get them myself! But some measure of the difficulty may be indicated by an inconsistency between your results for object IV-94 (which you call a globular cluster) and those of Mould, Oke and Nemec (who call it a background galaxy, with a velocity ~2900 km/sec.

HUCHRA: Yes. I'd love to see their spectrum, but can't really comment on the difference until I do.

CAYREL: Is M 87 actually globular cluster rich per unit mass, in comparison with other less massive elliptical galaxies?

HUCHRA: I haven't looked at that correlation at all but I can say that it is probably the case that computing the frequency per unit mass will decrease the scatter. M 87 is much more massive (4x) than M 49 and has 3 or 4 times the number clusters although the same luminosity.

HARRIS: The fact that the specific frequency (number per unit light) is roughly constant for E galaxies over a factor of more than 1000 in luminosity might itself be interesting. The reason is that, if small elliptical galaxies have a lower M/L than large ellipticals, it means that the small ellipticals were actually more efficient at forming clusters than the big galaxies were.

HUCHRA: I think that the constancy of M/L for galaxies as a function of L over large ranges of L is actually a very, very important point. This is seen in things like the Tully-Fisher relation for spirals and Faber-Jackson relation for ellipticals where the slopes of those relations indicate that the M/L for such galaxies varies by less than a factor of 2 over several orders of magnitude in luminosity.

ZINNECKER: As far as the Hubble constant is concerned: when you correct the Hubble constant derived from the peak of the M 87 globular cluster luminosity function for the infall velocity of the galaxy towards Virgo, it becomes uncomfortably high to accommodate the ages of the galactic globular clusters. This begs the question: how could one reduce the globular cluster ages (to ~10 Gyr say)?

HUCHRA: Ah yes. The Hubble constant. My belief is that we're really not in trouble yet - the lowest globular cluster ages and the longest expansion times from H_o just about overlap. I think the value of H_o is still pretty soft, we've seen in the last decade several major changes in fundamental calibrators - revision of the Hyades distance modulus, a change in the galaxy scale from 10 to 8 kpc, etc. More such things are bound to come, especially with ST, and each one can change H_o by 20 or 30%. I believe that it is also the case that cluster ages may change. When I first got into the H_o game and got (with Marc

Aaronson and Jeremy Mould) a high value, in conflict with the cluster ages, Al Cameron, a pundit on stellar interiors, said to me "don't worry too much just yet, after all we still can't get the structure of the Sun - the solar neutrino flux - right".

LILLER: You mentioned that the background of the galaxy gave problems. Could you say a little about how this component was subtracted from your spectra?

HUCHRA: We used a two aperture (sky subtracting) spectrograph and both switched the object between aperture and rotated (you get that free of charge with the MMT) to sample the sky.

NEMEC: Mould, Oke and Nemec (1986 preprint) find a mean metal abundance of the 27 globular clusters in the sample $<[Fe/H]> = -1.2 \pm 0.2$ i.e. no more than a factor of 2 more metal rich than the galactic globular cluster system. There is no evidence for the existence of young luminous LMC-like clusters in the M 87 system.

HUCHRA: Sorry, I'd only heard third party that you guys had gotten [Fe/H] ~ -1. I also agree, Jean and I found no objects which could be called young clusters.

Chapter IV

Review Papers

Evolution of Globular Clusters

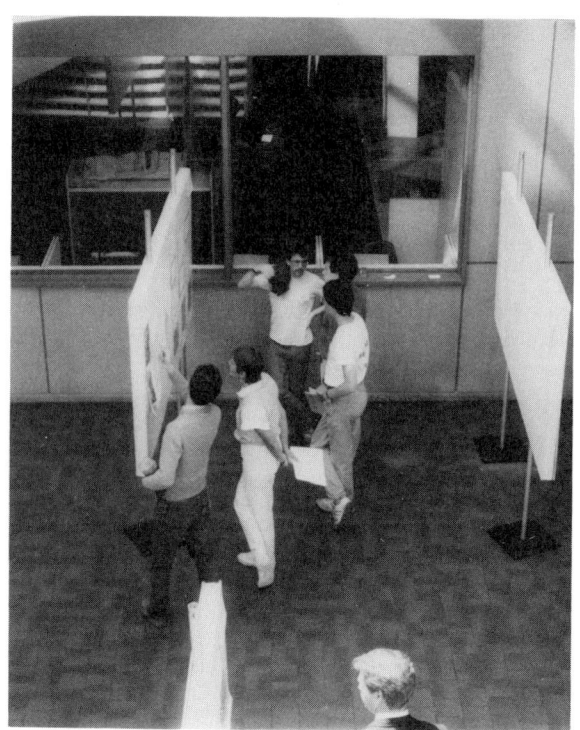
Poster papers were also set up downstairs

Paul Hodge checking in

THE EVOLUTION OF THE SYSTEM OF GLOBULAR CLUSTERS

Jeremiah P. Ostriker

Princeton University Observatory

1. INTRODUCTION

Globular clusters have a sufficiently distinct character that we can treat the <u>system of globular clusters</u>, as a distinct object which has a size, mass and angular momentum, but also, an internal density distribution, velocity distribution, and, in general, a detailed interior structure. Shapley, of course, pioneered in the effort to model, that is to describe, the existing state of the system from our vantage point. At the present time we probably have an inventory of clusters which is largely complete, and for the majority of the identified systems we have good knowledge (cf for example Webbink, 1985) of positions in the galaxy, one component of galactic velocity and a sufficiently detailed picture of the interior dynamical state to characterize each cluster by a luminosity and two radii: a core radius r_c where the surface density has fallen to half the central value, and a tidal radius r_t determined by a fit to the King (1966) truncated isothermal profiles, which is thought to represent the radius beyond which the tidal force due to the galaxy effectively exceeds the force due to the clusters own gravity.

Treated as a dynamical system, can we go beyond description to the conditions which establish physical equilibrium? The answer is yes, but satisfying equilibrium conditions does not convey very much information. In an axisymmetric galactic potential (surely true to first order), any distribution in phase space which is a function of the integrals (J_z, E) will, due to Jeans' theorem, satisfy the Vlasov equation. It is quite likely that for most galactic orbits there exist three integrals which, consistent with our imperfect knowledge of the kinematic quantities, can be approximated by (J^2, J_z, E), total angular momentum, angular momentum about the galactic rotation axis and energy. Several authors have constructed such equilibrium solutions, with best fit indications (cf Frenk and White, 1980) that the degree of rotation (dependence on J_z) is small overall, although not perhaps small for the subset of "disk" systems (Zinn, 1985), the degree of velocity anisotropy (dependence on J^2) also small, and velocity distribution approximately gaussian with no significant variation over the galaxy. Thus to a remarkable degree the system can be described, at the present time, as a simple isotropic isothermal distribution $f(p,q) = \text{Const} \exp(-E/kT)$ with a temperature $3kT \equiv \langle v^2 \rangle$ smaller than that of the underlying mass distribution by a factor 4/7. This would allow the total mass density of the galaxy to fall as $\rho(r) \propto r^{-2}$ between r = 3 kpc and r = 30 kpc and the globular cluster distribution fall off approximately as $r^{-3.5}$ in that region (cf Zinn, 1985), with both laws steeper exterior to this region and flatter interior to it. Mathematically put, if both the total density ρ_t and the density of clusters ρ_{cl} have, over some part of space, constant velocity dispersions ($\langle v^2 \rangle_t$, $\langle v^2 \rangle_{cl}$) and exist in equilibrium in the same potential ϕ_t, then

$$\frac{1}{\rho_{cl}} \vec{\nabla} P_{cl} = \frac{1}{\rho_t} \vec{\nabla} P_t = \vec{\nabla} \phi_t \quad , \tag{1}$$

or

$$\langle v^2 \rangle_{cl} \vec{\nabla} \ln \rho_{cl} = \langle v^2 \rangle_t \vec{\nabla} \ln \rho_t \quad . \tag{2}$$

Since $P = \frac{1}{3} \rho \langle v^2 \rangle$ for each component, we find that

$$\rho_{cl}(\vec{r}) = \text{Const } \rho_t(\vec{r})^{(\langle v_t^2 \rangle / \langle v_{cl}^2 \rangle)} = r^{-2(\langle v_t^2 \rangle / \langle v_{cl}^2 \rangle)} \quad , \tag{3}$$

where the last equation assumes a flat rotation curve.

The result, although simple, is quite surprising, since the stars comprising the globular clusters are in most respects similar to those in the spheroidal population of the galaxy. Thus, despite the fact that the chemical composition (including gradients) and the age of the two populations are broadly similar, the spheroid is, in contrast to the cluster system, locally quite anisotropic (axis ratios)~(2:1:1) and has a much smaller core radius (0.1 kpc vs 1 kpc).

To understand the present state of the globular cluster system we must go further than a consideration of equilibrium and treat evolution, which hopefully will specify more uniquely the physical state of the system. Here it is useful to bring in a metaphor from the study of the internal structure of stars. It was possible to construct equilibrium models in the fashion of Emden and Chandrasekhar before there was any accurate knowledge of atomic or nuclear physics, but specification of even main sequence models required understanding the atomic sources of opacity. To determine the unique characteristics of a given star with a specific age required modeling stellar evolution with enough understanding of the microphysics to calculate energy generation rates. Similarly, a knowledge of the microphysics of the globular cluster system, which is an understanding of the physical processes affecting individual clusters, is required to treat

the evolution (but not the equilibrium) of the globular cluster
system. Our inventory of relevant physical processes has been growing
and it is quite possibly still incomplete in essential ways, but the
obvious lacunae, which existed until just a few years ago have now
been filled. As an example, it was known (Antonov, 1962, Lynden-Bell
and Eggleton, 1980; Cohn, 1980) that after of order 15 half-mass
relaxation times, a typical cluster would undergo a process called
"core collapse" during which the central values of density and
velocity would -according to existing theories -approach infinity in a
finite time. It was obvious that real clusters could pass through
this phase (which only affected the centers) successfully, but how
they did so was unknown. Now (cf IAU Symposium #113) we know of
several processes which are potentially able to produce a "bounce" and
reexpansion. Thus, we are now in a position to start assembling the
pieces, putting together the microphysics with the equations for
statistical equilibrium and evolution, to describe evolutionary paths
which the system may have taken.

The first questions to be answered are obvious: can we account
for the differences between spheroid and globular cluster system
kinematics as due to evolution of the latter? A further more radical
question suggests itself: can the spheroid itself be the result of
the destruction of an initially much more populous system of clusters?
In Section 2 a catalog of physical processes will be presented and in
Section 3 I will briefly summarize my understanding of our present
preliminary knowledge of the system evolution.

2. PHYSICAL PROCESSES AFFECTING CLUSTER EVOLUTION

2.1 Evolution of Internal Structure

2.1.1 <u>Evaporation, Core Collapse and Expansion</u>. As is well known,
the rms escape velocity from a self-gravitating system is only a
factor of two larger than the rms velocity itself. Thus, if the
velocity distribution were everywhere locally a Maxwellian, 0.74% of
the stars would have energies sufficient to excape. Hence, we are led
to define a quantity ξ_e, the fractional evaporation per unit
half-mass relaxation time (Spitzer 1987),

$$t_{rh} \equiv 0.135 \frac{N^{1/2} r_h^{3/2}}{m^{1/2} G^{1/2} \ln \Lambda} = 1.8 \times 10^5 \frac{r_h(pc)^{3/2} N^{1/2}}{(m/m_o)^{1/2}} \text{ yrs}, \quad (4)$$

where N is the number of stars in the cluster and m is the mean
stellar mass. Detailed numerical treatments of the evolution of
isolated clusters summarized by Spitzer (1987) indicate that for
isolated, single component, precollapse systems, an appropriate
estimate for ξ_e is

$$\xi_e = -\frac{1}{N}\frac{dN}{dt} t_{rh} = 4 \times 10^{-3} \quad , \tag{5}$$

about half the value of the naive estimate given earlier. If this process were to act alone, the cluster would evaporate totally in a time which may be estimated as $t_{ev} \approx t_{rh}(0)/4\xi_e \approx 60\, t_{rh}$ with the loss in the a time interval $0 < t < 15\, t_{rh}$ of one quarter of the total mass.

However, after the central density exceeds the half-mass density by more than a factor of 177, the core contracts rapidly. The process may be understood as due to the negative heat capacity of self-gravitating systems. The temperature gradiant in the cluster carries heat outward, causing the core to contract, which in turn forces its temperature higher in an unstable fashion. The process stops and reverses when some source of heat in the cluster appears which can balance the conductive flux. Current theory indicates that either binaries formed by two body (Fabian, Pringle and Rees, 1975; Lee and Ostriker, 1986) or three body (Hut, 1985) processes will be the vehicle; or stellar collisions will either directly or indirectly (through production of high mass stars which explode) cause mass loss from the center which acts as an effective heat source.

In any case for $t > 15 t_{rho}$ the clusters should follow the evolution prescribed by Henon (1961, 1965, see also Goodman, 1984) within which energy generation in the core approximately balances conductively carried flux at the half-mass point and the cluster steadily expands. For an isolated cluster in this phase $r_h \propto t^{2/3}$ and $v_h \propto t^{-1/3}$.

But for a tidally limited cluster the energy input drives mass over the tidal boundary causing the total mass to linearly approach zero. Calculations by Lee and Ostriker (1987) show that, in the post core collapse stage, the rate is approximately

$$\dot{N} = -20 \ln(0.4 N_i)/t_{tid} \quad , \tag{6}$$

where t_{tid} is given in terms of the mean density within the tidal radius ρ_t

as
$$t_{tid} \equiv 2\pi \left(\frac{4\pi}{3} G\rho_t\right)^{-1/2} \quad . \tag{7}$$

This typically will cause the cluster to disintegrate totally in a time t_{ev} which can be conveniently expressed in units of its orbital galactic period

$$t_{ev} = 1.3 \left(\frac{N_c}{10^5}\right)\left(\frac{R}{kpc}\right)\left[\frac{V_c}{220 km/s}\right]^{-1} \times 10^{10} \text{ yrs} \quad , \tag{8}$$

where N_c is the number of stars in the system (~3/4 of the initial number) at the time of core collapse.

2.1.2 <u>Tidal Shocks.</u> When a cluster passes through the galactic plane, there is a short period during which matter in the disc acts to compress the cluster (Ostriker, Spitzer and Chevalier, 1975). A similar tidal force acts on clusters passing near the inner parts of the galactic spheroid. The impulsive energy input causes the cluster to expand making it still more subject to tidal shocks. If this process were the only one acting (and allowing for its acceleration) the time to destruction would be

$$t_{sh,d} = \frac{GM_c P_c V_c^2}{20 g_m^2 r_h^3} \qquad (9)$$

(cf Spitzer 1987), where g_m is the maximum z acceleration in the galactic disc and P_c is the orbital period.

If one were to simply model the bulge shocking by the same method, one would merely replace g_m by the appropriate tidal accelleration which, for a cluster perastion of R_{per} in a galaxy with a flat rotation curve, gives

$$g_m \rightarrow a_m = \frac{V_{cir}^2}{R_{per}} \qquad . \qquad (10)$$

Spitzer (1987) discusses the combined (and at first paradoxical) effects of evaporation and shocks for isolated clusters. Depending on cluster characteristics and position in the galaxy, evaporation, tidal overflow or either of the two shock processes may be the dominant destructive force. All lead ultimately to small, low surface density clusters which might be difficult to detect and recognize.

2.1.3 <u>Interaction with other massive components of the Galaxy.</u> There are various possibilities for interactions of clusters with other components of the galaxy that are more speculative, where the effects might be important but the physical process assumes a component whose existance is in doubt.

2.1.3.1 <u>Massive Black Holes.</u> Wielen (1987) has analyzed the effects on globular clusters of a hypothetical halo of massive ($\sim 10^{6.3} M_\odot$) black holes that has been postulated by Lacey and Ostriker (1985) and by Ipser and Semenzato (1985). He finds that substantial depletion would have occurred due to tidal interactions between the clusters and the comparable mass black holes but that, within the uncertainties of our knowledge of the original cluster distribution, it is not possible to dismiss this conjecture that such a halo does exist.

2.1.3.2 <u>Molecular Clouds</u>. Grindlay (1985) has examined the effects of disruptive collisions with molecular clouds. If there were a substantial population of clouds with internal mass density larger than those of the clusters, this effect could also be important. Chernoff et al (1986) showed that for some clusters evolution would be significantly accelerated.

2.1.3.3 <u>Cluster Cluster Interactions</u>. Currently, the cluster number density is so small that this process is quite unimportant, but if one were to speculate that most of the galactic spheroid component had been initially in globular clusters, the total number would have been ~100 times greater than at present and cluster-cluster interactions, especially in the inner parts of the galaxy could have been devastating. One can estimate the destruction rates simply by scaling to the calculations of Wielen (1987), since the cluster and black hole masses and velocities are similar.

2.1.3.4 <u>Triaxiality</u>. Lastly, it may be interesting to mention a relatively newly discussed mechanism treated in other contexts by Norman et al (1985) and Binney and Ostriker (1987). In an axisymmetric galaxy, the perigalacticon distance of a cluster orbit is fixed even as the orbit processes, but if the galactic potential is triaxial, another possibility exists. If the orbit is loop-like, it is qualitatively similar to those in an axisymmetric galaxy. But some fraction of orbits will be box-like. These, ultimately fill the space available to them, and in particular, will come arbitrarily close to the central point in a manner qualitatively like orbits in a three dimensional harmonic oscillator. Thus, destructive tidal interactions either of the Roche overflow type or of the shock type will always occur given enough time. In work currently underway at Princeton, it appears that even a very modest degree of triaxiality in the galaxy (i.e. enough to produce the "expanding" 3 kpc arm as an artifact of noncircular motions) could result in very effective destruction of low density clusters in the inner few kiloparsecs of the galaxy.

2.2 Evolution of Orbital Parameters

2.2.1 <u>Dynamical Friction</u>. This is by now a relatively well studied phenomenon. Massive objects will spiral towards the center of the galaxy on a time scale $t_{df} \equiv (-dR/Rdt)^{-1} \propto f(v_{rms})/\rho_{gal}M_{cl}$. In a first assay of the problem Tremaine, Ostriker and Spitzer (1975) indicated that this would deplete the inner part of the galaxy (R<1kpc) of massive and dense clusters, but would not greatly effect other parts of the system. Further work, (cf for example Chernoff et al 1986) has left that conclusion intact, but emphasized still more strongly the dependence of the result on orbital parameters. If the galaxy is triaxial, the first effect mentioned, bulge shocking could cooperate with dynamical friction. Since almost all of the drag occurs for orbits which spend time in the denser parts of the Galaxy, those on box-like orbits may eventually wander into regions where the drag will be large even if they are initially not subject to this process.

Drag due to the interaction which the galactic disc does not seem to have been investigated. One may anticipate that for the small fraction of prograde orbits which have moderate eccentricity and small z velocity, the effect could be large.

2.2.2 <u>Scattering by Massive Subcomponents</u>. All of the massive components such as massive black holes, other clusters, or molecular clouds could in principle scatter globular cluster orbits. Since the scattering time is of order N periods where (1/N) is the fraction of the mass in the total system in the other component, all of these potential sources of scattering might be important. It is conceivable that the velocity distribution has been isotropized to some degree by these processes. If there is a loss cone due to destruction of clusters on highly radial orbits, then such scattering processes may be important in refilling the loss cone (or repopulating the box orbits) and thus preventing the shutdown of destructive processes.

2.2.3 <u>Galactic Neighbors</u>. Finally we must consider the interactions with nearby systems. Cluster swapping (Muzzio, 1986) may occur and account for some of the objects with strange orbits or metallicity. But also close passages of the Magellenic clouds (and other equivalent galaxies like M32 for M31) could have a major stirring effect. With a mass $\epsilon = 10^{-2}$ of the galaxy its orbit carves out a volume on each close passage at distance R_p which is of order $\pi \epsilon^{2/3}$ of the total volume within R_p within which the tidal force is large and orbits would be scattered. Since ϵ is of order 10^{-2} this could be a large effect if the period is not too small compared to the Hubble time.

3. OVERVIEW OF SYSTEM EVOLUTION

At a given place in the galaxy one can consider a plane in which the tidal radius and mass of local clusters are plotted. Upper and lower bounds on the mass are given by the processes of dynamical friction and tidal overflow, whereas upper and lower bounds on the radius are determined by shock and evaporation processes. Thus, there is an allowed area in this plane for clusters which have survived for a given length of time. This area shrinks as time proceeds and shrinks as one considers distances closer and closer to the galactic center. Closer than 1 kpc it is difficult for any cluster to survive for very long.

In work underway (Aguilar, Hut and Ostriker 1987) a detailed examination of the various destructive processes is presented for best estimates of the properties of the real cluster system. The reader is referred to that work for further information. An oversimplified summary of the results might state that destruction is effective

1. for all clusters with apogalacticon <2 kpc,

2. for clusters with apogalacticon <10 kpc and very elongated orbits, or unusually low density or unusually high mass,

3. for clusters apogalacticon <50 kpc with extremely low density (like the Palomar clusters).

It is premature to say at present whether the observed cluster system represents a large fraction or a small remnant of the initial cluster system.

This work was supported in part by NASA Grant #NAGW-765, and NSF Grant #AST83-17118.

REFERENCES

Antonov, V. A. 1962 Vest. Leningrad Univ. 7, 135, (cf English translation in 1985 in IAU Symposium No. 113, Dynamics of Star Clusters, J. Goodman and P. Hut, eds., Reidel, Dordrecht, p. 525).
Aguilar, L., Hut, P. and Ostriker, J. P. 1987, in preparation.
Binney, J. and Ostriker, J. P. 1987, in preparation.
Chernoff, D. F., Kochanek, C. S. and Shapiro, S. S. 1986 Astrophys. J. 309, 183.
Cohn, H. 1980 Astrophys. J. 242, 765.
Fabian, A. C., Pringle, J. E. and Rees, M. J. 1975 Monthly Notices Roy. Astron. Soc. 172, 15p.
Frenk, C. and White, S. D. M. 1980 Monthly Notices Roy. Astron. Soc. 193, 295.
Goodman, J. 1984 Astrophys. J. 280, 298.
Grindlay, J. E. 1985 in IAU Symposium No. 113, Dynamics of Star Cllusters, J. Goodman and P. Hut, eds., Reidel, Dordrecht, p. 43.
Henon, M. 1961 Annal. d'Astrophys. 24, 179.
Henon, M. 1965 Annal. d'Astrophys. 28, 62.
Hut, P. 1985 in IAU Symposium No. 113, Dynamics of Star Clusters, J. Goodman and P. Hut, eds., Reidel, Dordrecht, p. 231.
Isper, J. R. and Semenzato, R. 1985 Astron. Astrophys. 149, 408.
King, I. 1966 Astron. J. 71, 64.
Lacey, C. and Ostriker, J. P. 1985 Astrophys. J. 299, 633.
Lee, H. M. and Ostriker, J. P. 1986 Astrophys. J. 310, 176.
Lee, H. M. and Ostriker, J. R. 1987 Astrophys. J., in press.
Lynden-Bell, D. and Eggleton, P. 1980 Monthly Notices Roy. Astron. Soc. 191, 483.
Muzzio, J. C. 1986 Astrophys. J. 306, 44.
Norman, C. A., May, A. and van Albada, T. S. 1985 Astrophys. J. 296, 20.

Ostriker, J. P., Spitzer, L. and Chevalier, R. A. 1975 Astrophys. J. Letters **176**, L51.
Spitzer, L. 1987 *Dynamical Evolution of Globular Clusters*, Princeton University Press, Princeton.
Tremaine, S., Ostriker, J. P. and Spitzer 1975 Astrophys. J. **196**, 407.
Webbink, R. F. 1985 in *IAU Symposium No. 113, Dynamics of Star Clusters*, J. Goodman and P. Hut, eds., Reidel, Dordrecht, p. 541.
Wielen, R. 1987 in *IAU Symposium No. 126, Globular Cluster Systems in Galaxies*, J. E. Grindlay and A. G. D. Philip, eds., Reidel, Dordrecht, p. 393.
Zinn, R. 1985 Astrophys. J. **293**, 424.

DISCUSSION

KING: Do your calculations give a figure for how many clusters have already undergone core collapse?

OSTRIKER: As of the present, we are unable to make such a calculation for lack of information concerning the initial distribution of cluster parameters. It is quite possible that the vast majority were destroyed.

NEMEC: Concerning the question of blue straggler formation through mergers, the 50 blue stragglers found in NGC 5466 (which has a relaxation time ~ a few billion years) would seem to be primordial, if one believes the collision probabilities given by Hills and Day, which imply only ~ a few collisions over 15 Gyr.

OSTRIKER: The required calculation would use a modification of the tidal capture cross-section which is considerably larger than the physical collision cross-section. But, if the observed blue stragglers are slowly rotating, then most were not made by mergers.

HARRIS: Your model shows that almost no cluster destruction occurs outside of 10 kpc, which oddly enough I find more interesting than the region inside where the action occurs. The _observational_ statement that the clusters follow a wider space distribution than the halo light (as in M 87) relies more on the _wide_ field region far outside 10 kpc than it does on the inner few kpc. Would you then agree that we are still left with the need to have a primordial distribution of clusters that is different from the halo stars all the way from the start?

OSTRIKER: I am not sure that I understand your remark. In M 87 the star and cluster distributions are parallel (on a log-log plot); in the outer regions with arbitrary offset. Thus a decrease in the cluster density in the inner parts will be translated to an apparent increase in the "size" of the cluster system.

CAYREL: If the globular cluster system in our Galaxy formed before the collapse of protogalactic gas into a disk, what kind of effect on the specifications and homogeneity of the system, had the strong gravitational potential variation associated with collapse?

OSTRIKER: In an examination of the related problem of the compression in the Z direction of the dark halo, Binney and I found only a small flattening at large radii. However, an increase in density of ~ 30% in the galactic plane is quite possible.

GRINDLAY: As you know, I have published a series of papers in the past few years suggesting that globular clusters with orbits that have low inclinations to the disk can be disrupted by GMCs at R < 4 kpc, thereby "solving" the galactic bulge X-ray source problem, In this case, your statement about cluster "survivors" being those on circular orbits needs modification since those near the plane will not survive (cf. Chernoff and Shapiro). This means, in turn, the number of disk globular clusters (cf. Zinn) must have originally been greater than now observed.

OSTRIKER: I agree that those clusters on circular orbits in the plane of the galaxy will be preferentially destroyed.

RICHER: A comment: Many observers recently have become interested in dynamical processes in globular clusters. However, many of the theoretical calculations have not been easy to interpret into things that observers can actually measure. For example, in a given globular cluster an observer can measure the luminosity function as a function of radius (the stars at a given mass). It has not been easy to convert evolutionary dynamical models of globular clusters into this kind of data. This is just a plea from one observer to a theoretician to try and keep in mind what observers can actually measure directly.

OSTRIKER: It is only in the last few years that the theoretical calculations have approached the sophistication required for comparison with nature to be sensible.

TIDAL HEATING OF GLOBULAR CLUSTERS

David F. Chernoff and Stuart L. Shapiro

Center for Radiophysics and Space Research
Cornell University

ABSTRACT: The influence of tidal heating on the evolution of globular clusters (GC's) in circular orbits about the Galactic center is studied. Giant Molecular Clouds (GMC's) stretch a globular cluster in a direction transverse to its orbit through the disk. The variation in acceleration with height in the disk compresses the cluster in a longititudinal direction. Numerical and analytic calculations of heating and mass loss for GC's, represented by King models, show that disk heating dominates. We apply the results to calculate GC evolution prior to core collapse or tidal disruption using a three parameter (energy, mass, and tidal radius) sequence of King models. The changes in the parameters are calculated for tidal perturbations, relaxation and evaporation. Clusters close to the Galactic center (less than 3 kpc) undergo core collapse in a Hubble time. The effect of tidal perturbations on energy and mass loss of the cluster is strongest between 3 and 5 kpc where it can substantially effect the evolution of the cluster. Here, depending upon their initial concentration, clusters are either tidally heated and dissolved, or forced towards a gravothermal catastrophe in times that are a fraction of a Hubble time. These inner regions of the Galaxy should be fertile territory for the search for post-collapsed clusters.

1. INTRODUCTION

In this paper we report calculations of the heating and mass loss suffered by a Globular Cluster (GC) on its passage through the Galactic disk. We illustrate the qualitative effect on GC evolution with schematic calculations which incorporate two-body relaxation, mass loss across a tidal boundary and heating and mass loss by orbital passage through the Galactic disk. In the past, accurate treatments of the evolution of isolated GC's have employed the Fokker-Planck equation, in which the cluster is modeled as a sequence of instantaneous solutions to the collisionless Vlasov equation (for reviews see Spitzer 1975, 1985, Lightman and Shapiro 1978, Shapiro 1985 and Cohn 1985). Monte-Carlo and finite difference techniques have led to some understanding of the basic physics of *two-body relaxation* of an evolving, self-gravitating system. It is clear that a variety of other processes will be important in the evolutionary history of a GC. In the early stages, *stellar evolution* (SE) will significantly affect the energy and mass budget of the cluster. Once core collapse is well underway, the finite size of the stars may become important. Even when SE is no longer important and core collapse not very far advanced, we find that *tidal effects* may be important. In a recent paper (Chernoff, Kochanek and Shapiro 1986; hereafter CKS) we focused on the tidal interaction of GC's with giant molecular clouds (GMC's) and with the Galactic disk during crossings. This study was motivated in part by the suggestions of Grindlay (1984, 1985, 1986, Grindlay and Hertz 1985) that galactic bulge burst sources may come from disrupted GC's. We presented simplified (i.e. *not* Fokker-Planck) evolutionary calculations employing King models. In addition to the perturbation to the cluster during its disk passages, we incorporated the effect of the Galaxy's large-scale tidal field. Here we present a global Galactic survey of GC evolution based on that model. Elsewhere (Chernoff and Shapiro 1986), we will deal with the mass loss by stellar evolution which dominates the first 5×10^9 years of life of a GC. We conclude that it is essential to include the effects of tidal shocks in modeling clusters within 8 kpc of the center of the Galaxy. Our results also suggest that core collapsed clusters will be found preferentially near the center of the Galaxy.

2. GALAXY MODEL

We parameterize the effects of external physical forces in terms of the Galactocentric location of the GC. To this end, we adopt the Bahcall, Schmidt and Soneira (1982; hereafter BS&S) model of the Galaxy. The potential of that model allows nearly circular orbits for point masses with arbitrary angles of inclination with respect to the disk. We restrict ourselves to circular (or nearly circular) orbits because they are simplest to treat theoretically -- the tidal strength in the BS&S model is nearly constant, except during disk passages, so that the tidal boundary condition for the GC varies only on the relaxation timescale, not on the GC's orbital timescale as it would for an eccentric orbit. To treat the disk passages, we combined the disk surface density of the BS&S model with Bahcall's (1984) determination of the local acceleration above the disk. Together with the assumption of constant scale heights (as observed in edge-on spirals, Van der Kruit and Searle 1982), these Galactic parameters then give the acceleration perpendicular to the disk $K_z(R_g, z)$ at the galactocentric radius R_g and height z. Two components with scale heights 175 pc and 550 pc contribute to K_z. We also included a population of GMC's uniformly distributed, with an average surface density, as a function of R_g, matching the observations of Sanders, Solomon and Scoville (1984). The GMC's heat the GC during close encounters.

3. GC MODEL FOR EVOLUTION

The GC's are represented by King models, which provide a simple theoretical description agreeing remarkably well with the observational data (King 1966). Each model is specified by total mass, M, energy, E, and tidal radius r_t. (These quantities imply the concentration c of the GC and the value of W_0, the fundamental parameter of the King sequence, when we restrict our consideration to the set of models with $0 \leq W_0 \leq 8.5$.) In our study we have included two physical effects:
(1) Two-body relaxation causes the concentration of the cluster to increase so that eventually the core collapses. At the same time, mass is shed across the tidal boundary. Early treatments (the evaporation picture) have been superceded by detailed calculations which show that single-component core collapse proceeds essentially independently of the outer envelope, *once the density contrast is great enough*. Recent Fokker-Planck calculations of GC's in an external tidal field (Wiyanto, Kato and Inagaki 1986) compared the evolution an (initial) King model configuration to the King sequence, in which evolution is estimated via the evaporation picture. Until $W_0 \geq 7.4$ the central density as well as the escape rate matched the simple estimates. Extremely condensed clusters with $W_0 \geq 7.4$ are considered to have undergone total core collapse. We treat relaxation and tidal mass loss with a single-component model (whose particle mass is the average mass per particle) in the evaporation picture. Significant improvements on this treatment have been made (Apple-

gate 1986, Stodolkiewitz 1985, Chernoff, Weinberg and Shapiro 1986) but are considerably more complicated.

(2) Disk passages perturb the cluster, heating it and causing it to shed mass. The essential physical effects are the transverse and longitudinal perturbations created by passing masses, both of which have been well-studied before. Spitzer (1958) treated the disruption of Galactic clusters by interstellar H I clouds (the transverse effect) and Wielen (1985) has extended the theory to impulsive disruption of open clusters mediated by GMC's. Ostriker, Spitzer and Chevalier (1972) and Spitzer and Chevalier (1973) studied the longitudinal effect for GC's -- an extended body ''feels'' a gradient in the force; the center of mass is freely falling but the extremities of the object are accelerated at slightly different rates. Since the internal gravity binds the matter at the edge, the tidal force does work on the cluster on each passage through the disk.

4. QUALITATIVE EVOLUTION OF KING MODELS

The key equations for the evolution of the King model may be written schematically

$$K : \begin{bmatrix} E \\ M \\ r_t \end{bmatrix} \rightarrow \begin{bmatrix} E' \\ M' \\ r'_t \end{bmatrix}, \qquad (1)$$

where the tidal boundary condition is

$$r_t = \left(\frac{M}{3M_g}\right)^{1/3} R_g, \qquad (2)$$

where M, E, r_t are the GC mass, energy and tidal radius and R_g is the radius of the hoop-like-orbit and M_g is the mass interior to the orbit in the BS&S model. (The term dM_g/dR_g in the tidal force has been ignored.) The tidal condition links the evolution of the three variables together, viz. $\delta E = E' - E$ and $\delta M = M' - M$ completely determine the cluster evolution.

To illustrate the qualitative trend in GC concentration we have extended King's (1966) model of cluster relaxation for general δE and δM. Let T be the kinetic energy and V the potential energy. The King models are solutions to the time-independent Vlasov equation and are in virial equilibrium

$$2T + V = 0. \qquad (3)$$

Define ν

$$\nu = \frac{E}{GM^2/r_t}, \qquad (4)$$

a simple monotonic function of W_0 over the range $0 \le W_0 \le 8.5$, ranging from -0.6 (exactly) to -2.1 (approximately). Using the tidal condition and the definition above, for small changes we have

$$\frac{\delta E}{E} = \frac{\delta \nu}{\nu} - \frac{2\delta M}{M} - \frac{\delta r_t}{r_t} = \frac{\delta \nu}{\nu} - \frac{5\delta M}{3M}. \qquad (5)$$

The sizes of δE and δM are dictated by the specific processes which perturb the King model. However, the qualitative tendency of evolution depends only upon the ratio of δE to δM which we can express as

$$f \equiv \frac{\delta E}{(-GM\delta M/r_t)}, \qquad (6)$$

to give

$$\frac{\delta E}{E} = -\left(\frac{f}{\nu}\right)\frac{\delta M}{M}. \qquad (7)$$

Using eqn. (5) and (7) we have

$$\frac{\delta \nu}{\nu} = -\left(\frac{5}{3} + \frac{f}{\nu}\right)\frac{\delta M}{M}. \qquad (8)$$

The nature of GC evolution is determined by the sign of $f + (5\nu/3)$. Since the mass is always lost, i.e. $\delta M/\delta t < 0$, the sign of $(f + (5\nu/3))$ is the sign of $\delta \nu$. ν decreases as W_0 (or concentration) increases, hence if $f + (5\nu/3) > 0$ the relative concentration of the cluster decreases, if $f + (5\nu/3) < 0$ it increases. The average cluster density is fixed by the tidal condition, so that the central density decreases in the first case and increases in the second.

5. CALCULATION OF δE AND δM

5.1 TWO-BODY RELAXATION

In the evaporation picture, marginally bound stars are lost from the GC by two-body gravitational encounters with small velocity changes. The energy change due to the loss of a star with mass m is $\delta E = mGM/r_t$. Hence, $f = 1$, which implies, since $\nu \leq -0.6$, that the tendency of relaxation alone is to always force the GC towards greater concentration. The rate of relaxation is given by King (1966) who integrated the local escape rate over the King model, using the local energy threshold for escape. For GC's on orbits with zero eccentricity, King's formula becomes

$$\frac{dM}{dt} = -\frac{27}{8}\left(\frac{G}{2\pi}\right)^{1/2}\left(\frac{3M_g}{R_g^3}\right)^{1/2}\langle m\rangle\ln(N_c/2)R(W_0) \qquad (9)$$

where N_c is the number of stars within a homogenous core of radius r_c and number density $\rho(0)/\langle m\rangle$. $R(W_0)$ is a pure function of W_0 which King tabulated for $2.5 \leq W_0 \leq 10$; we have assumed it constant for $W_0 \leq 2.5$. It varies only by about a factor of 2 over the entire range King gives. $\langle m\rangle$ is the average stellar mass. King's formula gives a nearly constant rate of mass loss over the GC lifetime.

It is important to point out that this prescription for evaporation does not apply to a young, isolated cluster without a halo extending to r_t. Such a cluster undergoes core collapse in a few hundred *central* relaxation times, generating a halo as it does so. Only when the halo reaches r_t can the cluster be considered, even approximately, to be a member of the King sequence with a ''starting'' value of W_0 determined by the concentration at that time. A global theory of GC formation (such as that of Fall and Rees, 1985) would determine whether clusters form at high concentrations or are ''born'' filling their tidal surface. Our treatment of evaporation allows us to make a comparison between tidal heating and collisional evaporation only for those clusters which extend to r_t.

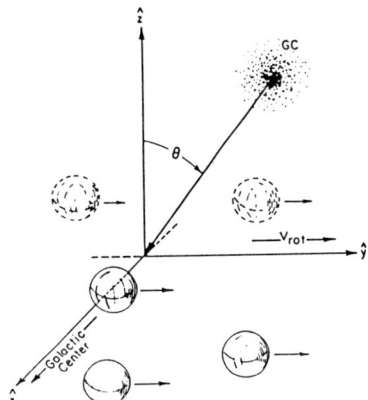

Fig. 1. GMC-globular cluster encounter geometry.

5.2 GMC HEATING OF GC's

The tidal heating during disk passage was studied in detail by CKS who found that in virtually all cases $f = 1.5 - 2.0$. When $f = 1.5$ this implies that under the influence of tidal heating alone, GC's with $W_0 > 4$ collapse and those with $W_0 < 4$ dissolve. Figure 1 illustrates the encounter geometry of the GC with the disk.

Clouds in the disk orbit the Galaxy (in the BS&S model) and the GC's trajectory pierces the disk with some fixed angle of inclination. The rate of heating by clouds was calculated as follows. Construct a GC by choosing a large number (2000) of points from the distribution function describing the King model. Next generate a random distribution of clouds with the appropriate surface density (4.8 $M_\odot$/pc^2 in the solar neighborhood and 15 $M_\odot$/pc^2 maximum near $R_g = 5$ kpc). Our canonical clouds have radii, $r_{cl} = 20$ pc, and masses, $M_{cl} = 10^5$ $M_\odot$ and are homogenous spheres, confined to the plane. The impulse approximation is used to calculate individual velocity perturbations of stars in the GC due to the gravitational pull of a GMC. Penetrating encounters are accurately treated. The entire cluster is scattered in the encounter. Using the cluster's new center of momentum, we find the number of unbound particles (δM) and change in cluster energy (δE), including the work done by the tidal field as material is removed to infinity. For the particles which remain bound, we find the change in kinetic energy. These results are combined to give the overall change in mass and energy and, hence, the new cluster (i.e. King) model. In the simulations, the procedure is repeated, using the current King model parameters, for each disk passage.

Let $\epsilon = 0.5(1 + \sin(\theta))$ where θ is the angle of the GC orbit with respect to the normal to the disk, at the point of entry (see Figure 1). θ is taken as positive for a retrograde orientation ($\epsilon \to 1$) and negative for a prograde orientation ($\epsilon \to 0$). Although the numerical results include the effect of interpenetrations, the finite number density of GMC's in the plane (which cause nonlinear terms in the surface density to enter the heating rate), and the effect of particle loss, we find that the rate of heating is well approximated by an average over impact parameters of the tidal heating

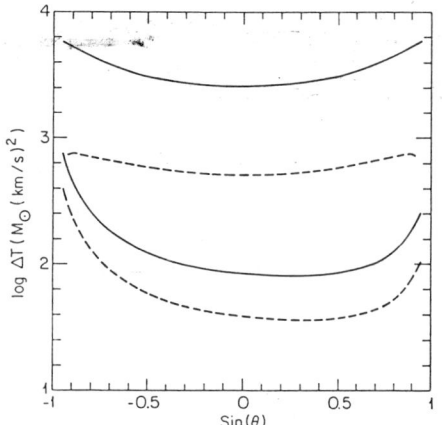

Fig. 2. Heating by GMC's (lower two lines) and tidal shocks (upper two lines) during disk passages as a function of the angle of incidence for $M = 10^5 \, M_\odot$ at $R_g = 8 \, \text{kpc}$ for $W_0 = 1$ (dashed) and $W_0 = 7$ (solid).

by single particles. We find

$$\Delta T \approx (200 \, M_\odot \, \text{km}^2 \text{s}^{-2}) \left[\frac{\langle r^2 \rangle}{p_{min}^{*2}}\right] \left[\frac{1}{4\epsilon(1-\epsilon)^{1/2}}\right] \quad (10)$$

$$\left[\frac{M}{10^5 \, M_\odot}\right] \left[\frac{\sigma_{cl}}{4.8 \, M_\odot/\text{pc}^2}\right] \left[\frac{M_{cl}}{10^5 \, M_\odot}\right] \left[\frac{200 \, \text{km s}^{-1}}{v_c}\right]^2.$$

where ΔT is the change in kinetic energy, $\langle r^2 \rangle$ is the mean square radius of a star in the GC, σ_{cl} is the cloud surface density, M_{cl} is the cloud mass, v_c is the local rotation velocity and p_{min}^* is the effective minimum impact parameter for the GC-GMC collision. CKS solved for p_{min}^* based on their numerical work, with results that were crudely $p_{min}^{*2} \approx 0.5(r_{cl}^2 + r_t^2)$. Figure 2 illustrates the numerical results for a cluster with $W_0 = 1$ (lower dashed line) and $W_0 = 7$ (lower solid line)), $M = 10^5 \, M_\odot$ as a function of the angle of incidence.

We tested the scaling of relation (10) above for $2 \times 10^4 < M_{cl} < 5 \times 10^5 \, M_\odot$ and fixed σ_{cl}. We found it to be accurate to about 50%; particle loss for the least concentrated cluster and the highest M_{cl} led to some suppression of the direct heating. However, indirect heating via the change in potential energy compensates for that suppression somewhat. Scaling with r_{cl} is somewhat less accurate, i.e. p_{min}^* is not well approximated by the expression above. The heating rate is greater than that given by eqn (10) by (at most) 2.5 when $r_{cl} = 4 \, \text{pc}$ and by (at most) 2.0 when $r_{cl} = 100 \, \text{pc}$.

We can now address the question whether disruption of GC's by GMC's can ever take place. The GMC density is a maximum at $R_g \approx 5 \, \text{kpc}$ with $\sigma_{cl} \approx 15 \, M_\odot/\text{pc}^2$. Using eqns. (10) and (4) we can write the fractional change per disk crossing, due to GMC perturbations, as

$$\frac{\Delta T}{E} = \frac{7.4 \times 10^{-5}}{\nu \epsilon \sqrt{1-\epsilon}} \left[\frac{\langle r^2 \rangle}{p_{min}^{*2}}\right] \left[\frac{\sigma_{cl}}{15 \, M_\odot \, \text{pc}^2}\right] \left[\frac{M_{cl}}{10^5 \, M_\odot}\right] \left[\frac{R_g}{5 \, \text{kpc}}\right]^2 \left[\frac{10^5 \, M_\odot}{M}\right]^{2/3} \quad (11)$$

In 10^{10} years There are at most 200 disk passages for an orbit at $R_g = 5\,\text{kpc}$. Accordingly, *GMC's cannot appreciably affect cluster evolution unless ϵ is very near 0 or 1, i.e. disruption requires extreme cluster orbit orientations.* To examine the situation in slightly more detail, we calculate ϵ^* such that $\Delta T/E = 1$ after 200 passages by eqn. (10) above. For $W_0 = 1, 3, 4, 7$ we find $\epsilon^* = 1.6 \times 10^{-2}, 1.1 \times 10^{-2}, 8.2 \times 10^{-3}, 2.5 \times 10^{-3}$ for prograde orbits or, in angles, $-75°, -78°, -80°, -84°$. Although the validity of the impulse approximation puts a lower limit on θ (such that the disk passage is sudden for most of the particles in the GC) these disruption angles all lie close to, but above that limit. Therefore, for the range $1 \leq W_0 \leq 7$, we conclude that disruption is energetically feasible for prograde orbits. However, the parameter space for disrupted orbits (even among purely circular orbits) is quite small: it is confined to the peak of the function $\sigma_{cl} R_g$ (near $R_g = 5\,\text{kpc}$) and to a narrow angular range.

Eqn. (10) shows that disruption of retrograde orbits requires such small $1 - \epsilon$ that the GC would never leave the disk. Expressing these results in terms of angles, *GC's on circular orbits with $-75° < \Theta < 90°$ are not disrupted by GMC's alone.* The physical interpretation of these results is clear: for a uniform distribution in Θ of cluster orbits, most orbits are not disrupted by the GMC's in the disk. However, a population of GC's confined to the disk will quite likely be destroyed. Therefore, if the galactic bulge burst sources are on nearly prograde orbits they may have come from destroyed GC's as suggested by Grindlay and collaborators (1984, 1985, 1986, Grindlay and Hertz 1985). Ironically, small- and large-angle orbits through the disk, lying outside the narrow ''parameter space'' described above, may be the safest haven for GC's within about 8 kpc of the Galactic center. These orbits do not suffer strong gravitational shocks which can be much more effective than the GMC's at heating the GC.

5.3 TIDAL SHOCKS

Our numerical treatment of tidal shocks is an extension of that of Ostriker, Spitzer and Chevalier (1972) in that it includes a realistic disk potential, an adiabatic cutoff and particle loss. We find that all three effects are important in determining the magnitude of the heating, which turns out to be large enough to significantly alter the concentration of a GC at small Galactocentric distance.

In the impulsive limit the GC heating rate per passage is given, without particle loss, by

$$\Delta T = \frac{2}{3} M (\frac{d\phi}{dz}|_{z_1})^2 \frac{\langle r^2 \rangle}{v^2 \cos^2 \Theta}. \tag{12}$$

where the effective disk thickness z_1 must be chosen so that the encounter is ''impulsive'' ($\lim_{z_1 \to \infty} d\phi/dz = 2\pi G \sigma_d$.). However, this purely impulsive limit is rarely realized, so we will use this equation only to estimate the *scaling* of the energy transfer with the Galactic and GC parameters. For convenience, we fix the normalization using the value of ΔT at $\Theta = 0, R_g = 8\,\text{kpc}, W_0 = 1$, and $M = 10^5\,M_\odot$, which we will refer to as $\Delta T|_{fix}$ and determine numerically. Expressing the dimensionless mean square radius, $\langle r^2 \rangle / r_t^2 = s(W_0)$, eqns. (2) and (12) give

$$\Delta T = \Delta T|_{fix} \left[\frac{M}{10^5\,M_\odot}\right]^{5/3} \left[\frac{s(W_0)}{s(1)}\right] \left[\frac{1}{\cos^2 \Theta}\right] \chi(R_g) \tag{13}$$

where

$$\chi(R_g) = \left[\frac{\sigma_d(R_g)}{\sigma_d(8\,\text{kpc})}\right]^2 \left[\frac{R_g}{8\,\text{kpc}}\right]^2 \left[\frac{M_g(8\,\text{kpc})}{M_g}\right]^{2/3} \left[\frac{v_c(8\,\text{kpc})}{v_c(R_g)}\right]^2. \tag{14}$$

$\chi(R_g)$ describes the scaling of the heating with Galactic position. Assuming constant v_c (with constant M_g/R_g) we have

$$\chi(R_g) \approx \left[\frac{\sigma_d(R_g)}{\sigma_d(8\,\text{kpc})}\right]^2 \left[\frac{R_g}{8\,\text{kpc}}\right]^{4/3}. \tag{15}$$

For an exponential disk surface density with a characteristic length of 3.5 kpc, the scaling of χ with R_g implies that the maximum heating per plane passage, for a fixed GC mass and W_0, occurs at about $R_g = 2.3$ kpc. When the frequency of passage is folded in (proportional to $1/R_g$) the maximum rate of heating per unit time due to tidal shocks occurs at about $R_g = 0.5$ kpc. To summarize the increase in surface density towards the center of the disk, means that GC heating becomes stronger as the orbital radius decreases. However, for fixed M, the tidal radius decreases as $R_g \to 0$ so that eventually the heating must decrease and this leads to a maximum heating rate at a finite R_g. The scaling for χ ignores particle loss and the adiabatic cutoff.

To illustrate the new qualitative features of the numerical calculation we show the heating as function of angle of incidence for two values of W_0 and for two values of R_g. The GC had a mass of $M = 10^5\,M_\odot$. Several runs were performed to illustrate the effects of the frequency cutoff, of the scale height, and of particle loss. In all the runs the potential was determined by the law

$$K_z(R_g, z) = K_z(R_0, z) \left(\frac{\sigma_d(R_g)}{\sigma_d(R_0)}\right) \sum_i K_{0i} \tanh(z/z_{0i}), \tag{16}$$

where R_0 is the local position and σ_d is the disk surface density in the BS&S model. We define the ``standard model'' as

$$K_{01} = 3.47 \times 10^{-9} cm/s^2,$$

$$z_{01} = 175\,\text{pc};$$

$$K_{02} = 3.96 \times 10^{-9} cm/s^2,$$

$$z_{02} = 550\,\text{pc}. \tag{17}$$

and z is in pc. This is a smooth fit to Bahcall's local acceleration data. The ``one component'' model is given by K_{01} above and a scale-height assumption. To illustrate the magnitude of the different effects which enter, we have plotted the heating rate as a function of angle of incidence under four different sets of assumptions. To begin with, we do not allow any particle escape. Consider first the four upper lines in the Figure 3a which all relate to the heating of a $W_0 = 1$ model at $R_g = 3$ kpc.

The line of short dashes is the ``one component'' model with $z_{01} = 0$, i.e. impulsive heating by the low-scale height surface density. Note the steep rise as $\cos\Theta \to 0$ (consistent with the scaling of eqn. (12)). The line of long dashes is the ``one component'' model with $z_{01} = 175$ pc; there is a substantial suppression in the heating due to the adiabatic cutoff. The dash-dot line is the standard model given above (two components with separate adiabatic cutoffs) and the solid line is the same model but with particle loss allowed. The last two lines illustrate that there is some heating by the large scale-height component and some inefficiency on account of particle escape. A completely analogous set of curves is given in figure 3a for a $W_0 = 7$ model,

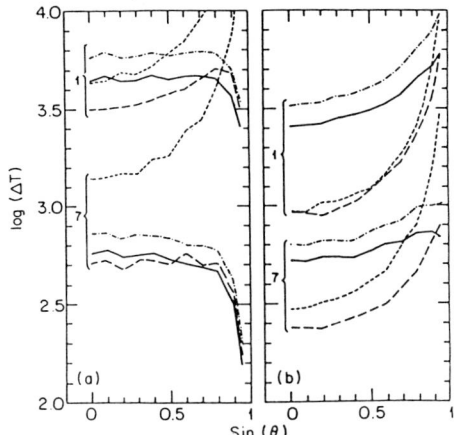

Fig. 3a and 3b. Comparison of disk heating of $10^5 \, M_\odot$ GC as a function of angle of incidence at (a) $R_g = 3$ kpc and (b) 8 kpc for two values of $W_0 = 1$ and 7. Four lines for each (R_g, W_0) pair are shown. The first three have no particle loss: they are (short dashes) impulsive heating with disk acceleration $K_{01} = 3.47 \times 10^{-9} \, \mathrm{cm/s^2}$, (long dashes) heating by same model with finite scale height ($z_{01} = 175$ pc) and (dash-dot line) standard model. The standard model with particle loss included is illustrated by the solid line.

which is much more centrally condensed, illustrating the significant role of the impulse approximation. Figure 3b illustrates heating of the $W_0 = 1$ and $W_0 = 7$ models at the local position ($R_g = 8$ kpc). Comparing the curves, the main effect in 3b is the extra heating associated with K_{02}. In the solar neighborhood, the adiabatic cutoff is not as important as at 3 kpc because the tidal radius is larger with longer dynamic timescales; meanwhile the disk passage timescale remains constant. Hence, the impulsive approximation becomes more accurate at large radii. Figure 2 illustrates, the final heating rate (upper two lines) as a function of angle of inclination of the GC orbit, for $W_0 = 1$ (dashed line) and $W_0 = 7$ (solid line) at 8 kpc. It is interesting that despite the reductions in magnitude of ΔT by the various corrections, *the disk heating still exceeds the heating by clouds by about a factor of 10.* This is true at all R_g and for all W_0.

CKS parameterized their numerical results in terms of a multiplicative correction factor, γ, to the simple expression for χ, given by eqn. (15), over the range $1 \, \mathrm{kpc} < R_g < 24.5 \, \mathrm{kpc}$ and $\Theta = 0.0, 0.5, 0.9$. The large range of the correction ($0.00 < \gamma < 2.3$) is noteworthy and illustrates the effects of particle loss and the adiabatic cutoff on the simple scaling arguments. *The maximum total heating per passage occurs at about 5 kpc for $W_0 = 1, 3$ and at about 7 kpc for $W_0 = 4, 7$, both at small $\cos \Theta$; the maximum heating per unit time occurs between 3 kpc and 5 kpc at small $\sin \Theta$.* We have also investigated the rate of heating for orbits which cross the disk with velocity different from v_c; typically, the variation with v is much more gradual than the $1/v^2$ dependence implied by eqn. (12).

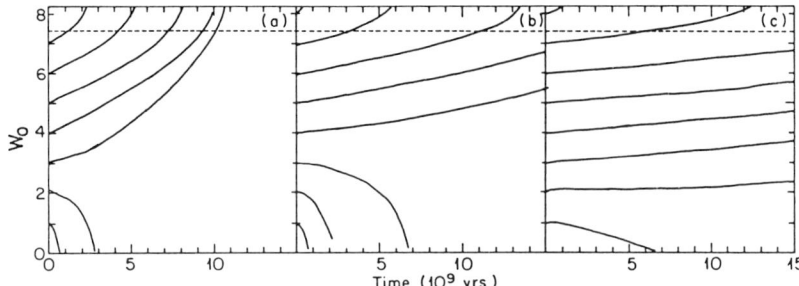

Fig. 4. Cluster evolution of concentration: $W_0(t)$ for a family of initial conditions ($M = 10^5 \, M_\odot$) at $R_g = 3$ kpc (a), 8 kpc (b) and 14 kpc (c). Relaxation causes the concentration to increase (higher W_0) and tidal shocking causes the concentration to decrease. The dashed line marks $W_0 = 7.4$, core collapse. $W_0 = 0$ is interperted at tidal dissolution.

6 EVOLUTIONARY CALCULATIONS

We found that heating and mass loss from tidal shocks is more important that of interactions with GMC's for all King models with $10^5 \, M_\odot < M < 10^6 \, M_\odot$ on orbits with $2 \, \text{kpc} < R_g < 14 \, \text{kpc}$ for all angles of incidence, with the exception of nearly prograde GC orbits with very small angles of inclination. The rate of heating by tidal shocks at fixed R_g is only weakly dependent on the angle of incidence of the cluster orbit, on account of the suppression of energy transfer by the adiabatic cutoff. Hence, we will present results for fixed angle of inclination, $\theta = 0$. Figure 4 illustrates the evolution of W_0 of a $10^5 \, M_\odot$ GC without SE, at $R_g = 3, 8,$ and 14 kpc for a family of initial cluster concentrations.

GC evolution proceeds most quickly near the center of the Galaxy. There is an unstable tendency to either collapse or dissolve, i.e. there exists a critical value of W_0 at each R_g such that more concentrated clusters collapse and less concentrated ones dissolve. An entirely analagous effect was originally pointed out by King (1958) in his treatment of Galactic clusters. Mass loss by relaxation and by tidal shocks is comparable at the critical W_0. According to our previous discussion, we interpert $W_0 \geq 7.4$ as meaning that core collapse has occurred, while $W_0 = 0$ implies that the cluster has dissolved. Comparison of runs with and without tidal shocking shows that the time until collapse of a given initial mass *decreases* for most initial concentrations when the shocking is included. Apparently if the cluster collapses, the increased mass loss and decreased relaxation time offset the shock's tendency to dissolve the cluster.

Calculations similar to those shown in Figure 4 have been performed for $2 \, \text{kpc} < R_g < 15 \, \text{kpc}$. Figure 5 summarizes the global trends in GC evolution under the environmental effects of Galactic position.

Contours of W_0 after an evolution of 10^{10} years are plotted. The graph's axes are

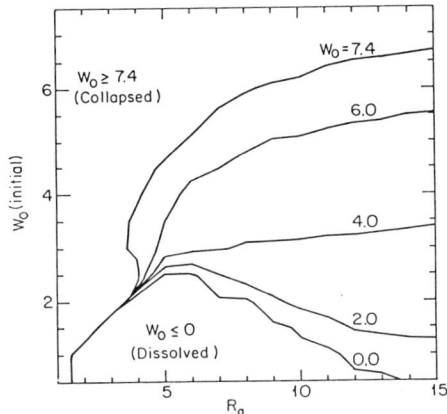

Fig. 5. Contours of cluster concentration after 10^{10} years evolution for a cluster with galactocentric position R_g (abscissa), initial concentration W_0 (ordinate), and initial mass $10^5 \, M_\odot$. Clusters lying below the lowest contour, $W_0 = 0$, have dissolved; those above the topmost contour, $W_0 = 7.4$, have collapsed.

the initial value of W_0 (ordinate) and R_g (abscissa, assumed constant throughout the evolution; dynamical friction is insignificant for $R_g > 2 \, \text{kpc}$ and $M < 2 \times 10^6 \, M_\odot$, Tremaine, Ostriker and Spitzer 1975). The contour of $W_0 = 0$ is the limit of clusters which have ''dissolved,'' while $W_0 = 7.4$ is the limit of those that have undergone core-collapse. Clusters at large R_g are unaffected by external forces and closely resemble their initial structure, meaning loosely speaking, their structure when SE ''turned off.'' At the other extreme, most clusters at small R_g have already been destroyed by tidal heating or have undergone core collapse. Our analysis of course cannot be extended beyond core collapse. However, if post core-collapsed clusters are distinguishable from pre core-collapsed clusters, one expects a definite trend: *more collapsed objects near the Galactic center.* Such a trend may have been observed (Djorgovski 1986). It is highly likely that tidal shocking continues to be important during the life of all clusters at small R_g, including those that have undergone core collapse.

As a corrollary to these results, the detection of a GC with collapsed core at *large* R_g would be profoundly signficant. The density of a King model is controlled by the strength of the tidal field. If a GC is dense enough to core collapse in a Hubble time at large R_g then the initial conditions for that GC must be denser than that allowed by the King model. That is, the initial mass must be well concentrated within the allowed (i.e. tidal) radius. A non-gravitational force must then be at work in creating such proto-cluster conditions, e.g. the cooling instability studied by Fall and Rees (1985).

This research was supported in part by NSF Grant No. AST-84-15162 at Cornell University.

REFERENCES

Applegate, J. H. 1986 Astrophys. J. 301, 132.
Bahcall, J. N., Schmidt, M. and Soneira, R. M. 1982 Astrophys. J. Letters 258, L23.
Bahcall, J. N. 1984 Astrophys. J. 287, 926.
Blitz, L. 1979 Astrophys. J. Letters 231, L115.
Cohn, H. 1985 in IAU Symposium No. 113, Dynamics of Star Clusters, J. Goodman and P. Hut, eds., Reidel, Dordrecht, p 161.
Chernoff, D. and Shapiro, S. 1986 submitted.
Chernoff, D., Kochanek, C. and Shapiro, S. 1986 in press.
Chernoff, D., Weinberg, M. and Shapiro, S. 1986 in preparation.
Djorgovski, G. 1986 personal communication.
Fall, S. M. and Rees, M. J. 1985 Astrophys. J. 298, 18.
Grindlay, J. E. 1984 Adv. Space Res. 3, 19.
Grindlay, J. E. 1985 in Proc. Japan-US Seminar in Galactic and Extragalactic Compact X-Ray Sources, Y. Tanaka and W. Lewin, eds., ISAS, Tokyo, p. 215.
Grindlay, J. E. 1986 in Evolution of Galactic X-Ray Binaries, J. Trumper, W. Lewin and W. Brinkman, eds., NATO ASI series, p. 25.
Grindlay, J. E. and Hertz, P. 1985 in Cataclysmic Variables and Low Mass X-Ray Binaries, D. Q. Lamb and J. Patterson, eds., Reidel, Dordrecht, p. 79.
King, I. R. 1958 Astron. J. 63, 465.
King, I. R. 1966 Astron. J. 71, 64.
Lightman, A. P. and Shapiro, S. L. 1978 Rev. Modern Phys., 50, 437.
Ostriker, J. P., Spitzer, L. and Chevalier, R. A. 1972 Astrophys. J. Letters 176, L51.
Ostriker, J. P. 1985 in IAU Sysmposium No. 113, Dynamics of Star Clusters, J. Goodman and P. Hut, eds., Reidel, Dordrecht, p. 347.
Sanders, D. B., Solomon, P. M. and Scoville, N. Z. 1984 Astrophys. J. 276, 182.
Shapiro, S. L. 1985 in IAU Symposium No. 113, Dynamics of Star Clusters, J. Goodman and P. Hut, eds., Reidel, Dordrecht, p. 373.
Soloman, P. M., Sanders, D. B. and Scoville, N. Z. 1979 in IAU Symposium No. 84, Large Scale Characteristics of the Galaxy, W. B. Burton, ed., Reidel, Dordrecht, p. 35.
Spitzer, L. 1985 in IAU Symposium No. 113, Dynamics of Star Clusters, J. Goodman and P. Hut., eds., Reidel, Dordrecht p. 109.
Spitzer, L. 1975 in IAU Symposium No. XXX, Dynamics of Stellar Systems, A. Hayli, ed., Reidel, Dordrecht, p. xxx
Spitzer, L. 1958 Astrophys. J. 127, 17.
Spitzer, L. and Chevalier, R. 1973 Astrophys. J. 183, 565.
Stodolkiewicz, J. S. 1985 in IAU Symposium No. 113, Dynamics of Star Clusters, J. Goodman and P. Hut, ed., Reidel, Dordrecht, p. 361.
Tremaine, S., Ostriker, J. and Spitzer, L. 1975, Astrophys. J.

196, 407.
Wielen, R. W.   1985   in <u>IAU</u> <u>Symposium</u> <u>No.</u> <u>113,</u> <u>Dynamics</u> <u>of</u> <u>Star</u> <u>Clusters</u>, J. Goodman and P. Hut, eds.,   Reidel, Dordrecht, p. 449.
Wiyanto, P., Kato, S. and Inagaki, S.   1985   <u>Publ.</u> <u>Astron.</u> <u>Soc.</u> <u>Pacific</u> **37**, 715.
van der Kruit, P. C. and Searle, L.   1982   <u>Astron.</u> <u>Astrophys.</u> **110**, 61.

DISCUSSION

GNEDIN: There is a point of view that the low-mass X-ray binaries in the bulge originated as a result of collisions between globular clusters and giant molecular clouds. What do you think of this?

CHERNOFF: Disruption appears to be energetically feasible for prograde orbits with very large angles of incidence with respect to the disk; i.e., orbits nearly in the disk. [I would also like to point out that individual GMC's not complexes of GMC's were used to estimate the destruction. The larger complexes will enhance the distinction.]

ZINNECKER: This is a question which is also addressed to Dr. Ostriker. Are there destruction processes for globular clusters which operate preferentially in spirals or ellipticals? If so, this would seem to have some bearing on the specific frequency of globular clusters in spirals versus ellipticals?

CHERNOFF: The tidal shocking by disk and GMC's will be absent in present-day ellipticals. The average strengths of the tidal field will still enter into the collapse rate.

INNANEN: Could you apply your results to old open clusters like M 67 and NGC 188, in view of the fact that your orbits are nearly circular?

CHERNOFF: I would be hesitant to apply the general method, to open clusters because these systems are small N systems; i.e., the separation of dynamic and relaxation timescales is not as clear.

GOODMAN: Both your own and Ostriker's analysis of relaxation have assumed equal mass stars. But we know that in the presence of a mixture of stellar masses, the collapse time is greatly reduced, perhaps by a factor of five.

CHERNOFF: I agree with you. It may be that stellar evolution narrows the range of the mass function sufficiently, and (additionally) pumps enough energy into the cluster by mass loss to counteract these effects. In any case, these effects cannot be studied with this sort of model.

CLUSTER SWAPPING

Juan C. Muzzio

National University of La Plata

ABSTRACT. The investigation of globular cluster swapping in clusters of galaxies has resulted in some interesting theoretical findings and, at the same time, it offers a promising field for observers. Numerical simulations of galaxy clusters where the galaxies have swarms of test particles around them showed that, in addition to tidal stripping, tidal accretion plays an important role in the dynamical evolution of clusters of galaxies; it also turns out that, even in clusters where the gravitational field is dominated by a massive background, the galaxy-galaxy attraction cannot be ignored when estimating the outcome of collisions. Cluster swapping is just an example of tidal accretion and, taking the globulars as probes of halo material, it might offer an opportunity to observe some consequences of that effect; besides, although the difficulties look formidable at present, the study of globulars lost through tidal stripping is a possibility that should not be neglected. Tidal stripping and accretion processes are very sensitive to the ratio of galactic mass to total mass, so that observations related to the cluster swapping phenomena may provide a new means to investigate the missing mass problem.

1. INTRODUCTION

What happens to the globular cluster system of a galaxy that suffers an encounter with another galaxy? If the especial conditions required for a merger (low relative speed and small impact parameter) are not fullfilled, the system will be modified in three different ways: 1) through the capture of globulars from the other galaxy (cluster swapping), 2) through the loss of globulars to the other galaxy or to the intergalactic medium (tidal stripping), and 3) through changes in the orbits of the globulars that are kept by the system (Muzzio 1986a, hereafter Paper II).

Galaxies in groups and clusters suffer several encounters with other galaxies during their lifetime and, besides, they are also affected by their interaction with the cluster as a whole (Miller 1984);

thus, it seems to be an interesting problem to investigate how their globular cluster systems evolve. Obviously, the stars of the galaxy will also be affected, but there are some advantages in concentrating in the evolution of the globular cluster systems because, although we can only study the integrated effect of the stellar light from distant galaxies, their globular clusters can be recognized as separate units. Accordingly while the interpretation of surface brightness observations should take into account the stellar luminosity function and its possible variation, it is much simpler to interpret the observed surface densities of globulars. Besides, the possibility is open to the study of individual globulars and to the derivation of metallicities and radial velocities that, again, are not weighted by luminosity. Therefore, the study of globular cluster systems of galaxies that belong to clusters, in addition to be of interest by itself, may also provide useful information about the dynamical processes in clusters of galaxies.

2. ENCOUNTERS OF GALAXIES

The two most cited investigations about encounters of galaxies that do not lead to mergers are probably those of Richstone (1975) and of Dekel et al. (1980); interestingly enough, they did not take into account the possibility of swapping (or tidal accretion) and they just considered the changes due to tidal stripping and to the redistribution of the galactic material (but see Figure 4 of Dekel et al. and its caption). The omission had an unfortunate consequence because, since most authors that later on made numerical simulations of clusters of galaxies used the results of those two investigations, the currently available cluster models (Merritt 1983; Miller 1983; Richstone and Malumuth 1983; Malumuth and Richstone 1984) do not include the effect of tidal accretion.

Alternatively, something can be learned about swaps and escapes from the work of Hills (1975), Hills and Fullerton (1980), Fullerton and Hills (1982) and Hills (1984) who investigated encounters between a binary star (or a star-planet system) and a stellar intruder. Since their experiments were planned for other purposes, they have some limitations for the study of the present subject, but they are well supplemented by the results of Muzzio, Rabolli and Martínez (1984) who used similar methods to investigate encounters of a galaxy-globular cluster pair with another galaxy. Several interesting facts were revealed by all these studies of collisions: a) essentially the same results are obtained using either the exact formula for the attraction law between two Schuster's spheres or the asymptotic approximation of Muzzio and Martínez (1982); b) the probabilities of swap and of escape are strongly dependent on the ratio between the two galactic masses but, provided that such ratio is not altered, the extension of the galaxies does not influence so much the results of the encounter; c) the probability of swap for head-on collisions increases steeply as the mass of the interloper is increased from one-half to twice the mass of the parent galaxy, but for larger mass ratios the change is much

gentler up to, at least, a mass ratio of 64; the probability of
swap is maximum for low relative velocities, and smoothly falls
to zero as the speed increases; d) for low relative velocities
the probability of escape for head-on encounters increases
steeply as the mass ratio is increased from one-half to four and
it virtually does not change for subsequent increases of that
ratio; the probability of escape is low for small relative velocities
and, for mass ratios up to four, decreases with increasing relative
velocity immediately after reaching a maximum, while for larger mass
ratios the probability remains at its maximum possible value of 100%
over a large range of speeds and falls off only for very large relative
velocities; e) the velocity parameter that best characterizes the
collision is the ratio between the relative velocity of the galaxies at
the time of impact and the orbital velocity of the globular; f) the
highest probability of escape is obtained for a value of the impact
parameter that increases with decreasing relative velocity; since the
decrease of the probability of swap as the impact parameter increases
is less steep, swaps may outnumber escapes for large impact parameters.

Muzzio et al. (1987, hereafter Paper IV) confirmed that the most
adequate velocity parameter is the ratio of the relative velocity of
the galaxies at the time of impact to the internal velocity of the
galaxy and showed that, since the former velocity depends on the
relative orbit of the galaxies which in turn depends on their mutual
attraction, the forces that the galaxies exert on each other cannot
be ignored even in cluster models dominated by a massive background.

3. THE TEST PARTICLE APPROACH

As indicated above, current models of galaxy clusters do not
include the effect of tidal accretion and they are not suitable for
the study of cluster swapping. Thus, in order to investigate in detail
the swapping process, my coworkers and me (Muzzio, Martínez and
Rabolli 1984, hereafter Paper I; Paper II; Muzzio 1986b, hereafter
Paper III; Paper IV; Rabolli 1986) followed the approach pioneered
by Forte et al. (1982) and we investigated models of clusters of
galaxies where the galaxies had swarms of test particles around them.
The advantage of using test particles, rather than full N-body codes
(e.g. Carnevali et al. 1981), is that the gravitational fields of the
galaxies and of the intracluster medium (ICM) are independent of the
number of particles that belong to them; thus, there is no problem in
having a galaxy with just a few test particles around it, so that one
can use numbers of test particles proportional to the galactic mass
and still be able to use a rather broad range of galactic masses.
Besides, the computing time is approximately proportional to the number
of galaxies times the number of galaxies plus test particles, and is
shorter than in full N-body codes where it is proportional to the
square of the total number of particles. Finally, since the test
particles do not interact with each other, relaxation effects pose
no problem and one can use softening parameters that realistically

represent the sizes of galaxies. The main disadvantage of the test particle approach is that galaxies are not self-consistent; therefore, the size, mass and gravitational field of a galaxy remain always the same, despite the changes that the galaxy suffers. Thus, our method cannot compete with the methods used by the authors cited above to simulate the evolution of whole clusters of galaxies; alternatively, it is able to provide a realistic cluster environment for the galaxies, and it is better than the other methods to investigate what happens to the galaxies themselves when they move within that environment.

Muzzio and Vergne (1986) have developed a new code that might offer the advantages of both the full N→body and the test particle methods without suffering their shortcomings. We use test particles but the masses and sizes of the galaxies and of the background are determined by the number and distribution of the particles that belong to them; in other words, the particles are <u>individually</u> regarded as massless, but when they act <u>collectively</u> they behave as massive particles whose distribution determines the general gravitational field. We have been able to preserve in this way the advantages of the test particle approach attaining, at the same time, such a high degree of self-consistency that we have been able to produce mergers with this new code.

4. RESULTS

Following Miller (1983), a given set of parameters and initial conditions will be referred to as a model and a particular realization of these parameters as a run. Between three and eight runs (obtained using different seed numbers for the random number generator) enter in each model.

All the models investigated thus far had adopted virialized initial conditions and, in view of the results of Miller (1983), there is an urgent need for models that start from non-equilibrium conditions. M.M. Vergne and myself have just begun to investigate such models and our, still very preliminary, results suggest that the cluster swapping processes are intensified during the pre-virialization stages of the cluster evolution.

4.1 Number of galaxies and halo extension

The number of galaxies in the Virgo-like cluster models of Papers I and III was limited to the 61 ("bright" models) or 111 ("faint" models) brightest, and most massive, galaxies. No significant differences arose between the results of both kinds of models, however, so that in other investigations "bright" models were preferred because they allow the inclusion of a larger number of test particles and, thus, the obtainment of more meaningful results.

Since we suspected that the extension of the globular cluster systems could be a critical parameter of the tidal stripping and swapping processes, both "concentrated" and "extended" models were investigated in Papers I and III. The former imply a limiting radius of 420 kpc for a galaxy like M87, while the corresponding value for the latter is 1050 kpc; these radii are proportional to the square root of the galactic masses. It is important to emphasize that these are <u>limiting</u> values and that, due to the steep radial distribution of globulars, the distances to the corresponding galactic centers are less than one-fourth of those radii for 91.5% of the globulars. The results of the models showed that the differences were much less important than we had suspected, and that reasonable inaccuracies in the extension of the more realistic "concentrated" models could not seriously affect their results. It is worth noting that Miller (1983) found that the mass-radius relationship was of little relevance for the tidal stripping results of his models.

4.2 Influence of location

Rabolli (1986) studied the influence of the galaxy location on the cluster swapping processes. She grouped the galaxies in radial distance bins according to two different criteria: a) their final position after 10^{10} years of dynamical evolution, and b) their time-averaged radial distance over the same interval. The results obtained from the two different samplings are essentially the same, and in both cases it clearly turns out that galaxies in the central region of a cluster are strongly affected while galaxies outside that region suffer smaller or negligible effects.

Table I presents the global results obtained by Rabolli when the galaxies were grouped according to their final positions.

TABLE I
Influence of location on cluster swapping.

Radius (kpc)	Lost by capture (%)	Lost by escape (%)	Captured (%)
0.0- 412.5	23.3	21.3	24.2
412.5- 825.0	8.7	11.7	7.4
825.0-1650.0	5.7	5.1	5.1
≥1650.0	0.0	1.2	0.6

4.3 Influence of the distribution of missing mass

The results of Merritt (1983) and of Richstone and Malumuth (1983) suggested that the conclusions of Papers I, II and III could be altered if not all the cluster mass lay in the galaxies as had been assumed in our investigations. The suspicion was confirmed by the results of Paper IV which showed that when part of the cluster mass was assigned to a smoothly distributed background the previous conclusions were appreciably altered. If all the missing mass belongs to the background, the cluster swapping processes turn out to be irrelevant; under the more likely assumption that half of the total mass lies in the galaxies and half in the background, those processes are damped but they should still yield observable effects.

4.4 Mass dependence and massive central galaxies

One of the most interesting consequences of the cluster swapping process is its strong dependence on the masses of the galaxies involved, already noted by Forte et al. (1982). Figures 3 and 4 of Paper I present the histograms of the percentage of captured globulars versus the absolute magnitudes of the capturing galaxy, while Figure 5 of the same Paper presents the corresponding histograms for the globulars that are lost, due either to escape or to capture by other galaxies.

We emphasized in Paper I that even the least massive galaxies capture globulars, albeit a small percentage, but it is also important to emphasize that the percentage of lost clusters decreases with decreasing mass of the parent galaxy. The reason to insist on this point is that the idea exposed in the paper by Forte et al. (1982), that giant galaxies capture globular from dwarf galaxies, is still entertained by other researchers (cf. Harris 1986) and that idea is wrong according to the more recent evidence. Since the number of globulars captured by other galaxies decreases with decreasing mass of the parent galaxy, and since the total number of available globulars in a cluster like Virgo has a maximum near $M_V=-21$ mag and decreases rather sharply toward faint magnitudes (Figure 2 of Paper I), the number of swapped globulars that come from dwarf galaxies is, in fact, fairly small.

As a check, we examined the absolute magnitudes of the parent galaxies of the globulars captured by the massive central galaxies of all the runs used in Paper II. The mean difference between the visual absolute magnitude of the parent galaxy and that of the capturing galaxy turned out to be:

$$\Delta M_V = -1.41 \pm 0.13 \text{ mag} \quad (N = 53, \sigma = 0.92 \text{ mag})$$

and only 2, out of 53, test particles came from galaxies more than 3 mag fainter than the capturing galaxy. This result strengthens our assertion that the bulk of the captured globulars does not come from dwarf galaxies. Cluster swapping is a game reserved to the big shots!

Another interesting question is the one raised by Forte et al. (1982): Could M87 have obtained its globular cluster system capturing them from other galaxies? Figure 1 shows the net (i.e., captures minus losses) percent variation of the number of globular clusters of the brightest galaxy versus the visual absolute magnitude of that galaxy for all the runs investigated thus far. Several facts

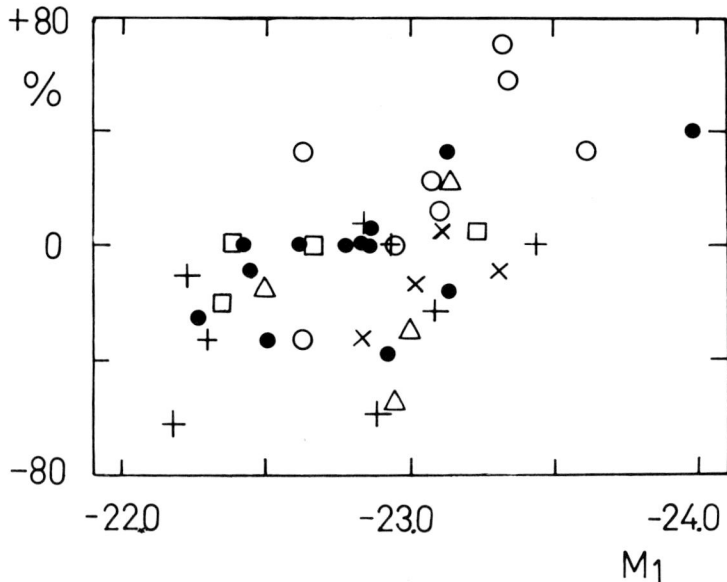

Fig. 1. Net percent variation of the number of globulars vs. absolute magnitude of first-ranked galaxies. Full circles: "concentrated" (Paper I and Rabolli 1986); open circles: "extended" (Paper I); plus signs: Paper II; triangles, crosses and squares: "a", "b" and "c", respectively, from Paper IV.

are clear from the Figure: 1) no model allows an increase larger than 70%, or 40% if we consider only the "concentrated" models; 2) an absolute magnitude brighter than about -23.5 mag seems to rule out the possibility of a decrease, while galaxies fainter than $M_V=-22.5$ mag have no chance to increase their globular cluster population at the expense of others; notice that the masses implied by these absolute magnitudes are about $6.1\ 10^{13}\ M_\odot$ and $2.4\ 10^{13}\ M_\odot$, respectively; 3) the influence of the central location of the most massive galaxy is confirmed by the generally lower percentages yielded by the runs of Paper II as compared with the results of Paper I; 4) the influence of the galactic- to background-mass ratio is also confirmed by the smaller percentages (be they positive or negative) as one goes from model "a", through "b", to "c".

Unfortunately, we cannot yet be sure of which is the precise parameter that governs the net gain (or loss) of globulars because, for example, the magnitude difference between the first- and the second-ranked galaxy yields a correlation similar to that shown in Figure 1. V.H. Dessaunet and myself are now preparing a new set of models that might help to clarify this situation; for the time being, it just seems possible for very massive galaxies to moderately increase their globular cluster population taking them from other galaxies.

4.5. Changes in the cluster distribution

The relatively small net changes in the total population of a massive galaxy result from much larger gains and losses that partly compensate each other (Papers I and II); a central galaxy suffers quite a beating, and the structure of its globular cluster system becomes much more extended than that of an isolated galaxy. This scenario is strongly modified if most of the missing mass lies in the background and not in the galaxies (Paper IV) but if, as seems likely (Forman et al. 1985), the galaxies themselves contain an appreciable amount of missing mass, there will still be a substantial modification of the distribution of globular clusters around a massive central galaxy.

The results of Papers I, II and IV showing that virtually all the outer halo (say, globulars beyond 210 kpc from the center of the galaxy) is lost due to tidal interactions merely confirmed what one could have reasonably expected. What was rather surprising was the result that even globulars belonging to the inner parts of the halo are also lost in large quantities. Since the idea that tidal stripping only affects the outer parts of galaxies seems to be prevalent (e.g. Richstone and Malumuth 1983; Miller 1983; van den Bergh 1984) it is important to emphasize that, due to the central concentration of globulars (and matter in general) in galaxies, losses from the inner halo outweight those from the outer halo. If we consider the results of models "a" and "b" from Paper IV, which do include a reasonable amount of background mass, we notice that for a galaxy like M87 the globulars lost from the region within a 50 kpc radius almost double those lost from the rest of the galaxy; such radius corresponds to an angular radius of about 10', i.e., well within the region where globulars have been studied in M87.

The globulars that are captured by a galaxy also tend to prefer the central regions, and their final radial distribution seems to be fairly independent of the original extension of the cluster systems (Paper I). The number of captures depends strongly on the ratio of galactic- to background-mass, however, and although model "a" of Paper IV still shows a substantial number of captured globulars, that number is very small for model "b".

The radial distribution of the globulars that remain with their parent galaxy becomes much more extended as a consequence of the tidal interactions. This result is strongly related to the amount of galaxy-galaxy interaction: it was negligible for the non-central galaxies of Paper I, very important for the central galaxies of Paper II, and varying from very important to completely negligible as the cluster dynamics become more and more dominated by the background (models "a", through "b", to "c" of Paper IV).

4.6. Lost globulars

Following the ideas of Merritt (1983, 1984), Harris (1986) made the interesting suggestion that, as a result of tidal stripping, systems of globular clusters not related to any obvious central galaxy could sit at the center of clusters of galaxies. That possibility was investigated in detail in Paper III, and some related results were also given in Papers II and IV. The conclusion was that such systems are indeed possible but, unfortunately, the predicted surface density of globulars is very low and, therefore, very difficult to confirm observationally. The number of tramps decreases steadily as the background- to galactic-mass ratio increases (Paper IV).

5. OBSERVATIONAL POSSIBILITIES

If there were observable effects of the cluster swapping processes, it would be possible to use the observational results to investigate some poorly known properties of clusters of galaxies (e.g. the distribution of missing mass) and to investigate individual globulars that had suffered the effect of those processes.

Following the ideas of Forte et al. (1982), Harris (1983) and van den Bergh (1984) used metallicity arguments to explore some consequences of cluster swapping. Unfortunately, if the bulk of the captured globulars does not come from dwarf galaxies but from galaxies only slightly less massive than the capturing galaxy, there is not much hope to get an observable metallicity difference. Nevertheless, the metallicity gradients within the globular cluster systems themselves (Strom et al. 1981) might allow the use of metallicity arguments in a different form: since the whole original outer halos are lost due to tidal stripping and replenished by cluster swapping with globulars that come mainly from the inner halos, we can expect that the metallicity of the outer halos will be higher for the galaxies that have suffered strong tidal stripping and accretion than for those that have had more peaceful lives. Accordingly, metallicity investigations of the globulars that lie beyond about 50 kpc from the center of galaxies like M87 may provide an interesting observational test of the consequences of cluster swapping.

An even simpler test may be offered by the radial distribution of the globular clusters, because the results of Papers I and II and of Rabolli (1986) show that a galaxy located in the central region of a cluster should have a more extended distribution of globulars than a galaxy which lies farther from that center.

Harris (1986) derived revised radial distributions for the surface density of the globular clusters in both M87, located near the center of the Virgo cluster, and M49, located about 4° from that center. If Harris's Figures 4 and 7 are superimposed and displaced 0.25 units along the vertical axis to account for the difference between the population

of globulars in each galaxy, it turns out that for very large
radii the surface density of globulars in M87 exceeds that in M49
at the same radius. The observational results are somewhat
doubtful both for the innermost and the outermost regions, but
it is interesting to note that, although small, the effect is in
qualitative agreement with the theoretical predictions. We can
only hope that the observational results will be improved in a near
future, through the use of CCD photometry for the inner regions
(Grillmair et al. 1986), and of large field telescopes for the outer
regions (Harris 1986).

It is very important to emphasize that a comparison between only
two galaxies is not of much relevance, and that many systems should be
observed and compared. The reason is that the theoretical predictions
are of a statistical nature: they include averages of several runs,
averages over several galaxies, averages over several globulars, and
so on. Not only are the dispersions large, but the average results can
be unapplicable to particular cases. For example, although we agree
with Harris (1986) in that M32, wich has no globulars, is an excellent
example of a dwarf galaxy that has been tidally stripped, we must
emphasize that, as a satellite of M31, the situation of M32 is very
peculiar; in fact, as the results of Paper I show, dwarf galaxies are
much less affected than giant galaxies by tidal stripping!

Since tidal accretion must be always accompanied by tidal
stripping, the search for lost globulars seems to offer interesting
possibilities, particularly considering that they might also offer a
test for some theories of galaxy formation (e.g. van den Bergh 1980).
Unfortunately, such a search presents formidable observational
difficulties due to the very low surface density of the tramps: how to
recognize them against the much larger surface density of foreground
stars and background galaxies, how to be sure that they do not belong
to the extended halo of a galaxy, and so on. Nevertheless, considering
the interesting opportunities that their discovery would offer
(systems of globular clusters without central galaxy, the end products
of tidal stripping, and so on), and the rapid progress of the
observational techniques, their study might soon become possible.

6. ACKNOWLEDGEMENTS

The technical assistance of Mrs. S.D. Abal de Rocha, M.C. Fanjul
de Correbo and M.E. Mac Williams de Carstens is gratefully
acknowledged. The author is a member of the Carrera del Investigador
Científico del Consejo Nacional de Investigaciones Científicas y
Técnicas de la República Argentina (CONICET). This work was supported
by grants from the CONICET, the Comisión de Investigaciones
Científicas de la Provincia de Buenos Aires and the Secretaría de
Estado de Ciencia y Tecnología.

REFERENCES

Carnevali, P., Cavaliere, A. and Santangelo, P. 1981 Astrophys. J. 249, 449.
Dekel, A., Lecar, M. and Shaham, J. 1980 Astrophys. J. 241, 946.
Forman, W., Jones, C. and De Faccio, M. 1985 in Proceedings of the ESO Workshop on the Virgo Cluster, O.-G. Richter and B. Binggeli, eds., European Southern Observatory, Garching, p. 323.
Forte, J. C., Martinez, R. E. and Muzzio, J. C. 1982 Astron. J. 87, 1465.
Fullerton, L. W. and Hills, J. G. 1982 Astron. J. 87, 175.
Grillmair, C., Pritchet, C. and van den Bergh, S. 1986 Astron. J. 91, 1328.
Harris, W. E. 1983 Publ. Astron. Soc. Pacific 95, 21.
Harris, W. E. 1986 Astron. J. 91, 822.
Hills, J. G. 1975 Astron. J. 80, 809.
Hills, J. G. 1984 Astron. J. 89, 1559.
Hills, J. G. and Fullerton, L. W. 1980 Astron. J. 85, 1281.
Malumuth, E. M. and Richstone, D. O. 1984 Astrophys. J. 276, 413.
Merritt, D. 1983 Astrophys. J. 264, 24.
Merritt, D. 1984 Astrophys. J. 276, 26.
Miller, G. E. 1983 Astrophys. J. 268, 495.
Miller, R. H. 1984 European Southern Observatory Scientific Preprint No. 343.
Muzzio, J. C. 1986a Astrophys. J. 301, 23 (Paper II).
Muzzio, J. C. 1986b Astrophys. J., in press (Paper III).
Muzzio, J. C., Dessaunet, V. H. and Vergne, M. M. 1987 Astrophys. J. in press (Paper IV).
Muzzio, J. C. and Martinez, R. E. 1982 Obs. Astron. Univ. Nac. La Plata Serie Astron. XLI.
Muzzio, J. C., Martinez, R. E. and Rabolli, M. 1984 Astrophys. J. 285, 7 (Paper I).
Muzzio, J. C., Rabolli, M. and Martinez, R. E. 1984 unpublished.
Muzzio, J. C. and Vergne, M. M. 1986 in preparation.
Rabolli, M. 1986 Univ. Nac. La Plata Ph. D. Thesis in preparation.
Richstone, D. O. 1975 Astrophys. J. 200, 535.
Richstone, D. O. and Malumuth, E. M. 1983 Astrophys. J. 268, 30.
Strom, S. E., Forte, J. C., Harris, W. E., Strom, K. M., Wells, D. C. and Smith, M. G. 1981 Astrophys. J. 245, 416.
van den Bergh, S. 1980 in Globular Clusters, D. Hanes and B. Madore, eds., Cambridge University Press, Cambridge, p. 175.
van den Bergh, S. 1984 Publ. Astron. Soc. Pacific 96, 329.

DISCUSSION

HARRIS: I think this simulation work is extremely interesting and something that we have needed to have for a long time. I would like to ask whether this work could be extended to small groups of galaxies as well as to richer environments like Coma. In the small groups of galaxies the encounter velocities are much lower, and you might end up with qualitatively different results. For example, in the fraction of escaped clusters or those that go through exchanges of any kind. Do you have any comments about the expected direction of the effects?

MUZZIO: I fully agree with you in that it will certainly be of interest to do models of small groups but since, as you say, the encounter velocities are lower, mergers may be more important there. Thus, I would rather investigate small groups with my new code, after having it thoroughly tested, because we will need self-consistency for that study.

OSTRIKER: I wonder if you will ever change the ratio of globular clusters to stars since when you find clusters are swapped, then stars will be swapped in the same ratio. Similarly the colors of the stars and the clusters are changed by the same amount. What are your views?

MUZZIO: I no longer entertain the idea of our original paper by Forte et al. (1982) that only the globulars are swapped; I now agree with you in that both globular clusters and stars are swapped and that is why I now prefer to call this effect tidal accretion rather than cluster swapping.

AGUILAR: The number of swapped clusters that you find seems very large. Simon White and I have studied tidal stripping using N-body experiments with a self consistent target galaxy and found that particle swapping is negligible in most cases, and becomes important only when the encounter velocity at infinity is comparable to the internal rms velocity of the interacting galaxies. Since these encounters are very inelastic most of these encounters may lead to mergers in which case there is no swapping.

WALLERSTEIN: Is it possible to estimate what fraction of the 150 globular clusters in our Galaxy have been captured?

MUZZIO: Thank you very much for that question, because it allows me to emphasize once again that our results are average values. We did run some models of the local group, but the swapping activity turned out to be low there and dominated by random factors (say, the close passage of one galaxy near another), or, as Richstone and Malumuth (1983) very accurately put it, by "luck".

HANES: Your overlay of the Harris counts for M 87 and M 49 suggested that the outer parts of the M 87 cluster system is somewhat more distended. You should remember that the error bars are from size-of-sample statistics but that the counts in the outer parts are systematically very sensitive to the correct adoption of the local field contribution.

MUZZIO: I am fully aware of the uncertainty of the values in the outer regions and, besides, since our emphasis in on average values, a comparison of just two galaxies carrier little weight. I just wanted to show that M 49 and M 87 may be not identical and that the difference, although small, is in the sense predicted by the models; we certainly need more accurate observational results.

GALAXY FORMATION AND CLUSTER FORMATION

Richard B. Larson

Yale Astronomy Department

1. INTRODUCTION

A primary motivation for studying globular clusters is that, as the oldest known galactic fossils, they trace the earliest stages of galactic evolution; indeed, they may hold the key to understanding galaxy formation. Thus it is clearly of great importance to learn how to read the fossil record. To do this, we need to understand something about how the globular clusters themselves formed. Were they the first bound objects to form, or did they form in larger pre-existing systems of which they are just small surviving fragments? If the latter, what were the prehistoric cluster-forming systems like? And how did they manage to produce objects like globular clusters?

2. CLUES TO CLUSTER FORMATION

Several pieces of evidence indicate that most globular clusters formed very early. The ages of the well-studied galactic globular clusters (VandenBerg, this symposium; Demarque, this symposium) imply that these clusters formed within at most the first few Gyr of the history of the universe. In addition, as has been noted by Harris (1986; this symposium), the globular clusters in elliptical galaxies generally have a more extended spatial distribution and a flatter radial density profile than the background stars; in the brightest ellipticals the density distribution of the globular clusters follows approximately $r^{-2.5}$, compared with $r^{-3.0}$ for the stars. The globular clusters are also generally bluer and thus presumably more metal-poor than the field stars at the same distance from the center in elliptical galaxies (Harris 1983). These properties suggest that the globular clusters formed earlier than the field stars in these galaxies, since in any picture of galaxy formation involving progressive condensation and enrichment of the gas, a population that forms earlier has both a lower metallicity and a flatter density profile (Larson 1969).

However, there is also evidence that globular clusters are not primordial objects but formed only after considerable star formation activity had already taken place. The radial distribution of the globular clusters in elliptical galaxies, while flatter than that of the stars, is steeper than the r^{-2} profile usually assumed for the dark matter in galactic halos, suggesting that the dark halos formed first; if the dark matter is in faint stars or stellar remnants that originated in an early Population III, the globular clusters formed after the Population III objects. More compelling evidence that globular clusters are not primordial is provided by the fact that, with very few exceptions, they are internally very homogeneous in metallicity: the dispersion in measured stellar [Fe/H] values in a cluster is less than 0.15 (Norris, this symposium), and the narrowness of the giant branch in M3 implies a dispersion in [Fe/H] less than 0.04 (Buonanno et al., this symposium). This is a much smaller spread in metallicity than has been found in any system of galactic size, including the dwarf spheroidal galaxies (Da Costa, this symposium). The gas from which a cluster formed must therefore have been thoroughly mixed, presumably by turbulence generated by energy input from massive stars, subsequent to its chemical enrichment but before the formation of the observed stars. This implies that the pre-cluster gas was retained in a bound star-forming system long enough for turbulence to homogenize it before the cluster formed. In order to retain the gas, such a system must almost certainly have been much more massive than the cluster itself, implying that globular clusters formed only as small parts of much larger star-forming systems (e.g. Searle 1977, Searle and Zinn 1978).

If the formation of globular clusters is at all similar to the formation of open clusters in our Galaxy, observations of nearby regions of star formation also suggest that bound clusters form only as small parts of much larger complexes of gas and young stars. For example, in Orion there is a small and very dense grouping of young stars immediately surrounding the Trapezium that will probably survive as a bound open cluster (Herbig and Terndrup 1986), but it contains only a few hundred solar masses, compared with a total of several thousand solar masses of young stars and several times 10^5 solar masses of gas in the whole Orion region of star formation. Thus only about 10^{-3} of the mass in this region seems likely to end up in a bound star cluster. If a similarly low efficiency applies to the formation of globular clusters, this suggests that they originated in star-forming systems or complexes with masses exceeding 10^8 solar masses. It may be relevant that the smallest galaxy known to contain globular clusters, the Fornax dwarf spheroidal galaxy, has a present mass of about 2×10^7 solar masses, and it may well have had a much larger initial mass.

3. SITES OF CLUSTER FORMATION

What did the primitive cluster-forming systems look like? The above considerations suggest that they typically had the masses of dwarf galaxies, and some of them may have been much larger. Indeed, since there is a population of globular clusters associated with the old disk of our Galaxy (Zinn 1985; this symposium), globular clusters can evidently form in the disks of large spiral galaxies, at least early in their evolution. Thus globular clusters may have formed in systems with masses anywhere between perhaps 10^8 and 10^{11} solar masses. Most presently observed gas-rich, actively star-forming galaxies in this mass range are rotationally flattened disks, so many globular clusters may actually have formed in disks. Most of these early disk systems must since have been destroyed by collisions or mass loss that left them unbound and dispersed their stellar contents.

Since the field stars in elliptical galaxies may well also have formed in disk systems that later merged (Toomre 1977), and since most of the stars in elliptical galaxies are probably quite old, the early disk systems must have formed their stars quite rapidly, probably mostly within the first few Gyr. They must therefore have been more compact than typical present-day irregular or spiral galaxies, since a higher star formation rate requires a higher mean density (Larson 1977). Typical irregular or spiral galaxies also do not make clusters as massive as typical globular clusters, so they cannot serve as suitable prototypes for the birthplaces of globular clusters. The original star-forming systems must have been both more gas-rich and more compact than present-day galaxies, and they must have formed stars more vigorously and in larger complexes. If observed today, such a system would probably be called a blue compact or starburst galaxy.

A similar picture is suggested by the available evidence on the circumstances in which very massive star clusters form. The disk of our Galaxy evidently stopped forming globular clusters early in its history, and now forms only much less massive open clusters. The Large Magellanic Cloud is currently forming more massive clusters, as is M33 (Christian, this symposium), but these clusters are still perhaps an order of magnitude less massive than typical galactic globular clusters. As in our Galaxy, the young clusters in these galaxies are less massive than the old ones, and only the oldest LMC clusters approach typical galactic globular clusters in size. The most luminous young cluster in the Local Group, located in the bright LMC HII region 30 Doradus, has a total luminosity of about 4×10^7 suns (Jones et al. 1986) and contains about three hundred O stars (Moffat et al. 1985); with a conventional IMF, this implies a total stellar mass of about 10^5 solar masses. However, nothing is known directly about low-mass stars in 30 Dor, so it cannot be predicted how much mass will remain in a bound cluster; probably it will not exceed a few times 10^4 solar masses. The resulting cluster may then have up to a hundred times the mass of the Trapezium cluster, but it will still fall at least an order of magnitude short of being a good candidate for a young globular cluster, which should have at least a thousand times the mass of the Trapezium cluster. The total mass of gas in the 30 Dor region is about 5×10^7 solar masses (Rohlfs et al. 1984), also roughly a hundred times the total mass in the Orion region (Maddalena et al. 1986) rather than the three orders of magnitude more that may be required to form a globular cluster.

Starburst galaxies may be more promising sites for the formation of globular clusters. The nearby prototype M82, which is currently experiencing a vigorous burst of star formation in a compact nuclear gas disk, contains a number of bright stellar knots with luminosities exceeding 10^8 suns (van den Bergh 1971) that could represent, or contain, young globular clusters. Again, however, nothing is known about the low-mass stellar content or the spatial structure of these knots. Another system of interest is I Zw 18, which contains a very bright HII region that has a luminosity of order 10^8 suns, an estimated stellar mass of 10^6 solar masses, and a metal abundance [Fe/H] of -1.5 (Lequeux and Viallefond 1980). The most massive young cluster known may be the "super star cluster" in NGC 1705 (Melnick et al. 1986), which has a luminosity of 2×10^8 suns but contains no O stars, implying an age of at least 10^7 years. Its inferred mass is then about 5×10^6 solar masses, assuming a conventional IMF, although this assumption may not be valid in starburst galaxies (Larson 1987). In any case it would appear that the conditions required for globular cluster formation are at least approached, if not met, in some starburst galaxies.

Similar conditions may exist in some of the giant HII regions in the outer parts of the giant spiral galaxy M101, which typically contain several million solar masses of young stars (Viallefond et al. 1981). If sufficiently massive bound clusters exist in these luminous HII regions, the conditions required for globular cluster formation may be met or approached even in some late-type giant spirals. If so, it may be that the essential requirement for forming globular clusters is simply a very gas-rich system with active star formation occurring in very massive complexes.

4. THE MECHANISM OF CLUSTER FORMATION

How, then, do globular clusters form in the larger, flattened star-forming systems that have been discussed? And can one account for the fact that very massive clusters seem to form only in youthful, gas-rich, vigorously star-forming systems, while the typical mass of the clusters that form in a galaxy decreases with time as the galaxy ages?

The massive gas complexes in which star clusters are born must be self-gravitating, and their formation almost certainly involves gravity acting on large scales. A mechanism that is probably of very general importance for star formation is the gravitational instability of a galactic gas layer or disk; this can play a role not only in generating spiral structure in galaxies (Goldreich and Lynden-Bell

1965, Toomre 1981) but also in initiating or organizing star formation in them (Larson 1977, 1983). In a typical spiral galaxy like our own, gravitational instability of the gas layer produces structures with typical sizes of a few kpc and masses of several times 10^7 solar masses, comparable to a major spiral arm segment or to the 30 Dor complex in the LMC. Thus gravitational instability can account for at least the large-scale structure of major star-forming regions in our Galaxy and others (Elmegreen and Elmegreen 1983). Further gravitational condensation processes on smaller scales, which are not yet well understood, are evidently required to form the small dense cloud cores in which star clusters and ultimately individual stars are born. The development and fragmentation of flattened or filamentary structures may play an important role also in these later stages of cloud condensation (Larson 1985).

The stability of a rotating gas disk is determined by the parameter $Q = c\kappa/\pi G\mu$, where c is the velocity dispersion of the gas, μ is its surface density, and κ is the epicyclic frequency. In a thin disk, violent instability occurs when $Q < 1$; limited growth of spiral density perturbations occurs for $1 < Q < \sim 2$ (Larson 1984); and stability occurs for $Q > \sim 2$. Numerical simulations of disks containing gas (e.g. Sellwood and Carlberg 1984) show that, while dissipation tends to reduce Q and make the disk more unstable, the ensuing instabilities tend to raise the velocity dispersion again and maintain Q at a nearly constant value of about 2.

If the gas layer in a disk galaxy is thus maintained in a state of marginal stability throughout its evolution, the typical size and mass of any structures formed by gravitational instability will decrease with time as the gas is depleted by star formation. This is because, if the epicyclic frequency κ does not change much, a constant value of Q implies that the velocity dispersion c of the gas decreases as the gas surface density μ decreases, and this results in smaller length scales for instability (e.g. Carlberg and Freedman 1985). In the solar neighborhood, for example, the gas surface density may have decreased by as much as an order of magnitude from the time of disk formation to the present, implying a similar order-or-magnitude decrease in the velocity dispersion of the gas (if the contribution of stars to the instability of the gas layer is neglected.) The typical mass of structures formed by gravitational instability varies as c^4/μ (Larson 1985), so if both μ and c decrease by an order of magnitude, the predicted mass scale decreases by three orders of magnitude. This is just the factor by which the Trapezium cluster (which is relatively small) must be scaled up in mass to make a typical globular cluster. Thus it may be possible to understand in this way why globular clusters formed only during the early evolution of the Milky Way disk, and why the presently forming clusters are much less massive.

This conclusion probably does not depend crucially on the assumption that the gas layer remains marginally stable at all times. Even if the velocity dispersion of the gas is much too small to maintain stability, there is still a maximum length scale on which instability can operate that is proportional to μ/κ^2. If this maximum length scale is identified with the size of the largest star-forming complexes that develop, which are probably the birthplaces of the most massive clusters, then the mass of these largest complexes is predicted to vary as μ^3, just as is predicted by the above argument for "typical" structures.

Although not intended to represent the formation of star clusters, several simulations of the fragmentation of gas disks (Larson 1978, Wood 1982, Miyama et al. 1984) show results that are at least qualitatively consistent with the predictions of gravitational instability theory as outlined above. In particular, they show that when the velocity dispersion (temperature) of the gas is decreased, the disk becomes thinner and the typical size and mass of the fragments formed is reduced. The simulations of Larson (1978) also show that very dense cores can form and will eventually accrete most of the mass if no disruptive effects operate. Thus, in addition to gravitational instability, accretional growth of condensations in a galactic disk may play a role in the development of massive clouds and of dense cluster-forming cores within them (see also Larson 1983). Effects such as ionization by massive stars may then limit how massive these cluster-forming cores can become before they self-destruct.

Some evidence that settling of the galactic gas layer occurred early in the evolution of the Milky Way disk and was accompanied by a decrease in the size of the clusters formed is provided by the fact that the luminosities of the disk globular clusters in our Galaxy appear to increase with distance from the galactic plane (Zinn 1985); for example, the disk clusters more than 500 pc from the plane are on the average about 1.3 mag more luminous than those closer to the plane.

5. STAR FORMATION IN GLOBULAR CLUSTERS

In order for a bound globular cluster to be produced, a cluster-forming cloud must form low-mass stars with high efficiency. Also, if the protocluster gas is not to be ionized and dispersed by massive stars before most of its has condensed into low-mass stars, it is probably necessary for star formation to be completed in a time shorter than stellar evolution timescales, i.e. shorter than 10^6 years. Both of these requirements may be satisfied if cluster-forming cloud cores are dense enough. Very high densities must indeed have existed, since the observed densities of globular clusters, allowing for some expansion due to stellar mass loss, imply that their birth clouds must have had remarkably high average densities of at least 10^5 molecules per cm^3.

It has often been suggested that the earliest star formation in galaxies produced mostly or exclusively massive stars: for example, the initial enrichment of halo stars and their high oxygen-to-iron ratio have been attributed to an early generation of massive stars, and it has been suggested that the early formation of mostly massive stars is to be expected because of the relatively high gas temperature (Larson 1986). Also, a variety of cosmological phenomena may be explainable by postulating large numbers of pregalactic very massive stars (Carr et al. 1984). If early star formation favored massive stars, then globular clusters must have been exceptional objects, since star formation in them evidently produced mostly low-mass stars.

The fragmentation process and the resulting stellar IMF in a star-forming cloud depend critically on the temperature-density relation: fragmentation can continue as long as the gas temperature decreases with increasing density, but it becomes less likely and may stop altogether when the temperature stops decreasing and begins to increase with increasing density (Larson 1985). In a region of very active star formation, the dust is heated by the strong infrared radiation field present (Falgarone and Puget 1985), and this causes the gas temperature to turn upward with increasing density when the critical mass for fragmentation is approximately 5 solar masses (Larson 1987). This suggests that in starburst regions the formation of stars less massive than several solar masses is inhibited, as has also been suggested on observational grounds. However, the formation of low-mass stars can still occur in exceptionally dense cloud cores where the density exceeds 10^6 molecules per cm^3; in such regions the critical mass becomes less than one solar mass despite the elevated temperature (Larson 1987). Thus a globular cluster can form in a massive cloud core in which most of the gas has a density of 10^6 molecules per cm^3 or more. In nearby molecular clouds, only very small regions with masses of several tens or hundreds of solar masses have such high densities, and this suggests again that the formation of globular clusters requires much more massive clouds.

If a protocluster cloud core has an average density of 10^6 molecules per cm^3, its free-fall time is only 3×10^4 years, so it is possible that star formation is largely completed within 10^5 years, i.e. well before any of the stars begin to evolve and before ionization has eroded away much of the gas. However, most of the gas will probably have been ionized and dispersed by the time supernovae begin to explode in the cluster after about 10^7 years (Tenorio-Tagle et al. 1986). For this reason, chemical self-enrichment of globular clusters seems unlikely to occur.

Another implication of the required very high initial density of globular clusters is that at this density, the dissolution time for a globular cluster of typical mass is of the same order as the Hubble time. For a fixed density, the dissolution time of a cluster is proportional to its mass; thus globular clusters with much smaller masses should not have survived to the present time, and this may explain why such clusters are not now observed.

Clearly many properties of globular clusters remain to be explained, including most of those discussed at this symposium. Not only cluster formation processes but also destruction mechanisms will clearly be important in accounting for the properties of globular clusters and for their correlations with the environment. Given the current rudimentary state of theorizing on the formation of globular clusters, it will be difficult to make any very quantitative predictions about how they began; indeed, it seems enough of a challenge to understand the mere existence of these remarkable objects.

REFERENCES

Carlberg, R. G. and Freedman, W. L. 1986 Astrophys. J. 298, 486.
Carr, B. J., Bond, J. R. and Arnett, W. D. 1984 Astrophys. J. 277, 445.
Elmegreen, B. G. and Elmegreen, d. M. 1983 Monthly Notices Roy. Astron. Soc. 203, 31.
Falgarone, E. and Puget, J. L. 1985 Astron. Astrophys. 142, 157.
Goldreich, P. and Lynden-Bell, D. 1965 Monthly Notices Roy. Astron. Soc. 130, 125.
Harris, W. E. 1983 Publ. Astron. Soc. Pacific 95, 21.
Harris, W. E. 1986 Astron. J. 91, 822.
Herbig, G. H. and Terndrup, D. M. 1986 Astrophys. J. 307, 609.
Jones, T. J., Hyland, A. R., Straw, S., Harvey, P. M., Wilking, B. A., Joy, M., Gatley, I. and Thomas, J. A. 1986 Monthly Notices Roy. Astron. Soc. 219, 603.
Larson, R. B. 1969 Monthly Notices Roy. Astron. Soc. 145, 405.
Larson, R. B. 1977 in The Evolution of Galaxies and Stellar Populations B. M. Tinsley and R. B. Larson, eds., Yale University Observatory, New Haven, p. 97.
Larson, R. B. 1978 Monthly Notices Roy. Astron. Soc. 184, 69.
Larson, R. B. 1983 Highlights of Astronomy 6, 191.
Larson, R. B. 1984 Monthly Notices Roy. Astron. Soc. 206, 197.
Larson, R. B. 1985 Monthly Notices Roy. Astron. Soc. 214, 379.
Larson, R. B. 1986 Monthly Notices Roy. Astron. Soc. 218, 409.
Larson, R. B. 1987 in Stellar Populations C. A. Norman, A. Renzini, and M. Tosi, eds., Cambridge University Press, in press.
Lequeux, J. and Viallefond, F. 1980 Astron. Astrophys. 91, 269.
Maddalena, R. J., Morris, M., Moscowitz, J. and Thaddeus, P. 1986 Astrophys. J. 303, 375.
Melnick, J., Moles, M. and Terlevich, R. 1985 Astron. Astrophys. 149, L24.
Moffat, A. F. J., Seggewiss, W. and Shara, M. M. 1985 Astrophys. J. 295, 109.
Miyama, S. M., Hayashi, C. and Narita, S. 1984 Astrophys. J. 279, 621.
Rohlfs, K., Kreitschmann, J., Siegman, B. C. and Feitzinger, J. V. 1984 Astron. Astrophys. 137, 343.

Searle, L. 1977 in The Evolution of Galaxies and Stellar Populations B. M. Tinsley and R. B. Larson, eds., Yale University Observatory, New Haven, p. 219.
Searle, L. and Zinn, R. 1978 Astrophys. J. 225, 357.
Sellwood, J. A. and Carlberg, R. G. 1984 Astrophys. J. 282, 61.
Tenorio-Tagle, G., Bodenheimer, P., Lin, D. N. C. and Noriega-Crespo, A. 1986 Monthly Notices Roy. Astron. Soc. 221, 635.
Toomre, A. 1977 in The Evolution of Galaxies and Stellar Populations B. M. Tinsley and R. B. Larson, eds., Yale University Observatory, New Haven, p. 401.
Toomre, A. 1981 in The Structure and Evolution of Normal Galaxies S. M. Fall and D. Lynden-Bell, eds, Cambridge University Press, p. 111.
van den Bergh, S. 1971 Astron. Astrophys. 12, 474.
Viallefond, F., Allen, R. J. and Goss, W. M. 1981 Astron. Astrophys. 104, 127.
Wood, D. 1982 Monthly Notices Roy. Astron. Soc. 199, 331.
Zinn, R. J. 1985 Astrophys. J. 293, 424.

DISCUSSION

GRINDLAY: If globular clusters are indeed formed in gaseous disks by instabilities, how do you account for the clusters systems in giant ellipticals such as M 87? What are the implications for formation of the ellipticals themselves?

LARSON: I would suggest that giant ellipticals form by mergers of disk systems of various sizes. There is now good enough observational evidence to make a convincing case that at least some elliptical galaxies form or are built up by mergers of disk galaxies. The globular clusters in ellipticals probably form as a result of starburst activity during the early evolution of such disks, while the field stars mostly form somewhat later in these disks when the metallicity has increased and globular clusters have stopped forming.

OSTRIKER: If globular clusters are all made via gravitational instabilities in disks, why is the system so spherical now? If they were made in preexisting disk systems that merged, then how did the clusters know how to arrange themselves so that the more metal-rich clusters became more centrally concentrated in our Galaxy?

LARSON: I imagine that spherical globular cluster systems form from subsystems that merge with little total angular momentum, although the subsystems may individually be flattened and rotating. The entire proctology probably had some overall organization, perhaps provided by the potential well of a dark halo, and leftover gas continued to dissipate and settle toward the center as it became chemically enriched. I am suggesting that, in order to form globular clusters, the protogalaxy contained internal structure consisting of flattened star-forming subsystems.

MATHIEU: I have two questions regarding star formation in globular clusters. First, do you think that the apparent paucity of binaries in globular clusters indicates differences in the star formation processes relative to those of Pop I stars and Pop II field stars and do you have any conjecture as to why these differences might arise? And second, how many supernovae would be required to remove the parent gas of a globular cluster? Does the presence of hundreds of O stars suggest greater coevality (< 10^6 yr) than that seen in newly star forming regions (~ 10^7 yr)?

LARSON: I know of no reason why globular clusters should have formed relatively few binaries; maybe most of the original binaries in globular clusters were destroyed. The chances for destruction might be increased if the clusters were initially much denser. Supernovae are probably not a very important mechanism of cloud destruction because ionization by O stars is likely to have done even greater damage before supernovae began to explode, after ~ 10^7 years. Of course one wants to avoid destruction by ionization before most of the low-mass stars have formed, but at the densities of > 10^6 cm^{-3} that I suggest for proto-globular clusters the free-fall time is only a few times 10^4 years. So star formation may indeed occur within only a very short time interval.

VAN DEN BERGH: It seems to me that the correlations of metallicity, cluster radius and the second parameter of globular clusters with galactocentric distance suggest that the proto-galaxy must already have been quite highly organized at the time clusters were formed. There is also a qualitative difference between the formation of clusters in the Magellanic Clouds and in the protogalaxy: (1) clusters in the clouds have larger ξ than their galactic counterparts and (2) the mass spectrum and globular clusters differs from that with which clusters are formed in the clouds.

LARSON: I agree that there must have been some overall organization, as I indicated in my answer to Ostriker. I would also reiterate that the star-forming, subsystems that made the globular clusters could not have been just like typical present irregular or spiral galaxies, which are too loose in structure and form stars at too leisurely a pace. Thus the Magellanic Clouds are not a good prototype. More compact, gas-rich, and actively star forming systems containing exceedingly massive gas complexes are required, and the closest present analogies are more likely to be starburst galaxies.

BHATIA: How do you explain the formation of sparse clusters like Pal 14 which exist in the halo of the Galaxy?

LARSON: I don't pretend to be able to predict the structure of globular clusters when they form. Maybe a great many sparse clusters were formed initially, and only those at large galactocentric distances have survived. A possible way of making sparse clusters from initially dense ones would be as a result of stellar mass loss in a system whose initial mass function contained a large proportion of massive stars. Maybe the dwarf spheroidal galaxies are very loose because they lost a lot of mass in this way.

RICHER: Can you form Jupiter's (few thousandths of a M_V) in your star formation picture in globular clusters?

LARSON: My feeling is that it would be difficult to make many Jupiters in globular clusters. To make Jupiters you need very low temperatures and/or exceedingly high densities. I think that fragmentation to smaller masses continues effectively as long as the temperature continues to decrease with increasing density, but becomes much less likely when the temperature stops decreasing and begins to rise again. In nearby cold dark clouds, this allows one to understand the formation of objects as small as a few tenths of a solar mass, but suggests that Jupiters are much harder to form. Maybe they only form in dense pre-planetary disks around stars like the Sun.

THE ORIGIN OF GLOBULAR CLUSTERS

S. Michael Fall

Space Telescope Science Institute and
Johns Hopkins University

Martin J. Rees

Institute of Astronomy, Cambridge

1. INTRODUCTION

The purpose of this article is to review some recent attempts to understand the origin of globular clusters. To put this in perspective, it may help to recall the analogous problem of the origin of galaxies. This splits into two parts. First, given a proto-galaxy with a specified mass and radius, how does it collapse, form stars and settle into a state of dynamical equilibrium? Richard Larson explored these topics in an important series of numerical simulations in the 1970s. Progress in this area brings into sharper focus a second set of questions that really has precedence over the first. Why did proto-galaxies have properties like the initial conditions in the collapse calculations and what distinguishes galaxies from structures on much larger and much smaller scales? Similar questions face us when we consider the origin of globular clusters. First, how did stars form in a proto-cluster, what was the efficiency, the initial mass function and so forth? It is appropriate that Larson has discussed these topics in the preceding article but here we are mainly concerned with the second kind of question: What is special about objects with masses of order 10^5-10^6 $M_\odot$ and dimensions of a few tens of parsecs?

2. DISRUPTION

One possible answer to the last question is that star clusters formed with a wide range of properties and that only those with a much narrower range of properties survived to the present. In this spirit, we once emphasized the gradual disruption of clusters by dynamical friction, tidal shocks and internal relaxation (Fall and Rees 1977). When the time scales for these processes are set equal to a Hubble time, they define a "survival triangle" in the mass-radius plane. Most globular clusters lie inside the triangle but any objects that formed well outside it would have been destroyed or severely damaged by now (This was also noticed by J. P. Ostriker, unpublished). Disruption is not a complete answer because dynamical friction sets an upper limit on the masses that increases with galactocentric distance whereas the observed luminosities of globular clusters show no such dependence (Caputo and Castellani 1984). Moreover, tidal shocks, which occur as clusters pass through a massive disk, would not have any effect in elliptical galaxies. Finally, there is some doubt as to whether internal relaxation leads to the complete disruption of clusters. Thus, although these stellar dynamical processes may have played some role in restricting the range of sub-galactic structures, they cannot by themselves account for the special properties of globular clusters.

Two other disruptive effects, tidal limitation and star formation are potentially more important than the previous ones and act on shorter time scales. To remain bound, a cluster must have a mean density that exceeds a value set by the tidal field of the parent galaxy. As the orbit of the cluster carries it closer to the galactic center, it will experience a stronger tidal field, and consequently, shed some stars. A cluster on a nearly radial orbit might even be destroyed, releasing all its stars into a field population. The expulsion of gas from a proto-cluster during star formation can lead to disruption in either of two ways. If more than half the total mass is removed quickly (i.e. in a time shorter than the internal crossing time), the proto-cluster, including the stars that formed in it, cannot remain bound. Alternatively, if any amount of mass is removed slowly, the proto-cluster will expand, and in the presence of a tidal field, release its least bound gas and stars. The expulsion of gas by stellar winds, HII regions and supernovae is thought to be important in star-forming regions in the galactic disk today. Its importance during the formation of globular clusters, however, is hard to quantify because we know almost nothing about the number of massive stars that were produced.

The discussion of disruptive effects is necessarily rather vague but it does raise two issues worth emphasizing at this point. First, globular clusters today may bear only a loose resemblance to their progenitors. This should be kept in mind when comparing any predictions of the initial masses and densities with observations. Second, many field stars in the spheroidal components of galaxies may be the debris of disrupted or failed globular clusters. The traditional view is that, if field stars and globular clusters share a common origin, they should, as populations, have the same space distributions, kinematics and chemical compositions. We must not, however, insist on complete similarity in all these respects because the likelihood of a cluster being disrupted depends on its position, orbital motion, stellar mass function and so forth. Oort (1965) estimated that the intial number of globular clusters was at least an order of magnitude larger than the present number. He supposed that the stars liberated from disrupted clusters would be strung out along families of tube orbits and those passing through the solar neighborhood would appear as "moving groups". Unfortunately, since there is some doubt as to the reality of the moving groups considered by Oort, the initial number of globular clusters must be regarded as a free parameter.

3. PRIMARY FORMATION

Theories in this subject can be classified as primary, secondary or tertiary depending on whether globular clusters are assumed to form before, during or after the collapse of proto-galaxies. We discuss each of these possibilities in turn. Primary formation, first suggested by Peebles and Dicke (1968), relies on the fact that the baryonic Jeans mass just after recombination is of order 10^5-10^6 $M_\odot$. This defines the smallest objects that can form by gravity alone in some cosmological pictures. In others, perturbations on small scales are damped out and the first objects to form are galaxies or clusters of galaxies. Globular clusters may have a primary origin in a universe dominated by weakly interacting particles with small random velocities, i.e., "cold dark matter" (Peebles 1984). In this picture, the initial spectrum of perturbations is a decreasing function of mass with negative curvature. The development of structure is roughly hierarchical on large scales but more complicated on small scales. Luminous objects are assumed to form by the dissipative collapse of baryons in the potential wells provided by the dark matter. The collapse occurs at redshifts of 2-4 on galactic scales and at redshifts of up to 10-20 on smaller scales. If globular clusters formed in this way, they would, at least initially, be surrounded by dark halos with masses of order 10^7-10^8 $M_\odot$.

There are several objections to the idea that globular clusters formed before the collapse of proto-galaxies. Each has a counter argument that may or may not seem convincing. First, galaxies contain very few objects with masses in the range above 10^6 $M_\odot$ where a continuum of structures might be expected. It is, however, conceivable that many of the objects more massive and therefore less dense than globular clusters were tidally disrupted. Second, globular clusters are more concentrated toward the centers of galaxies than the dark matter. A corrollary is that intergalactic clusters are extremely rare. These problems are alleviated by "biasing", which ensures that the perturbations destined to form globular clusters are located preferentially but not exclusively inside the perturbations destined to form galaxies. Third, globular clusters have significant abundances of heavy elements rather than primordial compositions. Self-enrichment is a possible solution although this is severely constrained by the narrow spread in the metallicities of the stars within most globular clusters. One must therefore postulate that all the low-mass stars observed today formed after the proto-clusters were enriched by high-mass stars. Another problem is that the metallicities of globular clusters are correlated with their positions, which is hard to understand if they formed before the collapse of proto-galaxies.

4. SECONDARY FORMATION

The idea that globular clusters formed during the collapse of proto-galaxies has been advocated by many authors. One argument in favor of secondary formation, mainly emphasized by observers, is based on the overall similarity between globular clusters and field stars in the spheroidal components of galaxies (see, for example, Searle 1977 and Searle and Zinn 1978). Such comparisons are most natural when restricted to the "halo" clusters (Zinn 1985). Another line of reasoning, which we emphasize here, is based on physical plausibility (Gunn 1980, McCrea 1982, Fall and Rees 1985). Our starting point is the widely held view that fragmentation and star formation should occur in a proto-galaxy when it can cool in a free-fall time (Binney 1977, Rees and Ostriker 1977, Silk 1977). This condition picks out a mass of order 10^{12} $M_\odot$ and a radius of order 10^2 kpc. The cooling arguments have been extended to a picture in which the luminous components of galaxies form by the collapse of gas in dark halos that cluster hierarchically from small perturbations in the early universe (White and Rees 1978). The latest version of this story is the one with cold dark matter (Blumenthal, Faber, Primack and Rees 1984). Our theory for the origin of globular clusters is motivated in part by these ideas but the general features should apply in a much wider range of cosmological pictures.

For a proto-galaxy to collapse in free fall, the radiative cooling must remain at least as efficient as the gravitational heating. If the gas is lumpy, as expected in any realistic proto-galaxy, the overdense regions will cool more rapidly than the underdense regions. This process--a thermal instability--will produce a two-phase medium, i.e., cold dense clouds embedded in and confined by hot diffuse gas. Now there are two characteristic temperatures in the problem. One, the temperature of the hot gas, can be expressed as $T_h \approx (\mu_h/3k) V_{gal}^2$, where V_{gal} is a typical velocity for large-scale, gravitationally-induced motions and $\mu_h \approx 0.6$ m_p is the mean mass per particle of ionized gas. The other characteristic temperature, $T_c \approx 10^4$ K, is where hydrogen recombines and the cooling rate drops precipitously. We assume for the moment that the clouds do not cool to lower temperatures and justify this later. The densities of the two phases, once they reach pressure balance, are related by $\rho_c/\rho_h = (\mu_c/\mu_h)(T_h/T_c)$, where $\mu_c \approx 1.2$ m_p is the mean mass per particle of neutral gas. For $V_{gal} = 300$ km s^{-1}, a value appropriate to the Milky Way, we find $T_h \approx 2 \times 10^6$ K and therefore $\rho_c/\rho_h \approx 400$. Our detailed calculations show that this state is reached during the collapse of the proto-galaxy if the initial amplitudes of the perturbations giving rise to the clouds are of order 10%. Perturbations with larger amplitudes grow even more rapidly.

Any clouds with masses greater than some critical value will be gravitationally unstable and will collapse. The standard formula for an isothermal sphere confined by an external pressure p_h is

$$M_{crit} = 1.2(kT_c/\mu_c)^2 G^{-3/2} p_h^{-1/2}. \tag{1}$$

This can be simplified by noting that the hot gas as a whole remains near the threshold for gravitational instability while it collapses. Combining (1) with a similar expression for a proto-galaxy of mass M_{gal} then gives

$$M_{crit} \approx \frac{1}{4} (T_c/T_h)^2 f_h^{-1/2} M_{gal}, \tag{2}$$

where f_h is the fraction of the mass in the hot phase. In general, we expect f_h to be near unity when the first clouds form and to decrease thereafter. The exact value is not crucial, however, because f_h enters

(2) only through a square root. Another way to estimate p_h and hence M_{crit} is by assuming that the cooling time of the hot gas is comparable to the free-fall time, as in the arguments that lead to a preferred scale for proto-galaxies. (This is what was done in our 1985 paper.) For $T_h \approx 2 \times 10^6$ K and $M_{gal} \approx 3 \times 10^{11} M_\odot$, we find $M_{crit} \approx 2 \times 10^6 M_\odot$. This is somewhat higher than but reasonably close to the masses of globular clusters. Since T_h scales roughly as $M_{gal}^{1/2}$ (the Faber-Jackson and Tully-Fisher relations), the critical mass, as given by (2), should have little variation from one galaxy to another.

The clouds produced by a thermal instability can have a wide range of masses. In the absence of magnetic fields, thermal conduction sets a lower limit, which, unless the clouds are highly flattened or filamentary, is well below M_{crit}. If there is a tangled magnetic field with a strength of only 10^{-15} G, conduction is suppressed and the lower limit on the masses is even smaller. Most of the clouds will therefore be gravitationally stable. They will persist in pressure balance with the hot gas until collisions produce agglomerations that are massive enough to collapse. We therefore expect the proto-clusters to have a narrow range of masses near M_{crit}. The value $M_{crit} \sim 10^6 M_\odot$ is, however, special only if the temperature of the cold gas "hangs up" at 10^4 K. A necessary condition for this to occur is that the cooling times of the clouds be comparable to or longer than their internal free-fall times so that they contract quasi-statically. If this condition were not satisfied, the gas would cool rapidly through 10^4 K and M_{crit} would be drastically reduced. Some of the smaller clouds might eventually reach temperatures low enough to become gravitationally unstable, but if the cooling time is large in comparison with the free-fall time, a feature near $10^6 M_\odot$ will still be imprinted in the mass spectrum.

In gas with a primordial composition, the only significant cooling at temperatures just below 10^4 K is caused by molecular hydrogen. This would spoil our theory were it not that H_2 can be destroyed by radiation just longward of the Lyman limit (Stecher and Williams 1967). Even the hot gas in a proto-galaxy emits enough ultraviolet photons to keep molecular cooling at modest levels and this could be reduced further by radiation from massive stars or an active galactic nucleus. Once heavy elements are produced and dispersed within a proto-galaxy, they provide another source of cooling. In an idealized model with no heat input, we find that the temperatures of the clouds would remain near 10^4 K as long as the metallicity is less than or of order $10^{-2} Z_\odot$. This estimate is in reasonable agreement with the abundance of heavy elements in many globular clusters. A completely realistic treatment would include heating mechanisms and might therefore be compatible with the higher metallicities of some clusters. There are several possibilities: (a) heating by supernovae, stellar winds, etc. within the proto-clusters, (b) photo-ionization by massive stars elsewhere in the proto-galaxy, (c) heating by cosmic rays, (d) photo-ionization by an active galactic nucleus. Any of these effects could raise the metallicity at which cooling becomes important but none of them can be calculated without additional assumptions.

A consequence of the previous arguments is that the first generation of stars would form in clouds with masses of order $10^6 M_\odot$.

As the proto-galaxy is progressively enriched in heavy elements, cooling becomes more important and clouds with smaller masses can collapse. These objects would be more susceptible to disruption (they would lie outside the survival triangle) and the stars that formed in them would be released into a field population. The metallicity at which the transition occurs is a little vague because of the uncertainties in the heating mechanisms and the possibility of some self-enrichment in the proto-clusters. Moreover, some of the field stars with very low metallicities may have formed in globular clusters that were later disrupted. Nevertheless, we do expect the field stars, on average, to be slightly younger and to have higher metallicities than the globular clusters. The field stars, by forming later in the collapse of a proto-galaxy, should also have a space distribution more centrally concentrated than that of the globular clusters. As the result of various selection biases, these suggestions are not easy to test for the Milky Way but they are consistent with the available data for other galaxies (Forte, Strom and Strom 1981, Harris 1986, Mould 1986, Mould, Oke and Nemec 1986).

Gunn (1980) and McCrea (1982) pointed out that globular clusters might form in the compressed gas behind strong shocks in proto-galaxies. This could be especially important in collisions between sub-galactic fragments. To show the connection with our work, we consider two streams or fragments, each with a density ρ_0, that collide supersonically with a velocity V_{rel}. The resulting shocks propagate away from the center of mass with a velocity $V_{rel}/6$, leaving the layer of hot gas between them at rest. Just behind the shocks, the density and temperature are $\rho_h = 4 \rho_0$ and $T_h = (\mu_h/12k) V_{rel}^2$. After a cooling time, a layer of cold gas forms, sandwiched between two layers of hot gas. Since, to a good approximation, all the gas between the shocks is isobaric, the densities and temperatures of the two phases are related by $\rho_c/\rho_h \approx (\mu_c/\mu_h)(T_h/T_c)$. The critical mass for gravitational instability is given by an expression that differs from (1) only in numerical factors of order unity. If the density of the fragments ρ_0 is comparable to the mean density within the proto-galaxy, (2) should also be a valid approximation. For $V_{rel} \approx 2 V_{gal}$ and $T_c \approx 10^4$ K, we obtain roughly the same result as before, $M_{crit} \sim 10^6 M_\odot$. Thus, as regards the formation of globular clusters, it probably makes little difference whether the two-phase medium is produced by shocks or by a thermal instability. What is crucial is that the gas not cool to temperatures much below 10^4 K.

5. TERTIARY FORMATION

There are several ways in which globular clusters might form after the collapse of proto-galaxies. One suggestion is based on the fact that the central members of some clusters of galaxies with X-ray cooling flows have unusually large populations of globular clusters (Fabian. Nulsen and Canizares 1984). M87, at the center of the Virgo cluster, provides an interesting example. The pressure in the hot gas, which can be inferred directly from X-ray observations, is $p_h \approx 1 \times 10^{-9}$ $(R/kpc)^{-1}$ dyne cm^{-2} over the radial range 1 kpc $\lesssim R \lesssim$ 30 kpc (Stewart, Canizares, Fabian and Nulsen 1984). Furthermore, the optical filaments indicate that some of the gas is relatively cold, with a temperature near 10^4 K, and may be the result of a thermal instability. Under these conditions, the critical mass, $M_{crit} \approx 4 \times 10^5$ $(R/kpc)^{1/2}$ $M_\odot$, given by (1) is comparable to the masses of globular clusters. However, since the metallicity of the gas is nearly solar and since there are no strong sources of heat, any clouds can cool through 10^4 K in a time of order 10^{-2} of their internal free-fall times (Fall 1986). Thus, a characteristic mass of order 10^6 $M_\odot$ cannot have been imprinted in the recent past. The globular clusters must have formed when the metallicity was much lower or the heating rate much higher, perhaps at the time M87 itself formed. These arguments are consistent with some recent spectroscopic observations, which show that the metallicities of the globular clusters in M87 are similar to those of the globular clusters in the Milky Way (Mould, Oke, and Nemec 1986).

In another version of the tertiary hypothesis, globular clusters are assumed to form in disks and those now in spheroids are assumed to have got there by the merging of smaller galaxies or proto-galaxies (Rogers and Paltoglou 1984, Larson 1986). This is motivated in part by Zinn's (1985) observation that the globular clusters in our galaxy more metal rich than [Fe/H] $\approx$ -0.8 have the kinematics and space distribution of a thick disk. Moreover, the Magellanic Clouds and other late-type galaxies have many rich clusters of young and intermediate ages associated with the disk populations (Freeman, Illingworth and Oemler 1983). Although these objects are often referred to as globular clusters, they have many properties in common with the open clusters in the Milky Way, including a luminosity function with no preferred scale (Elson and Fall 1985a, b). When galaxies merge, they could hardly avoid adding clusters to a spheroidal component. However, if this process was ever important in our galaxy, it must have ended fairly early (within a few $\times 10^9$ yr) because all the halo clusters appear to be old. The distinction between secondary and tertiary formation then becomes very blurred and some of the arguments about colliding fragments may apply.

REFERENCES

Binney, J. 1977 Astrophys. J. 215, 483.
Blumenthal, G. R., Faber, S. M., Primack, J. R. and Rees, M. J. 1984 Nature 311, 517.
Caputo, F. and Castellani, V. 1984 Monthly Notices Roy. Astron. Soc. 207, 185.
Elson, R. A. W. and Fall, S. M. 1985a Publ. Astron. Soc. Pacific 97, 692.
Elson, R. A. W. and Fall, S. M. 1985b Astrophys. J. 299, 211.
Fabian, A. C., Nulsen, P. E. J. and Canizares, C. R. 1984 Nature 310, 733.
Fall, S. M. 1986 in preparation.
Fall, S. M. and Rees, M. J. 1977 Monthly Notices Roy. Astron. Soc. 181, 37p.
Fall, S. M. and Rees, M. J. 1985 Astrophys. J. 298, 18.
Forte, J. C., Strom, S. E. and Strom, K. M. 1981 Astrophys. J.

Letters **245**, L9.
Freeman, K. C., Illingworth, G. and Oemler, A. 1983 Astrophys. J. **272**, 488.
Gunn, J. E. 1980 in Globular Clusters D. Hanes and B. Madore, eds., Cambridge University Press, Cambridge, p. 301.
Harris, W. E. 1986 Astrophys. J. **91**, 822.
Larson, R. B. 1986 in Nearly Normal Galaxies: From the Planck Time to the Present S. M. Faber, ed., Springer-Verlag, New York, in press.
McCrea, W. H. 1982 in Progress in Cosmology A. W. Wolfendale, ed., Reidel, Dordrecht, p. 239.
Mould, J. 1986 in Stellar Populations C. A. Norman, A. Renzini and M. Tosi, Cambridge University Press, Cambridge, in press.
Mould, J. R., Oke, J. B. and Nemec, J. M. 1986 preprint.
Oort, J. H. 1965 in Stars and Stellar Systems, Vol. 5, Galactic Structure A. Blaau and M. Schmidt, eds, University of Chicago Press, Chicago, p. 455.
Peebles, P. J. E. 1984 Astrophys. J. **277**, 470.
Peebles, P. J. E. and Dicke, R. H. 1968 Astrophys. J. **154**, 891.
Rees, M. J. and Ostriker, J. P. 1977 Monthly Notices Roy. Astron. Soc. **179**, 541.
Rogers, A. W. and Paltoglou, G. 1984 Astrophys. J. Letters **283**, L5.
Searle, L. 1977 in The Evolution of Galaxies and Stellar Populations B. M. Tinsley and R. B. Larson, eds, Yale University Observatory, New Haven, p. 219.
Searle, L. and Zinn, R. 1978 Astrophys. J. **225**, 357.
Silk, J. 1977 Astrophys. J. **211**, 638.
Stecher, T. P. and Williams, D. A. 1967 Astrophys. J. Letters **149**, L29.
Stewart, G. C., Canizares, C. R., Fabian, A. C. and Nulsen, P. E. J. 1984 Astrophys. J. **278**, 536.
White, S. D. M. and Rees, M. J. 1978 Monthly Notices Roy. Astron. Soc. **183**, 341.
Zinn, R. 1985 Astrophys. J. **293**, 424.

DISCUSSION

GNEDIN: My question concerns the cooling mechanism. You consider thermal bremsstrahlung only. What about a magnetic field? I believe it is very important for the cooling process.

REES: I don't think the magnetic field is important for cooling by a thermal plasma. However, even a weak field affects the conductivity, and thereby determines the minimum size of cool clouds embedded in a hot medium, as well as their likely shapes (sheets? filaments? etc.)

OSTRIKER: Martin, if I understand the Fall-Rees picture, it would predict a definite relation between the characteristic mass (at the peak of the luminosity function) of globular clusters and the mass of the parent galaxy (perhaps $M_{cl} \propto M_{gal}^{1/2}$). How well does the predicted relation accord with observation?

REES: Idealized versions of the model do indeed predict a slow dependence of cluster mass on galaxy mass - and indeed on galactocentric distance within a given galaxy. However I don't think too much should be made of these, because the efficiency of star formation and mass retention within each forming cluster is likely to depend on environment (e.g. external pressure).

PRYOR: Globular clusters in Zinn's disk look very similar to clusters in the halo. Could you comment on how this similarity arises in your model?

REES: Disk clusters probably formed in a qualitatively similar fashion to the halo clusters. The young "globular clusters" in, for instance, the LMC may not, however, form pressure-confined clouds in the same way.

COHEN: It appears that dark matter may not be necessary to stabilize the disk of the Milky Way and that there may not be a missing mass problem locally. Could you tell us your views on dark matter, particularly non-baryonic dark matter?

REES: There seems little doubt that there some kind of dark matter exists in halos and in clusters of galaxies. I'm personally agnostic about whether this is baryonic or not - but it's impressive how well the so-called "cold dark matter" cosmology has stood up to two or three years of intense scrutiny. As far as globular clusters are concerned, there is no firm evidence that they contain dark matter. However if the CDM cosmology is correct and globular clusters are pregalactic, then they would be surrounded by non-baryonic mini-halos.

ZINNECKER: May I inject a word of caution about your cooling curve below $T \sim 10^4$. I believe molecular hydrogen cooling is likely to be more efficient than calculated in the Fall and Rees (1985) Ap. J. paper which would pose a threat to your globular cluster formation theory. In this paper, you do not consider all the channels for H_2-formation (for example, the route via H_2^+ is not included). Moreover, H_2-formation is a tricky business involving non-equilibrium ionization, non-LTE level population, shielding etc. I wonder whether you could comment on these points?

REES: I agree that the thermal history of the 10^4 K clouds is important, because non-equilibrium processes are involved. However, a sufficiently intense UV background can unquestionably prevent H_2 formation - the most detailed calculations so far being those of Kang and Shapiro - though it is unclear how plausible it is that a protogalaxy generates this background at the appropriate stage.

OZERNOY: Martin, could you describe within the framework of your scheme, as a particular example, the differences in possible evolutionary ways of globular cluster formation in our Galaxy as compared with that in the Magellanic Clouds.

REES: I'm honestly quite unclear whether the Magellanic Cloud clusters are the same kind of beast at all. The work of Elson and Fall suggests that they formed continuously over the entire lifetime of the LM. Moreover, their mass function extends down to low values more typical of open clusters. (This is, unmistakably, a constraint on theories which attribute the globular clusters in our Galaxy to mergers with small disks.)

WEBBINK: Is there any difficulty posed to either the primordial or secondary scenarios for cluster formation in importing sufficient angular momentum to the condensing globular cluster to avoid strongly radial orbits and destruction by the galactic tidal field?

REES: Protogalaxies probably acquired their overall angular momentum via tidal torques. Gas that starts off at ~ 100 kpc would typically acquire the angular momentum appropriate to an orbit with perigalacticon at ~10 kpc. Of course the orbits of individual clusters would, in the Fall-Rees model, be influenced by random motions in the protogalactic gas. A better-developed theory than we yet have should be able to say something about the distribution of orbital eccentricities and hence tidal effects.

SURFACE PHOTOMETRY OF GLOBULAR CLUSTERS

S. Djorgovski

Harvard-Smithsonian Center for Astrophysics

ABSTRACT. Much of what we know about the structure, dynamics, and evolution of globular clusters derives from their observed density profiles, and their interpretations. In this review, I will briefly describe the problems and techniques specific to the surface photometry of globular clusters, show some new results, and offer suggestions for future ground-based work.

Globular clusters are our main testing ground for the dynamics of stellar systems. All manner of interesting processes takes place in these systems, on the time scales generally shorter than the Hubble time: two-body relaxation, core collapse and its reversal, tidal shocks, possibly even gravothermal oscillations, equipartition, etc. The main channel for testing our theories are the projected *density* profiles of clusters, which in the absence of a strong mass segregation (a well-justified assumption) we can almost always identify with their *surface brightness* profiles. Therein lies the importance of surface photometry. Substantial progress has been achieved in the last few years in obtaining more complete, and better quality data. An important stimulus was provided by the vigorous theoretical activity related to the problems of core collapse, and the post-collapse dynamical evolution: the volume edited by Goodman & Hut (1985) contains several excellent reviews. In this paper, I will first describe the modern techniques used in surface photometry of globular clusters, summarize some of the new results, and finally suggest some possible directions for the future work. For the earlier work in this field, and more complete accounts of the relation between the density profiles and the underlying stellar dynamics, the reader should consult the reviews by King (1975, 1980, 1981, 1985) and Spitzer (1984), and the references therein.

Good to excellent surface brightness and/or star counts profiles now exist for almost all known Galactic globulars, that is, some 130 clusters. Published surveys include King *et al.* (1968), Illingworth & Illingworth (1976), Peterson (1976), Djorgovski & King (1984, 1986), Kron, Hewitt & Wasserman (1984), Djorgovski & Penner (1985), Hertz & Grindlay (1985), Lugger, Cohn & Grindlay (1985), Lugger *et al.* (this conference), etc. The ellipticities and ellipticity gradients were measured by White & Shawl (1987). Much of the new data has not been systematically analysed yet, but in a year or so we should have a new compilation of dynamical and structural parameters for globular clusters, which would replace the classical, but now obsolete study of Peterson & King (1975), or the somewhat heterogeneous compilation by Webbink (1985).

Most of the difficulties and problems associated with surface photometry of globulars are caused by their "bumpy" nature: clusters are composed of finite numbers of stars, and a large fraction of the total light is contributed by a relatively small number of giants. There is not much that one can do about this, except to use the bluest bandpass possible: in the U band, for example, the HB and the RGB stars have approximatelly the same luminosity, so that a relatively large number of stars contributes most of the light, and thus the $1/\sqrt{n}$ fluctuations are smaller. This rule was often emphasized by Ivan King (e.g., in his 1985 review). The second important source of difficulties is the presence of foreground stars, in particular at low Galactic latitudes or in the Bulge, where most of the clusters are. This difficulty can be coped with if one uses an imaging detector, as will be briefly described below, but it is much more detrimental for a concentric-aperture photometry work.

There are several techniques which can be used in measuring the surface brightness profiles of clusters: First, there is "real" surface photometry, done with an imaging detector, such as a CCD, or a photographic plate. This is the best way, but it is usually good only for the inner regions of clusters (which are dynamically the most interesting, anyway). Second, there are star counts, which are currently the only way of measuring profiles in the tidal cutoff region. With the HST, we should be able to do star counts in the cores, and that should prove to be very interesting, as it may give us the first solid evidence for mass segregation or population gradients in clusters. Third, one may use concentric-aperture photoelectric photometry. That technique suffers from centering and foreground difficulties much more than the imaging work, and is completely insensitive to the core structure. It should not be trusted for radii less than ~ 10 arcsec, and it really works only in the intermediate regions, and only if the foreground is not too heavy. Finally, there are techniques not worth serious attention, such as one-dimensional scans (which are noisy, and need to be inverted...), etc. I will concentrate here on the imaging surface photometry work; King (1986) gives further discussion and comparisons of various methods and techniques.

Generally, the foreground stars need to be removed before any of the photometry is performed. This can be done interactively (by hand and a cursor), simply by editing out or flagging all pixels suspected of being polluted with the foreground stars, or uncorrected detector defects (bad CCD columns, etc.). Flagged pixels or areas are then ignored by the profile-computing routines. A much better approach to this task is allowed by the modern digital stellar photometry software, such as the famed *DAOPHOT*. One can form a color-magnitude diagram of all resolved stars in the frame, or simply find all stars sufficiently brighter than the obvious cluster giants, and then remove them by point-spread function (PSF) fitting and subtraction. The process is illustrated here in Fig. 1. It may be necessary to flag the central pixels from underneath the removed stars, just to guard against the imperfect PSF subtraction. This technique has the advantage of being more objective and more automatic than the simple star flagging by hand, and it is more reliable, especially if there is color information available.

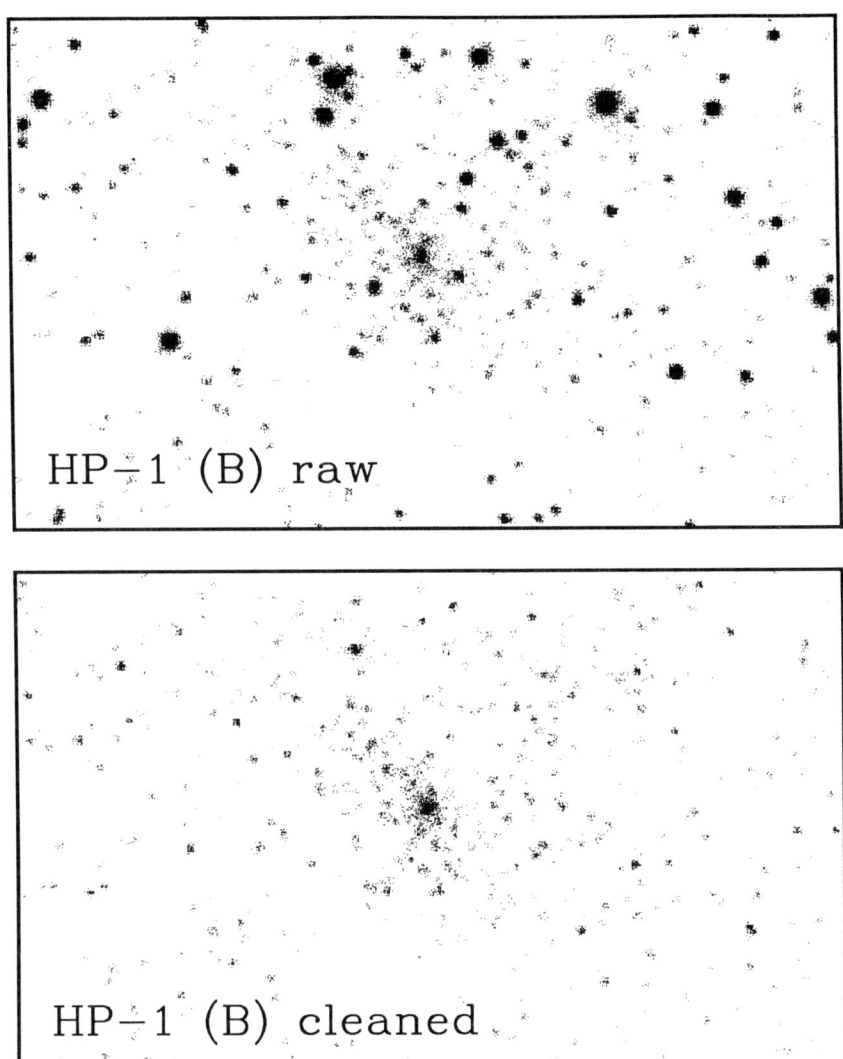

Fig. 1. An example of *DAOPHOT* foreground cleaning. The technique is particularly effective for the heavily obscured clusters at low Galactic latitudes, like the HP-1 shown here, and for the bluer bandpasses. A color-magnitude array of all well-resolved stars was formed; there is no need for the zero-point or color calibrations. The cluster giant branch was evident, and all brighter stars which were clearly outside the cluster sequence were then removed by the PSF fitting and subtraction. Even if the color information is not available, one can remove all stars brighter than the obvious cluster giants.

The next problem that one encounters is the cluster centering. Missing the true cluster center would always artificially flatten the profile, and may hide a possible post-collapse core. One centering method, developed by Ivan King, is the mirror-autocorrelation technique, illustrated in Fig. 2. This method is very robust, and it uses the full two-dimensional information present in the image. An alternative technique, the maximum symmetry method, was developed by Hertz & Grindlay (1985), and it employs separate centering in X and Y projections.

One then proceedes with the profile derivation. The image is divided into annular, concentric pseudo-apertures, each of which is divided in a number of sectors (most authors use 8 sectors, as that is an easy number to implement, and it is an almost optimal one). This process is illustrated in Fig. 3. The use of sectors to determine the internal error-bars is essential: most of the errors are due to the discreteness of light distribution in the cluster, and all other sources of

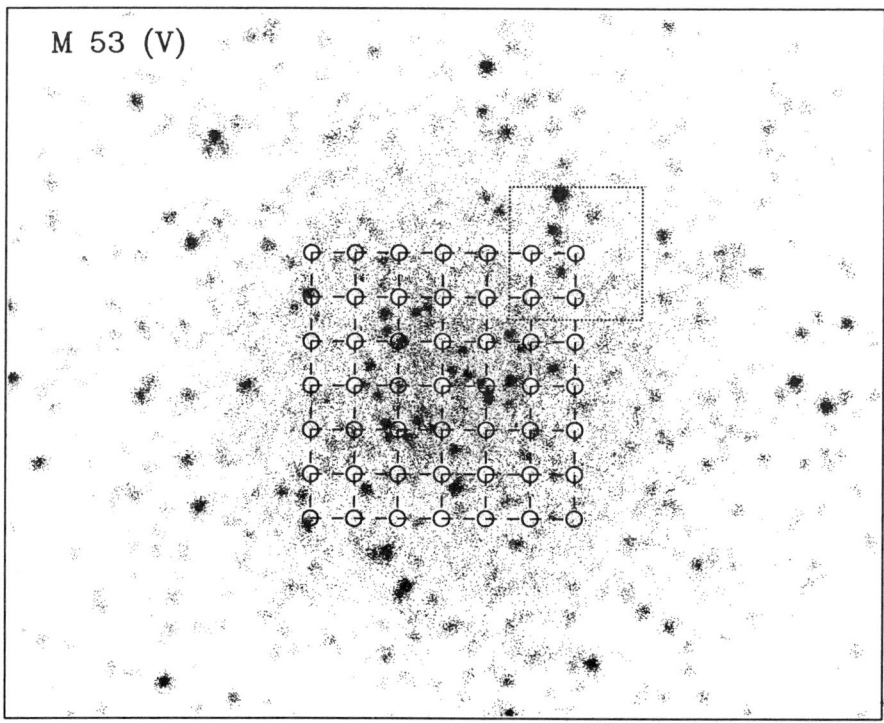

Fig. 2.   Mirror-autocorrelation centering is performed as follows: A grid of test centers (circles) is set to cover the cluster core. At each test center, a square window sub-array (dotted lines) is extracted, and amplitude of the autocorrelation of this data sample with its mirror image is computed. Thus, there is one number (autocorrelation amplitude) at each test center, and it is maximized for the most symmetric sample. Finally, a paraboloid is fitted to this grid of amplitudes. The vertex of the paraboloid is the optimal cluster center.

errors tend to be negligible in comparison. The error-bars also signal a presence of azimuthal asymmetries, or "bumps": if there is a lump of stars, or an unremoved foreground star somewhere in the aperture, the corresponding data point will be artificially high, but the error-bar will be appropriatelly increased as well. In particularly difficult cases, it may be advantageous to use the *median* of the mean surface brightness in sectors in each annulus, rather than the mean; this is a more robust way, but it is not flux-conserving, since we are dealing here with highly asymmetric, non-gaussian noise. A more complete discussion of various sources of errors is given by Illingworth & Illingworth (1976) and Newell & O'Neil (1978).

Finally, there is the problem of sky determination, which can be accute in the case of small-field CCD's. A practical way of doing it is from the mode or the median of a sky histogram, which is compiled from the pixels as far from the cluster center as possible. This sky estimate will necessarily be polluted by the unresolved cluster light, but the situation is much better here than it is in the case

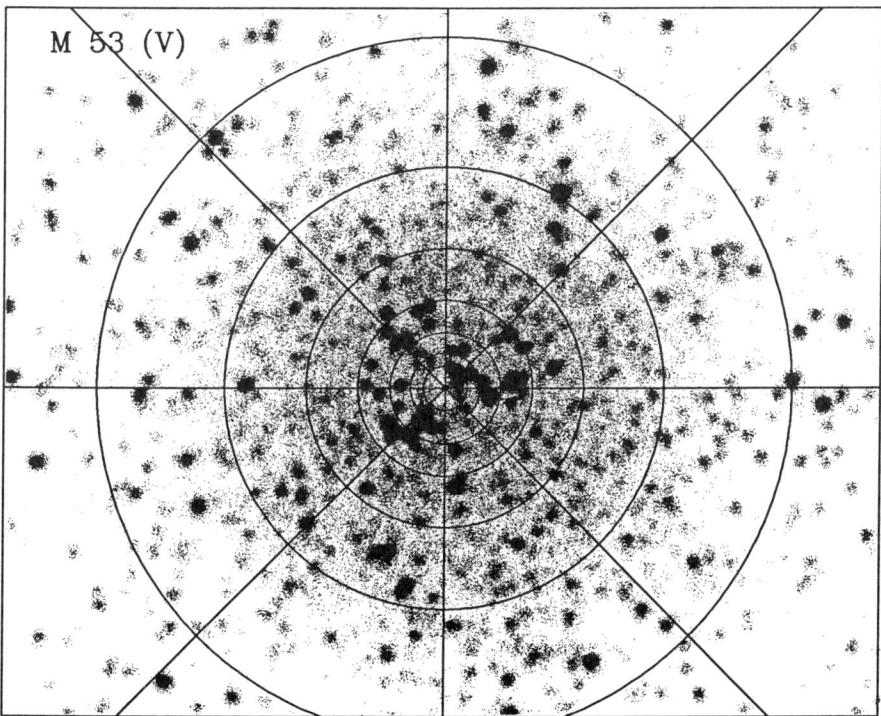

Fig. 3. Aperture grid for the photometry. Concentric circular annular apertures are centered on the optimal cluster center. The aperture spacing is typically chosen to be equidistant logarithmic. Each annulus is subdivided into eight sectors. Mean surface brightness of all non-flagged pixels in each sector is computed. Mean and the sigma of the values for the eight sectors are adopted as the mean surface brightness for the annulus and its internal error.

of surface photometry of galaxies: most of the light in clusters comes from the resolved giants, and their pixels do not affect the mode or median of the sky histogram very much. The small residual pollution would make the sky too bright; however, there is also an opposing effect: the unremoved background/foreground of faint stars and galaxies would add to the signal in the central regions, which is treated as being from the cluster alone. Thus, the uncertainty of sky determination in the CCD work may be important only for the outermost point or two of a cluster profile.

The final step in surface photometry is the determination of the photometric zero-pont for a profile, if that is needed (it is certainly unimportant for the morphological studies). One major advantage of imaging detectors is that the data can be taken in non-photometric conditions, and calibrated later. This is a point in which the photoelectric aperture work can make its most useful contribution.

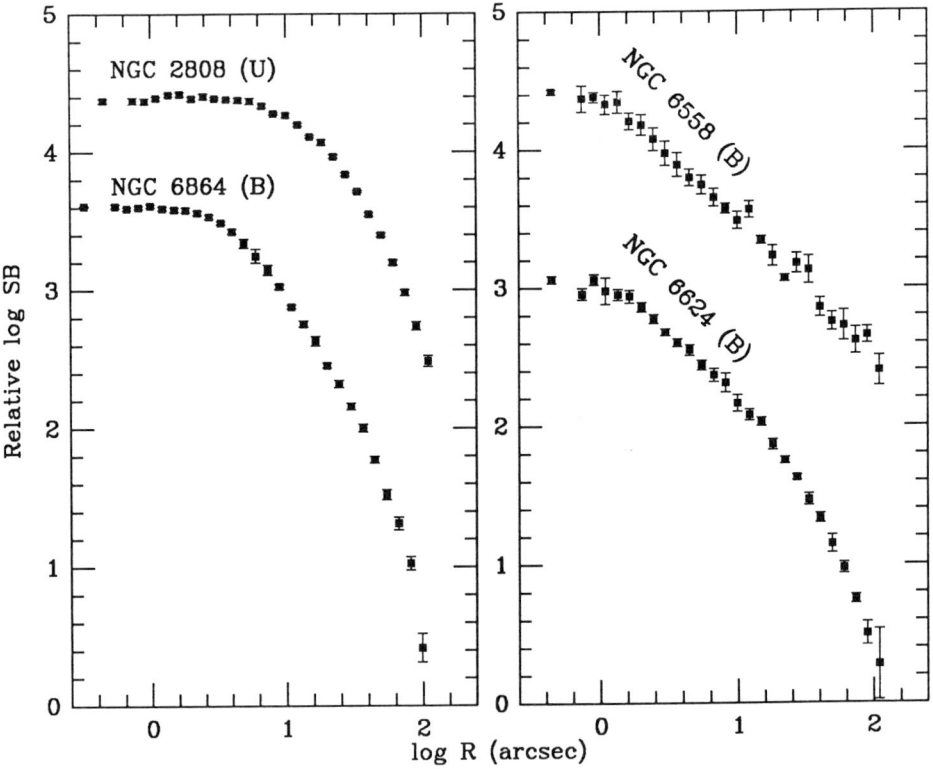

Fig. 4. Profiles of clusters with different core morphology, plotted on a 1:1 log-log scale. On the left, two clusters described well by the King (1966) models; they are distinguished by flat cores and steep envelopes. On the right, two clusters with the post-core-collapse morphology; their central parts are relatively shallow, slope $\simeq$ -1 power-law cusps, going into the seeing disk.

The core morphology of globular clusters shows the existence of two distinct types: the well-known family of King (1966) models, characterized by flat cores and steep envelopes, and the power-law cusps with slopes $\sim$ -1, approximating singular isothermal spheres, which gradually roll off into a King-model envelope, or a tidal cutoff. We now believe that the two families represent different stages of dynamical evolution of globular clusters: the clusters with cusps are those which underwent the core collapse, and its reversal. Their existence was first predicted in the pioneering work by Hénon (1961). Examples of clusters with different morphology are shown in Fig. 4. The characteristic which distinguishes the post-core-collapse (PCC) clusters is the existence of a shallow (slope $\simeq$ -1) power-law section of the profile in their central parts, going into the seeing disk. This can be checked by subtracting a -1 power-law from the data. However, it is a mistake to think of PCC centers as something occuring at very small angular scales, and hiding in the seeing limit: these power-laws typically cover tens of arcseconds. It is also inappropriate to think of PCC clusters as King models with an "extra" spike or a cusp in the middle; the structure of these whole clusters is fundamenally different, although they may be regarded as infinite-concentration limiting cases of the King sequence. There is always a fundamental dynamical difference: the centers of PCC clusters are thought to contain close binaries, which serve as sources of energy, stabilizing the cluster, and generating a positive radial flow of energy which gradually expands the cluster envelope. There is nothing like that in the King models, whose phase space distribution is a steady-state, lowered Maxwellian, without any energy flow. In both cases, however, there is a loss of stars due to evaporation in the tidal cutoff regions.

Given enough time, every cluster should go through core collapse, unless it evaporates away first. The collapse time can be as short as a few half-mass relaxation times, and it is shortened substantially if there is a mixture of stars of different masses (Inagaki & Saslaw 1985). The collapse and the recovery are thought to be very rapid, and once a cluster reaches the PCC state, it stays that way until it evaporates away, that is, many billions of years (Goodman 1984). Core oscillations may occur, due to the ejection and replacement of central binaries (McMillan & Lightman 1984), or due to an as-yet poorly understood gravothermal instability mechanism (Bettwieser & Sugimoto 1984). Thus, the PCC clusters are not an anomaly, but rather a natural dynamical state for evolved clusters. Indeed, in the Berkeley globular cluster surface photometry survey (Djorgovski & King 1986, Djorgovski et al. 1986), it was found that $\sim 1/5^{th}$ of all known Galactic globulars shows the characteristic PCC morphology. The problem is actually the other way round: there are some highly concentrated King-model-like clusters, which had enough time to collapse, but apparently did not. One possible explanation is that they *do* have collapsed cores composed of dark stellar remnants, whose dynamical evolution was too rapid for ordinary visible stars to follow (Larson 1984). Another possibility, that the clusters recover from the collapse in a *much* shorter time than what we think, seems unlikely for the reasons which will be explained below.

The results from the Berkeley survey were used to investigate relations between the cluster morphology and other properties. It was immediately noticed that the PCC clusters are more concentrated towards the Galactic center than the King-model clusters. Furthermore, the high-concentration King-model cluster are more concentrated towards the Galactic center than the low-concentration ones. By the same token, the PCC and high-concentration clusters are also more concentrated towards the Galactic plane. This effect was predicted by Chernoff, Kochanek & Shapiro (1986) before they knew about our results. They investigated the influence of tidal shocks from disk passages on cluster evolution, and found that the shocks selectively accelerate dynamical evolution: at a given initial cluster mass and central concentration, there is a critical galactocentric radius within which all clusters collapse, and outside of which all clusters dissolve. At any given moment the distribution will depend on the initial conditions, but the trend will be to have more PCC or highly concentrated survivors at lower galactocentric radii, just as is observed. Moreover, there may be a very weak trend that at a fixed galactocentric radius, more concentrated clusters tend to have higher Z-distances from the plane, which would mean more inclined orbits. In any case, the core collapse and its aftermath are not driven by the internal Antonov-Hénon and Spitzer instabilities alone. Another observed trend is that the more concentrated King-model clusters tend to have higher luminosities (or masses), but that trend is reversed for the PCC clusters: they are less massive than the high-concentration King-model clusters. This may reflect the fact that smaller-N systems have shorter dynamical time scales, and so collapse first, and/or that the PCC clusters evaporate faster because of their internal energy sources and ensuing envelope expansion. These trends are illustrated in Fig. 5. We found no significant correlation between the dynamical and the chemical properties of clusters.

The fact that there are good correlations between the cluster morphology and global variables, such as their distribution in the Galaxy, or mass, indicate that the core collapse and its reversal are a "once in a lifetime" affair, and that the clusters do not recover back to a King-model state within Hubble time. Unless, that is, if the relative durations of King-model and PCC phases in such tentative collapse-and-recovery cycles depend on the tidal shocks and mass in a suitable way, but that seems to be too contrived.

Finally, the power-law slopes in the cusps are generally close to -1, but tend to be shallower, and can be as low as -0.7 or so (the measured value depends somewhat on the radial range used: the slopes are shallower at the lower radii because of the seeing, and steeper at the larger radii, where the King-model envelope or the tidal cutoff begins). Djorgovski & King (1986) find the median value for the slopes to be around -0.9, but with a large scatter which is real. This may reflect real differences in mass spectrum between the cusps, which in turn may reflect IMF differences between the clusters.

Thus, we are now getting a good handle on the dynamical structure of Galactic globular clusters. We could use more kinematical information, but as for the surface photometry, it is unlikely that we can do much better from the ground. Some seeing-compensation schemes are worth a try, but the final word in high resolution observations of cores will come from the Hubble Space Telescope. The HST should enable us to do star counts deep into the cores, where detectable mass segregation may exist; we may even detect the true core radii or unusual central objects in the PCC clusters. Even more interesting would be to look in the cores of those puzzling high-concentration King-model clusters which should have collapsed by now. So much about the cores, but there are other interesting projects which can be done from the ground.

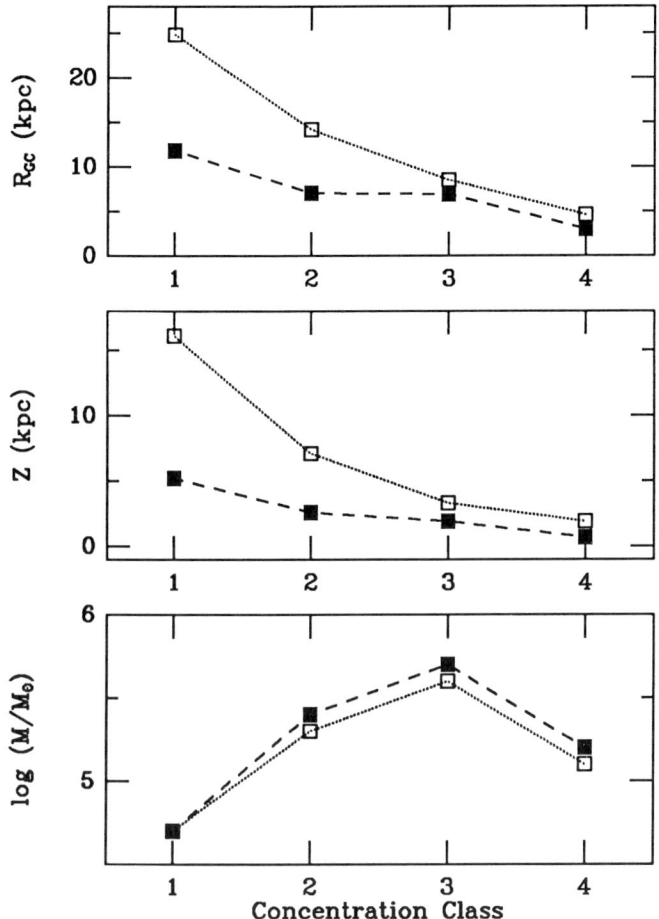

Fig. 5. Dependence of the galactocentric radius (top), distance from the Galactic plane (middle), and the cluster mass (bottom), as functions of cluster concentration. The sample of Djorgovski & King (1986) was divided in four approximately equal groups ($\sim$ 30 clusters in each group), in increasing order of concentration: class 1 are the King-model clusters with $c \leq 1.2$, class 2 are with $1.2 < c < 1.7$, class 3 with $c \geq 1.7$, and class 4 are the PCC and possible PCC clusters. Solid squares and dashed lines indicate the median values for the groups, and open squares and dotted lines the mean values. More concentrated clusters tend to be closer to the Galactic center and plane, and tend to have higher masses, except that for the PCC clusters the mass is lower again. The cluster masses were computed from their extinction-corrected visual luminosities by assuming a universal $M/L = 3$.

First, we can move on to other galaxies: Magellanic Clouds, M 31, and the dwarf spheroidals near our Galaxy. The Magellanic clusters should be easy, they are only a few times farther than the typical Galactic globulars, and should be well resolved. They present a nice complement to the Galactic system: they have a wide spectrum of ages, they are generally less massive, and thus have shorter dynamical time scales and faster evolution, and they do not suffer the strong tidal shocks like the Galactic globulars. A good census of PCC clusters and the cluster morphology in the Magellanic Clouds may gain us some new, valuable insights in the dynamical evolution of clusters in general, and in particular solve the problem of collapse-and-recovery *vs.* the collapse-only-once. Some surveys have already started (Mateo & Hodge, or Papenhausen & Schommer, this conference; Meylan & Djorgovski, in preparation).

The M 31 clusters present a more difficult challenge, because they are much less resolved. Here the problem becomes more similar to the surface photometry of galaxies, and we can use the corresponding software and methods: measure the ellipticity and position angles of isophotes easier than what we can do for the Galactic clusters, etc. The M 31 globulars are bright, and we can obtain sufficiently high S/N data as to attempt some seeing deconvolutions. As a family, they seem to be somewhat different in their stellar population properties from the Galactic globulars; comparing their dynamical properties would be very interesting. By the way, in the terms of the sampling and angular resolution, observing M 31 globulars from the ground is practically equivalent to observing globulars in the Virgo cluster with the HST. An example of surface photometry of a bright M 31 cluster Mayall II is shown in Fig. 6.

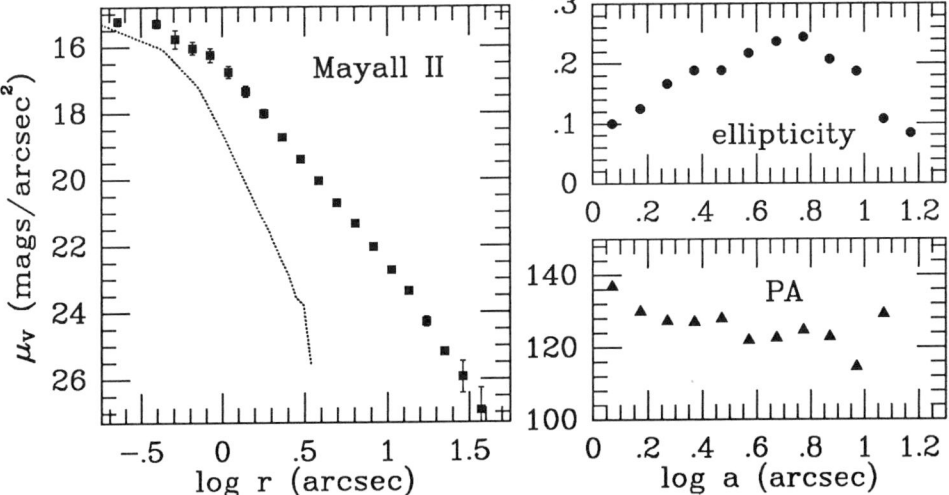

Fig. 6. Left: Circularly averaged surface brightness profile of the M 31 cluster Mayall II, plotted as a function of radius, in 2:1 log-log scale. The dotted line indicates the PSF. Right: Ellipticity and major axis position angle profiles, plotted as functions of the semi-major axis. The cluster giants are marginally resolved in a good seeing. The data were obtained with the KPNO 4-m telescope, and the exposures lasted less than 1 minute. Seeing deconvolution work on this and other M 31 clusters is now in progress (Bendinelli *et al.*, in preparation).

One disreputable project on which we should try our luck again, but with modern data and technology, are the color (or stellar mass) gradients. They could be caused by equipartition, but most clusters may still not be old enough for the effect to be appreciable. The PCC clusters may be more hopeful targets. The early reports of color gradients by Chun & Freeman (1979) are now widely regarded as being spurious, and caused by a preferred centering of photoelectric apertures on chance lumps of red giants near the cluster centers; the effects reported by Scaria & Bappu (1981) probably have a similar cause. However, several authors at this conference reported radial dependence in the fraction of blue stragglers... The subject is still wide open. The color gradients can be measured in the same way as described above for the surface brightness profiles, but by reducing two different bandpass frames at once: assure that the same cluster center is used for both frames, and compute the mean colors in the corresponding annuli and sectors. Since the same stars cause profile fluctuations in both frames, this procedure would assure that there is a good match, and provide the correct error-bars. Alternatively, one can smear both frames to have the identical PSF's, carefully register them, and produce a color frame, from which the color profile can be extracted.

There is another approach, which we may call "star stripping". Good CCD imaging data and PSF subtraction software afford us this new opportunity. First, it would be interesting to separate the giants from the unresolved background, and see if both groups have the same density distributions, and whether the star counts and the surface brightness profile (=luminosity-weighted star counts) have the same shape. It is essential to have a good handle on the completeness of star subtraction, both as a function of magnitude, and the distance from the cluster center. This is easily doable with *DAOPHOT* and similar programs, by inserting artificial stars in the data, and recovering them in repeated analysis. Comparing the "giants" and the "dwarfs" profiles may show some indications of a mass segregation. Or, one can isolate the stars by their color, or both color and magnitude (e.g., blue HB stars). The simplest thing would be just to remove the brightest giants, which cause most of the profile fluctuations, and get the color profile of the remainder in one of the ways described above.

Finally, one neglected aspect of globular-cluster structure are the tidal radii, which are measurable only through star counts. There is much space and need for improvement here: most of the existing measurements date from the old work by King and collaborators, when the counts were done by eye (thus the lack of enthusiasm for follow-up work). There is no reason why the star counts should not be done automatically, and better. We already have the necessary plates, the scanners, and the software. Some pioneering attempts were already done by Herzog & Illingworth (1977), and Irwin & Trimble (1984). A good way to do it would be to remove most of the foreground stars by color-magnitude selection. This approach should be much more powerful and efficient than the counts by eye. We should be able to measure or constrain the tidal radii for a much larger number of clusters than heretofore available. In the cleanest cases (rich clusters at high latitudes), we may even be able to examine the *azimuthal* structure of tidal cutoff regions, and probe directly the shapes of their Roche surfaces.

I would like to thank to my good collaborators, on whose work most of this review was based, and in particular Ivan King, Abe Oren, Howard Penner, and

Carl Vuosalo. The data for our work were obtained at Cerro Tololo and Lick observatories, whose staff provided invaluable help in numerous observing runs. Discussions with David Chernoff, Haldan Cohn, Jeremy Goodman, Josh Grindlay, Piet Hut, Shogo Inagaki, Phylis Lugger, Georges Meylan, and many others were most valuable and stimulating. Partial support from Harvard University is also gratefully acknowledged.

REFERENCES

Bettwieser, E. and Sugimoto, D. 1984 Monthly Notices Roy. Astron. Soc. 208, 493.
Chernoff, D., Kochanek, C. and Shapiro, S. 1986 preprint.
Chun, M. and Freeman, K. 1979 Astrophys. J. 227, 93.
Djorgovski, S. and King, I. R. 1984 Astrophys. J. Letters 277, L49.
Djorgovski, S. and Pennner, H. 1985 in IAU Symposium No. 113, Dynamics of Star Clusters, J. Goodman and P. Hut, eds., Reidel, Dordrecht, p. 73.
Djorgovski, S. and King, I. R. 1986 Astrophys. J. Letters 305, L61.
Djorgovski, S., King, I. R., Vuosalo, C., Oren, A. and Pennner, H. 1986 in IAU Symposium 118, Instrumentation and Research Programmes for Small Telescopes, J. Hearnshaw and P. Cottrell, eds., Reidel, Dordrecht, p. XXX.
Goodman, J. and Hut, P. 1985 IAU Symposium 113, Dynamics of Star Clusters Reidel, Dordrecht, (GH85).
Goodman, J. 1984 Astrophys. J. 280, 298.
Henon, M. 1961 Ann. d'Astrophys. 24, 369.
Hertz, P. and Grindlay, J. 1985 Astrophys. J. 298, 95.
Herzog, A. and Illingworth, G. 1977 Astrophys. J. Suppl. 33, 55.
Illingworth, G. and Illingworth, W. 1976 Astrophys. J. Suppl. 30, 227.
Inagaki, S. and Saslaw, W. 1985 Astrophys. J. 292, 339.
Irwin, M. and Trimble, V. 1984 Astron. J. 89, 83.
King, I. R. 1966 Astron. J. 71, 64.
King, I. R., Hedemann, E., Hodge, S. and White, R. 1968 Astron. J. 73, 456.
King, I. R. 1975 in IAU Symposium 69, Dynamics of Stellar Systems, A. Hayli, ed., Reidel, Dordrecht, p. 99.
King, I. R. 1980 in IAU Symposium 85, Star Clusters J. Hesser, ed., Reidel, Dordrecht, p. 139.
King, I. R. 1981 Quar. Journ. Roy. Astron. Soc. 22, 227.
King, I. R. 1985 in IAU Symposium No. 113, Dynamics of Star Clusters, J. Goodman and P. Hut, eds., Reidel, Dordrecht, p. 1.
King, I. R. 1986 in Proceedings of the International Workshop on Data Analysis in Astronomy L. Scarsi and V. Di Gesu, eds., Plenum, New York, in press.
Kron, G., Hewitt, A. and Wasserman, L. 1984 Publ. Astron. Soc. Pacific 96, 198.

Larson, R. 1984 Monthly Notices Roy. Astron. Soc. 210, 763.
Lugger, P., Cohn, H. and Grindlay, J. 1985 in IAU Symposium No. 113, Dynamics of Star Clusters, J. Goodman and P. Hut, eds., Reidel, Dordrecht, p. 89.
McMillan, S. and Lightman, A. 1984 Astrophys. J. 283, 813.
Newell, E. and O'Neil, E. 1978 Astrophys. J. Suppl. 37, 27.
Peterson, C. and King, I. R. 1975 Astron. J. 80, 427.
Peterson, C. 1976 Astron. J. 81, 617.
Scaria, K. and Bappu, M. 1981 J. Astrophys. Astron. 2, 215.
Spitzer, L. 1984 Science 225, 465.
Webbink, R. 1985 in IAU Symposium No. 113, Dynamics of Star Clusters, J. Goodman and P. Hut, eds., Reidel, Dordrecht, p. 541.
White, R. and Shawl, S. 1987 Astrophys. J., in press.

DISCUSSION

BAUM: We derived core radii for 13 globular clusters in M 31, which would be of interest to compare with your M 31 data. Our results were obtained with a TI 800x800 CCD on the 1.8 meter Perkins telescope and were reported at the June 1983 AAS meeting. Observed profiles were matched with King functions convolved with the PSF.

KING: Even when an ultraviolet image is not available, one can be synthesized through pixel-by-pixel combination of B and R images. It is essential to make the point-spread functions identical, but then the synthesis works quite well and makes a smoother image with the individual red giants nicely suppressed.

COHEN: Perhaps we could use a very small telescope with poor spatial resolution but large spatial coverage to measure a new set of tidal radii for clusters in relatively uncrowded fields. Also there are new tidal radii for NGC 6229 and NGC 7006 in my recent AJ paper.

DJORGOVSKI: I do not think that CCD imaging with a wide head telescope would work because of the heavy foreground pollution. Namely, you will end up with at least one foreground star in each pixel. Steve Kent recently did such imaging of M 31, which is an even easier case and he ran into this problem.

BAILYN: We have tried to do "star stripping", of the kind you suggest, on NGC 6712 using DAOPHOT, and we run into bigger problems than one might expect, due primarily to the radial dependence of the background light. Current software is not yet able to deal with this.

DJORGOVSKI: That is an important remark, but the problem should be curable.

ALTNER: I want to point out that IUE has shown color gradients where the central regions of several clusters (M 15, M 92) have shown an excess of blue objects. This is probably a selection effect.

COHN: Lugger, Grindlay, Bailyn, Hertz and I have found that a number of post-core collapse clusters have central surface brightness profiles with slopes significantly flatter than -1. We determine these slopes by fitting seeing-convolved power laws to the central parts of the profiles. M 15. for example, has a central slope of -0.64, which happens to agree exactly with a post-collapse evolution model, reported by Y. M. Lee at this conference, that includes nonluminous white dwarfs. We take our slopes as evidence for nonluminous remnants, in cluster cores, that are somewhat more massive than the stars that dominate the luminosity profile.

MENDEZ: We have also found that the nuclei of some clusters (NGC 6266, 7099, and others) are bluer than the whole cluster, as Bruce Altner pointed out.

RICHER: Didn't Peterson in a paper earlier this year say something about color gradients in clusters?

PETERSON, C.: Yes. The conclusion that was stressed from the study of concentric aperture photometric colors was that apparent color gradients are produced by the random spatial distribution of these bright giant stars or even field stars. The concentric aperture photometry has not produced evidence for real radial variations in color due to radial variations in the stellar population.

KRON: Electronic camera photometry is as good as counts in the outer regions of globular clusters.

FUSI PECCI: In a paper published in 1981 (Buonanno et al., Astron. Astrophys.) we have applied to M 5 the technique here suggested by Djorgovski's information on possible color gradients. In particular, using photographic plates, we have obtained both the CMO and the integrated magnitudes and colors over spots and annuli to simulate the observations made by Chun and Freeman (1979) which led them to claim the existence of a strong color gradient within that cluster. Then by "taking off" star-by-star (in a sequence, starting with the brightest) we have shown that the effect they found was due to sampling rather than to intrinsic properties of the cluster.

KING: I am sorry to contradict Gerry Kron but surface photometry in the outer parts of clusters is statistically very much inferior to star counts.

X-RAY BINARIES IN GLOBULAR CLUSTERS

Jonathan E. Grindlay

Harvard-Smithsonian Center for Astrophysics

ABSTRACT: X-ray binaries in globular clusters provide a powerful tool for the exploration of the evolution of compact binaries and their host globular clusters. Recent x-ray and optical studies of these systems have yielded long-sought binary periods and fundamental properties for two sources (in NGC 6624 and M 15). It appears that tidal capture formation of compact binaries in globular clusters can proceed by several different routes and lead to exotic systems such as the white dwarf-neutron star binary with an 11-minute period recently discovered in NGC 6624. Combined with previously reported long-term periods for several globular cluster (and field) x-ray sources, this suggests again that many of these systems may in fact be hierarchical triple systems. The prospects for forming these in the dense cores of clusters undergoing core collapse is discussed, and searches for color gradients in the cores of globular clusters showing cusps in their central surface brightness distribution are presented. A program to test for the high central density of binaries (and triples) expected in cusp clusters by searching for diffuse line emission from their constituent cataclysmic variables is briefly described. Finally, the case for globular cluster disruption and the formation of galactic x-ray burst source is reviewed in light of recent developments.

1. INTRODUCTION

The study of compact x-ray sources in globular clusters continues to provide important constraints on both the nature and physics of short-period (and ultra-short period) binaries containing neutron stars, the physics of accretion onto these objects, and the evolution of globular clusters themselves. In this review, we shall summarize a number of developments which bear on each of these topics. We shall build on the material presented in our recent reviews of these problems (Grindlay 1985a,b, 1986, 1987) as well as the recent discoveries, both observational and theoretical, which have triggered new developments in the field.

We begin with an update on the process of tidal capture for the formation of binaries in globular clusters since several recent calculations have indicated this is more complex than previously recognized. The tidal capture of a triple companion by a compact binary is then briefly discussed. We then turn to the recent discoveries of the binary periods for the first two globular cluster x-ray sources: those in NGC 6624 and in M 15. These apparently disparate systems, with periods of 11.4 minutes and 8.5 hours (respectively), may in fact be linked in a common evolutionary scenario which we outline. In an attempt to identify more globular cluster x-ray binaries, both low and high luminosity, we are carrying out extensive optical imaging and spectroscopic studies. A status report on this work is briefly described. Similarly, a status report on our ongoing study of the central surface brightness of clusters is presented. We describe preliminary results on the colors of the cusps in NGC 6624 and M 15 which suggest they are not significantly bluer than their surrounding clusters, and we present plans for follow-up studies with narrow-band imaging to search for diffuse line emission from the cataclysmic variables (CVs) expected to have been formed. Finally, we return to our often-stated hypothesis that the galactic bulge x-ray sources may have been created in globular clusters since disrupted and consider this, vs. the alternative scenario of binary ejection, in light of the work discussed here and in progress. Our conclusions and directions for future work follow.

2. TIDAL CAPTURE FORMATION

The formation of compact binaries in globular clusters can proceed by at least two different mechanisms: three-body interactions or tidal capture (see Cohn 1987). Since the two-body tidal capture process must become important as the stellar density in a cluster core increases (e.g. during core collapse) before the three-body mechanism sets in, we shall focus on the tidal capture process. It can, or should, be operative for both the formation of binaries and "higher order binaries", or hierarchical triples (cf. Grindlay 1985a,b, 1986). We note, however, that three body interactions may be especially important for the formation of binaries containing degenerate components if the central densities (e.g. in cluster core collapse) are high and are dominated by relatively massive degenerate stars (Lee 1987).

2.1 Binaries

The approximate maximum closest approach distance for tidal capture was found to be ~ 3 stellar radii by dimensional arguments in the original discussion of Fabian et al. (1975) and in the detailed calculations of Press and Teukolsky (1977). Recently, these calculations have been extended, and a (small) numerical error corrected, by Lee and Ostriker (1986). New calculations by McMillan, McDermott, and Taam (1987) and by Antia et al. (1987) have also included a more realistic treatment of the effects on the structure of the interacting stars by the tidal dissipation of energy in their encounter. In spite of these improvements, the maximum tidal capture impact parameter is still found to be approximately 3.3 stellar radii (McMillan et al. 1987), so that the various inferences drawn for the number of neutron stars required in globular clusters to produce the numbers of sources observed (e.g. Lightman and Grindlay 1982, Grindlay 1987) are essentially unaffected by the new capture cross section results, though they are affected by other considerations (cf. sections 3.3, 5, and 7 below).

The new results for the effects of tidal heating during the tidal capture process (e.g McMillan et al. 1987) have shown that the captured star on which the tidal bulge is raised will expand and be heated significantly by the dissipative effects of the encounter. This means that for tidal capture between two normal (i.e. non-degenerate) stars, the expansion of each could lead to a common envelope binary and coalescence of the binary. (We note that tidal capture of main-sequence stars by cluster giants may thus be even more effective than main-sequence star - white dwarf collisions in creating the centrally-concentrated faint blue horizontal-branch star systems discussed by Bailyn et al. (1987)). Tidal capture binaries thus coalesced would therefore be unavailable to interact with other stars in the cluster, suggesting that such binaries formed from main-sequence (or perhaps also giant) stars in the cluster may not be (McMillan 1986) the effective sources of dynamical heating previously suspected (Ostriker 1985). In this case, cluster core collapse might proceed to still higher densities and be halted only by the formation then of three-body binaries. However, tidal capture binaries involving a white dwarf or a neutron star as one member of the pair are probably not so affected (Taam 1986), since the expansion of the radius of the captured star (main sequence or giant) is then not great enough to engulf the degenerate companion causing it to spiral in. Tidal capture would thus remain the most efficient source (at intermediate densities) of binary formation for compact x-ray sources involving either white dwarfs (Hertz and Grindlay 1983, Hertz and Wood 1985) or neutron stars (Lightman and Grindlay 1982).

2.2 Triples

A compact binary (however created) should be an effective target for the tidal capture of a third star, as suggested by Grindlay (1985a,b, 1986). This process has now been partially calculated by McMillan (1986) and is being studied in greater detail by Bailyn, Grindlay and McMillan (1987). Preliminary results suggest that at least 3 % of the collisions of a third star with a compact binary at a minimum separation of at least three times the binary semi-major axis result in a stable hierarchical triple system. Such systems must have a ratio of inner to outer semi-major axes of at least ~ 3 for dynamical stability. In this case, and in fact out to separation ratios of ~ 30, the outer binary is still "hard" in the frame of the globular cluster (i.e. its orbital velocity exceeds the central velocity dispersion of the cluster). In actual fact, separation ratios of ~ 10 are probably not exceeded since this is also limited by the tidal capture cross section, or radius, of the third star. Thus the intrinsically stable hierarchical triples created by tidal capture in a dense globular cluster core are not disrupted as "soft" cluster binaries.

3. STUDIES OF COMPACT BINARIES IN GLOBULAR CLUSTERS

The study of x-ray binaries in globular clusters has taken a great quantum leap forward with the recent discoveries of the first binary periods for members of this class of objects. An x-ray periodicity of 11.4 minutes was found (Stella, Priedhorsky and White 1986) for the archetype globular cluster source 4U1820-30 in NGC 6624 using the EXOSAT x-ray observatory. Within two months, an 8.5 hour period was announced from the optical photometry (Ilovaisky et al. 1986) and absorption line spectrum (Naylor et al. 1986) of the proposed (Aurière et al. 1984) optical counterpart, AC211, for the bright x-ray source 4U2127+12 in M 15. The period and the optical counterpart, the first for any globular cluster source, was confirmed by a detection of the periodicity in archived x-ray data for this source (Hertz 1986). We have incorporated these discoveries into a model (Bailyn and Grindlay 1987a; hereafter referred to as BG) for the formation and evolution of a class of tidal capture x-ray binaries in globular clusters which is summarized below (section 3.3).

3.1 The 11 Minute Binary in NGC 6624

The 11.4 minute period discovered for the bright x-ray source 4U1820-30 in NGC 6624 (Stella et al. 1986) was also found in archival data extending over the past 10 years and showing that the period remains constant to within ~ 0.1 msec per year (Morgan et al. 1986). This greatly strengthened the original claim of Stella et al. (1986) that the period must be orbital, not rotational (which, given the large mass transfer rate inferred from the x- ray luminosity of ~ 5×10^{37} erg/sec, would lead to a much larger rate of change of period as the neutron star spins up). The ultra-short binary period, which is shorter than any binary period known, requires the mass- losing star be a degenerate dwarf to fit in so small an orbit. A helium white dwarf of mass 0.06 $M_\odot$ orbiting a 1.4 $M_\odot$ neutron star could supply the necessary mass transfer by Roche lobe overflow at a rate of 1×10^{-8} $M_\odot$/yr (as required by the x-ray luminosity) by virtue of the angular momentum losses expected from the gravitational radiation from such a system (Stella et al. 1986, BG).

Such a relatively well constrained system begs for an explanation. Verbunt (1986) suggested that it is the result of a collision of a red giant (containing a white dwarf core) and a neutron star in the cluster, but BG show that this is relatively less likely than an alternative mechanism (described below). Red giant collisions may also be subject to a possible mass transfer instability which would cause the white dwarf and neutron star to coalesce. The system parameters imply that the x-ray lifetime for the system in its current (second) mass transfer phase (cf. section 3.3 below) has been relatively short (~ 10^7 years). This suggests that the total number of such systems in a dormant (i.e. not transferring mass) state must be large and that the corresponding number of capture binaries involving neutron stars is also large.

We note also that the previously discovered long-term period of 176 days for the source (Priedhorsky and Terrell 1984) may be due to a hierarchical triple companion (cf. Grindlay 1986). If the 176 day period is due to precession of the 11.4 minute orbit, an outer orbit with a period of ~ 15 hours is required (cf. Bailyn and Grindlay 1987b). However, as discussed in the evolutionary scenario of BG, the binary period for 4U1820-30 may have been originally ~ 9 hours in which case a ~ 15 hour triple period would not be stable; the resolution may be (Bailyn 1987) that the longer (i.e. ≥ 40 hour) periods required for stability of the hierarchical triple companion to a ~ 9 hour binary could still give rise to the 176 day precession period when the binary has shrunk to 11.4 minutes if the system is not co-planar (as was assumed in Bailyn and Grindlay 1987b). Indeed, non-coplanarity would be expected for a tidally captured triple companion.

3.2 The 8.5 Hour Binary in M 15: AC211

As mentioned above, the x-ray source 4U2127+12 in M 15 has been found to have an ~ 8.5 hour period in both its optical and x-ray flux. Although a best estimate period of 8.53 hours is given by Ilovaisky et al., the period is still not well determined due to aliasing with finite observing time windows. In any case, the ~ 8.5 hour period is almost certainly the orbital period as it is appropriate to a 0.8 $M_\odot$ terminal age main-sequence star in an orbit such that it overflows its Roche lobe onto a 1.2 $M_\odot$ neutron star (BG). These parameters are, in turn, those expected by our evolutionary model which can account for both the 8.5 hour and 11.4 minute binaries.

It is likely (e.g. Grindlay 1986) that 4U2127+12 is much more luminous than suggested by its relatively modest x-ray flux (~ 5-10 UFU), which itself would indicate a luminosity of ~ $10^{36.4}$ erg/sec. Rather, its luminosity is probably in excess of 10^{38} erg/sec and the source is surrounded by a so-called accretion disk corona (ADC) which prevents us from seeing the compact object directly. The x-ray flux observed is then scattered to us by the corona, which has only modest optical depth and intercepts only a fraction of the emitted flux. Such ADC sources are usually the highest luminosity objects, such as 4.8 hour binary Cyg X-3, but they can include somewhat lower luminosity systems such as the 5.6 hour binary 4U2127+49 from which an x-ray burst was recently discovered (Garcia and Grindlay 1987). The high luminosity is indicated for 4U2127+12 by the apparent lack of burst activity for this source (indicating it is probably more luminous than, say, 4U2127+49), by the anomalously low x-ray to optical flux ratio (only ~ 20 vs. ~ 10^3 for typical low mass x-ray binaries), and by the unusual variable low energy x-ray variability of the source which suggests variable absorption around the system. The latter variability was discovered for the M 15 source by Hertz and Grindlay (1983) (cf. Fig. 3 in their paper) but has been found to be generally present in a re-analysis of the bulk of the Einstein observations of M 15. In Fig 1 we show the general correlation evident between the low energy absorption (expressed as the equivalent neutral column density N_H) and the hard x-ray flux from the source which measures the overall x-ray luminosity: when the luminosity increases, the apparent low energy absorption decreases. This indicates that there may be substantial material surrounding the binary system which can be partially ionized, and made more transparent, when the x-ray flux increases. This material is expected if the system is in fact entering a common envelope stage.

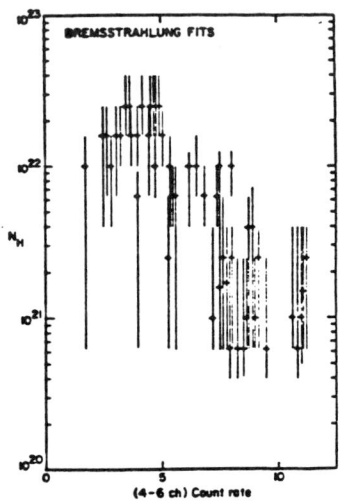

Fig. 1. Low energy absorption column vs. hard x-ray flux for M 15.

If the system is accreting at or near the Eddington limit to produce the inferred high luminosity, and is surrounded by the inferred circumsource material, then mass loss from the system in an outflowing wind is expected. This is the likely explanation for the systemic velocity of ~ -150 km/sec in the λ 14471 He I absorption line reported by Naylor et al. (1986), rather than their suggestion of ejection of the binary from the cluster core since this would occur in only $\sim 10^3$ years. A marginal confirmation of the blueshifted helium line velocity was obtained by Grindlay and Huchra (1987) in an average of three short exposure MMT spectra obtained on three successive nights in September 1986. It is encouraging that the possible velocity of -150 km/sec is close to what might be expected from around the Roche surface of the binary. The velocity variation of ~ 45 km/sec at the orbital period (Naylor et al. 1986) might then be due to the variable potential gradient, and thus velocity of the wind, around the Roche surface as it rotates at the binary period relative to our line of sight. A detailed treatment of the mass loss expected and its expected velocity variation is needed.

The MMT spectra (both recent and previous - cf. Grindlay 1985b, 1986) also showed that the Balmer line (absorption) spectrum from the (vicinity of) AC211 is blueshifted by only -30 ± 10 km/sec from the cluster mean velocity of -107 km/sec. The discrepancy with the He line velocity is probably because the Balmer lines, which are much narrower than those from M 15 horizontal-branch stars (e.g. the velocity standard HB star I-51), arise from the typical metal-poor cluster K-giants (R. Peterson, private communication) in the core (or cusp) of M 15 and not from the much hotter object AC211, which shows such a pronounced uv-excess relative to a HB star. This can be seen by comparing the MMT spectra of AC211 vs. I-51 in Fig. 2.

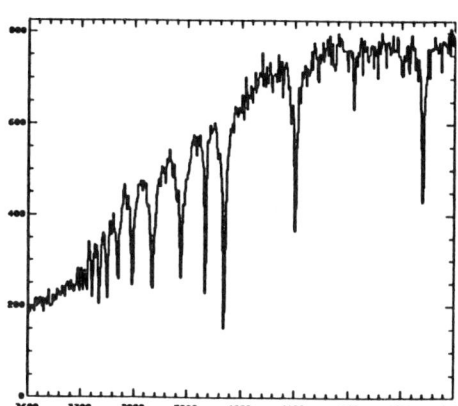 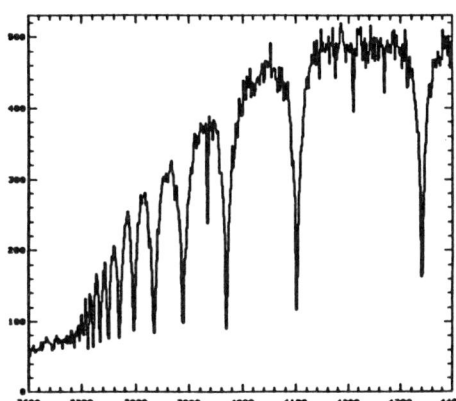

Fig. 2. MMT spectra of (a) AC211 and (b) the HB star I-51 in M 15.

We note that although we have never detected the Balmer emission features (e.g. Hα), reported as variable by Charles et al. (1986), we are confident we are observing AC211 since i) the object coordinates we have used (obtained originally from Aurière 1984 and updated by Aurière, private communication) to position the 1 arcsec MMT spectrograph aperture to ~ 0.3 arcsec setting accuracy (relative to nearby SAO stars) have been confirmed (to within ~ 0.5 arcsec) by our own astrometry of AC211 as detected on our CCD study of the core of M 15 (Bailyn et al. 1987), ii) the object observed clearly has a uv excess relative to a HB star, and iii) the Balmer absorption lines (sharp component) observed by Charles et al. are also not as significantly blueshifted as the He absorption spectrum (Charles, private communication).

3.3 Evolutionary Model

The tidal capture formation of compact binaries in globular clusters does not necessarily mean production at the outset of a mass transfer binary. Instead, if the binary is formed by an encounter with closest approach less than the maximum (~ $3.3R_*$) for capture but greater than the maximum for Roche lobe overflow of the secondary (with radius R_*), then a "dormant" x-ray binary will be formed. Mass transfer and x-ray emission will only begin when the captured star evolves off the main sequence and expands to fill its Roche lobe. This is the key idea in the evolutionary scenario of BG for the 11.4 minute binary system and possibly also the source in M 15. In this case, a 0.8 $M_\odot$ main-sequence star is captured into an orbit which circularizes at a separation $\geq 3R_{TAMS}$ (i.e. it encountered the neutron star with a separation of half this value), where R_{TAMS} is the radius of the star when it (later) has evolved to the terminal age main sequence and is perhaps ~ 1.5-2 times the radius R_* of the star on the main sequence. Hence the required range of closest approach distances is ~ 2.3-3.3 R_*, so that "dormant" tidal capture might occur for ~ 0.3-0.5 of all tidal capture encounters (since the cross section scales linearly with closest approach distance due to gravitational focusing).

When mass transfer finally begins, it will be unstable if the mass ratio $q = m_2/m_1 \geq 0.67$, where the subscripts 1 and 2 denote the primary and secondary masses, and if the accretion proceeds via an accretion disk (BG). This critical mass ratio would be exceeded for stars more massive than the present 0.8 $M_\odot$ turnoff mass which are captured by 1.2 $M_\odot$ neutron stars, where such a neutron star mass is expected if the neutron stars in globular clusters are predominantly the result of the accretion induced collapse of (massive) white dwarfs (Taam and Van den Heuval 1986, Grindlay 1987). The unstable mass transfer will result in accretion onto the neutron star at approximately the Eddington limit as well as accumulation of the excess mass (which cannot accrete) in the disk and in a surrounding circumstellar shell. This will lead to the formation of a common envelope and an evolution of the binary which is very difficult to predict in detail or in timescale. It is likely, however, that ultimately the circumstellar envelope will cause enough frictional drag on the secondary that it will spiral into the primary (probably finally on a rapid timescale) to a separation where the orbital energy of the binary exceeds the binding energy of the common envelope and atmosphere of the secondary exterior to the remaining core of the secondary. Since the secondary star is assumed to have (just) evolved off the main sequence, it will have developed a small (~ 0.1 $M_\odot$) white dwarf core. Thus the common envelope phase is expected to end with a detached white dwarf-neutron star binary in a ~ 31 minute orbit (BG). Gravitational radiation will then spiral the detached low mass white dwarf towards the neutron star primary on a timescale of 10^7 years after which it will begin a second stage of mass transfer onto the primary and a luminous (~ 10^{38} erg/sec) x-ray source will again be present. This is the stage at which the 11.4 minute binary is now found; the 8.5 hour binary in M 15 may well be the first phase (i.e. post main-sequence star secondary, entering a common envelope stage) of this generic tidal capture evolution.

4. OPTICAL SEARCHES FOR X-RAY BINARIES IN GLOBULAR CLUSTERS

We now turn to a brief update on our ongoing optical studies of the cores of globular clusters. We are carrying out both CCD photometry and spectroscopic investigations in an effort to both identify x-ray binaries and study the effects of binaries on globular cluster evolution.

4.1 The CCD Surveys

A massive program of UBV(RI) surface photometry of globular cluster cores has been carried out from CTIO, KPNO and the CFHT. While the primary goals have been to study the cusps in the central surface brightness profiles recently found in the cores of a number of clusters (Djorgovski and King 1984, 1986; Hertz and Grindlay 1985; Lugger et al. 1985, 1987a and 1987b); this work has also allowed a search for optical counterparts of both low and high luminosity x-ray binaries in clusters. Optical counterparts are sought as uv-excess (i.e. (U-B) $\leq$ -0.5) objects with absolute magnitudes of ~ 2-6 and ~ 1-3 for the low and high luminosity objects, which are expected to be predominantly CVs and low mass x-ray binaries, respectively (Hertz and Grindlay 1983).

Identification of optical counterparts for the low luminosity sources, which are generally located only to ~ 30 arcsec by the Einstein IPC, is only possible (if at all) for sources well removed from the densest central regions of the cluster. Of the two globular clusters (ω Cen and M 22) with (several) non-central low luminosity point sources detected in the Einstein IPC survey (Hertz and Grindlay 1983), UBV frames have been obtained for the first (M 22 being too reddened). (Although 47 Tuc has at least two sources well removed from its central source, they did not meet the angular displacement criteria of the IPC survey and they have thus far not been included in our CCD survey). One additional cluster, NGC 5824, with a low luminosity source well resolved from the central core and also located to only ~ 10 arcsec uncertainty by the Einstein HRI detector, has also been observed in UBV.

A particularly promising 21st magnitude candidate has been found (Grindlay 1986) in the small error circle for the source in NGC 5824; spectroscopic confirmation of its suspected identity as a CV was clouded out in our 1986 CTIO run but will be attempted again in 1987. Several uv objects were found in the error circles of the ω Cen sources, and spectra for the brighter ($m_v \simeq$ 17-18) of these were obtained under marginal conditions in May 1986. They were found to be members of the class of faint blue horizontal-branch (FBHB) stars which we have found to themselves be of special interest by virtue of their apparent central concentration in ω Cen (Bailyn et al. 1987). Although this latter study has indicated these objects may therefore be the result of binary mergers (of white dwarfs and main-sequence stars), there is no reason to suppose that they are in fact counterparts for the low luminosity x-ray sources. Although each IPC error circle (~ 30 arcsec radius) contains at least one of these objects, others of comparable brightness and uv excess are found outside the x-ray source positions, and their lack of emission lines would suggest the sources were (if CVs) in a perpetual outburst state. Additional spectra, which include both greater spectral coverage (e.g. Hα) and higher resolution are planned.

For the high luminosity sources, our major efforts have been to identify additional candidates beyond the AC211 counterpart for the M 15 source. For each source, now located to within ~ 3 arcsec by the Einstein HRI (Grindlay et al. 1984), we have searched for uv excess objects and can rule out objects as bright and blue as AC211 in each of the clusters NGC 1851, 6441, 6624 and 6712 (the remaining high luminosity clusters are heavily reddened and were thus not observed in U). Individual magnitude vs. color limits as well as color maps for the cores of each of these clusters (as well as M 15), are presented by Bailyn et al. 1987. Although analysis is still in progress, thus far the only other high luminosity source cluster with a possible optical candidate is NGC 6712.

4.2 Studies of NGC 6712

Within the HRI error circle of the high luminosity (~ $10^{36.4}$ erg/sec) source in NGC 6712 we have found a possible uv-excess object (Bailyn et al. 1987). Both the magnitude ($\geq$ 20) and color (U-B) $\leq$ -0.5) of this object are very uncertain because of the severe crowding effects (only ~ 0.2 core radii from the cluster center). However, in the divided (U/B) image, the pixels containing the candidate stand out as among the bluest in the core. Spectra will be attempted at CTIO in May 1987.

The search for uv candidates and the color analysis of cluster cores in and around x-ray positions has required that the data be sky subtracted first to remove the bias otherwise present in the study of radial color gradients in cusps (see section 5.1). Sky subtraction is also important for determining the true surface brightness profile and the possible departures from a King model for the relatively diffuse cluster NGC 6712. All previous measurements of the structural properties (e.g. core radius) of this important x-ray cluster have used photographic data (cf. Hertz and Grindlay 1985). An improved surface brightness profile is particularly important for NGC 6712 because of the conjecture (Grindlay 1985a) that this cluster may be in an advanced state of post core collapse in which it has expanded to its present apparently diffuse form. The tidal capture binary systems (e.g. the high luminosity source) in NGC 6712 would then have formed during an earlier epoch of higher densities. To carry out such a sky-subtracted surface brightness profile and dynamical study of this cluster, we have recently obtained B and V frames of the adjacent surrounding fields out to ~ 10 arcmin radii. Analysis of these is in progress, and will also be used to extend (using DAOPHOT) the stellar photometry of the cluster to derive a color magnitude diagram at larger radii (than the central frames) for use in cluster membership studies.

The latter is necessary to extend our study (Grindlay et al. 1987) of the velocity dispersion (and its radial variation) of the cluster to larger radii. The present totals of some 50 stars out to ~ 3 core radii will each be at least doubled. This will enable a search for departures from isothermality of the measured central velocity dispersion value of 4.2 km/sec and the implied low M/L ~ 0.6 (Grindlay et al. 1987).

5. CUSPS (OF BINARIES ?) IN GLOBULAR CLUSTERS

It is interesting that the two x-ray binaries which may be representative of the dormant tidal capture model and consequent (relatively) short x-ray emission lifetime (cf. section 3) are in the two most conspicuous globular clusters with central cusps in their surface brightness profiles: M 15 and NGC 6624. These are the two "original" cusp clusters, as the measurements and references to earlier work in Hertz and Grindlay (1985) make clear. Our recent detailed CCD study of each of these clusters (Lugger et al. 1987b), together with the "control" cluster NGC 6388, shows that the cusps are well described by power laws with indices of approximately 0.7 for both clusters. This is flatter than the profile slope found for post-core collapse cluster models and indicates the cusps may be dominated by heavy remnants (Cohn 1987). Thus the fact that both cusp clusters contain luminous and relatively short-lived x-ray binaries suggests that the number of binaries containing neutron stars in cusps may be significantly enhanced, as expected.

5.1 Blue Colors ?

In general, globular cluster binaries should be blue. This seemingly glib statement reflects the fact that binaries are expected to be predominantly tidal capture products, especially in the dense cores of cusp clusters, and that these will undergo (either initially or eventually) mass transfer. If the binary components are both main-sequence stars, as in most cases, the products may resemble the cluster blue stragglers (cf. Nemec 1987). If the binary components are (initially) a white dwarf and main-sequence star, then either FBHB stars may form (Bailyn et al. 1987a) or CVs and low luminosity x-ray sources are the expected result (Hertz and Grindlay 1983). Finally, if the components are a neutron star and a main-sequence star, the products are either the "normal" globular cluster x-ray sources (Lightman and Grindlay 1982) or the "dormant" cluster sources (BG). In either of the last two cases, the optical counterparts are expected to be blue by virtue of x-ray heating and/or the energy released directly in a high temperature accretion disk.

If the cusp clusters are indeed rich in binaries, as suspected above and as predicted if they are indeed post core collapse signatures with binary fractions approaching 50 % (Statler et al. 1986), then they should be bluer than their surrounding clusters. We have searched for this by analyzing the U vs. B images of the central regions of both NGC 6624 and M 15 in a novel way (Bailyn et al. 1987b). The usual method would be to compare the surface brightness profiles in each band, or the profile of the divided U/B image, which would be sensitive to the fluctuations in brightness due to individual bright giants in or out of a given annulus in a radial distribution. Instead, we have measured the distribution of the number of pixels at each value in the divided (U/B) image, or equivalently the relative number of pixels at each (U-B) color index. This is relatively less sensitive to the presence of individual bright stars although they enter by crowding effects and are also still relatively more important at small annuli where the available number of pixels from which the distribution is drawn is small. To simulate the color effects of crowding, we co-add enough cluster pixels at large radii to match the measured intensity (counts/pixel) at the small radii of interest in the cusp. To remove the slight color gradients which could arise from increasing sky contribution at large radii from the cluster centers, all frames are sky subtracted first using the technique of Djorgovski (1987). In practice, we have used 90 x 90 pixel (0.6 arcsec/pixel) to derive the cluster "background" color and to simulate crowding effects and have compared these color distributions with those obtained for 30 x 30 and 10 x 10 pixel regions about the cluster center.

Our preliminary results on NGC 6624 and M 15 using this technique indicate that the two central regions are not significantly bluer than the outer comparison region of the clusters. This is shown in Fig. 3 for M 15. This result is surprising, since the central core of M 15 appears to be relatively blue on short exposure images. This is probably due to the effects of crowding of relatively blue horizontal-branch stars and does not seem to require the underlying cusp be significantly bluer than the surrounding cluster. However, the limited spatial resolution and relatively large pixel size severely limit the significance of this result; higher resolution data have been obtained on the CFHT and will be analyzed for this effect.

5.2 Diffuse Line Emission from CVs ?

Although we have not yet apparently detected the additional blue continuum from an underlying population of a binary population in the cusp of M 15 or NGC 6624, it is possible that the diffuse line emission (e.g. Hα) from the component of the binary population which are CVs has been detected. Possible detections of diffuse Hα for the core of NGC 6624 were reported by Grindlay and Liller (1977) and for the core of

M 15 by Peterson (1976). Whereas the first result was obtained with

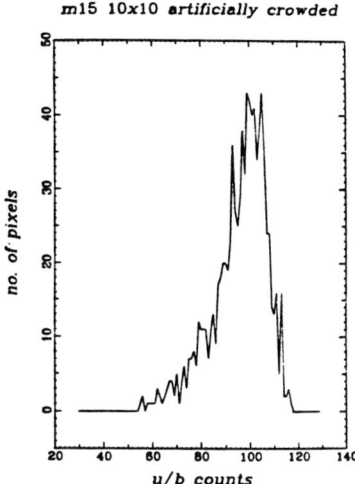
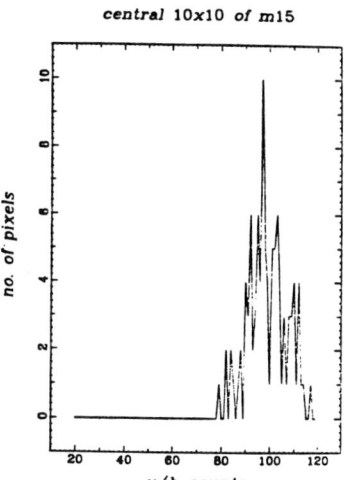

Fig. 3. Comparison of U/B pixel distributions for M 15 regions

concentric aperture photometry, and therefore is subject to the uncertainties of centering and the effects of individual red giants in the aperture (which, by undergoing mass loss, might be Hα emitters, as noted by Grindlay and Liller), the second result was photographic and obtained with a pre-filtered (10 Å) slitless spectrograph. The Hα image was noted as being smaller than the (central region) of the cluster and thus could be the central portion of the cusp instead of a new planetary nebula (much fainter than the known planetary in M 15) as suggested by the author. Once again, we have obtained (August 1986), but not yet been able to analyze, several test interference band (Hα) frames at the CFHT; we intend to extend this search to much deeper limits with future observations. For a cusp with only a ~ 15% fraction of CVs, which might be expected from the post-core collapse binary frations of Statler et al. (1987), it should be possible to detect a 10% excess in a 10 Å wide band centered on Hα relative to the neighboring continuum. This estimate assumes a "typical" CV with Hα emission equivalent width of ~ 10 Å vs. cluster giants (which dominate the light) with Hα absorption equivalent widths of ~ 1 Å.

6. THE CASE FOR CLUSTER DISRUPTION

The possibility that many of the x-ray burst sources in the galactic bulge were formed by tidal capture in globular clusters, which have since been disrupted (Grindlay 1984, 1985a,b, 1986), is made somewhat more plausible by several recent developments. First, the "dormant" tidal capture model (BG) would help to solve the otherwise questionable lifetime problem: bursters or compact low mass x-ray binaries with emission lifetimes of ~ 10^9 years would imply cluster disruption on a similar (or shorter) timescale (Grindlay 1986). However, if a significant fraction of the tidal captures are initially dormant until the secondary has evolved, cluster disruption can occur over a more plausible ~ 10^{10} years. The 50-minute binary 4U1916-05, which we have pointed to as a prime candidate for origin in a cluster, is also a prime contender for a product of dormant tidal capture since its short period implies an evolved secondary (cf. Swank et al. 1984).

Secondly, the recent work on tidal disruption of globular clusters (Chernoff et al. 1986, Chernoff and Shapiro 1987) by the galactic field and encounters with individual giant molecular clouds, which we had suggested could disrupt globular clusters, has shown that indeed clusters within ~ 4 kpc of the galactic center can be disrupted. This work also bears out our hypothesis that only globular clusters on a rather narrow loss cone of orbital inclinations (with respect to the galactic plane) will be disrupted by GMCs and it reinforces the association (Grindlay 1984) of galactic bulge x- ray sources (with their flattened distribution) with a subset of bulge globular clusters (those with orbital inclinations initially near the galactic plane).

Finally, Cowley et al. (1986) have reported preliminary evidence that the velocity dispersion and distribution of the bulge x-ray sources (primarily bursters in their sample) is similar to that of the bulge x-ray globular clusters. They interpret this as kinematic evidence for the Pop II nature of the bulge x-ray sources, but it is also consistent with the globular cluster disruption picture for these objects. This is a promising area for further study, and many more velocities (than in their limited sample) are needed.

7. CONCLUSIONS

The study of x-ray binaries in both the x-ray and optical regimes has yielded important new results for the origin and evolution of compact binaries in globular clusters. The recent discoveries of the first binary periods for these objects has allowed fundamental constraints to be derived for the nature of the secondary stars supplying the accretion onto the neutron star primaries. It is surprising, but perhaps not coincidental, that these first binary identifications were made for the two prototype "cusp clusters": NGC 6624 and M 15. This is consistent with their having produced a significant enhancement in their populations of compact binaries if the cusps are indeed signatures of post-core collapse. This interpretation is also supported by the model described here and in BG whereby both the NGC 6624 and M 15 binaries are the two stages expected for a common channel of binary formation and evolution: the "dormant" tidal capture model. The relatively short x-ray lifetimes expected in this model, particularly for the second phase of x-ray emission due to accretion of a low mass white dwarf onto the neutron star primary (as in NGC 6624), may imply a total number of binaries in the cluster core ~ 10 times greater than previously suspected (e.g. by Lightman and Grindlay 1982).

Optical identification of the x-ray binaries in other globular clusters will allow the relative frequency of "dormant" tidal capture systems vs. the "normal" tidal capture binaries, with immediate production of lower luminosity but longer lasting x-ray emission. A possible optical candidate has been found for the source in the peculiar (diffuse) x-ray globular cluster NGC 6712. It may be possible to derive magnitudes and colors for this object from high resolution images (e.g. from the CFHT) but this may require HST. Our current estimates would allow either type of tidal capture binary, with absolute magnitudes expected to be ~ 1-2 (normal type) vs. ~ 4-6 (second stage dormant type), although a (first stage dormant type) system such as the M 15 source can already be eliminated. Our optical studies (Bailyn et al. 1987) also rule out M 15-type optical counterparts for any of the other four bright x-ray globular clusters (NGC 1851, 6441, 6624 and 6712) for which we have U-band CCD images, although higher spatial resolution is once again needed to make these limits more restrictive than, typically, ~ 2-3 magnitudes fainter than the AC211 optical counterpart in M 15.

Higher spatial resolution is also needed to properly study the expected blue color gradients in cusps if they are indeed dominated by binaries. Our preliminary results do not show significant differences in the (U-B) colors in and out of cusps, but we are limited by pixel size and crowding effects to regions outside the central ~ 6 arcsec. Searches for diffuse line emission (e.g. Hα) expected from a central population of CVs in cusps may be more productive, and once again it is tantalizing that the two prototype cusp clusters both have reported Hα emission "cores". Higher resolution narrow-band visible (from CFHT) and uv (with HST) observations are needed, and are planned, to properly carry out these studies.

I thank my various collaborators in this work, particularly Charles Bailyn. This work was supported in part by NSF grant AST-84-17846 and by NASA grants NAGW-624 and NAS8-30751.

REFERENCES

Antia, H. M., Kembhavi, A. K. and Ray, A. 1987 in IAU Symposium No. 126, Globular Cluster Systems in Galaxies, J. E. Grindlay and A. G. D. Philip, eds., Reidel, Dordrecht, p. 671.
Bailyn, C. 1987, in preparation.
Bailyn, C. and Grindlay, J. 1987a, Astrophys. J. Letters, in press (BG).
Bailyn, C. and Grindlay, J. 1987b, Astrophys. J., 312, 748.
Bailyn, C., Grindlay, J. E., Cohn, H. N. and Lugger, P. M. 1987a in IAU Symposium No. 126, Globular Cluster Systems in Galaxies, J. E. Grindlay and A. G. D. Philip, eds., Reidel, Dordrecht,

p. 679.
Bailyn, C., Grindlay, J., Cohn, H. and Lugger, P. 1987b, in preparation.
Bailyn, C., Grindlay, J. and McMillan, S. 1987, in preparation.
Charles, P., Jones, D. and Naylor, T. 1986 Nature, 323, 426.
Chernoff, D., Kochanek,C. and Shapiro, S. 1986 Astrophys. J., 309, 183.
Chernoff, D. and Shapiro, S. 1987 in IAU Symposium No. 126, Globular Cluster Systems in Galaxies, J. E. Grindlay and A. G. D. Philip, eds., Reidel, Dordrecht, p. 673.
Cohn, H. 1987 in IAU Symposium No. 126, Globular Cluster Systems in Galaxies, J. E. Grindlay and A. G. D. Philip, eds., Reidel, Dordrecht, p. 379.
Cowley, A., Hutchings, J., Crampton, D. and Hartwick, F. 1986, Astrophys. J., in press.
Djorgovski, S. 1987 in IAU Symposium No. 126, Globular Cluster Systems in Galaxies, J. E. Grindlay and A. G. D. Philip, eds., Reidel, Dordrecht, p. 333.
Djorgovski, S. and King, I. 1984 Astrophys. J. Letters 277, L49.
Djorgovski, S. and King, I. 1986 Astrophys. J. Letters 305, L61.
Fabian, A., Pringle, J. and Rees, M. 1975 Monthly Notices Roy. Astron. Soc. 172, 15P.
Garcia, M. and Grindlay, J. 1987 Astrophys. J. Letters, in press.
Grindlay, J. E. 1984 Adv. Space Res., 3, No. 10, 19.
Grindlay, J. E. 1985a in IAU Symposium No. 113, Dynamics of Star Clusters, J. Goodman and P. Hut, eds., Reidel, Dordrecht, p. 43.
Grindlay, J. E. 1985b in Proc. US-Japan Seminar on Galactic and Extragalactic Compact X-ray Sources, Y. Tanaka and W. Lewin, eds., ISAS, Tokyo, p. 215.
Grindlay, J. E. 1986 in The Evolution of Galactic X-Ray Binaries, J. Trumper, W. Lewin and W. Brinkman, eds., NATO ASI Series, Vol. 167, p. 25.
Grindlay, J. E. 1987 in IAU Symposium No. 125, Origin and Evolution of Neutron Stars, D. Helfand and J. Huang, eds., Reidel, Dordrecht, in press.
Grindlay, J. E., Bailyn, C., Mathieu, R. D. and Latham, D. W. 1987 in IAU Symposium No. 126, Globular Cluster Systems in Galaxies, J. E. Grindlay and A. G. D. Philip, eds., Reidel, Dordrecht, p. 659.
Grindlay, J. E., Hertz, P., Steiner, J. E., Murray, S S. and Lightman, A. P. 1984 Astrophys. J. Letters 282, L13.
Grindlay, J. E. and Huchra, J. 1987, in preparation
Grindlay, J. and Liller, W. 1977 Astrophys. J. Letters 216, L105.
Hertz, P. 1986 IAU Circular No. 4272.
Hertz, P. and Grindlay, J. 1983 Astrophys. J. 275, 105.
Hertz, P. and Grindlay, J. 1985 Astrophys. J. 298, 95.
Hertz, P. and Wood, K. 1985 Astrophys. J. 290, 171.
Ilovaisky, S., Chevalier, C., Aurière, M. and Angebault, P. 1986 IAU Circular No. 4263.
Lee, H. M. 1987 in IAU Symposium No. 126, Globular Cluster Systems in Galaxies, J. E. Grindlay and A. G. D. Philip, eds., Reidel,

Dordrecht, p. 665.
Lee, H. and Ostriker, J. 1986, preprint.
Lightman, A. P. and Grindlay, J. E. 1982 Astrophys. J. 262, 145.
Lugger, P., Cohn, H. and Grindlay, J. E. 1985 in IAU Symposium No. 113, Dynamics of Star Clusters, J. Goodman and P. Hut, eds., Reidel, Dordrecht, p. 89.
Lugger, P., Cohn, H., Grindlay, J., Bailyn, C. and Hertz, P. 1987a in 1987 in IAU Symposium No. 126, Globular Cluster Systems in Galaxies, J. E. Grindlay and A. G. D. Philip, eds., Reidel, Dordrecht, p. 657.
Lugger, P., Cohn, H., Grindlay, J., Bailyn, C. and Hertz, P. 1987b Astrophys. J., in press.
McMillan, S. 1986 Astrophys. J. 306, 552.
McMillan, S., McDermott, P. and Taam, R. 1987, preprint.
Morgan, E., Remillard, R, Stella, L., White, N. and Garcia, M. 1986 IAU Circular No. 4261.
Nemec, J. 1987 in IAU Symposium No. 126, Globular Cluster Systems in Galaxies, J. E. Grindlay and A. G. D. Philip, eds., Reidel, Dordrecht, p. 677.
Ostriker, J. P. 1985 in IAU Symposium No. 113, Dynamics of Star Clusters, J. Goodman and P. Hut, eds., Reidel, Dordrecht, p. 347.
Peterson, A. W. 1976 Astron. and Astrophys. 53, 441.
Priedhorsky, W. and Terrell, J. 1984 Astrophys. J. Letters 284, L17.
Press, W. and Teukolsky, S. 1977 Astrophys. J. 213, 183.
Statler, T., Ostriker, J. and Cohn, H. 1986 Astrophys. J., in press.
Statler, T., Ostriker, J. and Cohn, H. 1987 Astrophys. J., in press.
Stella, L., Priedhorsky, W. and White, N. 1986 Astrophys. J. Letters, 312, L17.
Swank, J., Taam, R. and White, N. 1984 Astrophys J. 277, 274.
Taam, R. 1986, private communication.
Taam, R. and Van den Heuvel, E. 1986 Astrophys. J. 305, 235
Verbunt, F. 1986 Astrophys. J. Letters 316. L 23.

DISCUSSION

GNEDIN: Is there any evidence for extended X-ray sources? I am referring to your paper with Hartwick and Cowley. What is the present situation?

GRINDLAY: There are no further data available from the Einstein Observatory beyond what we published. Further interpretation of these data is contained, however, in my paper in the I.A.U. 113 Proceedings. Upper limits for diffuse X-ray emission in ω Cen are reported by Koch-Miramond and Aurière at this meeting which do not contradict the Einstein results.

MCMILLAN: I have two remarks concerning tidal binaries. First, both the formation process and subsequent interactions with third stars are very likely to lead to stellar merges, significantly reducing the fraction of simple binary systems. Second, in my simulations of tidal binary interactions, I do indeed find a substantial number of hierarchial triples being formed.

DJORGOVSKI: I wonder how wise it is to present PCC clusters as a sum of a broad-core King model, and a cusp. They are single, unified dynamical systems, with a structure profoundly different from the King models. And another remark: Shri Kulkarni and I looked for a possible optical counterpart of the tentative pulsar in M 28, using our UBVRI CCD data; we did not find anything striking.

GRINDLAY: I agree PCC clusters are not composite dynamical systems. However, our fitting procedure is motivated by current models which indicate power law profiles (cusps) merging at larger radii with an apparently undistorted King model.

OSTRIKER: The close encounters which produce tidal binaries will also destroy them through their ejection or physical collisions. In the Princeton work we found a typical lifetime less than 10^9 years for tidal binaries at the relevant stage of cluster evolution. This increases the required rate of production.

GRINDLAY: I agree. This means either the neutron star fraction or the total density must be larger than the values assumed by Lightman and me (1982) for a "typical" cluster (i.e., not a post-core collapse cluster, since present indications are that these account for at most ~ 25% of the clusters).

KING: I have not yet done enough modeling to speak with certainty but it is my impression that if the neutron-star mass is as much as 10^{-3} of the total, it should have a strong effect on the central distribution of light. (White dwarfs have almost no effect, because they have a mass so similar to that of the bright visible stars.)

GRINDLAY: A neutron star mass fraction of 10^{-3} (total) or perhaps 1-10% (in the core alone), may indeed have a pronounced effect on the light distribution and probably contributes to the flatter slopes (than -1) for the PCC cluster cusps. The much larger mass fraction in white dwarfs (e.g. 35% in 47 Tuc claimed by Da Costa and Freeman) may also significantly affect the light profile, however, since many of these WDs are probably relatively massive (~1 $M_\odot$) remnants from the more massive stars in the cluster IMF than the stars producing ~0.6 $M_\odot$ white dwarfs currently.

OZERNOY: Have you considered the formation of binaries in the very central stellar core of the Galactic nucleus? What could you tell us about X-ray sources in the vicinity of the Galactic center?

GRINDLAY: That's a very interesting question which I have not considered. A dense stellar cluster at the Galactic nucleus should indeed be capable of producing (by tidal capture) a compact X-ray binary, and it is interesting then that the X-ray luminosity (~ 2-10 kev) of our own galactic center is ~ 10^{36} erg/sec, or about the same as a globular cluster source.

REES: If neutron star binaries form via tidal capture, there should be a comparable number of events when the neutron star undergoes a physical merging ending up inside an ordinary star. The resultant "Thorne-Zytkow Object" would be conspicuous, like a supergiant, with a lifetime that would (when mass loss was rapid) be at least several times 10^7 yrs. Have any optical observers found objects answering this description? And, if not, is it a constraint on anything?

GRINDLAY: Another interesting question. I question, however, whether a "Thorne-Zytkow Object" would really form (in a direct collision) without losing most, if not all, of the stellar envelope. In this case, it is not obvious how optically luminous such an object should be. Partial loss of the envelope would presumably cause the object to appear very blue (and luminous), so it is interesting to speculate whether the uv-luminous stars found by Zinn could be such objects and not Post-AGB stars as usually assumed. Bailyn and I (this conference) have also suggested a possible analogue object class for white dwarfs - star collisions; the faint blue horizontal branch stars.

PRYOR: Our velocity dispersion data for NGC 6712 shows a low central M/L. This central M/L suggests that there are not very many heavy remnants in the center of this cluster.

GRINDLAY: Our results for NGC 6712 (this conference) also suggest a low M/L ~ 0.6. We have commented this may be related to the possible expansion of this cluster.

COHN: I would like to say a few words in defense of our profile

fitting procedure that has been commented on by Djorgovski and King.
Our primary motivation is to determine the central slope of the cusp;
thus we fit seeing-convolved power laws to the inner part of the
profile. Our fitting King models to the outer part of the profile is
motivated by the result from numerical simulations that while power law
structure develops in the central region, the outer regions continue to
resemble King models. Thus our empirical model fits both the simulated
and real cluster profiles.

PRECOLLAPSE EVOLUTION OF GLOBULAR CLUSTERS

Shogo Inagaki

University of Kyoto

ABSTRACT: Density profiles of most globular clusters are well fitted by a King (1966) model. The evolution of a King model in the tidal gravitational field of the Galaxy is first discussed. If the concentration parameter c (= $\log(r_t/r_c)$) is small enough, the evolution is nearly along the King model sequence and c becomes larger. When c becomes large enough (about 2.1), gravothermal instability sets in. The basic properties of gravothermal instability is next discussed. The stability criterion and its interpretation are given. Globular clusters consist of stars with disparate masses, so that finally the evolution of multi-component clusters is discussed. Acceleration of evolution in multi-component clusters and equipartition of the kinetic energies among components are discussed, and conclusions and future problems are given.

1. INTRODUCTION

Spitzer (1985) made an excellent review of this field at IAU Symposium No. 113 at Princeton and relatively little progress has been made in this field since then. In this paper I will review precollapse evolution from a somewhat different point of view.

Globular clusters are bounded by the gravitational field of the galaxy. Their radii are determined by the balance of the gravitational force of the Galaxy and their own gravitational force. Most clusters are well fitted by a King (1966) model. The model is characterized by a parameter c (= $\log(r_t/r_c)$), where r_t is the tidal radius and r_c is the core radius.

If c is smaller than the critical value (about 2.1), clusters evolve due to evaporation of stars from the tidal radii and if c is larger than the critical value, clusters evolve due to gravothermal instability (Katz 1980, Wiyanto et al. 1985). In the former case the central density becomes higher and higher and c becomes larger. In this stage the evolution is nearly along the King sequence. At late-time epochs the evolution is due to gravothermal instability.

2. EVOLUTION DUE TO EVAPORATION OF STARS FROM THE TIDAL RADIUS

If c is smaller than the critical value, the evolution is

governed by the escape of stars from the cluster. Wiyanto et al. (1985) calculated the evolution of a King model by numerical integration of isotropized Fokker-Planck equation and found that evolution is nearly along a King sequence if c is small enough (Fig. 1). When c becomes about 2.1, gravothermal instability takes place and evolution is accelerated. They also show that the rate of evaporation of stars is constant in time and in good agreement with King's (1966) prediction.

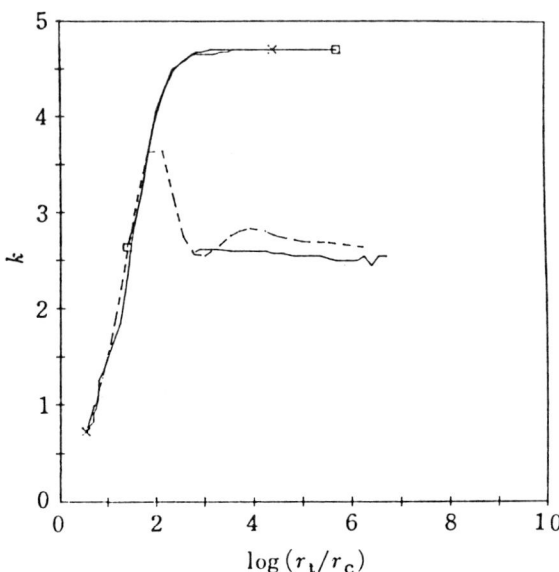

Fig. 1. Comparison of the k-(r_t/r_c) relation of the simulations of Wiyanto et al. (1985) to that of King's (1966) sequence, where k is defined by $E = - (k+1/2)GM^2/r_t$ and E is the total energy of the cluster. The broken curve denotes King's sequence. Solid curves denote the evolutionary sequence obtained by the simulations. The curve between the two cross marks is for model A, which is started from a stable configuration. The curve between the two square marks is for model B, which is started from a nearly marginally stable configuration. A nearly horizontal curve in the middle of the figure is for model C, which is started from an unstable configuration.

Chernoff, Kochanek and Shapiro (1986) investigated the effect of tidal heating due to giant molecular clouds and to passage through the galactic plane. They showed that passage through the galactic plane is dominant and that small concentrated clusters are disrupted and those with large concentrations are accelerated toward a gravothermal instability. Both processes may occur within the Hubble time for any concentration parameter. For further details, see Chernoff and Shapiro (1987).

3. EVOLUTION DUE TO GRAVOTHERMAL INSTABILITY

If c is larger than 2.1 (Katz 1980), globular clusters evolve due to gravothermal instability. Originally gravothermal instability was investigated under the assumption that clusters are strictly isothermal and therefore bounded by spherical walls (Antonov 1962, Lynden-Bell and Wood 1968, Hachisu and Sugimoto 1978). Lynden-Bell and Wood (1968) examined the gravothermal instability by considering linear series (Fig. 2). The character of instability changes at the turning point of E-v-diagram, where v is the dimensionless central potential. The validity of a linear series to investigate stability is verified by Inagaki and Hachisu (1978), Yoshizawa et al. (1978) and Katz (1978). Hachisu and Sugimoto (1978) formulated the problem of stability by using a Green's function. They expressed the variation of the temperature δT in terms of the variation of the entropy:

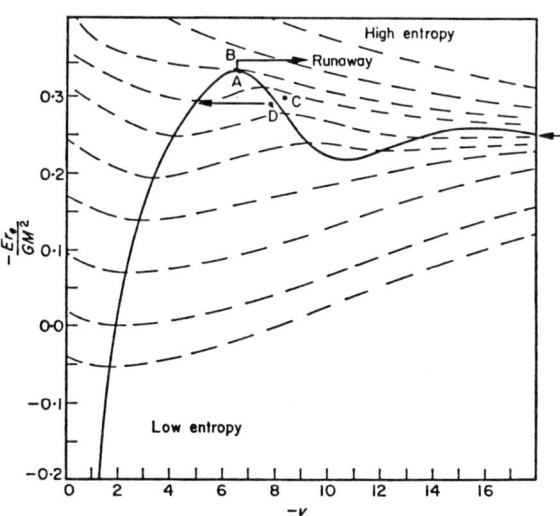

Fig. 2. Energy-v relation for isothermal clusters. The central potential v has a one-to-one relation with the density contrast. The first turning point corresponds to the density contrast of 709. The cluster is unstable for larger values of v.

$$\delta \ln T(M_r) = \int_0^M F(M_r, M_r') \delta s(M_r') dM_r' \quad (1)$$

where δs is the variation of the specific entropy. Therefore F corresponds to the inverse specific heat. The meaning of equation (1) is as follows: If we transport heat inside a cluster, it is represented by the variation of the entropy distribution. The cluster may reach then a new hydrostatic equilibrium and the temperature distribution is changed, and F may be calculated. The second order variation of the entropy is expressed as

$$\delta^2 S = - \int_0^M dM_r \int_0^M dM_r' F(M_r, M_r') \delta s(M_r) \delta s(M_r') \quad (2)$$

Equation (2) shows that if the inverse specific heat tensor F is negative, the second order variation of the entropy is positive, i.e., instability. The Green's function F is shown for the stable case, marginally stable case and for the unstable case in Fig. 3. In the stable case the region of negative specific heat is small but it grows as the cluster becomes more unstable. Hachisu and Sugimoto (1978) also calculated the eigenvalues and eigenfunctions which maximize $\delta^2 S$. They are shown in Figs. 4 and 5, respectively. Fig. 4 shows that the fundamental mode becomes unstable at the first turning point of Fig. 2 and the second mode becomes unstable at the second turning point and so on. Thus the cluster is unstable if the density contrast is larger than 709. Fig. 5 shows that if the cluster is unstable, δS and δT take the opposite sign, which is consistent with equations (1) and (2).

Evolution after gravothermal instability was first examined by Larson (1970) by using moment equations of the Fokker-Planck equation. Detailed numerical calculation by numerical integration of the isotropized Fokker-Planck equation was done by Cohn (1980), who showed that evolution takes place homologously (Fig. 6). Homologous evolution after gravothermal instability was examined in detail by Lynden-Bell and Eggleton (1980). They adopted a conductive gas model following Hachisu et al. (1978), by using modified conductivity:

$$K = 6GC \log N \rho/v$$

The usual expression for conductivity for plasmas is inadequate for stellar systems because it increases as the relaxation time becomes larger. Thus Lynden-Bell and Eggleton adopted a expression for conductivity such that the conductivity decreases as the relaxation time increases. According to their results, the central quantities of the cluster change as follows:

$$\rho_c \propto (1 - t_{cc}/t)^{-1.17} \quad (3)$$

$$v_c \propto (1 - t_{cc}/t)^{-0.055} \quad (4)$$

$$R_c \propto (1 - t_{cc}/t)^{0.52} \quad (5)$$

$$M_c \propto (1 - t_{cc}/t)^{0.42} \quad (6)$$

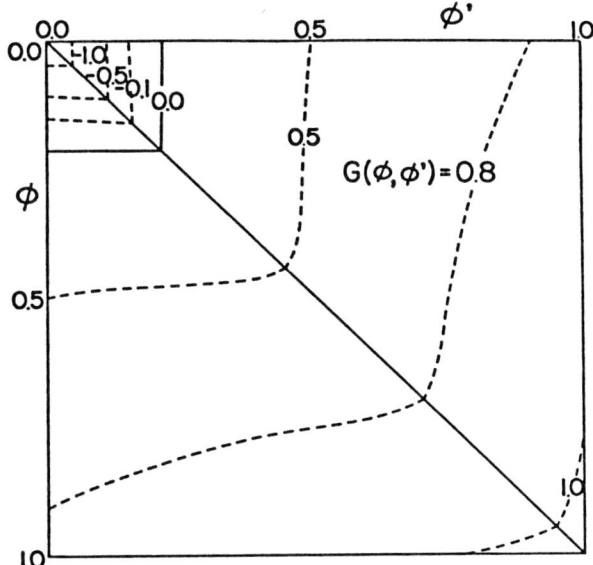

Fig. 3a. Contour map of the off-diagonal part of the inverse tensor specific heat $F(\phi,\phi')$ where $\phi = M_r/M$. This is for a stable system with a small density contrast, 2.5.

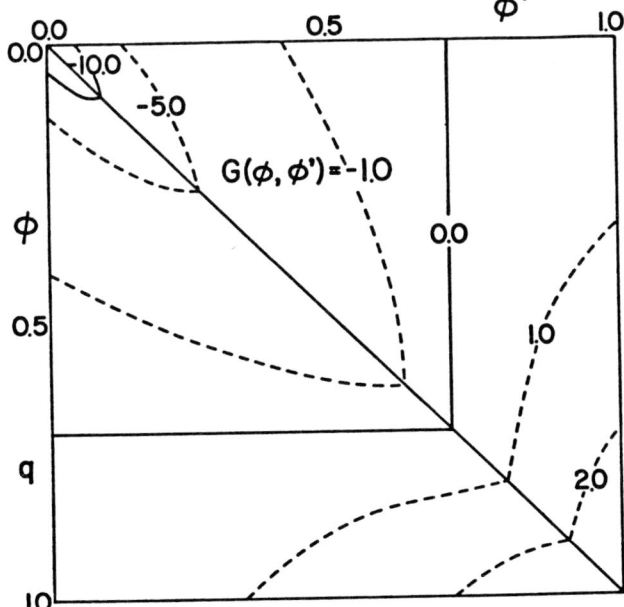

Fig. 3b. The same as Fig. 3a but for a marginally stable system with a density contrast of 709.

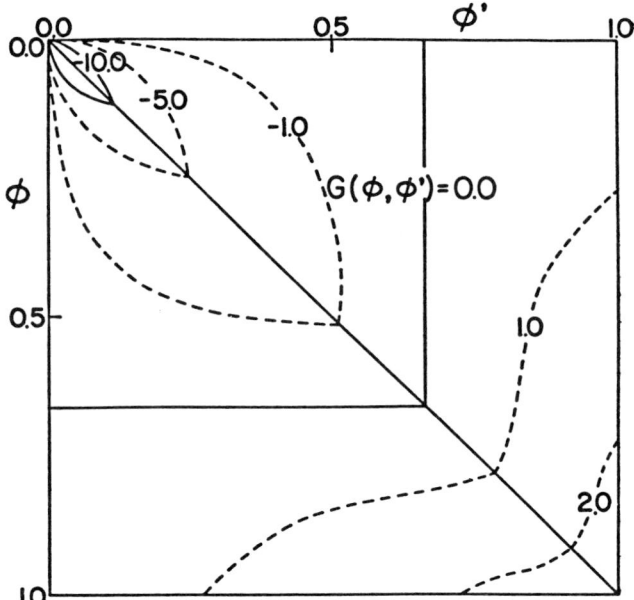

Fig. 3c. The same as Fig. 3a but for a strongly unstable system with a density contrast of 3.16×10^7.

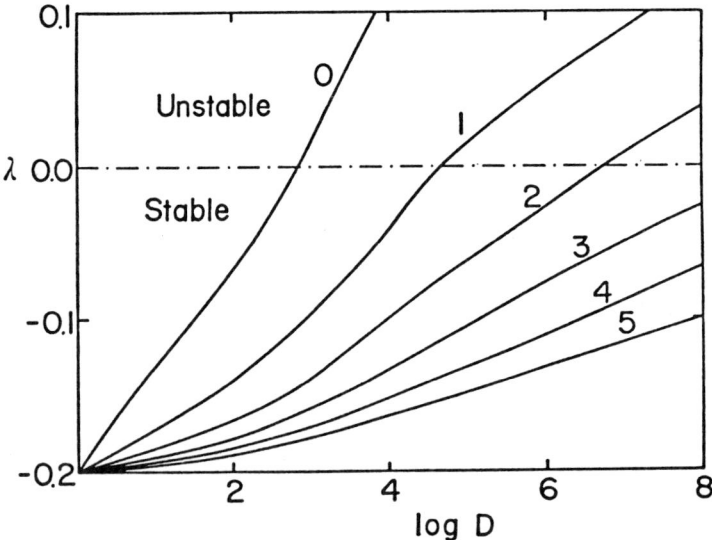

Fig 4. The eigen value λ is shown against the density contrast D for the fundamental mode (0) as well as higher harmonics (1, 2, ...). The condition for gravothermal instability is $\lambda > 0$.

where ρ_c is the central density, v_c the central velocity dispersion, r_c the core radius and M_c is the core mass. From equation (3) we see that the central density diverges when $t=t_{cc}$. Another remarkable point is that the logarithmic density gradient dln ρr/dln r is -2.21 in Lynden-Bell and Eggleton's model, which is very close to Cohn's (1980) value -2.23, though they adopted a conductive gas model.

4. EVOLUTION OF MULTI-COMPONENT CLUSTERS

Globular clusters consist of stars with different masses. In this subsection, we consider the effect of disparate masses in globular clusters. The main extent of disparate masses is to greatly accelerate the cluster evolution. Let us consider the simplest multi-component cluster, i.e., two-component cluster. Inagaki (1985) carried out numerical integrations of the one-dimensional Fokker-Planck equation and showed that the evolution time scale is about ten times faster than the single component cluster (Table I).

Another feature of the two-component cluster is destabilization due to mass-segregation instability (Spitzer 1969).

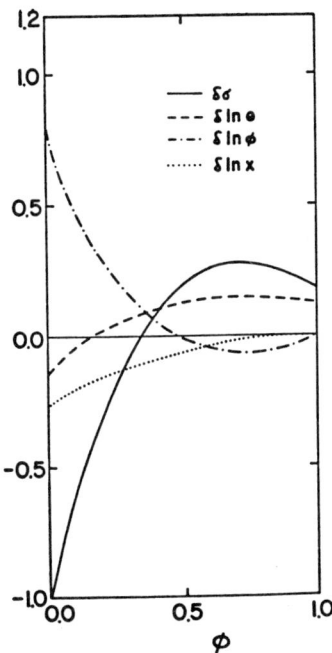

Fig. 5a. Eigenfunctions of the fundamental mode are shown in arbitrary units against the Lagrangian mass coordinate $\phi = M_r/M$ for the case of a stable system with D = 24.2 and λ = -0.11. σ, θ, and x are dimensionless entropy, temperature and radius, respectively.

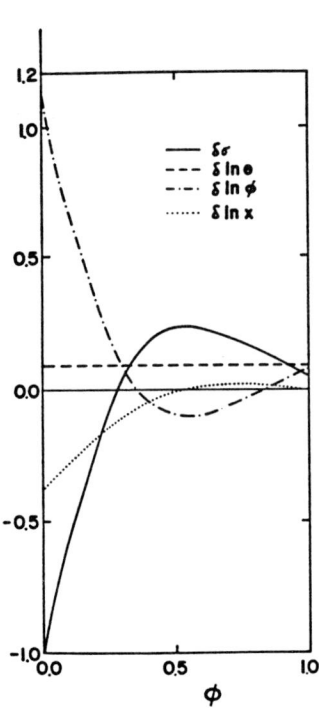 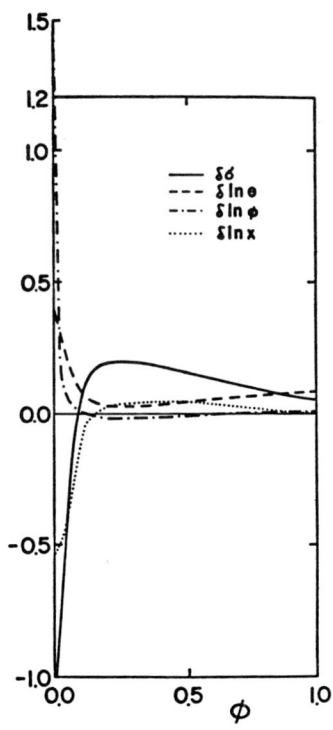

Fig. 5b. The same as Fig. 5a but for the marginally stable system with D = 709 and λ = 0

Fig 5c. The same as Fig. 5a but for the case of an unstable system with D = 1.41 x 10^6 and λ = 0.2.

TABLE I
The time (in the unit of t_{rh}) required for the complete collapse in two-component clusters with $m_1/m_2 = 5$.

m_1/m_2	0.001	0.005	0.014	0.072	0.30
t_{cc}	14.8	10.8	4.2	1.7	1.9

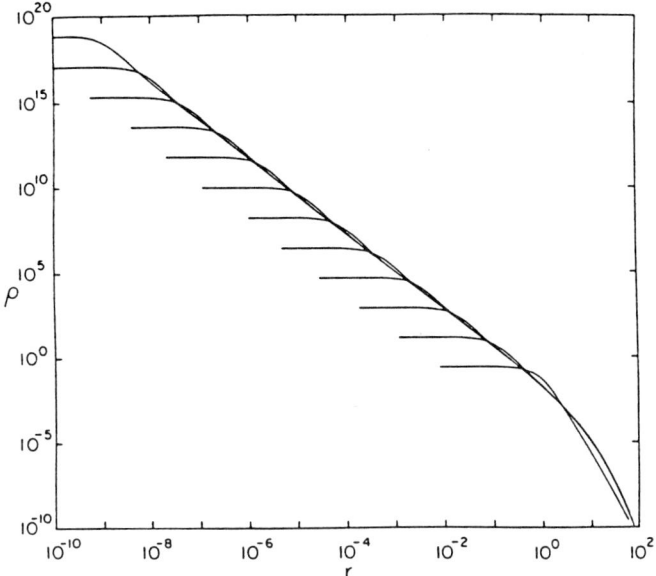

Fig.6. Cohn's (1980) calculation of a single component cluster. Homologous evolution of the cluster is seen from the figure.

Spitzer (1969) showed that if

$$S = (M_2/M_1)(m_2/m_1)^{-1.5} \qquad (7)$$

is larger than 0.16, equipartition between the more massive component and the less massive component is impossible. The stability of isothermal two-component clusters was analyzed by Yoshizawa et al. (1978). The curve of marginal stability is shown in Fig. 7. They showed that the states of marginal stability lie for large range of the density contrast: the cluster can be unstable even if the density contrast is 19 which is much smaller than 709 for a single component system. In Fig. 7 some stabilization effect is seen, i.e., the isothermal cluster with $m_2 = 10$ and $M_2/M_1 = 0.003$ is stable if $\rho_c/\rho_b <$ 5012. This stabilization can be understood as follows. Fig. 7 shows that $\rho_2(0)/\rho_1(0)$ is constant (≈ 8) along the marginally stable states near this model. This means that the development of the halo does not affect the stability or that the stability is determined by the state of the core. In other words, the stability is caused by the exchange of energy between components in the core. However, if the halo becomes too extended (for example, density contrast of the less massive massive component exceeds 709), the less massive component becomes unstable as a single component cluster. The density contrast at this stage is about $\rho_2(0)/\rho_1(0)$ times 709 so that it is significantly larger than 709. In this sense, the stabilization in a two-component cluster is deceptive.

Another interesting point is whether equipartition of kinetic energies between different components may be achieved. Inagaki and Wiyanto (1984) and Inagaki (1985) made numerical integration of isotropized Fokker-Planck equation and confirmed Spitzer's prediction. Inagaki and Saslaw (1985) further made simulations of fifteen component clusters and found that equipartition is impossible if the mass spectrum is shallower than $dM \propto m^{-6} dm$. This conclusion should be compared with Vishniac's (1978) result that equipartition is possible if the mass spectrum is steeper than $dM \propto m^{-2.5} dm$, although this assumed a homologous density profile for each mass component, which is never realized.

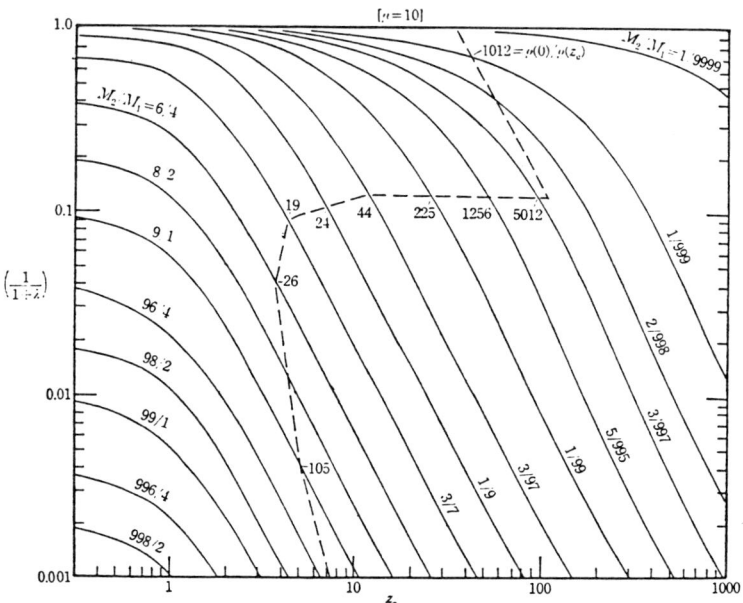

Fig. 7. Curves of fixed M_2/M_1 (solid curves) and of the marginal stability (dashed curve) in the $(1 + \lambda)^{-1}$-z_0 diagram for the case of $\mu = 10$, where $\lambda = \rho_2(0)/\rho_1(0)$ and $\mu = m_2/m_1$. Note that not only each curve of a fixed M_2/M_1 but also each line of a constant λ crosses the curve of the marginal stability criterion only once. The left side of the marginal stability curve is the stable region. The density contrast $\rho(0)/\rho(z_0)$ between the center and the surface is monotonically increasing with increasing z_0 along the curve of a fixed M_2/M_1. The value of the critical density contrast at which the cluster becomes marginally stable is shown along the curve of the marginal stability.

5. CONCLUSIONS AND FUTURE PROBLEMS

The basic processes in precollapse phase are:
1) Overflow from the tidal radius
2) Tidal shock
3) Gravothermal instability
4) Multi-mass effects, i.e., acceleration of evolution, mass-segregation instability, and lack of equipartition.

All these processes have now been investigated in detail. The primary future problem is to construct realistic models of globular clusters including the above mentioned effects and to see whether most clusters collapse in Hubble time and are currently in post-collapse phase.

REFERENCES

Antonov, V. A. 1962 Vestnik Lenigrad Univ. 7, 135.
Chernoff, D. F., Kochanek, C. S. and Shapiro, S. L. 1986 preprint.
Cohn, H. 1980 Astrophys. J. 242, 765.
Hachisu, I., Nakada, Y., Nomoto, K. and Sugimoto, D. 1978 Prog. Theor. Phys., Kyoto 60, 393.
Hachisu, I. and Sugimoto, D. 1978 Prog. Theor. Phys., Kyoto 60, 123.
Inagaki, S. 1985 in IAU Symposium No. 113, Dynamics of Star Clusters, J. Goodman and P. Hut, eds., Reidel, Dordrecht, p. 189.
Inagaki, S. and Hachisu, I. 1978 Publ. Astron. Soc. Japan 30, 39.
Inagaki, S. and Saslaw, W. C. 1985 Astrophys. J. 292, 339.
Inagaki, S. and Wiyanto, P. 1984 Publ. Astron. Soc. Japan 36, 391.
Katz, J. 1978 Monthly Notices Roy. Astron. Soc. 183, 765.
Katz, J. 1980 Monthly Notices Roy. Astron. Soc. 190, 497.
King, I. R. 1966 Astron. J. 71, 64.
Larson, R. B. 1970 Monthly Notices Roy. Astron. Soc. 147, 323.
Lynden-Bell, D. and Eggleton, P. P. 1980 Monthly Notices Roy. Astron. Soc. 191, 483.
Lynden-Bell, D. and Wood, R. 1968 Monthly Notices Roy. Astron. Soc. 138, 495.
Chernoff, D. F. and Shapiro, S. L. 1987 in IAU Symposium No. 126, Globular Cluster Systems in Galaxies, J. E. Grindlay and A. G. D. Philip, eds., Reidel, Dordrecht, p. XXX.
Spitzer, L., Jr. 1969 Astrophys. J. Letters 158, L139.
Spitzer, L., Jr. 1985 in IAU Symposium No. 113, Dynamics of Star Clusters, J. Goodman and P. Hut, eds., Reidel, Dordrecht, p. 119.
Vishniac, E. T. 1978 Astrophys. J. 223, 986.
Wiyanto, P., Kato, S. and Inagaki, S. 1985 Publ. Astron. Soc. Japan 37, 715.
Yosizawa, M., Inagaki, S., Nishida, M. T., Kato, S., Tanaka, Y. and Watanabe, Y. 1978 Publ. Astron. Soc. Japan 30, 279.

DISCUSSION

DJORGOVSKI: Since the mass contrast accelerates the collapse so much, it is possible to contemplate core collapse in dwarf ellipticals; for example, M 32 has the surface brightness profile which is a slope -1.2 power law for radii < 30". Do you think that it is possible that M 32 is collapsing now, or that it is a PCC galaxy?

INAGAKI: If the half-mass relaxation time of M 32 is smaller than several billion years, it is quite possible that M 32 is a PCC galaxy. However, a tentative calculation shows that the half-mass relaxtion time of M 32 is of the order of the Hubble time. Therefore it is unlikely that M 32 is a PCC galaxy. The power-law density profile of M 32 may be created by some other mechanism.

LEE, H. M.: How much mass is lost due to the high velocity tail of Maxwellian velocity distributions (estimated); and how much is lost due to the conductive flux through the tidal boundary?

INAGAKI: Since we did not take account of close encounters and put the value of distribution function zero at the energy corresponding to the tidal radius, no mass is lost due to high velocity tail of Maxwellian distribution.

AFTER CORE COLLAPSE. WHAT?

Haldan Cohn

Indiana University

ABSTRACT. As our understanding of core collapse in globular clusters has improved through detailed computer simulations, attention has naturally turned to dynamical evolution of globular clusters after core collapse. The results of recent simulations of post-collapse cluster evolution are reviewed. An assessment is given of progress towards the goal of developing astrophysically realistic models that cover all phases of globular cluster evolution. A focus of this review is the stability of the post-collapse expansion phase to the large amplitude core oscillations first observed in the simulations of Sugimoto and Bettwieser and now confirmed by several other studies. The implications of core oscillations for the observation of post-collapse clusters are discussed.

1. INTRODUCTION

Detailed computer simulations of globular cluster dynamical evolution in the past 15 years have firmly established the inevitability of the core collapse process (Inagaki, this volume) which drives cluster core radii to very small values ($< 10^{-2}$ pc) while the central density increases to extremely high values ($> 10^8 \, M_\odot \, pc^{-3}$). At the same time, the central surface brightness cusps expected to result from core collapse have been convincingly detected in a substantial number of clusters (Djorgovski and King 1986; Djorgovski, this volume; Lugger et al. 1987a and this volume). It is also becoming clear that core collapse has important implications for stellar evolution in globular clusters (Bailyn et al., this volume; Lee 1986a). Following up a line of inquiry initiated by Hénon 25 years ago, a number of investigations of post-collapse evolution have been carried out in the past three years. Under the dense conditions that result from core collapse, rates of binary formation by two-star tidal interactions and three-star interactions are substantially enhanced. It is generally thought that core collapse is halted and cluster cores undergo a post-collapse expansion due to energy release from hard binaries that interact with single stars and each other (Hut 1985). Since it seems likely that as many as 25% of all Galactic globular cluster have already undergone core collapse based on theoretical (Lightman 1982, Cohn and Hut 1984) and observational (Djorgovski and King 1986) grounds, the dynamical evolution of globular clusters in this phase is of great interest.

Heggie (1985) presented an excellent summary of the status of studies of dynamical evolution of globular clusters after core collapse at IAU Symposium No. 113, two years ago. The major uncertainty at that time was the *stability* of the post-collapse phase. Sugimoto and Bettweiser (1983) had observed the development of large-amplitude nonlinear core oscillations in their fluid dynamical simulations of cluster evolution that had not yet been seen in any other study. Two important developments have since occurred concerning these 'gravothermal' oscillations. First, core oscillations of quite similar character have now been observed in direct Fokker-Planck simulations of cluster evolution (Cohn and Hut 1985; Cohn et al. 1986; Cohn, Hut, and Wise 1987) and fluid dynamical simulations based on the Fokker-Planck equation (Heggie and Ramamani 1986). Second, Goodman (1987) has performed a stability analysis which convincingly demonstrates that the core oscillation instability is intrinsic to the Fokker-Planck model for large-N systems with three-body binary heating.

Core oscillations greatly complicate the study of cluster evolution since the computational time steps that must be used to track oscillations are $\sim 10^4 - 10^5$ times smaller than would otherwise be required. Consequently, there are many uncertainties and unanswered questions concerning core oscillations. This paper will review recent progress towards understanding post-collapse evolution and the major areas in which more work is needed.

2. BINARIES AND CORE COLLAPSE

The pioneering work of Hénon (1961, 1965, 1975) indicated that a central energy source could halt core collapse and drive a subsequent expansion. Two possible energy sources have been suggested: a massive central black hole (Shapiro 1977) and a centrally concentrated population of hard binaries (Heggie 1975, Hills 1975). The presence of massive black holes at the centers of globular clusters appeared, at one time, to present an attractive explanation of observed X-ray emission from clusters (Silk and Arons 1975, Bahcall and Ostriker 1975). However, since there is now strong evidence that the 10 known high luminosity globular cluster X-ray sources are neutron star binaries (Grindlay 1981), there is little evidence for massive black holes in clusters. Instead the X-ray data provide direct evidence for the presence of binaries (Grindlay, this volume).

Three possible mechanisms have been considered for formation of binaries in clusters: primordial formation, two-body tidal capture (Fabian, Pringle, and Rees 1975; Press and Teukolsky 1977), and three-body interactions (see Hut 1985). While most simulations of cluster evolution with binaries have concentrated on only one of these mechanisms, the comprehensive, though highly idealized, simulations of Stodólkiewicz (1985) included binary formation by both tidal capture and three-body interactions. The investigations of the evolution of clusters of *identical main sequence* stars by Ostriker (1985) and Hut and Inagaki (1985) suggest that binaries produced by tidal capture should reverse core collapse before the core density becomes high enough to form binaries by three-body interactions. However Stodólkiewicz (1985) and Lee (1986b and this volume) find that three-body binary formation will occur at significant levels when degenerate remnants (either white dwarfs or neutron stars) are present. In the latter study, which included a primordial degenerate component, three-body interactions are found to be the *dominant* binary formation mechanism when the degenerate population is sufficiently large. Thus it is important to carry out detailed simulations that consider both tidal capture and three-body binary formation.

3. RECENT RESULTS

3.1 Goodman's Stability Analysis

Goodman (1987) has performed a stability analysis of an idealized model of a cluster of identical stars undergoing core expansion due to energy input by binaries produced by three-body interactions. He finds that for total star number $N < 7 \times 10^3$, the model is stable, while for $N > 4 \times 10^4$, the model is unstable. In between these two limits, the model is overstable. Since N substantially exceeds 4×10^4 for globular clusters, they are predicted to be unstable. Goodman's results support Bettwieser's (1985) interpretation of the oscillations as being a manifestation of the same 'gravothermal' instability that accounts for the original core collapse event. Even with a central energy source, clusters with too high a degree of central concentration are unstable.

Goodman's (1987) work, which is essentially analytic and quite rigorous in nature, lays to rest any lingering suspicions that the core oscillations found in cluster evolution simulations might merely be due to instabilities in the *numerical methods* used. A more fundamental question — and thus even more difficult to answer — is whether the fluid dynamical and direct Fokker-Planck models, for which the core oscillation instability is intrinsic, accurately represent real globular clusters. This issue is returned to in §3.3 below.

3.2 Direct Fokker Planck Calculations

3.2.1 The Method

As discussed at IAU Symposium No. 113, the direct Fokker-Planck method is well suited for simulating post-collapse dynamical evolution (Cohn 1985; Ostriker 1985). This method works with continuous stellar distribution functions in energy-space and thus does not treat the discreteness of individual stars. Binary formation must be explicitly included; three-body binary formation does not naturally occur as in a direct N-body simulation (c.f. McMillan and Lightman 1984a,b). The direct Fokker-Planck method has been extended to include three-body binaries by Cohn, Hut, and Wise (1987) and to include tidal-capture binaries by Statler, Ostriker, and Cohn (1986 and this volume). Three-body binaries are treated by adding a heating term to the Fokker-Planck equation which represents the energy input due to superelastic scatterings of singles by hard binaries. An instantaneous binary 'burning' approximation is adopted, i.e. binary formation is assumed to be immediately followed by energy input into the cluster and ejection of the binary from the cluster. The treatment of tidal capture binaries is much more complex and includes a separately tracked binary component, detailed tidal capture binary formation rates, and binary-single and binary-binary interaction rates.

3.2.2 Time Step Size and Oscillations

Initially, the codes were run with a time step selection algorithm that limited the fractional change in central density per time step to a small value (e.g. 5%). This resulted in time steps nearly 10^5 times longer than the central relaxation time t_{ro}, in the simulations of cluster evolution with three-body binaries reported by Cohn (1985). Smooth, monotonic reexpansion of the cluster core was found for both three-body binaries (Cohn 1985) and tidal capture binaries (Ostriker 1985). Following a suggestion by E. Bettweiser, the three-body and tidal capture codes were run with time steps limited in size to t_{ro}. In both cases, large amplitude core oscillations are observed (Cohn et al. 1986; Cohn, Hut, and Wise 1987; Statler, Ostriker, and Cohn 1986). Some representative results from these studies are included here.

3.2.3 Results for Three-Body Binaries

Figure 1 illustrates the time evolution of the central density from a simulation of a cluster of identical stars including three-body binary heating (Cohn, Hut, and Wise 1987). Time is measured in units of the initial half-mass relaxation time, t_{rh}, which is of order $10^8 - 10^9$ yr for the more centrally concentrated Galactic globular clusters. Central density is measured in units which correspond to approximately $10^4 \, M_\odot \, pc^{-3}$. Since three-body binaries only form when extremely high central densities ($\sim 10^{10} \, M_\odot \, pc^{-3}$) are achieved, the pre-collapse evolution is the same as that of a cluster not containing binaries. Energy input from the binaries halts the core collapse at $t = 15.7 \, t_{rh}$ and a brief expansion phase begins. However the expansion is highly unstable, as predicted by Goodman's (1987) analysis and nonlinear oscillations rapidly develop (Fig. 1a,b). The oscillations are nonsinusoidal in character with more of the time spent near the density minima than the maxima. The expansion phases are considerably faster than is the case when the instability is artificially damped using large time steps (Cohn 1985). For example, the density minimum which occurs near $t = 18\ t_{rh}$ is over a factor of 100 lower than the corresponding value in the large time step run. During the last oscillation cycle of the simulation, the cluster spends most of the time near the density minimum. Although the core repeatedly undergoes brief collapse phases during which the central density returns the large value achieved during the first core collapse, the fraction of time spent at these high densities decreases in time. Thus on a time-averaged basis, the *effective* core expansion rate is greatly enhanced over the artificially smoothed result for the large time step run.

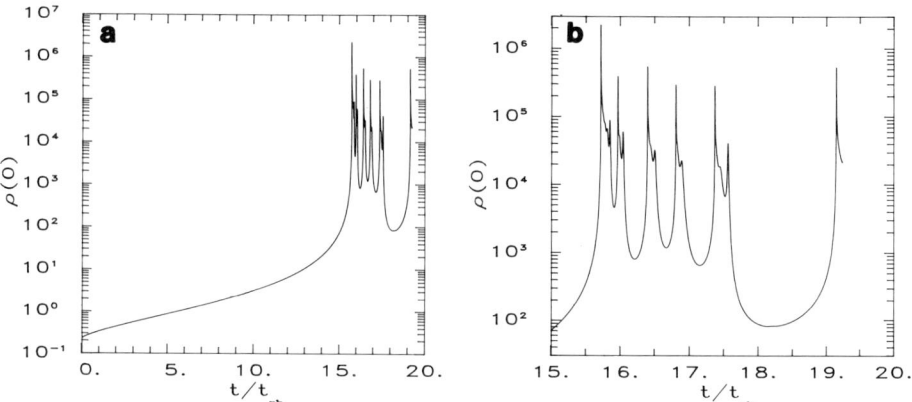

Fig. 1. Evolution of central density from a simulation including binaries produced by three-body interactions. (a) The entire simulation. (b) The core oscillation phase.

Some additional insight into the nature of the oscillations is provided by replacing physical time by a timelike parameter τ that measures the elapsed number of central relaxation times since the start of the simulation, $\tau \equiv \int_0^t dt'/t_{ro}(t')$. The parameter τ is analogous to optical depth in radiative transfer theory. Figure 2 illustrates the evolution of central density as a function of τ. The oscillation appears very much more regular and symmetric with respect to the mean density than when plotted using physical time. This regularity reflects the fact that the time scale of the oscillation is determined by the instantaneous value of the central relaxation time t_{ro}. Since t_{ro} varies inversely with central density, the fraction of an oscillation cycle during which the cluster hovers near the density minimum increases as the amplitude of the oscillation increases as can be seen from Figure 1b.

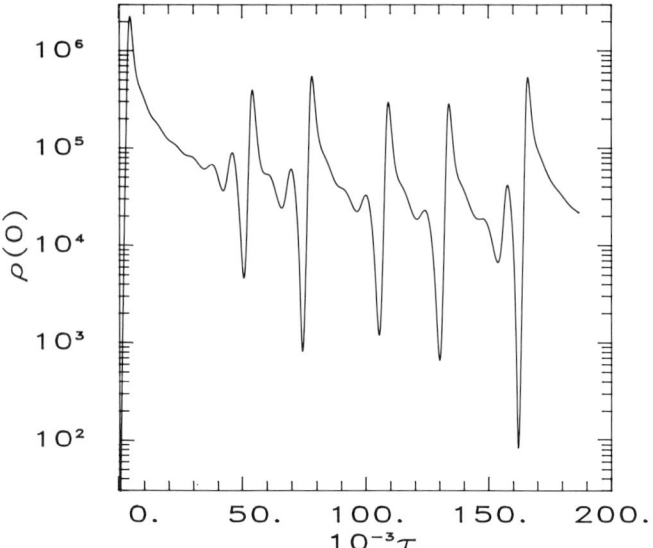

Fig. 2. Regularized core oscillations for the same simulation as in Fig. 1. Central density is plotted versus elapsed number of central relaxation times, τ.

Figure 3 illustrates the time evolution of the core radius in addition to radii containing fixed fractions of the cluster mass. The unit of radius is approximately 1 pc. The outermost radius shown, which contains 10% of the cluster mass, is clearly affected by core oscillations. In contrast, the half-mass radius increases smoothly and monotonically during this period.

Goodman (1987) has predicted that core oscillations will saturate when the core contains 1-2% of the total cluster mass at maximum expansion with a ratio of core radius to half-mass radius of 0.02. Indeed, at maximum expansion during the final oscillation shown in Figure 3, the core radius contains exactly 1% of the cluster mass and $r_c/r_h = 0.02$. While further integrations are necessary to determine whether the core oscillation amplitude has in fact saturated, the apparent agreement between the direct Fokker-Planck calculations and Goodman's (1987) analytic analysis, which is based on a fluid dynamical model, is strikingly good. Thus, there is a high degree of confidence that the oscillations, observed in both direct Fokker-Planck and fluid dynamical calculations of cluster evolution with three-body binary formation, are intrinsic to the models and not a numerical artifact.

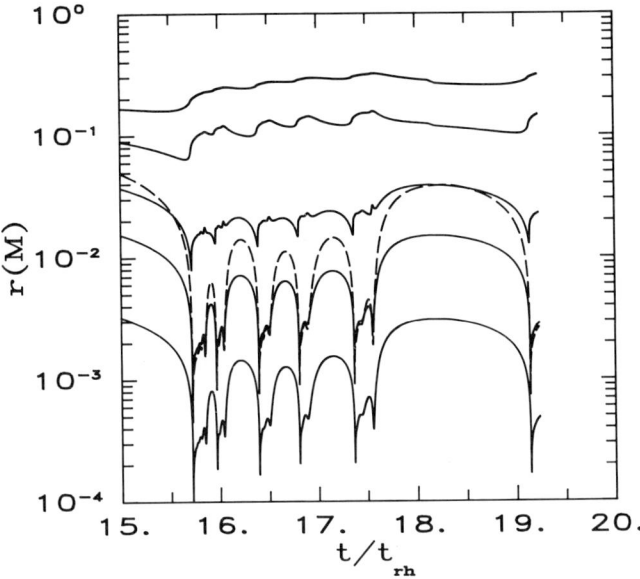

Fig. 3. Evolution of core radius (dashed line) and radii containing 0.001%, 0.1%, 1%, 5%, and 10% of the total cluster mass (solid lines), for the same simulation as in Fig. 1.

3.2.4 Results for Tidal Capture Binaries

While Goodman (1987) only explicitly considered the case of binary formation by three-body interactions (for technical reasons) he argues that his results do not depend on the specific energy input mechanism but instead on the structure of the cluster. He predicts that any expanding equilibrium with $r_c/r_h \lesssim 10^{-2}$ will be unstable to core oscillations. Figure 4 illustrates the development of core oscillations in a cluster simulation including binaries formed by the two-body tidal capture process, bearing out Goodman's (1987) prediction. Statler, Ostriker, and Cohn (1986 and this volume) have reported the results of large time step runs ($\gg t_{ro}$) with this code; the results shown here are for small time steps ($\sim t_{ro}$) which permit core oscillations to develop. There is an initial damping of the instability immediately after the reversal of core collapse, but this is followed by a linear growth phase leading to nonlinear oscillations. During the immediate post-collapse period, the ratio of core radius to half-mass radius hovers near the critical value of 10^{-2}. As can be seen in Fig. 1 of Statler, Ostriker, and Cohn (this volume) r_c/r_h *decreases* following core collapse in the tidal capture binary formation case, so that even if a cluster starts the post-collapse phase on the stable side of the stability criterion, it is expected to cross over to the unstable side. As for the three-body binary formation case, the density hovers near the minimum value for most of an oscillation cycle. Thus, the effective expansion rate is again faster than for smoothed reexpansion.

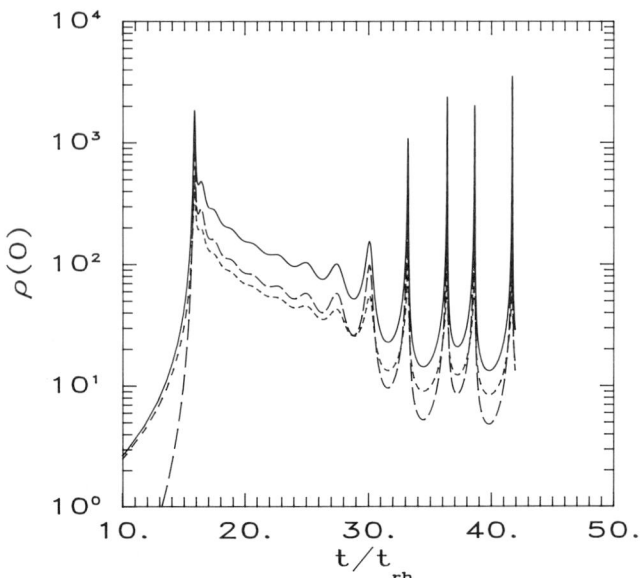

Fig. 4. Evolution of central density from a simulation including binaries formed by tidal capture. Results are shown for total mass density (solid curve), single stars (short dashes), and binaries (long dashes).

3.3 Other Recent Results

Heggie and Ramamani (1986 and private communication) have recently obtained results using a fluid dynamical code that are quite similar in character to the large amplitude oscillations obtained with the direct Fokker-Planck method (Fig. 1a,b). While the the agreement is not exact, it is extremely good. This provides another indication that at least for clusters of identical stars, fluid dynamical analogues are able to reproduce the results of the more exact (and thus more computationally expensive) direct Fokker-Planck method. A common property of these two approaches is a lack of the discreteness present in both real globular clusters and direct N-body simulations.

Neither Inagaki's (1986) direct N-body simulations of cluster evolution nor McMillan's (1986) hybrid N-body/Fokker-Planck simulations

4.2 Observation of Post-Collapse Clusters

As discussed in §3.2, gravothermal oscillations accelerate the *effective* expansion rate following core collapse. In the case of binaries produced by three-body interactions, the core radius reaches a value of nearly 0.04 pc at maximum expansion (Fig. 3) which subtends an angle of 0.8" for a typical cluster near the Galactic center and is thus marginally resolvable by ground-based observations. Djorgovski and King (1986) find that NGC 6397, one of the nearest globular clusters, has a collapsed core. At the distance of this cluster (2.2 kpc) 0.04 pc subtends an angle of 3.8" and is easily resolvable. In fact, Lugger et al. (1987b) find that the central power law profile of this cluster flattens out at a radius of order 2". Observations of the central structure of globular clusters, with the highest angular resolution possible are needed to test the predicted effective size of post-collapse cluster cores. While additional ground-based studies — particularly those that take advantage of sites with excellent seeing — are needed, Hubble Space Telescope observations of cluster cores will be of critical importance.

If the most massive stars in a cluster are nonluminous remnants while most of the luminosity in due to less massive giants or horizontal branch stars, then the slope of the central surface brightness cusp in a post-collapse cluster can be significantly flatter than the value of $d \ln S/d \ln r = -1$ expected in a cluster of identical stars. This point is illustrated by the simulations of Lee (1986b), for which a ratio of the mass of a nonluminous star to that of a luminous star of 1.9 produces a slope of -0.6. This slope is quite close to that measured for the central surface brightness cusp in M15 by Lugger et al. (1987a and this volume). There is considerable potential for useful interaction between realistic dynamical models and the improved data base of cluster structure observations becoming available.

4.3 Prospects for Future Work

Ideally, an evolving, dynamical model for a globular cluster would include all of the following physical properties and mechanisms:

- realistic stellar mass spectrum
- stellar evolution and mass loss
- tidal capture binary formation
- physical collisions between stars and mergers
- 3-body binary formation
- binary-single and binary-binary interactions
- velocity distribution anisotropy
- cluster rotation
- large angle gravitational scattering
- discreteness of the stellar distribution
- tidal cutoff
- tidal shocks
- and make sure the time steps are small!

show the large amplitude 'gravothermal' oscillations seen in the direct Fokker-Planck and fluid dynamical simulations. However, small amplitude core oscillations, attributable to particular binary formation and ejection events, are seen. These authors interpret their findings as indicating that the presence of discreteness prevents the development of gravothermal oscillations. Goodman (1987) notes that Inagaki (1986) included at most 3000 stars in his simulations, which is below the critical number for instability, and that McMillan's (1986) hybrid simulations only extend for a short period following core collapse which corresponds to the linear growth phase of the instability. Thus, Goodman argues that large amplitude oscillations would not have been *expected* in either simulation. Recently, Makino, Tanekusa, and Sugimoto (1986) have reported gravothermal oscillations in an N-body simulation. Further investigation of this issue is clearly needed.

As noted in §2 above, Lee (1986b and this volume) has carried out direct Fokker-Planck simulations of globular cluster evolution including a primordial degenerate remnant population. This work may be regarded as a first step towards a simulation of cluster evolution beyond core collapse that includes both an astrophysically realistic stellar mass spectrum and stellar evolution. These simulations use a large time step and thus do not allow for the possibility of gravothermal oscillations. Lee (1986a) has also carried out direct Fokker-Planck simulations in which successive stellar mergers take place, as a result of close encounters, leading to the production of massive stars (up to 22 $M_\odot$).

Subsequent stellar evolution mass loss provides an energy input into the cluster which plays a role similar to binary heating. Heating by binaries formed by three-body interactions is also included, and it is energy input from massive binaries that ultimately reverses core collapse. The post-collapse phase is, however, driven by energy input from stellar evolution mass loss.

4. DISCUSSION

4.1 Gravothermal Oscillations

'Gravothermal' oscillations of globular cluster cores undergoing binary driven expansion are now well established by direct Fokker-Planck and fluid dynamical simulations, as well as by a rigorous stability analysis based on the fluid dynamical model. Some lingering uncertainty about the role that discreteness might play in suppressing the instability remains to be cleared up by further investigation. The general convergence of the results from different investigations, since IAU Symposium No. 113, is an encouraging sign that progress is being made towards a full understanding of the post-collapse life of globular clusters.

Each of these items has been considered in a particular cluster simulation and some simulations (most notably that of Stodólkiewicz 1985) have included a good number — but not all — of these. That this statement can be made indicates substantial progress. However, it is often the case in the study of a complex physical system like a star cluster that the joint effect of two physical processes is quite different than that of either individual process. In the present age of supercomputers and large-scale computation it does not seem unreasonable to expect a fully inclusive simulation of globular cluster evolution in the not too distant future.

REFERENCES

Bahcall, J. N. and Ostriker, J. P. 1975 Nature 256, 23.
Bettwieser, E., Fricke, K. J. and Spurzem, R. 1985 in IAU Symposium No. 113, Dynamics of Star Clusters, J. Goodman and P. Hut, eds, Reidel, Dordrecht, p. 219.
Cohn, H. 1960 Astrophys. J. 242, 765.
Cohn, H. 1985 in IAU Symposium No. 113, Dynamics of Star Clusters, J. Goodman and P. Hut, eds, Reidel, Dordrecht, p. 161.
Cohn, H. and Hut, P. 1984 Astrophys. J. Letters 277, L45.
Cohn, H. and Hut, P. 1985 Bull. Am. Astron. Soc. 17, 603.
Cohn, H., Hut, P. and Wise, M. 1987 Astrophys. J., in press
Cohn, H., Wise, M. W., Yoon, T. S., Statler, T. S., Ostriker, J. P. and Hut, P. 1986 in Proceedings of the Use of Supercomputers in Stellar Dynamics, P. Hut and S. McMillan, eds., Springer, in press.
Djorgovski, S. and King, I. R. 1986 Astrophys. J. Letters 305, L61.
Fabian, A. C., Pringle, J. E. and Rees, M. J. 1975 Monthly Notices Roy. Astron. Soc. 172, 15P.
Goodman, J. 1987 Astrophys. J. 313, in press.
Grindlay, J. E. 1981 in X-Ray Astronomy with the Einstein Satellite, R. Giacconi, ed., Reidel, Dordrecht, p. XXX.
Grindlay, J. E. 1985 in IAU Symposium No. 113, Dynamics of Star Clusters, J. Goodman and P. Hut, eds., Reidel, Dordrecht, p. 43.
Heggie, D. C. 1975 Monthly Notices Roy. Astron. Soc. 173, 729.
Heggie, D. C. 1985 in IAU Symposium No. 113, Dynamics of Star Clusters, J. Goodman and P. Hut, eds., Reidel, Dordrecht, p. 139.
Heggie, D. C. and Ramamani, N. 1986 in preparation.
Henon, M. 1961 Ann. D'Astrophys. 24, 369.
Henon, M. 1965 Ann. D'Astrophys. 28, 62.
Henon, M. 1975 in IAU Symposium No. 69, Dynamics of Stellar Systems, A. Hayli, ed., Reidel, Dordrecht, p. XXX.
Hills, J. G. 1975 Astron. J. 80, 809.
Hut, P. 1985 in IAU Symposium No. 113, Dynamics of Star Clusters, J. Goodman and P. Hut, eds., Reidel, Dordrecht, p. 231.
Hut, P. and Inagaki, S. 1985 Astrophys. J. 298, 502.
Inagaki, S. 1986 Publ. Astron. Soc. Japan 38, XXX.
Lee, Y. M. 1986a preprint.

Lee, Y. M. 1986b preprint.
Lightman, A. P. 1982 Astrophys. J. Letters 263, L19.
Lugger, P. M., Cohn, H., Grindlay, J. E., Bailyn, C. D. and Hertz, P. 1987a, Astrophys. J., in press.
Lugger, P. M., Cohn, H., Grindlay, J. E., Bailyn, C. D. and Hertz, P. 1987b, in preparation.
Makino, J., Tanekusa, J. and Sugimoto, D. 1986 preprint.
McMillan, S. L. W. 1986 Astrophys. J. 307, 126.
McMillan, S. L. W. and Lightman, A. P. 1984a Astrophys. J. 283, 801.
McMillan, S. L. W. and Lightman, A. P. 1984b Astrophys. J. 283, 813.
Ostriker, J. P. 1985 in IAU Symposium No. 113, Dynamics of Star Clusters, J. Goodman and P. Hut, eds., Reidel, Dordrecht, p. 347,
Press, W. H. and Teukolsky, S. A. 1977 Astrophys. J. 213, 183.
Shapiro, S. L. 1977 Astrophys. J. 217, 281.
Silk, J. and Arons, J. 1975 Astrophys. J. Letters 20, L131.
Statler, T. P., Ostriker, J. P. and Cohn, H. 1986 Astrophys. J. in press.
Stodolkiewicz, J. S. 1985 in IAU Symposium No. 113, Dynamics of Star Clusters, J. Goodman and P. Hut, eds., Reidel, Dordrecht, p. 361.
Sugimoto, D. 1985 in IAU Symposium No. 113, Dynamics of Star Clusters J. Goodman and P. Hut, eds., Reidel, Dordrecht, p. 207.
Sugimoto, D. and Bettwieser, E. 1983 Monthly Notices Roy. Astron. Soc. 204, 19P.

DISCUSSION

KING: Some years ago Retterer provided, in the Kolmogorov-Feller equation, a mechanism for dealing with large-angle scattering. Why isn't it used?

COHN: Jeremy Goodman's treatment of large angle scattering in an energy-space formulation can be regarded as a follow-up to Retterer's very important work. I'll let Jeremy comment on this.

GOODMAN: I have done collapse calculations with the KF equation to describe large-angle encounters. I was disappointed to find that my results were very similar to those obtained by Cohn with the Fokker-Planck equation.

LEE, H. M.: Could you make a short comment on the reason why we expect gravothermal oscillations for highly concentrated clusters?

COHN: I believe that it is for very much the same reason that highly concentrated clusters without central energy sources are subject to the gravothermal instability leading to core collapse: the cores of these concentrated clusters have negative specific heat.

DJORGOVSKI: Can you tell us what is the behavior of the profile of an oscillating PCC [post core collapse] cluster? Does it ever look like a very concentrated King model?

COHN: This is something that I will have to check. While we did not save radial structure information for the tidal capture binary runs, we do have it for the three-body binary runs. When I analyzed the profiles for the large time step runs I presented at IAU Symposium No. 113, I found that the late time profiles did resemble very extended King models with $W_o \sim 17$. I suspect that this will also be true for the oscillating case.

STATLER: I'd like to amplify one point that you touched on. It's important that the different calculations that now show oscillations have "zero-order" solutions with quite different physics. In those with 3-body binaries, heat is deposited locally in the core and conducted outward, while in those with tidally captured binaries, much of the energy is deposited in the vicinity of the half-mass point and conducted inward. So the presence of the oscillations is independent not only of the microphysics of the heating, but also on the global characteristics of the background solutions

COHN: To amplify on your amplification, I think that it is important to run simulations for as many different treatments of binaries as possible, using small time steps, in order to determine exactly what conditions lead to oscillations. I suspect that Jeremy Goodman's criterion that oscillations occur when the ratio of core radius to half-mass radius is less than $\sim 10^{-2}$ is quite generally valid.

BLECHA: How do you form the binaries? Is it really necessary to go to a core radius of 10^{-6} pc before the energy input from binaries prevents the further collapse?

COHN: The results that I presented were from two sets of simulations that separately treated binaries formed by three-body interactions and binaries formed by tidal captures. In the three-body case, the core radius drops to less than 10^{-3} pc before core collapse is reversed. In the tidal capture case, core collapse is reversed at a core radius of about 10^{-2} pc.

MCMILLAN: The results of runs with my current hybrid code cannot be compared directly with Jeremy Goodman's criterion for gravothermal oscillations, as N is not defined in my work. There are other possible reasons why I have not yet seen gravothermal oscillations in my runs. If they really do exist, and are not just a consequence of the continuous description of the cluster employed in the fluid and Fokker-Planck approaches, it is possible that I did not run my code for long enough to see them. It may also be that my resolution of the outer regions of the cluster (outside fifty or so core radii) was insufficient to distinguish the effect. Both these possibilities are being investigated.

COHN: Thank you for the clarification. I am looking forward to learning of the results of your current simulations.

NEMEC: Hills and Day (~1978) predict that only a few collisions occur during the lifetime of a loose globular cluster. It follows that the 50 blue stragglers, which are relatively massive and therefore possibly binaries, found in NGC 5466 are primordial in their origin. What is your opinion on the possibility that massive blue stragglers are formed by collisions in low central concentration globular clusters?

COHN: A loose globular cluster does not provide a favorable environment for forming binaries by the tidal capture process.

INAGAKI: Your result shows that the mass fraction inside the core radius is $\sim 10^{-4} - 10^{-5}$ in the case of three-body binaries. This means that the number of stars contained in the core is 10 - 100. Do you think that the Fokker-Planck approximation is still valid?

COHN: Of course the Fokker-Planck approximation, which is based on large-N asumptotics, breaks down for sufficiently small N. Steve McMillan has found that the self-similar core collapse solution given by direct Fokker-Planck calculations is valid until about 30 stars remain in the core. At this point, a three-body interaction results in the formation of a binary in Steve's simulations. We have modified the direct Fokker-Planck approach to model explicitly the energy input into the core due to three-body formation and interaction. Further comparisons of our results with those of direct N-body and hybrid N-body/Fokker-Planck simulations will be necessary to establish the validity of our modified Fokker-Planck approach (in an ensemble-averaged sense) when the number of stars in the core becomes small.

DISOLUTION OF STAR CLUSTERS IN GALAXIES

Roland Wielen

Astronomisches Rechen-Institut

ABSTRACT

We present a procedure which allows to predict the dissolution times of star clusters in a simple way. The dissolution time of a cluster depends mainly on its total mass, its median radius and its galactic environment (galactic tidal field and passing massive objects). As an example, we discuss the lifetimes of LMC clusters. Finally, we show that massive black holes, which have been proposed as the major constituent of the dark coronae of galaxies, are very effective in destroying globular clusters.

1. INTRODUCTION

Star clusters in galaxies provide often more detailed information than individual stars: Clusters are usually brighter, the sample of clusters is relatively more complete, and, most importantly, individual ages of clusters are usually of higher accuracy than those of single stars, and the ages of observable clusters cover the whole range between the youngest and oldest objects in galaxies. Due to internal or external dynamical effects, star clusters often dissolve during a period shorter than the Hubble time. The dynamical dissolution of clusters can therefore strongly determine the distribution of presently observable properties of clusters. It is then more difficult to deduce the distribution of the original properties of clusters at birth. On the other hand, the observable age distribution of open clusters contains the most direct information on the dynamical dissolution of star clusters in a galaxy.

2. THEORETICAL PREDICTIONS FOR THE LIFETIMES OF STAR CLUSTERS

There are many mechanisms which are causing the dissolution of star clusters. The most important internal mechanism is the evaporation of stars, due to the exchange of orbital energy during close encounters between the cluster members. The nearly stationary tidal field of the parent galaxy can increase the evaporation rate dramatically by lowering the limiting energy for bound stars. Internal evolution of the individual stars causes a mass loss from the cluster, thus weakening the binding of the cluster. Tidal shocks due to passing objects, such as giant molecular clouds, are often a very effective external mechanisms for the partial or complete destruction of clusters.

2.1 Dissolution Times of Clusters Predicted from N-Body Simulations and Theoretical Extrapolations

The most direct way to determine theoretically the dissolution times of star clusters are provided by N-body simulations. The results of such N-body simulations have been extensively reviewed (Aarseth 1985, Aarseth and Lecar 1975, Wielen 1974, 1975, 1985). The most recent N-body experiments have been carried out by E.Terlevich (1980, 1983, 1986). It is now possible to follow the dynamical evolution of a star cluster essentially until its final dissolution. The main limitation for N-body experiments is set by the initial number N of stars in a cluster which can be handled on the available computers within reasonable computing times. For star clusters, this limit is presently of the order of N = 1000 stars. In order to extrapolate the N-body results to higher values of N, we have to rely on theoretical extrapolations.

We consider first the dissolution of a cluster due to the combined effects of the internal relaxation and the stationary tidal field of the galaxy alone. Mass loss of evolving stars is neglected, since it does not seem to alter the results dramatically. Tidal shocks due to passing giant molecular clouds or massive black holes are discussed in Section 2.2.

2.1.1 Isolated Clusters

For isolated clusters, we use the theory developed by Hénon (1960, 1969) in order to extrapolate the results of N-body simulations to higher values of N. Hénon's theory is based on the assumption that close encounters between cluster stars are the most important source of escaping stars. This assumption has been nicely confirmed by the N-body simulations, which showed that most of the escapers gain the necessary energy indeed during close encounters with single stars or binaries. This result is not in contradiction to another result of N-body simulations (Aarseth, Hénon and Wielen, 1974), namely that the overall dynamical evolution of a cluster is rather accurately described by a Fokker-Planck equation based on the predominance of more distant, weak encounters. It means only that the rare escape events are not due to the otherwise dominant diffusion of stars in phase space.

Using Hénon's theory, we predict for the evaporation time T,

$$T = -N/(dN/dt) = NT_{cr}/\phi_0 \ . \tag{1}$$

The crossing time T_{cr} for a Plummer model is given by

$$T_{cr} = (32/(3\pi))^{3/2}(Gm_c/r_c^3)^{-1/2} \ , \tag{2}$$

where m_c is the total mass of the cluster, r_c is the median radius of the cluster (containing half of m_c in projection), and G is the gravitational constant. The quantity ϕ_0 in Eq. 1 is best determined from N-body experiments. For a cluster with a realistic distribution of stellar masses, similar to Salpeter's law, we find for the number of stars which escape during one crossing time:

$$\phi_0 = 0.56 \ . \tag{3}$$

Eqs. 1, 2 and 3 allow to predict the dissolution time of an isolated star cluster for any initial value of N, m_c and r_c, provided that our assumptions are also valid outside the experimentally explored range of N.

2.1.2 Tidally Limited Clusters

The theory presented in Section 2.1.1 can be extended to tidally limited clusters. The N-body experiments show that even for tidally limited clusters, close encounters produce still the overwhelming fraction of escapers. It can be shown that, to a first approximation, the number of escapers per crossing time depends on the ratio between the radius r_c and the tidal radius r_t of the cluster only. This leads to the following prediction for the dissolution time T of tidally limited clusters:

$$T = NT_{cr}/(\phi_0 f(r_c/r_t)) \ . \tag{4}$$

ϕ_0 is given by Eq. 3, and the function $f(r_c/r_t)$ has been derived by fitting the results of N-body simulations of tidally limited clusters (Fig. 1). The fitting curve corresponds to

$$f(r_c/r_t) = (1 + (43r_c/r_t)^3)^{1/2}. \tag{5}$$

This expression has the desired properties: f approaches 1 for small radii r_c (isolated clusters), and T becomes independent of r_c for large radii r_c (strongly tidally limited clusters). The tidal radius r_t is given by

$$r_t = (Gm_c/(4\omega^2-\kappa^2))^{1/3} \ , \tag{6}$$

where the local rotational frequency ω and the local epicyclic frequency κ have to be derived from the rotation curve of the parent galaxy. If we introduce a 'tidal time' T_t,

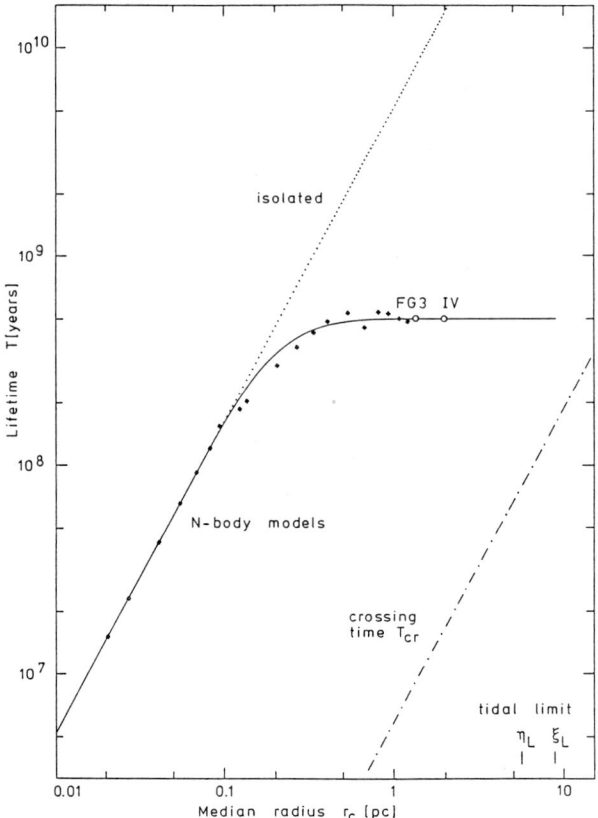

Fig. 1. Fit of the results of N-body simulations for the dissolution time T of a cluster with N = 500 stars and m_c = 250 $m_\odot$ as a function of the median radius r_c of the cluster. N-body models from Wielen (1975, full squares and FG3) and Terlevich (1983, IV). Solid curve: T from Eqs. 4 and 5.

$$T_t = (r_t/(43 r_c))^{3/2} T_{cr}$$
$$= (32/(3\pi \cdot 43))^{3/2} (4\omega^2 - \kappa^2)^{-1/2}$$
$$= 0.022 \, (4\omega^2 - \kappa^2)^{-1/2} , \qquad (7)$$

then the function f can be expressed in terms of the ratio between the crossing time T_{cr} of the cluster and the tidal time T_t,

$$f(r_c/r_t) = f(T_{cr}/T_t) = (1 + (T_{cr}/T_t)^2)^{1/2} . \qquad (8)$$

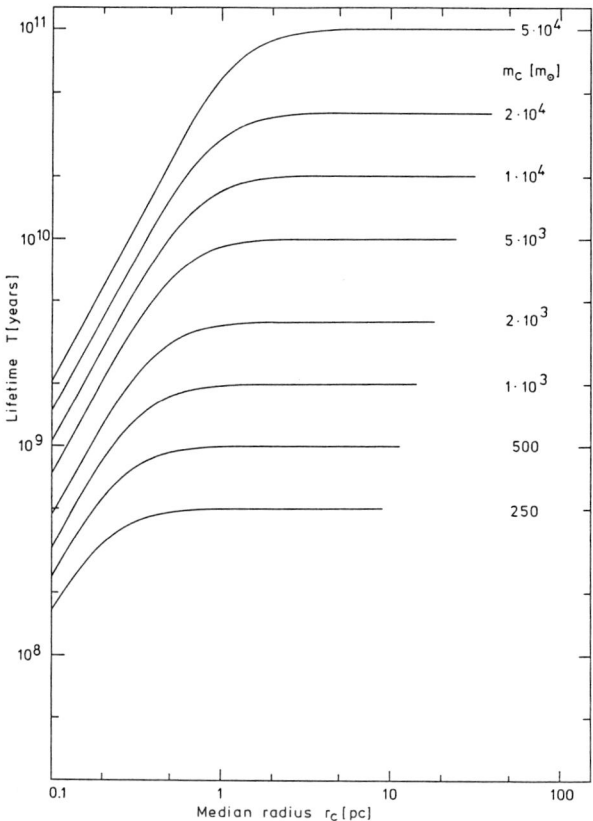

Fig. 2. Dissolution time T of a tidally limited star cluster in the solar neighbourhood as a function of the median radius r_c of the cluster for various total masses m_c, based on Eqs. 4 and 5.

Eqs. 4 and 5 or 8, together with the expression for T_{cr}, r_t, T_t and ϕ_0 = 0.56, provide now the necessary instruments for predicting the dissolution time of a tidally limited star cluster with any value of N, m_c, r_c and r_t.

In Fig. 2, we present results based on our procedure, for star clusters in the solar neighbourhood with a realistic spectrum of stellar masses and a mean stellar mass of $m_c/N = 0.5\ m_\odot$. The dissolution time T is plotted as a function of r_c for various values of m_c.

In the limit of large cluster radii r_c, i.e. for strongly tidally limited clusters in which r_c is not much smaller than r_t, we get the following simple expression for the dissolution time:

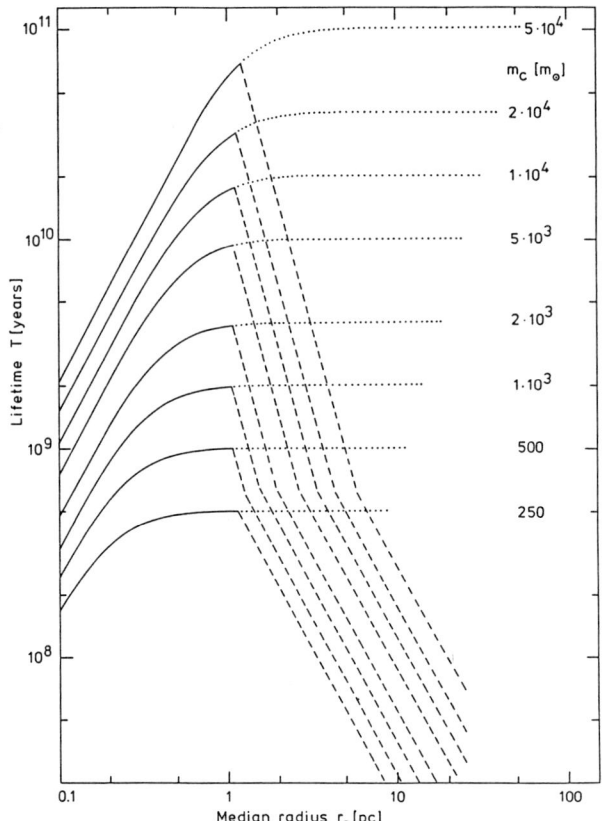

Fig. 3. Dissolution time T of a star cluster in the solar neighbourhood. The full (and dotted) curves are taken from Fig. 2. The dashed lines give the dissolution times due to giant molecular clouds (GMCs).

$$T_{lim} = NT_t/\phi_0 \quad . \tag{9}$$

The dissolution time T_{lim} is independent of r_c and proportional to N, or to m_c for a fixed value of the mean stellar mass (m_c/N). In the solar neighbourhood, the tidal time T_t is about $0.56 \cdot 10^6$ years. This leads to $T_{lim} = N \cdot 10^6$ years or, for example, to $T_{lim} = 10^{10}$ years for N = 10000 stars ($m_c = 5000\ m_\odot$ for $m_c/N = 0.5\ m_\odot$). Clusters with N much higher than 10^4 do not dissolve in a Hubble time, if we consider only internal relaxation and a stationary galactic tidal field. Hence our extrapolation from N $\leq$ 1000, covered by N-body simulations, to the highest values of real interest, N $\sim 10^4$, is by one order of magnitude only. I would consider that as not too risky.

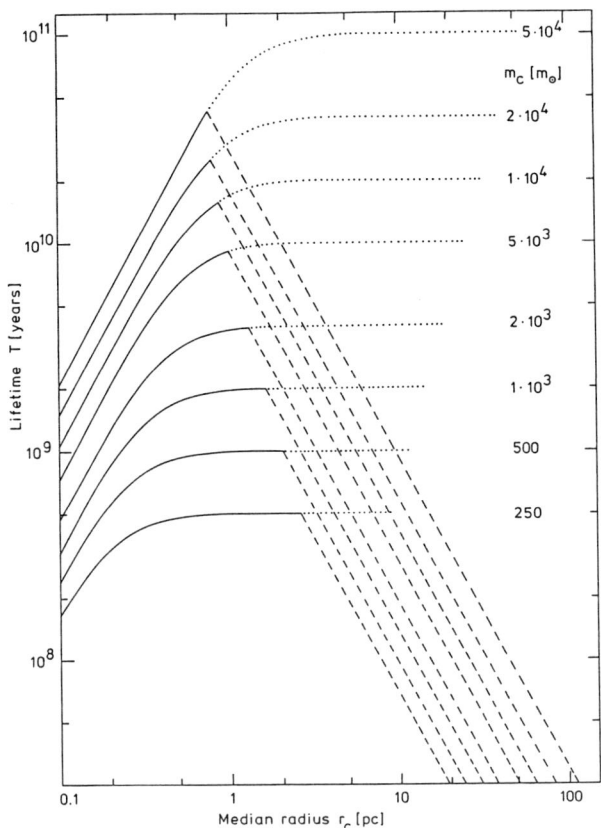

Fig. 4. Dissolution time T of a star cluster in the solar neighbourhood. The full (and dotted) curves are taken from Fig. 2. The dashed lines give the dissolution times due to massive black holes (MBHs).

2.2 Dissolution Times of Clusters due to Passing Massive Objects

Star clusters can also be dissolved by passing massive objects. Important effects can be caused especially by giant molecular clouds (GMC) or by massive black holes (MBH). For a detailed discussion we refer to an earlier paper (Wielen, 1985). For completeness we repeat here the final results for the dissolution times T_n of star clusters due to passing massive objects. A massive object is characterized by its mass m_n and its radius r_n; ν_n is the number density of the massive objects in the neighbourhood of the cluster; $\rho_{an} = m_n \nu_n$ is the overall mass density of the objects; V is the relative velocity between the object and the cluster; and αr_c is the effective radius of the cluster for tidal shocks ($\alpha = 1$ may be used as a first guess).

Two different cases for the dissolution of clusters by passing objects occur. They are characterized by the relative size of the limiting impact parameter p_0,

$$p_0 = \max(r_n, r_c) ,\qquad (10)$$

with respect to the critical impact parameter p_1,

$$p_1 = 4(\alpha/3)^{1/2}(G/\pi)^{1/4} m_n^{1/2} V^{-1/2} (m_c/r_c^3)^{-1/4} . \qquad (11)$$

In the case of $p_1 > p_0$, the cluster can be destroyed completely by a single encounter, if the massive object passes with an impact parameter $p < p_1$. In this case ($p_1 > p_0$), the mean dissolution time $T_{n,1}$ of a cluster is given by

$$T_{n,1} = (3/(32\alpha))\pi^{-1/2} (Gm_c/r_c^3)^{1/2}/(G\rho_{an}) . \qquad (12)$$

It is remarkable that $T_{n,1}$ depends only on the overall mass density ρ_{an}, but not on the individual values of m_c, r_c or V of the passing objects.

In the case $p_0 > p_1$, a number of successive encounters are necessary for disrupting the cluster. The resulting dissolution time $T_{n,0}$ of a cluster is given by

$$T_{n,0} = (9/(512\alpha^2)) \, G^{-1} m_n^{-1} \rho_{an}^{-1} V \, m_c r_c^{-3} p_0^2 , \qquad (13)$$

where p_0 is either r_n or r_c according to Eq. 10.

Both expressions, $T_{n,0}$ and $T_{n,1}$, are only valid if the encounter between the cluster and the object is impulsive (Spitzer, 1958), i.e. if $p/V < T_{cr}$. This is essentially fulfilled if

$$\max(p_0, p_1) < V \, T_{cr} . \qquad (14)$$

In order to calculate T_n for a given cluster, one has firstly to determine p_0 and p_1 from Eqs. 10 and 11, secondly to check the impulsive approximation according to Eq. 14, and then finally to compute T_n either from Eq. 12 ($T_{n,1}$ for $p_1 > p_0$) or from Eq. 13 ($T_{n,0}$ for $p_0 > p_1$). The resulting dissolution time T_n neglects the effect of the galactic tidal field, which would tend to reduce the values for T_n, because it again lowers the limiting energy for bound stars.

In Figs. 3 and 4, we have plotted the dissolution time T_n of clusters as dashed lines, either due to giant molecular clouds (Fig. 3) or massive black holes (Fig. 4). The parameters used are $m_n = 5 \cdot 10^5 \, m_\odot$, $r_n = 25$ pc, $\rho_{an} = 0.02 \, m_\odot/pc^3$ and $V = 10$ km/s for GMCs, and $m_n = 3 \cdot 10^6 \, m_\odot$, $r_n = 0$, $\rho_{an} = 0.006 \, m_\odot/pc^3$ and $V = 250$ km/s for MBHs (see Wielen, 1985).

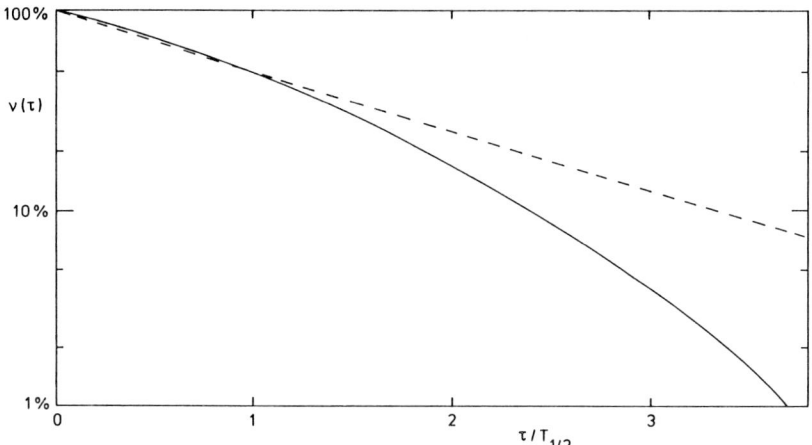

Fig. 5. Survival rate ν of clusters with the same initial values of N, m_c and r_c as a function of the age τ (scaled by the median value $T_{1/2}$ of the dissolution times). Disruption of clusters due to GMCs (or MBHs) in a case with $p_1 \gg p_0$. Evaporation due to internal relaxation neglected here.

If a type of clusters may be destroyed completely by a single encounter with a passing object, i.e. if $p_1 > p_0$, then there is a wide spread in the actual lifetimes of individual clusters, caused by the stochastic occurence of such encounters. Hence the dissolution time T_n derived above has to be interpreted as an expectation value (mean value) for the lifetime only. Kuhrau and Wielen (unpublished) have carried out Monte-Carlo simulations on the disruption of clusters due to passing objects. A typical result of these studies is shown in Fig. 5. The full curve gives the survival rate $\nu(\tau)$, i.e. the percentage of clusters which live longer than τ, for a case with $p_1 \gg p_0$. All clusters had initially the same total mass and radius. The initial decline of $\nu(\tau)$ is roughly exponential (dashed line), as expected for random close encounters. For larger ages τ, the accumulating number of distant encounters causes a steeper decline of ν.

2.3 Total Dissolution Times

In order to derive the dissolution time T_{total}, due to the combined effect of internal relaxation and passing objects, we add the corresponding escape rates, $1/T$:

$$T_{total}^{-1} = T_{N-body}^{-1} + T_{GMC}^{-1} + T_{MBH}^{-1} \ . \qquad (15)$$

Here, T_{N-body} is to be taken from Section 2.1, while T_{GMC} and T_{MBH} have to be derived from Section 2.3. In Figs. 3 and 4, we have not

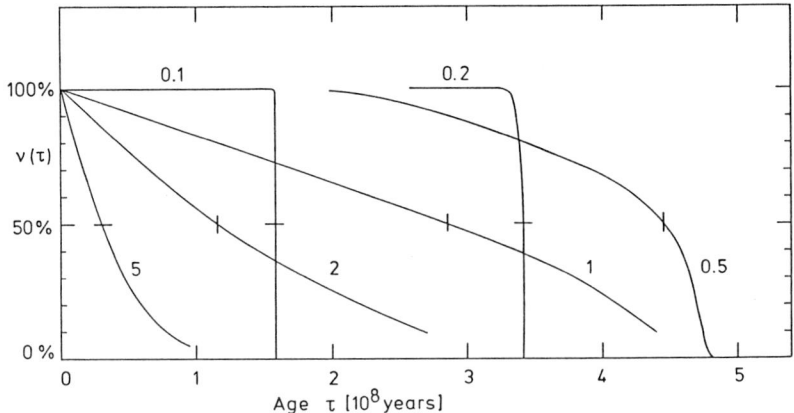

Fig. 6. Survival rate ν of clusters with the same initial values of $m_c = 250\ m_\odot$, $N = 500$ stars and various values of r_c (in pc) as a function of age τ. Dissolution of clusters due to internal relaxation, galactic tidal field and GMCs.

plotted the total dissolution time, since this would have obscured the picture. An example of a curve for the total dissolution time is shown in Fig. 1 of Wielen (1985). At the 'crossing points', the total dissolution time is half of $T_{N-body} = T_{GMC}$, while for small (large) radii, T_{total} approaches T_{N-body} (T_{GMC}).

As discussed in Section 2.2, close encounters with passing massive objects produce a wide distribution of actual lifetimes, even for clusters with the same initial values of m_c, r_c and r_t. To the contrary, the internal relaxation leads to rather well-defined dissolution times. The width of the distribution function of total lifetimes depends therefore on the relative importance of internal relaxation and close encounters with GMCs or MBHs. An example for the survival rate $\nu(\tau)$ of clusters with $m_c = 250\ m_\odot$, $N = 500$ stars and various values of r_c is shown in Fig. 6 (Kuhrau and Wielen, unpublished).

3. STAR CLUSTERS IN THE LARGE MAGELLANIC CLOUD (LMC)

We will now apply the procedure presented in Section 2 to star clusters in the LMC. For deriving the tidal time T_t (Eq.7) as a function of the distance R from the center of the LMC, we need the run of $4\omega^2-\kappa^2$ based on the rotation curve of the LMC. We use the results of Elson and Freemann (1986). The tidal time in the LMC increases from $0.4 \cdot 10^6$ years at $R \sim 1°$ to $1.1 \cdot 10^6$ years at $R \sim 6°$. At $R \sim 3°$, the tidal time in the LMC is just about equal to the value in the solar neighbourhood, $0.6 \cdot 10^6$ years, so that Fig. 2 can be directly used for LMC clusters at $R \sim 3°$.

If we consider only the dissolution of clusters due to internal relaxation combined with the LMC tidal field, then all clusters with median radii r_c larger than about 1 pc and with total masses m_c above 7500 $m_\odot$ at R $\sim$ 3°, above 9000 $m_\odot$ at R $\sim$ 1°, and above 4000 $m_\odot$ at R $\sim$ 6° should have survived over $15 \cdot 10^9$ years. Clusters with smaller total masses would live shorter, in direct proportion to their value of m_c.

The overall density of molecular gas in the LMC is still rather uncertain (Israel, 1984, Cohen et al., 1984). Hence it is essentially impossible to estimate at present the effect of molecular clouds on LMC clusters. Since the molecular gas seems to be less abundant in the LMC than in our Galaxy, its effect on LMC clusters is correspondingly smaller than for galactic clusters. For LMC clusters with large median radii, however, large molecular clouds may still be the dominant source of cluster dissolution.

Elson and Fall (1985) have derived the age distribution for rich star clusters in the LMC from observational data. They find a decrease of the number of clusters with increasing age similar to that of open clusters in our Galaxy (Wielen, 1971) for ages up to $3 \cdot 10^8$ years, and a flatter tail for older LMC clusters. The indicated short median dissolution time of about $2 \cdot 10^8$ years for these rich LMC clusters with $m_c > 10^3$ $m_\odot$ would require strong effects from passing massive objects such as giant molecular clouds. However, due to the various assumptions which enter into the derivation of the 'observed' age distribution of LMC clusters, this conclusion is quite uncertain at present. Mateo (1986) has presented an age distribution for less massive LMC clusters which differs strongly from the results of Elson and Fall. Mateo's data indicate a median dissolution time of about $3 \cdot 10^9$ years for a constant formation rate. The discrepancy may be, at least partially, due to the different locations of the cluster samples (typically R $\sim$ 2° for Elson and Fall, and R $\sim$ 6° for Mateo). Mateo's results can be explained, if the effect of GMCs are negligible in the LMC at R $\sim$ 6°. For this distance, we would then predict a dissolution time of $2 \cdot 10^9$ years for a cluster with $m_c = 500$ $m_\odot$ and $r_c > 1$ pc.

4. THE EFFECT OF MASSIVE BLACK HOLES ON GLOBULAR CLUSTERS

Massive black holes (MBHs) have been proposed by Lacey and Ostriker (1985) and Ipser and Semenzato (1985) as the major constituents of dark galactic coronae. These MBHs would nicely explain the increase in the velocity dispersion of disk stars as a function of age as derived by Wielen (1977). If the MBHs really exist, they would be extremely efficient in dissolving globular clusters.

In Table 1, we give the dissolution times for globular clusters of various total masses and median radii, at the solar distance from the galactic center, due to MBHs. For the MBHs, we use the same

Table I
Disruption time T_n (in 10^9 years) for globular clusters due to massive black holes

mass m_c (in $m_\odot$) of the cluster	median radius r_c (in pc) of the cluster				
	1	2	5	10	20
$2 \cdot 10^6$	**188**	69	28	14	6.9
$1 \cdot 10^6$	**133**	**47**	14	6.9	3.5
$5 \cdot 10^5$	**94**	**33**	**8.4**	3.5	1.7
$2 \cdot 10^5$	**59**	**21**	**5.3**	**1.9**	**0.7**
$1 \cdot 10^5$	**42**	**15**	**3.8**	**1.3**	**0.5**
$5 \cdot 10^4$	**30**	**11**	**2.7**	**0.9**	**0.3**
$2 \cdot 10^4$	**19**	**6.6**	**1.7**	**0.6**	**0.2**
$1 \cdot 10^4$	**13**	**4.7**	**1.2**	**0.4**	**0.1**

bold face : $T_{n,1}$ ($p_1 > r_c$)
normal : $T_{n,0}$ ($p_1 < r_c$)

Table II
Survival rate ν (in %) for globular clusters of age $\tau = 15 \cdot 10^9$ years in the presence of massive black holes

mass m_c (in $m_\odot$) of the cluster	median radius r_c (in pc) of the cluster				
	1	2	5	10	20
$2 \cdot 10^6$	96	89	(71)	((47))	((19))
$1 \cdot 10^6$	94	83	47	(19)	((1))
$5 \cdot 10^5$	92	75	24	1	0
$2 \cdot 10^5$	87	62	8	0	0
$1 \cdot 10^5$	80	50	1	0	0
$5 \cdot 10^4$	73	36	0	0	0
$2 \cdot 10^4$	59	15	0	0	0
$1 \cdot 10^4$	44	5	0	0	0

parameters as in Section 2.2, namely $m_n = 3 \cdot 10^6$ $m_\odot$ and $\rho_{an} = 0.006$ $m_\odot/pc^3$. For the relative velocity between MBHs and globular clusters, we assume $V = 2^{1/2} \cdot 220$ km/s = 311 km/s. A typical globular cluster with $m_c = 2 \cdot 10^5$ $m_\odot$ and $r_c = 5$ pc has a mean lifetime T_n of only $5 \cdot 10^9$ years, i.e. about one third of the Hubble time. Clusters with larger radii or smaller masses are even more vulnerable by MBHs.

The dissolution times of globular clusters are inversely proportional to the overall mass density ρ_{an} of the MBHs in the corona. Therefore, the dissolution time T_n of a globular cluster depends strongly on its (mean) galactocentric distance R, T_n being shorter in the inner parts of the Galaxy and longer in the outer regions. The decrease of ρ_{an} at larger distances follows the relation $\rho_{an} \propto R^{-2}$ for a flat rotation curve. This would imply $T_n \propto R^2$. The increase of ρ_{an} towards the galactic center is not well-known quantitatively. According to Lacey and Ostriker (1985), the central density ρ_{an} of MBHs may be higher than their local density by a factor of 2 or more. The dissolution times of globular clusters due to MBHs in the central region of the Galaxy are then correspondingly shorter than the values given in Table I.

It may seem at first that dissolution times significantly shorter than the Hubble time would be a strong argument against the existence of MBHs, since the observed globular clusters have ages of about $15 \cdot 10^9$ years. One should remember, however, that the disruption of globular clusters by MBHs is a highly stochastic process, leading to a wide spread in the lifetimes of clusters (Section 2.2). We have therefore derived the percentage ν of globular clusters which survive over a period of $15 \cdot 10^9$ years as a function of m_c and r_c. The survival rates ν presented in Table 2 are based on the dissolution times given in Table 1, and on a relative survival function similar to that shown in Fig. 5. From Table 2, we see, for example, that 8 % of globular clusters with an initial total mass of $m_c = 2 \cdot 10^5$ $m_\odot$ and a median radius of $r_c = 5$ pc would survive over $15 \cdot 10^9$ years at the solar distance R_o. This means that in order to explain the presently observed number of such clusters, the initial population of clusters of that type had to be larger than now by a factor of $1/0.08 = 12.5$. The general conclusion is that if the proposed massive black holes do exist, the initial number of globular clusters was much higher than now. It may even be that all halo field stars were formerly members of now dissolved clusters.

A possible argument against the existence of MBHs would come from a universal distribution function for the total luminosities of globular clusters. It is clear from Tables I and II that MBHs would strongly modify any initial distribution of total masses of globular clusters and would probably even determine the location of the peak in the luminosity function of globular clusters and the shape of this function for low cluster luminosities. In galaxies like our one, the dependence of T_n on R would produce a strong radial change in the luminosity function, which does not seem to be observed. From galaxy to galaxy, one would expect a variation in the location of the maximum of the luminosity function of globular clusters, if there is not a kind of universal coupling between the distribution of globular clusters and the density distribution of MBHs in dark coronae of galaxies.

REFERENCES

Aarseth, S. J. 1985 in Multiple Time Scales (Computational Techniques), J. U. Brackbill and B. I. Cohen, eds., Academic Press, Orlando, p. XXX.
Aarseth, S. J. and Lecar, M. 1975 Ann. Rev. Astron. Astrophys. 13, 1.
Aarseth, S. J., Henon, M. and Wielen, R. 1974 Astron. Astrophys. 37, 183.
Caldwell, J. A. R. and Ostriker, J. P. 1981 Astrophys. J. 251, 61.
Cohen, R., Montani, J. and Rubio, M. 1984 in IAU Symposium No. 108, Structure and Evolution of the Magellanic Clouds S. van den Bergh and K. S. de Boer, eds., D. Reidel Publ. Co., Dordrecht, p. 401.
Elson, R. A. W. and Fall, S. M. 1985 Astrophys. J. 299, 211.
Elson, R. A. W. and Freeman, K. C. 1986 preprint.
Henon, M. 1960 Ann. Astrophys. 23, 668.
Henon, M. 1969 Astron. Astrophys. 2, 151.
Ipser, J. R. and Semenzato, R. 1985 Astron. Astrophys. 149, 408.
Israel, F. P. 1984 in IAU Symposium No. 108, Structure and Evolution of the Magellanic Clouds S. van den Bergh and K. S. de Boer, eds., D. Reidel Publ. Co., Dordrecht, p. 319.
Lacey, C. G. and Ostriker, J. P. 1985 Astrophys. J. 299, 633.
Mateo, M. 1987 in IAU Symposium No. 126, Globular Cluster Systems in Galaxies, J. E. Grindlay and A. G. D. Philip, eds., Reidel, Dordrecht, p. 557.
Spitzer, L. 1958 Astrophys. J. 127, 17.
Terlevich, E. 1980 in IAU Symposium No. 85, Star Clusters, J. E. Hesser, ed., D. Reidel Publ. Co., Dordrecht, p. 165.
Terlevich, E. 1983 Ph.D. Dissertation, Univ. Cambridge, England.
Terlevich, E. 1986 Preprint.
Wielen, R. 1971 Astron. Astrophys. 13, 309.
Wielen, R. 1974 in Proceedings of the First European Astronomical Meeting, Vol. 2, Stars and the Milky Way System L. N. Mavridis, ed., Springer-Verlag, Berlin, p. 326.
Wielen, R. 1975 in IAU Symposium No. 69, Dynamics of Stellar Systems A. Hayli, ed., D. Reidel Publ. Co., Dordrecht, p. 119.
Wielen, R. 1977 Astron. Astrophys. 60, 263.
Wielen, R. 1985 in IAU Symposium No. 113, Dynamics of Star Clusters J. Goodman and P. Hut, eds., D. Reidel Publ. Co., Dordrecht, p. 449.

Note added in proof: Elson and Freeman 1986 should be replaced by Elson, R. A. W., Fall, S. M. and Freeman, K. C. 1987, preprint.

DISCUSSION

INAGAKI: Does the disruption time become significantly shorter if you take into account the post-collapse expansion?

WIELEN: The N-body simulations of cluster dissolution cover in many cases the 'post-collapse' phase. There is no indication for a significant change of the escape rate with time. In the case of an encounter between a cluster and a passing massive object, the actual radius of the cluster at that time determines the fate of the cluster. This effective radius of the cluster may be, of course, affected by post-collapse expansion.

SCHOMMER: Two comments on your LMC comparison. It would be relatively easy for someone to measure tidal radii, using CCD frames from small telescopes, of many LMC clusters. Secondly, while I find the Elson and Fall analysis impressive, I am worried about their sample completeness. In particular, they need to correct the Van den Bergh compilation by counting from Hodge Atlas Plates, by factors of 4-5 for the faintest, oldest clusters, precisely those with $t > 10^9$ yrs, which lie away from your galactic line; these correction factors may vary from region to region, which could easily bring their age distribution up to that of Mateo.

COHEN: What effect do those proposed massive black holes have on the velocity dispersion of halo stars?

WIELEN: The relative increase in the velocity dispersion of halo stars, due to the proposed massive black holes, is rather small at the solar distance. To get a first guess, one may just add quadratically the velocity dispersion of the oldest disk stars (~ 80 km/s) to the total velocity dispersion of halo stars (say 270 km/s). This leads to about 280 km/s, or an increase by 4%. In the inner part of the Galaxy, the increase in the velocity dispersion of halo stars by black holes is larger, but probably still not overwhelming, depending mainly on the assumed increase in the number density of the black holes towards the galactic center.

BHATIA: What is the effect of the globular/open clusters on the molecular clouds?

WIELEN: The passing clusters will increase the internal energy of the molecular clouds by tidal heating. Since most open clusters have much smaller masses than giant molecular clouds, the effect is probably small in most cases.

INNANEN: There is a relative absence of globular clusters between R ~ 20 kpc and R ~ 70 kpc. Could the proposed black holes selectively destroy clusters that once might have been there?

WIELEN: Probably not. The disruption time for a given type of globular clusters is mainly proportional to the overall density of the black holes. This density is probably decreasing monotonically with increasing galactocentric distance R. There is also no indication that the initial masses or radii of globular clusters vary in a dramatical way with R.

KING: Can you estimate how much your results would change if it turned out that in the richer clusters the dominant escapes are due to Fokker-Planck diffusion?

WIELEN: In the case of a Fokker-Planck diffusion, the evaporation time of a cluster would increase only as $N/\log_{10}(0.4N)$ instead of N. For example, extrapolating from $N = 500$ to 5000 stars, the resulting evaporation time for $N = 5000$ would have to be decreased by a factor of $\log(0.4 \times 500)/\log(0.4 \times 5000) = 0.7$ relative to my results based on escapers by close encounters. Even for a globular cluster with $N = 10^6$ stars, the reduction factor would be only about 0.4.

Chapter V

Review Papers

Globular Clusters

as Tracers

and HST

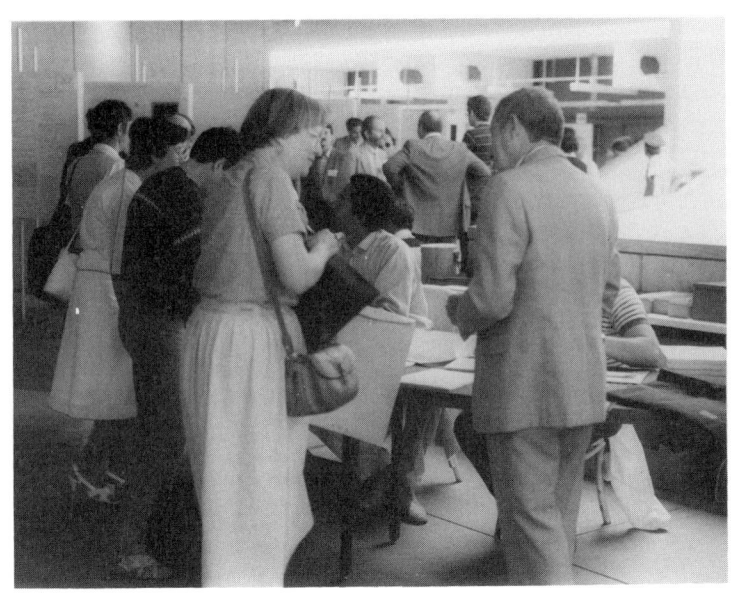

Martha Hazen and Morton Roberts in discussion

The crowd awaits the opening remarks

INTERSTELLAR MATTER IN GLOBULAR CLUSTERS

Morton S. Roberts

National Radio Astronomy Observatory

> *"Is there any point to which you would wish to draw my attention?"*
> *"To the curious incident of the dog in the night-time."*
> *"The dog did nothing in the night-time."*
> *"That was the curious incident,"* remarked Sherlock Holmes.
>
> Memoirs of Sherlock Holmes

I. INTRODUCTION

The source for intracluster matter is seen in various mass loss processes ongoing within clusters and is supported by the theoretical need for mass loss to explain the morphology of cluster color-magnitude diagrams. A variety of techniques ranging from X-ray to radio wavelengths have been employed to search for such matter but with few exceptions has not been found. The amount of material expected to collect between cleansing passages through the galactic plane has variously been estimated at between $\sim 10^2$ and $\sim 10^3$ $M_\odot$. In contrast, observed upper limits for many clusters are well below these values, often < 1 $M_\odot$. The few detections are at levels of $\lesssim 10^{-2}$ $M_\odot$.

This may not be a dilemma for there are a number of proposed mechanisms which will remove diffuse matter from a cluster. Such proposals invoke various assumptions whose acceptance can, in some instances, be guided by observation. Specifically the range of cluster escape velocities and galactocentric distances must be more fully observed before the conflict between prediction and observation is fully resolved.

2. MASS LOSS - PREDICTED AND OBSERVED

The need for mass loss by evolving low mass stars follows most simply from their mass at turn-off, ~ 0.85 $M_\odot$, and the observed mean mass of (field) white dwarfs of ~ 0.55 $M_\odot$. A similar value derives from a detailed modeling of the evolution of Population II stars which requires ~0.2 $M_\odot$ loss prior to the horizontal branch phase as well as an additional ~0.1 $M_\odot$ loss in the asymptotic giant branch phase (see, e.g., Renzini 1979 and references therein). For a population of 10^3 stars past turn-off, we expect ~300 $M_\odot$ of material to be fed into intracluster space; this within a period of a few 10^8 years. More quantitative calculations of the cumulative mass loss between passages through the galactic plane give values of 10^2-10^3 $M_\odot$ (e.g., Roberts 1960; Hills and Klein 1973; Knapp, Rose and Kerr 1973; Taylor and Wood 1975). Intracluster matter of this amount in the form of atomic hydrogen, either neutral or ionized, or in molecular form, is easily detectable.

The presence of at least one planetary nebula and two, possibly three, novae within globular clusters tells us that at least some mass loss is ongoing. A more significant source is indicated by the spectra of red giant stars where $H\alpha$ emission wings are often seen adjacent to the normal $H\alpha$ absorption line (Cohen 1976; Mallia and Pagel 1978, Peterson 1981; Gratton 1983; Cacciari and Freeman 1983). The emission appears to the red of the absorption line, as in P Cygni stars, to the blue, and in both locations; the last case is reminiscent of Type III P Cygni stars (Beals 1950). In some instances the $H\alpha$ emission is seen to vary on a time scale of days (Cacciari and Freeman 1983). Measured with respect to the absorption line, the emission features have velocities that are typically ± 50 km s^{-1} and reach as high as ± 80 km s^{-1}.

Cohen (1976) as well as all of the other observers noted above propose that these $H\alpha$ features arise from a circumstellar shell due to mass loss from the parent star. She derives a minimum mass loss rate ~2×10^{-9} $M_\odot$ yr^{-1} which in the lifetime of a red giant, ~10^8 yrs, will result in a total loss of 0.2 $M_\odot$. Others, e.g., Mallia and Pagel (1978), derive significantly higher mass loss rates, up to 6×10^{-8} $M_\odot$/yr. The difference is related to the adoption of substantially larger radii for the parent stars; values derived from luminosities and effective temperatures. Clearly this latter, larger value of mass loss cannot be continuous over the entire red giant phase. The variability of the $H\alpha$ emission and its apparent dependence on luminosity (e.g., Cacciari and Freeman 1983) support this conclusion. The detailed dependence is unknown but a total mass loss of order 0.2 $M_\odot$ appears reasonable.

Peterson (1981) has shown that the NaD and $H\alpha$ lines in several, but not all, globular cluster giants that she studied at high resolution show a core shift to the blue; 11 km s^{-1} for NaD and 6 km s^{-1} for $H\alpha$. Others have also reported a blue core shift for $H\alpha$ in some giants.

The data on red giants are extensive; about two hundred have been observed with over a third showing $H\alpha$ emission features. However, the proposal that these features are indicative of a circumstellar shell produced by mass loss is not universally accepted. Several concerns have been raised (Reimers 1981, Dupree et al. 1984, Jura 1986). These include:

- (1) The appearance in some instances of only a blue wing.
- (2) The short time scale of Hα variability.
- (3) The large range in Hα emission for stars of similar L and T_e.
- (4) The (unknown) mechanism of mass loss.

Dupree et al. (1984), using semiempirical atmospheric models, find Hα emission wings arising from static chromospheres. An extended atmosphere model yielding a mass loss much less than 2×10^{-9} $M_\odot$ yr^{-1} also yields line parameters similar in many respects to those observed. They propose observational tests for this chromospheric model and also note that their models do not rule out massive cool winds or transient events that could give rise to substantial mass loss.

As with variants of P-Cygni type stars and spectral-line variations in Be stars, we find the picture for globular cluster red giants complex. The details are not understood but all three categories of stars do show a signature of mass loss.

A further sign of mass loss is suggested by the different spatial distributions of red giants and horizontal branch stars (Oort and van Herk 1959; Woolf 1964); with the latter appearing more extensive. However, the time available, $\sim 10^8$ years, for the apparent redistribution of mass appears short relative to the expected relaxation time (King 1972).

3. REMOVAL OF INTERSTELLAR MATTER FROM GLOBULAR CLUSTERS

The discrepancy between the predicted amount of intracluster matter and the low upper limits derived from observation has prompted a number of viable explanations. These are separable into two categories: "intrinsic," the cleansing mechanism is internal to and part of the cluster, and extrinsic. A basic requirement for both follows from the steady-state nature of the production of intracluster matter: the removal mechanism must clearly also be so.

Several of the intrinsic models involve cluster winds. Scott and Rose (1975) propose that the wind energy arise from stellar ultraviolet radiation. The authors point out that for massive, concentrated clusters (i.e., large, central escape velocity, V_{esc}) such as NGC 6388, a large, significant amount of gas could accumulate in the center. Observations of such high V_{esc} clusters yield the conflicting result of upper limits of ionized hydrogen below those predicted.

Wind models invoking high gas input velocities are proposed by Burke (1968), Faulkner and Freeman (1977), and VandenBerg and Faulkner (1977). Their calculations are consistent with observations but require injection velocities $\gtrsim 100$ km s^{-1} for extreme cases. Such velocities are significantly greater than red giant emission features imply but not the (low $\dot{M}$) solar wind velocities. VandenBerg (1978) combines both a UV field and a high initial input velocity.

Other intrinsic suggestions include nova-driven winds (Scott and Durisen 1978), winds driven by flare-stars (Coleman and Worden 1977), condensation into stars (Roberts 1960, Johnson 1975), and accretion processes drawing upon a central gas reservoir (e.g., Faulkner and Coleman 1984).

The only extrinsic mechanism proposed (Frank and Gisler 1976, Lea and De Young 1976) invokes continuous sweeping of clusters by the gaseous medium of the galactic halo. For halo densities of 10^{-3} atoms cm^{-3} and relative velocities of $> 10^2$ km s^{-1} such stripping is effective even for large V_{esc}.

All of the methods proposed are strongly dependent on either the assumptions invoked or on poorly known parameters, a situation clearly stated by many of the authors referenced above. Tests of these proposals are possible (e.g., Frank and Gisler 1976, Faulkner and Freeman 1977) and several are made below. A convincing case for any mechanism remains to be made.

4. THE SEARCH FOR INTERSTELLAR MATTER WITHIN GLOBULAR CLUSTERS

There are over two dozen observational searches for intracluster gas and dust; a bibliographic listing is given in Table 1. These cover neutral hydrogen via 21 cm; ionized hydrogen using Fabry-Perot techniques, slit spectra, narrow-band photometry and radio continuum measurements for free-free transitions; searches for distributed OH and for OH and H$_2$O masers; searches for infrared radiation and for extended X-ray sources. With few exceptions (see Table III) all yield negative results. Upper limits in some instances are less than a solar mass of neutral or ionized hydrogen, M_{gas}. For M15, Conklin (1986) derives an upper limit of 0.1 $M_\odot$ for neutral hydrogen after a 7-hour integration with the Arecibo telescope. Clusters with $1 \leq M_{gas}/M_\odot < 5$ and $M_{gas}/M_\odot < 1$ are listed in Table II.

Such limits raise the question as to whether the best globular clusters have been observed for testing removal mechanisms. Specifically clusters with high central escape velocities are best suited to retain material if an intrinsic mechanism is the correct explanation while large galactocentric distances as well as high escape velocity would favor retention in the extrinsic case.

Data on central escape velocity and galactocentric distance for close to 150 globular clusters are tabulated by Webbink (1985) and displayed in Figure 1 as open histograms. Also shown are the distributions in these two parameters of those clusters which have upper limits noted in Table II.

Table I.
Bibliography on Observational Searches
for Diffuse Matter in Globular Clusters

HI - 21 cm

1. 1959   M. S. Roberts, Nature, 184, 1555.
2. 1964   S. J. Goldstein, Jr., Astrophys. J., 140, 802.
3. 1966   C. Heiles and R. C. Henry, Astrophys. J., 146, 953.
4. 1967   B. J. Robinson, Astrophys. Letts., 1, 21.
5. 1972   F. J. Kerr and G. R. Knapp, Astron. J., 77, 573.
6. 1973   G. R. Knapp, W. K. Rose, and F. J. Kerr, Astrophys. J., 186, 831.
7. 1974   E. K. Conklin and R. A. Kimble, Bull. Amer. Astron. Soc., 6, 468.
8. 1975   E. K. Conklin, private communication in 1986.
9. 1979   P. F. Bowers, et al., Astrophys. J., 233, 553.
10. 1983 M. Birkinshaw, P. T. P. Ho, and B. Baud, Astron. Astrophys., 125, 271.

HII - Hα

11. 1976 M. G. Smith, J. E. Hesser, and S. J. Shawl, Astrophys. J., 206, 66.
12. 1977 D. J. Faulkner and K. C. Freeman, Astrophys. J., 211, 77.
13. 1977 J. E. Grindlay and W. Liller, Astrophys. J. (Letters), 216, L105.
14. 1977 J. E. Hesser and S. J. Shawl, Astrophys. J. (Letters), 217, L143.

HII - Radio continuum radiation (free-free)

15. 1973 J. G. Hill and M. J. Klein, Astrophys. Letts., 13, 65.
16. 1975 J. W. Erkes and A. G. Davis Philips, Astrophys. J., 197, 533.
17. 1976 M. J. Klein, Astrophys. Letts., 18, 25.

Molecules (and masers): CO, OH, H_2O - radio

18. 1973 G. R. Knapp and F. J. Kerr, Astron. J., 78, 458.
19. 1978 T. H. Troland, J. E. Hesser, and C. Heiles, Astrophys. J., 219, 873.
20. 1978 M. H. Schneps, et al., Astrophys. J., 225, 808.
21. 1979 N. L. Cohen and M. A. Malkan, Astron. J., 84, 74.
22. 1980 J. M. Dickey and M. A. Malkan, Astron. J., 85, 145.
          Also see No. 9 above.

Infra-red observations

23. 1973 A. D. MacGregor, J. P. Phillips, and M. J. Selby, M.N.R.A.S., 164, 31P.
24. 1975 O. L. Hansen and J. E. Hesser, Nature, 257, 568.
25. 1984 F. C. Gillett, et al., Bull. Amer. Ast. Soc., 16, 526.
26. 1984 F. C. Gillett, et al., Bull. Amer. Ast. Soc., 16, 948.

Extended X-ray

27. 1982 F.D.A. Hartwick, A. P. Cowley, and J. E. Grindlay, Astrophys. J. (Letters), 254, L11.
28. 1985 J. E. Grindlay in Dynamics of Star Clusters, ed. J. Goodman and P. Hut, p. 43.

Dust (references after 1958)

29. 1959 G. M. Idlis and G. M. Nikol'skii, Soviet Astron. (AJ), 3, 652 (1960).
30. 1959 H. S. Hogg, Astron. J., 64, 425.
31. 1960 P. A. Hodge, Publ. Ast. Soc. Pac., 72, 308.
32. 1960 M. S. Roberts, Astron. J., 65, 457.
33. 1978 S. P. Kanagy and S. P. Wyatt, Astron. J., 83, 779.
34. 1981 P. G. Martin and S. J. Shawl, Astrophys. J., 251, 108.

Table II.
Clusters with Low Upper Limits of Interstellar Gas

(a) Ionized Hydrogen from Slit and Fabry-Perot Spectroscopy

$1 \leq M_{HII} \leq 5$		$M_{HII} < 1$	
NGC 104	47 Tuc	NGC 1851	
NGC 362		NGC 5824	
NGC 2808		NGC 5904	M5
NGC 3201		NGC 6093	M80
NGC 4147		NGC 6388	
NGC 5286		NGC 6397	
NGC 5694		NGC 6441	
NGC 5927		NGC 6522	
NGC 6121	M4	NGC 6541	
NGC 6218	M12	NGC 6624	
NGC 6266	M62	NGC 6681	M70
NGC 6333	M9	NGC 6864	M75
NGC 6715	M54	NGC 7078	M15
NGC 6752		NGC 7089	M9
		NGC 7099	M30

(b) Neutral Atomic Hydrogen from 21-cm Spectroscopy

$1 \leq M_{HI} \leq 5$		$M_{HI} < 1$	
NGC 6341	M92	NGC 5272	M3
NGC 6712		NGC 5904	M5
NGC 6809	M55	NGC 6121	M4
		NGC 6205	M13
		NGC 6656	M22
		NGC 7078	M15

See Table I for references. HI data use Webbink (1985) distances.
Units: Solar masses.

Table III.
Reported Detections of Diffuse Matter in Globular Clusters

GAS			
1.	NGC 1851		Hα detection via narrow band photometry. Inferred mass of ionized hydrogen ~0.02 $M_\odot$. Ref.: Table I, No. 13.
	NGC 5286		
	NGC 5824		
	NGC 6266	M62	
	NGC 6441		
	NGC 6624		
	NGC 7078	M15	
2.	NGC 104	47 Tuc	Extended X-ray emission. Ref.: Table I, No. 27, 28.
	NGC 5139	ω Cen	
	NGC 6656	M22	
DUST			
1.	NGC 5272	M3	Multicolor photometry of dark regions Avg. cloud ~0.003 $M_\odot$. Ref.: Table I, No. 33.
	NGC 6205	M13	
	NGC 7078	M15	
2.	NGC 7078	M15	Tentative detection of polarization in dark region 0.22% ± 0.06%. Ref.: Table I, No. 34.
3.	NGC 104	47 Tuc	IRAS observations. Extended emission at 12, 25, 60 μ · ~ 10^{-6} $M_\odot$ dust in central region. Ref.: Table I, No. 25.

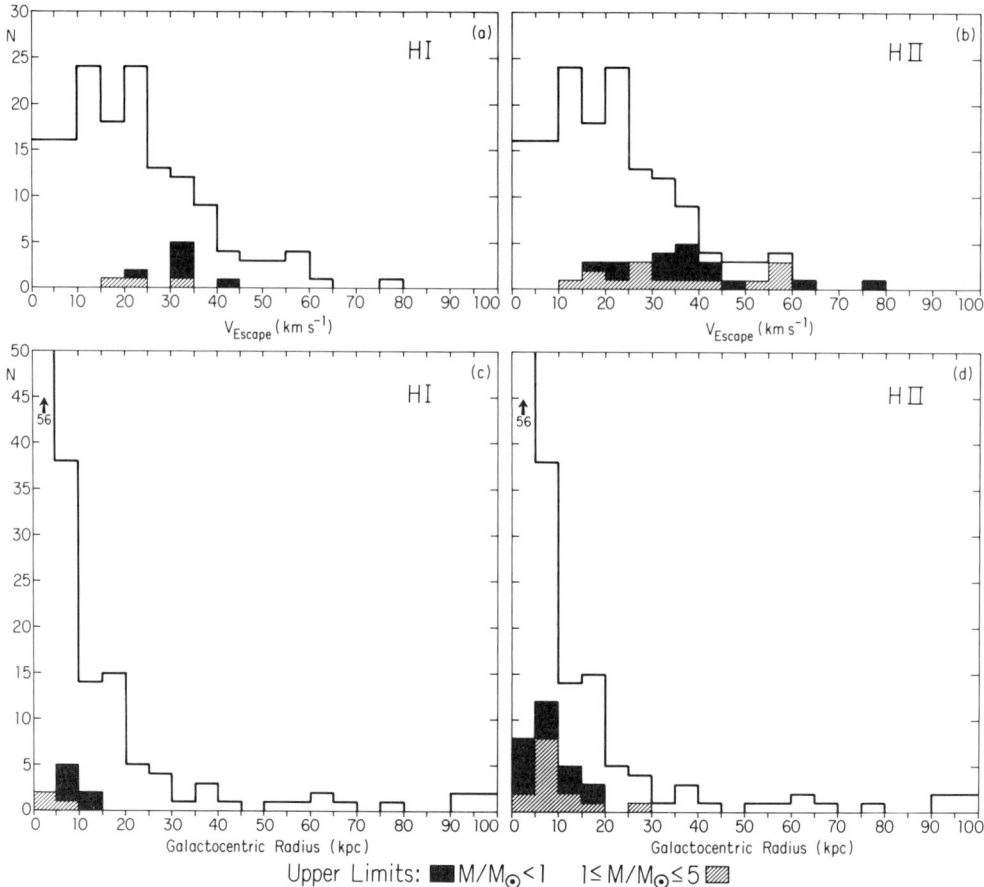

Fig. 1. Open histograms display the distribution of central escape velocities, V_{esc}, and galactocentric distances for globular clusters as tabulated by Webbink (1985). Panel (a) shows as shaded boxes those clusters with low upper limits of neutral hydrogen, HI, while Panel (b) is for ionized hydrogen, HII. Panels (c) and (d) repeat these data as function of galactocentric radius.

Except for HII measurements which do cover the highest V_{esc} values known, the observational data for both HI and HII poorly sample V_{esc} and galactocentric radius. Measurements of extreme cases of these parameters are needed and would effectively limit the various models proposed. Other tests are possible, e.g., measurement of the ultraviolet flux of those clusters with particularly stringent upper limits; further X-ray measurements, etc.

The latter example is particularly important in light of the extended X-ray sources reported by Hartwick et al. (1982) for 47 Tuc, ωCen and M22. A possible interpretation has the X-rays originating from a bow shock in the interaction between a diffuse intracluster medium and the hot gaseous galactic halo (Grindlay 1985).

Dark regions, presumably dust concentrations, are seen towards a number of globular clusters (Idlis and Nikolski 1959, Hogg 1959, Roberts 1960). A drawing of M13 by the Earl of Rosse (1861) based on observations with his 6-foot telescope shows one of the best such examples, a feature confirmed many times over by photographic studies. Monte Carlo simulations (Roberts 1960) and an approximate inverse wavelength dependence of the extinction within these features (Kanagy and Wyatt 1978) rule out as an explanation the chance arrangement of stars within these clusters. One such dark region in M15 shows possible polarization, 0.22% ± 0.06% (Martin and Shawl 1981). Whether the obscuring matter is in or foreground to the cluster is less certain. Either case is interesting.

5. CONCLUSIONS

The low limits, often < 1 $M_\odot$, on the diffuse matter within globular clusters require an efficient, steady-state cleansing mechanism for removal. Several mechanisms have been proposed. The existence of extended X-ray sources in the three clusters studied for such sources favors sweeping by the hot gaseous galactic halo. Observations of clusters at large galactocentric distances and with large escape velocities will aid in testing this and other proposed mechanisms. The explanation for the (mostly) missing diffuse matter in globular clusters will open new understanding to the evolution of these clusters.

I am particularly pleased to acknowledge aid in the preparation of this review from W. P. Bidelman, J. E. Bregman, E. K. Conklin, P. S. Conti, D. E. Hogg, I. R. King, and R. T. Rood. The National Radio Astronomy Observatory is operated by Associated Universities, Inc., under contract with the National Science Foundation.

REFERENCES

Beals, C. S. 1950 Publ. Dominion Astrophys. Ob. **9**, 1.
Burke, J. N. 1968 Monthly Notices Roy. Astron. Soc. **140**, 241.
Cacciari, C. and Freeman, K. C. 1983 Astrophys. J. **268**, 185.
Cohen, J. C. 1976 Astrophys. J. **203**, L127.
Coleman, G. D. and Worder, S. P. 1977 Astrophys. J. **218**, 792.
Conklin, E. K. 1986 private communication.
Dupree, A. K., Harman, L. and Avrett, E. H. 1984 Astrophys. J. Letters **281**, L37.
Faulkner, D. J. and Freeman, K. C. 1977 Astrophys. J. **211**, 77.
Faulkner, D. J. and Coleman, C. S. 1984 Monthly Notices Roy. Astron. Soc. **206**, 121.
Frank, J. and Gisler, G. 1976 Monthly Notices Roy. Astron. Soc. **176**, 633.
Gratton, R. G. 1983 Astrophys. J. **264**, 223.
Grindlay, J. E. 1985 in IAU Symposium No. 113, Dynamics of Star Clusters J. Goodman and P. Hut, eds., D. Reidel, Dordrecht, p. 43.
Hartwick, F. D. A., Cowley, A. P. and Grindlay, J. E. 1982 Astrophys. J. Letters **254**, L11.
Hills, J. G. and Klein, M. J. 1973 Astrophys. Letters **13**, L65.
Hogg, H. S. 1959 Astron. J. **64**, 425.
Idlis, G. M. and Nikol'skii 1960 Soviet Astron. J. **3**, 652.
Johnson, H. M. 1975 Astrophys. J. **202**, 490.
Jura, M. 1986 Astrophys. J. **301**, 624.
Kanagy, S. P. and Wyatt, S. P. 1978 Astron. J. **83**, 779.
King, I. R. 1972 in Evolution of Population II Stars A. G. D. Philip, ed., Dudley Obs., Albany, p. 31.
Knapp, G. R., Rose, K. W. and Kerr, F. J. 1973 Astrophys. J. **186**, 831.
Lea, S. M. and De Young, D. S. 1976 Astrophys. J. **210**, 647.
Mallia, E. A. and Pagel, B. E. J. 1978 Monthly Notices Roy. Astron. Soc. **184**, 55P.
Martin, P. G. and Shawl, S. J. 1982 Astrophys. J. **251**, 108.
Oort, J. H. and van Herk, G. 1959 Bull. Astron. Inst. Netherland **14**, 299.
Peterson, R. C. 1981 Astrophys. J. Letters **248**, L31.
Reimers, D. 1981 in Physical Processes in Red Giants I. Iben, Jr. and A. Renzini, eds., D. Reidel, Dordrecht, p. 269.
Renzini, A. 1979 in Stars And Star Systems B. E. Westerlund, ed., D. Reidel, Dordrecht, p. 155.
Roberts, M. S. 1960 Astron. J. **65**, 457.
Rosse, Earl 1861 Phil. Trans. Roc. Soc. London, 151, plate 28.
Scott, E. H. and Rose, W. K. 1975 Astrophys. J. **197**, 147.
Scott, E. H. and Durisen, R. H. 1978 Astrophys. J. **222**, 612.
Taylor, R. J. and Wood, P. R. 1975 Monthly Notices Roy. Astron. Soc. **171**, 467.
VandenBergh, D. A. 1978 Astrophys. J. **224**, 394.
VandenBergh, D. A. and Faulkner, D. J. 1977 Astrophys. J. **218**, 415.

Webbink, R. F. 1985 in IAU Symposium No. 113, Dynamics of Star Clusters J. Goodman and P. Hut, eds, D. Reidel, Dordrecht, p. 541.
Woolf, N. J. 1964 Astrophys. J. **139**, 1081.

DISCUSSION

GRINDLAY: The possible Hα emission reported in the Grindlay and Liller paper was _not_ claimed as necessarily due to diffuse gas ($\sim 0.02 M_o$) but instead perhaps due to individual red giants, as found at about the same time by Cohen. As for the extended X-ray emission, the derived total mass of the hot gas is uncertain because of the uncertain geometry of the emitting X-ray shell (cf. discussion in my paper in the IAU 113 Proceedings).

ROBERTS: Because of the obvious uncertainty in the mass estimate from the X-ray data I do not quote the value given in IAU 113 Proceedings in my table of detections.

VAN DEN BERGH: 2 or 3 novae have been discovered in globular clusters indicating a nova rate of $\sim$ 2 per century. Since globular clusters contain $\sim$ 1% of the halo population this suggests a rate of $\sim$ 200 galactic halo novae per century. Do you think this is reasonable or might the nova frequency in globular clusters be unusually high?

ROBERTS: It does appear high but the globular cluster numbers are so few that I hesitate to draw a meaningful comparison.

GNEDIN: In my opinion there are two sensitive methods for a search for hot gas: 1. The change of the position of radio sources behind a cluster due to refraction. 2. The change of the polarization plane due to the Faraday effect. Both effects are noticeable especially in the decameter radio range of the spectrum.

KING: Continuous sweeping by halo gas is more attractive than sweeping by the Galactic plane. If a cluster has to wait for passages through the plane, the gas is likely to condense centrally to a density where it will be retained.

SCHOMMER: Lake and Schommer looked at HI in low luminosity ellipticals several years ago. They also looked at two dwarf spheroidals (Leo I & II). They detected no HI, although failed to publish that result. As I recall the upper limits are <50 M_o, but someone should harass me if they wish the actual numbers. Some earlier searches did not look at the correct velocity range, I believe.

ROBERTS: The sequence of globular clusters, dwarf spheroidals and elliptical galaxies, all of which have essentially negative results (except for a few apparently special case ellipticals) must have an important clue on the cleansing mechanism(s).

RENZINI: A straightforward calculation shows that when clusters were $\sim$ 10^8 yrs old, the rate of mass return from stars was $\sim$1000 times higher than it is today. The cooling rate in the gas was then about one million times higher, and the conditions were much more favorable for retaining gas in clusters. This may have left some observable consequences (e.f. composition anomalies??).

MATHIEU: Since Bob has begun confessing of unpublished data, Leo Blitz and I searched several of the "dust" lanes in several clusters for CO emission. We found no emission to sensitive limits. You presented a reference to a similar study. Could you say something more about that study; what limit did they place on the presence of H_2?

ROBERTS: Troland, Hesser and Heiles in Ap. J. 219, 873, 1978 looked for CO in some of the more prominent dust lanes. They found only upper limits but did not estimate an upper limit to the H_2.

COHEN: Can EXOSAT confirm the rather tentative detection by Einstein of extended X-ray emission around 3 globular clusters?

ROBERTS: Somebody said that EXOSAT is not sensitive enough.

CUDWORTH: The proper motion of M 22 is in the wrong direction for the extended X-ray source to be a bow shock. In this line of sight, however, the association of the source with the cluster must be rather uncertain.

ROBERTS: Hartwick et al. (Ap. J. 254, L11, 1982) note that the proper motion for ω Cen is appropriate for a bow shock explanation.

ADUR: Your data seem to be interesting but I remember a similar search was carried out by Faulkner et al. (1977) where his estimates were lower than the ones you have mentioned. Could you take the same value, or do you intend to give any upper limits for the values.

ROBERTS: I took Faulkner, et al. data. They are listed under the category $< 1 M_\odot$ for H II.

GLOBULAR CLUSTERS AS TRACERS OF THE GALAXY MASS DISTRIBUTION

K. A. Innanen

York University and CITA, Toronto

This review will consist of two parts: (a) a brief description of a new method to determine V_o, the circular speed of the local standard of rest, and which, of course, plays a fundamental role in the mass distribution of the Galaxy, and (b) a review of the globular clusters as tracers of the mass distribution in the Galaxy.

1. A NEW ESTIMATE OF THE CIRCULAR SPEED V_o OF THE LOCAL STANDARD OF REST.

V_o is poorly known, with contemporary values ranging from 180 to 280 km/sec (Mihalas and Binney 1981). The consequent uncertainty in the mass distribution is highly unsatisfactory. The new method to estimate V_o is straightforward to explain, and, in principle, yields a relatively accurate result (Carlberg and Innanen, 1986). The Galaxy is believed to contain a nucleus with a substantial mass that manifests itself as a strong rise in the rotation curve inside 2 kpc, peaking at about 250 km/sec, at around 500 pc. Stars with very eccentric orbits that pass through the nuclear region generally cannot be confined to a flattened disk distribution; their orbits are chaotic, i.e., they do not have third "integrals", and consequently they spend most of their time in the halo. In a local sample of old disk stars, there will be accordingly an apparent deficiency of stars with low angular momentum, because they are in a much higher scale height distribution. The deficiency should be observed in galactocentric tangential velocities, as a gap centered on the circular velocity as reflected in the motion of the LSR. The value obtained in this way thus is independent of R_o or the mass model; it depends only on the presence of a gap. The expected size and velocity width of the gap has been calibrated with a new mass model. Using the compilations of Woolley et al. (1970), and Gliese (1969), evidence is presented that the expected deficiency of low angular momentum stars does exist in the local stars. The strength of the conclusion is limited by the size of the sample of appropriate intermediate population stars, i.e., those that are both in a strongly flattened distribution and possess significant numbers of members at low angular momentum.

The available data favor a scale and model-free value of V_o in the range 225-245 km/sec with a most probable value of 235 km/sec. The implication of this result for the system of globular clusters in the Galaxy is that the latter system must rotate with a mean speed of 235-170 = 65 km/sec, with an estimated uncertainty of 25 km/sec. This value is in reasonable accord with the value of approximately 80 km/sec obtained by Huchra et al. (1982) for the M31 globular cluster system. It would be most useful to have similar radial velocity data for globular cluster systems in other nearby disk systems.

2. GLOBULAR CLUSTERS, TIDAL RADII AND THE MASS OF THE GALAXY

The theory for the tidal truncation of a satellite stellar system (i.e, a globular cluster) moving in the gravitational field of a much more massive galaxy rests on a theory dating back to the time of Roche. In one form, this theory states that, in order to survive tidal disruption, the mean density of the satellite system must exceed by some fixed ratio the mean density of the galaxy within the perigalactic circle of the satellite:

$$m/r^3 > kM/R^3 \quad (1)$$

where
- m = mass of the globular cluster
- r = tidal radius of the globular cluster
- R = perigalactic distance of the globular cluster
- M = mass of the galaxy interior to R
- k = a constant, the appropriate value of which is somewhat controversial.

The hope is that knowledge of m, r, R and k will produce the galaxy mass inside R. The realization of this hope has proven elusive; a fairly complete discussion of the associated problems can be found in the review of Innanen, Harris and Webbink (1983) (hereafter IHW). What follows is a brief review of the problems associated with the use of Eq. (1).

(i) The first point to be noted is that the mass M varies as the ratio of the cubes of two linear distances, so that errors in the latter quantities produce masses which suffer triply. This is self evident but a point that nevertheless deserves repetition.

(ii) r is usually referred to as the "observed" value of the tidal radius of the cluster. It is not normally observed at all in the classical sense, but rather is an empirical extrapolation of the run of surface density of stars in the central parts of the cluster to a zero value which invariably is buried well out in the field of the galaxy. Most of the globular clusters in the Galaxy are rather round and there is reasonable accord in various estimates of this "observational" quantity for about 66 clusters (IHW). There are, however, some well known exceptions. Very discordant data exist for such clusters as NGC 362, NGC 1851, M 53 and M 3 and must be discarded. Although special scale plates may be required, Irwin and Trimble (1984) have demonstrated with M 55 that automated equipment can match the results of tedious, human star-counting work.

(iii) It is normal practice next to assume that the observational value of r in (ii) is equal to a tidal radius caused by the galaxy and which can be evaluated by a simple theory. This has turned out to be a significant oversimplification. It is clear that those clusters spending most of their time in shorter period in the inner part of the Galaxy suffer more vigorous tidal perturbations than the more remote outriders. Thus the casual application of the same value of "k" in Eq. (1) to all clusters may be unrealistic. A gross example is to try to apply Eq. (1) with the King (1962) value of k = 3 + e (e is the eccentricity of the cluster orbit) to Jupiter's outer, direct satellite to measure the Sun's mass. Seitzer and Freeman (see Freeman and Norris, 1981) have advocated the use of the King formula, whereas Keenan (1981a,b) and IHW have favored larger values for k. Both viewpoints may well be based on inadequate physics and/or oversimplified numerical experiments. As is now well known, many globular clusters have at their centers gravitational "machines" which continuously populate their envelopes with stellar ejecta from their cores. The rate at which the galactic tide can prune this envelope and thereby eventually establish a nominal equilibrium tidal radius is not known. It is interesting to note that in the much softer tide of the LMC, the younger, bluer clusters have very elliptical shapes.

(iv) The mass of the cluster is estimated from its integrated luminosity through the assumption of a constant mass to luminosity ratio m/l. Although m/l(visual) = 1.7 is the most common value, there is some evidence (IHW) that it could be as high as 3 or 5. Such high values imply the existence of dark matter in globular clusters (Peebles, 1984).

(v) In order to relate the cluster's present galactocentric distance to the perigalactic distance, it is necessary to invoke certain orbit- averaging methods. These methods (IHW) give the "most probable" distance which can be considered to be the present distance. Of course this requires an a priori assumption of the kind of potential the cluster moves in. Analytic expressions are available for both Keplerian orbits and for orbits in a logarithmic potential (flat rotation curve). The outcome of this exercise favors the flat rotation model (IHW), but it is not at all clear that the rotation curve of the Galaxy inside R_o is flat.

A fair summary of the above list (i)-(v) is that the basic problem is not yet well enough posed in the physical sense to warrant detailed analysis. This was the basic, disappointing conclusion reached by IHW. Despite this weakness, the available data indicate that the orbits of the clusters are more commonly round than elongated.

To escape this conclusion, either the theoretical tidal radii are too small by a factor of 2, or the mass-luminosity ratio of the clusters is too small by a factor of 2 to 3, or a combination of both of these.

The data for the galactic globular cluster system produces reasonable statistics out to a galactocentric distance of approximately 20 kpc. Beyond that value there are very few clusters. IHW used $R_0 = 8.5$ kpc and $V_0 = 236$ km/sec. The latter value is fortunately essentially the same as the value advocated above. Consequently, one may still use with confidence their derived mass distribution

$$M(R) = M_0(R/R_0)^{1.27+0.18} \qquad (2)$$

where $M_\odot = 1.1 \times 10^{11}$ solar masses
so that $M(20) = 3.26 \times 10^{11}$ solar masses.

The large step from 20 to 100 kpc must be accomplished by using the radial velocities of a dozen or so distant globular clusters and dwarf spheroidals and by assuming that the virial theorem may be applied, together with an additional assumption such as isotropy. Hartwick and Sargent (1978) first performed this analysis and obtained a value of $8 - 10 \times 10^{11}$ solar masses inside R = 100 kpc. A reexamination of the problem using more recent radial velocities (Olszewski et al. 1986) has been performed by Tremaine (1986) who finds a mass of $2 - 4 \times 10^{11}$ solar masses. This mass confirms a similar value obtained by Lynden-Bell (1983). Evidently the outcome depends strongly on the accuracy of the sparse data, as well as the assumption of isotropy. These lower values of the coronal mass imply that the galactic rotation curve must decline beyond R = 25 kpc. It may well be the case that some of the most useful information about the outer mass distribution in the Galaxy will come from studies of the kinematics of relatively nearby high-velocity stars, as can be seen from the work of Carney and Latham (1986; also this symposium).

It is a pleasure to acknowledge useful conversations with Ron Webbink, Bill Harris, John Huchra, Sidney van den Bergh, Ruth Peterson, Ray Carlberg, Jean Brodie and Scott Tremaine. This work has been supported, in part, by grants from NSERC (Canada).

REFERENCES

Carney, B. W. and Latham, D. W. 1986 Astron. J. 92, 60.
Carlberg, R. G. and Innanen, K. A. 1986 Astron. J. (submitted).
Freeman, K. C. and Norris, J. 1981 Ann. Rev. Astron. Astrophys. 19, 319.
Gliese, W. 1969 Veroff. Astr. Rech. Inst. Heidelberg, No. 22.
Huchra, J., Stauffer, J. and Van Speybroeck, L. 1982 Astrophys. J. Letters 259, L57.
Innanen, K. A., Harris, W. E. and Webbink, R. F. 1983 Astron. J. 88, 338.

Irwin, M. J. and Trimble, V. 1984 Astron. J. **89**, 83.
Mihalas, D. and Binney, J. 1981 Galactic Astronomy, Freeman, San Francisco.
Keenan, D. W. 1981a Astron. Astrophys. **95**, 334.
Keenan, D. W. 1981b ibid **95**, 340.
King, I.R. 1962 Astron. J. **67**, 471.
Lynden-Bell, D. 1983 Monthly Notices Roy. Astron. Soc. **204**, 87P.
Olszewski, E. W., Peterson, R. C., and Aaronson, M. 1986 Astrophys. J. Letters **302**, L45.
Peebles, P. J. E. 1984 Astrophys. J. **277**, 470.
Tremaine, S. D. 1986 Contributed paper at Santa Cruz Workshop on Galaxies.
Woolley, R., Epps, E. A., Penston, M. J., and Pocock, S.B. 1970 Roy. Obs. Ann. No. 5.

DISCUSSION

GRINDLAY: First, an additional comment for the last talk. When I said in response to Cudworth's remark about the M 22 proper motion not agreeing with an X-ray bow shock, I should have pointed out that in our paper we explicitly mentioned M 22 could be a doubtful case because of the more likely contamination by other extended X-ray sources (e.g., uncatalogued supernova remnants) near the galactic plane. Now my question: How did you actually do orbit averaging to derive R_p values?

INNANEN: This can be done analytically for both "Keplerian" and logarithmic potentials (i.e., flat rotation curve potentials). The details are in Innanen, Harris and Webbink, A.J. __88__, 338, 1983).

CARNEY: The field stars are germane to determining the Galaxy's total mass, too. There are two results in particular, Hawkins (1983) claimed to find an RR Lyrae variable with V_{rad} = -465 km/sec at R_{gc} = 59 kpc. This immediately indicates a Galactic mass of $1.4 \times 10^{12} M_\odot$, if the star is bound. Second, the survey Dave Latham and I have been doing has revealed a few dozen nearby stars with Galactic-frame velocities of over 400 km/sec, and what looks like a power law distribution extending to 550 km/sec. As you said, the implied total mass depends are V_o, but it appears the Galaxy's mass exceeds that interior to the solar orbit by a factor of at least five.

INNANEN: I agree entirely that such extreme velocity stars provide a invaluable probe as test particles of the Galaxy potential. It would seem extremely unlikely that we should see even a single intergalactic "tramp" star in the solar neighborhood.

OSTRIKER: 1) Lee and I recently found that clusters may be larger by a factor of 1.5 times the nominal tidal radius due to relatively slow loss rate outside the tidal radius. 2) It would be interesting to combine your analysis with that of Tremaine (who used radial velocities). Then, the orbital anisotropy could be estimated rather than assumed. Have you considered combining the two approaches?

INNANEN: Not yet, but the idea is worth pursuing.

WHITE: Of the four clusters with "terrible tidal radii", the first two, NGC 362 and 1851, have fields contaminated by the SMC so background sky contributions are 1) variable and 2) uncertain. The other two, NGC 5024 and 5272 (M 53 and M 3, respectively) lie in the direction of the NGP; don't know what went wrong there.

INNANEN: Yes, some clusters are intrinsically very difficult, but it's still disappointing to not be able to include them.

COHEN: What is the expected effect of the Magellanic Clouds on the dynamics of the outer clusters in our galaxy?

INNANEN: They undoubtedly play a role in the range $40 < R < 70$ kpc with capture and exchange as one possibility. I have no quantitative data to offer.

CUDWORTH: Anyone in North America who may be considering a massive plate scanning project to re-derive tidal radii should contact Roberta Humphreys at Minnesota regarding use of the automated plate scanner. This machine is much faster than a PDS microdensitometer with slightly poorer photometric precision, which should be adequate for such a project.

ARMANDROFF: Since the Olszewski et al. study, Gary DaCosta and I have redetermined the velocity of Sculptor based on 16 K-giants. The value has increased from +20 km/s adopted by Olszewski et al. to +107 km/s. This increases their mass estimate by ~10%. The last word has probably not been said about the velocities of many of the outer halo systems.

PETERSON, R.: Since the Olszewski et al. paper appeared, I have remeasured the space velocity of Pal 15 with MMT echelle values for 4 stars. The resulting systematic galactocentric radial velocity is as large as the largest values tabulated there. If its distance is correct (cf. Seitzer and Carney), it implies substantial matter at large galactocentric distance.

CARNEY: Pat Seitzer and I have obtained a color-magnitude diagram for Palomar 15, and the cluster appears to be severely reddened. If so, its distance declines by a factor of about three, and so it becomes less interesting (at least for this issue).

OLSZEWSKI: I pointed out that detector technology has vastly improved the quality of velocities and distances of distant globular clusters. Secondly, I pointed out that we used Lynden Bell's analysis -- the assumption of isotropy requires some objects on radial orbits -- this is a problem given the fragility of many of the outer systems. Using Lynden Bell's favorite correction the quoted mass estimate from our paper goes up ~ a factor of 2. We need <u>measured</u> orbits. Finally, I'll bet that the velocity of Hawkin's RR Lyrae is wrong by at least 100 km/s.

GLOBULAR CLUSTERS AND PRIMORDIAL COMPOSITION

Roger Cayrel

Observatory of Paris

1. INTRODUCTION

So much has been learned already, during these last four days, about globulars that I shall start by addressing the second part of my topic first: the primordial chemical composition. Then I shall review the abundance of the primordial elements in globulars, or more generally in Population II. Actually if the globulars are the oldest stars ever observed in the Universe one could hope that the subject is over and that there is no need to speak about elements synthesized in stellar interiors for globulars. Unfortunately these rascals show always some amount of stellar nucleosynthesis products and this may very well be the central problem raised by their chemical composition. We shall discuss in turn the amount of the stellar synthesized elements in globulars and then the clues given by the peculiarities of the abundance ratios, which may help in understanding their origin. The last section summarized the present knowledge of the subject.

2. THE PRIMORDIAL COMPOSITION

Let us now turn our attention towards a refreshingly simple physical system (when compared to a globular cluster): the Universe as a whole.

Table I summarizes the main events which we believe to have occured in the Universe. The period of interest for the building up the primordial composition is between 0.1s and 1000s. At the time 0.1s the Universe was indeed a very simple system: almost perfect uniformity in temperature and density, interactions in the domain of 1 Mev with $T \simeq 10^{10} K$. No departure from LTE, no molecules, and every species extraordinarily close to thermodynamic equilibrium. One could think that the word 'local' does not sound right when one is speaking of the whole Universe but at this time the 'causal' horizon was encompassing a mass of the order of $10^{-3} m_\odot$ only, so the nucleosynthesis has been more local than what could be thought a first view. The physical basis for the study of the elaboration of the primordial composition have already been

TABLE I

Time-Table of the Big-Bang

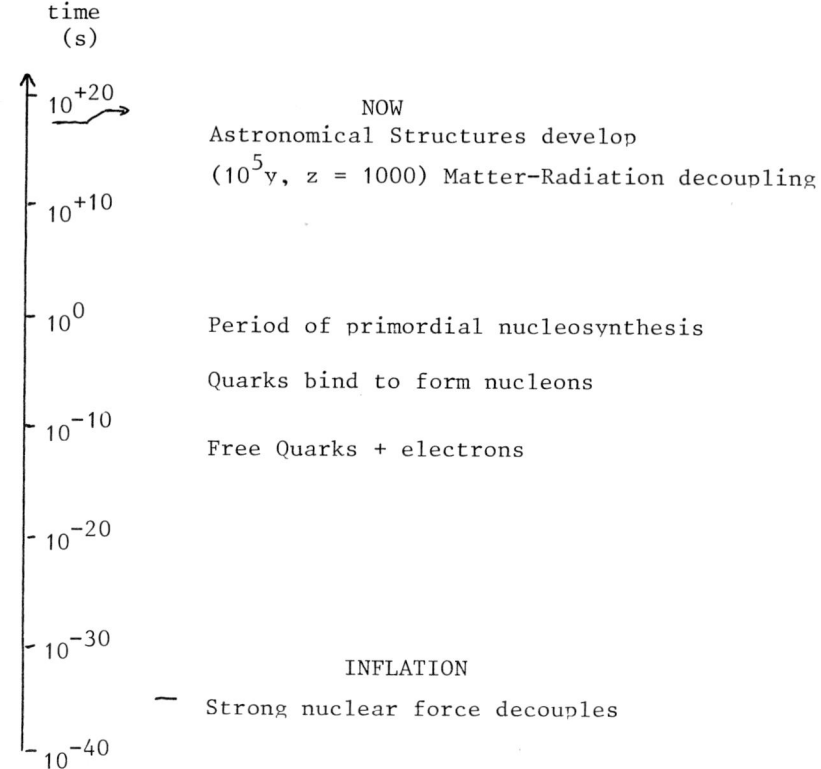

Gravitation decouples from the other forces
NO ADEQUATE PHYSICS

established by Wagoner 1973, but the basic recent paper on the subject is by Yang and al. 1984, and a nice review of both the theoretical aspect and the observational evidence concerning the primordial elements is given by Boesgaard and Steigman 1985.

What has driven some action in the physical conditions of the Universe between the time 0.1s and 1000s is of course the expansion. The semi-thermodynamical equilibrium existing at the time 0.1s has not been able to survive because of the cooling of the Universe and of the density decrease, both factors leading to a decrease of all the reaction rates down to practically zero at the time 1000s ('freeze out' of the nuclear composition). All the processes do not freeze out at the same time, because rates do not decline all equally. A close up of the main events in the period of primordial nucleosynthesis is given in table II.

TABLE II

Important events during Nucleosynthesis period

Time (s)	Event
0,1	neutrinos decouple, proton and neutron in statistical equilibrium
1	protons and neutrons decouple
10	e^+, e^- pair production stops N(nucleons)/N(photons) freezes out. Deuterium production begins
30	^{3}He and ^{4}He production starts
300	Li production starts
1000	all reaction rates become unsignificant, Big-Bang primordial composition is set.

The predicted frozen composition does depend upon a single parameter η, the ratio between the number of nucleons and the number of photons (conserved later during the expansion until, of course, non primeval photons have been generated bu nuclear burning in stars). Fig.1 shows the predicted abundances as a function of η, under the usual assumptions that the metrics was of the Robertson-Walker type, and that the expansion rate was controlled by the equations of General Relativity. To this, one must add that the number of Relativistic particles existing was known (3 mass-loss neutrinos are usually assumed). Within the frame of the Grand Unified Theory other particles may have existed (photinos, gravitinos, axions) and if this is the case their contribution to the density may have to be included, and the expansion rate would be affected. For a discussion of this point see Boesgaard and Steigman 1985 or Audouze 1986). We shall only be concerned today with the so called 'standard' Big-Bang for which these new particules are not taken into account.

The direct or (indirect) best evidence for the initial abundances of the primordial elements is shown in Fig.1. It is very very **remarkable** that they agree more or less, and are not orders of magnitude apart from each other. The most accurate evidence for the primordial abundance of ^{4}He does not come from old stars but from the ^{4}He/H ratio in extragalactic H II regions of low metallicity(Kunth and Sargent 1983, Peimbert 1986). The case of ^{3}He and D is more involved as these elements have been subjected to some amount of astration which has to be estimated.

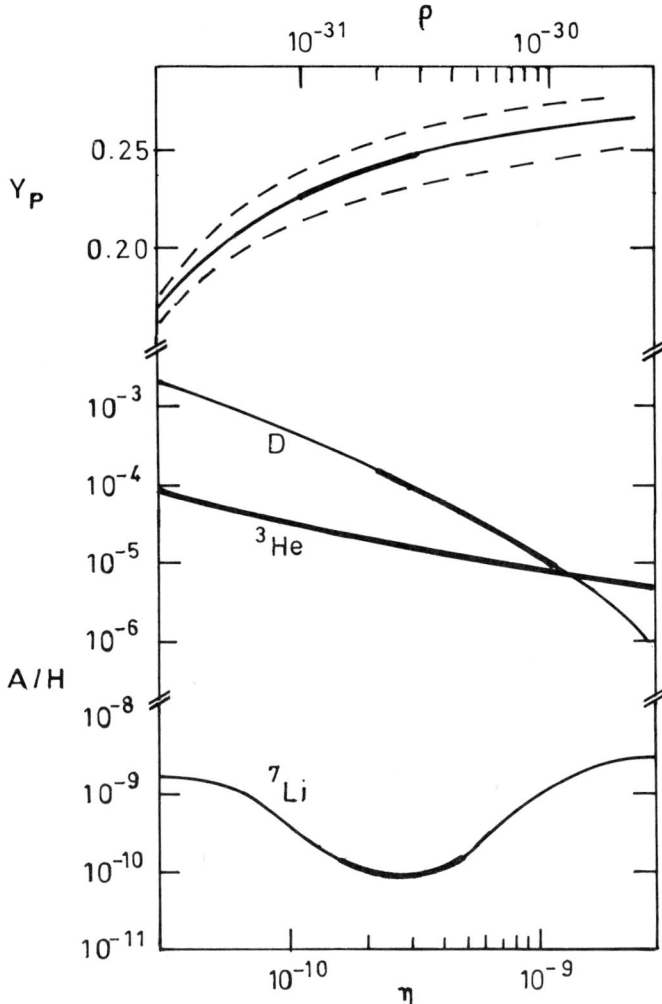

Fig. 1. Predicted primordial abundances in the 'standard' Big-Bang, according to Yang et al.. Note that in the upper part the ^{4}He abundance Y_p is by mass, whereas for D, ^{3}He and ^{7}Li the abundances are by number of particles, relative to H. The abscissa at the lower edge of the frame is the number ratio of nucleons to photons, whereas the abscissa at the upper edge is the present baryonic mean density of the Universe in gcm^{-3}. On each curve the values compatible with observational evidence are reinforced. The three curves for ^{4}He are respectively for a number of neutrons species equal to 2, 3 and 4, from bottom to top. The assumed neutron life-time is 10.6 minutes.

3. THE PRIMORDIAL ELEMENTS IN GLOBULARS

3.1 Helium

If the extrapolation of the helium abundance to zero O/H ratio in oxygen-poor H II regions represents the primordial helium abundance how does this value compare to the helium abundance in globulars? This last one can be determined by several methods : the position of the blue edge of the instability trip in the color-luminosity diagram, the position of the main sequence in the theoretical HR diagram, the magnitude difference between the horizontal branch (HB) and a point of the unevolved main sequence of given effective temperature, the ratio of the number of stars on the HB and on the red giant branch (RB), etc... These methods have been reviewed recently by Caputo and Castellani (1983) and Cole et al. (1983). The most accurate method is probably the last one mentioned. Buzzoni et al. (1983) give:

$$Y_{GC} = 0.23 \pm 0.02$$

in agreement, within the error bars, with the primordial abundance Y_p:

$$Y_p = 0.24 \pm 0.01$$

derived from the study of oxygen poor extragalactic H II regions (Kunth 1986). The accuracy of such determinations has been discussed recently (Davidson and Kinman 1985).

3.2 Deuterium and ^{3}He

These elements are not observable in low mass star Population II.

3.3 Lithium

By far this is the most interesting element. By an incredible luck, metal-poor stars have a much shallower convective zone than population I stars. Lithium, which is burnt in all Population I unevolved stars as old as the sun, has survived in Population II main sequence stars with metallicities lower than 1/20 solar, and with an effective temperature larger than 5500 K. (Spite^2 1982, Spite et al. 1984, Spite^2 1986). Globular cluster main-sequence stars are too faint to allow a good spectroscopic determination of lithium abundance but the parent population of field subdwarfs has been well studied by the Spites, who have found the results shown in Fig.2 from a nearly complete sample of subdwarfs brighter than V = 9.5. Lithium is the only element which does not show any correlation with Fe/H , and this constitutes the most direct evidence that this element is primordial. All other elements show a strong correlation with Fe/H , with a slope 1, when one does consider a pure halo star sample (François 1986). The only other element which has been claimed to show a slope significantly lower than 1 is nickel (Luck and Bond 1983), but further work by other authors has not confirmed this result. One should note that the abundance found for lithium in popu-

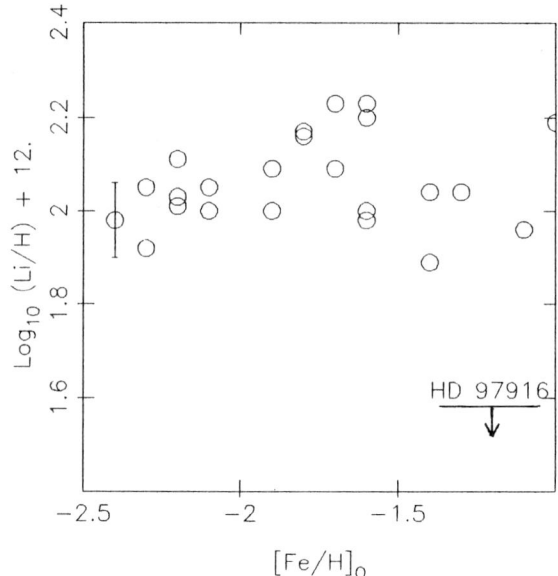

Fig. 2. Lithium abundance found by the Spites in subdwarfs as a function of [Fe/H]. Note the independence of the lithium abundance vs. [Fe/H]. This is a unique case among all other elements.

lation II subdwarfs is 10 times smaller than the so-called cosmic abundance (Li/H = 10^{-9}) found in young population I stars and in the today interstellar matter.

4. THE METALLICITY OF GLOBULAR CLUSTERS.

During a couple of decades the metallicity of globulars was thought to be the result of progressive enrichment of the Galaxy in elements synthesized in stellar interiors and rejected into the interstellar medium, mostly by supernovae explosions (Eggen, Lynden, Bell, Sandage 1962).

However, if GC are the oldest objects in our Galaxy and in other galaxies as well, how is it that none of them has the primordial composition which was the only possibility prior to enrichment? This problem which was shown by Bond 1981, to occur as well in the field population II, is the so-called population III problem. Because we are more concerned in this symposium by GC than by field population II, and also because the metallicity distribution of field population II is still a matter of controversy (see Poster n° 108), we shall discuss mostly the metallicity of globular clusters. Zinn 1985 has shown that the histogram of the metallicity of globular clusters has a double peak when the logarithm of the metallicity is used for abscissa.

The metal rich group has kinematical properties resembling a thick disk population rather than a spheroidal population. If, in order to discuss the evolution of the metallicity according to current galaxy enrichment models we use the metallicity instead its logarithm, we found the results shown in Fig.3 and 4. On a coarse scale (Fig.3) the first peak appears as an intense 'flash' of production of heavy elements at the very beginning of the life of the Galaxy, whereas the second peak is smeared out and has the aspect of an 'on-going' activity at a much lower level. We believe that the progressive enrichment model of Eggen, Lyndel-Bell and Sandage does describe this second family of GC, but that the initial flash shown at an enlarged scale on Fig.4 may have little to do with anything 'progressive'. An important step was accomplished by Searle 1977 who postulated that GC were born of 'fragments' of interstellar matter orbiting freely in the galactic gravitational field prior to the formation of the disk. This assumption allows to understand the fact that the metallicity of GC appears to be independent of galactocentric distance, as shown already in this meeting. In fact the metallicity depends only on what has happened in each of these fragments. However this model keeps the progressive enrichment model in each of the fragments, and for this reason has the almost horizontal start inherent to the so-called simple model, for which the amount of heavy elements produced up to time t, is proportional to the number of low mass stars produced within the same laps of time. Let us note that a purely stochastic law (log-normal) fits as well, and indeed better, the observed histogram.

So the basic question is to understand, why even if we extend our sample to extragalactic GC, there is no GC with metallicity significantly lower than 1/200 of the solar metallicity. Two types of explanations have been proposed to explain this fact:

i) There was a pregalactic stellar generation, with an initial mass function truncated below $0.8\, m_\odot$, which has produced heavy elements but no shining star, only invisible remnants.

ii) The star formation in a collapsing primordial cloud has suffered some amount of self-pollution, and low mass stars have been polluted by type II Supernovae resulting of fast evolving O stars. Explanation 1) has been first advocated by Truran and Cameron 1971. A variant of it by Carr, Bond and Arnett 1984 proposes that the first (pregalactic' generation is made of very massive objects (VMOs), with masses in the range 100 to 1000 $m_\odot$. The remnants of such objects would be of course invisible now.

The physical reason justifying a cut-off of the initial mass function below $0.8\, m_\odot$ in a zero metal environment is the absence of radiative mechanism otherwise available in present day interstellar matter. However when one includes H_2 formation (Palla 1983) the Jeans mass may well be as small as $0.1\, m_\odot$ even in a metal free medium.

Explanation ii) has been originally proposed by Peebles and Dicke 1968. The existence of the large structures of the Universe makes unli-

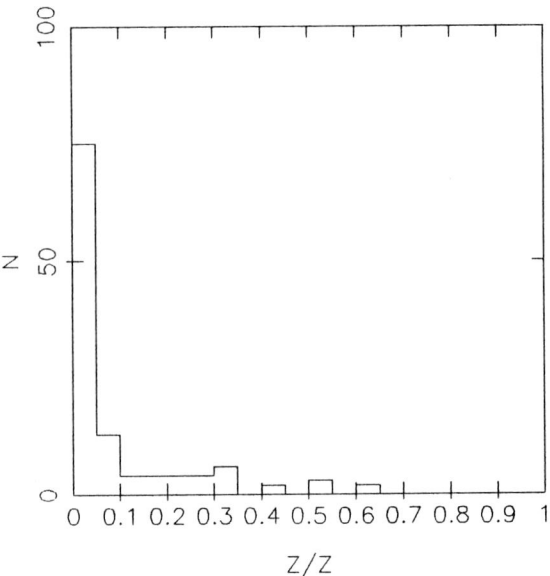

Fig. 3. The histogram of the metallicity of GC with bins on a linear scale in Z. (coarse scale). The source is Zinn 1985.

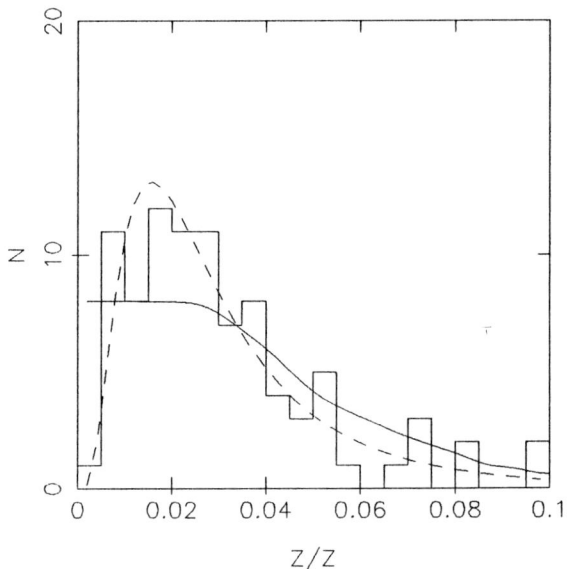

Fig. 4. The histogram of the metallicity of GC with bins on a linear scale in Z (finer scale, showing the low metallicity part). The full curve is according to Searle 1977, the dashed curve is a fitted lognormal law with [Fe/H] = -1.6 and dispersion σ = 0.3 dex.

kely that GC did form first as in this 'bottom-up' cosmological model, but recently the author of this review has developped the idea that even if the early galaxy was made of primordial clouds, the gravitational collapse of such clouds may have formed first a central dense nucleus rich in very massive stars which, in becoming SN II, have polluted the cloud itself before low mass stars had time to develop (Cayrel 1986).

At the present time there is no definite proof on which one of these two explanations holds more truth.

5. ABUNDANCE RATIOS IN GLOBULAR CLUSTER STARS

If it is well known that population II is metal poor, the question of how do differ the abundances ratios in population II and in population I has not been very clear until recent work. The recommended references on the subject are Spite[2] 1985, and François 1986. An older reference, but dealing more specifically with cluster stars, is Freeman and Norris 1981. The basic fact which has emerged from recent work is that there is a clear indication that abundance ratios are stable within population I and within population II, with some differences between both, and a transition occuring in the intermediate population with metallicities between $1/10\ Z_\odot$ and $1/3\ Z_\odot$. The most well established difference between the two populations is the enhancement of oxygen in population II versus the iron abundance by a factor of about 3 (Clegg et al. 1981). The case of nitrogen is very peculiar. This element does follow the abundance of iron most of the time, but is sometimes very overabundant with respect to iron in a few subdwarfs. Magnesium and silicium have moderate overabundances with respect to iron in population II (0.5 and 0.3 dex respectively). Heavy elements, produced by the s or the r processes, either show some overdeficiency with respect to iron on follow the iron abundance. Except for nitrogen, for which a primary mechanism of production is still conjectural, these differences are explainable by recent models of galactic evolution (Greggio and Matteucci 1985, Matteucci 1986). The predicted halo composition agrees fairly well with the observed one if one assumes that the halo composition is set only by SN II explosions whereas the chemical composition of population I is set by slower processes as well (SN I, planetary nebulae ejectae). Fig.5 and 6 illustrate this point. The success of this interpretation makes more likely that elements produced after the Big-Bang, found in population II, have been produced by normal SN II with O stars progenitors rather than in VMO's, as in Carr et al. 1984 proposal. This, however, leaves the choice between the Truran and Cameron 1971 proposal, and the self-pollution explanation.

6. CONCLUSION

In globulars, and more generally in population II :
i) helium and lithium have their primordial abundances
ii) element products of stellar nucleosynthesis are always present at some level, with a mean abundance of about $1/30^{th}$ of the solar value an a more or less log-normal distribution with σ = 0.3 dex.

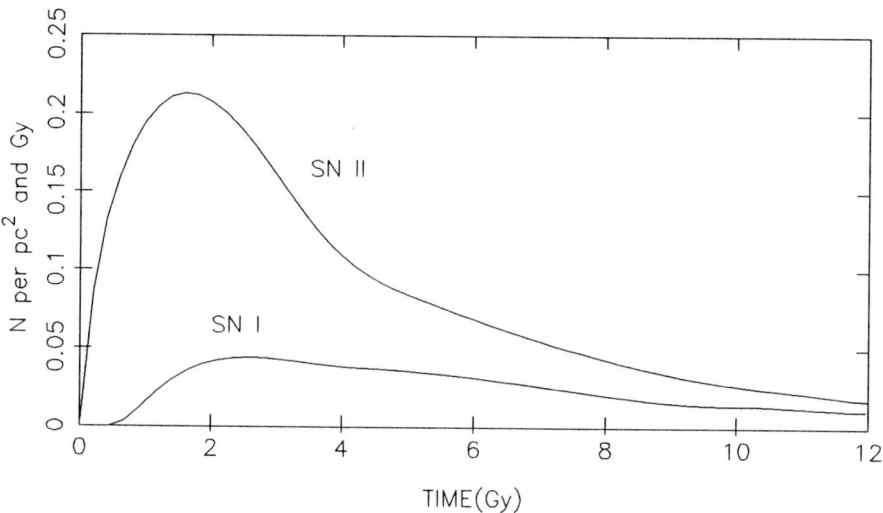

Fig. 5. Occurence of supernovae of type I and II in the life-time of the Galaxy according to Greggio and Matteucci 1985. Note that only SN II can account for the initial enrichment of the Galaxy.

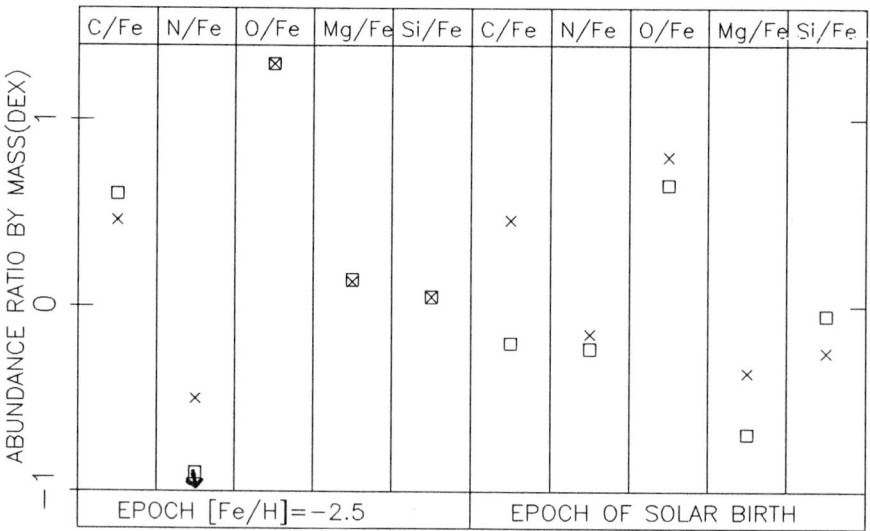

Fig. 6. Predicted (□) and observed (✗) ratios of some elements i) early in the galactic life   ii) at the time of solar birth. Note the good fit in case i) (according to Matteucci 1986, with new $^{12}C\,(\alpha,\gamma)^{16}O$ reaction rate).

iii) Abundance ratios are slightly different from those in population I and suggest strongly that the heavy elements content of population II comes exclusively from massive SN II explosions.

iv) The absence of metal free objects may either be due to a truncation of the IMF in a first stellar generation having left no visible remnants, or to the early pollution of primordial clouds by SN II in the galaxy, before they have generated the low-mass stars we see in GC and in the halo field.

REFERENCES

Audouze, J. 1986 in Workshop on Dark Matter in the Universe, Reidel, Dordrecht, p. XXX.
Boesgaard, A. M. and Steigman, G. 1985 Ann. Rev. Astron. Astrophys. 23, 319.
Bond, H. E. 1980 Astrophys. J. Suppl. 44, 517.
Buzzoni, A., Fusi Pecci, F., Buonanno, R. and Corsi, C. E. 1983 ESO Workhop on Primordial Helium P. A. Staver, D. Kunth and K. Kjar, eds., ESO, Garching, p. 231.
Caputo, F. and Castellani, V. 1983 ESO Workhop on Primordial Helium P. A. Staver, D. Kunth and K. Kjar, eds., ESO, Garching, p. 213.
Carr, B. J., Bond, J. R. and Arnett, W. D. 1984 Astrophys. J. 277, 445.
Cayrel, R. 1986 Astron. Astrophys, in press.
Clegg, R. E. S., Lambert, D. and Tomkin, J. 1981 Astrophys. J. 250, 262.
Cole, P. W., Demarque, P. and Green, E. M. 1983 ESO Workhop on Primordial Helium P. A. Staver, D. Kunth and K. Kjar, eds., ESO, Garching, p. 235.
Davidson, K. and Kinman, T. D. 1985 Astrophys. J. Suppl. 58, 321.
Eggen, O. J., Lynden-Bell, D. and Sandage, A. R. 1962 Astrophys. J. 136, 748.
Francois, P. 1986 Astron. Astrophys. 160, 264.
Greggio, L. and Matteucci, F. 1986 Astron. Astrophys. 154, 279.
Kunth, D. 1986 Publ. Astron. Soc. Pacific in press. (Also preprint no. 143 of the Institut d'Astrophysique de Paris).
Kunth, D. and Sargent, W. L. W. 1983 Astrophys. J. 273, 81.
Luck, R. E. and Bond, H. E. 1983 Astrophys. J. Letters 271, L75.
Matteucci, F. 1986 ESO preprint no. 434, The Effect of the New $12(\alpha,\gamma)$ 16-O Rate on the Chemical Evolution of the Solar Neighbourhood.
Palla, F. 1983 Mem Soc. Astron. Italiana 54, 235.
Peebles, P. J. E. and Dicke, R. H. 1968 Astrophys. J. 154, 891.
Peimbert, M. 1986 in Star Forming Dwarf-Galaxies and Related Objects Workshop held at the I.A.P., D. Kunth, and T. X. Thuan, eds., J. Tran Thank Van. Editions Frontieres, p. 403.
Searle, L. 1977 in The Evolution of Galaxies and Stellar Populations B. M. Tinsley and R. B. Larson, eds., Yale University Observatory, New Haven, p. 219.
Spite, M., Maillard, J. P. and Spite, F. 1984 Astron. Astrophys. 141, 56.

Spite, M. and Spite, F.   1982   Astron. Astrophys. 115, 357.
Spite, F. and Spite, M.   1986   Astron. Astrophys. 163, 140.
Truran, J. W. and Cameron, A. G. W.   1971   Astrophys. Space Sc. 14, 179.
Wagoner, R. V.   1973   Astrophys. J. 179, 343.
Yang, J., Turner, M., Steigman, G., Schramm, D. N. and Olive, K. A. 1984 Astrophys. J. 281, 493.
Zinn, R.   1985   Astrophys. J. 293, 424.

DISCUSSION

OSTRIKER: If I understand you correctly, the current lithium abundance is much higher than that in old stars. Is it possible that whatever the process which has increased Li also acted before the old stars were born so that the Li seen in these stars is not, in fact, primordial.

WALLERSTEIN: Currently lithium is reasonably well understood in terms of spallation of interstellar heavy elements by cosmic rays. If this is correct then it would not have worked when there were no heavy elements.

CARNEY: The search for and study of extremely metal-poor stars is critical to understanding our Galaxy's history (and that of others by inference). The existence or lack of such stars remains unsettled, but Beers, Preston and Shectman (1986) and Laird, Latham and I have claimed such stars exist in numbers consistent with simple models of Galactic chemical evolution.

CAYREL: It is true that the properties of extremely metal-poor stars in the field are quite important, especially because the diffuse Pop II represents 99% of the mass of the halo. However, the statistics of these very metal-poor stars still relies on very small numbers and is not well established yet.

ZINNECKER: 1) Can you explain in more detail how you solve the problem of the chemical homogeneity of globular clusters in your supernova self-enrichment scenario? 2) Is the supernova supposed to trigger the formation of the low-mass stars of the cluster? If so, you would not expect the same chemical pollution of the protostellar fragments throughout the proto-cluster cloud.

CAYREL: Concerning your first question, the level of pollution in a single burst of star formation is set by the ratio between the number of polluting supernovae and the mass of gas polluted by them. It seems reasonable to assume that these two numbers are roughly proportional, so the resulting metallicity is more or less unique. Concerning your second question, the answer is yes. Although you are right in saying that one does not expect exactly the same pollution in all protostellar fragments throughout the cloud, my answer to the first question shows that you do not expect wide variations either. But I agree with you that the most difficult point in the self-pollution mechanism is the apparent chemical homogeneity of most globular clusters.

STELLAR EVOLUTION IN GLOBULAR CLUSTERS AND HST

Alvio Renzini

Department of Astronomy, Bologna

1. INTRODUCTION

Globular clusters (GC) have been regarded among the most obvious targets for the Hubble Space Telescope since the very first conception of this project, and some observational programs have been exemplified in publications concerning the future use of HST (e.g. Westphal 1982 and Macchetto 1982, in the "Patras Book", Bahcall 1985, see also the 1985 Report of the STScI working group on Stars and Star Clusters).

In this talk I will focus on a few issues concerning the evolution of stars in globular clusters, for which HST observations are expected to provide crucial insight. I will not touch upon problems in cluster dynamics, which certainly will also take advantage from HST observations, (although, to some degree, stellar evolution and cluster dynamics may not be completely decoupled subjects).

A discussion of this general subject can start by considering that there are six evolutionary phases to be investigated (MS, RGB, HB, AGB, POST-AGB, WD), six instruments on board of HST (FGS, HSP, FOS, HRS, FOC, WF/PC) and six GC families that HST can resolve (MW, FORNAX, LMC, SMC, M31, M33), for a total of $6^3 = 216$ possible combinations (!). While, by good fortune, most *matrix elements* are just nonsensical (e.g. HSP observations of WDs in M31 globulars), still there is a fairly large number of sensible combinations. I will then further restrict the discussion to studies of cluster color-magnitude diagrams (CMD) which will become feasible with HST, a choice largely dictated by my own personal taste. I will then exemplify the case with three hypothetical projects making use of HST imaging capabilities, each concerning respectively globulars in the Milky Way, in the Magellanic Clouds, and in M31.

2. GALACTIC GLOBULARS: FOR COMPLETENESS SAKE

There are basically two ways of using HST imaging capabilities in connection with the subject of stellar evolution in galactic globulars: either one tries to go deep (faint) or to go *complete*. The only really faint evolved stars in MW globulars are white dwarfs, and much has already been said about them as targets for HST. The other evolved stars (i.e. stars in post-main sequence evolutionary stages) are so bright that have not attracted as much the attention of the potential HST astronomer. However, HST is basically a photon-starving telescope, as John Bahcall (this volume) will emphasize, and it is then particularly suited for the observation of relatively bright stars located in crowded fields. Therefore, what HST will easily make possible, thanks only to its superior angular resolution, is the photometry of ALL the evolved stars in a GC in a fairly short observing time. In other words, it allows the construction of really complete CMDs. Conversely, the observation of very faint GC stars (e.g. WDs) requires to push HST to its limits in a partly innatural way.

With few exceptions, the advantages of *going complete* in at least a well defined portion of a GC have not been fully appreciated. Quite often only the isolated stellar images are measured, with the result of obtaining CMDs where magnitudes and colors are (internally) very accurate, but where the luminosity function (LF) is virtually useless because of uncontrolled selection effects. However, as emphasized by Paczynski (1984), the LF provides a much harder information, compared to the shapes of cluster loci and isochrones, which are so much undermined by color/temperature uncertainties (e.g. Iben and Renzini 1984).

2.1 Luminosity function, ages, and model testing

The LF of a globular cluster, in a broad sense, is the (number) distribution of stars along the various branches of the CMD. For several reasons the LF can hardly be used to derive a cluster age (cf. Renzini 1986a,b), but can be used in a more subtle, fundamental way that I shall try to illustrate.

Indeed, dating GCs is a three step procedure: 1) a number of observable quantities are measured via astronomical observations of a Globular cluster (e.g. ΔM_{TO}^{RR}, [Fe/H], etc.), 2) these observables are fed into a theoretical machinery that we call the clock (i.e. a theoretically established relation between some observables and age), and 3) one faithfully sorts an age. Certainly, it is important to improve the accuracy with which the observables are measured, as this will reduce the internal error in the age determination. But what about the accuracy of the clock itself? i.e. what about the reliability of theoretical evolutionary models?

In strict connection with these questions, it appears that accurate (i.e. complete) cluster LFs provide the best possible way of testing the evolutionary models, and then of assessing the reliability of the clock. When an extensive and complete photometric survey of cluster stars becomes available, a whole variety of meaningful checks becomes feasible. For example: i) checking the contribution of the various evolutionary stages to the total cluster light, in the mood of Figure 5 in Renzini and Buzzoni (1986), ii) checking the duration of the various evolutionary stages, according to Eq. (1) below, and iii) checking the composition stratification inside stars, using the LF of the RGB.

Concerning points i) and ii), the first attempt in these directions is part of the archetipal study of M3 by Buonanno et al (BBCFS, this volume), based on the complete and accurate photometry of 10,000 cluster stars. Still, this study covers only $\sim 30,000$ $L_\odot$ of cluster light, or $\sim 5\%$ of the total, and does not allow but a first, rough study of the advanced evolutionary stages. For example, when inserting this figure into the number/luminosity/time relation (Renzini and Buzzoni 1986):

$$N_j = B(t) L_T t_j \qquad (1)$$

one infers the presence of ~ 10 early AGB stars in the sample ($B = 2\ 10^{-11} L_\odot^{-1} yr^{-1}$, $t_{AGB} = 1.5\ 10^7 yr$). While the theoretical prediction is in excellent agreement with the actual number of observed AGB stars (just 10 !), still the whole comparison is limited by the small number statistics.

The extension of the BBCFS study to the whole cluster ($\sim 6\ 10^5 L_\odot$) would in fact make available some 200 AGB stars, with an obvious improvement in the statistical accuracy, and even opening the possibility of studying the LF of the AGB in some detail. This is however impossible using ground based observations, as severe crowding prevents a complete and accurate photometric survey from being extended over a significantly larger fraction of the cluster (light). Conversely, HST will allow such a survey to be extended over the whole cluster, including the central regions!

Point iii) above is the one most intimately connected with the problem of cluster dating. Indeed, the clock itself is ultimately a relation between isochrone turnoff luminosity and age and then only main sequence, core hydrogen burning models are involved. These are therefore the models most worth testing, if one is primarily interested in GC ages. Moreover, in this like in many other cases, the accuracy of models of a certain evolutionary phase is best ascertained by looking at the immediate progeny of such models.

In this specific case, the main sequence lifetime is controlled by the rate of hydrogen burning, and therefore there is close connection with the actual composition stratification inside the star, as established in the course of the core hydrogen burning phase. In turn, such a composition profile controls the rate of evolution during the subsequent RGB phase. For example, a hypothetical mixing of the central regions would both prolong the MS lifetime (by providing fresh new fuel) and produce a shallower composition profile. Seemingly, a hypothetical diffusion of helium relative to hydrogen (cf. Stringfellow et al 1983) would both reduce the MS lifetime and produce a steeper composition profile.

While such, or other processes are hypothetically operating (during the MS stage) they may have little immediately observable consequences. But they would concur in establishing the final composition profile, through which the hydrogen burning shell will later eat its way during the subsequent RGB phase. Ultimately, there is therefore close connection between the composition profile at the end of the MS stage, and the actual LF of the RGB. Moreover, during the RGB phase there exists a relation between the mass of the stellar core (i.e. the mass coordinate of the hydrogen burning shell) and the stellar surface luminosity. Putting all this together, one can easily realize that the LF of the RGB allows to study the composition stratification inside the stars, and therefore to test the reliability of the models which provide the age calibration (clock).

However, this kind of "evolutionary stratigraphy" requires big, complete stellar samples (cf. Rood and Croker 1985, for illuminating simulations). From Eq. (1) we see that the larger the sample (larger sampled luminosity L_T), the larger the number of stars per unit duration, and then the larger the TIME resolution along an evolutionary sequence. In turn, thanks to the link between core mass and luminosity, the larger L_T, the larger the MASS resolution in the stratigraphic approach.

In principle, these considerations apply equally well to the AGB LF as a tool for probing the C-O-He stratification, as established during the core helium burning phase. The main difference is just that, compared to the RGB, the AGB is intrinsically less populated and the helium-burning shell is somewhat thicker, thus reducing the mass resolution of the evolutionary stratigraphy. Anyway, one can anticipate that good AGB luminosity functions will certainly provide the ultimate test concerning the nature and extension of the mixing processes active during the HB phase.

2.2 The luminosity function and HST planning

As emphasized by Bahcall (1985), the knowledge of the cluster LF can be extremely useful for planning HST observations of globular clusters, in particular of white dwarf stars. For the bright part the LF used by Bahcall is the same as that obtained by Da Costa (1982) for the cluster 47 Tuc. I have recently realized that this LF is in severe conflict with what one would expect from theoretical evolutionary models (Renzini 1986b). In brief, from Da Costa LF one can infer a subgiant branch (SGB) to horizontal branch (HB) number ratio $N_{SGB}/N_{HB} \simeq 6$, the SGB being defined as the portion between turnoff and the base of the RGB. By adopting a HB lifetime of 0.1 Gyr, one then infers an SGB duration $t_{SGB} = 0.6$ Gyr.

This compares to $t_{SGB} \simeq 3.5$ Gyr, as predicted theoretically by evolutionary models (Mengel et al 1979) with the appropriate composition, [Fe/H]= -0.7. There is therefore a discrepancy by at least a factor of 5, even when allowance is made for the error in reading the LF directly from Bahcall's Figure 1. On the contrary, the LF obtained by BBCFS for M3 does not disagree with theoretical expectations, and I conclude that Da Costa LF is probably incomplete by about a factor of 5 at the level of the SGB. This conclusion is not based on a faithful belief in the theory, but rather on the appreciation that star counts down to a plate limit is the most dangerous way of deriving a LF. Note also that an indication of severe incompleteness in the old LF is already evident when looking at the more recent 47 Tuc LF obtained by King, Da Costa and Demarque (1985, their Figure 3). In particular, when normalizing to the integral of the bright part (rather than at one arbitrary bin) already at V=17 the new LF is roughly a factor of 3 above the old one.

Concerning the anticipated number of white dwarfs in globular clusters, these findings will change slightly some of the numbers given in Renzini (1985). Having in general:

$$N_{WD} = a \, 10^{-11} L_V t_{cooling}, \qquad (2)$$

I obtained $a \simeq 4.7$ using Bahcall/Da Costa LF, and one now finds $a = 3.0 \pm 0.2$ using the LF of BBCFS. Moreover, the BBCFS visual to bolometric conversion factor for M3 is 1.43 (i.e. $L_{bol} = 1.43 L_V$, rather than $2L_V$ as adopted in Renzini, 1985). Therefore, also the fully theoretical approach gives $a \simeq 3$, in very good agreement with the semiempirical approach. The conclusion seems to be that in Bahcall (1985) both the number of WDs, and the number of lower main sequence contaminants have been underestimated by about a factor of 5 (this latter statement rests on the assumption that Bahcall's adopted lower main sequence LF is correct). Correspondingly, the number of pixels available per stellar image has probably been overestimated in Renzini (1985).

Finally, one important aspect of cluster LFs needs to be strongly emphasized. Obviously enough, most of the cluster light comes from few stars, while most stars provide very little light. So, the LF for the bright part of the M3 CMD obtained by BBCFS covers most of the stars contributing to the integrated light of the studied portion of the cluster, but does not extend faint enough to include the bulk of the cluster stars, which are still fainter than the completeness limit. Conversely, other deep LFs obtained with CCDs over small cluster portions (small sampled L_V) involve mainly low mass stars which contribute little or negligible light to the cluster luminosity. On the contrary, the luminosity to number conversion function (cf. Bahcall 1985) requires a LF (correctly) encompassing all stars, i.e. extending from the top to the bottom of the CMD. Such a LF is not yet available, and then it appears appropriate to give very high priority to attempts at properly, carefully matching the bright and faint portions of the LF in at least a few clusters. This is really important for the most efficient planning of deep HST observations of MW globular clusters.

2.3 Unanticipated applications (a curious example)

I would like to exemplify now how the knowledge of the LF for whole, popolous globular clusters might have useful applications far beyond those one can currently anticipate, some of which have been mentioned in the previous sections. I will illustrate the case by mentioning a recent, unexpected application of stellar counts on GC color-magnitude diagrams (Renzini 1987).

It has been recently suggested that perhaps "the solutions to the neutrino problem in the Sun and the missing mass problem in the Galaxy are one and the same" (Press and Spergel 1985, see also Gilliland et al 1986 and references therein). In brief, in this scenario weakly interacting massive particles (WIMPs, also called cosmions), while providing the missing mass, are continuously collected by the Sun, where they isothermalize the inner, neutrino producing core, thus reducing the solar neutrino flux below the observed ~ 2 SNUs.

However, if the solar core is isothermalized by WIMPs, so it would happen also to the inner core of HB stars, and core convection would then be suppressed during the core helium burning stage. This would have major, dramatic consequences on HB and AGB stars; in particular the HB lifetime would be drastically reduced by the suppression of the continuous fuel replenishment otherwise ensured by convection. On the contrary, the AGB lifetime would be considerably lengthened, because shell helium burning would now start from a point much closer in mass to the stellar center. Correspondingly, the lifetime ratio t_{AGB}/t_{HB} would be increased by almost one order of magnitude over the canonical value, ~ 0.14. Since star counts over complete samples for 15 globulars give $N_{AGB}/N_{HB} \simeq 300/2000 = 0.15 \pm 0.01$ (Buzzoni et al 1983), one can safely infer that the core of HB stars cannot be kept isothermal by pervasive WIMPs.

The conclusion is that, most likely, the WIMPs idea is not a viable solution for the solar neutrino problem, unless the WIMP annihilation cross section is tuned to within a permitted range of ± 1.5 dex, in such a way as to ensure WIMPs survival in the Sun, but their digestion and destruction prior to the HB phase. The WIMP supporter could then maintain that star counts in GCs provide a measure of the WIMP annihilation cross section....

3. THE GLOBULAR CLUSTERS IN THE MAGELLANIC CLOUDS

Globular clusters in the Magellanic Clouds span an age from ~ 0.01 to ~ 10 Gyr. The construction of complete CMDs for whole clusters will correspondingly allow the test of stellar models in a mass range from $\sim 1 M_\odot$ up to $\sim 10 M_\odot$. Moreover, the combined study of CMDs, integrated colors and spectra of MC clusters represents a fundamental step for population synthesis investigations, as it provides the "template stellar populations" for the test and calibration of synthesis codes and their ingredients (cf. Renzini and Buzzoni 1986, Renzini 1986c). In this regard, of particular interest are clusters ~ 0.5 to a few Gyr old, as most of their light is provided by stars around $2M_\odot$, just as in the case of the high redshift galaxies which are either within reach now, or will become so with the next generation of Very Large Telescopes.

For some time I had the impression that the greatest advantage of using HST on MC clusters would have been in reducing the field contamination, the argument being that the field-to-clusters ratio is obviously minimum in the crowded central regions of the clusters. However, it turns out that in excellent seeing conditions many clusters can be thoroughly resolved from the ground, and CCD photometry is possible for virtually all the stars around turnoff or beyond. The photometric accuracy is however seriously degraded, and it appears that accurate and complete CMDs for whole GCs can best be obtained with HST, which will also allow near-UV photometry (note that in clusters 1 Gyr old the turnoff temperature is $\sim 10,000K$). The emphasis has then shifted from field contamination (which may remain a problem) to accuracy and completeness. In this regard, an application of Eq. (1) to MC clusters indicates the necessity of using virtually all the clusters in crucial age bins, if one wants to ensure a statistically significant coverage of all the relevant evolutionary stages.

In conclusion, also the HST study of stellar evolution in MC globulars does not require to go really deep (at least in young and intermediate age clusters), but rather "to go complete". Again, the specific characteristic of HST which will be exploited is its angular resolution, which will allow accurate photometry even in very dense fields.

4. THE M31 CLUSTER FAMILY

It is widely recognized that GCs play a very important role in our attempts at understanding the origin and early evolution of the MW galaxy. Having said this, it becomes immediately clear that it is of the highest possible interest for the general problem of galaxy formation and evolution to extend such studies to another giant spiral galaxy, such as M31.

Indeed, HST will make possible the construction of M31 cluster CMDs extending 1 to 2 magnitudes below the HB level (which in Andromeda is at $<B> \simeq 25.5$, cf. Pritchet 1986), and simulations of FOC observations show that fairly accurate photometry can be obtained (Bragaglia et al 1986). Among the many possible uses of such CMDs two are particularly worth noting: 1) a plot of V_{HB} vs the metallicity indicator $(B-V)_{o,g}$, and 2) a plot of any HB morphology indicator vs $(B-V)_{o,g}$. We can correspondingly refer to a "vertical" and a "horizontal" study of the HB in M31 globulars.

The "vertical" plot is of crucial importance for both the distance scale problem, and for the age of GCs, as it allows a direct assessment of the metallicity dependence of the HB luminosity (cf. Sandage 1982; Renzini 1986a; Buonanno 1986; Sweigart, Renzini and Tornambe' 1987).

The "horizontal" plot will allow the study of the "second parameter" problem in another GC family, with all its meaningful aspects (e.g. comparison with the MW family, trends with M31 galactocentric distance, and so forth, cf. Fusi Pecci's and Zinn's reviews in this volume). This application of the CMDs of M31 globulars is also of great interest for the study of the integrated light of stellar populations, as clearly demonstrated by the still uncertain interpretation of the integrated spectra of these clusters (Burstein et al 1984; Rose 1985; Renzini 1986c; Fusi Pecci, this volume). Indeed, it will be possible to directly assess as to whether the famous $Mg_2 - H\beta$ anomaly is produced by a systematically different metallicity dependence of the HB morphology (e.g. Cacciari et al 1982), or if other causes need to be envisaged.

5. CONCLUSIONS

Obviously enough, one cannot offer any conclusion concerning HST observations of globular clusters, but there are two points worth making, which both concern the impact on this field of the two-year delay in the planned launch of HST. First, there are now far more cluster CCD data still on tape, than have been elaborated and presented at this meeting. The exciting results we have seen here represent just the tip of the CCD iceberg, while by 1988 an impressive body of first quality cluster CMDs and LFs will likely be available. It is hard to say whether this situation will make advisable to update some of the GTO projects. But certainly it would be important to dispose of complete and deep cluster LFs before GO projects are completely finalized (cf. Section 2.2).

The second point concerns the HST photometric systems (cf. Koornneef et al 1986), which are significantly different from those used so far in ground based observations of GCs. I will then conclude by just asking the following question to the audience: would it be useful to soon start adopting the HST photometric systems also for ground based observations?

I would like to express my gratitude to the Space Telescope Science Institute for its hospitality, and to STScI staff members Ralph Bohlin, Carla Cacciari, Holland Ford, Jan Koornneef, Duccio Macchetto, and Francesco Paresce, for useful conversations about the HST operations and capabilities. I would also like to thank the BBCFS group members for their patience in day by day satisfying my curiosity for their work on M3, and to Franceso Ferraro for computing at my request several LF integrals.

REFERENCES

Bahcall, J. N. 1985 Dynamics of Star Clusters J. Goodman and P. Hut, eds., Reidel, Dordrecht, p. 481.
Bragaglia, A., et al. 1986 Astronet 1984-1985 G. Sedmak, ed., Osservatorio Astronomico, Trieste, p. 431.
Buonanno, R. 1986 Mem Soc. Astron. Italiana xxx.
Burstein, D., Faber, S. M., Gaskell, C. M. and Krumm, M. 1984 Astrophys. J. 287, 586.
Buzzoni, A., Fusi Pecci, F., Buonanno, R. and Corsi, C. E. 1983 Astron. Astrophys. 128, 94.
Cacciari, C., Cassatella, A., Bianchi, L., Fusi Pecci, F. and Kron, R. G. 1982 Astrophys. J. 261, 77.
Da Costa, G. S. 1982 Astron. J. 87, 990.
Gilliland, R. L., Faulkner, J., Press, W. H. and Spergel, D. N. 1986 Astrophys. J. 306, 703.
Iben, I. Jr. and Renzini, A. 1984 Physics Reports 105, 329.
King, C. R., Da Costa, G. S. and Demarque, P. 1985 Astrophys. J. 299, 674.
Koornneef, J., Bohlin, R., Buser, R., Horne, K. and Turnshek, D. 1986 Highlights of Astronomy J.-P. Swings, ed., Reidel, Dordrecht., p. 833.
Macchetto, F. 1982 The Space Telescope Observatory D. N. B. Hall, ed., NASA CP-2244, p. 40.
Mengel, J. G., Sweigart, A. V., Demarque, P. and Gross, P. G. 1979 Astrophys. J. Suppl. 40, 733.
Paczynski, B. 1984 Astrophys. J. 240, 670.
Press, W. H. and Spergel, D. N. 1985 Astrophys. J. 296, 679.
Pritchet, C. J. 1986 Galaxy Distances and Deviations from Universal Expansion B. F. Madore and R. B. Tully, eds., Reidel, Dordrecht, p. 35.
Renzini, A. 1985 Astronomy Express 1, 127.
Renzini, A. 1986a Galaxy Distances and Deviations from Universal Expansion B. F. Madore and R. B. Tully, eds., Reidel, Dordrecht, p. 177.
Renzini, A. 1986b Mem. Soc. Astron. Italiana, in press.
Renzini, A. 1986c Stellar Populations C. A. Norman, A. Renzini and M. Tosi, eds., Cambridge University Press, p. 213.
Renzini, A. 1987 Astron. Astrophys., in press
Renzini, A. and Buzzoni, A. 1986 Spectral Evolution of Galaxies C. Chiosi and A. Renzini, eds., Reidel, Dordrecht, p. 135.
Rood, R. T. and Crocker, D. A. 1985 Production and Distribution of CNO Elements I. J. Danziger, F. Matteucci, and K. Kjar, eds., ESO, Garching, p. 61.
Rose, J. A. 1985 Astron. J. 90, 1927.
Sandage, A. 1982 Astrophys. J. 252, 553.
Stringfellow, G. S., Bodenheimer, P., Noerdlinger, P. D. and Arigo, R. J. 1983 Astrophys. J. 264, 228.
Sweigart, A. V., Renzini, A. and Tornambe', A. 1987 Astrophys. J. in press.

Westphal, J. A. 1982 The Space Telescope Observatory D. N. B. Hall, ed., NASA CP-2244, p. 28.

DISCUSSION

GNEDIN: There is an interesting idea, by Prof. Okun' from the Moscow Inst. of Theo. Phys., concerning the solar neutrino problem. The interaction between the spin of a neutrino and the solar magnetic field may reverse the spin and convert antineutrinos into neutrinos. Therefore one could expect that the antineutrino flux would decrease. As the strength of the magnetic field depends on the solar cycle, one could expect the antineutrino flux to vary with the solar cycle. There is some evidence of this effect.

REES: A comment about WIMPS, in case anyone here takes them seriously. If the Dark matter is in the form of WIMPS, and if globular clusters formed by a "primary" (pregalactic) mechanism, each cluster would be embedded in a "mini-hole" of WIMPS gravitationally bound to it. The stars in globular clusters would then have capture ~ 100 times as many WIMPS as the Sun -- primarily because the capture cross section (taking gravitational focusing into account) goes inversely with the WIMPS random velocities, which would be < 10 km/s for those trapped in clusters, rather than > 200 km/s for those filling the halo.

RENZINI: Yes, so the N_{AGB}/N_{HB} argument may tell something about dark matter very hypothetically associated to globular clusters.

TRIMBLE: Have you given any thought to the effect on estimated ages of globular clusters. Faulkner thought it might be important.

RENZINI: I heard that John has made this suggestion, but I've not seen the preprint.

TRIMBLE: I think the effect will be small, but no exact calculations have been done and, as Martin Rees says, it depends on the number of WIMPS.

RENZINI: It must certainly depend on how many WIMPS one is willing to dump.

ALCAINO: Due to the fact that with the HST we will reach one magnitude below the horizontal branch level of globular clusters in M 31, the HB position will be a useful check to the current distance determination of Andromeda via the Cepheids.

RENZINI: Yes, I agree.

KING: Your suggestion of using HST photometric systems from the ground is a good one. Using synthetic magnitudes (**from** spectrophotometric curves and sensitivities), I have tried to convert one UBV system to another, and so far I have failed.

DA COSTA: The 47 Tuc luminosity function data of Da Costa (1982) agrees well with the original work of Hesser and Hartwick in the subgiant region, and with the newer data of Hesser and Harris at fainter magnitudes (V < 21,5). I do not think that it is possible that a factor of 5 would have been missed.

RENZINI: I think the contrary, and your more recent luminosity function of 47 Tuc (King, Da Costa and Demarque 1985) already indicates that the 1982 luminosity function was severely incomplete at the level of the subgiant branch.

HESSER: The luminosity functions, of which I'm aware, for the 47 Tuc subgiant branch are **heavily** weighted towards photographic data. Our recent CCD study in two small fields at large radii do not sample well the subgiant region, a shortcoming we hope to overcome.

RENZINI: So the point is that there is a factor of 5 disagreement between theory and stellar counts for the subgiant branch of 47 Tuc. I bet for the theory.

BOND: In regard to using the M 31 globular clusters to get the dependence of horizontal-branch luminosity upon metallicity, won't the depth of the M 31 system along the line of sight smear out any relationship?

RENZINI: No, the dispersion in distance modulus is of the order of five hundredths of magnitude, considerably smaller than the effect one is seeking to see.

OZERNOY: After all, what is the best estimate for the age of the oldest globular clusters and what is the uncertainty of the estimate.

RENZINI: As Don VandenBerg mentioned some days ago, the best estimates give 16 billion years with an uncertainty of ~ 25%

SIMULATIONS OF HST OBSERVATIONS OF GLOBULAR CLUSTERS

John N. Bahcall and Donald P. Schneider

Institute for Advanced Study

1. INTRODUCTORY REMARKS

The high angular resolution of the Hubble Space Telescope will provide opportunities for many fundamental observations of globular clusters, most of which have been extensively discussed in the literature. We have therefore chosen to devote our time (and pages) to a presentation of what HST observations may reveal about some aspects of galactic globular clusters. To avoid infringing upon programs that others may propose, we have limited ourselves to simulations of observations that are part of our Guaranteed Time Observations. [The complete catalog of GTO observations has published by the Space Telescope Science Institute and is available upon request.]

The science programs described here are: 1) studies of the cores of nearby galactic globular clusters (exemplified in what follows by pictures of M13); and 2) stellar population studies of a nearby relatively unobscured globular cluster (exemplified in what follows by pictures of NGC 6397). The purposes of the first program include the determination of the stellar density distributions and luminosity functions in the innermost regions of the globular clusters. The second project is a joint investigation with the WF/PC instrument development team and is aimed at detecting white dwarfs and the faint red end of the cluster luminosity function.

2. WHAT IS INCLUDED IN A SIMULATION?

The simulations we present were created using the image processing code developed by one of us (DPS) and refer to the Wide Field/Planetary Camera (Westphal et al. 1982) operating in the Wide Field mode. The Wide Field Camera (WFC) has a field of 160" on a side and an image scale of 0.1" per pixel. Because of hardware limitations, we display only a 500 pixel by 500 pixel subregion of a WFC frame. This is not an important constraint for the purposes of this paper, since our goal is to present a qualitative demonstration of some of the ways that HST observations can supplement ground-based studies.

The recently updated instrument and telescope parameters (Westphal et al. 1986) were used to construct the HST pictures. The instrumental parameters required for the simulations include the point spread function, the quantum efficiency, the filters, the sky background, and the instrumental noise. For the clusters, we need the stellar luminosity function (taken from Bahcall 1985), a distribution of stellar types with absolute magnitude (adapted from Bahcall and Soneira 1984), the central surface brightness, the core radius, the distance, and the reddening. The observational parameters were taken from Peterson and King (1979) and Harris and Racine (1979). We have purposely chosen not to use the latest and most detailed parameterizations from the ground based data, in order that our simulations be properly regarded as only illustrative. Since the dynamic range of the detectors in the WFC far exceeds that of a photograph, the visual appearance of the pictures depends strongly upon what range of intensities (the "stretch") of the data are selected to be displayed. We have made pictures with a variety of stretches and present here the images that we believe are the most representative and revealing.

3. THE PICTURES

The only real data we show is a picture of M13 (distance 6.1 kpc) which is displayed in Figure 1 (and again in zoom-format in Figure 2); this image was obtained with a 10 second exposure using the "4-Shooter" camera (Gunn et al. 1984) and the r filter on the Palomar 5 meter telescope. The image displays the inner regions of this bright globular cluster; the figure is 250" on a side. The exposure was acquired in excellent seeing; the full-width at half-maximum of the stellar images is 0.8". The limiting visual magnitude corresponds to $m_V \sim 22.5^m$.

In order to have a ground-based image that is on the same scale as the HST simulations we will show later, we display in Figure 2 the inner $50'' \times 50''$ of Figure 1. Notice that the images appear to be almost merging blobs, even with excellent seeing conditions.

The first "HST" image is shown in Figure 3. This simulation represents a 10 second exposure of the core of M13 taken with the (wide visual) F555W filter. There are approximately 10^3 stars visible down to the limiting magnitude of $m_V \sim 22.5^m$.

How would the field of Figure 3 appear if it had been observed in 0.8" seeing with the same filter? Figure 4 shows the simulated ground-based image. Note that enormously fewer stars are visible and that many of the images that are clearly separated in Figure 3 have merged to form amorphous blobs in Figure 4. Many of the bright "stars" in Figure 4 are really small stellar groups (cf. Figure 3), which suggests caution is required in interpreting ground-based observations of dense cluster cores. The simulated ground-based picture shown in Figure 4 is similar to what has been obtained under good conditions by observers who have studied the cores of globular clusters.

HST observations of the inner regions of globular clusters will make important contributions to the observational understanding of core collapse and mass segregation.

Figures 5 and 6 show HST images of NGC 6397 (distance 2.1 kpc). The ultraviolet exposure is shown in Figure 5 and represents a 45 minute (full-orbit) image taken with the F336M filter (effective wavelength about 3360 Å). The limiting visual magnitude is about $m_V \sim 27^m$. The I-band picture is shown in Figure 6 and represents a 7.5 minute exposure with F785LP; it reaches to about $m_I \sim 26^m$. Saturation of the CCD charge storage capacity by bright stars creates the ugly streaks that are so prominent in Figure 6 and which, unfortunately, will be a common occurence of HST images of globular clusters. It will not be easy to perform accurate photometry with such data, especially when one considers that the images are severely undersampled.

We hope to observe $\sim 10^2$ white dwarfs on the full WFC image. A rather different estimate has been given by Renzini (this conference), who based his results on theoretical stellar evolution calculations. With the many uncertainties (theoretical and observational, including cooling; see Bahcall 1985) that enter such predictions, an uncertainty in the estimated number of white dwarfs of order a factor of 5 or even 10 is not surprising. The first HST observations of white dwarfs may provide important constraints on calculations of the advanced stages of stellar evolution.

We also expect to observe $\sim 10^4$ faint red dwarfs (on the full WFC image) in an effort to determine the low-mass end of the cluster luminosity function, and with luck, obtain some insight into the initial mass function of globular clusters.

The images shown in Figures 1-6 give what we believe is a representative impression of the potentialities of the HST for studies of the cores of galactic globular clusters and for measuring the faint ends of the cluster luminosity functions.

This work was supported in part by NASA grants NAS8-32902 and NAS5-29225.

REFERENCES

Bahcall, N. N. and Soneira, R. M. 1984 Astrophys. J. 55, 67.
Bahcall, J. N. and Soneira, R. M. 1985 in IAU Symposium No. 113, Dynamics of Star Clusters, J. Goodman and P. Hut, eds., Reidel, Dordrecht, p. 481.
Gunn, J. E. 1984 Bull. A. A. S. 16, 477.
Harris, W. E. and Racine, R. 1979 Ann. Rev. 17, 241.
Peterson, C. J. and King. I. R. 1975 Astron. J. 80, 427.
Westphal, J. A. and WF/PC IDT Team 1982 in The Space Telescope Observatory, IAU 18th General Assembly, D. B. Hall, ed., NASA CP 2244, Washington, p. 28.
Westphal, J. A. and WF/PC IDT Team 1986, private communication.

Fig. 1. A Palomar 5 meter Image of the Inner Region of M13. The exposure was obtained with excellent seeing using an r filter.

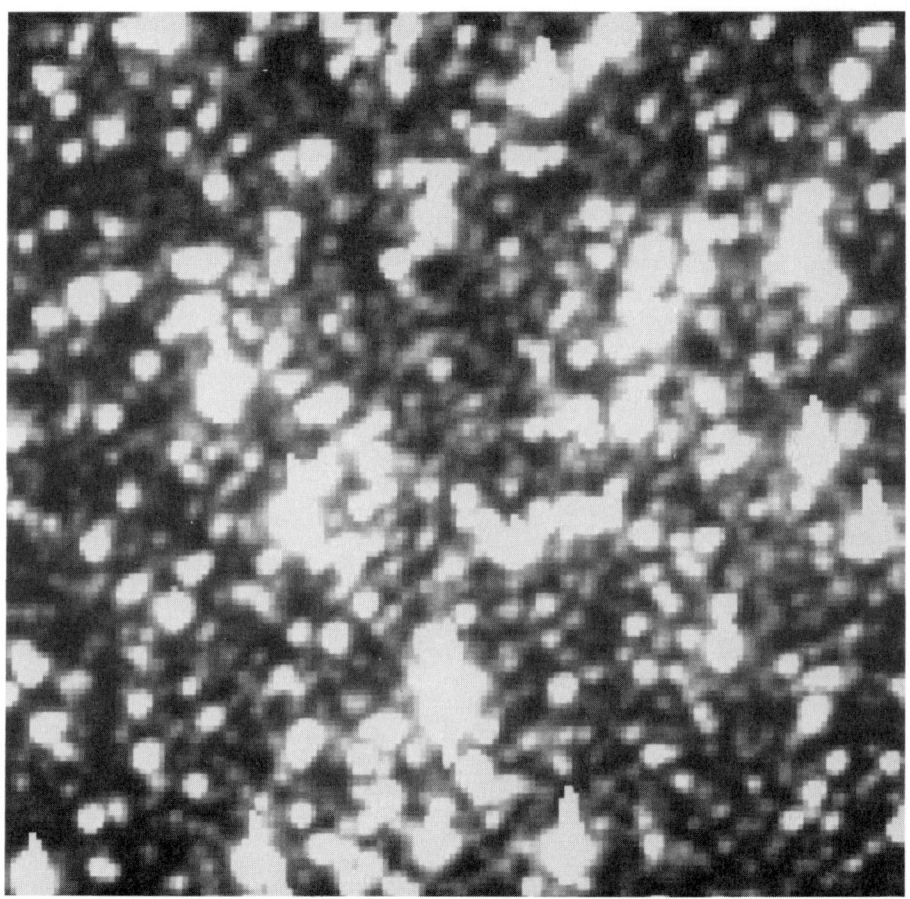

Fig. 2. The Innermost 50″ × 50″ of Figure 1.

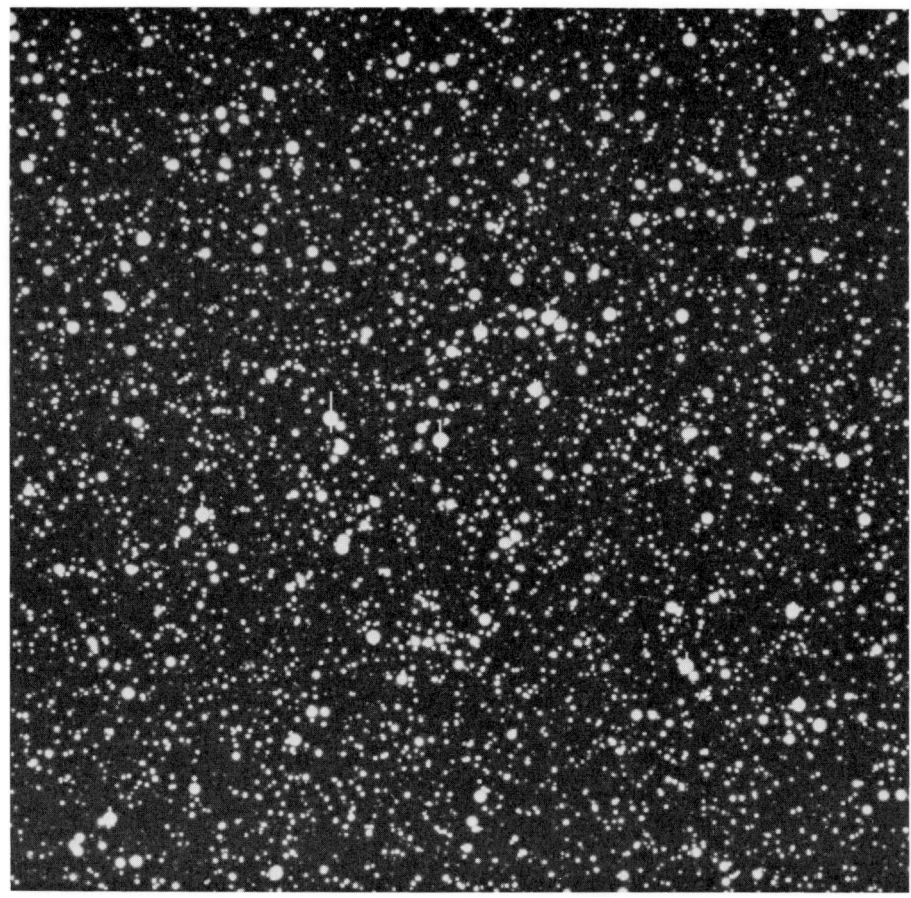

Fig. 3. A Simulated HST Picture of the Core of M13 with a Visual Filter. The field is $50'' \times 50''$.

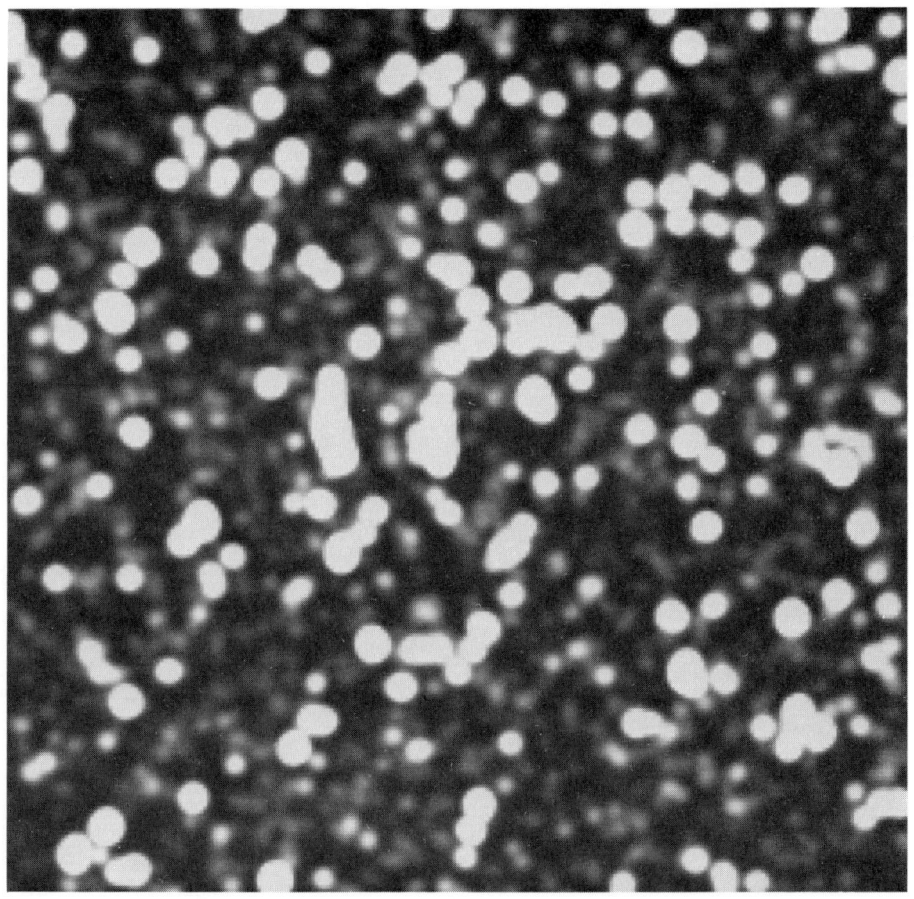

Fig. 4. The Field of Figure 3 "Observed" with 0.8″ Resolution.

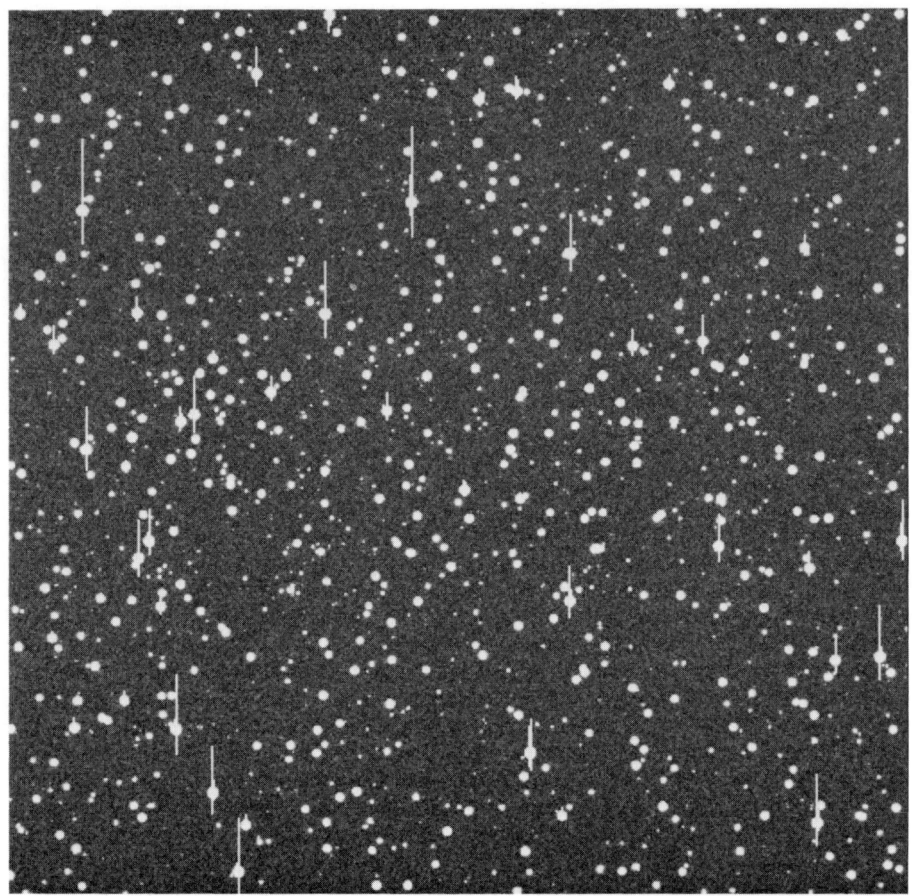

Fig. 5. A Simulated Ultraviolet HST Exposure of the Core of NGC 6397.

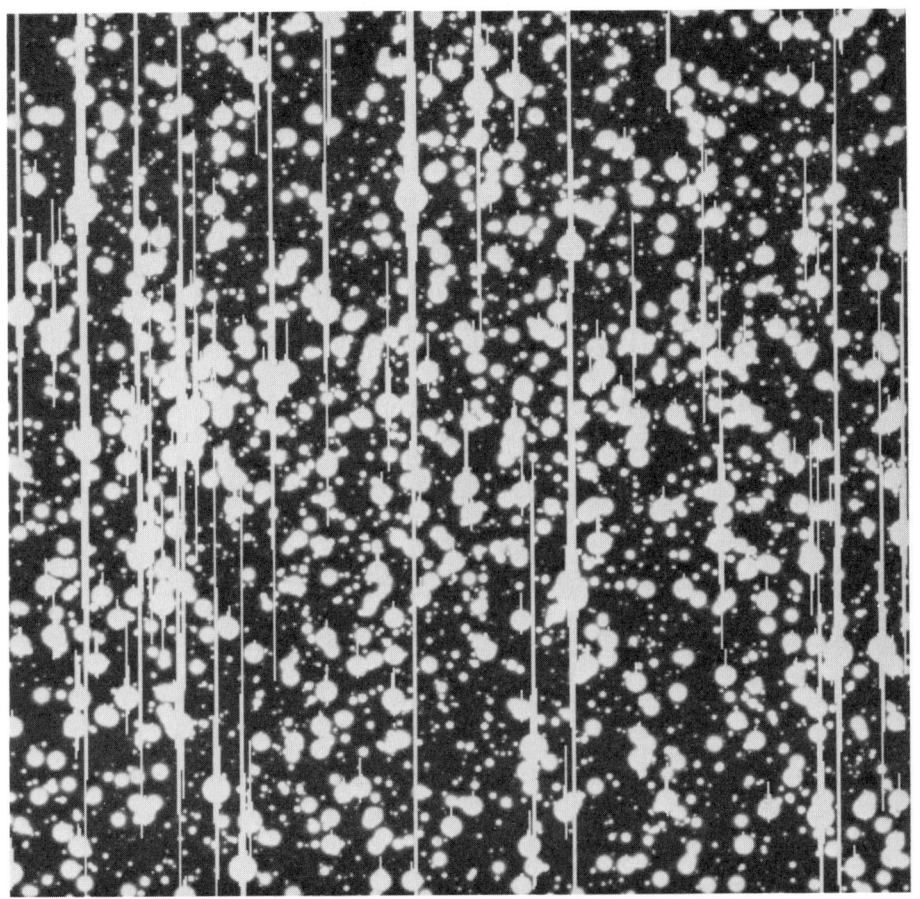

Fig. 6. A Simulated I-band HST Exposure of the Core of NGC 6397.

DISCUSSION

CHRISTIAN: While it is appreciated that we should carefully consider the expected performance of the HST, and that comparison of existing ground based data to HST data will be interesting if not problematic. I think the capabilities of ground based facilities have been underestimated in this talk. At CFHT we have obtained images with FWHM of 0.25 arc sec. If we had available the HST budget for 2 months we could commission such facilities such as adaptive optics, tip-tilt mirrors and fast shutters: then we would be able to achieve very high resolution (spatial) routinely in the next few years before HST is launched.

COHEN: The 200" inch mirror is only good to 0.6", based on Hartmann tests. I agree with everything Carol Christian said. The base atmosphere above a high site has a PSF with about 0.3" FWHM. All new ground-based telescopes will be built to take advantage of the best ground-based seeing. Image profiles on existing ground-based telescopes could be significantly improved if funds were available.

KING: May I briefly describe the FOC Guaranteed Time program? In five globular clusters (ω Cen, 47 Tuc, NGC 6397, 6624 and 6752) we will take one orbit B and V exposures at the center, 1 r_c and 3 r_c. The outer parts should be done from the ground. We will also look at the central profiles of about a dozen collapsed-core clusters. In M 31 we have 14 globular clusters on the program.

Chapter VI

Review Papers

Summary

Josh Grindlay thanks attendees for not being in China

Another view of the Symposium participants

GLOBULAR CLUSTER SYSTEMS IN GALAXIES: MAIN TRENDS AND FUTURE DIRECTIONS

Sidney van den Bergh

Dominion Astrophysical Observatory
Hertzberg Institute of Astrophysics

This was a really exciting conference! The main problem in summarizing it is that so many new and important results were presented, making it impossible to mention each one individually. In reviewing this conference I shall first discuss new results, then problem areas revealed by papers and during discussions and finally desiderata and trends for future work.

1. NEW RESULTS:

Perhaps the most impressive feature of this conference has been the tremendous increase in the observational database on Galactic and extra-Galactic globular clusters. Mainly due to the advent of CCD detectors numerous new precision color-magnitude diagrams (Peterson 1986) have recently become available. Hesser, Shawl and Meyer (1986) have provided a uniform database on the radial velocities of globular clusters and Peterson and Reed (1987) and Djorgovski (1987a) have provided much new information on the structural parameters of globulars. Finally Harris (1987) has given us an excellent review of recent progress on the study of globular clusters in extra-Galactic systems.

2. At a number of other meetings on globular clusters (e.g. IAU Colloquium No. 68) and Population II during the last few years the abundance scale for globular clusters had been a subject of lively controversy. This problem now appears to have been settled to almost everyone's satisfaction. The source of problems with the [Fe/H] scale derived from echelle spectra seems to have been traced to difficulties with the placement of the stellar continuum.

3. One of the most exciting results presented at this conference consisted of the discovery by McClure et al. (1987) that the slopes of the luminosity functions of globular clusters correlate with their metallicity in the sense that metal-poor clusters have steeper mass spectra than do metal-rich ones. If this correlation is confirmed by subsequent observation it might provide valuable new insights into the

mechanisms by which globular clusters form and subsequently fragment into stars. A slight extrapolation of the observations to higher masses and luminosity suggests that flat spectrum (high metallicity) clusters will contain a larger number of neutron stars than metal-poor clusters, as discussed by Grindlay (1987).

4. Don VandenBerg (1987) showed us that models for metal-poor stars, that are enriched in oxygen, appear to give a slightly better fit to the observations than do models in which the mix of heavy elements is essentially solar. This result is, perhaps, not unexpected since supernovae, such as Cassiopeia A, (which had a very massive progenitor), are observed to eject large quantities of oxygen. With this enhanced oxygen abundance globular cluster ages are reduced to 13 to 14 Gyr. Finally VandenBerg showed that presently available color-magnitude diagrams for globular clusters are consistent with the hypothesis that any age differences between Galactic globular clusters are < 1.5 Gyr. It would, however, clearly be very worthwhile (and difficult!) to extend main sequence photometry to globular clusters situated very close to the Galactic center and to clusters in the outer Galactic halo to see whether cluster formation took place synchronously over the entire proto-Galaxy. A number of Symposium speakers emphasized the fact that populous clusters in the LMC and SMC exhibit a wide age range in contradistinction to the Galactic globulars which all appear to be approximately coeval.

5. Harris (1987) showed that the Coma galaxy NGC 4874 has a high specific globular cluster frequency. This observation strengthens the conjecture, previously based on observations of only 3 galaxy clusters, that the central galaxies in rich clusters are special. In fact such central galaxies must be special <u>ab initio</u> if the globular clusters in such systems are older than their parent galaxies. Considerable support for this idea was provided by new observations by Judy Cohen (1987) which support similar results obtained previously by Forte, Strom and Strom (1981).

6. We now understand that spheroidal galaxies (Wirth and Gallagher 1984, Kormendy 1985) have structural parameters which show that they differ from ellipticals. At this meeting Gary Da Costa (1987) showed us that the stellar content of dwarf spheroidal galaxies differs substantially from that encountered in Galactic globular clusters. It is important to remember that the spheroidal galaxy "form family" comprises both relatively luminous objects like NGC 147, NGC 185 and NGC 205 and low luminosity dwarf objects such as And I, And II, And III near M31 and the "seven dwarfs" associated with our own galaxy.

7. The radial density distribution of globular clusters appears to be intermediate between that of elliptical galaxies, for which $\rho \propto r^{-3}$, and that of dark matter which has $\rho \propto R^{-2}$. In particular the observations by Mould et al. (1986) and Huchra (1987) demonstrate that the globular clusters associated with M87 exhibit kinematical properties which show that the mass distribution derived from globular

clusters smoothly joins that derived from integrated starlight (Sargent et al. 1978) and that calculated from X-ray emission of hot gas by Fabricant and Gorenstein (1983).

8. Tremendous progress (Inagaki 1987, Cohn 1987) has recently been made in our understanding of the pre- and post core collapse phases of cluster evolution. These studies show that clusters will expand on a timescale comparable to the Hubble time after core collapse i.e. there is life after core collapse! The importance of these theoretical studies is enhanced by recent observations of cusps in clusters cores by Djorgovski and King (1986). John Bahcall (1987) provided some stunning simulations of Space Telescope data which show how much information ST may be expected to provide on the crowded central regions of globular clusters. In the meantime studies of the same problem are proceeding at somewhat lower resolution with the Canada-France-Hawaii Telescope.

9. Innanen (1987) pointed out that Population II stars with low angular momentum orbits will be scattered into "chaotic orbits" by the mass concentration near the nucleus of our galaxy. As a result there should be a marked deficiency of such stars in the vicinity of the Sun. From inspection of orbital data in the Gliese and Woolley Catalogs Innanen very tentatively concludes that the circular velocity near the Sun is 235 ± 10 km s^{-1}. It is particularly interesting to note that this value does not depend on any assumption regarding R_0. It would be very interesting to strengthen these results by looking at the velocity components of the Lowell proper motion stars.

10. In his review Renzini (1987) pointed out that observations of globular clusters may already rule out the existence of weakly interacting massive particles (WIMPs) via the effect that these objects have on the computed lifetimes of globular cluster horizontal branch stars. With semi-convection, $\tau_{AGB}/\tau_{HB} = 0.14$ compared to $\tau_{ASB}/\tau_{HB} > 1.5$ for WIMP models. The observed value $\tau_{AGB}/\tau_{HB} = 0.15$ seems to rule out the presence of WIMPs.

PROBLEMS:

1. More than twenty years ago (van den Bergh 1965) I stumbled on the "second parameter" problem. At that time I suggested that this second parameter might be due to either age differences between clusters or to different relative abundances of elements heavier than hydrogen. During the last two decades much new observational evidence has been found for the existence of a second parameter but our understanding of its nature has not really improved. In fact the problem may have become somewhat worse. In his review, Jim Hesser (1987) showed us that modern high-quality color-magnitude diagrams of such clusters as M13, M92 and NGC 5466 show subtle but distinct differences among blue horizontal branch clusters suggesting that more than one parameter, in addition to metallicity, may be required to describe cluster color-magnitude diagrams.

2. Our understanding of the Oosterhoff (1939) classes (van Albada and Baker 1972) has not really improved in recent years nor has our knowledge of the metallicity dependence of the absolute magnitudes of RR Lyrae stars (Sandage 1981). At the present meeting papers like those of Cacciari, Clementini and Prevot (1987) and Jones et al. (1987) show no evidence for the expected metallicity dependence of the absolute magnitudes of RR Lyrae stars. Recent observations by Pritchet and van den Bergh (1987) show that the halo of M31 appears to resemble an Oosterhoff Type I population. For 24 RR Lyrae stars $\langle P_{ab} \rangle$ = 0.548 while the variables with largest amplitudes have 0.45 < P < 0.50 days.

3. One of the most hotly debated issues at this meeting was the question as to whether globular clusters represent fair samples of halo populations. From the similarity of the kinematics of Galactic high velocity stars and globular clusters the answer to this question appears to be "yes". On the other hand Harris (1987) emphasized that there are important differences between globular clusters and associated halo populations in external galaxies:
(a) In elliptical galaxies the distribution of globular clusters is more extended than that of the background galaxy light.
(b) Isopleths of globular clusters are found to be rounder than the galactic isophotes.
(c) Globular clusters are bluer (and hence presumably metal-poorer) (Cohen 1987, Forte, Strom and Strom 1981) than the galaxy background on which they are superimposed.
(d) Globular cluster systems appear to be dynamically hotter (Mould, Oke and Nemec 1986, Huchra 1987, Lauer and Kormendy 1986 and Grillmair, Pritchet and van den Bergh 1986) than their parent galaxies.

Ostriker (1987) cautioned that the presently observed distribution of globular clusters might have been affected by tidal friction, disk shocks, etc. The observation by Grillmair et al. (1986) that there is no noticeable radial gradient in the luminosity function of the M87 globular clusters does, however, place constraints on the importance of dynamical friction as a contributor to the difference in the core size of M87 and its globular cluster system. Detailed numerical simulations by Muzzio (1987) have shown that swapping of globular clusters may be an important factor for cluster galaxies. His computations do, however, show that stripping of normal cluster galaxies cannot account for the huge excess cluster populations associated with central galaxies of rich clusters.

4. Pritchet and van den Bergh (1987) have studied a 2 x 3 arcmin field situated along the minor axis of M31 at a distance of 40 arcmin (9 kpc) from the nucleus. In this field they detected 32 variables of which 30 are either confirmed or probable RR Lyrae stars. From the frequency of multiple discoveries of individual variables these authors estimate that their data are only ~25% complete. The total number of RR Lyrae stars in their field is therefore probably of the order of 120. For comparison it is noted that this M31 field would contain only ~2 variables if its stellar population were similar to that of the

cluster 47 Tucanae and ~200 variables if it were like M3. From the study of the red giant stars in a field adjacent to that of Pritchet and van den Bergh, Mould and Kristian (1986) estimate a mean field star metallicity <[Fe/H]> ~ -0.6 i.e. a value similar to that of 47 Tucanae for which Gratton, Quarta and Ortolani (1987) find [Fe/H] = -0.8. The most likely explanation for the observation that the M31 halo is both variable-rich and relatively metal-rich is that the halo of the Andromeda nebula contains a population of old metal-rich stars with a strongly developed horizontal branch. In this respect the halo of the Andromeda nebula would then resemble the M31 globular clusters which exhibit strong metallic lines and strong Hδ simultaneously (Burstein et al. 1984). Confirmation of the hypothesis that M31 globular clusters have unexpectedly strong blue horizontal branches is provided by the IUE observations of the cluster Bo 158 (Cacciari et al. 1982). The most straightforward interpretation of these observations is that the M31 globular clusters and the halo stars in the Andromeda nebula exhibit similar "family traits" which differ systematically from those of halo stars and globular clusters in the Galaxy. A special search for RR Lyrae stars in metal-rich Galactic globular clusters (Hazen 1987) shows a low specific frequency of RR Lyrae stars in the clusters NGC 6388 and NGC 6652. A specific frequency of cluster type variables, which is only a factor a 3 or 4 lower than that in the M31 halo field studied by Pritchet and van den Bergh (1987), is found in NGC 6569. It would be interesting to see whether the relatively high RR Lyrae star frequency in the latter cluster is unusual or whether Zinn and West's (1984) metallicity estimate [Fe/H] = -0.86 might be too high.

5. Recently Lacey and Ostriker (1985) have suggested that the missing mass in galactic halos might be in the form of ~10^6 $M_\odot$ black holes. Wielen (1987) examined the effects which the presence of ~10^5 such "Nemeses" would have on Galactic globular clusters. His calculations showed that a large population of black holes might devastate parts of the Galactic globular cluster system. More detailed computations will be required to see which constraints the observed distribution and luminosity function of Galactic globular clusters places on such hypothetical massive black holes in the Galactic halo.

FUTURE:

1. All indications are that high-resolution ground-based observations, and eventually observations from Space Telescope, will contribute tremendously to our understanding of density cusps and collapsed cores in globular clusters (Bahcall 1987). Such observations should also provide information on gradients of blue stars (Nemec and Harris 1987), the hypothesized dwarf enrichment in some cluster cores (Rose and Tripicco 1986, 1987), X-ray sources etc.

The discovery of highly obscured globular clusters (mainly near the Galactic center) would be important so that we can study a complete sample of Galactic globular clusters. Djorgovski's (1987b) use of the IRAS database to search for obscured clusters near the Galactic center is a good first step in this direction.

3. It was somewhat disappointing to me that so little attention was paid to the flattening of globular clusters at this meeting. We now know from kinematical studies of a number of globular clusters that this flattening is due to rotation. Flattening of clusters therefore provides information on the angular momentum of globular clusters. A major mystery is why galactic open clusters and globular clusters are almost spherical whereas the young and old clusters in the Magellanic Clouds are much more flattened (van den Bergh 1984). Good first steps in the study of cluster ellipticities are provided by White and Shawl (1987), by Djorgovski (1987a) and by Spassova, Staneva and Golev (1987) who have studied the flattenings of clusters in M31.

4. The new observations reported by Harris (1987) greatly strengthen the hypothesis that the central galaxies in rich clusters are unusual ab initio. Since cD galaxies are central galaxies par excellence it would be very interesting to study the frequency and distribution of globular clusters in cD galaxies. The nearest example of a cD galaxy in a rich cluster is NGC 6166 with V_0 = 9075 km s^{-1}. If this object is similar to M87 one would expect it to contain ~140 globular clusters brighter than B = 25 and ~900 globulars with B < 26. These objects would appear to be within range of CFHT and could be studied in great detail with the Hubble Space Telescope.

REFERENCES

Bahcall, J. 1987 IAU Symposium No. 126, Globular Cluster Systems in Galaxies, J. E. Grindlay A. G. D. Philip, eds., Reidel, Dordrecht, p. 455.

Burstein, D., Faber, S. M., Gaskell, C. M. and Krumm, N. 1985 Astrophys. J., 287, 586.

Cacciari, C., Cassatella, A., Bianchi, L., Fusi Pecci, F. and Kron, R. G. 1982 Astrophys. J., 261, 77.

Cacciari, C., Clementini, G. and Prevot, L. 1987 in IAU Symposium No. 126, Globular Cluster Systems in Galaxies, J. E. Grindlay and A. G. D. Philip, eds., Reidel, Dordrecht, p. 587.

Cohen, J. 1987 in IAU Symposium No. 126, Globular Cluster Systems in Galaxies, J. E. Grindlay and A. G. D. Philip, eds., Reidel, Dordrecht, p. 605.

Cohn, H. 1987 in IAU Symposium No. 126, Globular Cluster Systems in Galaxies, J. E. Grindlay and A. G. D. Philip, eds., Reidel, Dordrecht, p. 379.

Da Costa, G. S. 1987 in IAU Sympsium No. 126, Globular Cluster Systems in Galaxies, J. E. Grindlay and A. G. D. Philip, eds., Reidel, Dordrecht, p. 217.

Djorgovski, S. G. 1987a in IAU Symposium No. 126, Globular Cluster

Systems in Galaxies, J. E. Grindlay and A. G. D. Philip, eds., Reidel, Dordrecht, p. 333.
Djorgovski, S. G. 1987b in IAU Symposium No. 126, Globular Cluster Systems in Galaxies, J. E. Grindlay and A. G. D. Philip, eds., Reidel, Dordrecht, p. 527.
Djorgovski, S. and King. I. R. 1986 Astrophys. J. Letters, 305, L61.
Fabricant, D. and Gorenstein, P. 1983 Astrophys. J., 267, 535.
Forte, J. C., Strom, S. E. and Strom, K. E. 1981 Astrophys. J. Letters, 245, L9.
Gratton, R. G., Quarta, M. L. and Ortolani, S. 1987 in IAU Symposium No. 126, Globular Cluster Systems in Galaxies, J. E. Grindlay and A. G. D. Philip, eds., Reidel, Dordrecht, p. 495.
Grillmair, C., Pritchet, C. and van den Bergh, S. 1986 Astron. J., 91, 1328.
Harris, W. E. 1987 in IAU Symposium No. 126, Globular Cluster Systems in Galaxies, J. E. Grindlay and A. G. D. Philip, eds., Reidel, Dordrecht, p. 237.
Hazen, M. L. 1987 in IAU Symposium No. 126, Globular Cluster Systems in Galaxies, J. E. Grindlay and A. G. D. Philip, eds., Reidel, Dordrecht, p. 593.
Hesser, J. E. 1987 in IAU Symposium No. 126, Globular Cluster Systems in Galaxies, J. E. Grindlay and A. G. D. Philip, eds., Reidel, Dordrecht, p. 61.
Hesser, J. E., Shawl, S. J. and Meyer, J. E. 1986 Publ. Astron. Soc. Pacific, 98, 403.
Huchra, J. 1987 in IAU Symposium No. 126, Globular Cluster Systems in Galaxies, J. E. Grindlay and A. G. D. Philip, eds., Reidel, Dordrecht, p. 255.
Inagaki, S. 1987 in IAU Symposium No. 126, Globular Cluster Systems in Galaxies, J. E. Grindlay and A. G. D. Philip, eds., Reidel, Dordrecht, p. 367.
Innanen, K. A. 1987 in IAU Symposium No. 126, Globular Cluster Systems in Galaxies, J. E. Grindlay and A. G. D. Philip, eds., Reidel, Dordrecht, p. 423.
Jones, R. V., Carney, B. W., Latham, D. W. and Kurucz, R. L. 1987 in IAU Symposium No. 126, Globular Cluster Systems in Galaxies, J. E. Grindlay and A. G. D. Philip, eds., Reidel, Dordrecht, p. 589.
Kormendy, J. 1985 Astrophys. J., 295, 73.
Lacey, C. G. and Ostriker, J. P. 1985 Astrophys. J., 299, 633.
Lauer, T. R. and Kormendy, J. 1986 Astrophys. J. Letters, 303, L1.
McClure, R. D., Stetson, P. B. Hesser, J. E.Smith, G. H. and VandenBerg, D. A. 1987 in IAU Symposium No. 126, Globular Cluster Systems in Galaxies, J. E. Grindlay and A. G. D. Philip, eds., Reidel, Dordrecht, p. 485.
Mould, J. and Kristian, J. 1986 Astrophys. J., 305, 591.
Mould, J. R., Oke, J. B. and Nemec, J. M. 1986, preprint.

Muzzio, J. C. 1987 in IAU Symposium No. 126, Globular Cluster Systems in Galaxies, J. E. Grindlay and A. G. D. Philip, eds., Reidel, Dordrecht, p. 297.
Nemec, J. M. and Harris, H. C. 1987 in IAU Symposium No. 126, Globular Cluster Systems in Galaxies, J. E. Grindlay and A. G. D. Philip, eds., Reidel, Dordrecht, p. 677.
Oosterhoff, P. T. 1939 Observatory, 62, 104.
Ostriker, J. P. 1987 in IAU Symposium No. 126, Globular Cluster Systems in Galaxies, J. E. Grindlay and A. G. D. Philip, eds., Reidel, Dordrecht, p. 271.
Peterson, C. J. 1986 (preprint).
Peterson, C. J. and Reed, B. R. 1987 in IAU Symposium No. 126, Globular Cluster Systems in Galaxies, J. E. Grindlay and A. G. D. Philip, eds., Reidel, Dordrecht, p. 489.
Pritchet, C. J. and van den Bergh, S. 1987, preprint.
Renzini, A. 1987 in IAU Symposium No. 126, Globular Cluster Systems in Galaxies, J. E. Grindlay and A. G. D. Philip, eds., Reidel, Dordrecht, p. 443.
Rose, J. A. and Tripicco, M. J. 1986 Astron. J., 92, 610.
Rose, J. A. and Tripicco, M. J. 1987 in IAU Symposium No. 126, Globular Cluster Systems in Galaxies, J. E. Grindlay and A. G. D. Philip, eds., Reidel, Dordrecht, p. 499.
Sandage, A. 1981 Astrophys. J., 248, 161.
Sargent, W. L. W., Young, P. J., Boksenberg, A., Shortridge, K., Lynds, C. R. and Hartwick, F. D. A. 1978 Astrophys. J., 221, 731.
Spassova, N. M., Staneva, A. V. and Golev, V. K. 1987 in IAU Symposium No. 126, Globular Cluster Systems in Galaxies, J. E. Grindlay and A. G. D. Philip, eds., Reidel, Dordrecht, p. 569.
van Albada, T. S. and Baker, N. H. 1972 in The Evolution of Population II Stars, A. G. D. Philip, ed., Dudley Observatory Report No. 4, Albany, p. 193.
VandenBerg, D. A. 1987 in IAU Symposium No. 126, Globular Cluster Systems in Galaxies, J. E. Grindlay and A. G. D. Philip, eds., Reidel, Dordrecht, p. 107.
Van den Bergh, S. 1965 Journ. Roy. Astron. Soc., 59, 151.
Van den Bergh, S. 1984 in IAU Symposium No. 108, Structure and Evolution of the Magellanic Clouds, S. van den Bergh and K. S. de Boer, eds., Reidel, Dordrecht, p. 1.
White, R. E. and Shawl, S. J. 1987 in IAU Symposium No. 126, Globular Cluster Systems in Galaxies, J. E. Grindlay and A. G. D. Philip, eds., Reidel, Dordrecht, p. 491.
Wielen, R. 1987 in IAU Symposium No. 126, Globular Cluster Systems in Galaxies, J. E. Grindlay and A. G. D. Philip, eds., Reidel, Dordrecht, p. 393.
Wirth, A. and Gallagher, J. S. 1984 Astrophys. J., 282, 85.
Zinn, R. and West, M. J. 1984 Astrophys. J. Suppl., 55, 45.

Chapter VII

Poster Papers

Harlow Shapley

Globular Clusters in the Milky Way

Drs. McCarthy, Gingerich, Hazen, and Alcaino examine historic plate

Bob McClure explains his pivotal poster to Rene Racine, Juan Forte and others

HARLOW SHAPLEY: A VIEW FROM THE HARVARD ARCHIVES

Barbara L. Welther

Harvard-Smithsonian Center for Astrophysics

ABSTRACT: This exhibit featured facsimiles of some letters that Shapley exchanged with George Ellery Hale, Henry Norris Russell, and Heber Doust Curtis from 1917 when he was at Mount Wilson working on globular clusters to 1921 when he became Director of Harvard College Observatory.

The Harvard University Archives holds a rich collection of Shapley's correspondence and memorabilia that spans seven decades of his life: from his diary begun in 1902 when he was 16 to letters received in 1972 when he died. Facsimiles of his correspondence with Hale, Russell, and Curtis were selected to reveal his aspirations for the directorship of Harvard College Observatory, as well as his intensity in studying the clusters at Mount Wilson from 1917 to 1920. His work, which revolutionized previous notions about the scale of the universe, led Hale to invite Shapley to present his novel views before the National Academy of Sciences. To enliven the presentation, Hale also invited Curtis to state his more conservative views. Engaged in nebular photography at Lick, Curtis was skeptical of the "startling if true" results from Mount Wilson: Shapley's calibratrion of the Period-Luminosity diagram and Van Maanen's rotational velocities in spirals.

Among the facsimiles exhibited was a proposal from Shapley to Hale in 1917 to publish a series of papers in both the *Astrophysical Journal* and the *Contributions from Mount Wilson Solar Observatory* on the colors and magnitudes in globular clusters. In reply Hale wrote to Shapley very supportively:

> *I have always felt that a comprehensive investigation of star clusters should prove exceedingly valuable, and it is a pleasure to see that you are fully realizing my expextations* [sic] *in this regard.*

Also supportive of Shapley was his former mentor at Princeton University, Henry Norris Russell, with whom Shapley kept up a lifelong friendship and correspondence. However, in February 1919, a few days after Pickering's death, when Shapley wrote to Russell that he aspired to become the next director of Harvard College Observatory, Russell rebuked his ambition:

> ... I would be very glad to see you in a good position at Harvard, free from executive cares,... under a sympathetic director,... But I would not recommend you for Pickering's place, and I believe that you would make the mistake of your life if you tried to fill it.

Less than two years later when Shapley did, indeed, receive a Harvard appointment, Russell did an about-face in his letter of January 24, 1921:

> Glory, glory, hallelujah! I "did my possible" ... to induce President Lowell to appoint you....
>
> I am delighted to think of you at Harvard.... It is a man-size job, but you can swing it, ... you can get good advice if you want it ... Bailey ... Schlesinger ... myself ... I have learned a good deal in the last three years, — principally what a fool I used to be. But I was no fool when I tried to get you at Harvard.

Meanwhile, before Shapley received his Harvard appointment, he had to debate Curtis in 1920. In February of that year, Hale sent him a telegram:

PROGRAM COMMITTEE NATIONAL ACADEMY PROPOSES DEBATE BETWEEN YOU AND HEBER CURTIS ON SUBJECT SCALE OF UNIVERSE ...

Hale's request posed a dilemma for Shapley. Because Hale had been so supportive of him, Shapley felt obliged to honor the invitation to debate Curtis. However, because he feared the debate could jeopardize his chances for Harvard, Shapley did not wish to appear in a scientific forum against Curtis, a much more experienced speaker than himself. Letters from Curtis did not allay Shapley's anxiety:

> I agree with you that it should not be made a formal "debate", but I am sure that we could be just as good friends if we did go at each other "hammer and tongs",... For my part, I am quite willing that you should attack the island universe theory and the smaller dimensions for the galaxy to the limit, provided you will let me support my side and attack yours, also to the limit....

Ultimately, Shapley need not have worried about the outcome of the debate. He soon stepped into the directorship of Harvard College Observatory, where he spent the next three decades garnering many prestigious honors, gold medals, and scientific positions.

In addition to the facsimile letters, the exhibit contained copies of Shapley's papers on the clusters and the debate, newspaper clippings of his Harvard appointment, Hale's telegram, and period photographs of Curtis, Hale, Russell and Shapley.

ACKNOWLEDGMENTS

I wish to thank the Harvard University Archives for help in locating materials in the Shapley collection and permission to quote from them. Thanks also to Owen Gingerich for suggesting this exhibit and giving invaluable criticisms and comments.

HARLOW SHAPLEY AND THE UNIVERSITY OF MISSOURI

Charles J. Peterson

University of Missouri - Columbia

The Laws Observatory of the University of Missouri is an example of the small university observatory that under lean financial circumstances succeeded for a number of years (1890-1920) as a successful research institution. Productive research appears to have been due to the temperament of the individual astronomers who had the ability to find projects which could be accomplished with small telescopes, minimal equipment, and few other resources. In these circumstances, a handful of students began careers in astronomy at the University of Missouri.

The Laws Observatory began in 1880 with the acquisition of a 7.5 inch Merz und Sohne refractor originally manufactured in 1848. Initially used only for instructional purposes, this changed with establishment of a Department of Astronomy in 1893 under the direction of Milton Updegraff, the first professionally trained astronomer on the faculty. Both the size and age of the Merz telescope were handicaps to Updegraff; however, many observatories with newer and larger telescopes were much less productive in research. Lack of funds for repairs and to modernize the Observatory equipment were not all that troubled him; appended to an "Observatory Report" (Updegraff 1894, Observatory 17, 246), the editors noted, "We hope to hear more of Mr. Updegraff, although he is somewhat hampered with his University duties." Updegraff resigned in 1899 and was soon replaced by Frederick H. Seares.

Although many of the problems about which Updegraff had so persistantly complained continued through the tenure of Seares, Seares seems to have been more able to work under the conditions that prevailed. His requests for renovation of Observatory facilities tended to be less expensive and with improvement in University finances [these years would actually be long remembered as a Golden Age according to later University historians (F.F. Stephens 1962, A History of the University of Missouri)], limited resources became available for new equipment and other needs. As Seares would later write (19 October 1911) to his successor: "I hope that you may have as much pleasure in the work there as I had during the years that I held the position. As you say, the observatory is not what one would call modern." In reference to the main telescope, "I am afraid the driving clock of the 7 1/2 inch is hopeless. I never attempted to use it for any purpose except to keep the instrument under way in a rough kind of fashion when visitors were using the instrument." Otherwise, "you will find any amount of polite interest in the affairs of the institution, but it is extremely difficult to carry the matter beyond that point, - at least such was my experience. The University authorities themselves, however, I always found to be most generously inclined within the limits of their financial abilities, which, of course, were necessarily very limited...." Thus in an era in which research at the University was not considered a highly important part of the duties of a professor, Seares was able to build a reputation for research.

Actually Seares was able to make considerable additions to the Observatory equipment. Of particular relevance to his observations of variable stars were acquisitions of Pickering photometer (which used a calibrated absorbing wedge to dim an artificial light source to match the brightness of the star being observed) and a Zöllner-Müller photometer (which used polarizing prisms to vary the intensity of the artificial comparison light). These instruments were mounted on a new 4.5 inch refractor permitting work on fifth to tenth magnitude stars. He also acquired by gift a significant addition of books and journals for the Observatory library and initiated the Laws Observatory Bulletin.

In this environment, the young Harlow Shapley was introduced to astronomy. The astronomy program was small and Shapley was the only astronomy major at the time. In 1908 Seares requested and received funds ($300 per annum) for a student assistant a) to work as a stenographer to care for the "burdensome" Observatory correspondence, b) to assist in the reduction of observations, and c) to take care of miscellaneous duties such as winding clocks. Thus after only two years as a student, Shapley essentially became a staff member sharing responsibilities with Professor Seares and Instructor Eli S. Haynes who had also studied under Seares. This was in part due to the flexible structure of required credits. Although a number of astronomy courses were listed in the University Catalogue, there were really only two, an introductory course which Shapley ended up teaching (as part of his miscellaneous duties?) and a course in practical astronomy using a text that Seares had written. The other courses were given more in the manner of private discussion with Seares. As for research, Shapley joined in the observation of variable stars, noting years later that "we got some of our education by finding faults in the instruments." Nevertheless, Shapley could still say that the Laws Observatory was a complete observatory. With the two small telescopes, Shapley, Seares, and Haynes accomplished a fair amount of observational work, primarily on eclipsing variable stars, which well prepared Shapley for his subsequent doctoral research at Princeton under the supervision of Henry Norris Russell. By then, however, Seares had left for better opportunities at Mt. Wilson Observatory and Haynes had gone to Lick Observatory for additional study. Robert H. Baker assumed the Professorship of Astronomy at the University of Missouri in 1911, but after nearly a decade of frustration and inability to obtain a more modern observational facility, he resigned to take a position at the University of Illinois. With his departure, the Laws Observatory ceased to be a research institution.

HARLOW SHAPLEY AND RED GIANT STARS

Martin F. McCarthy S.J.

Vatican Observatory

In five papers written between 1951 and 1955 Shapley considers the topic of red giant stars and reddish variable stars in the Magellanic Clouds. These works coincide with Shapley's final year as Director at Harvard and the first years of his retirement which extended a full score of years before his death in 1972. They include the following: Magellanic Clouds II (Supergiants/Red Variable Stars in the Small Cloud; January 1951); Magellanic Clouds IV (On Period Frequency Anomalies; February 1952); Magellanic Clouds VII (Star Colors and Luminosities in Five Constellations; March 1953); Magellanic Clouds VIII (On the Populations Characteristics of the Two Clouds; October 1953); and Magellanic Clouds XVI (Infrared Stars and Stellar Evolution; July 1955). These five papers, which appeared originally in the Proceedings of the National Academy of Science, may be found in the Harvard Reprint Series I as numbers 346, 360, 373, 376, and 425.

Here Shapley explores the brightness and the size of the stars in the Clouds, then realizes that the domain of the main sequence and much of the giant branches must remain projects for future studies and limits himself to a first survey and examination of the stars of high luminosity. Bounds were imposed on Shapley's research on giants in the Clouds because he did not have available sufficient material for spectral classification and for radial velocity measures of the faint stars. Lacking these he could not with certainty separate Cloud members from foreground stars. He used what he had: preliminary magnitudes and colors plus measures of the variations of the red variables, where he built upon the measures of C.P. and S. Gaposchkin as well as the photometric measures of Nail; he was aware of the importance of interpolating magnitudes on photographic plates from reliable photoelectric photometry and he used the new sequences developed by Uco van Wyck. Shapley obtained the best available distance moduli to the Clouds and argued from a comparison of star fields outside the Clouds at comparably high galactic latitudes to a discrimination between Cloud members and superposed stars of the galaxy. He estimated absorption effects as carefully as possible but realizes he has no substitute for actual colors derived from normal colors based on spectral studies.

While these studies are overshadowed by his earlier and greater works on the galactic center, Shapley stands for us as the Gatekeeper to the realms of the Magellanic Clouds. He yearned to explore these in detail but was limited in his access and knew he could not enter these Promised Lands fully. When he came to the end, he found that it was only the beginning. Today those who understand so much more about the size, structure and development of objects in the Clouds than Harlow Shapley could thank him for affording to them something of his vision and his giant's shoulders.

The text of this paper will be published fully elsewhere.

THE DEVELOPMENT OF A RED-GIANT BRANCH IN LOW TO INTERMEDIATE MASS STARS

A. V. Sweigart

NASA Goddard Space Flight Center

L. Greggio and A. Renzini

University of Bologna

ABSTRACT: A new grid of evolutionary sequences has been computed for the main sequence and first red-giant branch (RGB) phases of low to intermediate mass stars. From these sequences we have obtained new intermediate age isochrones.

1. MOTIVATION

The evolution of low to intermediate mass stars is of theoretical and observational interest for a number of reasons, including:

a) The study of the phase transition from degenerate to nondegenerate helium-core ignition.

Low mass stars develop a degenerate helium core following the main sequence phase and, as a result, have an extended RGB phase prior to helium ignition, as found in galactic globular clusters. In contrast, intermediate mass stars do not develop a degenerate helium core and therefore do not have an extended RGB phase prior to helium ignition. Thus the transition between low and intermediate mass stars represents a transition in both the interior structure and observable morphology in the HR diagram.

b) The need for intermediate age isochrones covering the evolutionary phases up to helium-core ignition.

Such isochrones are necessary, for example, in order to understand the HR diagrams of intermediate age globular clusters in the Magellanic Clouds (cf. Poster 210 by Renzini et al. in this symposium).

c) The interpretation of the integrated properties of stellar populations.

Intermediate age theoretical sequences are needed to determine the contributions of the different evolutionary phases to the integrated luminosities and colors as well as to determine the dependence of the integrated properties on chemical composition and age.

2. COMPUTATIONS

One hundred canonical evolutionary sequences consisting in total of $\sim 10^5$ models have been constructed for the evolution of low to intermediate mass stars from the zero-age main sequence to helium-core ignition. These sequences, obtained with a modified version of the evolution code described in Sweigart and Gross (1978), have been computed for each combination of the following main sequence helium and heavy element abundances: Y_{MS} = 0.20 and 0.30 and Z = 0.004, 0.01 and 0.04, and for stellar masses between 1.4 and 3.4 $M_\odot$. A small mass spacing of only 0.05 $M_\odot$ was used around the RGB phase transition in order to map this transition very precisely.

The present sequences have been used to construct isochrones for ages ranging from 0.3 to 1.75 Gyr. These isochrones, which include the first RGB phase, have been transformed into the observational M_V versus B-V plane for comparison with the HR diagrams of the intermediate age globular clusters in the Magellanic Clouds. These isochrones can also be used to construct model stellar populations for high red-shift galaxies, when the dominant stellar population consists of ~ 1 Gyr old stars.

3. RESULTS

The objective of the present work has been to study the phase transition between low mass stars with degenerate helium cores and prominent RGB's and intermediate mass stars without these evolutionary characteristics. From our results we conclude:

a) The phase transition occurs abruptly with increasing mass. Typically this transition covers a mass range of ~ 0.4 $M_\odot$.

b) The mass at the phase transition varies from ~ 2 to ~ 3 $M_\odot$, depending on the composition. The transition mass increases with either decreasing Y_{MS} or increasing Z.

c) During the phase transition the luminosity at the tip of the RGB changes by ~ 2.5 mag.

d) The average age at the phase transition is 6×10^8 yr. Most importantly, this age varies by only $\pm 10^8$ yr for $0.20 \leq Y_{MS} \leq 0.30$ and $0.004 \leq Z \leq 0.04$ and thus is insensitive to the composition.

REFERENCE

Sweigart, A, V, and Gross, P. G. 1978 Astrophys. J. Suppl. 36, 405.

NEW MAIN-SEQUENCE LUMINOSITY FUNCTIONS FOR GLOBULAR CLUSTERS

Robert D. McClure, Peter B. Stetson, James E. Hesser and
Graham H. Smith

Dominion Astrophysical Observatory

William E. Harris

McMaster University

Don A. VandenBerg

University of Victoria

We report new results from a program which is aimed at obtaining deep CCD photometry for a sample of relatively nearby globular clusters having a wide range of metallicities. The CCD cameras on the CFHT 3.6 m, CTIO 4 m and KPNO 4 m telescopes have been used over the past 4 years to obtain deep exposures in regions of a number of clusters. In order to avoid the severest crowding, all of our observations have been obtained at distances of greater than ~ 5 core radii from the cluster centers. The images have been analysed by using the DAOPHOT point-spread-function fitting routines.

The primary result of these investigations has been the determination of cluster V, B–V color magnitude diagrams, a number of which have now been published. Typical total integration times per filter are $1 - 1.5^{hr}$. One of the main uses to which these data have been put involves comparisons with the theoretical isochrones of VandenBerg and Bell (1985). This aspect of our program is described in the review by VandenBerg in this symposium.

A further feature of this deep photometry which has yielded surprising results is the main sequence luminosity functions (LF's) obtained. Our earliest results have been discussed by McClure et al. (1986) [see also the review by Hesser, this symposium], who compared LF's for deep CCD CMD's both from their own work and from the literature. That study suggested that the power-law index x of the main-sequence mass function correlates with cluster metallicity, with the most metal rich clusters having the flattest LFs.

In this paper we add new data for M 92, M 12, and NGC 6362 to those of McClure et al. (1986). The LF's for these clusters are shown in Fig. 1 as solid lines, with the cluster metallicity displayed in brackets. The M 15 and 47 Tuc LF's are shown for comparison by dashed lines. [The [M/H] values are means of those values compiled by Pilachowski (1984), Zinn and West (1984), and Webbink (1985).] Fig. 2 is a plot of the power-law x index versus metallicity [M/H]. The clusters discussed by McClure et al. (1986) are shown as filled circles, the new data as open circles. Whereas McClure et al. (1986) derived x values by simple eye comparison of the data with isochrone predictions in LF plots, we have adopted a more quantitative approach in the present work. We have measured the number of observed stars in the magnitude bins $M_V = 4.25 - 5.75$ and $5.75 - 8.25$, taken their number ratio, and compared them with the predictions of a simple power-law mass spectrum. (VandenBerg and Bell's isochrones were used to determine the appropriate mass intervals corresponding to the adopted M_V bins.) Consequently, the x values appearing in Fig. 2 are somewhat different from those in McClure et al. (1986), typical differences being 0.5, the estimated uncertainty in the determination. As can be seen, the correlation between x and [M/H] is still clearly evident among the cluster sample available to McClure et al. (1986). The addition of the three new clusters in the present work still preserves a correlation, but with some scatter. Considering the errors in both the x and [M/H] determinations, however, we cannot say whether the scatter has physical significance.

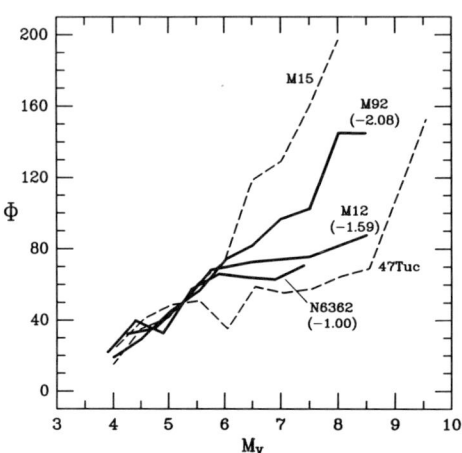 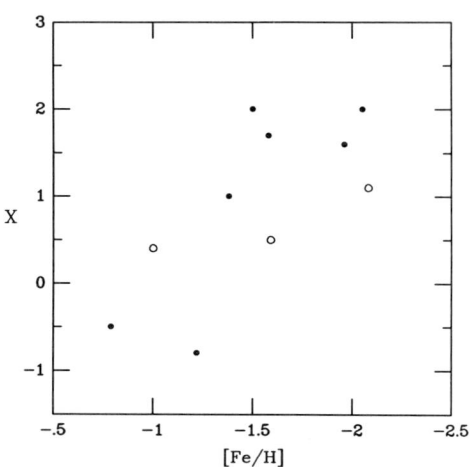

Fig. 1. LF's from three unpublished CMD's are compared with those for M 15 and 47 Tuc as displayed by McClure et al. (1986). For the data points illustrated, corrections have been made for field star contamination and incompleteness. The latter is always less than $\sim 30\%$.

Fig. 2. The power-law index, x ($dN \propto m^{-(1+x)}dm$), computed as described in the text, is shown versus metallicity.

REFERENCES

McClure, R. D., VandenBerg, D. A., Smith, G. H., Fahlman, G. C.
   Richer, H. B., Hesser, J. E., Harris, W. E., Stetson, P. B.
   and Bell, R. A. 1986 Astrophys. J. Letters 307, L49.
Pilachowski, C. A. 1984 Astrophys. J. 281, 614.
VandenBerg, D. A. and Bell, R. A. 1985 Astrophys. J. Suppl.
   58, 561.
Webbink, R. F. 1985 in IAU Symposium No. 113, Dynamics of Star
   Clusters, J. Goodman and P. Hut, eds., Reidel, Dorgrecht, p. 541.
Zinn, R. and West, M. J. 1984 Astrophys. J. Suppl. 55, 45.

GLOBAL VERSUS LOCAL MASS FUNCTIONS

Ivan R. King

University of California, Berkeley

ABSTRACT: Multimass dynamical models are used to study the difference between a mass function determined in an outer field and the global mass function of the cluster. The differences are small.

McClure et al. (*Astrophys. J. Letters*, 15 Aug. 1986) have recently suggested that the mass functions of globular clusters correlate with their metallicities, citing seven examples for which they give the exponent x in the mass-function formula
$$N(m)\,dm = m^{-(1+x)}dm.$$
With decreasing metallicity their values of x range from -0.5 to $+2.5$.

Their mass functions are derived from observed luminosity functions in a single field in each cluster. Since the stars of different mass are distributed differently, however, a local mass function does not correctly reflect the global mass function of the cluster. The latter will not be directly observable until the era of Hubble Space Telescope, but it can be deduced reasonably well by fitting dynamical models to the observations in each cluster. The present note shows how great the differences should be, using models that fit the general characteristics of each cluster.

In the models, each stellar type has a lowered Maxwellian distribution with a modulus of precision that corresponds to its mass, the whole being in mutual dynamical equilibrium. The main groups have masses of 0.75, 0.60, 0.45, 0.30, and 0.15 $M_\odot$ respectively, with stars of 0.775 $M_\odot$ added to represent the red giants. In addition, the models contain white dwarfs, and small numbers of brown dwarfs and neutron stars, all of which have an inconsequential effect on the dynamics.

For each cluster the proportions of the groups from 0.15 to 0.775 $M_\odot$ were chosen according to a power law with the exponent given for that cluster by McClure et al. These proportions were imposed on the projected density distribution at the radial distance at which the luminosity function had been studied. The central concentration of each model was chosen to agree with the value given for that cluster.

The total number of stars of each type was integrated in each model, and was compared with the local mass function. The results are displayed in the figure, in which the circles represent the local mass function and the connected crosses the global one. The vertical zero point of each set of points is arbitrary; what matters is the difference in slope and curvature.

The models had isotropic velocity distributions. To investigate the effect of anisotropy, a model was calculated for M13 with anisotropy radius $4r_c$, near the value found by Lupton, Gunn, and Griffin (preprint); it is shown as M13A.

The following conclusions can be drawn. (1) Global mass functions tend to be less steep than local ones in the outer parts, but the differences are far too small to destroy the correlation shown by McClure *et al.*. (2) The difference is larger in a cluster where the central concentration is high and the observations are taken far out than in a cluster of lower concentration with observations closer to the center. (3) For a given outer mass function, anisotropic models imply a flatter global mass function than do isotropic models. (4) In general, the differences found here tend to bend the global mass function up from a power law, at the high-mass end.

The support of contract NASA5-28086 is acknowledged.

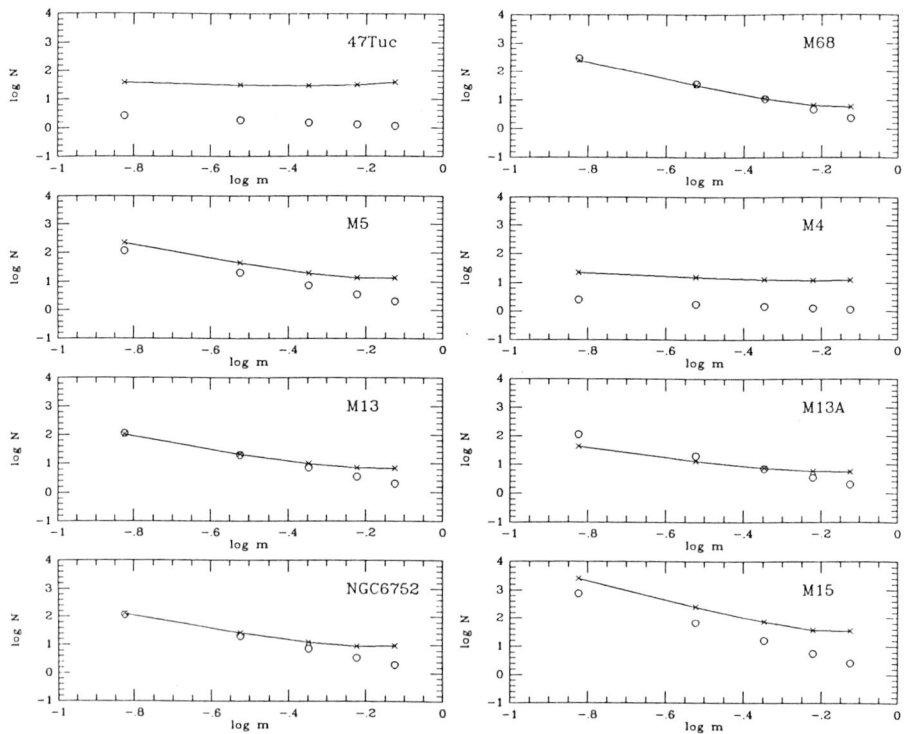

A NEW SURVEY OF GLOBULAR CLUSTER STRUCTURAL AND LUMINOSITY PARAMETERS

B. Cameron Reed

St. Mary's University, Halifax

Charles J. Peterson

University of Missouri

We have made an analysis of the visual photometric data contained in the <u>Catalogue of Concentric Aperture UBVRI Photoelectric Photometry of Globular Clusters</u> (Peterson 1986). Structural parameters have been obtained by use of the Simplex algorithm of Caceci and Cacheris (1984) to fit the model curves of King (1966) to the run of cluster luminosity with radius. We find that concentric aperture photometry alone can be used to determine globular cluster core radii and central surface brigtnesses reliably. Application of this techique, however, is limited to about two-thirds of the known clusters of the Galaxy because no or inadequate numbers of photometric measurements exist for the remaining clusters. Accurate determination of cluster concentration classes still requires use of other types of data, such as star counts.

Comparison of our derived cluster structural and luminosity parameters with those determined in other studies gives the following conclusions:

a) The core radii given by Peterson and King (1975) and Peterson (1976) appear to be systematically too large by approximately 0.1 dex in $\log (r'_c)$.

b) Our core radii are in good agreement with the values that Kron, Hewitt, and Wassermann (1983) obtained from surface photometry of electronic camera exposures. The estimated r.m.s. scatter about an equality relationship is 0.08 dex, implying an intrinsic error of 0.06 dex in our results if equal weight is given to the data of the two studies.

c) From comparison to our core radii, the smallest core radii obtained by Webbink (1984) appear to be underestimated in size; his larger core radii appear to be slightly overestimated in size. Of the 69 clusters studied by Kron, Hewitt, and Wassermann, their values of log (r'_c) for 53 clusters are identical with the values tabulated by Webbink; thus, a systematic problem appears to exist only in those values of core radii obtained by Webbink from consideration of other data (either the concentric aperture photometry or star counts).

d) Our analysis (as well as that of Kron, Hewitt, and Wassermann) does not obtain values of cluster concentration class $c = \log(r_t/r_c)$ that are of high statistical accuracy, although our current results appear to be in best agreement with the studies of Peterson and King (1975) and Peterson (1976).

e) Our results for central surface brightness appear well-behaved and agree well with the prior values obtained by Peterson and King, Peterson, and by Webbink over a range of 5 to 14 visual magnitudes per square minute of arc. The scatter about equality is significantly smaller for the comparison with Webbink.

f) Although concentration class appears to be poorly defined in our calculations, total magnitudes which are strongly dependent upon the adopted value of c are reasonably well-behaved. The majority of our values agree with those in the Peterson and King (1975) and Peterson (1976) studies and with those of Webbink, though the r.m.s. scatter about equality appears larger in the comparison with Webbink. In a number of cases in which the concentric aperture data span an insufficient radial range to define c precisely, our method appears to overestimate cluster concentration and hence overestimates cluster brightness.

One of us (C.J.P.) wishes to acknowledge the support and hospitality of the Berkeley Astronomy Department and the Dominion Astrophysical Observatory where this work was performed during a sabbatical leave from the University of Missouri. B.C.R. is supported in part by a grant from the Natural Sciences and Engineering Research Council of Canada.

REFERENCES

Cacec, M. S. and Cacheris, W. P. 1984 Byte 9, #5, 340.
King, I. R. 1966   Astron. J. 71, 64.
Kron, G. E., Hewitt, A. V. and Wassermann, L. H. 1983 Publ. Astron. Soc. Pacific 96, 198.
Peterson, C. J. and King, I. R. 1975 Astron. J. 80, 427.
Peterson, C. J. 1976 Astron. J. 81, 617.
Peterson, C. J. 1986 Astron. Data Center Bull. 1, in press.
Webbink, R. F. 1984 in I.A.U. Symposium 115, Dynamics of Star Clusters, J. Goodman and P. Hut, eds., Reidel, Dordrecht, p. 541.

AXIAL RATIOS AND ORIENTATIONS FOR 100 GALACTIC GLOBULAR STAR CLUSTERS

Raymond E. White

Steward Observatory, University of Arizona

Stephen J. Shawl

Clyde W. Tombaugh Observatory, University of Kansas

INTRODUCTION

The non-spherical appearance of globular clusters was first noted by Pease and Shapley (1917) and discussed in some detail by Shapley (1930) who analyzed cluster shapes determined from star counts made by Helen Sawyer on the Franklin-Adams star charts. This classic work has provided most of the data set used in all subsequent discussions of cluster shapes. A number of studies reporting cluster shapes have appeared in the years since we began this project, the most recent of which include Geyer, Hopp and Nelles (1983), Frenk and Fall (1982), and Kadla et al. (1976, 1977).

The observational basis of our study has been the availability of a complete sample of photographic material and the availability of digital image processing techniques. The present reinvestigation of the shapes of the galactic globular clusters provides a large sample of cluster ellipticities on a uniform system.

METHODOLOGY AND DISCUSSION

The ellipticity analysis was performed on digitized images which were smoothed to resemble the smooth distribution of an elliptical galaxy (see Shawl and White 1980, 1986 for details). We have satisfied ourselves in two ways that the blurring process does no irreparable damage to the image geometry: first, we produced artificial star clusters of known axial ratio and subjected them to the identical procedures used on the actual clusters; second, our results compare favorably with both those of Geyer et al. (1983) and with that of Kadla et al. (1976, 1977).

The analysis provides us with values for a cluster's axial ratio, major axis orientation, and relative intensity, all as functions of the radial distance from the cluster center. Considerations of the correlation length within the blurred images, together with the typical error-value in (b/a) of ± 0.03, lead us to the conclusion that there are no significant variations of (b/a) with cluster radial distance within the ranges we have been able to analyze. The axial ratios determined in this work have been compared with those from Shapley (1930), Kholopov (1953), Frenk and Fall (1982) and Geyer (1983). In all four comparisions, the average differences are zero within 1 standard error of the mean. The comparisions are of *average* values, and are not, necessarily, made at the same distance from each cluster's center. Comparisons of our axial ratios as a function of radial distance from the cluster center with those of Geyer (1983) and Kadla *et al.* (1977) show excellent agreement. For 99 clusters (NGC 6273 is neglected because of nonuniform interstellar extinction) the average axial ratio is 0.93 ± 0.01. We find that 32% have (b/a)-values <0.90, while only 5% are flatter than (b/a)<0.80; hence, the clusters are systematically quite spherical. Following the example of van den Bergh (1984), we divided the cluster sample into two distinct groups on the basis of the total visual absorption A_V. A two-sample Kolmogorov–Smirnov test failed to demonstrate a clear distinction between the two groups of clusters. Our null hypothesis, that there is statistically no difference between the two subsamples may be rejected at a significance level no better than 20%. Such a value is insufficiently strong to be able to state categorically that the apparent cluster ellipticities are due solely to the effects of the interstellar absorption. Thus, our larger, more homogeneous, data set only weakly confirms the conclusion by van den Bergh (1984) that, because of the effects of foreground absorption, the *intrinsic* shapes of the clusters are even more spherical than previously thought. However, our data show that there are tendencies for greater cluster eccentricities both in the direction of the galactic plane and towards the Galactic Center.

From an analysis of the orientations relative to the Galaxy, we conclude that the observed distribution is consistent with that expected from randomly oriented clusters and that tidal effects do not dominate *within the range of radial distances from the cluster center we have been able to consider.*

A more detailed manuscript has been submitted to *The Astrophysical Journal.*

REFERENCES

Frenk, C. S. and Fall, S. M. 1982 Monthly Notices Roy. Astron. Soc. **199**, 565.
Geyer, E. H., Hopp, U. and Nelles, B. 1983 Astron. Astrophys. **125**, 359
Kadla, Z. I., Richter, N. Strugatskaya, A.A. and Hogner, W. 1977 Soviet Astron. J., **20**, 49.
Kadla, Z. I. Richter, N., Hogner, W. and Strugatskaya, A. A. 1977 Izv. Glav. Astron. Obs. Pulkova No. 195, Astrofiz. Astrometr. 74
Kholopov, P. N. 1952 Astron. Zh. **29**, 671.
Kholopov, P. N. 1953 Publ. Astron. Sternberg Inst. **23**, 250.
King, I. R. 1961 Astron. J. **66**, 68.
Pease, F. G. and Shapley, H. 1917 Contrib. Mt. Wilson Obs. Nr. 129.
Shawl, S. J. and White, R. E. 1980 Astrophys. J. Letters **239**, L61.
Shawl, S. J. and White, R. E. 1986 Astron. J. **91**, 312.

ABUNDANCES IN STARS IN GLOBULAR CLUSTERS FROM PALOMAR CCD SPECTRA

E. Myckki Leep and George Wallerstein

University of Washington

J. B. Oke

California Institute of Technology

We have completed abundance analyses of stars in three globular clusters: M71, M4, and M22. Spectra of resolution 0.3 and 0.6 (two pixel) resolution have been obtained with the Palomar coudé spectrograph and a TI CCD. The analysis was carried out with model atmospheres and f-values derived from three sources: absolute f-values derived by theory for the 6300 line of OI and for CN bands, laboratory f-values for lines that are too weak in the sun to be useful, and solar f-values. The last introduce an uncertainty of about 0.25 dex because solar f-values derived via the Holweger-Muller model differ from those derived via the BEGN model.

Resulting abundances of iron are as follows: for M71, which is important as a calibrator of strong-lined globular clusters, we find that [Fe/H] lies between -0.6 and -1.0, depending on which model is used for solar f-values and which wavelength region is used. For M4 we find [Fe/H] to lie between -1.4 and -1.2, which is similar to photometric determinations. For M22, which has been reported to be inhomogeneous in composition, we find star III-3 to be richer in iron by 0.25 dex, as compared to star IV-102. This difference is similar to prior findings and confirms a small inhomogeneity in M22.

For CNO abundances we find the following: in M71 [O/H] = -0.6, which becomes a range from 0.0 to $+0.4$ for [O/Fe], depending on the iron abundances. For M4 we did not observe oxygen, but Geisler (Ph.D. thesis, 1983) found [O/Fe] = $+0.9$ from two stars in M4.

Our analysis of the 2-0 vibrational band of the red CN system yields a line in the (C/H, N/H) plane. A search for the $\lambda 8727$ line of CI in two clusters was not successful.

For M71 an analysis of CN in three stars yields similar lines in the (C/H, N/H) plane, which can be understood if C/N is between 1 and 3, the main sequence values of [C/H] and [N/H] are near -1.0, and the expected CN cycling and mixing have modestly increased N at the expense of C.

For M4 two stars also show reasonable lines in the (C/H, N/H) plane, provided that their initial values of [C/H] and [N/H] were about -1.2.

For M22 there is a gross difference between the relatively metal-rich star III-3 and IV-102, with CN stronger in the former star by about a factor 10. This favors the idea that C and N follows Fe and are all more abundant by about a factor of 3 in III-3, as compared with IV-102. In star III-3 we have detected the clump of ^{13}CN lines near $\lambda 8005$ and find a ratio of $^{12}C/^{13}C$ near 4. At our resolution of only 0.6 Å this is very uncertain, despite the signal-to-noise of about 150. If correct, it indicates much deeper mixing than predicted by standard evolution and mixing theory.

Our results for various elements relative to iron are best shown in a table. We have rather little that is new for M22 and hence show relative abundances for the other two clusters in the following table.

TABLE I
[X/Fe] for Various Elements

Element	M71	M4 this work	M4 Geisler Thesis	Element	M71	M4 this work	M4 Geisler Thesis
O	+0.2		+0.8	Ca	+0.6	+0.6	+0.4
Na	+0.5	+0.4	+0.5	Sc	+0.3		+0.1
Mg		+1.1	+0.6	Ti	+0.4	+0.3	+0.6
Al	+0.6	+0.9	+1.4	Fe peak			+0.1
Si	+0.3	+0.9	+0.8	s-process			0.0

It is clear from the table that both clusters show a very substantial excess of the light metals relative to iron. The observed effect is noticeable for all the light elements, not just the integral-α nuclei and includes titanium, but not scandium. In M4 the iron-peak elements go with iron, as do the four s-process elements observed by Geisler. We confirm his high sodium and aluminum abundances.

This research was supported by NSF Grant 84-15353 to G. Wallerstein and by NASA Grant NGL 05-002-134 to J. B. Oke. Computing time was granted by the National Center for Atmospheric Research.

THE METAL ABUNDANCE OF METAL-RICH GLOBULAR CLUSTERS

Raffaele Gratton

Astronomical Observatory of Rome

Maria Lucia Quarta

Institute of Space Physics, Frascati
Institute of Astronomy and Geophysics, Sao Paulo

Sergio Ortolani
Astrophysical Observatory, Asiago

We think that it is possible to find the correct scale of abundance for metal-rich globular clusters thanks to the new generation of spectrographs, equipped with CCD cameras. We analyzed giants in ten globular clusters and Arcturus using high dispersion spectra acquired through the CASPEC spectrograph at the 3.6 m telescope at La Silla. The detector was an RCA CCD. Stars cooler than 4150 K were avoided since their absorption spectrum is too strong. By a comparison with standard Arcturus spectra, we found a small trend to overestimate equivalent widths. This systematic error affects the derived abundances only marginally. However, too large equivalent widths must produce too large metal abundances. Abundances were derived following a standard procedure.

Since line analysis may be affected by systematic errors due to continuum tracing, we derived abundances also by a comparison with synthetic spectra, using a technique which avoids any continuum tracing. This comparison provides a weighted mean abundance of metals biased versus Fe. A total error of 0.15 dex may be given to the abundances derived from synthetic spectra. The derived abundances are presented in Table I. Line analysis probably overestimates the true metal abundance (due to the use of equivalent widths which are too large). Synthetic spectra probably underestimate the abundances (due to a possible zero point error in wavelengths). Synthetic spectra are completely independnet of continuum tracing. The agreement between line analysis and synthetic spectra gives us confidence in our abundances for [Fe/H] < -1.5. Synthetic spectra are inaccurate for very low metal abundance since the selected spectral region includes only quite weak lines. The comparison with previous high dispersion abundance determinations (Pilachowski et al., 1983, PSW) is quite poor. <[Fe/H]LA - [Fe/H]PSW> = +0.02 ± 0.11 (σ = 0.30). The comparison with other metal-abundance determinations (from photometric indices and low dispersion spectroscopy) is good.

Oxygen abundances were derived by means of synthetic spectra of the region including the 6300.31 Å [OI] line. The use of relatively hot stars makes CO formation of less importance than in previous studies. Errors are about 0.2 dex. The comparison between our oxygen abundances and those by PSW is poor <[O/H]US - [O/H]PSW)> = +0.22 ±0.13 (σ = 0.32). The comparison with R(CO) (Caputo 1985) is good. If O > C, the correlation of R(CO) with oxygen abundances indicates that carbon and oxygen abundances are correlated. The cluster to cluster scatter in [O/Fe] is within the uncertainties in the analysis. There is no evidence for an underabundance of oxygen for NGC 288, while NGC 6752 provided a low oxygen abundance (significant at a 2σ level). In the mean, they do not differ from the other clusters. The mean oxygen overabundance in the observed globular clusters is <[O/Fe]> = +0.4 ±0.1. Oxygen is overabundant as in field halo stars (see e.g. Sneden et al. 1979). α-elements (Mg, Si, Ca and Ti) are overabundant by <[α/Fe]> = +0.3 ±0.1. From our abundances, the metal enrichment process of globular cluster material is indistinguishable from that of halo stars. The simplest interpretation is that there was no important self pollution.

REFERENCES

Caputo, F. 1985 Astron. Astrophys. 147, 317.
Pilachowski, C. A., Sneden, C., and Wallerstein, G. 1983 Astrophys. J. Suppl. 52, 241 (PSW).
Sneden, C., Lambert, D. L., and Whitaker, R. W. 1979 Astrophys. J. 231, 762.

TABLE I

Abundances

Cluster	[Fe/H] LA	[α/Fe]	[Fe/H] SS	[O/H]
Arcturus	-0.54+0.04	+0.20+0.09	-0.47+0.05	-0.34+0.07
47 Tuc	-0.82+0.03	+0.18+0.08	-0.88+0.08	...
NGC 288	-1.31+0.03	+0.34+0.08	-1.35+0.09	-0.76+0.07
NGC 362	-1.18+0.03	+0.22+0.05	-1.10+0.09	-0.96+0.18
NGC 5897	-1.84+0.01	+0.22+0.04	-2.36+0.16	-1.44+0.11
M5	-1.42+0.05	+0.29+0.04	-1.44+0.04	-0.93+0.05
M4	-1.32+0.03	+0.36+0.13	-1.48+0.04	-0.99+0.06
NGC 6352	-0.79+0.06	+0.23+0.06	-0.75+0.04	-0.49+0.04
NGC 6362	-1.04+0.06	+0.33+0.06	-1.05+0.07	...
NGC 6752	-1.53+0.05	+0.31+0.06	-1.68+0.06	-1.37+0.09
M71	-0.81+0.04	+0.28+0.05	-0.89+0.04	-0.16+0.06

THE COMPOSITION OF WARM GIANTS IN M 71 AND M 5

Catherine A. Pilachowski

National Optical Astronomy Observatories*
Kitt Peak National Observatory

Christopher Sneden

McDonald Observatory, University of Texas

In 1979 a disturbing controversy arose in the field of globular cluster research when Cohen (1980) and Pilachowski, Canterna, and Wallerstein (1980) announced the results of the first high dispersion studies of the composition of giants in the globular clusters M 71 and 47 Tucanae. In contrast to earlier studies, which found metallicities of typically -0.3 and -0.5 dex, these investigators obtained values of -1.3 and -1.1. Since then, many have attempted to redetermine the abundances of M 71 and 47 Tuc to explain the discrepant results. These efforts have all suffered from the absence of high signal-to-noise, high resolution spectra of stars with temperatures above 4300 K.

Improvements to the 4m echelle spectrograph in the last 5 years have allowed us to get better spectra than previously possible. These improvements include a new, fast focal length camera, new coatings, better gratings, and CCD detectors. Furthermore, Cudworth (1985) has also provided a proper motion membership survey of stars in M71, so that members can be selected unambiguously.

New CCD spectra of two warm giants in M 71 and 3 giants in M 5 were obtained in May, 1986, with the Kitt Peak National Observatory 4m Telescope and echelle spectrograph equipped with the new UV Fast Camera and a TI CCD detector. The spectra have a resolution of 28,000, and a signal-to-noise of 50 or greater for the M 71 stars and 30 or greater for the M 5 stars, and are complete from 5000 to 7000 Å. Spectra of Arcturus and of the K0 III star ϵ Cygni were obtained as standards.

Model atmosphere effective temperatures were established based on
*Operated by the Association of Universities for Research in Astronomy, Inc., under contract with the National Science Foundation.

the Ridgway et al. (1981) calibration of (V-K) for the M 71 giants and the Böhm-Vitense (1981) calibration of (B-V) for the M 5 giants. The (V-K) colors for the two M 71 giants were kindly provided by M. Sitko. Surface gravities for the cluster giants were determined from the apparent magnitude, the cluster distance modulus, the temperature, bolometric corrections, and a stellar mass assumed to be 0.8 $M_\odot$. The adopted atmospheric parameters for the stars are given the Table I. The differential model atmosphere analysis followed procedures described by Pilachowski et al. (1983).

TABLE I
Astrophysical Parameters

Star	V	B-V	T(K)	log g	[Fe/H] (Arcturus)	[Fe/H] (eps Cygni)
M 71 - 56	13.25	1.37	4500	1.5	-0.2	-0.7
M 71 - 95	13.39	1.24	4600	1.6	-0.2	-0.7
M 5 - I-71	13.10	1.20	4340	1.2	-0.7	-1.2
M 5 - III-18	13.30	1.00	4580	1.4	-0.6	-1.1
M 5 - IV-34	13.06	1.20	4340	1.2	-0.7	-1.2

For M 71, we used E(B-V) = 0.25 and $(m-M)_v$ = 14.0; the photometry is from Arp and Hartwick (1971). For M 5, we used E(B-V) = 0.03 and $(m-M)_v$ = 14.51; the photometry is from Cudworth (1979).

The abundance of iron in our standard stars is given in Cayrel de Strobel and Bentolila (1983); in Arcturus, [Fe/H] = -0.5, and in ε Cygni, [Fe/H] = -0.06. With these values, we conclude that the abundance of iron in M 71 is [Fe/H] = -0.75 ± 0.15; M 5 is more metal poor, at [Fe/H] = -1.2 ± 0.15. In both clusters the abundances [Ca/Fe] and [Si/Fe] are near +0.2. We found no evidence these species are more enhanced in M 71 than in M 5.

REFERENCES

Arp, H. C. and Hartwick, F. D. A. 1971 Astrophys. J., **167**, 499.
Böhm-Vitense, E. 1981 Ann. Rev. Astron. Astrophys., **19**, 295.
Cayrel de Strobel, G. and Bentolila, C. 1983 Astron. and Astrophys., **119**, 1.
Cohen, J. 1980 Astrophys. J. **231**, 751.
Cudworth, K. M. 1979 Astron. J. **84**, 1866 (M 5).
Cudworth, K. M. 1985 Astron. J. **90**, 65 (M 71).
Pilachowski, C. A., Canterna, R. and Wallerstein, G. 1980 Astrophys. J. Letters, **235**, L21.
Pilachowski, C. A., Sneden, C. and Wallerstein, G. 1983 Astrophys. J. Suppl., **52**, 241.
Ridgway, S. T., Joyce, R. R., White, N. M. and Wing, R. F. 1980 Astrophys. J., **235**, 126.

THE INTEGRATED SPECTRA OF METAL-RICH GALACTIC GLOBULAR CLUSTERS: A TWO-PARAMETER FAMILY

James A. Rose

University of North Carolina

Michael J. Tripicco

Institute for Astronomy
University of Hawaii

ABSTRACT. Integrated photographic image-tube spectra have been obtained for the central regions of a number of metal-rich Galactic globular clusters with a wavelength resolution of 2.5 Å over the wavelength region $\lambda\lambda 3400-4500$ Å. The spectra have been analyzed using a variety of quantitative spectral indices that compare the strengths of neighboring absorption features. Our main result is that two parameters are needed to describe the integrated spectra of metal-rich globular clusters. The second parameter is manifested in two ways:

(1) In a diagnostic diagram sensitive to surface-gravity the metal-rich clusters do not form a well-defined linear sequence. Instead, we find large differences from one cluster to another in the mean surface gravity of the stars contributing to the integrated light at 4000 Å. It appears that the relative amounts of light contributed by dwarfs and giants varies considerably from one cluster to another.

(2) In a diagram that discriminates CN strength we find large differences from one cluster to another in the mean CN strengths of clusters having similar "spectral type". It is inferred that the mean CN anomaly varies from one cluster to another and that the anomalous CN strengths are present in main sequence stars as well as in giants.

The above two second parameter effects are shown to be strongly correlated in the sense that the "dwarf-dominated" clusters have the largest CN anomalies. Hence, our study indicates that no more than two independent parameters are required to explain the integrated spectra of metal-rich globular clusters.

Details of the above findings can be found in Rose and Tripicco (1986, _A.J._ 92, 1610).

IUE INVESTIGATIONS AT THE CORE OF M 79

Bruce Altner

Applied Research Corporation

As a result of their investigation of 27 galactic globular clusters with the ANS satellite van Albada, de Boer and Dickens (1981) classified M 79 (NGC 1904) as an "extremely blue" cluster. It was also found to be a low luminosity x-ray source based on data acquired with the Einstein X-ray Observatory (Grindlay 1981). In this brief paper we discuss the stellar spectra extracted from two short wavelength (SWP) IUE images acquired at the "center of light" of M 79. Discrete peaks in the "spatially resolved", cross-dispersion profile suggested the presence of at least three hot stars in the large aperture in both images (Fig. 1). The images in question are SWP 25303 and SWP 28936 for which it is important to note here that 1) the orientation of the large aperture on the plane of the sky differs for the two images by nearly 180° and 2) the target coordinates, determined from offset maneuvers with respect to a nearby SAO star, are nearly coincident. The cross-dispersion profile observed in the first spectrum is thus repeated in the second case, but is "flipped" left for right. Features immediately apparent in both images include a broad, asymmetric peak, almost certainly the blend of two or more components, and a second peak well separated from this blend and most likely a single star. In Fig. 2 we show the cross-dispersion profile of SWP 28936 and the best fit to that profile, assuming three components. From fits such as this and similar ones in other wavelength bins, we ultimately obtain from each image the entire SWP spectrum of each individual component.

The spectrum of the single source in each image was readily separated from the broader, blended profile. These two spectra are remarkably similar and it is most likely that they are spectra of the same star; we shall refer to this source as star "A". The mean spectrum of Star A is shown in Fig. 3, along with the Kurucz (1979) model atmosphere with T_{eff} = 13,000 K and log g = 4.5, and solar abundances. The IUE standard star HD 29335 (49 Eri; B7 V) is a close match to this spectrum. Not so easily resolved were the stars comprising the broad blend, due to the small separation of the peak centers (~5".5 in both images), but we finally were able to achieve reasonable results for the brighter component, which we refer to as star "B"; it is best fit with a Kurucz solar abundance model of T_{eff} 11,000 K and log g = 2.0. In this case we find that a B9 IV standard star (α Del) is a fairly close match to star B's IUE spectrum. Given the position angle and location of the aperture we tentatively identify star A as No. 214 and B as No. 210 from the list of Cordoni and Aurière (1983; V = 16.8, 16.0 and (B-V) = -0.2, 0.2, respectively). From the fits to the Kurucz models we derive $L/L_\odot$ = 2.1 and 2.5 for stars A and B respectively, indicating an above horizontal-branch location in the cluster color-magnitude diagram.

Investigation of Fig. 1 reveals a startling inconsistency in the above arguments, however it is clear that the separation between Star A and star B is not the same for both images. If real, such a shift would imply a tangential velocity nearly half of the speed of light, which is clearly unreasonable and suggests that perhaps these aren't the same stars at all. Yet, if one considers the total flux in all three components for each image (so that no errors due to the extraction procedure are involved) one finds remarkable agreement between the two (composite) spectra. Perhaps this close agreement is merely fortuitous; with only two images it is difficult to tell, but work continues on this and other questions in M 79 and other galactic globular clusters.

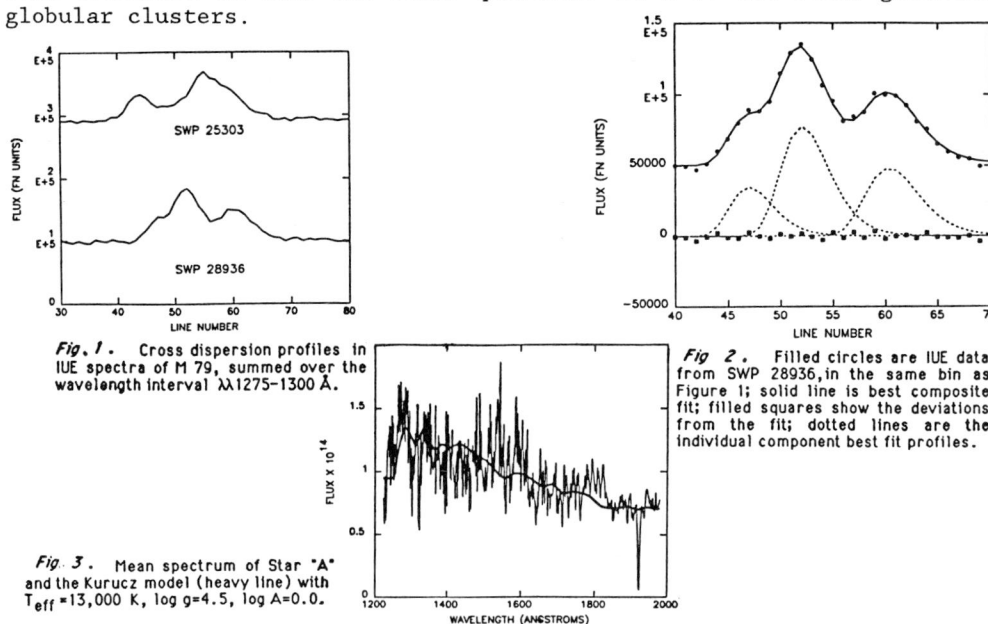

Fig. 1. Cross dispersion profiles in IUE spectra of M 79, summed over the wavelength interval λλ1275-1300 Å.

Fig 2. Filled circles are IUE data from SWP 28936, in the same bin as Figure 1; solid line is best composite fit; filled squares show the deviations from the fit; dotted lines are the individual component best fit profiles.

Fig. 3. Mean spectrum of Star "A" and the Kurucz model (heavy line) with T_{eff} =13,000 K, log g=4.5, log A=0.0.

REFERENCES

Cordoni, J. -P. and Aurière, M. 1983 Astron Astrophys. Supp. 54, 431.
Grindlay, J. E. 1981 in X-Ray Astronomy with the Einstein Satellite, R. Giacconi, ed., Reidel, Dordrecht, p. 79.
Kurucz, R. L. 1979 Astrophys. J. Suppl. 40, 1.
van Albada, T. S., de Boer, K. S. and Dickens, R. J. 1981 Monthly Notices Roy. Astron. Soc. 195, 591.

ONE MICRON PHOTOMETRY OF OMEGA CENTAURI GIANTS

Graeme H. Smith

Dominion Astrophysical Observatory
Herzberg Institute of Astrophysics

The globular cluster ω Centauri is known to be chemically inhomogeneous, a property that reveals itself via a wide giant branch in the V, B-V color-magnitude diagram (Cannon and Stobie 1973). Typically a color range of $\Delta(B-V) = 0.3 - 0.4$ exists among giants with $V < 13$. However, in the red R, R-I (Norris and Bessell 1975, Bessell and Norris 1976), and I, V-I (Lloyd Evans 1977) diagrams a much tighter giant branch is seen. This has lead to the suggestion (e.g. Bessell and Norris 1976) that molecular absorption in the bandpass of the B filter has significantly reddened the B-V color, and thereby produced an anomalously wide giant branch.

In order to further investigate the width of the giant branch in the near-infrared CMD, photometry has been obtained near 1μ for a sample of 29 ω Cen red giants. A filter having a central wavelength of 10,175 Å was used in conjunction with a standard V filter and an extended red response InGaAsP tube (Bessell 1979). Observations were made with the 1.0 and 0.6 m telescopes of the Siding Spring Observatory, during May, June, and July of 1979. A V-1μ color was measured, and reduced to an instrumental system defined by the observations obtained during the June run.

The V, V-1μ CMD is presented in Fig. 1. Also shown is the corresponding V, B-V diagram, the photometry being taken from Cannon and Stobie (1973), and the ROA catalog (Woolley 1966); corrections found by Cannon and Stobie being applied to the latter data. The two diagrams are morphologically similar. Color residuals relative to blue envelopes having the equations V-1μ = 4.20 - 0.3V and B-V = 4.90 - 0.3V were determined, and are plotted against each other in Fig. 2b. Similarly, δ(R-I) residuals were measured relative to a linear least squares fit to the R, R-I giant branch defined by those stars observed by Bessell and Norris (1976). Plots of δ(V-1μ) versus δ(R-I), and an analogous infrared residual R(V-K) from Persson et al. (1980), are shown in Figs. 2c,a. Some of the bluest stars in these diagrams, having δ(B-V) < 0.05, may be asymptotic branch giants.

These figures clearly demonstrate that ω Cen exhibits an

intrinsically wide giant branch. The δ(V-1μ) residual correlates well
with residuals in other colors. In particular it should be noted that a
correlation exists between the V-1μ and R-I residuals. This
demonstrates that although the dispersion in R-I may be small, it is
nonetheless real. The tight correlations between V-1μ, B-V, and V-K
residuals indicate that either the B-V color is not being significantly
affected by CN and CH molecular band blocking, or else such absorption
correlates closely with other heavy element line blanketing.

REFERENCES

Bessell, M. S. 1979 Publ. Astron. Soc. Pacific 91, 589.
Bessell, M. S. and Norris, J. 1976 Astrophys. J. 208, 369.
Cannon, R. D. and Stobie, R. S. 1973 Monthly Notices Roy. Astron.
    Soc. 162, 207.
Lloyd Evans, T. 1977 Monthly Notices Roy. Astron. Soc. 178, 345.
Persson, S. E., Frogel, J. A., Cohen, J. G., Aaronson, M. and
    Matthews, K. 1980 Astrophys. J. 235, 452.
Woolley, R. v. d. R. 1966 Ann. Roy. Obs., No. 2.

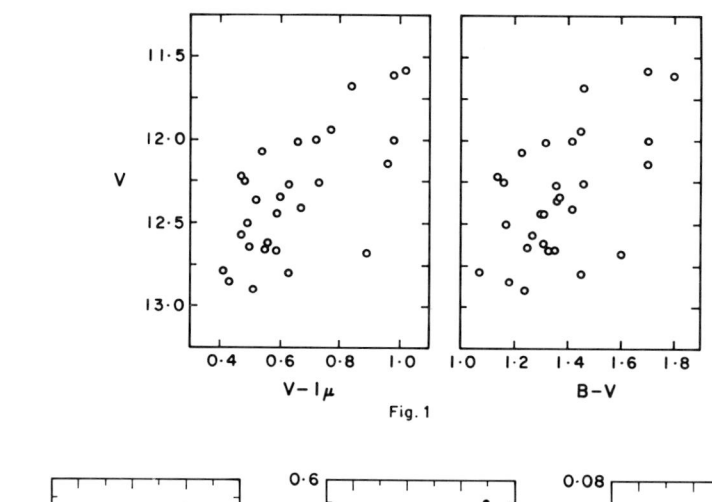

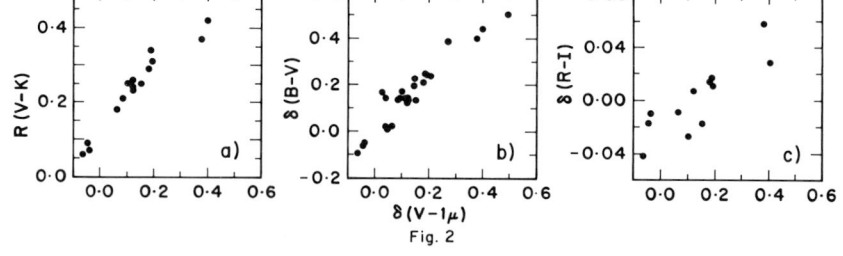

Fig. 1.   The V, V-1μ and V, B-V color-magnitude diagrams.
Fig. 2.   Color residuals, measured with respect to reference lines in
three different color-magnitude diagrams, are plotted against the
δ(V-1μ) residual for giants with 11.8 < V < 13.0.

ON THE BIMODAL DISTRIBUTIONS OF HORIZONTAL BRANCHES

Young-Wook Lee, Pierre Demarque and Robert Zinn

Yale University Observatory

New synthetic horizontal branch (HB) models are presented for some globular clusters known to have bimodal HB distributions. These models are based on new Yale HB evolutionary tracks for Y=0.25 and the core masses appropriate for the compositions. The distribution of stellar masses along the HB is given by a slightly modified version of Rood's (1973) function. Figure 1 compares the synthetic and the observed color-magnitude diagrams and the generalized histograms of the distribution of HB stars over (B-V)o (observational data from Alcaino & Liller 1984, Buonanno et al. 1981, Stetson 1981, and Menzies 1974 for clusters M4, M5, N1851, and N6723 respectively). Following Norris (1981), we have used the period-luminosity-color relations to estimate the colors of some RR Lyrae variables. Since none of the parameters in the models has a bimodal distribution, the excellent agreement between the color distributions of the models and the observations suggests that the observed bimodal distributions are a consequence of the evolution from the zero-age HB. Contrary to the suggestion of Norris (1981) and Smith & Norris (1983), there is no need to connect the bimodality of the HB with the observed bimodal CN distributions of the red giants in some of these clusters. The clusters in Figure 1 span a narrow range in metallicity ([Fe/H]=-1.40 to -1.09; Zinn 1985), and we are investigating other, less well observed, clusters in this range (e.g., N2808) to see if their bimodal HB distributions also have simple explanations.

This research was partially supported by NSF (AST-8304034) and NASA (NAGW-778) grants.

REFERENCES

Alcaino, G. and Liller, W. 1984 Astrophys. J. Suppl. 56, 19.
Buonanno, R., Corsi, C. and Fusi Pecci, F. 1981 Monthly Notices Roy. Astron. Soc. 196 435.
Menzies, J. 1974 Monthly Notices Roy. Astron. Soc. 168, 177.
Norris, J. 1981 Astrophys. J. 248, 177.
Rood, R. 1973 Astrophys. J. 184, 815.
Smith, G. and Norris, J. 1983 Astrophys. J. 264, 215.
Stetson, P. 1981 Astron. J. 86, 687.
Zinn, R. 1985 Astrophys. J. 293, 424.

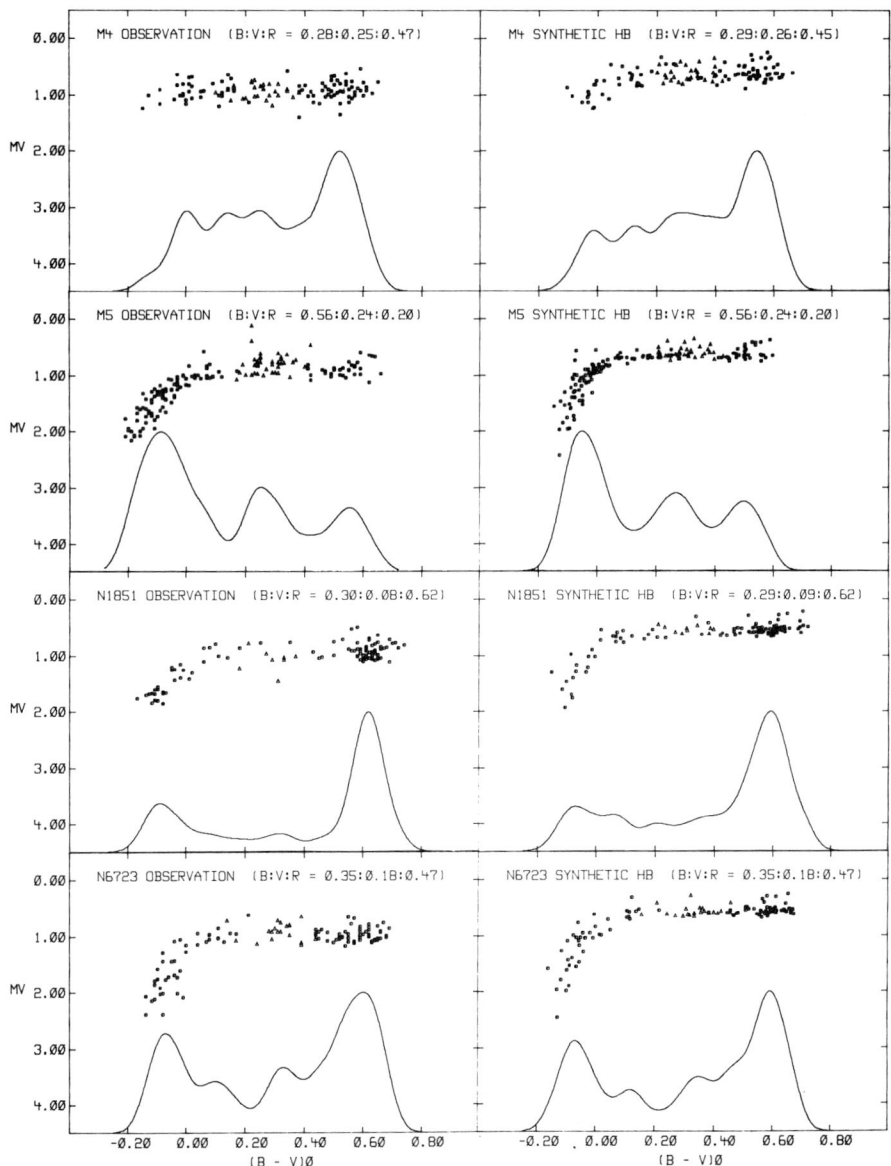

Fig. 1. Comparison of observations and synthetic HB's. Color-magnitude diagrams and generalized histograms (normalized to unity at the peak) are compared and proportions of BHB, RR Lyrae (triangles), and RHB stars (B:V:R) are indicated. The width of the variable strip is chosen to be 0.078 in log (T_{eff}). The small differences in luminosity between the observations and the synthetic HB's are due to the choices for the distance moduli of the clusters and have no effect on the comparisons of the distributions in color.

BIMODAL DISTRIBUTIONS ON THE HORIZONTAL BRANCH

Robert T. Rood and Deborah A. Crocker

University of Virginia

ABSTRACT. We present preliminary versions of log g - log T_{eff} diagrams for a number of clusters with gaps in their blue horizontal branches and for two normal clusters.

1. INTRODUCTION

A number of clusters show a bimodal distribution of stars along the horizontal branch (HB). In some cases there is actually a gap in the distribution; in others there is just a paucity of stars in a certain temperature range. The most dramatic examples are clusters with prominent gaps in very blue HB's (NGC 6752, NGC 288, M 15). There is no theoretical reason to expect such gaps. Unfortunately the distribution of stars along the HB is a sensitive function of many parameters. Possibilities include age, helium abundance, [Fe/H], [CNO/Fe], and parameters which affect mass loss which includes, at a minimum, rotation and magnetic fields. Bimodal distributions in any of these or other parameters could produce gaps.

As with the infamous "Second Parameter" problem a precise understanding of mass loss is required. Changes of a few percent in the mass loss parameter dramatically affect the distribution of stars on the HB. Since an understanding of mass loss at this level seems a rather remote possibility, we have begun a long term project to understand the HB without a preoccupation with its obvious feature --- the temperature distribution of the stars. Part of this project involves log g - log T_{eff} diagrams for a number of "blue gap" clusters as compared to a number of "normal" clusters. We report here on the preliminary results of that survey.

2. THE OBSERVATIONS

Spectra with a resolution of 4 - 7 Å were obtained using the IIDS on the 2.1 m telescope at KPNO for M 15, M 92, M 3 and M 5 and the 2D-Frutti on the 4 m at CTIO for NGC 288. Fits to Kurucz atmospheres gave gravities and temperatures. The gravities depend on the widths of Hβ, Hγ and Hδ, the temperatures on the Balmer jump and the slope of the continuum. From the internal consistency of Hβ, Hγ and Hδ the errors in gravity are generally less than 0.10. Errors in log T_{eff} are 0.02 or smaller.

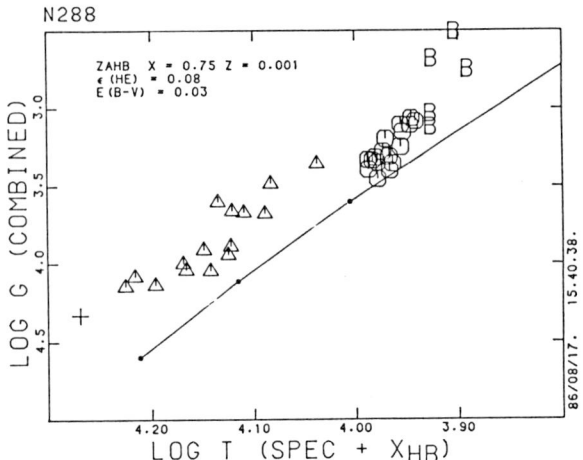

Fig. 1. The data for NGC288. The solid line is the ZAHB for the indicated composition. The same ZAHB fits the M 5 data well.

3. THE RESULTS

NGC 288 has a gap in the blue HB. Its metallicity is similar to M 5. It is a "too blue" 2nd parameter cluster. While the stars redward of the gap are close to the theoretical ZAHB, they do have systematically lower gravities than M 5. The stars blueward of the gap, the blue droop, are displaced significantly toward lower gravity. The blue droop does not seem to be an extension of the same ZAHB as the redder HB.

Coupled with our results for the other clusters we tentatively conclude: 1) There may be as much variety in cluster log g - log T_{eff} diagrams as in CM diagrams. 2) Some clusters (M 5 & M 92) fit the standard model pretty well. 3) In blue gap clusters the stars blueward of the gap, or blue droop, have (much?) lower gravities than if they lay on an extended HB. 4) The shift in the blue droop is in the direction which would result from rapid rotation or high helium abundances. The required change in either is very large. 5) Some clusters with fairly normal CM diagrams (M 3) show evidence for two overlapping populations in the log g - log T_{eff} diagram.

D.A.C. did most of the work reported here. We thank R. Buonanno, F. Fusi-Pecci, V. Caloi, and their colleagues for providing much pre-publication photometry, etc. We acknowledge the use of observing facilities at KPNO and CTIO.

HORIZONTAL-BRANCH STARS WITH STRONG HE LINES

Deborah A. Crocker and Robert T. Rood

University of Virginia

ABSTRACT. We discuss observations of He lines at λ 4026 Å in two stars in NGC 6752 and three in M 92. M 92 and NGC 6752 are moderate to extreme second parameter clusters. The presence of He lines in stars where diffusion should deplete atmospheric helium may be a clue to the identity of the second parameter. One such candidate which could easily inhibit diffusion is rotation.

1. INTRODUCTION

Observations of hot HB stars show that diffusion has altered the surface helium abundance (see Greenstein and Sargent 1974). So, in one of the few opportunities we have of directly measuring the He abundance in Pop II stars we are thwarted.

While making spectroscopic observations of blue cluster HB stars to investigate the origin of BHB gaps (see Rood and Crocker 1987) we detected He lines at λ 4026 Å in two stars in NGC 6752 and three in M 92. These lines are moderately strong, indicative of a Pop I helium abundance.

Fig. 1 shows the spectra of the three stars in M 92. The temperature and surface gravity of each star and the equivalent width of the He lines at λ 4026 Å are listed in the figures. Spectra of the stars in NGC 6752 with He lines are shown in Crocker, Rood, and O'Connell (1986 = CRO).

2. IMPLICATIONS

Heber et al. (1986) have reported measurements of He in NGC 6752 stars. One of the stars they observed was also observed by us (our 4104 = their 1083). They find reduced abundances (by factors of 1/2 to 3/4). We feel that neither our spectra, nor theirs are good enough to warrant a detailed analysis though. In addition, there are other observations of hot stars in M 3 and M 13 (Kadla, Gerashchenko, and Yablokova 1985) which show an apparently normal He abundance.

These four clusters (M 3, M 92, M 13 and NGC 6752) are all moderate to extreme second parameter clusters. In addition, M 92 and

NGC 6752 are gap clusters. The helium lines we observe may be related to a gap/second parameter candidate. For instance, rotation, in addition to producing a bluer HB, could induce meridional circulation which could inhibit diffusion. Atmospheric enrichment of helium during core flash would also produce blue stars.

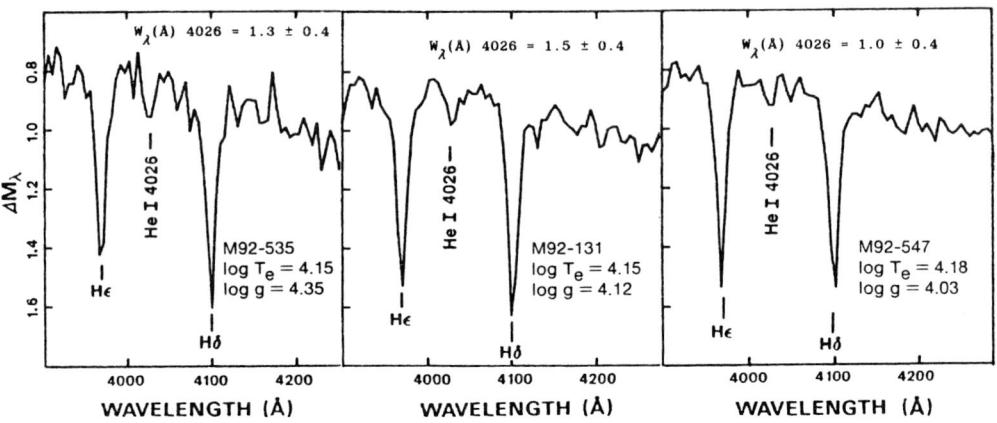

Fig. 1. Spectra of 3 stars in M 92

We do not detect He in observations of a fourth star in M 92 and one in M 3 at the same temperature and surface gravity as the stars with He lines. Any explanation for the appearance of the He lines we measure must also account for their absence in otherwise similar stars.

Further discussion on the implications of finding He lines in gap and second parameter clusters may be found in CRO.

3. AKNOWLEDGEMENTS

We thank R. Buonanno, F. Fusi Pecci, V. Caloi and their colleagues for providing NGC 6752 data in advance of publication. We also aknowledge use of the observing facilities at CTIO and KPNO.

REFERENCES

Crocker, D. A., Rood, R. T. and O'Connell, R. W. 1986 Astrophys. J. Lett., 309, in press.
Greenstein, J.L. and Sargent, A. I. 1974 Astrophys. J. Suppl., 28, 157.
Heber, U., Kudritski, R. P., Caloi, V., Castellani, V., Danziger, J. and Gilmozzi, R. 1986 Astron. Astrophys., 162, 171.
Kadla, Z. I., Gerashchenko, A. N., and Yablokova, N. V. 1985 Sov. Astron. Lett., 11, 142.
Rood, R. T. and Crocker, D. A. 1987 in IAU Symposium No. 126, Globular Cluster Systems in Galaxies, J. E. Grindlay and A. G. D. Philip, eds., Reidel, Dordrecht, p. 507.

SPECTRA OF BHB STARS IN M 3, M 13 AND M 92

A. G. Davis Philip

Van Vleck Observatory and Union College

N. N. Samus

Astronomical Council, Moscow

Spectra, at a dispersion of ~ 50 Å per millimeter, have been obtained of BHB stars in the globular clusters M 3, M 13 and M 92 with the TV scanner on the Soviet Union's Six Meter Telescope. The spectra cover a range of 700 Ångstroms in 500 channels in which counts were made of the intensity of the stellar spectrum. At this dispersion the hydrogen Balmer lines (γ, δ, ϵ, H8 - H12) can be seen as well as the Ca II line at λ = 3934.

The aim of this project is to measure the equivalent widths of the Ca II lines (on the basis that the strength of this line is a good indicator of the [Fe/H] value for these stars). The preliminary results show that the Ca II equivalent widths of the BHB stars in the three clusters scale well with their known [Fe/H] values, but there are some BHB stars which have much larger equivalent widths than the others. Strömgren four-color photometry has been obtained for some of the same stars for which spectra are available and it seems that the stars with the larger than normal Ca II equivalent widths are the stars which have larger than normal c_1 indices (~ 1.3 instead of ~ 1.2).

The spectrum of M 13 16 (a star with c_1 = 1.32) is shown on the left side of Fig. 1. It has the strongest Ca II line observed in any of the BHB stars investigated in this program. The Ca II line occurs at channel number 375, Hϵ is to the left and H8 is to the right. Just below this spectrum is one of M 13 18 (with a c_1 = 1.22) which has a negligible K line. The spectrum on the upper left is of the FHB star, HD 161817. The K line is fairly strong because this star is just on the blue edge of the instability strip. On the right side of Fig. 1. is the spectrum of another BHB star with a strong K line, M 3 182, which has a c_1 index of 1.31. M 3 IV 18 and M 92 II 26 are BHB stars with weak K lines.

Further investigations are planned to see if all the stars with the higher c_1 indices (> 1.3) have K lines with larger equivalent widths. The present data <u>suggest</u> that such a relation may be the case, but more observations are needed in order to confirm the hypothesis.

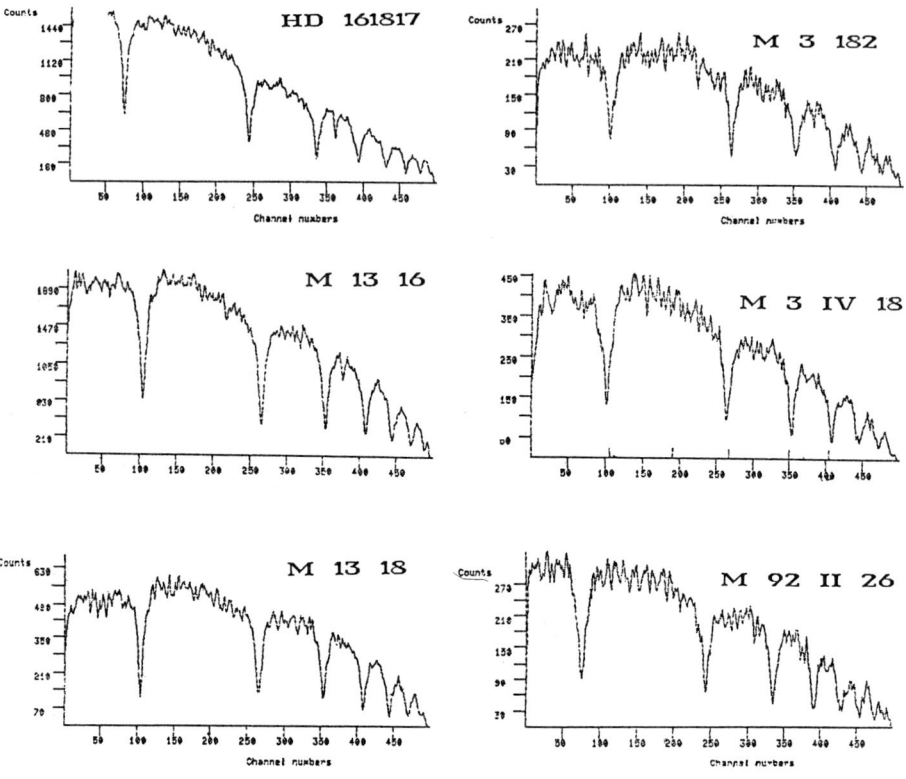

Fig. 1. Six Meter Spectra of Blue Horizontal-Branch A-Type Stars.

FOUR-COLOR MEASURES OF BHB STARS IN M 4, M 13 AND M 55

A. G. Davis Philip

Van Vleck Observatory and Union College

Strömgren four-color photometric measures have been made of blue horizontal-branch A stars in the globular clusters M 4, M 13 and M 55 with the Steward Observatory 90 inch telescope and with the 60 inch telescope at Cerro Tololo Inter-American Observatory. These stars are faint, ranging in V magnitude from 13.6 in M 4 to 15.5 in M 13 and the corresponding errors in the four-color indices are ± 0.04 to 0.06 in the c_1 index, for one observation. The error of the mean value of the c_1 indices is approximately ± 0.02 for most of the stars since they have been measured from 4 to 10 times each.

The distribution of the measured indices in the $(b-y)_0$, $(c_1)_0$ diagram is shown in Figs. 1a - c. The solid line indicates the position of the zero-age main sequence for Population I stars. The BHB stars scatter in the diagrams in the characteristic position for horizontal-branch stars 0.2 to 0.4 magnitudes above the main sequence. The vertical size of the triangles is approximately 0.02 mag. In each figure the majority of the points representing the four-color indices of the BHB stars fall about 0.15 magnitudes above the main sequence line. There are also stars in each diagram that fall ~ 0.25 magnitudes or more above the main sequence; these may be stars that are somewhat further along the HB evolutionary track than the stars with $\Delta c = 0.15$. When BHB stars in additional globular clusters of differing [Fe/H] values have been measured it will be interesting to see if a similar pattern is found in each cluster.

In February, 1986, CCD photometry in the four-color system was done at Cerro Tololo of three globular clusters, M 4, NGC 2808 and NGC 4833. In the latter two clusters, about 20 BHB stars per cluster were measured. Additional globular clusters will be investigated by means of CCD four-color photometry at Kitt Peak National Observatory. When these data are reduced then it will be possible to see if the relation derived for correcting surface gravities in M 4 is the same relation derived for the BHB stars in NGC 2808 and 4833. In Fig. 1 the distribution of BHB stars in the globular clusters M 4 and M 13 seem to be quite similar, but the c_1 indices for BHB stars in M 55 (the most metal-poor of the three clusters) seem to have average values somewhat larger than the c_1 indices in the other two clusters. If the relations for BHB stars are similar in the majority of globular clusters then one can have confidence in the derived surface gravity corrections. The BHB stars in the globular clusters in the observing program range in [Fe/H] from -1.3 to -2.2 so that if there is a difference as a function of metal abundance the nature of that relation can be determined.

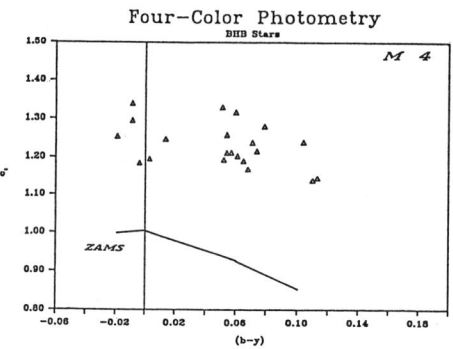

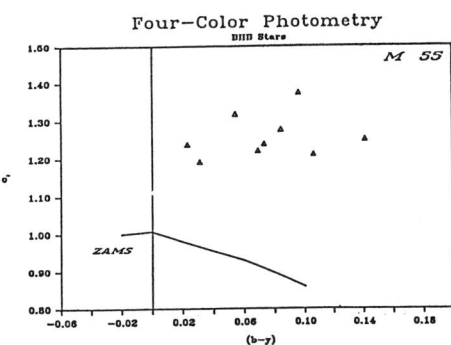

Fig. 1a, b, c.
Four-color photometry of BHB stars in the globular clusters M 4, M 13 and M 55. The solid line represents the location of the ZAMS.

GLOBULAR CLUSTERS IN THE VILNIUS PHOTOMETRIC SYSTEM

K. Zdanavičius

Institute of Physics, Vilnius

The Vilnius photometric system was developed for photometric two-dimensional classification of stars and for the determination of interstellar reddening. The system consists of seven magnitudes U, P, X, Y, Z, V and S with mean wavelengths of 345, 375, 405, 466, 516, 544 and 655 nm and half-widths of the order 20 - 30 nm (Straižys 1977). Later on the system was successfully used for the determination of temperatures and the metallicities of halo stars (Bartkevičius and Sperauskas 1983).

A research program concerning globular clusters in the Vilnius photometric system was started in 1978 with observations of their integrated color indices. Thirty nine globular clusters of the Galaxy were observed. The dependence of the integrated color indices with morphological types of the clusters and the interstellar extinction towards the observed globular clusters were studied (Zdanavičius 1983). After 1982, representative stars on the red giant, asymptotic giant and horizontal branches were observed. The metal abundances of the observed giant stars in the clusters, M 4, M 22, M 71 and M 92 were determined (Zdanavičius 1986) using the two-color and reddening-free Q,Q diagrams calibrated in metallicities (Bartkevičius 1983). The mean metal abundance and mean square errors are listed in the following table.

TABLE I

Cluster		Metallicity	Number of Stars
M 4	NGC 6121	-0.96 ±0.12	7
M 22	NGC 6656	-1.50 ±0.25	6
M 71	NGC 6838	-0.44 ±0.10	6
M 92	NGC 6341	-2.22 ±0.06	8

The analysis of different color-magnitude diagrams has shown that here is a possibility of separating the field stars from globular cluster stars which occupy the same area in a V, (B-V) diagram. For this several additional color-magnitude diagrams can be used. Various magnitudes respond differently to luminosity, metallicity, temperature and interstellar reddening and thus one can differentiate between the cluster members and the field stars. The separation is better with increasing photometric accuracy, decreasing metal abundance and with the use of ultraviolet magnitudes.

REFERENCES

Bartkevičius, A. and Sperauskas, J. 1983 Bull. Vilnius Obs. No. 63, 3.
Straižys, V. 1977 in Multicolor Stellar Photometry, Mokslas Publishers, Vilnius, p. 203.
Zdanavičius K. 1983 Astron. Zh. 60, 44.
Zdanavičius K. 1986 in Proceedings of Workshop on Stellar Clusters, May 19-23, 1986, Sverdlovsk (in press).

THE METALLICITY DISTRIBUTION FUNCTION OF HALO DWARFS AND GLOBULAR CLUSTERS

J. B. Laird, M. P. Rupin and B. W. Carney

University of North Carolina

D. W. Latham and R. L. Kurucz

Harvard-Smithsonian Center for Astrophysics

ABSTRACT: Metallicities have been determined for a chemically unbiased sample of field halo dwarf stars. Their metallicity distribution function is similar to the predictions of a simple model of chemical evolution, but somewhat different from that of globular clusters.

1. FIELD DWARFS

The Carney-Latham (1986) survey of high-proper motion stars has been used to identify a large sample of halo field dwarfs which is unbiased in metallicity. New, accurate metal abundances have been determined for these stars by comparing the spectra obtained for measuring radial velocities, having high resolution but poor signal-to-noise, to a grid of synthetic spectra. A sample of 124 stars, having retrograde orbits, was chosen from the survey to represent a true halo population.

Figure 1 compares the metallicity distribution function of these field stars to the predictions of a simple model of chemical evolution including mass loss (cf. Hartwick 1976, 1983; Searle and Zinn 1978; Bond 1981). The free parameter in the model is the effective yield, which equals $<Z>$ when the model goes to completion. The model having $\log(<Z>/Z_\odot) = -1.5$ is a fairly good fit, although it predicts slightly too many metal-poor stars, as noted in the past. For the first time the metal-rich end of the distributions can be compared, where too _few_ stars are predicted.

2. GLOBULAR CLUSTERS

Metallicity data for the globular clusters have been taken from Zinn's (1985) compilation. A limit was imposed in galactocentric radius of $R \geq 7$ kpc to exclude disk clusters, leaving 43 clusters. The same radius limit cannot be applied to the field stars, so a further restriction, rest frame velocity ≥ 300 km/s, was applied to the field

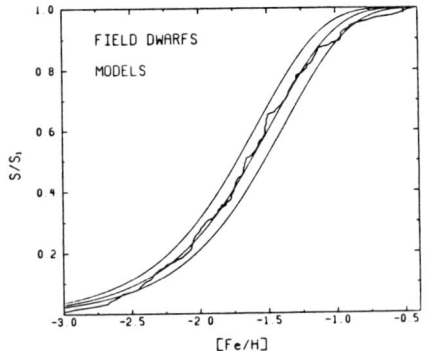

Fig. 1. Field dwarfs vs. the simple model. S/S_1 is the fraction of stars having that metallicity or less. The three models shown have $\log(<Z>/Z_\odot) = -1.4, -1.5,$ and -1.6.

stars in order to bias the sample toward outer halo stars. This smaller sample of 36 stars is compared to the clusters in Figure 2. The two distribution functions look substantially different, with the clusters lacking both metal-poor and metal-rich objects. A Kolmogorov-Smirnov test, however, shows that the difference is not significant, due in part to the small number of objects. The clusters were also compared to the predictions of the simple model. The models having $\log(<Z>/Z_\odot) = -1.5$ and -1.6 are significantly different from the cluster results at the 89% and 83% confidence levels, respectively, although a better fitting model will lie between these.

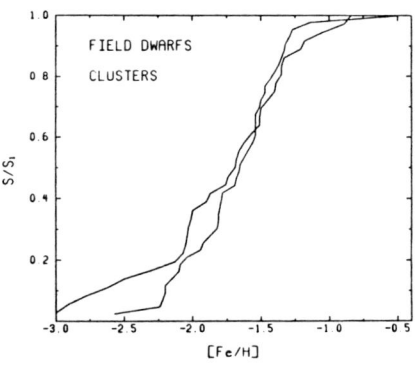

Fig. 2. Field dwarfs vs. globular clusters.

REFERENCES

Bond, H. E. 1981 Astrophys. J. 248, 606.
Carney, B. W. and Latham, D. W. 1986 submitted.
Hartwick, F. D. A. 1976 Astrophys. J. 209, 418.
Hartwick, F. D. A. 1983 Mem. Soc. Astr. Italia 54, 51.
Searle, L. and Zinn, R. 1978 Astrophys. J. 225, 357.
Zinn, R. 1985 Astrophys. J. 293, 424.

THE SIMILARITY OF THE HALO FIELD K GIANT POPULATION WITH THE GLOBULAR CLUSTER SYSTEM OF OUR GALAXY

Kavan U. Ratnatunga

Institute for Advanced Study, Princeton

Line-of-sight velocities and improved metal abundance estimates are available for a representative sample of 58 giants located by an objective prism survey (Ratnatunga and Freeman 1985), in a 20 square degree field near SA 127 ($l = 272$, $b = +39$). These in-situ K-giants of the outer regions of our galactic halo give a direct comparison of the field population with the globular cluster system. Fig. 1 illustrates the distribution of line-of-sight velocity with abundance for the sample of giant stars in SA 127. The mean and dispersion of the sample appears to be discontinuous at [FeH] ~ -0.8. Fig. 2 shows the distribution of [Fe/H] with distance from the Sun for the same stars. The metal stronger giants (filled symbols) represent a population of stars up to 6 kpc above the plane of the disk and have a velocity dispersion of about 50 km/s. In contrast, the metal weaker giants have a typical halo dispersion of about 120 km/s.

The field halo population appears to separate into two components with clearly different chemical and kinematical properties. (1) A metal-weak spheroidal halo component which is at most slowly rotating. (2) A metal-stronger thick disk-like component which is rotating with the disk and has a velocity dispersion of 50 km/s. Statistics for the spheroidal (SP) and thick-disk (TD) subsamples are given in Table I. The difference in both kinematic and chemical properties seems to be sufficient justification to separate the halo into two density components. This appears to be positive kinematic support for the presence of an intermediate scale height population, the existence of which cannot be convincingly proved by star counts alone (Bahcall et al. 1985). This result is very similar to the disk and halo sub-systems for the galactic globular clusters as shown by Zinn (1985). The change in kinematics is seen at about the same abundance.

REFERENCES

Bahcall, J. N., Ratnatunga, K. U., Buser, R., Fenkart, R. P. and Spaenhauer, A. 1985 Astrophys. J. 299, 616.
Ratnatunga, K. U. and Freeman, K. C. 1985 Astrophys. J. 291, 260.

Zinn, R.   1985   **Astrophys. J.** 293, 424.

TABLE I

MEAN KINEMATICS in SA127   (272.0   +38.6)

Sub sample	giants	<d> kpc	<z> kpc	<R> kpc	<Mv> mag.	<[Fe/H]> dex	<Vlos> km/s		σlos km/s	
SP	44	13.1	8.1	15.5	-0.5	-1.3	153	19	122	13
TD	14	4.9	3.1	9.3	1.2	-0.4	36	14	51	10

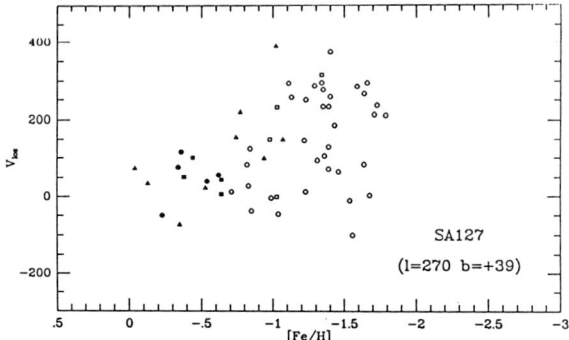

Fig. 1. Line of sight velocity of field halo K giants as a function of metal abundance for SA 127. Note the much smaller mean velocity and velocity dispersion of the metal stronger giants (filled symbols) in SA 127. I use circles, triangles and squares to identify observations made with the AAT:IPCS, MSO:B2PCA and MSO:R2PCA respectively.

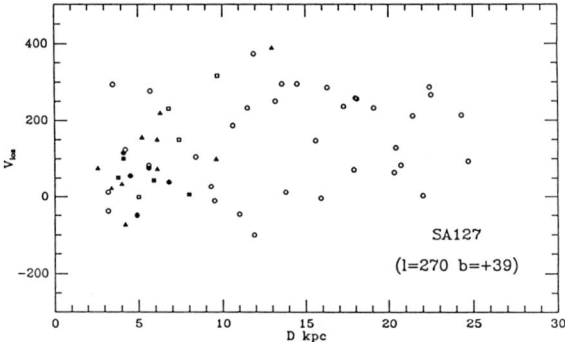

Fig. 2. Kinematics with distance from Sun of field halo K giants in SA 127 with same symbols as in Fig. 1.

APPARENT ROTATION OF THE GALACTIC GLOBULAR CLUSTER SYSTEM

J. Colin

Besançon Observatory, France

ABSTRACT: We use the radial velocities of galactic globular clusters, corrected for solar motion to calculate their apparent angular momentum. Comparing the apparent angular momentum with a numerical simulation we show, with 90% confidence, that the globular cluster system is in prograde rotation. The distribution of angular momentum with Fe/H reveals three distinct groups of clusters with different kind of rotation:
1) 19 clusters with Fe/H > -0.8 have a solid-body rotation in the direct sense.
2) 33 clusters with -1.75 < Fe/H < -0.8 do not exhibit any rotation (80%.confidence).
3) 27 clusters with Fe/H < -1.75 have a uniform rotational velocity in the direct sense (90% confidence).

1. SIMULATION OF THE GLOBAL SYSTEM

At the present time we have only the radial velocities of the globular clusters to determine the kinematics of the global system. Thus the results that we can obtain concern apparent motions. In this paper we calculate the apparent angular momentum of 99 clusters whose radial velocities are given by Zinn (1985) and the resulting apparent total momentum of the system. We find J_T = 7900 kpc km/s and we search to see if this value is sufficiently high to represent a global rotation. To do that we replace the radial velocity of each cluster by a random velocity whose value is between plus and minus 270 km/s, that is the maximum observed radial velocity corrected for the solar motion and we calculate the corresponding total angular momentum. This simulation is made 2000 times; the resulting moments J_S are always between ± 10,000 kpc km/s. Thus the observed value, J_T is inside these maxima and could be simply explained by the lack of tangential velocities. But the 2000 angular moments obtained by simulation are centered on zero and are distributed according to a Gaussian curve with a standard deviation T = + 5200 kpc km/s. The observed value is out of the standard deviation and the probability to obtain it is 10%. Thus, the probability that the system of globular clusters rotates is 90%.

2. ROTATION OF THE SUBSYSTEMS OF THE GLOBLAR CLUSTERS

The plot of the apparent angular moments against the galactocentric distance for each cluster is not very significant, but the angular moments against the metallicities divide the system of globular clusters into three groups (Fig. 1.):

1) Fe/H > -0.8 with only positive values of the angular moments.
2) -0.8 > Fe/H > -1.75 with positive and negative values.
3) -1.75 > Fe/H with positive values.

For each of these three subsystems we use the same simulation as that used for the global system. We find that the first group has a prograde rotation (95% confidence), the second group has no rotation (48% confidence) and the third group has a prograde rotation (92% confidence).

For each group, we investigated the angular velocity of a frame in which the sum of the residual angular moments is zero. We also investigated the constant rotational velocity V_{Rot} calculated in such a way that, if for each cluster we take out the partial angular momentum corresponding to V_{Rot}, the sum of the residual moments is zero. This is a way to define the global rotation.

Therefore, we find that the metal-rich clusters have a prograde solid-body rotation; the metal-intermediate group has no rotation (or perhaps a slow retrograde one), and the metal-poor clusters have a prograde differential rotation.

REFERENCE

Zinn, R. 1985 Astrophys. J. 293, 424.

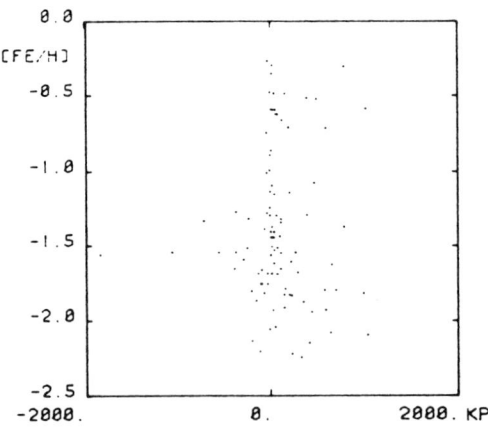

Fig.1. Distribution of Fe/H against angular moments for 99 globular clusters.

ASTROMETRIC DISTANCES OF GLOBULAR CLUSTERS

Kyle Cudworth

Yerkes Observatory, University of Chicago

Ruth C. Peterson

Whipple Observatory, Smithsonian Institution

ABSTRACT. With high-precision radial velocities and proper motions, one can equate the proper motion and radial velocity dispersions to obtain astrometric distances independent of any standard candles. We discuss the method and the small distance it yields to M 22.

1. INTRODUCTION

The distance scale of globular clusters has historically been based upon assuming an absolute magnitude for the RR Lyrae variables or the horizontal branch immediately redward or blueward. A more recent approach depends upon main-sequence fits of CCD photometry to magnitudes from ground-based parallaxes (e.g. Vilkki et al. 1986). Any such "standard candle" method is subject to uncertainties in the calibration of the candle and in reddening/extinction corrections.

These problems do not affect astrometric distances, in which the internal dispersion in radial velocities RV is matched to the apparent dispersion in proper motions PM. Cudworth (1979) and Lupton et al. (1985) have used Yerkes PMs and Palomar RVs to derive distances for M 3, M 13, and M 92. These agreed well with standard-candle distances, but the PM uncertainties were too large for definitive confirmation.

2. STATISTICAL PARALLAX METHOD

Recent PM measurements, such as those of Cudworth (1986) for M 22, have errors of 20 milliarcsec/century/star, if first-epoch plates are old enough and of sufficient number and quality. Such accuracy is essential, as is a proper error analysis, for even at 5 kpc a cluster with an RV dispersion of 5 km/s has a PM dispersion of 20 mas/cen.

The following problems must also be avoided. 1) In crowded fields, spuriously large PM values may be deduced for certain stars where subtle blending has arisen from plate-to-plate seeing and guiding differences. 2) The method is sensitive to the assumption that the velocity distribution is isotropic. Fortunately, this is usually the case in the inner regions of a well-populated cluster. Where departures from isotropy do occur in the outer regions, as a rule they are radial; their presence can be detected in the PMs alone by comparing the mean radial PM with the meant tangential PM. 3) Rotation, another form of anisotropy, must also be taken into account. Both PMs and RVs should show it. 4) RVs for the brightest giants show an extra dispersion ("jitter") of about 1 km/s, attributed to atmospheric motions; many such stars are photometric variables. (Giants in populous clusters are rarely spectroscopic binaries.)

3. APPLICATION TO M 22

Cudworth (1986) derived PMs with errors 20 mas/cen for more than 200 stars in M 22 with V < 14. Between 200" and 400", the internal dispersion was well-defined and radial and tangential PMs were equal. In July Peterson used the MMT echelle to obtain radial velocities good to ± 1 km/s (Latham 1985) for essentially all giants within this annnulus with 12.9 < V < 14, i.e. more than 1.5 mag from the tip. The RV dispersion of the 87 giants is 6.9 ± 0.5 km/s; peak-to-peak rotation is < 6 km/s; the PM dispersion is 61 ± 4 mas/cen yielding a distance of 2.4 kpc ± 10%, or $(m-M)_o$ = 11.9 ± 0.22. The distance derived taking E(B-V) = 0.32 ± 0.04 and M_V(blue HB) = +0.70 ± 0.20 is $(m-M)_o$ = 12.43 ± 0.24, or 3.06 ± 0.33 kpc, a significant discrepancy.

Support for the shorter distance is offered by the presence of three RV members whose total space motion would marginally exceed the cluster escape velocity if the larger distance were adopted. To be sure, our final analysis will include fitting a dynamical model to the data to fully account for dynamical effects.

Radial-velocity observations were obtained at the Multiple Mirror Telescope Observatory, a joint facility of the University of Arizona and the Smithsonian Institution. We thank the MMT schedulers for a generous time allotment, and the Harvard and Lick Observatories for the loan of old plates. This research was supported in part by NSF grants to the Yerkes astrometric program and to RCP.

REFERENCES

Cudworth, K. M.   1979   Astron. J. **84**, 1312.
Cudworth, K. M.   1986   Astron. J. **92**, 348.
Latham, D. W.   1985   in IAU Colloquium No. 88, Stellar Radial Velocities, A. G. D. Philip and D. W. Latham, eds., L. Davis Press, Schenectady, p. 21.
Lupton, R., Gunn, J. E. and Griffin, R. F.   1985   in IAU Symposium No. 113, Dynamics of Star Clusters, J. Goodman and p. Hut, eds., Reidel, Dordrecht, p. 19.
Vilkki, E., Welty, D. and Cudworth, K. M.   1986   Astron. J. **92**, in press.

ABSOLUTE PROPER MOTIONS AND SPACE VELOCITIES OF GLOBULAR CLUSTERS

H. -J. Tucholke[1,2], P. Brosche[1] and M. Geffert[1]

Observatorium Hoher List 1
University of Munster   2

ABSTRACT. Proper motions of globular clusters referred to extragalactic background objects provide absolute proper motions leading to space velocities. Some results from the Bonn Observatory proper motion program for galactic globular clusters are shown. Reference stars from the Lick program linked to galaxies are used. Low orbital angular momenta for NGC 4147, NGC 5466 and NGC 6218 were detected. In a further program, proper motions of 47 Tuc and NGC 362 are currently being measured relative to the background of the SMC.

1. INTRODUCTION

Absolute proper motions of globular clusters are important for the determination of the galactic halo dynamics, since, combined with radial velocities, they yield space velocities. In order to proceed from relative to absolute proper motions, one generally models the motions of the field stars by the superposition of solar motion and galactic rotation. The validity of this model has been questioned because of large unexplained components in the proper motions of fundamental stars (Brosche and Schwan 1981, 1986). An independent way to link globular cluster proper motions to an absolute reference frame is followed in Bonn: The cluster proper motions are referred to extragalactic objects, which should show zero proper motion within the measuring accuracy.

2. THE BONN OBSERVATORY PROPER MOTION PROGRAM

The following globular clusters are observed in the Bonn proper motion program: NGC 4147, NGC 5024, NGC 5272, NGC 5466, NGC 5904, NGC 6205, NGC 6218, NGC 6254, NGC 6341, NGC 6779, NGC 6934, NGC 7078 and NGC 7089. Old and new plates with epoch differences from 60 to 70 years were taken with the double refractor of the Bonner Sternwarte. The error of the cluster mean motion is $0\overset{''}{.}05$ to $0\overset{''}{.}10$/100 yr. For clusters at sufficiently high galactic latitude, stars from the Lick program with proper motions measured relative to galaxies are used as reference stars. In the future HIPPARCOS, whose system will be tied to an extragalactic inertial frame, will observe some reference stars in the field of each cluster.

Up to now results were published for NGC 5466 (Brosche et al. 1983) and NGC 4147 (Brosche et al. 1985). Below we list orbital parameters for both clusters along with preliminary results for NGC 6218. The solar motion and galactic rotation from Frenk et al. (1980) and a spherically symmetrical galactic mass distribution with $M(R) \sim R$ were used.

Cluster	$(\Pi,\Theta,Z)_{GSR}$	Eccentricity e	Perigalactic Distance (kpc)	Angular Momentum Parameter h
NGC 4147	(163,233,153)	0.62	12.2	0.62 ± 0.12
NGC 5466	(256,-61,223)	0.82	4.7	0.34 ± 0.11
NGC 6218	(-43,110,-130)	0.39	2.7	0.52 ± 0.16

Innanen et al. (1983) suggested a lack of high-eccentricity globular cluster orbits. They found a lower limit of 0.55 for the dimensionless angular momentum h. Our values for h are near or below this limit. In addition, we find high orbital eccentricities for the three clusters, whose common feature is their relative looseness. NGC 5466 is an example for a globular cluster in a retrograde orbit.

3. SPACE VELOCITIES FOR 47 TUC AND NGC 362

The galactic globular clusters 47 Tuc and NGC 362 appear projected onto the SMC. This provides the opportunity to relate directly the motions of the two clusters to that of the SMC, whose proper motion (predicted to be 0".2/100 yr by Lin et al. 1982) will be measured by HIPPARCOS. First-epoch plates taken with the 13"- and 24"-refractors of the Harvard Southern Station on loan from the Harvard Plate Collection were measured with the PDS 2020GM-Microdensitometer of the Münster Astronomical Institute. All objects in the cluster fields down to the limiting magnitude of 17^m were registered and are currently reduced.

ACKNOWLEDGEMENTS

We are very grateful to Dr. A.R. Klemola for providing us with results from the Lick program prior to publication. H.-J. Tucholke thanks Dr. M. Hazen for her cooperation in the use of the Harvard Plate Collection.

REFERENCES

Brosche, P., Geffert, M., Klemola, A. R. and Ninkovic, S. 1985 Astron. J. 90, 2033.
Brosche, P., Geffert, M. and Ninkovic, S. 1983 Publ. Astron. Inst. Czech. Acad. Sci. 56, 145.
Brosche, P. and Schwan, H. 1981 Astron. Astrophys. 99, 311.
Brosche, P. and Schwan, H. 1986 in IAU Symposium No. 109, Astrometric Techniques, H. Eichorn and R. J. Leacock, eds., Reidel, Dordrecht, p. 53.
Frenk, C. S. and White, S. D. M. 1980 Monthly Notices Roy. Astron. Soc. 193, 295.
Innanen, K. A., Harris, W. E. and Webbink, R. F. 1983 Astron. J. 88, 338.
Lin, D. N. C. and Lynden-Bell, E. 1982 Monthly Notices Roy. Astron. Soc. 198, 707.

A SEARCH FOR OBSCURED GLOBULAR CLUSTERS

S. Djorgovski

Harvard Smithsonian Center for Astrophysics

ABSTRACT. It has been estimated that up to several tens of globular clusters in our Galaxy remain undiscovered, because they are hidden by the dust. Most of those clusters are expected to be in the bulge area, but some may be anywhere in the galactic plane. This search is based on the use of $IRAS$ catalogs as candidate lists for the obscured globular clusters. It was found that the properties of detected known clusters are sufficiently distinct for a meaningful sifting through the PSC. A sequence of statistical "filters", described below, was applied on the PSC, until a couple of hundred most promising candidates were found in the bulge.

So far, only the Point-Source Catalog (PSC) has been analysed; it is expected that even better chances of finding clusters will be afforded by the use of Small Extended Sources Catalog. First, known globular clusters were identified in the PSC, dubious or confused sources rejected, and the statistics of their $IRAS$ properties established. The full PSC contains 245839 sources, and a selection procedure must be devised to extract a manageable number of cluster candidates.

The first step was to define the Galactic bulge and plane search area. The bulge area was defined as: $345° \geq l_{II} \leq 15°$, $-8° \geq b_{II} \leq 8°$. The plane area spans the rest of the Galactic equator, with $-2° \geq b_{II} \leq 2°$, except for the region $330° \geq l_{II} \leq 30°$, where the thickness is $-4° \geq b_{II} \leq 4°$. These two regions contain 71482 sources, out of which 60886 are unidentified, and 5956 are classified as multiple. Out of the unidentified, 1500 were subsequently identified with SAO stars. This leaves the total search sample of 60214 sources.

It was imediately noticed that all known globular clusters identified in the PSC are detected in the 12 μm band, and none are detected in the 60 and 100 μm bands. This is as expected, since we are expecting to find stars, rather than hot dust or gas. Applying the corresponding selection to the previous sample resulted in 36229 sources. At this point, it was decided to search the bulge area first, since that is where most of the missing clusters are expected to be, and to postpone the analysis of the galactic plane sample. The bulge sample as selected so far consists of 13837 sources.

Examination of log F_{12} histograms, for the PSC and the known bulge clusters indicated that the probability of finding a cluster is higher for the fluxes exceeding ~ 2 Jy in this band. This is as expected: the core of a globular cluster contains many giants, whereas most of the field PSC sources are probably individual stars.

This flux selection narrowed the sample down to 5018 sources. Examination of F_{25}/F_{12} histograms indicated that the clusters are never too cold, and an empirical threshold was imposed to this ratio, $F_{25}/F_{12} \leq 1.4$ for F_{25} detections, or ≤ 1.2 for F_{25} upper limits. This further reduced the sample to 4584 sources. From the log F_{25} histograms, it appears that the clusters avoid the range $F_{25} \sim 1.5 - 3$ Jy. This may be a spurious effect, but it is possible that the bimodality reflects a true distinction between clusters with and without much dust, or post-AGB stars. Application of this exclusion criterion lowered the sample to 2539 sources.

Some meager spatial resolution information is provided in the PSC in the form of correlation coefficients (CC) : 1=A=stellar, 2=B=slightly diffuse, etc. Not surprisingly, the clusters are much more often classified as diffuse (CC > 1). The contrast in CC distributions between the known clusters and the PSC sources increased after the rejection of "faint" and "cold" sources in the three steps above. Selecting only the sources with both CC(12μm) > 1, and CC(25μm) > 1, yielded the sample of primary candidates, consisting of 70 sources; the sample of objects where CC(12μm) > 1, or CC(25μm) > 1, but not both, is the sample of secondary candidates, consisting of 318 sources. If the F_{25} exclusion described above is relaxed, then there are additional 298 marginally extended candidates.

An examination of candidate fields on the ESO/SERC IR sky survey films is now in progress. A number of candidates can be immediately associated with obvious stars, and the remaining ones, mostly "blank fields", will be imaged with a CCD in R and I at CTIO.

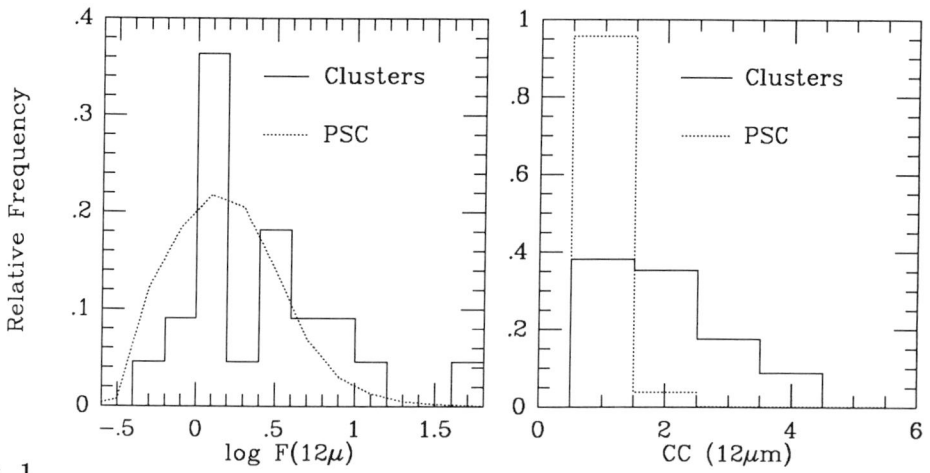

Fig. 1.
Left: Relative frequency distributions of log F_{12} for the bulge globular clusters detected by IRAS, and the "hot" PSC sources, in 0.2 log bins. Above $F_{12} \simeq 2$ Jy, the clusters dominate. Similar selection criteria can be devised based on the 25μm fluxes, and flux ratios.
Right: Distributions of angular resolution flags (correlation coefficients), given as ranks, 1 for a pointlike source, 2 for a slightly diffuse, etc., for the globular clusters, and the PSC sources, after the flux and flux ratio filters were applied. The clusters show a much higher tendency to be marginally resolved.

ON THE COLOR EXCESSES OF GLOBULAR CLUSTERS

V. Straižys and R. Janulis

Institute of Physics, Vilnius

The interstellar reddening of globular clusters of the Galaxy is still an important unresolved problem, especially for metal-rich objects that are found usually at low galactic latitudes in the general direction of the galactic center. Their color excesses are needed in order to correct their color-magnitude diagrams and to determine their intrinsic integrated color indices. For this we need some method which is not related to measures of the cluster stars. One such method is to use foreground field stars in the direction of the globular cluster to measure the interstellar reddening. Because most of the globular clusters lie outside the galactic plane, we need information about the reddening in all the layer of absorbing dust in different directions. This information can be obtained by investigating stars which are at different distances from the Sun up to the edge of the absorbing dust layer. On the other hand, these stars should be as close as possible to the position of the globular cluster to avoid possible variations in the interstellar reddening in the area of the cluster.

The Vilnius photometric system offers an excellent possibility of doing such research since it gives spectral and luminosity classes, absolute magnitudes and interstellar reddening for stars of all spectral types. The only problem is to achieve a deep enough limiting magnitude and to go far enough from the galactic plane. With this aim we have started a program for the investigation of interstellar reddening in the direction of a number of low latitude globular clusters using photoelectric photometry of stars in the Vilnius photometric system with the 1 m telescope of our institute placed on Maidanak Mountain Observatory in Uzbekistan. The work has been completed for three areas in the direction of the globular clusters M 56, M 71 and NGC 6712. In each area about 100 stars down to V = 13 have been investigated, all of them within 30 min. of the center of the cluster. The results for the cluster M 71 have been published (Janulis and Straizys 1984, Janulis 1984). The results for the cluster M 56 are in press (Janulis 1986). The results for NGC 6712 will be published later.

The investigation leads to the following conclusions:

1. In all the investigated areas the interstellar reddening of stars shows a considerable scattering of points in the plot of color excess vs. distance. This scattering for the cluster M 71 is between $E_{(B-V)} = 0.12$ and 0.32 and this exceeds the errors of the reddening determination. This means that the column density of the interstellar dust shows variations in different parts of the area by a factor of 3. Probably this is a result of the cloud structure of the interstellar dust. For a more exact determination of the cluster reddening a much small area around it should be studied.

2. The mean values of the interstellar reddening determined for the stars beyond the main bulk of the galactic dust layer are somewhat less than the reddening values determined by other methods, e.g. by the method of integrated color indices. This may mean that we encounter a selection effect caused by the limiting magnitude. To check this explanation we have to extend the measurements to fainter stars.

3. Both the problems listed above could be overcome by using CCD photometry in the Vilnius system down to V = 17 or 18 magnitude. This would give two-dimensional classification of stars and the interstellar reddening up to the edge of the Galaxy..

REFERENCES

Janulis, R. 1984 Bull. Vilnius Obs. No. 67, 18.
Janulis, R. 1986 Bull. Vilnius Obs. No. 75 (in press).
Janulis, R., and Straižys, V. 1984 Astrophys. Space Sci. 100, 95.

FIRST POSTER DISCUSSION

PILACHOWSKI: Welcome to our first poster discussion session on the topic of globular clusters in the Galaxy. I'd like to begin the discussion with the paper by Sweigart, Greggio, and Renzini. These authors have constructed new evolutionary sequences for low and intermediate mass stars evolving from the main sequence to the onset of core helium ignition in order to map the transition from intermediate mass stars without a prominent red giant phase to low mass stars which do undergo a prominent red giant phase. Their results show the RGB phase transition begins at an age of 500 to 600 million years, depending on composition. How well determined is this transition age from observations of star clusters? Do any star clusters with prominent red giant branches have ages less than a billion years?

WALLERSTEIN: M 11

PILACHOWSKI: It's young, all right, but it doesn't have much of a red giant branch above the clump. I can contrast two clusters, NGC 7789 and NGC 752, both of which have ages of about 1.5 Gyr. NGC 7789 has a very nice red giant branch; Saha and I have been searching for red giants in NGC 752, but we can find no stars above the luminosity of the clump.

RENZINI: The best observational test will come in the Magellanic Clouds, but we need more data for the clusters.

PILACHOWSKI: The paper by McClure, Stetson, Hesser, Smith, and VandenBerg presents luminosity functions for several clusters with a range of metallicity, and finds a trend between the power law exponent of the main sequence mass function and metallicity. King has added one additional paper to the discussion which suggests that mass functions determined from the outer parts of clusters are probably reliable - so there is no need to get all the way into the center. The determination of luminosity functions is a lot of work, but we really need to see more of them determined!

BELL: I'd like to emphasize that there are really two problems in determining luminosity functions; we need a better defined system than B and V, and we need broad bands to get to the faintest stars.

COHEN: Please tell us the magnitude of the crowding corrections applied to the observed star counts so we can decide how much credence to place on these results.

MCCLURE: As I remember the corrections are less than about 30% for the data appearing in the figures of this paper.

DA COSTA: Dynamical Evolution - the loss of low mass stars via tidal stripping, escape, etc - can mean that current mass functions are not necessarily related to INITIAL mass functions. For example, NGC 6397 is a very metal poor cluster with a current mass function that is deficient in low mass stars (alpha ~ 0, Da Costa 1982). But this is a very dynamically evolved cluster that may have had an initial mass function that supports the proposed correlation.

KING: It seems to me that the existence of a correlation between mass function and metallicity, undisturbed by dynamical properties, implies that there has not been any serious loss of low-mass stars.

PRYOR: R. McClure, G. Smith and myself have calculated mass segregation corrections for luminosity function power law exponents. Our results are similar to those in King's poster. The exponents become smaller by $\Delta X \sim 1$, but the trend with metallicity is preserved. More theoretical work is needed, though, on the amount of mass segregation expected at the very large radii where some of the luminosity functions are measured.

PILACHOWSKI: Let's move on to a discussion of the papers on the subject of cluster structure parameters. The first of these papers, by Peterson and Reed, offers new determinations of the core and tidal radii, and central relaxation times, densities, velocity dispersions, and escape velocities, among other parameters. The authors urge the use of these new values, which have corrected systematic errors in earlier results. The paper, by White and Shawl, tabulates axial ratios and major-axis orientations for 104 galactic globular clusters, and suggests that galactic tidal effects don't elongate clusters. I also recommend a peek at Figure 5 of their paper for a most interesting suggestion! People who want to correlate various composition parameters like CN with cluster flattening, etc. should probably look at these new data.

NORRIS: For the dependence of degree of CN strengthening on flattening I incorporate the data of White and Shawl!

PILACHOWSKI: Now we come to three papers which are dear to my heart! The papers by Leep, Wallerstein, and Oke; Gratton, Quarta, and Ortolani; and Pilachowksi and Sneden; all present the results of detailed analyses of high resolution CCD spectra of globular cluster giants, and M 71 giants in particular. All three papers follow similar techniques, and all three incorporate Arcturus as a standard. I believe that all the authors of these papers will agree with me that the introduction of new spectrographs, and particularly new detectors, in the last 7 years has provided a tremendous gain in our ability to obtain high-quality spectra of faint stars in globular clusters. We owe a great debt to the engineers who have made this possible. How nice it is that all these papers agree that the iron abundance in M 71 is near -0.75! The controversy can be laid to rest.

TRIMBLE: Your abstract doesn't mention the amount - what is the [Fe/H] of M 71?

PILACHOWSKI: -0.75 for 2 stars.

BELL: Is TiO contamination a possible resolution of the low echelle abundances? Indeed, aren't there TiO bands all over to worry about?

PILACHOWSKI: The stars analyzed here are too hot to form much TiO, and we see no evidence for it in the spectrum.

GEISLER: In the case of the cooler 47 Tuc giants in particular, there are regions where the TiO bands are so strong that they lower the "continuum" and can cause abundance underestimates by ~0.2 dex. In relative clean regions one does still need to worry about individual lines blending with TiO features.

PILACHOWSKI: Of course, we are careful to select lines that appear clean and unblended in the Arcturus atlas! It is also true that the line list used in the earlier analysis of cool 47 Tuc giants contained few lines in the regions Doug has shown to be contaminated with TiO.

PILACHOWSKI: I'd like to ask Rafaele Gratton what the prospects are for determining the abundance of oxygen in his 47 Tuc giants.

GRATTON: The geocentric radial velocity at the observing time was too low; therefore, the stellar line was blended with telluric emission. We could not derive reliable abundances.

PILACHOWSKI: The next paper by Rose and Tripicco reports on the integrated spectra of metal-rich galactic globular clusters. They find that the relative amounts of light contributed by dwarfs and giants varies from one cluster to another. Do you have any further results to add?

ROSE: We have very recently obtained integrated spectra of the core of 47 Tuc. Both its Sr II/Fe I and 3888/3859 indices are high, from which we infer that the core is dominated by dwarf light at 4000 Å, and that the mean core CN anomaly is very large. This result probably rules out the possibility that the observed spread in the Sr II/Fe I index between clusters is due to an age spread.

SCHOMMER: Which is your other cluster (beside 47 Tuc) which has a well-determined age and dwarf/giant ratio? I'm not sure any age is really reliable to 1 Gyr.

ROSE: Of the two clusters with well determined ages, only 47 Tuc shows an excess of dwarfs. The other one is M 71, and it does not show an excess of dwarfs.

KING: If you put an excess of dwarfs in the center of a well-relaxed cluster, you make life quite difficult for those of us who try to model the clusters dynamically.

COHEN: As I recall, the integrated light measurements of CO band strengths in globular clusters show no indication of dwarfs dominating the light. Are the samples disjoint or do I recollect incorrectly or are the CO measurements irrelevant?

ROSE: A factor of 2 variation in the relative amounts of dwarfs' vs. giants' light at 4000 Å, which is what we infer from the strontium index, will not necessarily show up at 2.2 microns, where the giant light will dominate except in extreme circumstances.

PILACHOWSKI: The paper by Altner discusses two mysterious broad absorption features at 1400 and 1600 Å in globular clusters. Bruce, would you like to tell us what you suspect the origin of these features might be?

ALTNER: There are 3 possibilities: the bands could be artifacts of the camera, the features could be real in individual stars, and then show up in the integrated spectra, or the short wavelength tails of the light of cooler stars may contribute a "saddle" effect.

PILACHOWSKI: Smith reports on V and 1.0 micron photometry of giants in ω Cen. He finds a broad giant branch even in the (V, (V - 1 μ)) diagram. Graeme, would you like to comment further on this paper?

SMITH, G.: (V - 1 μ) photometry was done of ω Cen giants, because John Norris and I wanted to obtain a color from which we could measure effective temperatures of stars for which no infrared photometry was available. We used (V and 1 μ) in an attempt to obtain a long baseline and avoid the problem with (R-I), which does not show a large spread on the ω Cen giant branch.

PILACHOWSKI: Now we come to the paper by Armandroff. I was impressed by the tight correlation between the calcium infrared triplet strength and [Fe/H]. Will you be applying this technique to more systems, reddened clusters in the direction of the galactic center? Radial velocities and metallicities are being derived from these spectra. It may be possible to apply this technique to extragalactic globulars. A parallel technique is being developed for clusters which are too diffuse for integrated light spectroscopy using individual-giant spectroscopy, again at the calcium triplet.

PETERSON: Are you planning to include galactic bulge giants in this program? These are basic ingredients for metal-rich extragalactic objects.

ARMANDROFF: It would be useful to include those giants.

NORRIS: Are you finding clusters with roughly solar abundance, similar to the most metal-rich objects found by Zinn among his disk globular cluster group?

DA COSTA: For very cool, luminous giants the triplet line strengths tend to saturate for abundances greater than [M/H] -1.0 dex, so metal abundance discrimination at high abundances in individual stars is reduced.

COHEN: There are some problems in extending the use of the IR calcium triplet to extragalactic globular clusters - you are limited by the atmospheric emission lines.

HESSER: Hugh Harris, Gretchen Harris, and I successfully obtained radial velocities for NGC 5128 globular clusters using the IR calcium triplet for clusters having V < ~ 19. We haven't yet tried to calculate abundance indices because, in part, we were not impressed by the difference in line strengths in the few giants we observed in clusters of widely differing metallicity.

PILACHOWSKI: Lee, Demarque, and Zinn have found that they can produce bimodal distributions of horizontal branch stars with a monomodal distribution of initial parameters. In your models, do you find it easier to create horizontal branch gaps at some effective temperatures than at others?

LEE: The positions of gaps depend on the [Fe/H] of the clusters. The combination of the shapes of the evolutionary tracks and the evolutionary speed along each track can naturally produce an apparent bimodal distribution of the horizontal branch, or gaps.

NORRIS: Would you predict that all horizontal branches which span the RR Lyrae gap should have bimodal distributions?

LEE: The horizontal-branch morphology strongly depends on the mean horizontal branch mass and dispersion factor.

PILACHOWSKI: Now we have two papers on horizontal-branch stars by Rood and Crocker, and Crocker and Rood. They find differences in surface gravity among very blue horizontal-branch stars which may be due to differences in the helium abundance or to rotation. In the second paper they detect a strong HE I 4026 line in two hot horizontal-branch stars in NGC 6752. The presence of strong helium lines may give us the opportunity of determining helium abundances in horizontal branch stars.

NORRIS: Given that one seems a range of helium abundance among blue horizontal-branch stars, how does one know which objects are telling you the truth about the Population II helium abundance?

CROCKER: It's important to observe a large enough sample of stars.

PILACHOWSKI: There are two papers by Philip and Samus and by Philip which discuss spectra and four-color photometry of blue horizontal-branch stars in several globular clusters. They find evidence for some stars with stronger calcium lines, and some stars with high Strömgren c_1 indices. They need more data to confirm that these two groups of stars are the same.

SMITH, H,: Do you see unusual abundances of any element but calcium? In particular, can you see anything unusual about lines which might be sensitive to diffusion?

PHILIP: The dispersion of these spectra is such that the K-line can be seen and measured. Weaker lines can not be resolved, and thus I have no information on other lines. It would take spectra at a higher dispersion to resolve such lines.

PILACHOWSKI: Turning to the question of the similarity of the cluster and field halo populations, we have two papers by Laird, Carney, Latham, and Kurucz and by Ratnatunga. The first of these examines the metallicity distribution function of halo dwarfs determined from low signal-to-noise, high resolution spectra using cross-correlation techniques. They find no compelling evidence that the metallicity distribution functions for the field stars and for the clusters are different.

LAIRD: I would like to clarify our result regarding the metallicity distribution functions of the field dwarfs and globular clusters. By a Kolmogorov-Smirnov test, which is a very conservative test, one cannot be sure that their distributions are different, but they certainly look different. The lack of certainty is largely due to the small number of objects.

NORRIS: What was the level of significance at which the cluster and field star distributions were different.

ZINN: The difference between the observed metallicity distribution of the clusters and the simple model may not be significant. The model predicts that only a very few additional metal poor clusters should exist, and it may be nothing more than small number statistics that these clusters are not in the present sample, which, by the way, is not a complete sample of all of the globular clusters in the Milky Way.

ZINNECKER: The result that halo field stars can be more metal-poor than the cluster stars may contradict the prediction of the theory of Fall and Rees (1985), the halo field stars should form only after the metallicity in the proto-galaxy is raised to 0.1 Z_o, or thereabouts.

ZINN: If my memory is correct, the work of Fall and Rees suggested that individual field stars could form only when the metal abundance of the halo was moderately high (i.e. [Fe/H]~-1.0). According to their picture, it is possible to form more metal-poor clusters of stars, and I believe that they suggest that the metal-poor halo stars originated in clusters that were disrupted.

PILACHOWSKI: Ratnatunga finds evidence for a thick disk in a sample of 58 giants located by his objective prism survey. His thick disk-like component is rotating with the disk, and has a velocity dispersion of 50 km/sec.

PILACHOWSKI: The next group of papers we will discuss concern the galaxy cluster system as a whole. The paper by Colin, suggests that the cluster system is in prograde rotation, and identifies groups of clusters with distinct rotation. The second paper by Tucholke, Brosche, and Geffert presents their efforts to determine absolute proper motions for clusters referred to extragalactic sources. Cudworth determines an astrometric distance to M 22, and Djorgovski describes his search of the IRAS data base to locate hidden globular clusters. I'd like to begin the discussion of these papers by asking Kyle how seriously we should take his new distance to M 22.

CUDWORTH: We want to move M 22 closer by ~0.5 magnitudes, but some of this is likely to be due to reddening problems in this cluster. Other papers at this meeting will claim fainter RR Lyrae absolute magnitudes, so there may not be a big discrepancy. We also do not currently understand M 22 dynamics well enough.

SCHOMMER: I have a question for Bob Zinn. Which is the outer halo blue horizontal-branch cluster without a known [Fe/H]?

ZINN: The cluster is Pal 15.

PILACHOWSKI: Finally, we have a paper dealing with the interstellar medium. Straizys and Janulis describe their study of the reddening toward M 71, M 56, and NGC 6712. They note that the reddening towards M 71 varies by a factor of 3 across the cluster. Zdanavicius also presents a paper on the use of the Vilnius photometric system to separate cluster and field stars.

WEBBINK: In compiling color excesses of globular clusters for the appendix to IAU Symposium 113, I was struck by the fact that in virtually every case of appreciable reddening (E(B-V) > 0.2), variations in reddening over the face of the cluster in question are quite apparent visually on blue sky survey plates. It is somewhat misleading to assign a single value to the reddening of these clusters.

PETERSON: Reddening of M 4 within the narrow area studied by Richer and Fahlman must be very constant to reproduce the narrow width of the main sequence turnoff.

RICHER: Although one field in M 4 with an area of 3' x 5' shows a tight turnoff, at least one other shows differential reddening of the indices of 0.03 - 0.05 in E(B-V).

PILACHOWSKI: This concludes the discussion. I'd like to thank the authors of all the papers in this session, and all of you for a stimulating discussion.

Chapter VIII

Poster Papers

Cluster Systems in Nearby Galaxies

Charles Peterson and Martha Hazen pinning up.

President Derek Bok officially welcomed the participants

A SEARCH FOR GLOBULAR CLUSTER CANDIDATES IN NGC 2403

M. L. Malagnini and P. Santin

Astronomical Observatory, Trieste

F. Bonoli, L. Frederici and F. Fusi Pecci

Department of Astronomy, Bologna

R. G. Kron

Yerkes Observatory

ABSTRACT: A complete search on KPNO 4-m plates has been undertaken in order to determine the whole observable part of the cluster luminosity function of the late-type spiral galaxy NGC 2403. Automatic procedures have been used for object detection and analysis, and then selection criteria, based on geometric and photometric properties, have been applied in order to define the sample of cluster candidates.

1. DATA ANALYSIS AND SAMPLE DEFINITION

The preliminary list of globular cluster candidates in NGC 2403 as derived from visual investigation of plates taken with the Loiano (Bologna) 152 cm telescope (Battistini et al., 1984) refers to objects brighter than $V = 20$. Since the cluster population of the galaxy is expected to contain members which are fainter than such a limit, a new and more complete search on deeper plates has been undertaken in order to determine the whole observable part of the cluster luminosity function. The photographic material, listed in Table I, was kindly lent to us by the observers.

TABLE I
KPNO 4-m Prime Focus plates

Plates	Date	Observer	Exposure	Emulsion	Filter
1879	1975 dec 30	Burkhead	45 min	IIIa-J	GG385
3698	1982 mar 23	Humphreys	75 min	IIa-O	UG2
3699	1982 mar 23	Humphreys	90 min	IIIa-F	RG610

The plates have been digitized with the Trieste PDS 1010A, by using a scanning aperture of 20x20 micron square and step of 20 microns. The digital images of 9000x9000 pixels have been processed with FODS (Faint Object Detection System) (Malagnini et al., 1985) at the Trieste Astronet Pole. Automatic procedures have been applied to the J image in order to detect the objects, and a main catalogue has been created with a total of 11,512 entries referring to "aggregates" with area in the range 10 ÷ 1000 pixels. Then the locations of the objects on the F and U images have been computed by means of coordinate transformations. To this purpose and for the calibration of machine magnitudes, the stars listed in Tammann and Sandage (1968) Table 1 have been used. For each object different parameters (area, magnitudes, ellipticity,...) have been computed and stored in the catalogue. Since the cluster population is expected in a limited magnitude range, a first selection produces the sample illustrated in Table II. By applying the conditions:
1) $16.5 \leqslant J \leqslant 21.5$, 2) $-0.2 \leqslant J-F \leqslant 1.4$, and 3) circular symmetry, the sample reduces to a subsample of 1001 "possible cluster candidates". Next step would be a supervised morphological classification of the objects of this subsample.

TABLE II.
Color-magnitude distribution of detected objects

J	J − F				J − F
	−0.5÷0.2	0.2÷0.9	0.9÷1.4	1.4÷2.8	−0.5÷2.8
16.5÷18.5	15	156	154	100	425
18.5÷20.5	47	350	268	291	956
20.5÷22.5	139	507	803	1385	2834
16.5÷22.5	201	1013	1225	1776	4215

REFERENCES

Battistini, P., Bonoli, F., Frederici, L., Fusi Pecci, F. and
    Kron, R. G. 1984 Astron. Astrophys. 130, 162.
Malagnini, M. L., Pasian, F., Pucillo, M. and Santin, P. 1985
    Astron. Astrophys. 144, 49.
Tammann, G. A. and Sandage, A. 1986 Astrophys. J. 151, 825.

SEARCH FOR GLOBULAR CLUSTERS IN NEARBY GALAXIES II. NGC 3109

A. Blecha

Geneva Observatory

ABSTRACT: We report on the search for globular clusters around NGC 3109, a SB(s)m nearby galaxy using observations taken with the wide field telescope at La Silla. Clusters are discriminated by using the advanced image processing software (MOAN). From 320 objects, 23 candidates are retained. Their luminosity function peaks at $m_v = 19.8$, thus giving the distance of the parent galaxy as 2.13 Mpc. The radial distribution follows the $D_p^{1/4}$ law well. The total number of clusters is estimated at 40 ± 25 and the specific frequency $S_v = 3$ clusters per $M_v = -15$.

1. OBSERVATIONS

The observations were done at La Silla in January, 1985, using the Mc Mullan electronographic camera (Mc Mullen and Powell 1976) attached to the f/8.6 focus of the Danish 1.54 Richey-Chrétien reflector. The usable field of the camera is 83 mm in diameter (scale 19"/mm) giving a total field of 26.2'.

2. DATA PROCESSING AND GEOMETRICAL ANALYSIS

All suspected non-stellar objects were numbered together with a regular net of well exposed stars. Objects were scanned with a 10 micron aperture on the Geneva Microdensitometer System (Blecha 1982). Scanned areas were centered on the object of interest, with 30 x 30 points spaced by 10 microns in both x and y directions (6 x 6").

Each field was processed in order to extract the geometrical parameters of the image. The MOAN software package (Blecha 1982, 1984) uses 3 shape parameters to describe the local true two-dimensional PFS, called a Gaussian Modified Profile (GMP). Two other parameters, C and D, are fixed for each plate in order to account for various seeing conditions and/or telescope adjustments. The parameters extracted are: x_0, y_0, the position of the centroid; e,f,w, shape parameters (ellipticity, orientation and width) and B_g, the local background.

The "net" of selected stars is used to "map" the PSF over the whole plate. The mapping is done by fitting a 3rd degree 2-D polynomial on the set of shape parameters extracted from 55 stars. As the variation of the PSF shape is slow, the mapped profiles match the image data for the stars almost perfectly. The method used here is that described by Blecha (1984, 1986).

The selection of clusters is based on the following criteria: Geometrical Aspects (object width > 0.5", residual ellipticity < 1.45, smooth objects, abnormally high RMS rejected) and Photometric Aspects; (only objects with 0 < (B-V) < 1.6)

3. GLOBULAR CLUSTER SYSTEM

From initially 320 objects, 23 remain in the list after the selection. We included four very large objects (FWHM > 2") because one of them has (B-V) = 0.8; such a globular cluster will be two to three times larger than the largest cluster in our galaxy (Blecha 1986). The luminosity function, though incomplete, peaks approximately at m_v = 19.8. Assuming A_v = 0.30 and the peak M_v = -7.2, we obtain the true distance modulus $(m_v - M_v)_o$ = 26.70, giving a distance of 2.18 (+0.5, -0.2) Mpc. Our distance is in good agreement with the H II region dimensions, but in strong contradiction with the distance of 1.17 Mpc given by Elias and Frogel (1985) based on infrared photometry of a few bright stars.

4. CLUSTER DISTRIBUTION AND THE TOTAL POPULATION

If a distance of 2.2 Mpc is assumed, the $r^{1/4}$ law fits well with the data;

$$Log(N) = 0.6 \pm 1.86 \; F_p^{1/4}.$$

By integrating the radial distribution and accounting for the missing part of the luminosity function we obtain a total population of 45 ± 25 globular clusters. Assuming the true distance modulus $(m_v - M_v)_{app}$ = 26.70, total apparent B magnitude BM_T = 10.27 (Carignan 1985), $(B-V)_T$ = 0.52 and A_B = 0.41, the specific frequency per M_v = -15 used by van den Bergh (1984) is 3.08.

REFERENCES

Blecha, A. 1982 Bull. CDS 22, 24.
Blecha, A. 1984 Astron. Astrophys. 135, 401.
Blecha, A. 1986 Astron. Astrophys. 154, 321.
Blecha, A. 1987 Astron. Astrophys., in prep.
Carignan, C. 1985 Astrophys. J. 299, 59.
Elias, J. H. and Frogel, J. A. 1985 Astrophys. J. 289, 141.
van den Bergh, S. 1984, DAO preprint.

A COMPLETE SAMPLE OF GLOBULAR CLUSTERS IN NGC 5128

Ray Sharples

Anglo-Australian Observatory

ABSTRACT. We present the results of an unbiased survey to search for globular clusters around the nearby elliptical galaxy NGC 5128. A total of 44 clusters has been identified on the basis of radial velocities alone. The cluster system appears to be flattened towards the major axis of the galaxy but does not show any dynamically significant rotation. There is a deficit of clusters in the core when compared with the power law density profile seen at large radii.

1. THE SAMPLE

Previous studies of the globular cluster system associated with NGC 5128 (Hesser, Harris & Harris 1986, and references therein) have relied primarily on selecting slightly resolved candidates for subsequent spectroscopic confirmation. However, only the brightest and largest clusters are predicted to be resolved, even on good seeing plates (Harris et al. 1984). Such samples are therefore limited in size and may possibly be biased in their properties. To determine the significance of any bias it is necessary to study complete magnitude-limited samples.

The present study is based on B, V photographic photometry from 4 m prime focus plates calibrated with a CCD sequence. After rejecting obvious galaxies, magnitudes and (B-V) colors have been derived for all objects $18.0 < V(mag) < 20.0$ between radial limits of $1.5 < R(arcmin) < 9.0$ from the nucleus of the galaxy. Spectroscopic observations were then obtained using the multi-object fiber spectrograph on the Anglo-Australian Telescope (Gray 1983) to acquire intermediate dispersion spectra of 227 of these candidates from which radial velocities have been derived using a standard cross-correlation technique.

2. RADIAL VELOCITIES

The systemic velocity of NGC 5128 has been accurately determined from studies of the kinematics of the stellar spheroid (Wilkinson et al. 1986) and the ionized gas in the dust lane (Bland 1985) to be 538 ± 10 kms^{-1}. Fig. 1 shows that, on the basis of radial velocities alone, 44 objects in the spectroscopic sample are probable globular clusters associated with NGC 5128. The remaining 183 are foreground stars, as expected from the low galactic latitude of the field.

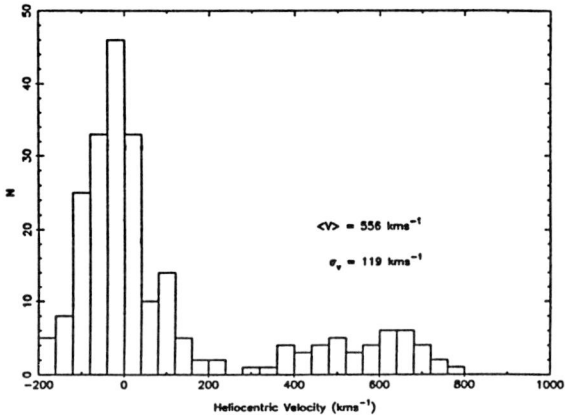

Fig. 1. Histogram of heliocentric velocities showing the clear separation into Galactic stars and globular clusters associated with NGC 5128.

3. SPATIAL DISTRIBUTION

The distribution of clusters is not symmetrical, but shows an excess in the NE and SW quadrants in which the major axis of the elliptical galaxy is also found (PA = 35°). If this distribution is modeled using a system of constant flattening, ellipticities > E2 are inferred for the cluster system i.e. flatter than the isophotes of the stellar spheroid. However, a search for global rotation of the cluster system about the minor axis shows only a marginal trend (NE: $< V > = 538 \pm 22$; SW: $< V > = 586 \pm 32$) in the same sense as that of the stellar component. This is insufficient to produce the inferred flattening of the cluster system.

The surface density profile falls as $R^{-1.8}$ for R > 3 arcmin, but is much flatter at smaller radii. The presence of a core in the radial distribution is similar to that found by Lauer & Kormendy (1986) for the M 87 cluster system, but with a smaller characteristic radius $R_c \sim 3(D/3)$ kpc, where D is the distance to NGC 5128 in Mpc.

REFERENCES

Bland, J. 1985 Ph.D. Thesis, University of Sussex.
Gray, P. 1983 Proc. SPIE 445, 57.
Harris, H. C., Harris, G. L. H., Hesser, J. E., and MacGillivray, H. T. 1984 Astrophys. J. 287, 185.
Hesser, J. E., Harris, H. C. and Harris, G. L. H. 1986 Astrophys. J. 303, L51.
Lauer, T. R. and Kormendy, J. 1986 Astrophys. J. 303, L1.
Wilkinson, A., Sharples, R. M., Fosbury, R. A. E. and Wallace, P. T. 1986 Monthly Notices Roy. Astron Soc. 218, 297.

ASTRONOMICAL CATALOGUES IN THE M 31 REGION

P. Battistini

Department of Astronomy, Bologna

In the course of work done in the Astronomy Department of Bologna University on "Search and Study of Globular Clusters in M 31", a remarkable quantity of data in the literature concerning optical, X-ray and radio surveys in the field of Andromeda Galaxy has been accumulated. This material is now presented organized in a data base. For each list/catalogue three files have been written:

1) The first one contains the data in their original form taken from the literature.

2) In the second file the coordinates of the objects have been reduced to a 1950.0 equinox, in order to have a direct and homogenous comparison between the different lists, and redundant data have been eliminated. In this file the format of the coordinates is HHMMSSDDD in integer form.

3) The third file is a documentation file containing:

a) Catalogue name
b) Keywords
c) Reference to the original note
d) Brief astronomical description
e) Description of the original table
f) Description of the data file
g) References

In the enclosed table we have reported the index of catalogues now available on magnetic tape at the Astronomy Department of Bologna University. The four columns represent: 1) identification keyword in the data beae; 2) type of object; 3) number of records; 4) reference.

CATALOGUES CONTAINED IN THE DATA BASE

36W	Radio Sources	403	Bystedt et al., 1984
37W	Radio Sources	249	Walterbos et al., 1985
5C3	Radio Sources	216	Pooley, 1969
BAR1	Emission Nebulae	686	Baade & Arp, 1964

BAR2	Extragal. Nebulae	68	Baade & Arp, 1964
BAR3	Globular Clusters	30	Baade & Arp, 1964
BAR4	Misc. Objects	40	Baade & Arp, 1964
BO1	Globular Clusters	288	Battistini et al., 1980
BO2	Misc. Objects	132	Battistini et al., 1980
BO3	Globular Clusters	353	Battistini et al., 1986
BO4	Globular Clusters	152	Battistini et al., 1986
BO5	Misc. Objects	218	Battistini et al., 1986
CSCC1	Globular Clusters	109	Crampton et al., 1984a
CSCC2	Globular Clusters	509	Crampton et al., 1984a
CRX	X-ray Sources	80	Crampton et al., 1984b
DOD	Supernova Remnants	19	D'Odorico et al, 1980
FJ	Planetory Nebulae	330	Ford & Jacobi, 1978, 1982
HO1	Open Clusters	403	Hodge, 1979 (unpublished)
HO2	Dark Nebulae	730	Hodge, 1980
M31C	Globular Clusters	355	Sargent et al., 1977
PEL	Emission Nebulae	981	Pellet et al., 1978
RIC1	Blue Objects	1131	Richter, 1976
RIC2	Radio Ident.	111	Richter et al., 1974
RO	Novae	90	Rosino, 1964, 1973
SH1	Globular Clusters	25	Sharov, 1973
SH2	Compact Galaxies	50	Sharov, 1974
VET	Globular Clusters	257	Vetesnik, 1962
WSB1	Misc. Objects	49	Wirth et al., 1985
WSB2	Misc. Objects	148	Wirth et al., 1985
SAO	Stars	133	SAO Star Catalogue
HEL22	Stars	384	Cat. Photographique du Ciel
HEL26	Stars	341	Cat. Photographique du Ciel
HEL30	Stars	340	Cat. Photographique du Ciel

THE BLUE STAR CLUSTERS OF M 31

Paul Hodge

University of Washington

ABSTRACT. CCD images in UBVR of 33 star clusters in the disk of M31 show them to be mainly bright and very young. Few cataloged clusters in the disk are likely to be as old as 10^9 years.

1. INTRODUCTION

Although the globular clusters of M31 have now been studied in considerable detail (Fusi Pecci 1986) and the young stellar associations in the disk (van den Bergh 1964; Efremov 1982) have been explored, little has been published about its disk clusters. A catalog of 404 clusters in the disk has been published (Hodge 1979), but no measurements of their properties have heretofore been given.

2. OBSERVATIONS AND RESULTS

Integrated UBVR photometry has been made for a selection of 33 disk star clusters in six regions of M31, using a CCD detector on the KPNO 4-m telescope. The brightest have absolute magnitudes of $M_v = -8.0$ and colors as blue as B-V = -0.3. Clusters in this sample also include objects as faint as $M_v = -3.7$.

The color distribution is similar to a composite of those of the LMC and M33; our sample includes a large fraction of luminous blue clusters, as in the LMC, as well as some intermediate color clusters like those in M33, but which are absent in the LMC.

The oldest disk clusters in the sample have ages, estimated from their BVR colors, of approximately 4×10^8 years, and the youngest are only a few million years old. Comparison of the cluster luminosity function with those of the LMC (van den Bergh 1981) and M33 (Christian and Schommer 1982) suggests that present sampling of M31 disk clusters is very incomplete. There are probably tens of thousands of such clusters still undetected

Figure 1 shows a comparison of the color-magnitude diagrams of M 31's disk clusters and its globular clusters. Even in this plane there is a fairly clear separation of cluster types. The figure also compares these data with that for clusters in the local group dwarf irregular galaxies NGC 6822 and IC 1613 (Hodge 1986), which have cluster populations like that of M 31's disk, but with fewer high-luminosity clusters, probably largely a population effect.

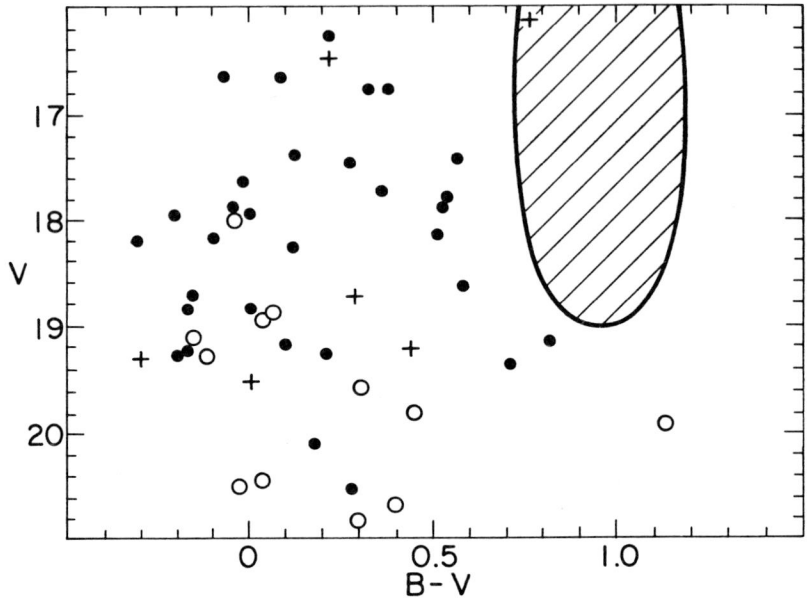

Fig. 1. Color-magnitude diagram for M 31 disk clusters (dots), compared with the area occupied by that galaxy's globulars (shaded) and with the clusters of NGC 6822 (crosses) and IC 1613 (open circles), both from Hodge (1986).

REFERENCES

Christian, C. and Schommer, R. 1982 Astrophys. J. Suppl. 49, 405.
Efremov, Yu. N. 1982 Astron. Zhur. 8, 585.
Fusi Pecci, F. 1987 in IAU Symposium No. 126, Globular Cluster Systems in Galaxies, J. E. Grindlay and A. G. D. Philip, eds., Reidel, Dordrecht, p. 173.
Hodge, P. 1979 Astron. J. 78, 959.
Hodge, P. 1986 Mem. Soc. Astron. Italiana in press.
van den Bergh, S. 1964 Astrophys. J. Suppl. 9, 65.
van den Bergh, S. 1981 Astron. Astrophys. Suppl. 46, 79.

SPATIAL DISTRIBUTION OF GLOBULAR CLUSTERS IN M 31

Jun-ichi Watanabe and Tomohiko Yamagata

Tokyo Astronomical Observatory

ABSTRACT: The spatial distribution of globular clusters in M 31 has been analyzed using the catalogues of Sargent et al. (1977) and Crampton et al. (1985). It is concluded that the globular clusters within the distance of 54' from the center of M 31 show an elliptical distribution aligned to the major axis of the disk. This is similar to the distribution of metal-rich clusters in our galaxy.

We re-examine whether the concentration of globular clusters on the major axis reflects incompleteness of the surveyed region or intrinsic distribution, using moment analysis up to the second order. We sample a region within 54' from the center which is the largest possible circle in the surveyed region of Sargent et al. (1977), and within 35' in that of Crampton et al. (1985). We regard the galactic nucleus as the center of the distribution. From the moments of the second order, we can reconstruct an ellipse of the distribution of globular clusters. If globular clusters concentrate onto the disk, the position angle of the major axis of this ellipse is close to that of the disk. Our calculations with four different radii of the circle are shown in Table I.

Globular clusters behind the nucleus may be hidden. This obscuration is limited within 15' (de Vaucouleurs and Buta, 1979). We calculated the second order moments without the clusters whose distances are less than 15' from the nucleus. These results are shown in Table II. These results suggest that the globular clusters within 54' from the center concentrate loosely onto the major axis of the disk.

In our Galaxy, globular clusters with high metallicity concentrate on the galactic nucleus (Harris 1976). Zinn (1985) also noted that a flattening of the spatial distribution of metal-rich clusters exists. If there is a metallicity gradient of the globular cluster system of M 31, then globular clusters of the inner region that we calculated, may be metal rich; since the globular cluster system in M 31 has the same characteristics as those in our Galaxy.

Table I.

Results of the calculation of the second order moments.

Radius	P.A.	Diff.	Axial ratio	N
Data from Sargent et al. 1977.				
25'	45.0	+7.3	0.73	154
35	39.1	+1.4	0.83	197
45	37.7	0.0	0.81	237
55	35.3	-2.4	0.76	273
Data from Crampton et al. 1985				
25	41.4	+3.7	0.73	197
35	35.5	-2.2	0.79	258
[45	36.9	-0.8	0.74	312]
[55	32.4	-5.3	0.68	370]

Table II.

Results without the central region.

Radius	P.A.	Diff.	Axial ratio	N
Data from Sargent et al. 1977.				
15-25'	46.0	+8.3	0.68	77
15-35	39.7	+2.0	0.83	120
15-45	38.0	+0.3	0.81	160
15-55	35.4	-2.3	0.76	196
Data from Crampton et al. 1985.				
15-25'	41.9	+4.2	0.68	92
15-35	35.7	-2.0	0.78	153
[15-45	36.9	-0.8	0.73	207]
[15-55	32.4	-5.3	0.67	265]

Disk   37.7   (de Vaucouleurs 1958)

Col. 1. Radius of the circle used for the calculations.
Col. 2. Position angle of the major axis of the fitted
    ellipse from the second order moments.
Col. 3. Displacement of the position angle of the major
    axis of the fitted ellipse from that of the disk.
Col. 4). Ratio of the major to the minor axis of the ellipse.
Col. 5. Number of globular clusters used for the calculations.

REFERENCES

Crampton, D. Cowley, A. P., Schade, D. and Chayer, P. 1985
    Astrophys. J. 288, 494.
de Vaucouleurs, G. 1958 Astrophys. J. 128, 465.
de Vaucouleurs, G. and Buta, R. 1978 Astron. J. 83,
    1383.
Harris, W. E. 1976 Astron. J. 81, 1095.
Sargent, W. L. E., Kowal, C., Hartwick, F. D. A. and van den S.
    1977 Astron. J. 82, 947.
Zinn, R. 1985 Astrophys. J. 293, 424.

FORMATION OF POPULOUS CLUSTERS FROM METAL-POOR GAS IN THE MAGELLANIC CLOUDS

T. Richtler and W. Seggewiss

Observatory of the University of Bonn

We present the results of low dispersion spectrophotometry (resolution about 5 Å, using IDS at ESO 3.6 m and 1.5 m telescopes) of 80 red supergiants in the Magellanic Clouds: 31 SMC field stars, 26 LMC field stars and 23 suspected members of young populous clusters. From these spectra, spectrophotometric indices have been derived which simulate Strömgren photometry and measure the strengths of the Ca I (4226) line and of the G-band. The indices are calibrated with respect to red giants and provide a set of metallicity indicators.

The results can be summarized as follows: The metallicities of SMC field stars cover a wide range from $-0.5 > [M/H] > -2$. The metallicity of NGC 330 turns out to be -1.4 dex. This low metallicity is in striking contrast to the usual idea of the metal enrichment of the young SMC population (it has been confirmed in a high dispersion study by Spite et. al 1986). The average metallicity of the LMC field stars is higher than that of the SMC stars, occupying the range $0 > [M/H] > -1.2$ (the lower limit may be shifted to a higher value, if individual reddenings for these stars become available). The young clusters show metallicities substancially below those expected for a young population. This leads us to propose that low metallicity is a condition to form globular clusters in the sense that high metallicity promotes the dissolution of previously bound protoclusters. The following possible scenario emerges.

As progenitors, we consider two cold HI-clouds with equal masses and different metallicities. If the clouds are assumed to be in thermal and ionisation equilibrium with the warm intercloud gas, the difference in metallicity corresponds to a difference in pressure (e.g. Talbot 1974). This pressure difference results mainly from an enhanced density for the metal poor cloud (rather than from an increased temperature). Accordingly, the radius of the (spherically symmetric) metal rich cloud is larger than the radius of the metal poor cloud (which we call the initial radii). If we assume that each cloud forms, after its collapse, a stellar cluster of a certain radius, the metal rich cluster obviously would show a larger ratio of the initial to its final radius.

This ratio is called the compression factor (Hills 1980) and it sets constraints for the stability of the system. Mass loss from the cluster (caused by radiation and stellar winds of luminous stars) also means a loss of binding energy and thus the total energy increases and eventually becomes larger than zero. Hills (1980) has derived a relation between the critical star formation efficiency, i.e. the minimum star formation efficiency to keep the system bound, and the compression factor: $\varepsilon_{min} = 1 - 1/f$ ($f = R_0/R_c$, R_0 and R_c are the initial radius and the radius, at which mass loss occurs, respectively; the mass loss is assumed to take a time short compared to the collapse time). Therefore, a metal poor (and thus dense) protocluster developes a larger stability against mass loss than a metal rich one, where only very little mass loss may be sufficient to dissolve it.

Furtheron, we argue that in the case of a dense and metal poor protocluster the mechanisms which increase the total energy of the system work less efficient than in a metal rich environment. Although the dependence of radiatively driven stellar winds on metal abundance is not yet exactly known, it is clear that the wind luminosity and therefore the energy input in the remaining protocluster gas drops with decreasing metallicity (e.g. Abott 1982). This is also true for the energy input by radiation so far as it is depictable by the concept of Stroemgren spheres. Then the mass which can be ionized by a luminous star is inversely proportional to the density of the ionized medium, so in a metal poor cloud (which is dense) the mass loss through expanding HII-regions is not so effective as in a metal-rich cloud. In reality, the associated stellar wind determines the density profile of the interstellar gas in the star's vicinity and the situation is modified by absorption of ionizing photons in the wind bubble. The principal effect, however, remains unchanged.

To summarize, protoclusters with low metal abundance are difficult to destroy by stellar winds and radiation and are thus expected to form populous, compact clusters like the young globular clusters in the Magellanic Clouds.

REFERENCES

Abbott, D. C. 1982 Astrophys. J. 259, 282.
Hills, J. G. 1980 Astrophys. J. 225, 986.
Spite, M., Cayrel, R., Francois, P., Richtler, T., and Spite, F. 1986 Astron. Astrophys., in press.

THE DEVELOPMENT OF THE RED GIANT BRANCH IN MAGELLANIC CLOUD CLUSTERS: PROGRESS REPORT

R. Buonanno and C. E. Corsi

Astronomical Observatory, Rome

F. Fusi Pecci, L. Greggio and A. Renzini

Department of Astronomy, Bologna

A. V. Sweigart

NASA Goddard Space Flight Center

   In close connection with a parallel theoretical study (see Sweigart, Greggio and Renzini 1987) CCD photometry in the B and V bands of star clusters in the LMC and SMC was started in 1983 using the ESO telescopes. To study the development of the red giant branch (RGB) we selected intermediate age clusters using the following criteria : 1) integrated (B-V) color between 0.3 and 0.6 and/or 2) membership in the type IV group of the Searle, Wilkinson and Bagnuolo (1980) classification (SWB). A few other clusters (e.g. NGC 121) were observed for calibration purposes. The observed clusters are NGC 121, 152, 1756, 1831, 1841, 1987, 2164, 2173, 2209, 2249, 1466, 1866, 2107, 2108 and 2134 and good or very good quality CM diagrams have been obtained for the first ten of them. Out of this list, those clusters with a well developed RGB are NGC 121, 152, 1841, 1987 and 2173. Using the CCD frames, integrated (B-V) colors have also been obtained for some of the clusters.

   A preliminary comparison of these CM diagrams with theoretical isochrones obtained from the models presented by Sweigart et al. (1987) gives the following indications: 1) Good agreement between the theoretical Main Sequence (MS) and the observations of the clusters is achieved, adopting a foreground reddening of 0.1 mag. and a true distance modulus of 18.5 mag. This result differs from that found at an early stage of this investigation (Buonanno et al. 1986) as our early CM diagrams were affected by a strong color equation. It appears that access to sufficiently blue standards is of crucial importance for settling the LMC distance issue (cf. Schommer, Olszewski and Aaronson 1984). 2) NGC 1756 exhibits an extended blue loop, indicating a fairly young age. From the MS fitting to theoretical isochrones one derives that this cluster is younger than 0.3 gyr. A fainter red clump is also present, which clearly belongs to the LMC field. This suggests that the intermediate color of 0.40 (van den Bergh 1981) reflects the field contamination, the cluster itself being intrinsically bluer. 3) An age between 0.3 and 0.5 gyr can be inferred for NGC 1831 and 2249. In these clusters no extended RGB is present. The integrated colors are respectively (B-V) = 0.34 and 0.43, suggesting SWB type III and IV for the two clusters. 4) For NGC 2209 the comparison with the isochrones gives an age of ~ 0.8 gyr, while the RGB is still only sparsely populated. This may indicate that the RGB develops at an age older than predicted by standard models, as first suggested by Barbaro and Pigatto (1984). The integrated (B-V) color turns out to be 0.52, taking into account all stars within 31" from the center, but drops to 0.46 when two very red carbon stars are taken out. 5) NGC 1987 is the youngest cluster in our sample which seems to have a well developed RGB. However, the CM diagram for the inner part (r < 31") of the cluster does not exhibit an extended RGB and the MS fitting yields a much younger age (~ 0.4 gyr) than the CM diagram for the outer part (~ 1 gyr). Therefore, this cluster turns out to be substantially contaminated by an older LMC field, in which the RGB has already developed. The integrated (B-V) colors are 0.54 and 0.65 for the inner and for the outer parts, respectively.

For a 62" 'aperture' we find (B-V) = 0.52 and 0.54 for NGC 2209 and 1987 respectively, in very good agreement with photoelectric values. This seems to indicate that the presence of the RGB has little effect on the integrated (B-V) colors of star clusters, as anticipated by Wyse (1985). This point is further illustrated by the following experiment: when subtracting RGB images from a CCD frame the (B-V) color for the outer part of NGC 1987 drops from 0.65 to 0.53. This shows that the presence of an extended RGB makes the cluster redder by only ~0.1 mag. For this reason, we are now inclined to exclude the development of the RGB from the list of possible origins of the dichotomy in the (B-V) distribution of Magellanic Cloud clusters (cf. Renzini and Buzzoni 1986). This dichotomy is instead more likely to be due to a degeneracy in the (B-V) of old and young clusters, coupled with the rapid fading of the clusters (cf. Elson and Fall 1985).

REFERENCES

Barbaro, G. and Pigatto, L. 1984 Astron Astrophys. 136, 365.
Buonanno, R., Corsi, C. E., Fusi Pecci, F., Greggio, L., Renzini, A. and Sweigart, A. V. 1986 Mem Soc. Astr. Italy, in press.
Elson, R. A. W. and Fall, S. M. 1985 Astrophys. J. 299, 211.
Renzini, A. and Buzzoni, A. 1986 in Spectral Evolution of Galaxies, C. Chiosi and A. Renzini, eds., Reidel, Dordrecht, p. 195.
Searle, L., Wilkinson, A. and Bagnuolo, W. G. 1980 Astrophys. J. 239, 803.
Schommer, R. A., Olszewski, E. W. and Aaronson, M. 1984 Astrophys. J. Lett. 285, L53.
Sweigart, A. V., Greggio, L. and Renzini, A. 1987 in IAU Symposium No. 126, Globular Cluster Systems in Galaxies, J. E. Grindlay A. G. D. Philip, eds., Reidel, Dordrecht, p. 483.
van den Berg, S. 1981 Astron. Astrophys. Suppl. 46, 79.
Wyse, R. F. G. 1985 Astrophys. J. 299, 593.

THE AGE DISTRIBUTION AND AGE-METALLICITY RELATION OF STAR CLUSTERS IN A NORTHERN REGION OF THE LMC

Mario Mateo

University of Washington

ABSTRACT. CCD images and UBV integrated photometry of 31 clusters comprising a magnitude limited sample in a remote northern region of the LMC have provided reliable ages and consistent abundances using isochrone 'fits' to deep color-magnitude (CM) diagrams for the 15 clusters with CCD data. Ages for the remaining clusters have been inferred from their integrated colors. The resulting LMC cluster age distribution is markedly different from the age distribution of Galactic clusters suggesting a significant 'burst' in the cluster formation rate in the outer LMC 2 - 4 Gyr ago. The age-metallicity relation (AMR) for our LMC cluster sample is also presented.

1. INTRODUCTION

Many studies have used LMC star clusters as probes of that galaxy's star formation and metal enrichment history (*e.g.*, Elson and Fall 1985). Such studies have suffered from large uncertainties in the ages of middle-aged and old clusters. With the advent of CCDs, the time is ripe for a re-examination of the LMC cluster age distribution (CAD) and age-metallicity relation (AMR). This paper describes such a study for a sample of remote LMC clusters. Analysis of this sample allows us to a) obtain a CAD and AMR subject to understandable selection effects, and b) address the puzzling questions of how and why the remote LMC clusters formed where there is now no gas or apparent star formation.

2. OBSERVATIONS AND PROCEDURES

The sample consists of 31 clusters located in a region centered about 6° NNE of the optical center of the LMC and is essentially complete to $V \sim 14.0$ ($M_V \sim -4.5$). The data include integrated UBV photometry of 29 clusters and BVR CCD photometry of 15 clusters using the CTIO 0.9m and 4m telescopes. All but one of the clusters are located beyond the limit of detected LMC HI (McGee and Milton 1966).

The analysis of these data include a) the subtraction of field stars from the cluster CM diagrams (Mateo and Hodge 1986), b) determining ages and metallicities via 'fits' of the CM diagrams to the isochrones of VandenBerg (1985), and c) using the integrated colors and ages of the clusters with CM diagrams to assign ages to the remaining clusters. Our analysis is 'self-contained' and involves no observations or age calibrations from other sources.

3. DISCUSSION

The normalized cluster frequency ($\equiv \frac{n}{\Delta t}$, where n is the number of clusters formed in the period Δt) is shown in Fig. 1. Comparison with the Galactic CAD (Lyngå 1982) suggests very different cluster formation histories in the two galaxies. A constant LMC cluster formation rate (CFR) implies excessively long cluster dissolution times (Wielen 1985) and implies that Fig. 1 is more consistent with a non-uniform CFR in the outer LMC with $CFR_{2Gyr}/CFR_{200Myr} \sim 20$. The AMR for the LMC clusters in our sample (Fig. 2) shows that a) the youngest clusters have abundances consistent with those of LMC Cepheids (Harris 1983), and b) the chemical enrichment of the LMC has been slower than in the Galaxy, but the present-day abundances of the two galaxies are quite similar.

Our data suggest that after forming the oldest clusters (*e.g.*, NGC 2010, H11), the LMC was relatively inactive. Then, about 2 – 4 Gyr ago, a sudden 'burst' in the CFR occurred throughout the LMC; the inner LMC is still forming stars. This picture is supported by a) the rarity of LMC clusters with accurate CCD CM diagrams with ages between 3.0 to 15 Gyr and b) the LMC field star data (Butcher 1977) suggesting a similar rise in the star formation rate 3 – 5 Gyr in many regions of the LMC.

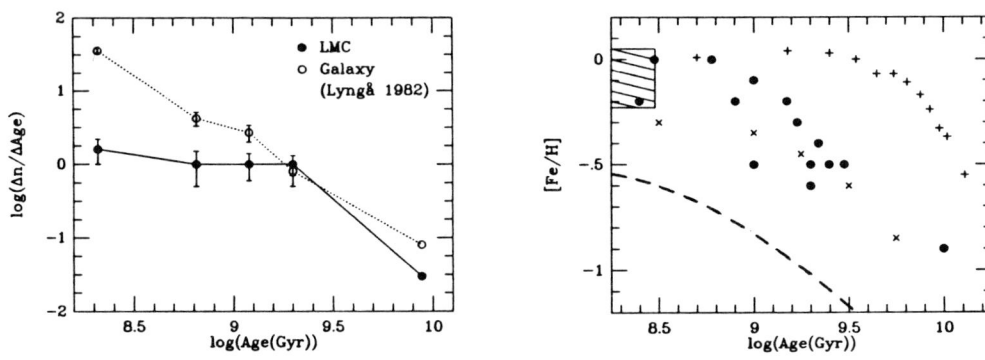

FIG. 1 (*left*) – The LMC and Galactic cluster age distributions. FIG. 2 (*right*) – The AMR for clusters in our sample (closed circles). Other AMRs – x's: Cohen (1982); line: Hodge (1981); +'s: Galactic (Twarog 1980); hatched: LMC Cepheids (Harris 1983).

REFERENCES

Butcher, H. R. 1977 Astrophys. J. 216, 372.
Cohen, J. G. 1982 Astrophys. J. 285, 143.
Elson, R. A. W. and Fall, S. M. 1985 Astrophys. J. 299, 211.
Harris, H. C. 1983 Astron. J. 88, 507. Hodge, P. 1981 in IAU Colloquium No. 68, Astrophysical Parameters for Globular Clusters,
    A. G. D. Philip and D. S. Hayes, eds., L. Davis Press,
    Schenectady, p. 205.
Lyngå, G. 1982 Astron. Astrophys. 109, 213.
Mateo, M and Hodge, P. 1986 Astrophys. J. Suppl. 60, 893.
McGee, R. X. and Milton, J. A. 1966 Australian J. Phys. 19, 343.
Twarog, B. 1980 Astrophys. J. 242, 242.
VandenBerg, D. A. 1985 Astrophys. J. Suppl. 58, 711.
Wielen, R. 1985 in IAU Symposium No. 113, Dynamics of Star Clusters,
    J. Goodman and P. Hut, eds., Reidel, Dordrecht, p. 449

PHOTOMETRIC MODELS FOR GLOBULAR CLUSTERS FROM POPULATION SYNTHESIS

R. Capuzzo Dolcetta

Astronomical Institute, Univ. of Rome

ABSTRACT: Integral fluxes (Bolometric and U, B, V) are computed in a completely theoretical frame in order to investigate the structural properties and stellar content of coeval stellar systems of various ages and metal abundance. Some results concerning the problem of the color gap in the distribution of the sample of Magellanic Cloud clusters are discussed.

1. THE SYNTHETIC EVOLUTIONARY MODELS

We computed a set of synthetic cluster models for bolometric and U, B, V fluxes for globular clusters of various ages and chemical compositions ($1.2 \times 10^7 \leq$ age $\leq 10^{10}$ (yrs); $X = 0.68$, $Y = 0.3$, $Z = 0.02$, 0.001) in the way described in detail by Battinelli and Capuzzo Dolcetta (1986, hereafter BCD).

2. AN APPLICATION TO MAGELLANIC CLOUD CLUSTERS

In BCD we investigated data on ages and masses of Magellanic Cloud (LMC and SMC) clusters. Here we report only on some considerations of the problem of the gap in the distribution of these clusters in the HR diagram, around $(B-V) = 0.5$ (Gascoigne 1980, van den Bergh 1981).

Following the line described in BCD and in Capuzzo Dolcetta (1986), we obtained theoretical color histograms and synthetic HR diagrams (to be compared with the corresponding observational ones). They show a clear gap in the $(B-V)_0$ color in the range of 0.3 to 0.5. One of the theoretical color-magnitude diagrams is shown in the left panel of Fig. 1. Fig. 1b shows the distribution of the LMC clusters.

What we can infer is that (being aware of some necessary approximations in the models) the occurrence of a gap in the synthetic HR diagram at roughly the same color as the gap in the observed distribution can be due to a maximum in the velocity along the synthetic evolutionary path for $(B-V)_0$ near 0.3 (for $Z = 0.001$). We do not need an ad hoc hypothesis of two bursts of cluster formation, only a reasonable choice of the cluster age distribution.

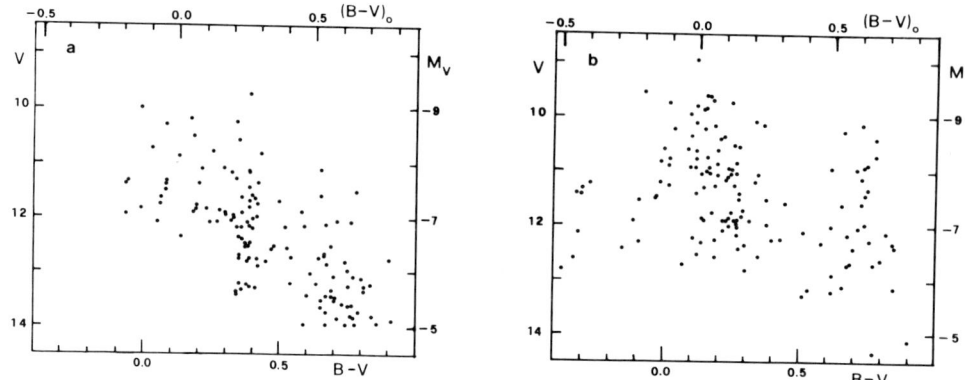

Fig. 1a. A theoretical CM diagram for a sample of clusters (in the range $-0.2 < (B-V)_o < 0.8$) to be compared with Fig. 1b. in which an observational diagram of LMC clusters is presented. (Data from van den Bergh 1981)

A more precise answer to the problem will be given by the introduction in our population synthesis code of a more detailed set of stellar evolutionary tracks with masses in the interval of 2 - 3 solar masses.

REFERENCES

Battinelli, P. and Capuzzo Dolcetta, R. 1986, in preparation.
Capuzzo Dolcetta, R. 1986 in Proc. of the Rome Astrophys. Coll., The Age of the Stellar Systems, in press
Gascoigne, S. B. C. 1980 in IAU Symposium No. 85, Star Clusters, J. E. Hesser, ed., Reidel, Dordrecht, p. 305.
van den Bergh, S. 1981 Astron. Astrophys. Suppl. 46, 79.

BVRI PHTOMETRY OF STAR CLUSTERS IN THE BOK REGION OF THE LARGE MAGELLANIC CLOUD

William Liller and Gonzalo Alcaino

Institute Isaac Newton

It is at present difficult from the literature to intercompare ages of clusters in the Magellanic Clouds owing to the variety of ways in which authors have interpreted the observational data and the several theoretical models used. With these considerations in mind we have embarked on a homogeneous investigation of clusters within a sky area of about 1 square degree. Because of its large concentration of clusters, we have chosen for study the so-called Bok region (Bok and Bok 1969) located in the northwestern part of the LMC bar. Photographic BVRI color-magnitude and color-color diagrams are provided for the 14 clusters listed in Table 1. None of these clusters have had previous stellar photometry in R, I and only four of them in the BV passbands.

The photographic plates have been obtained with the 1.5 m telescope at Tololo, the 2.5 m telescope at Las Campanas and the 3.6 m telescope at La Silla, and they have been calibrated with a BVRI photoelectric sequence of 15 stars in the magnitude range 9.4<V<15.3 (Alcaino and Liller 1982). This sequence has been extended using a Pickering-Racine wedge in the last two mentioned telescopes. Four global features are recognizable in the V vs (B-V) and V vs (B-R) CMDs: (1) the cluster main sequence population; (2) bright blue stars, probably evolved blue giants and supergiants, at $V\sim14$, $(B-V\sim0.0)$, $(B-R\sim0.0)$; (3) evolved luminous red giants to $(B-V\sim2.0)$, and $(B-R\sim3.0)$; (4) fainter cool giant branch stars extending from an upper red tip of $V\sim16.5$, $(B-V\sim2.0)$, $(B-R\sim3.0)$ to $V\sim19.0$, $(B-V\sim1.0)$, $(B-R\sim1.3)$. These last stars are known to belong to the field consisting of the older population abundantly present in this part of the LMC. Obviously the cluster features stand out much more clearly in well-populated clusters such as NGC 1850 and NGC 1854; they can hardly be discriminated in the stellar defficient clusters and NGC 1834, NGC 1860, SL 234 and SL 237.

In what follows, we have used the corrected distance modulus to the LMC of $(m-M)_o = 18.59$ which implies a distance of 52.2 kpc (Sandage and Tammann 1974). A value of $E(B-V)=0.06\pm0.02$ is generally adopted towards the LMC and the old and intermediate age star clusters, but the objects with ages $\sim 3\times10^8$ years are expected to be embedded in absorbing matter which should strongly increase the internal reddening. As our clusters are blue, and therefore young, individual reddening should be derived for each object. We have estimated the reddening by fitting the main sequence of our CMDs to an unreddened zero-age main sequence (Mermilliod 1981). The reddening derived in this way is listed in Table 1. In making a first estimation of the ages, we have followed the procedure used by Hodge (1983). To homogenize the age data, he assembled the pertinent CMDs and compared them with a single set of theoretical predictions. In order to age-date the younger clusters, Hodge used two indices, the magnitude of the MS turnoff and that of the brightest blue (B-V<0.4) evolved star. The mean values obtained by these two methods calibrated from Figure 1 in Hodge's paper are listed in Column 4 of Table 1. Another approach used to estimate the age has been fitting of the CMDs to the isochrones of Maeder and Mermilliod (1981). These isochrones are calculated with X=0.70, Z=0.03 and given in the M_v vs $(B-V)_o$ plane with ages ranging from 2.5×10^7 y to 6.3×10^9 y for models with $\alpha_c=0$. The isochrones have been superimposed in the V vs B-V diagrams, with the apparent distance modulus for each cluster listed in Table 1, Column 3. The weighted mean age for each object is listed in Table 1. It is seen from our samples that all the clusters are young with ages ranging from $15 \pm 5 \times 10^6$ y for NGC 1858 (B-V=-0.20) to $138 \pm 25 \times 10^6$ y for NGC 1860. They lie symetrically in the center of the age histogram of 245 galactic clusters (Lyngå 1982). Both distributions peak close to $\log t \sim 7.5$.

REFERENCES

Alcaino, G. and Liller, W. 1982 Astron. Astrophys. 114, 213.
Bok, J. B. and Bok, P. F. 1969 Astron. J. 74, 1125.
Hodge, P. W. 1983 Astrophys. J. 264, 470.
Lyngå, G. 1982 Astron. Astrophys. 109, 213.
Maeder, A. and Mermilliod, J. C. 1981 Astron. Astrophys. 93, 136.
Mermilliod, J. C. 1981 Astron. Astrophys. 97, 235.
Sandage, A. and Tammann, G. 1974 Astrophys. J. 190, 525.

TABLE I.

REDDENING AND AGES FOR THE CLUSTERS STUDIED

Cluster (NGC)	E(B-V)	$(m-M)_V$	MST+BBS 10^6 y	Isoch. 10^6 y	Weighted Mean Age 10^6 y
1834	0.10:	18.89	52:	43±7	47±8
1836	0.20:	19.19	31±5	37±8	34±8
1839	0.27	19.40	22±4	27±8	24±5
1847	0.25	19.34	18±4	18±4	18±4
1850	0.18	19.13	19±2	24±4	21±4
1854	0.20	19.19	20±1	25±5	22±4
1856	0.26	19.37	66±9	60±12	63±8
1858	0.15	19.04	10±3	19±6	15±5
1860	0.18:	19.13	115±28	160±25	138±25
1863	0.20	19.19	57:	55±10	56±10
1870	0.14	19.01	55±21	60±20	58±18
SL234	0.15:	19.04	57±11	40±15	48±12
SL237	0.17:	19.10	20±2	19±6	20±5
SL304	0.20	19.19	36±8	50±8	43±10

AGES AND METAL ABUNDANCES OF STAR CLUSTERS IN THE MAGELLANIC CLOUDS

Horace A. Smith

Michigan State University

Leonard Searle and Armando Manduca

Mt. Wilson and Las Campanas Observatories

It has long been apparent that the rich star clusters in the Magellanic Clouds differ widely in age, a circumstance which renders these clusters particularly useful in tracing the age-metallicity relations of their parent systems. We have attempted to exploit this potential by studying the integrated light of red globular clusters in the Large and Small Clouds.

We used the photon-counting spectrograph on the 2.5 m du Pont telescope to sample the integrated blue spectra of populous clusters. Two indices, h, an indicator of the strength of the Balmer lines, and m, a measure of the strength of the CaII H and K lines and the G-band, were determined for each cluster.

To calibrate h and m in terms of age and [Fe/H] we modeled h and m indices for clusters of given ages and metallicities. These models have been calculated with Bell's synthetic spectrum program and modified versions of the Yale isochrones. Similarly, synthetic Q(ugr) and Q(vgr) indices were calculated for comparison with the cluster photometry of Searle, Wilkinson, and Bagnuolo (1980). We do not propose that these calculations give definitive results, but they are useful in exploring the utility of the approach. Ages determined from the models are in good agreement with results from color-magnitude diagrams for older clusters, but may be slightly too old for clusters aged 1-2 Gyr. A semi-empirical correction has been applied to remedy this.

The age-metallicity relations we obtain are shown in figure 1. In interpreting this figure, it should be realized that for clusters aged 9-16 Gyr, the ages given by our approach could easily be in error by 2-4 Gyr. Averaged over the past 6-8 Gyr, the rate of chemical enrichment in the LMC appears greater than in the SMC or the solar neighborhood.

REFERENCES

Searle, L., Wikinson, A. and Bagnuolo, W. G. 1980 <u>Astrophys. J.</u> **239**, 603.
Twarog, B. A. <u>Astrophys. J.</u> 242, 242.

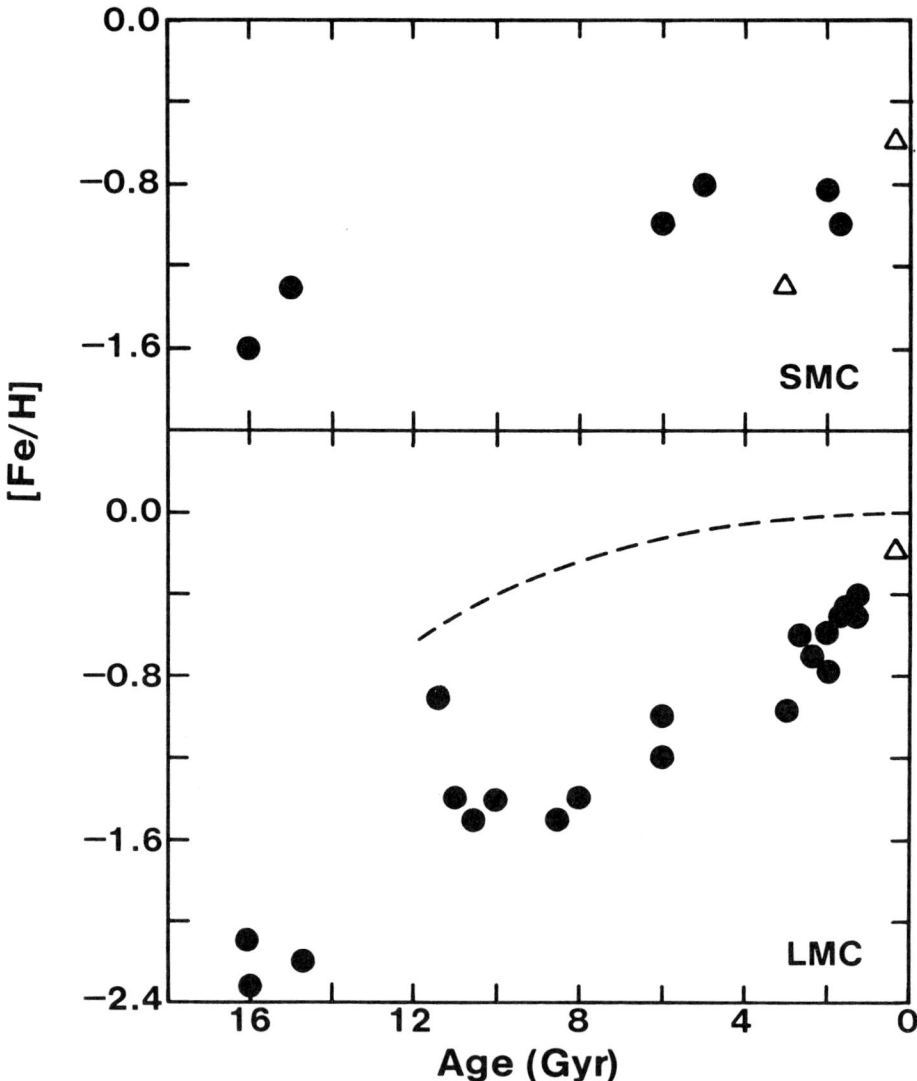

Fig. 1. Age-metallicity relations for the SMC and LMC. Closed circles represent star clusters. Triangles are recent results for variable stars in the field. The dashed line is the solar neighborhood relation according to Twarog (1980).

INTERNAL DYNAMICS OF MAGELLANIC CLOUD CLUSTERS

Paul Papenhausen and R.A. Schommer

Rutgers University

1. INTRODUCTION

It is believed that stellar systems become mass-segregated on a time scale comparable to the two-body relaxation time (Tr). Globular clusters in the Magellanic Clouds have an extensive range in age and richness. Thus, in the clouds, there are candidate clusters with ratios of Tr to age both > 1 and < 1. Our objective is to study the radial density structure of clusters as a function of mass (magnitude) and compare with expectations based upon Tr. We report here preliminary results on two such clusters at the age extremes in our sample: NGC 458 (SMC) and 121SC03 (LMC).

2. REDUCTION AND ANALYSIS

The frames (B & V) were obtained on the 4 m PFCCD Cerro Tololo for both program clusters and E region standard star fields. DAOPHOT was used to reduce the raw image data. CM diagrams are displayed in Fig. 1 and best fit Vandenberg (1985) isochrones are overlayed on the plots. Single mass, isotropic King models were fit to the brightness profiles of both clusters. Reasonably tight constraints on Wo and the tidal radius (Rt) served to determine a unique overall density profile for the clusters. No obvious segregation was evident from the B & V frame comparison for NGC 458 since these had identical best fit parameters. A simple tidal calculation for Rt, based upon a point mass SMC model, gave 146" for NGC 458 which agreed well with the model parameter. Table I displays best fit quantities for the clusters (including the isochronal ones).

High central crowding caused severe counting deficiencies, so a modified version of the ADDSTAR routine in DAOPHOT was used to find the correction factors to the radial stellar count profiles. Multiple runs were needed to prevent severe biasing in all cases. At ~ 24" (1.6 and 1.0 core radii respectively), reliable factors were obtainable for most of the M.S. of NGC 458 and all of 121SC03. Completeness correction factors for three radial regions are given in the first half of Table II, whereas the ratios of the number of stars in three magnitude bins after correction (the corrected luminosity functions) are given in the 2nd half of the table. From Spitzer & Hart (1971), the half-mass relaxation time (Trh) for NGC 458 & 121SC03 is: 260 Million, 415 Million yrs., respectively. Trh/Age = 1.3, 0.04 respectively.

3. DISCUSSION

The preceding calculation predicts that NGC 458 should be unrelaxed with no mass segregation, while 121SC03 should be segregated. Inspection of Table II shows little evidence for segregation in NGC 458. The uncertainties ranged from 0.03 to 0.16 (bright to faint bin) with the inner and outer annuli having identical ratios within 1 sigma. The middle radius has an excess of faint stars however. For 121SC03, the bright stars are more centrally concentrated, but this is a marginal result because only a 12% mass difference exists along the sample range. We will examine surface density profiles as a function of mass.

TABLE I.

Best Fit Parameters (both Isochrones & King fits)

	Dist. Mod.	Age(yrs)	Y	Z	Wo	Rt	Mv	Mass(solar)	CHI SQ.
NGC 458	18.80	300 Mill.	0.25	0.01	4.40	121"	-7.5	25,000	0.89
121SC03	18.20	10 Bill.	0.25	0.006	3.60	143"	-5.7	5,890	1.45

TABLE II.

Correction factors and Relative number of stars in three radial bins

	Mag. Range	24-48"	48-72"	72-120"	Mag. Range	24-48"	48-72"	72-120"
NGC 458	16.2-19.2	1.0	1.0	1.0	16.2-18.2	0.089	0.058	0.084
	19.2-20.2	1.38	1.0	1.0	18.2-20.2	0.668	0.427	0.626
	20.2-21.2	2.00	1.67	1.40	20.2-21.2	1.000	1.000	1.000
121SC03	18.4-22.0	1.0	1.0	1.0	18.4-20.8	0.120	0.129	0.099
	22.0-23.2	1.85	1.36	1.18	20.8-22.0	0.251	0.200	0.217
					22.0-23.2	1.000	1.000	1.000

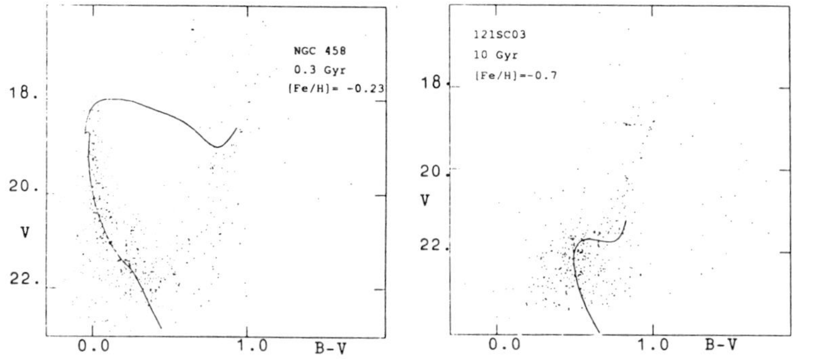

Fig. 1. CMD's of NGC 458 and 121SC03, with Vandenberg isochrones overlaid.

REFERENCES

King, I. 1966 Astron. J., 71, 64.
Spitzer, L. and Hart, M. H. 1971 Astrophys. J., 164, 400.
Vandenberg, D. A. 1985 Astrophys. J. Suppl., 58, 711.

DO BINARY CLUSTERS EXIST IN THE LARGE MAGELLANIC CLOUD?

D. Hatzidimitriou

University of Edinburgh

R. K. Bhatia

Royal Observatory, Edinburgh and Osmania University

The possible existence of binary clusters in our Galaxy (h and x Persei, Ocl 556) has been argued in the past, but it has never been a well established fact either in our Galaxy, or in external systems. An early speculation on the problem by Innanen et al (1972) has predicted a considerable degree of stability for binary clusters in low nuclear density galaxies, like the LMC.

A complete (down to the 17th visual magnitude) survey of the LMC cluster system on UKSTU plates has yielded a total of 69 double clusters with a centre-to-centre separation of less than 1.3 arcmin (~21 parsecs) (see e.g. Plate 1).

The large number of clusters in the field studied implies that a fraction of the double clusters observed may be due to projection effects. Taking into consideration the surface density distribution of the LMC cluster system we estimate that an upper limit on the expected number of double clusters due to projection effects is 31. Application of the Kolmogorov-Smirnov two-sample test has shown that the observed distribution of double clusters differs from the one predicted by chance at the 99.9% confidence level. Therefore, a significant number of the observed double clusters must be physically associated.

The dynamical stability of a binary cluster is a complicated problem that has to be treated in detail. To a first approximation, however, we can assume three main sources of instability for the system : (i) the tidal field of the parent galaxy, (ii) the mutual dynamical interaction of the members of the binary (leading either to coalescence, or to the disruption of the system) and (iii) the effect of passing giant molecular clouds (Alladin and Parthasarathy, 1978; Alladin et al. 1985; Bahcall et al. 1985). The timescales over which these processes affect significantly the binary cluster are functions of the initial separation of its members and of their mass and density ratios. Plausible values for these parameters give an upper limit of a few 10^8 yrs on the maximum possible age of a binary cluster.

It is important to note that the few clusters in our sample for which ages are available (van den Bergh, 1981; Elson and Fall, 1985) are all younger than a few 10^8 yrs, which is consistent with the above mentioned age limit. Moreover, the space distribution of the observed double clusters correlates (Spearman rank correlation test) very well with that of the very young and young clusters in the LMC, while it correlates relatively weakly with the intermediate and old clusters distribution .

Although further observations (especially on the clusters ages) are necessary to settle the question, the data presented here give considerable evidence in favour of the existence of binary clusters in the LMC, with obvious implications on the cluster formation (and evolution) processes in this galaxy and in Magellanic type irregulars in general .

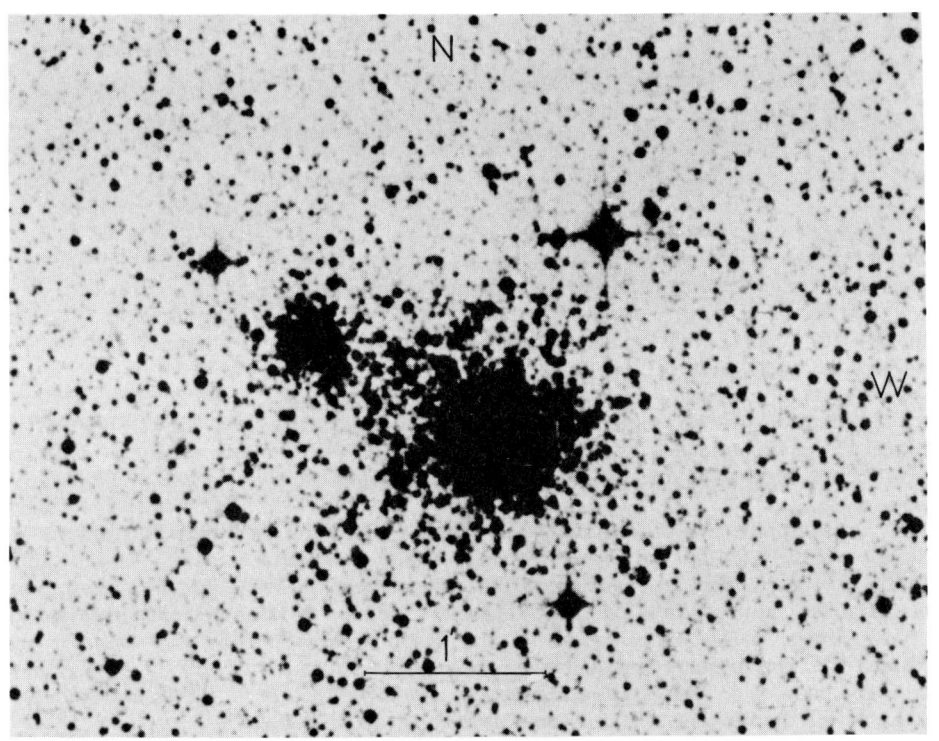

Fig. 1. A prominent LMC pair: NGC 2137 (left) and NGC 2136 (right).

REFERENCES

Alladin, S. M. and Parthasarathy, M. 1978 Monthly Notices Roy. Astron. 184, 871.
Alladin, S. M., Ramamani, N. and Sing, T. M. 1985 J. Astrophys. Astron., 6, 5.
Bahcall, J. N., Hut, P. and Tremaine, S. 1985 Astrophys. J. 290, 15.
Elson, R. A. W. and Fall, S. M. 1985 Astrophys. J. 299, 211.
Innanen, K. A., Wright, A. E., House, F. C. and Keenan, D. 1972 Monthly Notices Roy. Astron. Soc. 160, 249.
van den Bergh, S. 1981 Astron. Astrophys. Suppl. 46, 79.

ELLIPTICITIES OF GLOBULAR CLUSTERS IN THE ANDROMEDA GALAXY

Nedka M. Spassova

Bulgarian Academy of Sciences

Anelia V. Staneva and Valery K. Golev

Sofia University

A program of investigation of globular clusters' structural parameters in nearby galaxies has been initiated at the Rozhen National Observatory, Bulgaria (Spassova and Staneva 1984). Here we report on the projected ellipticities of 88 bright globular clusters in M 31, obtained on plates taken with the 2 m Ritchey-Chretien telescope at Rozhen (16 m focal length, 12"89/mm plate scale over a 1° x 1° field).

The cluster density distributions were obtained and digitized using the MDM-6 Joyce-Loebl microdensitometer. For each object four 32 x 32 raster data arrays (40 mcm pixel spot size and 50 mcm increment) were formed. For the ellipse parameter determination we applied the momentum analysis method (Stobie 1980) and the cluster's isodensity contour analysis (Geyer et al. 1953, Kadla et al. 1977). For each object we processed 30 - 50 different contours using the last method. Most of the clusters were measured on two or more plates. Table I lists the basic results of the investigation. The cluster number (Sargent et al 1977) is listed in column 1, the mean ellipticity e over all isodensities is listed in column 2. The analysis of the data in Table I yield the following conclusions:

1. The projected ellipticities of the Andromeda clusters range over 0.02 - 0.27 with a mean value of 0.097 ±0.04.

    a. The globular clusters in M 31 and in the Galaxy (e - 0.09 according to Frenk and Fall (1982); e = 0.07 according to Geyer et al. (1983) actually have identical mean ellipticities.

    b. The mean ellipticity obtained for M 31 clusters is similar to the one for LMC clusters in Frenk and Fall (1982) (e = 0.11) and in Geyer and Richtler (1981) (e = 0.13), but differs significantly from the value of 0.22 of Geisler and Hodge (1980).

    c. Another phenomenon needing explanation from a physical point of view is the obvious difference in the flatness of M 31 and SMC clusters (Kontizas et al. 1985).

TABLE I

The Ellipticities of the Globular Clusters in M 31

No.	e	No.	e	No.	e	No.	e	No.	e	No.	e
35	0.08	102	0.10	147	0.07	213	0.06	239	0.10	280	0.03
40	0.10	104	0.07	148	0.07	215	0.08	241	0.16	281	0.07
52	0.06	107	0.05	150	0.18	216	0.18	243	0.08	282	0.12
53	0.13	110	0.11	155	0.12	217	0.05	244	0.06	286	0.11
62	0.06	113	0.14	156	0.08	218	0.08	250	0.06	287	0.10
64	0.11	114	0.15	157	0.08	222	0.11	252	0.10	299	0.10
70	0.04	119	0.05	165	0.07	223	0.13	256	0.06	300	0.04
72	0.06	121	0.09	169	0.05	224	0.15	257	0.10	301	0.14
76	0.06	122	0.09	172	0.09	226	0.02	258	0.04	305	0.10
78	0.09	124	0.11	173	0.07	227	0.09	263	0.12	315	0.21
87	0.02	125	0.07	176	0.06	229	0.08	267	0.22	D83	0.14
90	0.08	127	0.08	183	0.07	230	0.12	272	0.08	D87	0.12
96	0.04	130	0.14	199	0.27	233	0.14	275	0.16	98	0.09
98	0.07	134	0.07	205	0.05	234	0.16	277	0.13		
101	0.12	144	0.05	212	0.17	235	0.11	279	0.12		

2. No explicit correlation between the visible ellipticities and stellar magnitudes, on the one hand, and the color indices on the other, could be established during the investigation. Although only preliminary, this conclusion seems interesting in view of the results of van den Bergh (1983) and Hesser et al. (1984), namely that the ellipticities of clusters in the Magellanic Clouds and in NGC 5128 correlate with their luminosities.

REFERENCES

Frenk, C. S and Fall, S. M. 1982 Monthly Notices Roy. Astron. Soc. 199, 565.
Geisler, D. and Hodge, P. 1980 Astrophys. J. 242, 66.
Geyer, E. and Richtler, T. 1981 in IAU Colloquium No. 68, Astrophysical Parameters for Globular Clusters, A. G. D. Philip and D. S. Hayes, eds., L. Davis Press, Schenectady, p. 239
Geyer, E., Hopp, U. and Nelles, B. 1983 Astron. Astrophys. 125, 359.
Hesser, J. E., Harris, H., van den Berg, S. and Harris, G. 1984 Astrophys. J. 276, 491.
Kadla, Z., Richter, N., Strugatzkaja, A. and Hagner, V. 1977 Pulkova Contr. 195, 74.
Kontizas, E., Dialetis, D. Prokakis, T. and Kontizas, M. 1985 Astron. Astrophys. 146, 293.
Sargent, W., Kowal, C., Hartwick, F. D. A. and van den Bergh, S. 1977 Astron. J. 82, 947.
Spassova, N. and Staneva, A. 1984 Sov. Astron. Lett. 10, 114.
Stobie, R. 1980 Journ. Brit. Interplanetary Soc. 33, 323.
van den Berg, S. 1983 Publ. Astron. Soc. Pacific 95, 839.

OBSERVED VARIATIONS IN THE DENSITY PROFILES OF STAR CLUSTERS IN THE LMC

M. Kontizas[1], D. Hatzidimitriou[2] and M. Metaxa[1]

Laboratory of Astrophysics, University of Athens   1

University of Edinburgh   2

Several dynamical theories have been developed in order to approach the dynamical evolution of stellar systems and explain the observational data. The observed density profiles of the clusters can be a valuable source of information towards the understanding of their dynamical properties. King in a series of papers has connected the established theories with the observed profiles in clusters of our own Galaxy (King, 1962, 1966; etc.). Density profiles can be obtained by means of star counts and/or by means of photometric photometry. So far the observations for clusters in our Galaxy and the MCs appear to fit well the so called King models and provide information of their tidal radii, total masses and concentration parameters (Kontizas, 1984).

Thirty eight remote LMC clusters randomly distributed around the LMC rotation centre were measured by means of star counts in plates taken with the 1.2 m U.K. Schmidt telescope to derive their tidal radii and total masses. For some regions I and J plates were available.

Seven of these clusters were found to violate the usual picture of a conventional density profile. Fig. 1 illustrates the number of stars per unit area versus the distance from the cluster centre for one of those clusters. A clear density fluctuation occurs at the outer regions of all seven clusters which is systematic and too large to be attributed to random background anomalies.

The radii r_a and r_b are defined as the distances from the centre of the clusters to the limits of the observed bumps (Fig. 1). If $w = r_b - r_a$, is the width of the fluctuation, the diagram of w versus r_b (Fig.2) shows that there is a linear correlation between these two quantities.

It was attempted to define a tidal radius r_t using the conventional methods usually applied to similar studies. Two values of the background density were used. Although it seems that the background is reached beyond the fluctuation, the tidal radius was not always possible to be found. For the background value corresponding to the density at

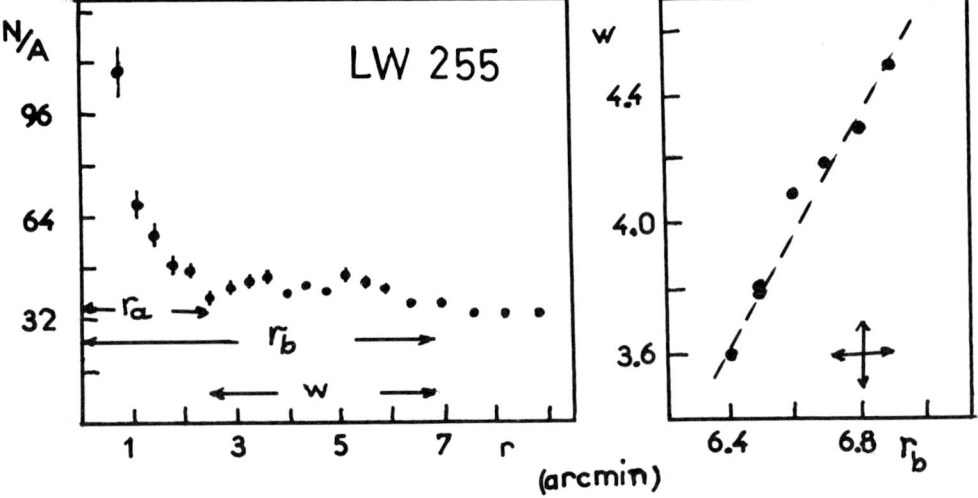

Fig. 1. Number of stars per unit area versus the distance from the cluster center.

Fig. 2. The width of the observed fluctuation versus the distance r_b.

r_a a tidal radius has been derived for all clusters. The so defined tidal radii show a clear correlation with the ratio r_b/r_t.

The phenomenon is extremely interesting and it has to be studied further both observationally and theoretically. Is it a stage of the cluster's dynamical evolution and how this is related to the gravitational attraction by the central mass of their parent galaxy? A possible explanationcan be given by the existence of a corona around the cluster due to the irregular forces of the star field surrounding it (Agekyan and Belozerova 1979).

ACKNOWLEDGEMENTS

We would like to express our sincere thanks to the U.K. Schmidt Unit for loan of the Observational material.

REFERENCES

Agekyan, T. A. and Belozerova, M. A. 1977 Soviet Astronomy 23, 4.
King, I. R. 1962 Astron. J. 67, 471.
King, I. R. 1966 Astron. J. 71, 64.
Kontizas, M. 1984 Astron. Astrophys. 131, 58.

RATIO OF EARLY TO LATE TYPE STARS IN SMC CLUSTERS

E. Kontizas [1], M. Kontizas [2],
A. Dapergolas [1], D. Hatzidmitriou [3]

Observatory of Athens.   1
University of Athens.   2
University of Edinburgh. 3

Spectral classification of stars in SMC clusters provide useful information on the evolutionary history of this galaxy and permit us to test the theory of stellar evolution.

For a large number of SMC clusters the bright stars were classified using the 1.2m U.K. Schmidt Telescope objective prism plates (Dapergolas et al. 1986; Kontizas et al. 1986). All classified stars are situated within the tidal radius of each cluster (Kontizas 1984). The plates and the classification criteria used for this investigation are described by Dapergolas et al. (1986). Stars in adjoining fields of each cluster were also studied for comparison . The field stars normalized to the cluster area were subtracted from the cluster stars and the observed ratios (R) of early (B+A) to late (K+M) spectral type stars were found for each cluster.

For a number of young and old clusters with known ages (Hodge 1982; Mould and Aaronson 1982) the derived ratios of early to late type stars are plotted versus their age in Fig. 1. The solid line in this diagram represents the best fit of the ratio of the blue supergiants plus the stars within 2.0 mag of the top of the main sequence to the red supergiants plus the rising giants as given by Schlesinger (1969, 1971) in his theoretical models, whereas the dashed line represents the best fit of the ratio of the blue supergiants to the red supergiants for the very young cluster.

The oldest SMC clusters with a small number of early type stars are located in the left upper part of the diagram. For the younger clusters there is a very good agreement with theory. In this diagram there are two exceptions, the cluster L83 and the cluster L54. L83 shows an unusually large number of early type stars for its adopted age based on the turn-off point of its CM diagram. It has been suggested that these early type stars are either blue stragglers or represent continuous star formation. For the cluster L54 with a photometric age of 1.2×10^7 yr (Carney et al. 1985) only the very bright stars are detected (blue and red supergiants) and the theoretical CM diagrams (Schlesinger 1969, 1971) suggest an age of the order $3-7 \times 10^7$ yrs for the observed ratio. Assuming this age the observed ratio coincides with the theoretical ones as can be seen in Fig. 1.

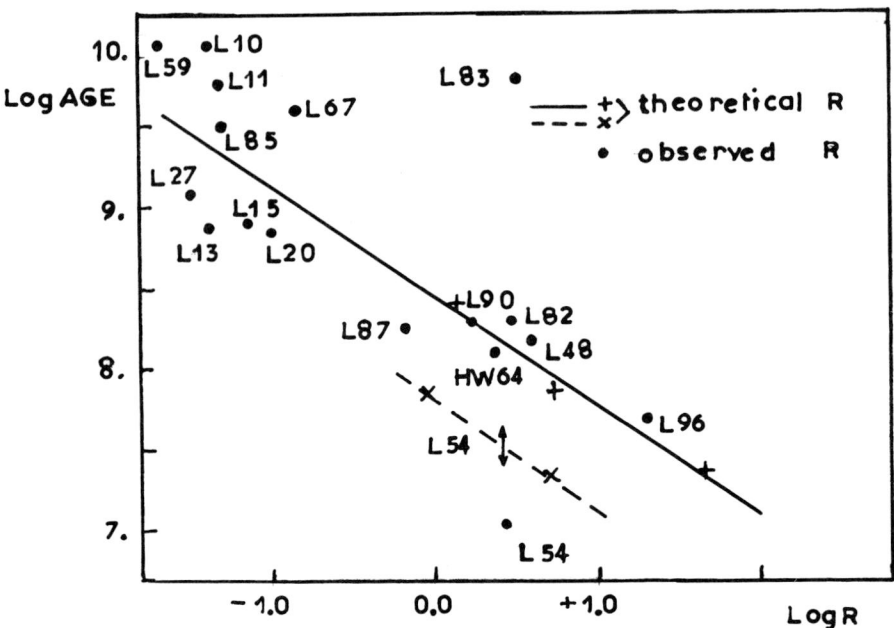

Fig. 1 Age versus ratio (R) of early to late type stars in SMC clusters. The arrow shows the new position of L54 assuming an age $3-7 \times 10^7$.

ACKNOWLEDGEMENTS

The authors would like to express their thanks to the 1.2m U.K. Schmidt Telescope Unit.

REFERENCES

Carney, B. W., Janes, K. A., Flower, P. J. 1985 Astron. J. 90, 7.
Dapergolas, A., Kontizas, E., Kontizas, M. 1986 Astron. Astrophys. Suppl., 65, 283.
Hodge, P. W. 1982 IAU Colloq. No. 68, p. 295.
Kontizas, E., Dapergolas, A., Kontizas, M. 1986 Astron. Astrophys. Suppl., 66, 51.
Kontizas, M. 1984 Astron. Astrophys. 131, 58.
Mould, J., Aaronson, M. 1982 Astrophys. J. 263,629.
Schlesinger, B. M. 1969 Astrophys. J. 157, 533.
Schlesinger, B. M. 1971 Astrophys. J. 166, 447.

THE SMC CLUSTER LINDSAY 11

J. Buttress[1], R. D. Cannon[2] and W. K. Griffiths[1]

University of Leeds   1

Royal Observatory, Edinburgh   2

This small cluster is situated in the western region of the SMC at $\alpha = 0^h\ 26^m\ 13^s$, $\delta = -73°$, 1', 20" (1950) and has been chosen for study in the initial post-launch period of the Hubble Space Telescope. This preliminary study was made using data obtained using a CCD camera on the SAAO 1 m telescope in October 1984.

The data consist of CCD exposures in B, V and R giving a total integration time of 2900s, 3000s and 1700s respectively.

The small size of the cluster and its crowded core would present severe problems but for the analytical routines using star profile-fitting techniques developed by A. J. Penny. The routines are very effective too in dealing with background luminosity gradients.

By concentrating on the stars within an annulus of 1.3' centred on the core, we ensure a cluster to field star ratio of at least 5:1. Field star contamination, a serious problem in earlier studies (Kontizas, 1980), is therefore minimized.

Our photometry is such that for 19<R<20 our rms error in R is 0.075 and in V-R is 0.12, these uncertainties increasing in the range 20<R<21 to 0.12 and 0.17 respectively.

We present here the CMDs for the cluster and the field in R vs V-R. The cluster, Figure 1, shows a strong RG branch and a clumpy horizontal branch. The HB of the cluster occurs at approximately V = 19.5 giving a distance modulus of 18.9 (assuming that M_V (HB) = 0.6). The integrated magnitude in V of this cluster within an annulus of 51" is 13.96. Grindlay (1978) obtained 14.04 photoelectrically using the same sized aperture.

We independently report the discovery of a presumed carbon star with R = 15.5 (V = 17.4) and V-R = 1.8 (with a corresponding B-V of 2.7). This was previously identified as a photometric carbon star by Mould and Aaronson (1982), and it lies just outside the core, ~ 10" to the south east.

The CMD of the field, Figure 2, has many similarities to that of the cluster. This, in part, will be due to the similar ages of both populations (van den Bergh, 1981; Hawkins and Bruck, 1984). The presence of genuine cluster stars in the surrounding field will, however, contribute to this similarity.

REFERENCES

Grindlay, J. E. 1978 *Astrophys. J. Letters* 224, L107.
Hawkins, M. R. S. and Bruck, M. T. 1982 *Monthly Notices Roy. Astron. Soc.* 198, 935.
Kontizas, M. 1980 *Astron. Astrophys. Suppl.* 40, 151.
Mould, J. and Aaronson, M. 1982 *Astrophys. J.* 263, 629.
van den Bergh, S. 1981 *Astron. Astrophys. Suppl.* 46, 79.

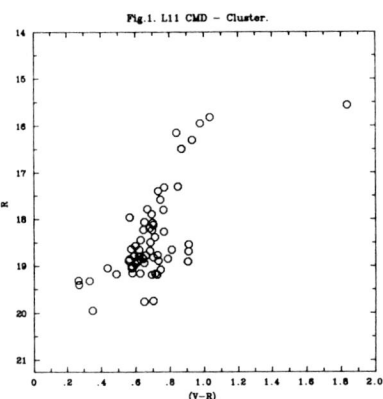
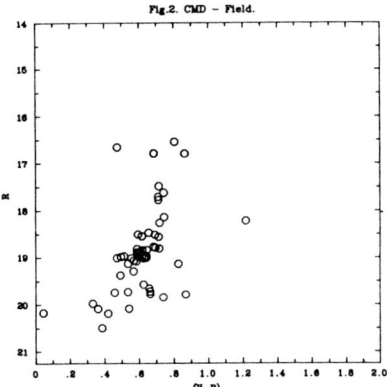

Fig. 1. CM Diagram for Lindsay 11.

Fig. 2. CM Diagram for field.

ABUNDANCES OF YOUNG LMC CLUSTERS

R.A. Schommer

Rutgers University

Doug Geisler

CTIO/NOAO

ABSTRACT: We present Washington system photoelectric photometry for stars in three young LMC clusters: NGC 1850, NGC 1866, and NGC 2164. We derive metal abundances [A/H] of -0.3, -0.4, and -0.6, respectively.

1. INTRODUCTION

While the ages of several dozen Magellanic Cloud clusters have now been determined by main sequence photometry, few direct abundance determinations have been made in these clusters since the pioneering work of Cohen (1982). This is in part due to the faintness of the stars, which prohibits high dispersion abundance analyses, and the relatively high abundances, which are often beyond the calibrations of low dispersion techniques. We present here Washington photometry of individual stars in three young LMC clusters.

2. THE DATA

Photometry, using the CTIO 1.5 meter telescope, was performed in the Washington system on individual stars in NGC 1850, NGC 1866 and NGC 2164. The mean errors for one observation were 0.025 mag in both $M-T_1$, and T_1-T_2. These rather large errors arise from the varying background in the crowded LMC fields. For this reason the clusters chosen are far from the LMC bar, and brighter giants outside the cluster cores were selected.

Table I lists the observed colors for the stars. The $\Delta(M-T_1)$ colors were then calculated and the corresponding abundances were derived using a new calibration of the system discussed in Canterna et al. (1986) and Geisler (1986). The LMC clusters are somewhat younger than the Galactic counterparts detailed in these calibrations, so some effects due to the different surface gravities are possible, but these are expected to be small in $M-T_1$ (Canterna and Harris 1979). The ages for these clusters from Hodge (1983) are also given.

TABLE I

STAR	$M-T_1$	T_1-T_2	$\Delta(M-T_1)$	n	E(B-V)	SWB	$<\Delta(M-T_1)>$	[A/H]	t(Gyr)
1850-34 ^1	0.647	0.461	0.060	2	0.08	II	0.048	-0.3	0.04
1850-48	0.944	0.614	0.035	2			±.007	±0.1	
1850-49	0.799	0.542	0.050	2					
1866-I-02 ^2	0.665	0.445		4	0.03	III	0.060	-0.4	0.08
1866-I-03	0.997	0.665	0.065	7			±.009	±0.1	
1866-III-09	0.747	0.532	0.099	3					
1866-IV-10	0.747	0.495	0.045	6					
1866-II-13	0.767	0.571	0.159	1					
1866-III-20	0.589	0.409	0.047	3					
1866-III-22	0.920	0.610	0.068	4					
2164-46 ^3	0.951	0.621	0.045	5	0.07	III	0.084	-0.6	0.05
2164-48	0.753	0.527	0.086	7			±0.013	±0.2	
2164-50	0.766	0.571	0.103	5					
2164-64	0.858	0.586	0.100	3					

1) Star identifications from Tifft and Connolly (1973), Table II SE.
2) Arp and Thackery (1967), inner annulus, except III-20, outer annulus
3) Star identifications in Hodge and Flower (1973).

3. DISCUSSION

The abundances are comparable with high dispersion analyses by Cohen (1982) for SWB type I-III clusters. We note that Nelles and Richtler (1984) derive a photometric abundance of -1.2 for NGC 1866, while Becker and Mathews (1983) analyzed the CMD of N1866 and found a solar metal abundance. It is premature to draw lengthy conclusions based upon the abundances in three LMC clusters. It appears that young LMC clusters are significantly metal-poor, by about 0.5 dex, compared with Galactic clusters of the same age (Canterna et al. 1986).

REFERENCES

Arp, H. C. and Thackeray, A.D. 1967 Astrophys. J., 149, 73.
Becker, S. A. and Mathews, G. J. 1983 Astrophys. J., 270, 155.
Canterna, R., Geisler, D., Harris, H., Olszewski, E.W. and Schommer, R.A. 1986 Astron. J., 92, 79.
Canterna, R. and Harris, H. C. 1979 in Problems of Calibration of Multicolor Photometric Systems, Dudley Obs. Rep. No. 14, A. G. D. Philip, ed., Dudley Obs., Schnectady, p. 199.
Cohen, J. G. 1982 Astrophys. J., 258, 143.
Geisler, D. 1986 Pub. Astron. Soc. Pacific, 98, in press.
Hodge, P. W. and Flower, P. J. 1973 Astrophys. J., 185, 829.
Hodge, P.W. 1983 Astrophys. J., 264, 470.
Nelles, B. and Richtler, T. 1984 in IAU Symposium No. 108, Structure and Evolution of the Magellanic Clouds, S. van den Bergh and K. S. de Boer, eds., Reidel, Dordrecht, p. 33.
Tifft, W. G. and Connolloy, L. 1973 Mon. Not. Roy. Astron. Soc., 163, 93.

THE ABUNDANCE OF THE LMC GLOBULAR CLUSTER NGC 2213

Doug Geisler

Cerro Tololo Inter-American Observatory

ABSTRACT: A new technique for determining accurate abundances in distant giants - Washington CCD photometry - has been applied to the intermediate-age LMC globular cluster NGC 2213. An abundance of -0.40 ± 0.15 was found from the analysis of 42 giants with V < 20, using data obtained with the 1.5 m telescope. Combined with published main-sequence photometry, the derived abundance indicates a true LMC distance modulus of 18.2 ± 0.2. A likely CN strong giant near the tip of the giant branch is identified. Abundances are also derived for a sample of 27 field giants. Results indicate that one could determine both the age and abundance of Magellanic Cloud clusters with high accuracy from Washington photometry using the 4 m in less than one hour of observing time per cluster.

1. INTRODUCTION

Despite the great progress made in recent years in obtaining high precision main-sequence photometry of Magellanic Cloud clusters, a nagging problem remains in our ability to use this information to determine accurate distances. Metal abundances in the clusters must be known to an accuracy of better than 0.2 dex before the distance modulus derived from main-sequence fitting can be trusted to better than 0.2 mag. Such accurate abundances have proven to be very difficult to achieve. Because of the faintness of the stars, we have been forced to rely on rather crude techniques: color-magnitude diagram morphology, integrated spectra or colors, or low resolution spectra of a few giants. These methods often yield uncertainties of 0.3 dex. It is therefore essential that our methods for determining metal abundance improve substantially before we can claim great confidence in distances derived from main-sequence fitting.

A technique that offers great promise is Washington CCD photometry. The filters are very broad (~ 1000Å) so that the system is very efficient. The system was designed to measure accurate temperatures of late-type giants, and to provide two independent abundance estimates. A recent recalibration of the Washington abundance indices demonstrates that the Fe abundance sensitivity of the $\Delta(C-M)$ index is comparable to or exceeds that of all other photometric or low resolution spectroscopic abundance indices at all metallicities. The system is also capable of detecting giants with strongly enhanced CN bands. By averaging together the abundance indices of many tens of giants per cluster, a mean abundance with an uncertainty as small as 0.1 dex should be attainable.

2. RESULTS

In order to investigate the ability of the system to obtain accurate abundances for Magellanic Cloud cluster giants, Washington CCD photometry of the LMC red globular cluster NGC 2213 was obtained on January 3, 1986 with the CTIO 1.5 m. The frames were reduced using DAOPHOT. The resulting photometry has internal errors less than 0.03 in C-M and < 0.02 in $M-T_1$, $T_1 - T_2$ and T_1 for most giants. These are sufficiently accurate for determining abundances.

After eliminating stars with large errors and those more distant than 1.5', 42 giants remained for the abundance analysis. One of these, a star near the tip of the giant branch, shows evidence for abnormally strong CN absorption. The remaining giants fall closely about the relation for an abundance of -0.5 dex in the $\Delta(C-M)$ index. The scatter is somewhat worse in the $\Delta(M-T_1)$ index, as expected from its lessened abundance sensitivity, and the mean abundance is 0.2 dex higher than that given by the $\Delta(C-M)$ index. The final mean abundance is -0.40 ± 0.15.

VandenBerg (1985) isochrones of the appropriate metal abundance were then compared to the deep color-magnitude diagram of Da Costa, Mould and Crawford (1985). The best fit was provided by the 1.5 Gyr isochrone using a distance modulus of 18.2. The experience gained from this study shows that both the abundance and age of a Magellanic Cloud cluster could be accurately determined from Washington photometry in less than 1 hour of 4 m time.

REFERENCES

Da Costa, G. S., Mould, J. R. and Crawford, M. D. 1985 Astrophys. J. 297, 582.
VandenBerg, D. A. 1985 Astrophys. J. Suppl. 58, 711.

DEEP PHOTOMETRY OF THE DRACO DWARF SPHEROIDAL GALAXY*

Bruce W. Carney

University of North Carolina

P. Seitzer

Kitt Peak National Observatory

ABSTRACT: We report the results of deep CCD photometry of two overlapping fields in the Draco dwarf spheroidal galaxy. We find $(B-V)_{o,g} = 0\overset{m}{.}69$, indicating a mean [Fe/H] = -2.0. The width of the giant branch below the horizontal branch is somewhat wider than our observational errors permit, from which we infer there is a spread in metallicities amongst the Draco giants, up to 0.8 dex, perhaps. Draco is found to contain numerous blue stragglers, like almost all other loosely bound halo systems. The number of such stars more massive than about 1.2 $M_\odot$ is roughly consistent with the number of anomalous cepheids discovered previously. It is not clear whether the blue stragglers represent a large number of mass transfer binaries or an intermediate age population, or both. A younger population component would help explain the galaxy's red horizontal branch stars in spite of its low mean metallicity. Signs of a younger population are seen in the color magnitude diagram for the entire sample, and especially from the higher quality data from the smaller area observed in the overlap of the two fields.

* Further details may be found in *Astronomical Journal*, **91, 23, 1986.**

CCD PHOTOMETRY IN THE CORE OF THE FORNAX DWARF GALAXY

Robert M. Light

P. Seitzer

Kitt Peak National Observatory/NOAO

1. INTRODUCTION

The present study is concerned with the examination of properties of stars in the core of the Fornax dwarf spheroidal galaxy. Previous studies have shown that Fornax has a very diverse stellar population. Four of the globular clusters associated with Fornax were found to have metallicities significantly lower than the mean metallicity of the field population of the galaxy (Buonanno et al. 1985); these clusters point out an older, metal-poor population. Also, there are a number of luminous carbon stars, which are indicative of a much younger population (see Mould and Aaronson 1986). Studies of the field population of Fornax (Demers, Kunkel, and Hardy, 1979; Buonanno et al., 1985) have shown a dispersion in metallicity. We have measured a large sample of giant branch stars, enabling a good determination of mean properties of the Fornax stellar population, as well as allowing a comparison of stars as a function of distance from the center of Fornax.

2. OBSERVATIONS AND ANALYSIS

The data for the color-magnitude diagrams are from 13 overlapping CCD frames, taken on the PFCCD at CTIO. They cover an area of approximately 145 square arcmin. around the geometrical center of Fornax. Each region was observed in B, V, and R, with exposure times of approximately 1 minute. They were reduced with Stetson's DAOPHOT program and put on the standard system using Graham's E region standards. The magnitudes of stars shared in the overlap regions were compared and no systematic offsets were found between frames. At $V = 18.4$, the errors in V, B-V, and V-R are 0.03, 0.04, and 0.05 mag., respectively. At $V = 20.8$, the errors are 0.08, 0.16, and 0.13. As a further comparison, one frame has stars in common with the Buonanno et al. (1985) region A1. Within the errors, our photometry is on the same system as theirs.

There is a noticeable dispersion in color among stars in the giant branch of Fornax. The bluest stars lie along the ridge lines of metal-poor globular clusters such as M92. The reddest giants seem to be limited by the 47 Tuc ridge line, assuming the M_V of the horizontal branch varies with metallicity as given by Sandage (1982). If, instead, $(M_V)_{HB}$

is defined to be +0.6, then there is a population of stars in Fornax more metal-rich than 47 Tuc. A mean giant branch ridge line was drawn for the color-magnitude diagram of the innermost stars, and we measure $\Delta V_{1.4} = 2.34$. Using Zinn and West's (1984) relation between $\Delta V_{1.4}$ and metallicity ([Fe/H] = $0.913 - 0.924 \Delta V_{1.4}$), the mean metallicity of Fornax is found to be [Fe/H] = -1.25, with an error of approximately 0.3 dex. This agrees with the value found by Buonanno et al. (1985) for the field of Fornax. The distribution of stars about this mean giant branch remains roughly constant as a function of radius from the center of Fornax.

A more detailed analysis of the dispersion in B-V shows that there is an intrinsic spread in color, in excess of that caused by errors in the photometry. Two subsets of data (19.2 < V < 20.0 and 20.0 < V < 20.5) were examined and, assuming that the error and intrinsic distributions in B-V are Gaussian, their intrinsic dispersions were measured. To convert this color spread to dispersion in metallicity, the difference in B-V between the Fornax ridge line and the giant branches of a number of globular clusters were plotted against [Fe/H]. Unfortunately, there is a large error associated with this calibration, possibly due to contamination of the giant branches by AGB stars. For both samples of data, σ([Fe/H]) is (very roughly) 0.10. Obviously, better measures of metallicity are needed for Fornax stars before a good value of the dispersion is known.

We would like to thank Jay Frogel for helping get the data for this study, and Bob Zinn for many useful discusions. One of us (R.M.L.) was supported by NSF grant AST-8304034.

REFERENCES

Buonanno, R., Corsi, C. E., Fusi Pecci, F. Hardy, E. and Zinn, R.
    1985 Astron. Astrophys. 152, 65.
Demers, S., Kunkel, W. E. and Hardy, E. 1979 Astrophys. J. 232, 84.
Mould, J. and Aaronson, M. 1986 Astrophys. J. 303, 10.
Sandage, A. 1982 Astrophys. J. 252, 553.
Zinn, R. and West, M. J. 1984 Astrophys. J. Suppl. 55, 45.

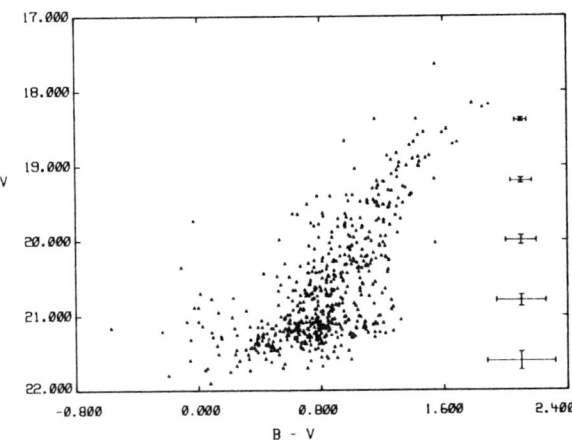

Fig. 1 CM diagram for one CCD field near the core of Fornax

A CANDIDATE FOR THE RECOVERED NOVA OF 1938 IN THE GLOBULAR CLUSTER M 14

Michael M. Shara and Michael Potter

Space Telescope Science Institute

Anthony F. J. Moffat

University of Montreal

Helen Sawyer Hogg

David Dunlap Observatory

Amelia Wehlau

University of Western Ontario

Although close binaries are believed to be of importance in the dynamical evolution of globular clusters, searches for such binaries have produced mostly negative results, aside from x-ray sources. Two dwarf novae which are possible cluster members are known (Margon and Downes 1983) and two classical nova candidates have been found. The crowded field around the nova observed in 1860 close to the center of M80 makes ground-based recovery of that star impossible with present techniques. Here we report on our attempt to recover the star which erupted in 1938 about 30" (0.8 core radii) from the center of M14.

In August, 1983 we imaged the nova field with the RCA CCD camera at the prime focus of the CTIO 4-m telescope using broadband R and B filters and an H_α filter. We again imaged the field in May, 1985 with the RCA CCD camera at the prime focus of the AAT using B, V and U filters. All subsequent reductions were done at the ST ScI. Using a position of the nova determined from the discovery plates of Hogg and Wehlau (1964) a faint star was identified on the CCD frames falling within half a pixel (0.3") of the position of the nova.

We have measured the brightness of this candidate and several hundred other stars in the CCD field using the photometry package DAOPHOT (Stetson 1984) with the zero point set by the photoelectric and photographic photometry in M14 of Kogan, Wehlau and Demers (1974). We find B = 20.2 ± 0.3 for our candidate with the rather large estimated error due to background and crowding. This translates into M_V = +2.7 ± 0.5 using the Harris and Racine (1979) values for reddening and distance modulus of the cluster. This is somewhat brighter than the mean value for old novae of 4.1 mag given by Patterson (1984) but within the brightness range which shows a FWHM of ~3 mag. However a calculation of the number of cluster stars in an error circle of 1" diameter centered on the 1938 nova position suggests we should expect to find 0.9 star with $M_V \leq 4$. Therefore our present candidate could be another star masking the fainter nova.

A check of the color of our candidate star can be done using the CMD shown in Figure 1. It can be seen that the nova candidate, designated by an "*", is about 0.2 mag redder than other stars of similar V. In the color-color diagram of M14 given in Figure 2 the candidate is seen to be about 0.7 mag brighter in U than other cluster stars of similar B-V. Its dereddened $(U-B)_0$ color of -0.3 ± 0.4 is similar to that of many old novae (Warner 1973). On the other hand its moderately red $(B-V)_0$ color of 0.8 ± 0.4 is somewhat unusual but could be due to an evolved secondary.

A spectrum of this star is needed before it can be confirmed or rejected as an old nova. A ground-based observation will be extremely difficult due to the star's faintness and the crowding of the field but such an observation would be possible with the Hubble Space Telescope.

REFERENCES

Harris, W. and Racine, R. 1979 Ann. Rev. Astron. Astrophys. 17, 241.
Hogg, H. S. and Wehlau, A. 1964 J. Roy. Astron. Soc. Canada 58, 163.
Kogan, C. S., Wehlau, A. and Demers, S. 1974 Astron. J. 79, 387.
Margon, B. and Downes, R. A. 1983 Astrophys. J. Letters, 274, L31.
Patterson, J. 1984 Astrophys. J. Suppl. 54, 443.
Sandage, A. 1970 Astrophys. J. 162, 841.
Stetson, P. 1984, private communication.
Warner, B. 1973 in IAU Symposium No. 73, Structure and Evolution of Close Binary Systems, P. Eggleton, ed., Reidel, Dordrecht, p. 85.

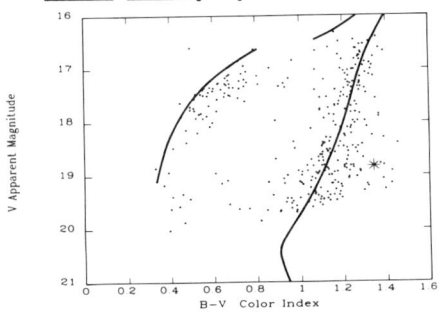

Fig. 1. C-M diagram of M14, derived from CCD camera images. The solid curve is taken from Sandage's (1970) C-M diagram of M13. The nova candidate is designated by an "*".

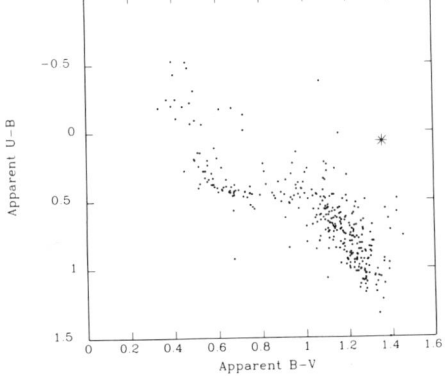

Fig. 2. Color-color diagram of M14 derived from CCD images and showing the candidate as an "*".

THE ABSOLUTE LUMINOSITY OF RR LYRAE VARIABLES

C. Cacciari[1,2], G. Clementini[1,2], L. Prevot[3]

Space Telescope Science Institute, Baltimore   1
Astronomical Observatory, Bologna   2
Astronomical Observatory, Marseille   3

We have taken UBVRI photoelectric photometry and CORAVEL radial velocities for 6 field RR Lyraes, i.e. SW And, YZ Cap, SW Dra, SS For, RV Phe and V440 Sgr, with the purpose of applying the Baade-Wesselink (B-W) method and determining their absolute magnitudes. The present improvements with respect to previous applications of the method are: a) the use of the (V-I) color, which shows the smallest sensitivity to gravity and metal abundance in the color-visual surface brightness plane. b) the use of a new grid of model atmospheres by Buser and Kurucz (1986), which include an improved treatment of opacity and convection. They provide a complete set of models for the relevant values of effective temperatures, gravities and metal abundances, thus avoiding the need of inaccurate interpolations or extrapolations.

The preliminary results of our analysis are summarized in Table I. The final results, along with a detailed discussion of the method and its individual aspects will be presented in two forthcoming papers.

A few remarks can be made: 1) For the stars affected by Blazhko effect (i.e. SW And and SS For) it was possible to match the light and radial velocity curves at the same phase in the Blazhko cycle. The effects of amplitude variations were then eliminated and the B-W method could be applied with some confidence. 2) Two stars (i.e. SW Dra and RV Phe) are affected by significant phase-lag between light and radial velocity curves. This prevents a correct application of the B-W method, in fact the values of M_v we have derived by a forced application of the method are unreasonably faint. Jones et al. (1987) have independently applied the B-W method to SW Dra, and found that the use of the (V-K) color and K magnitude minimizes the above mentioned phase-lag. The absolute magnitude they find for SW Dra is $M_v = 0.94 \pm 0.14$ mag. 3) For the c-type variable YZ Cap the method could not be applied, mainly because the scatter in the angular diameter curve derived from the photometry was too large.

Some conclusions can be drawn: 1) More stars need to be studied in order to obtain a statistically significant sample, since a fraction of them present problems in the application of the B-W method which cannot be easily foreseen. In particular more information is needed on metal-rich stars, if one wants to assess the dependence (if any) of the absolute magnitude M_v on metallicity. 2) Improvements to the method can still be made, and better accuracies in M_v determinations can be achieved, by using infrared magnitudes and colors, which are less affected by shock waves in the atmosphere, and by studying in detail the structure of the atmosphere during the pulsation cycle.

Table I

Data for RR Lyrae Stars

Star	Period	Fe/H	T_{eff}	R/R_o	M_v	Remarks
SW And	0.4423	-0.15	6650	4.3	1.0	Blazhko effect
YZ Cap	0.2735	-0.23				B-W method not applicable
SW Dra	0.5697	-0.70	6500	3.8	1.3:	$\Delta\varphi \sim 0.10$
SS For	0.4959	-1.50	6760	4.7	0.7	Blazhko effect
RV Phe	0.5964	-1.50	6460	4.1	1.3:	$\Delta\varphi \sim 0.10$
V440Sgr	0.4775	-1.35	6700	5.2	0.6	

REFERENCE

Jones, R. V., Carney, B. W., Lathamn, D. W. and Kurucz, R. L. 1987 in IAU Symposium No. 126, Globular Cluster Systems in Galaxies, J. E. Grindlay and A. G. D. Philip, eds., Reidel, Dordrecht, p. 589.

THE DISTANCES TO RR LYRAE VARIABLES

Rodney V. Jones and Bruce W. Carney

University of North Carolina

David W. Latham and Robert L. Kurucz

Harvard-Smithsonian Center for Astrophysics

Of all the different methods employed to estimate the mean absolute magnitude of RR Lyrae variables, only an analysis of the Baade-Wesselink type can determine this quantity directly. The distance to a globular cluster can therefore be measured by determining $<M_V>_{RR}$ for that cluster instead of being forced to assume that $<M_V>_{RR}$ is the same as that of the nearby field variables. This is important in that the field stars may have a different luminosity than cluster variables. In addition, since $<M_V>_{RR}$ should depend on the composition (especially helium) and history of mass loss of these stars, this quantity may vary from cluster to cluster.

Direct measures of the distances to the nearer globular clusters are now feasible with the Baade-Wesselink method due to the implementation of efficient spectrographs and detectors such as the digital speedometer on the MMT (Latham 1985; Wyatt 1985). We have successfully applied a version of this technique, the surface brightness method, to the nearby field variables X Ari ([Fe/H] ~ -2.2) and SW Dra ([Fe/H] ~ -0.7) utilizing simultaneous optical and infrared photometry and radial velocities with typical accuracies of 1 km sec^{-1} (Jones et al. 1986 a,b), and plan to extend the investigation to the nearer globular clusters such as M5.

Carney and Latham (1984) discovered a phasing problem in their analysis of VY Ser such that the phase of the radial velocities or the photometry had to be shifted in order for the Baade-Wesselink method to work. This problem also occurred for X Ari and SW Dra if optical color indices such as B-V or b-y were used to compute the effective temperatures needed for this type of analysis. The problem vanished when the V-K index was employed to calculate T_{eff}, and it was discovered that the optical colors yielded temperatures that were consistently hotter than those derived from V-K during the expansion phase of the pulsation cycle due to an excess of flux in the optical region, causing a distortion in the computed angular diameters which led to the apparent phase shift. The cause of this excess flux has not yet been determined, but it seems to be associated with the shock wave phenomenon.

We derived $\langle M_V \rangle = +0\overset{m}{.}88 \pm 0\overset{m}{.}15$ for X Ari and $\langle M_V \rangle = +0\overset{m}{.}94 \pm 0\overset{m}{.}15$ for SW Dra utilizing the V-K index and restricting the phase interval to exclude the shock waves. Since the major sources of error in the absolute magnitudes are systematic and affect both stars equally, the error in the magnitude difference is smaller, such that X Ari is only $0\overset{m}{.}06 \pm 0\overset{m}{.}10$ brighter than SW Dra despite the large difference in metallicity, $\Delta[Fe/H] = 1.5$. Sandage (1982) predicts that such a metallicity difference should produce a magnitude difference of $0\overset{m}{.}5$, and his period-luminosity-amplitude relation indicates that X Ari should be $0\overset{m}{.}27$ or $0\overset{m}{.}19$ brighter, depending on whether or not there is a metallicity dependence on horizontal branch star masses. Our results contradict these predictions; however, the Sandage relations were derived assuming that the stars were near their zero-age horizontal branch luminosities, which may not be valid here, since we cannot exclude the possibility that SW Dra is a well-evolved star that is not crossing the instability strip for the first time. The only sure way to adequately test the Sandage relations is to determine $\langle M_V \rangle_{RR}$ for globular cluster variables directly.

If the results for X Ari and SW Dra are valid for globular clusters of the appropriate metallicity, then the age of metal-poor clusters such as M15 and M92 derived from the luminosity of the main-sequence turnoff is 20×10^9 years for a helium abundance of $Y = 0.2$, with a slight dependence upon the helium abundance, while the age of metal-rich clusters such as 47 Tuc is 14×10^9 years for the same helium abundance. These results indicate that the metal-rich old disk clusters formed much later than the metal-poor halo clusters, but such a conclusion can be firmly established only by direct analysis of cluster variables.

REFERENCES

Carney, B. W. and Latham, D. W. 1984 Astrophys. J. 278, 241.
Jones, R. V., Carney, B. W., Latham, D. W. and Kurucz, R. L. 1986a Astrophys. J., in press.
Jones, R. V., Carney, B. W., Latham, D. W. and Kurucz, R. L. 1986b Astrophys. J., in press.
Latham, D. W. 1985 in IAU Colloquium No. 88, Stellar Radial Velocities, A. G. D. Philip and D. W. Latham, eds., L. Davis Press, Schenectady, p. 21.
Sandage, A. 1982 Astrophys. J. 252, 553.
Latham, D. W. 1985 in IAU Colloquium No. 88, Stellar Radial Velocities, A. G. D. Philip and D. W. Latham, eds., L. Davis Press, Schenectady, p. 123.

DOUBLE-MODE RR LYRAE STARS IN IC 4499

Christine M. Clement

David Dunlap Observatory

James M. Nemec

Palomar Observatory

Robert J. Dickens

Rutherford Appleton Laboratory

Elizabeth A. Bingham

Royal Greenwich Observatory

ABSTRACT: Thirteen double-mode RR Lyrae (RRd) stars, with mean magnitudes $<B>_c=18.30 \pm 0.10$ and $<V>_c=17.80 \pm 0.15$, have been identified in the variable-rich Oosterhoff type I globular cluster IC 4499. The stars have surprisingly uniform properties, and are considerably different from RRd stars found in Oo II systems. The mean first-overtone period (Fig.1) is $<P_1>=0.357^d \pm 0.005^d$ (cf. $<P_1>=0.40^d$ for Oo II RRd stars), and the mean ratio of the first-overtone period to the fundamental period is $<P_1/P_0>= 0.7443 \pm 0.0002$. The mean double-mode pulsation mass for the 13 stars, using the King Ia (Y=0.279, Z=0.001) mass calibration, is 0.535 ± 0.003 $M_\odot$. Such an average mass is 0.11 $M_\odot$ smaller (i.e. ~17% smaller) than that for RRd stars found in Oo II systems, and possibly ~0.01 $M_\odot$ smaller than the mean mass for the two RRd stars in M3 (it is important to note that the zero point of these mass determinations is uncertain by at least 15%). The metal abundances for the RRd stars, and for the system of RR Lyrae stars as a whole, are found to be consistent with $<[Fe/H]>=-1.38 \pm 0.20$, determined from ΔS spectroscopy. In the Peterson diagram (Fig.2), all known RRd stars now divide (apparently by mass) into two groups (split according to Oosterhoff type). With a reddening of $E_{B-V}=0.26 \pm 0.03$, the cluster distance modulus is $(m-M)_0=16.23^m \pm 0.23^m$.

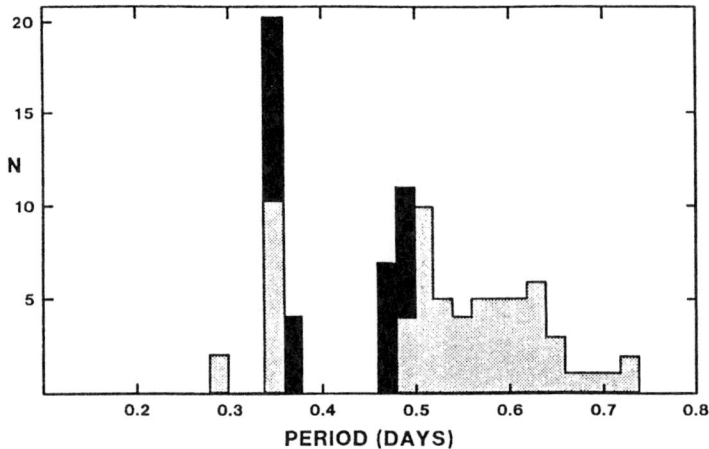

Fig.1 - Period histogram of the RR Lyrae stars in IC 4499. The black areas represent both period components of the double-mode RR Lyrae stars, and the shaded areas represent all other RR Lyrae stars.

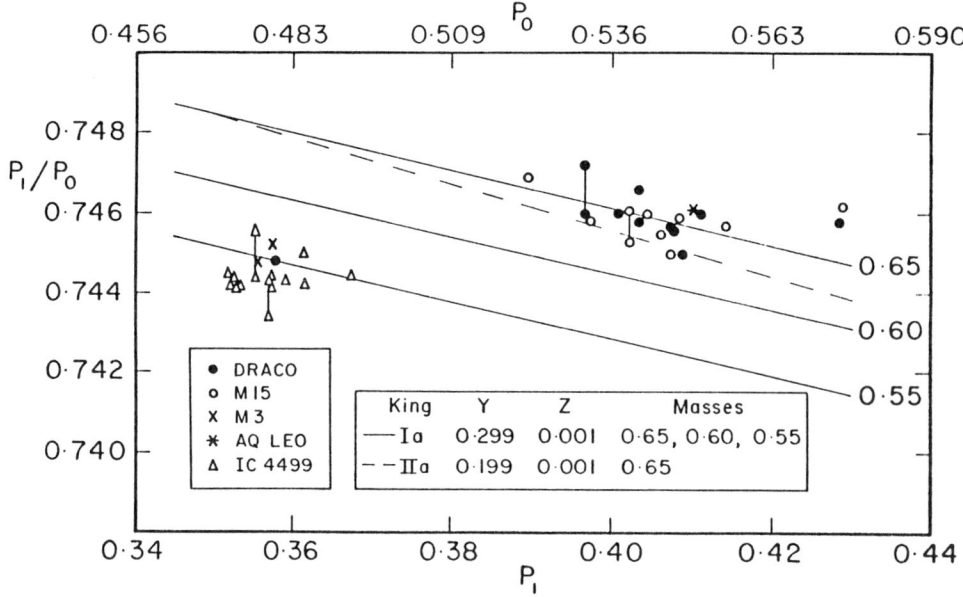

Fig.2 - Petersen diagram showing the positions of all known double-mode RR Lyrae stars. Note that the mean mass of the IC 4499 RRd stars is $\sim 0.535\ M_\odot$, and that the stars are virtually indistinguishable from the two RRd stars in M3 and the low mass RRd star in Draco.

SHORT-PERIOD VARIABLES IN GLOBULAR CLUSTERS OF MODERATE METALLICITY

Martha L. Hazen

Harvard-Smithsonian Center for Astrophysics

ABSTRACT. Three Galactic globular clusters of moderate metallicity, ($-0.7 >$ [Fe/H] > -0.9), have been searched for variable stars. NGC 6388 and NGC 6569 appear to contain RR Lyrae variables as members; NGC 6652 does not. NGC 6388 and NGC 6569 thus appear to have bluer horizontal branches than normally found at their metallicity.

As part of an ongoing program to study globular clusters that have not been extensively examined for variable stars, three Galactic globular clusters of moderate metallicity, ($-0.7 >$ [Fe/H] > -0.9 on the Zinn 1985 scale), have been searched for short-period variables. The clusters are listed in Table I along with their pertinent characteristics, including R, the distance from the Galactic center, and z, the distance above or below the Galactic plane.

Table I.
The clusters studied

NGC	[Fe/H]	R(kpc)	z(kpc)
6388	-0.74	3.8	-1.2
6569	-0.86	0.8:	-0.8:
6652	-0.89	5.2	-2.4:

NGC 6388 is found to contain four variable stars that are RR Lyrae type and probable members, and an additional two or more stars that are possible RR Lyrae members (Hazen and Hesser, 1986). NGC 6569 contains 8 RR Lyrae stars that are probable members and four that are possible members (Hazen-Liller, 1985). Preliminary results on NGC 6652 suggest that, although three stars lying within its tidal radius are RR Lyrae stars, none of them is a likely member.

NGC 6388 and NGC 6569 are the clusters of highest metallicity suspected to contain horizontal branch variable stars. The results imply that these two clusters have horizontal branches bluer than expected for their metallicities. The lack of RR Lyrae stars as possible members of NGC 6652 suggests that this cluster has a horizontal branch considered more typical of its metallicity.

REFERENCES

Hazen, M. L. and Hesser, B. H.   1986   Astron. J., in press.
Hazen-Liller, M. L.   1985   Astron. J. 90, 1807.
Zinn, R.   1985   Astrophys. J. 293, 424.

SECOND POSTER DISCUSSION

LYNGA: Let us then discuss the astronomical papers. A general comment: as in yesterday's poster session the majority of today's papers do not address the topic of our symposium; they do not take the holistic view of cluster systems in galaxies but the reductionist approach of discussing individual clusters and individual stars. I guess this is the way towards future knowledge about cluster systems.

We start by discussing three papers which treat systems of galaxies at such distances that images of clusters are almost stellar. Image analysis is thus very critical; advanced techniques of identifying clusters are given by Blecha. Sharples, in his work on NGC 5128, has used radial velocities as well, a very powerful technique. The Italian group has used methods already in print. Who wishes to comment?

COHEN: What is the most efficient way to pick out clusters?

GRAHAM: Gary Da Costa and I did quite well five years or so ago when we looked for slightly diffuse near-stellar objects against dark dust clouds in NGC 55 (thus removing background contamination). We found one young and two old clusters by this method. We were less successful with NGC 300 where we found all our candidates to be background galaxies.

SCHOMMER: Echoing the previous comments, Christian and I felt quite satisfied with our survey of M 33 (this session). On the other hand, Hugh Harris and I had candidates in M 81, which we followed up spectroscopically with John Huchra, with less success. Only 4-5 (out of 15 I believe) are true M 81 clusters.

HANES: I'd like to ask Ray Sharples if he agrees with Hesser et al. (1984) that the mean velocity of the NGC 5128 cluster system is not the same as the velocity of the galaxy as a whole; if they now agree, why?

SHARPLES: No. The systemic velocity of the sample, 556, is the same within the errors as those found by recent analyses of the stellar spheroid and ionized gas systems in NGC 5128 (545 ± 10).

LYNGA: There are three papers about M 31. Battistini has made a useful compilation of catalogues; Hodge has studied the young clusters of M 31 with CCD images and found them significantly bluer than those of M 33 or the LMC; Watanabe and Yamagata find that the globular cluster system of M 31 is flattened. One question to Paul Hodge would be whether he thinks that the limited (although very high) resolution would bias his selection in favor of blue clusters.

HODGE: The unusually large number of very young clusters studied is the result of selection effects. The available catalog of disk clusters is very incomplete because of the difficulty of recognizing small, old clusters against the complex, bright M31 disk.

LYNGA: It is agreed that there are fundamental differences between the cluster systems of our galaxy, the LMC, M 31 and M 33. To explain these we must surely seek to understand the evolution of the systems and of their clusters. Six papers address such topics.

Let us first discuss the new theory suggested by Richtler and Seggewiss. They find that compact star clusters preferentially are formed from metal-poor material. The efficiency of mass dispersion would be low here, the star forming efficiency high, and thus bound clusters would result.

SEGGEWISS: We derive fairly low metallicities for young populous clusters in the Magellanic Clouds (mean [M/H] = -1.0 dex). This is the observational basis for the outlined scenario that populous clusters need metal-poor interstellar material for their formation.

CAYREL: T. Richtler was anxious about the reliability of the low metallicity he has obtained for the young populous cluster NGC 330. I want to report that M. and F. Spite, P. François, T. Richtler and myself have observed one supergiant in NGC 330 at high resolution with the CASPIEC Spectrograph in ESO, with a S/N ratio of 100. We have confirmed the low metallicity of the cluster and detected emission on the spectrum. This may be the first such detection.

SCHOMMER: I am somewhat worried about these low abundances, not just because we get 0.4 to 0.5 dex higher values. H II regions and Cepheids in the Large Magellanic Cloud appear to be at $0.0 < [Fe/H] < -0.3$. I think there are problems if these clusters are 10 times under-abundant.

GRAHAM: What would one of these young metal-poor clusters look like in a galaxy if it is being formed now?

SEGGEWISS: One probably has to look for a clustering of far-infrared sources in a massive cloud. The central cluster NGC 2070 of the 30 Dor H II region may be a more evolved example of a very young populous cluster.

ZINNECKER: 30 Dor has been viewed as a proto-globular cluster, but we have to keep in mind that the mass in the gas is much larger than the mass in the stars, so even though 30 Dor is very compact it may well dissolve after the gas is dispersed.

SEGGEWISS: Surely; but one has to keep in mind that the central 30 Dor cluster NGC 2070 lies in a "cleaned" region with fairly low reddening.

HATZIDIMITRIOU: Have you considered the possibility that the formation of a cluster might trigger cluster formation in a nearby cloud and on what timescale that might happen?

SEGGEWISS: Up to now we have not taken into account the triggering of cluster formation by adjacent young clusters.

SMITH, H. The difference between the relatively high metal abundances of LMC Cepheids and the low abundances you find for young LMC clusters is surprising, particularly since there seems to be no spatial segregation of the two types of objects. Clearly this discrepancy requires resolution.

SEGGEWISS: We are aware of this situation. Even the histogram of field star metallicities as shown in our poster is shifted to the metal-richard end with respect to the cluster metallicities.

LYNGA: The paper by Smith, Searle and Manduca is based on comparisons between spectra for 20 LMC clusters and synthetic spectra for varying age and metallicity. The relation obtained between these parameters can be fitted with a model of chemical evolution of the LMC, if a starburst lasting 2 Gyr occurred 3 Gyr ago and involved 2/3 of the stars. One may ask to what extent these three parameters can be varied and still fit the data?

SMITH, H.: Neither the age-metallicity relations nor the simple models of chemical evolution can by themselves reveal the entire history of the clouds. However, we can look forward to stronger constraints on the models when deep luminosity functions are added to the age-metallicity data.

LYNGA: The question was partly prompted by the finding of Mateo that the formation of star clusters in a region of the LMC hardly could have been a uniform procedure. I take it that some sort of size distribution of molecular clouds has been used in the disruption calculations. Might different conditions be responsible for the fact that you find cluster longevities in the LMC ten times larger than in our galaxy?

MATEO: Even if the effects of molecular cloud disruption are ignored, the inferred dissolution timescales for the LMC clusters in our sample is too long by a factor of about 2 to 4. Of course, we don't expect any encounters with HI clouds in the outer LMC, but the encounters between the clusters themselves should be equivalent to this.

WIELEN: Could you comment on the discrepancy between your results on the age distribution of LMC clusters and the results obtained by Elson and Fall.

CAPUZZO-DOLCETTA: I think that it is possible to explain the excess of relatively young clusters in the LMC without invoking a burst of cluster formation. It would be sufficient to have a distribution function decreasing as $(age)^{-1}$ (see my paper in this symposium).

LYNGA: Two more papers deal with evolution in the Magellanic Clouds. Renzini et al. discuss the RGB development and Liller and Alcaino present photometry of 14 clusters in the LMC. May I ask Alvio Renzini to comment on the effect of the development of the RGB on the integrated colors?

RENZINI: Let me first emphasize that in order to have accurate CM diagrams for Magellanic Clusters it would be important to devote considerable effort to the update and expansion of standard photometric sequences, particularly in the blue - indeed, our preliminary CM diagrams were affected by a strong color equation, and we were concerned with the mismatch of cluster sequences and isochrones (cf. Buonanno et al. 1986, Mem. S. A. Italy, in press). Having corrected for this effect, we now find good agreement with a true modulus of 18.5 and a reddening 0.1. Concerning integrated colors, the effect seems small in (B-V), while it is (probably) fairly large in (B-R) or (V-K). This remains to be investigated in more detail.

RICHER: You now say that your results are consistent with a distant modulus of 18.5 to the LMC. I'd like to hear from Schommer or Da Costa or Olszewski who are proponents of the "short" distance modulus to the cloud as to their comments on this result.

Da COSTA: If instead of reddening the isochrones by 0.1 mag and fitting at a modulus of 18.5, is it not possible that you would fit at a shorter modulus with a smaller (or zero) color shift?

RENZINI: As I said, we are happy with 18.5, but owing to the problem of the standards I would not make a great case on differences of 2 or 3 tenths of a magnitude.

SCHOMMER: I don't think we have a problem with standards or transformations. Zero points and small color offsets at the 0.02 to 0.04 level may still be a problem in the magnitude transfers. However, I think the different distances here are mainly a result of different reddenings and abundances used. I stress the importance of independent abundance measures for the distance. And I agree at 0.2 to 0.3 mag level, that the distance is still uncertain.

LYNGA: Let us now discuss some papers on the dynamics of clusters, starting with the study by Papenhausen and Schommer of mass segregation in two clusters. They find slight evidence in ESO 121-SC03 where they expect an effect but none in NGC 458 where no effect is expected.

KING: It is quite difficult to test segregation in Magellanic Cloud clusters, because of the crowding and the faintness. The test needs to be made in galactic globular clusters that have long relaxation times, such as NGC 5053.

NEMEC: There is a color magnitude diagram by Nemec and Cohen (1986 in preparation) which shows 30 blue stragglers. These are similar to those in NGC 5466 and show evidence for mass segregation.

LYNGA: There are three more papers on dynamical effects: Hatzidimitriou and Bhatia find statistical evidence for binary clusters in the LMC; Spassova, Stenova and Golev study the ellipticity of globular clusters in M 31. Kontizas, Hatzidimitriou and Metaxa find an outer fluctuation in the radial density gradient for seven clusters. This is something quite new to me.

COHEN: Are you sure that this effect is not due to fluctuations in the background field?

HATZIDIMITRIOU: The fluctuations observed in the density profiles of clusters in remote regions of the LMC, cannot be explained by random fluctuations of the field density for the following reasons: (1) the fluctuations are observable (at the same positions and with the same widths) in all four quadrants of the grid used for the star counts, and in all colors U, V, J, I. (2) The correlations shown in the paper between the morphological characteristics of the fluctuations and the dynamical parameters of the clusters cannot be due to random field effects. (3) All the clusters measured were situated beyond 5 degrees from the central region of the LMC. Therefore there are no significant field-density gradients. (4) The observed fluctuations are well above the expected statistical fluctuations of the counts (better than 2σ). I would like to note that some galactic globular clusters show similar fluctuations (e.g. Pal 14, NGC 1960). The theoretical explanation of the effect is not at all clear yet, but consideration of the proposed region of semi-bound escapers has given encouraging results.

COHEN: On the variations in the outer parts of clusters in the LMC, I wonder whether the variations have any statistical significance.

BHATIA: These variations have been found not only in clusters, but also in clusters of galaxies, indicating that it may be a dynamical effect.

LYNGA: We shall now discuss some photometric papers: Buttress et al. have selected Lindsay 11 for a very careful examination. It is planned to be a target for the Hubble Space Telescope mission; the Washington system has been used for an accurate determination of metallicity in three clusters; the ratio of early/late type members of SMC clusters is strongly correlated with age. These papers are now open for discussion.

SMITH, H.: I used to be very skeptical of results from the Washington System. However, I find Geisler's new calibration of the system very convincing. Perhaps he would comment on this calibration and its potential?

GEISLER: The new calibrations show that a CM diagram is an extremely sensitive metallicity indicator over the full range of abundances. The system has now been set up for use with a CCD and results of the intermediate age globular cluster NGC 2213 in the LMC indicate the system can obtain very accurate mean metallicities and thus improve our knowledge of the distance.

LYNGA: Two papers by Seitzer et al. have discussed dwarf galaxies and they show population differences between inner and outer parts. One paper by Shara et al. gives a candidate for Nova 1938 as a 20^h magnitude star in a dense field in M 14. Two papers on RR Lyrae stars fail to show a dependence of $<M_v>$ on metallicity. Any comments on these papers?

BELL: How well do Coravel radial velocities compare with traditional radial velocities from high resolution spectra?

CACCIARI: They compare very well. For the variable SW And, the only one we have in common, we have compared our Coravel radial velocities with those derived by Preston and Paczynski (1964) using weak metal lines from high resolution spectra, and they superpose almost perfectly.

LATHAM: The radial velocity curves are very well determined and are not the limiting factor in our Baade-Wesselink determinations of the radii and distances of RR Lyrae variables. Instead the photometry and analysis set the limit. In particular we believe it is essential to use infrared light curves, because they are less sensitive to temperature changes than visual light curves. Thus the infrared curves track the change in radius more reliably.

CARNEY: Not only are we convinced the CfA echelle-reticon spectra provide accurate radial, hence pulsational, velocities, but we believed they are reliable for the (B-W) method. Oke, Giver, and Searle (1962) suggested that line and continuum forming regions vary in separation, hence systematic errors could arise in the analysis of photometric and spectroscopic radii. KPNO 4 m echelle spectra, however, show no change in radial velocities derived from lines that form at a wide variety of depths.

GRAHAM: How many dates do you like to have to get a period for a RR Lyrae variable?

HAZEN: Over a period of 8-10 days, I like to have 15-20 plates; with this many, some stars yield obvious periods, but others still may have uncertain or ambiguous periods.

LYNGA: That ends today's poster session. Thanks to all for a lively discussion.

Chapter IX

Poster Papers

Cluster Systems in Distant Galaxies

Deep Photometry, CM Diagrams

Jim Hesser and a well-known globular cluster captivate the crowd

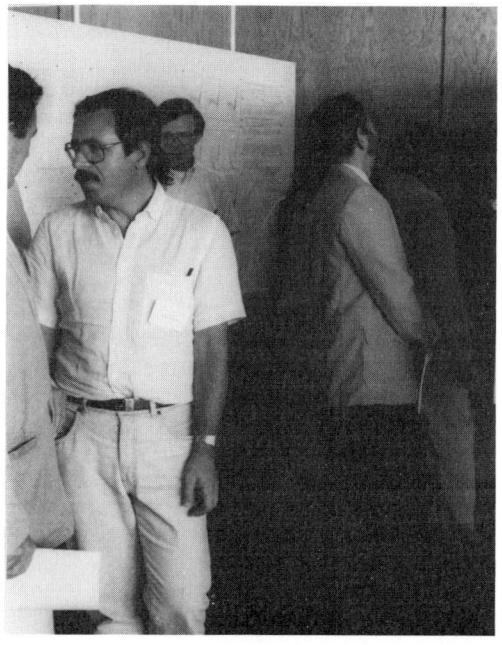

Roberto Buonanno with a cluster of globular luminaries

THE NUCLEI OF NUCLEATED DWARF ELLIPTICAL GALAXIES - ARE THEY GLOBULAR CLUSTERS?

H. Zinnecker[1], C. J. Keable[2], J. S. Dunlop[2]
R. D. Cannon[1] and W. K. Griffiths[3]

Royal Observatory, Edinburgh 1
University of Edinburgh 2
Leeds University 3

It came as a great surprise that many dwarf elliptical galaxies of very low surface brightness in the Virgo Cluster have conspicuous bright star-like nuclei (Reaves 1983, Binggeli, Sandage and Tammann 1985). These nuclei are at least a factor of 10 more luminous than the brightest globular clusters in the Local Group and comparable only to the very brightest globulars surrounding M87. They contain a considerable fraction (1 to 20%) of the total light of the parent galaxy (Binggeli, priv. commun.). Their physical nature and origin are a matter of debate (Zinnecker et al. 1985, van den Bergh 1985, Norman 1986, Zinnecker 1986) but optical spectroscopy for 3 objects indicates a stellar composition with a range similar to globular clusters (Bothun et al. 1985). It has been suggested that a central nucleus is formed when off-center bound star clusters migrate to the center as a consequence of dynamical friction (Norman 1986). Support for such a scenario comes from CCD observations of IC 3475 which reveal numerous knots near the center of this dwarf irregular galaxy (Vigroux et al. 1986). These knots have the same color as the parent galaxy and are interpreted as intermediate age star clusters.

We have obtained CCD observation in BRI at the 2.5 m Isaac Newton Telescope on La Palma for 3 nucleated dwarf ellipticals in the Virgo cluster (VCC 1185, VCC 1348, VCC 1539) in an attempt to study the colors and magnitudes of the nuclei. Exposure times were typically 5 min in each filter. The seeing was about 1".5 and the pixel size corresponded to 0".74. Although the frames were not taken under good photometric conditions, we could analyse the data to check for color differences between the nuclei and their parent galaxies. Wavelength-dependent seeing differences were the main complication in determining the color profile for small apertures. We have measured the flux from both the starlike nuclei and nearby reference stars, using a range of numerical apertures. The color (B-R) for both classes of object show the same variations with aperture diameters: for apertures 5"0 both star and nucleus redden by approximately the same amount, but with larger apertures the colors remain roughly constant (see Fig. 1). Therefore we conclude that to within 0.1 in (B-R) the star-like nuclei have the same color as the galaxies as a whole. This would be consistent with the IC 3475 observations and the above scenario for the origin of the nuclei. More work needs to be done to corroborate this conclusion.

Finally, we draw attention to the possibility that the nucleated dwarf ellipticals, when accreted and disrupted by a larger galaxy such a M 87, would contribute their naked nuclei as a population of globular clusters while the rest of the body of dwarf ellipticals would add to the halo stars of the giant galaxy.

More likely perhaps, dwarf spiral or dwarf irregular galaxies which contain knots (i.e. big star clusters) could, when swallowed by a larger galaxy, supply many if not all the globular clusters of that large galaxy directly (i.e. before dE nuclei are formed). In this way, not only would one avoid the accretion of over-massive clusters but also increase the number of accreted clusters. It is intriguing to realize that the ratio of mass in the knots to the mass in the bulk of some dwarf galaxies seems to be of the same order of magnitude (10^{-2} to 10^{-3}) as the ratio between the total mass comprised by globular clusters and the total mass of halo field stars in an ordinary large spiral or elliptical galaxy.

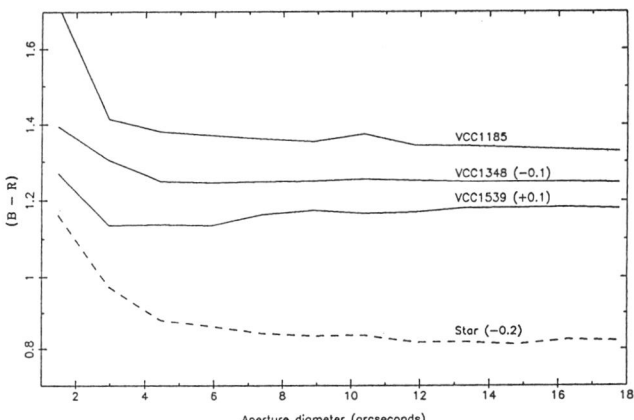

Fig. 1. Color profiles of the central regions of 3 Virgo dwarf elliptical galaxies. Comparison with the color profile of a reference star (dashed line) shows that the reddening of the innermost regions is a seeing artifact.

REFERENCES

Binggeli, B., Sandage, A. and Tammann, G. A. 1985 Astron. J. 90, 1681.
Bothun, G. D., Mould, J. R., Wirth, A. and Caldwell, N. 1985 Astron. 90, 697.
Norman, C. A. 1986 in Proc. IAP-Workshop, Star Forming Dwarf Galaxies and Related Objects, D. Kunth, T. X. Thuan and Tran Thanh Van, eds. editions Frontieres, p. 477.
Reaves, G. 1983 Astron. J. 53, 375.
van den Bergh, S. 1986 Astron. J. 91, 271.
Vigroux, L. Thuan, T. X., Vader, J. P. and Lachieze-Rey, M. 1986 Astron. J. 91, 70.
Zinnecker, H., Cannon, R. D., Hawarden, T. G. and MacGillivray, H. M. 1985 in Proc. ESO-Workshop on The Virgo Cluster, O. -G. Richter and B. Binggeli, eds., ESO, Garching, p. 150.
Zinnecker, H. 1986 in Proc. Rome Astrophysical Colloquium, Mem. Soc, Astron. Italia.

THE GLOBULAR CLUSTER SYSTEM OF M 87

Judith G. Cohen

Palomar Observatory

Long exposures with the 4-Shooter at the Cassegrain focus of the 200-inch telescope at Palomar Observatory have been obtained for M87 (and two other giant ellipticals in Virgo). Ellipse fitting with a code specially developed to reject point sources has been carried out to determine the surface brightness in various bandpasses of the underlying galaxy. The color gradients in the galaxy are quite small over the entire regime between 2 and 350 arc-sec from the nucleus of M87. Also I find that there is no difference between the ellipse parameters (position angle and eccentricity) derived in the various colors, i.e. the isochromes and the isophotes coincide. Details of the study of the halo of M87 are described in a paper submitted to the Astronomical Journal.

Subtraction of the best fit ellipses reveals the globular cluster system of M87 in its full glory. I added up the 7 best exposures (from a set of 15), each 800 seconds long, in the g filter of the Thuan-Gunn system after ellipse fitting had been carried out on each of them. Thus objects whose magnitude is near 26.5 are obvious on the final frame, and reliable photometry can be obtained down to g = 26 mag. Only 1 exposure each in the r and i bandpasses of the Thuan-Gunn system are available. Photometry down to about magnitude 23 is reliable in the r and i frames.

An automatic point source detection code was run on the summed g frames to find the globular clusters (plus small galaxies). Regions with chip defects or relatively large nearby resolved galaxies (of which there are several in the halo of M87) were not included. Then all the objects that are resolved (presumably more distant galaxies) were removed by hand. This eliminated about 70 objects in the 9x9 arc-minute field and effectively removed nearly all galaxies brighter than 23.5 mag. A total of about 7500 point sources remain after this in the M87 field. Most of these objects are globular clusters surrounding M87, although fainter than 23.5 mag there is substantial contamination by almost unresolved background galaxies.

The median and quartiles of the color distribution of the M87 globular clusters were found as a function of distance from the nucleus (expressed as bins in semi-major axis since beyond 1 arc-minute from

the nucleus M87 is noticeably elliptical). They reveal only a small
color gradient in the M87 cluster system itself, of size comparable to
that in the light of the underlying galaxy. It does appear that at any
given point, the median color of the globular cluster system is bluer
(by about 0.15 mag in (g-r) and by about 0.3 mag in (g-i)) than M87
itself.

Full details of this work and a similar analysis of the
globular cluster system of NGC 4472 and NGC 4406 will be presented in a
paper now being prepared for submission to the Astrophysical Journal.

THE CORE OF THE M 87 GLOBULAR CLUSTER SYSTEM

Tod R. Lauer

Princeton University Observatory

John Kormendy

Dominion Astrophysical Observatory

ABSTRACT: We have observed the central distribution of globular clusters in M 87. The core radius of the cluster system is an order of magnitude larger than that of the underlying galaxy.

DISCUSSION

This poster is a brief summary of work we have done on the central distribution of globular clusters in M 87; for a full description of our results see Lauer and Kormendy (1986). This work is based on high-resolution CCD observations taken at CFHT in sub-arcsecond (0"7 FWHM) seeing conditions. The light from M 87 was parameterized by a high resolution isophote fitting algorithm and then subtracted off. Globular clusters were visible in the residual image at all radii from the nucleus, including the core of the galaxy; 124 clusters with $m_B < 23.6$ have been detected within 60" of the M 87 nucleus.

Our most important finding is that the central distribution of globular clusters in M 87 is flat and has a core an order of magnitude larger than that of the underlying galaxy surface brightness distribution. The present data, when combined with those of other investigators, imply $r_C = 88 \pm 5$ and a central surface density of 72 ± 4 clusters arcmin^{-2} ($m_B < 23.6$) (see Fig. 1.). Further, the luminosity function of the central clusters is indistinguishable from that measured for clusters outside of 60" by van den Bergh, Pritchet and Grillmair (1985), which constrains any mechanisms operating near the center of M 87 that might selectively destroy or create clusters as a function of their mass. In particular, it appears that dynamical friction cannot produce a core this large from a system initially as centrally concentrated as the underlying galaxy; the large core of the cluster system may be a relic of the galaxy formation epoch.

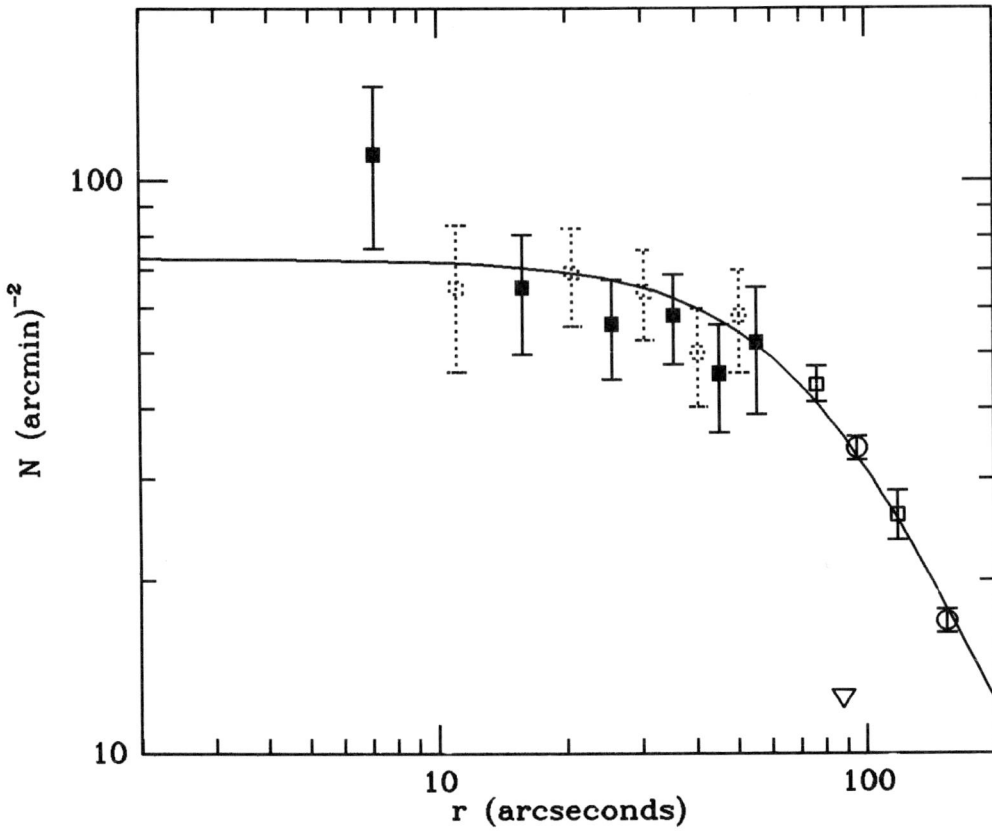

Fig. 1. Central surface density distribution of globular clusters in M.87 with $m_B < 23.6$. Data from the present study are shown in 10" radial bins both starting from the center (filled squares), and offset by 5" (dashed squares). Data outside of 60" are from van den Bergh, Pritchet and Grillmair (1985) (open squares), and Harris and Smith (1976) (open circles). The solid line is a concentration 2.5 King model fitted to the data. The location of the core radius is marked by an inverted triangle.

REFERENCES

Harris, W. E. and Smith M. G. 1976 <u>Astrophys. J.</u> 207, 1036.
Lauer, T. R. and Kormendy, J. 1986 <u>Astrophys. J. Lett.</u> 303, L1.
van den Berg, S., Pritchet, C. and Grillmair, C. 1985 <u>Astron. J.</u> 90, 595.

U PHOTOMETRY OF GLOBULAR CLUSTERS IN THE CENTRAL REGION OF M 87

E. V. Held

University of Padua

J. -L. Nieto

Observatoire du Pic du Midi

We present a photometric (U band) investigation of the globular clusters in the central region of M87 from deep calibrated photographic and electronographic material, obtained with the CFH telescope. The aim of this work is to complete previous photometric studies at longer wavelengths (Grillmair et al., 1986; Lauer and Kormendy, 1986).

The observational material was chosen from the deepest U photographic and electronographic data obtained during several CFH observing runs by one of us, for the study of the optical counterparts of the radio source Virgo A, notably the M87 jet (Nieto and Lelièvre, 1982; 1985).

In this preliminary report we have analyzed one prime focus 90 min photographic plate (IIaO baked) and two Cassegrain focus, respectively 120 min and 240 min electronographic plates, (obtained with the wide-field electronographic camera of Observatoire de Paris).

The fields investigated, scanned with the PDS of the Padua Observatory, are $6' \times 6'$ for the photographic plate and $3'30" \times 3'30"$ for the electronographic plates. The reduction was made using the IHAP and MIDAS packages developed by the European Southern Observatory. To remove the galactic background an unsharp mask was constructed and subtracted from the original image.

Only the photographic plate has objects in common with the list of photographic secondary UBV standards kindly communicated to us by R. Racine (1986). Our measurements show a good agreement with his in the range $19 < m_U < 22$ mag. We suspected that a deviation at $m_U \geq 22$ comes from a biased choice of noise–enhanced objects located near the detection limit of Racine's photographic material. This departure was confirmed by comparing our electronographic photometry with our photographic photometry in the central $3'30" \times 3'30"$. The zero point of the electronographic data was further adjusted to the photographic magnitudes. We found good internal consistency for our measurements down to the $m_U = 21.8$ photographic limit. We have adopted for each frame a magnitude limit brighter than the detection limit, in order to obtain a complete sample of globular clusters in the first phase of this study.

Figure 1 shows the globular cluster counts in 15" annuli. The inner 15" were not considered here because the central region deserves special treatment. A core is however visible that confirms Lauer and Kormendy's (1986) result. A preliminary differential luminosity function of the globular cluster candidates brighter than $m_U = 23$ was derived for each plate. Contributions of faint stars and background galaxies will be taken into account in a later stage of this study.

We thank René Racine for providing us with unpublished information, and G. Lelièvre, G. Wlérick and the Observatoire de Paris electronographic camera team for assistance during the observations.

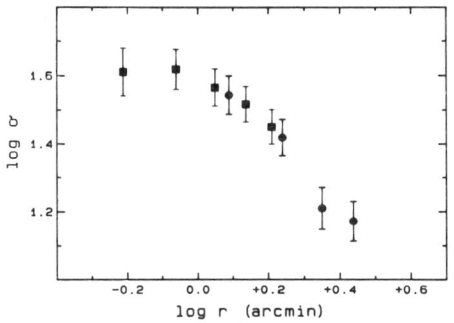

Fig. 1. Surface density of M87 globular clusters $(in\ arcmin^{-2})$ versus distance to the center. *Squares:* 120 min electronographic plate; *dots:* photographic plate. Error bars are computed as $\sqrt{n}$.

REFERENCES

Grillmair, C., Pritchet, C. and van den Bergh, S. 1986 Astron. J. 91, 1328.
Lauer, T. R. and Kormendy, J. 1986 Astrophys. J. Letters 303, L1.
Nieto, J. -L. and Levièvre, G. 1982 Astron. Astrophys. 109, 95.
Nieto, J. -L. and Levièvre, G. 1985 in ESO Workshop on the Virgo Cluster of Galaxies, O. -G. Richter and B. Binggeli, eds., ESO, Garching, p. 311.
Racine, R. 1986, private communication.

GLOBULAR CLUSTERS DETECTED IN THE COMA CLUSTER'S CENTRAL GIANT GALAXY NGC 4874

Laird A. Thompson

Institute for Astronomy, University of Hawaii

F. Valdes

National Optical Astronomy Observatories

We have used the Canada-France-Hawaii Telescope on Mauna Kea to obtain two deep CCD exposures in the central region of the Coma cluster, and we have used the new images to detect globular clusters in the central giant galaxy NGC 4874. One exposure with a total integration time of 2 hours lies in the halo of the giant galaxy. A second (comparison) exposure of the same integration time was taken within the cluster core but well away from any other bright galaxy. Our analysis, which includes both photometry and image classification with the automated routines called FOCAS, extends to a limiting B magnitude of 26.25. Image classification is a necessary part of this study because at the faintest limiting magnitudes globular clusters are easily confused with the rich population of faint background galaxies. We have been particularly successful in sorting stellar objects from galaxies because the image quality on our images is quite good: FWHM ~ 0.6 arcsec. The data were taken with the fast autoguiding instrument called ISIS (Thompson and Ryerson 1984).

In the near environs of NGC 4874, we find an excess of both stellar-like objects (the globular clusters) as well as an excess of extended galaxy-like objects (presumably dwarf galaxies). Because of the substantial population of true background galaxies, we can say very little about the radial distribution of dwarf galaxies relative to NGC 4874. However, once the galaxy-like objects are removed, the globular cluster population is relatively unaffected by background objects, and the corresponding radial distribution can easily be determined. As shown in Fig. 1, the globulars possess a very flat radial distribution when compared to the light of the underlying galaxy. This indicates that the globulars may be more closely associated with the cluster as a whole rather than being associated exclusively with the underlying starlight in NGC 4874.

By comparing the apparent luminosity functions of the globular cluster candidates in the Coma cluster with the well-studied globular cluster populations of M87 and our Galaxy, we can certify that we have, indeed, detected the globulars in Coma (see Fig. 2).

Unfortunately, the new Coma cluster data includes only the bright tail of the luminosity function, so there is no reliable way to use the new observations to determine the Coma cluster distance modulus. We can adopt the known distance modulus difference between Coma and Virgo and then use our new data to determine the total number of globular clusters in the small area of our study. This we combine with the luminosity of NGC 4874 contained in our data frame to determine the "specific frequency" of globular clusters. We find the value to be ~2.7, unusually low for a giant galaxy at the center of a rich cluster. Other giant galaxies have specific frequencies ~15 (Harris 1986). Our low value for NGC 4874 may be another indication that the globular clusters follow a very flat distribution relative to the galaxy light. In such situations, local values of the specific frequency should be lower than the norm in the central regions of the galaxy (where we observed) yet higher than the norm in the outer regions of the galaxy. This matter will remain unresolved until someone determines the specific frequency for NGC 4874 using images encompassing a wider field of view than ours.

REFERENCES

Harris, W. E. 1986 Astron. J. 91, 822.
Thompson, L. A. and Ryerson, H. R. 1984 Proc. S. P. I. E. 445, 560.

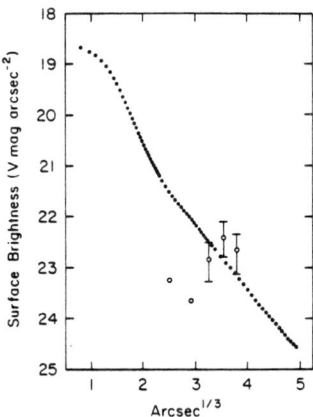

Fig. 1. Radial distribution of galaxy light (filled circles) and globular clusters (open circles).

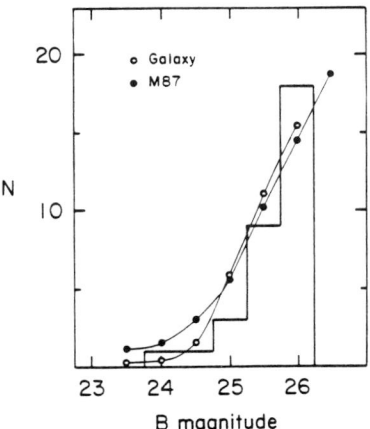

Fig. 2. Globular cluster luminosity functions NGC 4874 (histogram), M87 (filled circles), and the Galaxy (open circles).

GLOBULAR CLUSTERS IN DIFFERENT TYPES OF SPIRAL GALAXIES

Hugh G. Harris[†]
U. S. Naval Observatory

Gregory D. Bothun[†]
California Institute of Technology

James E. Hesser[†‡]
Dominion Astrophysical Observatory

ABSTRACT: Studies of globular clusters in spiral galaxies are still available for only a few galaxies with a limited range of galactic types. We have identified (or set upper limits on) the cluster systems in several edge-on spirals ranging from the early-type spiral NGC 7814 to the late type NGC 5907. We discuss the properties of the cluster systems and their relationship to the properties of the parent galaxies.

The data available on globular cluster systems in spiral galaxies are, in many respects, far more limited than data for ellipticals. Two reasons contribute to making spirals more difficult to study than ellipticals: the disks of spirals interfere through dust absorption and by cluttering images with disk objects, and spirals generally have fewer clusters (both in absolute numbers and relative to total light and bulge light). In addition, there is an added complication in interpreting cluster systems in spirals, caused by the several components contributing to the light and mass of spirals. We have observed several edge-on spirals of different types to try to expand the sample of spirals with useful data. We presently have images of the galaxies in Table I. Most are CCD images taken with the Canada-France-Hawaii Telescope, some are CCD images from Cerro Tololo InterAmerican Observatory, and a few are plates from CFHT. The data are in various stages of reduction, and results and notes are given in Table I.

Table I.
Spirals with New Data

Galaxy	Limits		Clusters		Notes
	Magnitude	Radii	Counted	Total	
NGC 2683	$J \leq 24.8$	$1.0 \leq R \leq 4.0$	100 ± 31	320 ± 110	Harris et al. 1985.
NGC 3717	$V \leq 24.5$				Detection marginal
NGC 5170	$V \leq 24.5$				Detection strong
NGC 5866	$V \leq 24.5$				Detection marginal
NGC 5907	$V \leq 24.5$	$0.3 \leq R \leq 3.6$	≤ 20	≤ 60	No detection
NGC 7814	$V \leq 24.5$	$0.8 \leq R \leq 6.6$	197 ± 28	720 ± 160	Detection strong

[†]Visiting astronomer, Canada-France-Hawaii Telescope.
[‡]Visiting astronomer, Cerro Tololo Inter-American Observatory.

In NGC 7814, the clusters show a significantly more extended radial distribution than the halo light, and also show a significantly flattened distribution, consistent with the flattening of the halo light. The flattening is perhaps surprising in view of the lack evidence for flattening for the cluster systems in other disk galaxies with flattened halos. Spherical distributions have been found in NGC 4594 and NGC 3115. There is evidence for flattening of the metal-rich subsystem of clusters in our Galaxy, but these clusters make up only a small fraction of the total Galactic system, and the shape of the Galactic halo for comparison is not well known. Previous kinematic studies indicate the flattening of the halo of NGC 7814 is a result of its rotation, not its disk mass which is not very significant. This result suggests that the cluster system is also rotating. The rotation of cluster systems in some galaxies with flattened halos like NGC 7814, and the lack of rotation in others like NGC 4594, might be explained under several different pictures for the origin of angular momentum and the relative time of cluster formation in disk galaxies.

The specific frequency S (the number of clusters normalized to total M_V of -15) for spirals is known to depend strongly on Hubble type. However, computing S using the spheroidal luminosity rather than the total luminosity gives similar values for the few spirals that have data. Table II gives the present status. The values of S(spheroid) are similar to those for elliptical galaxies lying outside rich clusters. For these cluster systems, however, the spatial distribution of the cluster systems differs (at least in some cases) from the spheroidal light, introducing the practical problem of how to compare the clusters with the light. Also, a difficulty sometimes occurs in separating the disk light, even when edge-on. In the late type spirals NGC 5170 and NGC 5907, there is no central bulge visible, and the minor axis light profile is dominated by the disk, probably to large enough radii to prevent detecting any spheroidal component. In several cases, the decomposition of an exponential disk, thick disk, bulge, and/or halo can be model-dependent. A comparison of the cluster population with galactic mass (within some radius) can be done through the galactic rotation velocity. The tendency is for late-type galaxies in Table II to have fewer clusters, in spite of their generally healthy rotation curves. The lack of clusters detectable in NGC 5907 is perhaps the clearest example of this trend. Hence a frequency relative to mass, S(mass), may depend on Hubble type in a manner similar to the frequency relative to total luminosity.

Table II.

Disk and S0 Galaxies with Cluster Data

Galaxy	Classification RSA	RC2	V_0 (km s^{-1})	W_{HI}	i (deg)	S(spheroid)	References
Galaxy	SbI-II	SXT4	...	450	...	2.2±0.5	Harris, W.E. 1981, Ap.J. **251**, 497.
M31	SbI-II	SAS3	-61	540	78	4.2±1.5	Harris, W.E. 1981, Ap.J. **251**, 497.
NGC 891	Sb	SAS3?/	706	490	90	≲0.1?	v.d. Bergh et al. 1981, A.J. **87**, 494.
NGC 2683	Sb	SAT3	373	454	82	9±3	Harris et al. 1985, A.J. **90**, 2495.
NGC 3115	S01(7)	L-/	472	...	~90?	2±1	Hanes et al. 1986, Ap.J. **304**, 599.
NGC 3717	Sb(s)	SA3:/	1477	433	83	...	This paper.
NGC 4565	Sb	SAS3?/	1122	524	90	0.9±0.5	v.d. Bergh et al. 1981, A.J. **87**, 494.
NGC 4594	Sa+/Sb-	SAS1/	963	750	84	3±1	Harris et al. 1984, A.J. **89**, 216.
NGC 5170	Sb:	SA5:/	1347	629	~88	...	This paper.
NGC 5866	S03(8)	LA+/	874	...	90	...	This paper.
NGC 5907	Sc	SAS5:/	780	491	~89	...	This paper.
NGC 7814	S(ab)	SA4:/	1249	(490)	90	...	This paper.

GLOBULAR CLUSTERS IN LENTICULAR GALAXIES: NGC 3115

E. V. Held and M. Capaccioli

Institute of Astronomy, University of Padua

We present a progress report for the ongoing study of the globular cluster (GC) system around the edge–on lenticular galaxy NGC 3115. This object is one of the best targets for this kind of investigation, due to its proximity ($\sim 10\ Mpc$) and favourable inclination (for a deep photometric study and a review of the properties of NGC 3115, see Capaccioli et al., 1986). Previously published data include a preliminary luminosity function of GC candidates, based on indirect calibration of KPNO photographs (Strom et al., 1977), and visual star counts on CFH and AAT deep plates (Hanes and Harris, 1986). Here we report on B–band photometry down to $m_B \sim 24$, based on CCD calibration.

We have obtained several CCD exposures for the central region (and for a few outer fields) of NGC 3115, taken with the ESO 2.2-m telescope in B, V and R. Three co–added 30 min exposures in B were reduced so far (covering an area of $2' \times 3'$ about the galaxy center), together with one frame of a comparison field $10'$ from the galaxy center. We also scanned a $6' \times 12'$ centered region, normal to the galaxy major axis, and a comparison field $22'$ from the center. This 50 min very high quality B plate of the CFH telescope was kindly lent us by Dr. J.-L. Nieto.

Both CCD frames and PDS scans were processed with the program INVENTORY (West and Kruszewski, INVENTORY ESO Manual), which provides a catalogue of parameters for all images in each field. The zero point of the CCD photometry was based upon standard stars (Landolt, 1983), while the photographic magnitudes were reduced to the CCD scale by direct comparison of unsaturated overlapping regions of the frames. No test for completeness has yet been made, but visual inspection of the frames suggests that the fraction of faint images lost by the search algorithm is negligible.

The radial distribution of all detected images is shown in Fig. 1. On the CCD frame the objects were counted in 10" wide annuli from $r = 30$" to 80" and, on the plate, in 30" annuli between $r = 60$" and 330" (the outermost being incomplete). The error bars are computed as $\sqrt{n}$. At this stage of the work, no attempt was made to separate stellar from non–stellar images.

Our results are compared in Fig. 1 with the deeper counts made by Hanes and Harris (1986) on their CFH plate only. Our data have been normalized by a shift in $\log \sigma$.

The apparent B–luminosity function (LF) for all the counted images from CCD frames is presented in Fig. 2. It increases steadily, and no maximum is seen to the detection threshold of $m_B \sim 24$. Our preliminary estimate is $m_B(max) > 23.8\ mag$, from which $(m - M) > 30.3\ mag$ (or $D > 11.5\ Mpc$) adopting an average

$B - V$ color of 0.8 mag together with $M_V(max) = -7.3$ mag (Harris and Racine, 1979). A better estimate will be derived taking into account the contamination by foreground stars and background galaxies.

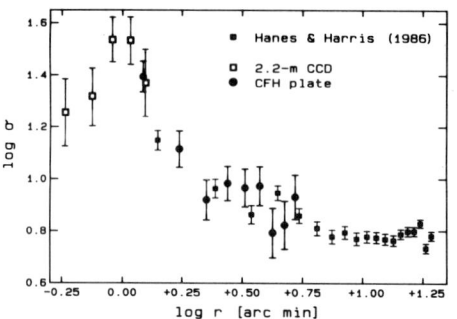

Fig. 1. Mean surface density, σ, of all objects found with the search algorithm, plotted vs. the radial distance from the center of NGC 3115. The *open squares* refer to the 90 min (total exposure) CCD frame, the *dots* to the central region of the galaxy in the CFH plate; *squares* are the counts from Hanes and Harris (1986).

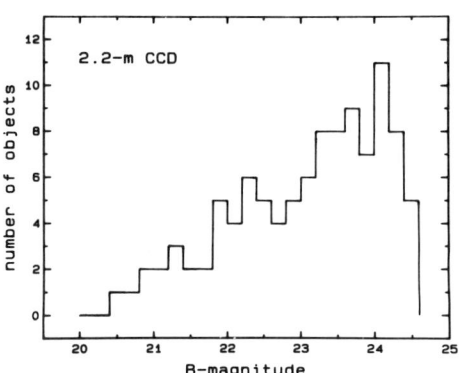

Fig. 2. Apparent luminosity function (in bins of 0.2 *mag*) for starlike objects around NGC 3115, from the 2.2-m CCD frames (central region of $2' \times 3'$).

REFERENCES

Capaccioli, M., Held, E. V. and Nieto, J. -L. 1986, preprint.
Hanes, D. A. and Harris, W. E. 1986 Astrophys. J. 304, 599.
Harris, W. E. and Racine, R. 1979 Ann. Rev. Astron. Astrophys. 17, 241.
Landolt, A. V. 1983 Astron. J. 88, 439.
Strom, K. M., Strom, S. E., Jensen, E. B., Moller, J., Thompson, L. A. and Thuan, T. X. 1977 Astrophys. J. 212, 335.

GLOBULAR CLUSTERS AS EXTRAGALACTIC DISTANCE INDICATORS: MAXIMUM LIKELIHOOD METHODS

David A Hanes and Donna G. Whittaker

Queen's University, Kingston

ABSTRACT

We have explored the use of maximum likelihood estimation techniques in the use of globular cluster luminosity functions (LFs) as distance indicators. In particular, we have tested size-of-sample effects through the analysis of Monte Carlo simulations of LFs drawn from an assumed universal population like that characterizing the globular clusters in the Local Group. Our working assumption, following others before us, is that the underlying LF is adequately well described by a Gaussian normal in a number vs. absolute magnitude representation.

For typically observable sample sizes in studies which are limited to the bright half of the LF, statistical limitations preclude a precise determination of the attributes which fully describe the LF, even in the absence of field object contamination. In particular, the intrinsic dispersion (the shape parameter of the LF) must be taken to be a universal constant, independent of galaxy type; only then may the turnover magnitude (which contains the distance information) be derived with good precision. Some data exist for nearby galaxies (including ellipticals) which permit an assessment of the universality of the intrinsic dispersion: they are not inconsistent with the hypothesis. However, it will be important to test this point in future as more data are secured.

Real globular clusters in remote galaxies are unresolved, and the samples are contaminated with foreground field stars and remote background objects. This contamination necessitates corrections which are statistical in nature, applicable to binned LFs. Through numerical simulations, we have tested the limitations imposed by realistic numbers of field objects in globular cluster LFs in remote galaxies, testing for systematic biases and assessing the attainable precision in derived distance as a function of the sample size and the limiting magnitude.

Our findings are that maximum likelihood methods are very robust. Distances precise to ± 10% are routinely derived (even in the presence of field object contamination) for moderately populous globular cluster samples within which photometry reaches nearly to the turnover: there is no need to strive for much deeper levels through extraordinary investments of telescope time. (Of course several very deep such studies will be wanted to test further the universality of the LF, which is also of interest as a diagnostic of cluster formation and disruption; thereafter, if our working assumption is indeed borne out, much more modest programs will yield the desired precision.) The implication is that globular clusters are potentially more far-reaching distance indicators than has previously been realized. Moreover, their previously-noted advantages make them doubly attractive: the method is insensitive to the Population I distance scale and calibrators, and permits the study of extremely remote systems in but a single step from the Milky Way.

Our complete results are in a paper submitted to the Astronomical Journal.

PHOTOMETRY OF FAINT STARS IN GLOBULAR CLUSTERS USING THE SIX METER TELESCOPE

N. Samus

Astronomical Council, USSR Academy of Science

Our program is based on photographic and photoelectric UBV photometry of globular cluster stars with the Soviet 6 m telescope. M 10 = NGC 6254 remains, regretfully, the only cluster for which we were able to gain photoelectric observational material sufficient for calibration of the photographic photometry for faint stars. Samus and Shugarov (1983) presented for M 10 a V, (B - V) diagram showing an unusually large magnitude difference between the main sequence turn-off point and the horizontal branch, $V(TO) - V(HB) \approx 3.8$ mag. It seemed of interest to compare M 10 and M 12 by their values of $V(TO) - V(HB)$. A preliminary calibration of the faint star photographic photometry in M 12 leads to $V(TO) - V(HB) \approx 4$ mag (Mironov et al., 1984). One may notice that in the classification introduced by Mironov and Samus (1974, 79), which is based mainly on the horizontal-branch morphology, both M 10 and M 12 belong to group I, presumably the older group. V, (B-V) diagrams were also published by us for NGC 288 and M 2 = NGC 7089 (Samus and Shugarov 1978, 79).

Two photographs of the globular cluster NGC 5053, one plate in B and one in V light, were taken on May 19/20 1980 at the primary focus of the 6 m telescope with the Pickering-Racine wedge. Unfortunately it turns out that the limiting magnitude of the plates, especially of the V plate, is far from the deepest possible for the telescope. The calibration is based on the photoelectric standards of Sandage et al. (1977) and on the electronographic observations of Walker et al. (1976). The main advantage of the present photometry in NGC 5053 is its practical completeness (in a round area of ~ 10' diameter) to 20^m V. The color of the turn-off region, somewhat too blue, cannot be attributed real significance because of the calibration difficulties at that magnitude level. There is no indication of $V(TO) - V(HB)$ being unusual. NGC 5053 belongs to Group II (presumably the younger one) in our classification. The shape of the subgiant branch is in agreement with VandenBerg's isochrones for $Y = 0.2$, $Z = 0.0001$, $T = 15$ and 18

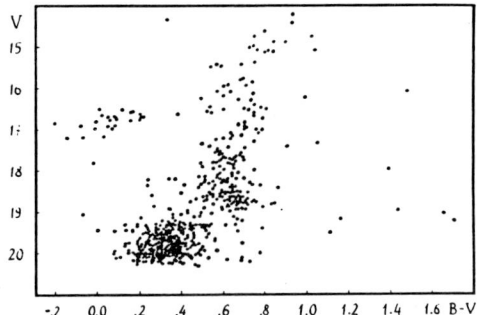
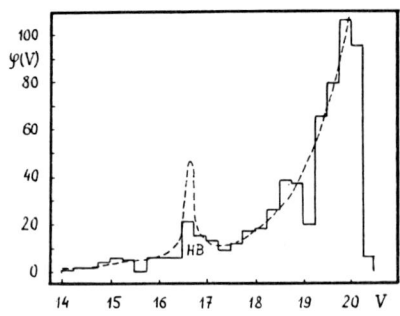

Fig.1. The V, (B-V) diagram for the globular cluster NGC 5053.

Fig. 2. The luminosity function for NGC 5053. No corrections of any kind have been applied. The dashed line shows the rescaled M 3 luminosity function (Sandage 1954).

billion years. The agreement of the luminosity function shape with that for M 3 (Sandage 1954) is surprisingly excellent.

Thanks are due to O. A. Baryshnikova and O. K. Silchenko who took part in the NGC 5053 studies.

REFERENCES

Mironov, A. V. and Samus, N. N. 1974, Peremennye Zvezdy, **19**, 337.
Mironov, A. V. and Samus, N. N. 1979 in Star Clusters UrGU, Sverdlovsk, p. 118.
Mironov, A. V., Samus, N. N., Shugarov, S. Yu. and Yuferov, A. O. 1984 Astron. Tsirkulyar, No. 1313.
Samus, N. N. and Shugarov, S. Yu. 1978 Astron. Tsirkulyar No. 1023.
Samus, N. N. and Shugarov, S. Yu. 1979 Astron. Zh. (USSR), **56**, 1323.
Samus, N. N. and Shugarov, S. Yu. 1983 Astron. Zh. (USSR), **60**, 1091.
A. R., Katem, B. and Johnson, H. L. 1977 Astron. J., **82**, 389.
VandenBerg, D. A. 1983 Astrophys. J. Suppl. **51**, 29.
Walker, M. F., Pike, C. D. and McGee, J. D. 1976 Monthly Notices Roy. Astron. Soc., **175**, 525.

HIGH PRECISION PHOTOMETRY OF 10,000 STARS IN M 3

R. Buonanno[1], A. Buzzoni[1], C. E. Corsi[1],
F. Fusi Pecci[2] and A. R. Sandage[3]

Astronomical Observatory, Rome.   1
Department of Astronomy, Bologna. 2
Mt. Wilson and Las Campanas Obs.   3

ABSTRACT: A new color-magnitude diagram for M 3 is presented. 10,000 stars have been measured down to V = 22 with an internal accuracy better than 0.03 mag to get complete and very accurate samples over well defined areas.

More than 10,000 stars have been measured down to V = 22 in two different areas. In the first, with 3.5 < r < 6.0 arcmin, photometric completeness has been achieved down to V = 21.5 and an algorithm to correct for losses due to unrecoverable crowding and blending has been experimentally computed. In the second, within a square field of 15 x 15 arcmin, completeness has been extended only to V = 18, well below the horizontal branch.

Many tests made on the data guarantee an internal photometric accuracy better than 0.03 mag at V = 21. Therefore, both the total population of each branch and the relative star-number ratios are "bona fide" representatives of the corresponding evolutionary time-scales. Here we simply present: 1) the color-magnitude diagram (see Fig. 1) obtained from the reduction of a wide collection of Palomar plates; 2) a table which presents the contribution of the various branches to the integrated cluster light; 3) the preliminary indication that, within the annulus we have considered, the blue stragglers seem to be slightly less centrally concentrated than the subgiants in the same magnitude interval.

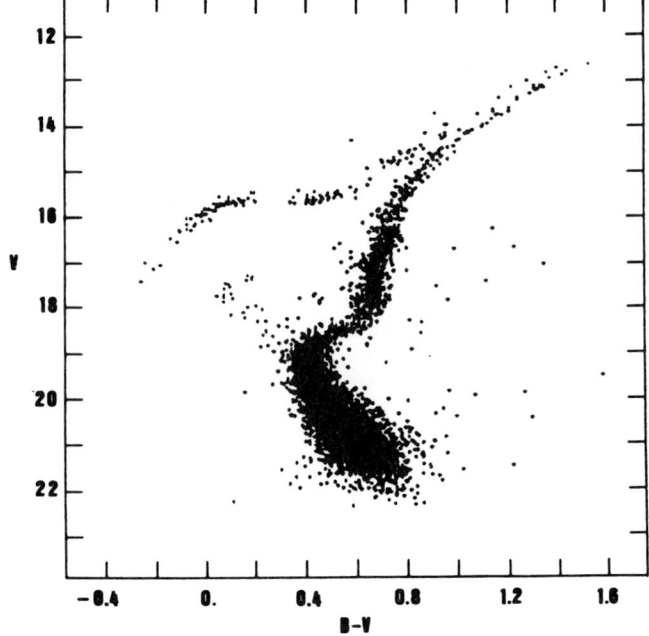

Fig. 1. CMD of M 3, composite sample.

Table I:

Contribution to the cluster integrated light of the various branches, computed over the complete samples

Phase	N(stars)	L/Lo	%Bol	% V	% B
AGB	10	5524	19	11	7
HB	82	4178	14	17	25
RGB	342	10739	37	36	29
SGB	764	2228	8	10	11
MS	8628	5908	21	25	28
BS	53	193	0.7	1.0	1.3

DEEP CCD PHOTOMETRY IN M 5

Harvey B. Richer and Gregory G. Fahlman

University of British Columbia

Deep UBV CCD imagery has been obtained in three fields of the galactic globular cluster M5. The locations of these fields are at distances of 8, 21, and 58 core radii. In the middle field, which overlaps substantially with the deep photometry field of Arp, the CCD photometry reaches fainter than V = 26. Color-magnitude diagrams constructed from stars in the inner two fields are identical, to within the errors, and can be used to set an upper limit of 4% to any metallicity difference between these two fields. A U, (U - V) color-magnitude diagram is also shown for the inner field and compared with that of a more metal rich and more metal poor cluster. Major differences in the morphology of these three diagrams are present as a function of metal abundance. From the color-color diagram the reddening in the direction of M5 is determined (E(B - V) = 0.02) as well as its metallicity ([M/H] = -1.13). The distance to M5 is then established from fitting local subdwarfs to the lower main sequence of the cluster. This yields $(m - M)_v$ = 14.3. Using the observationally determined parameters, an overlay of the appropriate VandenBerg and Bell isochrones yields an age estimate of 18 Gyrs for M5. Luminosity functions constructed from the three fields show excellent agreement through the range V = 17 to 23. Fainter than V = 23 there is some evidence for mass segregation effects due to dynamical relaxation.

Further details of this work can be found in Richer and Fahlman (1987).

REFERENCE

Richer, H. B. and Fahlman, G. G. 1987 Astrophys. J., in press.

PHOTOGRAPHIC PHTOTOMETRY OF 4500 STARS IN M 30

G. Piotto[1], M Capaccioli[1], S. Ortolani[2], L. Rosino[1], G. Alcaino[3] and W. Liller[3]

University of Padua 1
Asiago Astrophysical Observatory 2
Isaac Newton Institute, Santiago 3

We present B and V photometry of ~ 4500 stars (V < 21.3 mag) within the region 2'.1 < r < 10' centered on the southern globular cluster M 30 (NGC 7099), from two sets of four plates of the 4 m CTIO (CT) and the 2.5 m Dupont (LC) telescopes (Piotto 1986). Instrumental magnitudes (applying INVENTORY to PDS scans) are calibrated by four CCD exposures from the ESO 1.5 m Danish telescope. Our results removed the disagreement between previous photometry (Alcaino and Liller 1980) and the isochrones published by VandenBerg (1983).

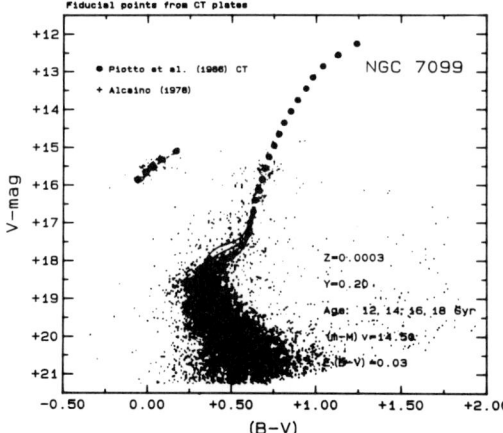

Fig. 1. CM diagram of M 30 showing all 4500 stars from CT plates. Fiducial points (full dots) are compared to the theoretical isochrones published by VandenBerg and Bell (1985).

The distance modulus $(m-M)_V = 14.5 \pm 0.1$, is computed assuming M_V (RR Lyrae) = 0.6 mag. From the CM diagram plotted in Fig. 1, we estimated that $m_{VTO} = 18.6 \pm 0.1$ mag and $(B-V)_{TO} = 0.36 \pm 0.06$ mag. The metallicity, [Fe/H] = -1.9 ± 0.3, is computed through the de-reddened color index (Webbink 1985) using $(B-V)_{o,g} = 0.71 \pm 0.03$ mag. With this value the isochrones (VandenBerg and Bell 1985) fit the data better than with [Fe/H] = -2.13 given by the index Q_{39} (Zinn 1985) providing an age of $(17 \pm 4) \times 10^9$ yr (model errors are included, as in Renzini, 1986).

The number of Milky Way stars contaminating the luminosity function (Fig. 2, left) is taken from Ratnatunga and Bahcall's (1985) tables. Crowding effects are estimated by determining the relative

J. E. Grindlay and A. G. Davis Philip (eds.),
The Harlow-Shapley Symposium on Globular Cluster Systems in Galaxies, 625–626.

number of stars per magnitude interval lost by INVENTORY among those randomly added to some fields sampling the cluster radially and by reproducing the star counts of King et al. (1968). The two methods give similar results, leading to a corrected luminosity function almost identical for the two sets of plates, in spite of their different resolution. This function, compared to other clusters after normalization to 50 stars between M_v = 5 and 5.5 mag, to avoid evolutionary effects (Fig. 2, left) runs between those of Ω Cen ([Fe/H] = -1.6) and M 15 ([Fe/H] = -2.1). McClure et al. (1986) have pointed out that the most metal-poor globular clusters have the steepest mass function. The theoretical luminosity function, obtained using VandenBerg and Bell's (1985) isochrones with Z = 0.0003 and a mass function $\xi(m) = \xi_0 m^{-(1+x)}$, is compared with the observations in the right panel of Fig. 2. To the limit of our photometry the theoretical curves do not allow discrimination of the value of the index x; there is only a marginal indication for the range 1.5 to 2.5.

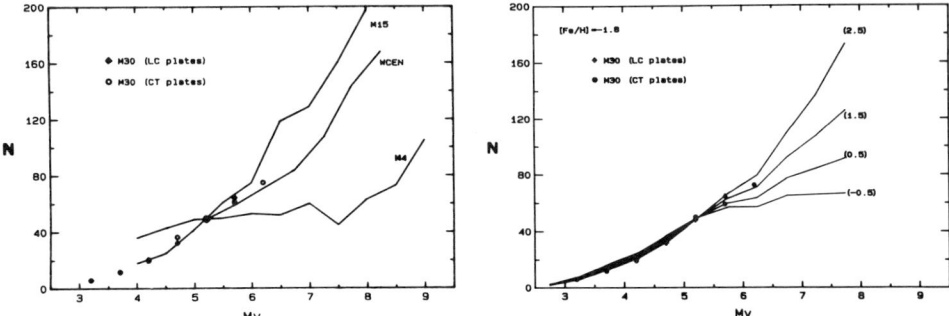

Fig. 2. Luminosity function of M 30 (left) compared with ω Cen (Ortolani 1986). M 15 and M 4 (McClure et al. 1986) and (right) with theoretical functions obtained form VandenBerg and Bell's (1985) isochrones (Z = 0.0003, Y = 0.2) using a power-law mass function with different values of the index x.

REFERENCES

Alcaino, G. 1978 Astron. Astrophys. Suppl. 33, 185.
Alcaino, G. and Liller, W. 1980 Astron. J. 85, 1330.
King, I. R., Hedemann, E. Jr. and Hodge, S. M. 1968 Astron. J. 73, 456.
McClure, R. D., VandenBerg, D. A., Smith, G. H., Fahlman, G. G., Richer, B. B., Hesser, J. E., Harris, W. E., Stetson, P. B. and Bell, R. A. 1986 Astrophys. J. Letters 307, L49.
Ortolani, S. 1986, private communication.
Piotto, G. 1986 Dissertation, University of Padua.
Ratnatunga, K. V. and Bahcall, J. N. Astrophys. J. Suppl. 59, 63.
Renzini, A. 1986, preprint.
VandenBerg, D. A. 1983 Astrophys. J. Suppl. 51, 29.
VandenBerg, D. A. and Bell, R. A. 1985 Astrophys. J. Suppl. 58, 561.
Webbink, R. F. 1985 in IAU Symposium No. 113, Dynamics of Star Clusters, J. Goodman and P. Hut, eds., Reidel, Dordrecht, p. 541.
Zinn, R. 1985 Astrophys. J. 293, 424.

AN AUTOMATED HR DIAGRAM FOR NGC 6809 (M 55)

Michael J. Irwin

Institute of Astronomy, Cambridge

Virginia Trimble

University of Maryland and University of California

ABSTRACT: For decades, star counts and HR diagrams extending below the main sequence turnoff in globular clusters meant the work of Sandage (1957). The advent of large CCD's at the foci of large telescopes has changed this (McClure et al. 1985, Harris & Hesser 1985, Christian & Heasley 1986, Heasley et al. 1986, Penny & Dickens 1986, Richer & Fahlman 1986, Smith et al. 1986) and made clear that clusters differ in the shapes of their luminosity functions and in the morphology of their HR diagrams. We return here to photographic methods, which can capture an order of magnitude more images and so possibly reveal new details.

J and R Anglo-Australian Observatory plates of the rich, open southern globular cluster M55 have been scanned and analyzed with the Automated Place Measurement facility of the Inst. of Astronomy and a new crowded-field algorithm (Irwin 1985). Almost 30,000 images were identifiable in both colors, of which slightly less than half represent cluster stars, extending nearly 3^m below the main sequence turnoff. Comparison of images in the two colors confirms earlier conclusions (Irwin & Trimble 1984) that (a) the luminosity function begins to flatten below $M_v = +5.5$ and (b) the radial profile, while generally well fit by a King model (with core and tidal radii of 2.2 and 26.6 pc at d = 4.8 kpc), shows several bumps and wiggles, of marginal statistical significance, that repeat from plate to plate.

The HR diagram (Fig. 1 is the raw data for the inner 16' or 22 pc radius) shows the usual features. The distribution of cluster stars around the ridge line of the main sequence is, at least, consistent with a sufficient population of close binaries to agree with a W UMa identification for the main sequence variables found in the earlier work. We have not yet attempted to fit evolutionary tracks or isochrones to the data, but the exceedingly sharp turnoff and a possible ridge-line job may prove interesting.

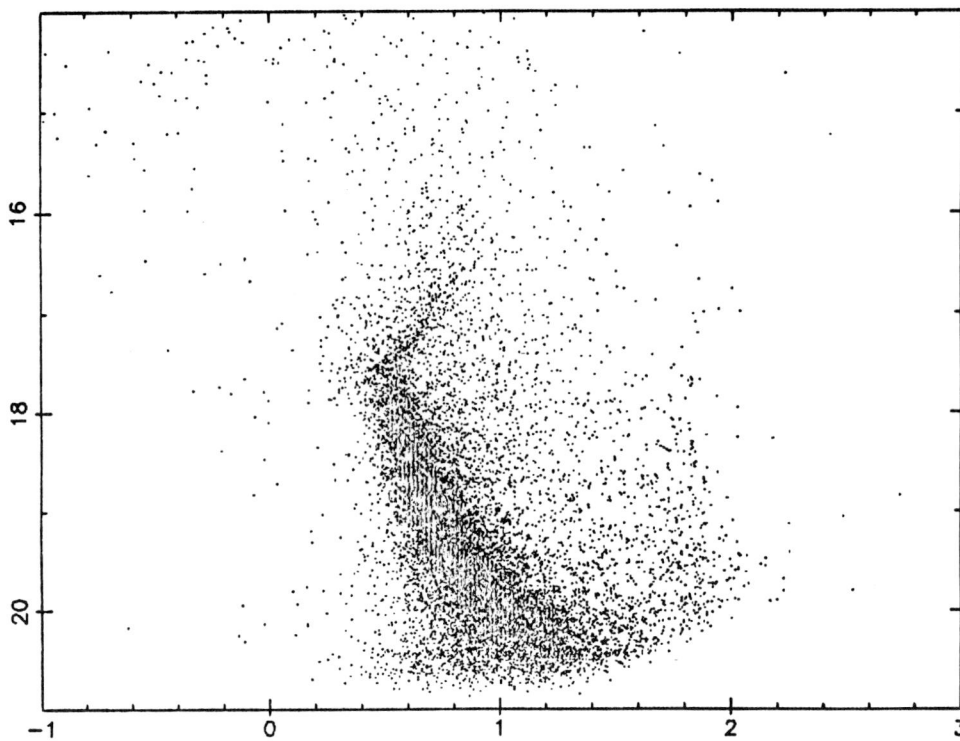

Fig.1. Automated HR diagram showing about 10,000 stars out to r=16' in the southern half of the cluster field. Axes are apparent magnitude R and color J-R. Diagonal bottom cutoff is the plate limit J=22 or M_J about 8.3 at the apparent distance modulus, 13.7, of M 55.

REFERENCES

Christian, C. A. and Heasley, J. N. 1986 Astrophys. J. 303, 216.
Harris, W. and Hesser, J.   1985   in IAU Symposium No. 113, Dynamics of Star Clusters, J. Goodman and P. Hut, eds., Reidel, Dordrecht, p. 81.
Heasley, J. N. Janes, K. A. and Christian, C. A. 1986 Astron. J. 91, 1108.
Irwin, M. J. 1985 Monthly Notices Roy. Astron. Soc. 214, 575/
Irwin, M. J. and Trimble, V. 1984 Astron. J. 89, 83.
McClure, R. D., Hesser, J. E., Stetson, P. B. and Stryker, L. L. 1985 Publ. Astron. Soc. Pacific 97, 665.
Penny, A. J. and Dickens, R. J. 1986 Monthly Notices Roy. Astron. Soc. 220, 845.
Richer, H. B. and Fahlman, G. G. 1986 Astrophys. J. 304, 273.
Sandage, A. R. 1957 Astrophys. J. 125, 422.
Smith, G. H., McClure, R. D., Stetson, P. B., Hesser, J. E. and Bell, R. A. 1986 Astron. J. 91, 842.

DEEP CCD PHOTOMETRY IN ω CEN AND NGC 3201[+]

S. Ortolani

Asiago Astrophysical Observatory

About 30 deep B and V CCD images, obtained at the 3.6 m ESO telescope, with the EFOSC in the focal reducer option, have been used to obtain color-magnitude diagrams and luminosity functions for the main sequence of the globular clusters ω Cen and NGC 3201. Two fields per cluster, at different distances from the center (25'- 30' for ω Cen and 7.5 - 9' for NGC 3201) were observed and separately reduced.

In Fig. 1-2 preliminary c-m diagrams of the inner fields are presented. Fig. 1 shows the ω Cen c-m diagram with VandenBerg and Bell (1985) isochrone superimposed (Y=0.2, [Fe/H]=-1.77, 16 Gyr). A reddening of 0.11 mag. and a distance modulus $(m-M)_V$=14.0 mag., as deduced from PLA method (Gratton, 1985), were assumed. The agreement is surprisingly good. The two lines on the lower blue side of the diagram are the mean lines for DA and DB white dwarfs assuming the same distance of the cluster (Sion and Liebert, 1977). The dots represent good stellar-shape images, possible white dwarf candidates of the cluster.

The c-m diagram of NGC 3201 is plotted in Fig. 2. A reddening of 0.21 mag. and a distance modulus $(m-M)_V$=14.15 mag. (Gratton, 1985, PLA method) were assumed to superimpose the VandenBerg and Bell isochrone with [Fe/H]=-1.3, Y=0.2 and 16 Gyr.

The luminosity functions, as derived from the c-m diagrams, corrected by crowding effects from experiments with artificial stars created with DAOPHOT, are represented in Fig. 3, compared with 47 Tuc (Ortolani, unpublished data) and with the theoretical predictions (McClure et al., 1986). The luminosity functions are corrected by field stars using the data from Ratnatunga and Bahcall's (1985) tables, after testing the consistency with the observed number of field stars on the red side of the main sequences.

They are consistent with an approximately Salpeter mass function index and much steeper than 47 Tuc luminosity function.

No appreciable variations have been found between the two fields.

(+) Based on observations obtained at the European Southern Observatory La Silla, Chile.

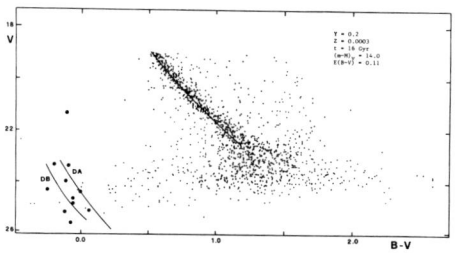

Fig. 1. c-m diagram of ωCen. The theoretical isochrone from VandenBerg and Bell (1985), is superimposed. The two lines on the blue side are DA and DB white dwarfs mean loci from Sion and Liebert (1977).

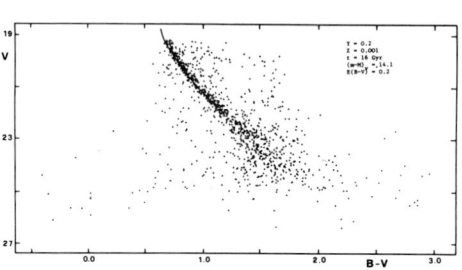

Fig. 2. c-m diagram of NGC 3201. The theoretical isochrone from VandenBerg and Bell is superimposed.

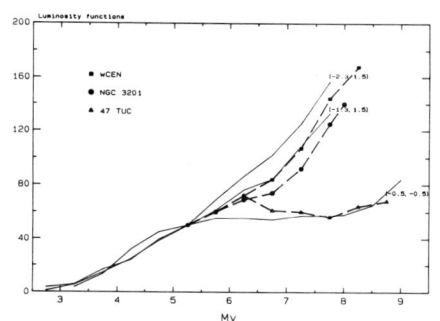

Fig. 3. Differential luminosity functions for the ω Cen, NCG 3201 and 47 Tuc main sequences, normalized at M_V=5.25. The continuous lines are the theoretical predictions from McClure et al. (1986): the numbers indicate metal abundance and power mass index.

REFERENCES

Gratton, R. 1985 Astron. Astrophys. 147, 169.
McClure, R. D., VandenBerg, D. A., Smith, G. H., Fahlman, G. G., Richer, B. B., Hesser, J. E., Harris, W. E., Stetson, P. B. and Bell, R. A. 1986 Astrophys. J. Letters 307, L49.
Ratnatunga, K. V. and Bahcall, J. N. 1985 Astrophys. J. Suppl. 59, 63.
VandenBerg, D. A. and Bell, R. A. 1985 Astrophys. J. Suppl. 58, 561.
Sion, E. M. and Liebert, J. 1977 Astrophys. J. 213, 468.

DEEP CCD PHOTOMETRY OF OMEGA CENTAURI

R. G. Noble[1], J. Buttress[1], W. K. Griffiths[1],
A. J. Penny[2], R. J. Dickens[2] and R. D. Cannon[3]

Physics Dept., University of Leeds   1
Rutherford Appleton Laboratory   2
Royal Observatory, Edinburgh   3

This paper presents a color magnitude diagram for the enigmatic cluster ω Centauri (NGC 5139 = C1328 - 472) tracing the main sequence down to V ~ 21.5. The spread in color on the upper main sequence is confirmed as intrinsic to the cluster. The CCD observations were made using the SAAO 1 m telescope with the UCL CCD camera and the RGO CCD camera at the prime focus of the AAT.

The fields studied lie approximately 20 arcmin. to the west of the cluster core (Cannon and Stewart 1981) and consist of three sets of observations: deep AAT exposures of one field with total exposure times of 1500 s in V and 1000 s in B, shorter AAT exposures overlapping the deep exposures, and a series of SAAO fields, with a range of exposure times. The analysis of the data was carried out using a profile fitting crowded field photometry program written by one of us (Penny and Dickens 1986). Fig. 1 shows the composite CM diagram for the region studied. The uncertainties in V and B, as derived from the scatter of the individual measures of the same star are approximately 0.02 at V = 18, increasing to 0.05 at V = 21.

Isochrones (VandenBerg and Bell 1985) corresponding to values of [Fe/H] of -1.75, -1.5 and -1.23 are plotted in Fig. 2. In Fig. 2a the isochrones have been shifted to represent a distance modulus of 13.9 and reddening of E(B-V) = 0.14. 18 Gyr isochrones are plotted. In Fig. 2b we show the same metallicities, but with an age of 16 Gyr. For these, E(B-V) = 0.17 and $(m-M)_V$ = 14.10.

REFERENCES

Cannon, R. D. and Stewart, N. J. 1981 Monthly Notices Roy. Astron. Soc. **195**, 15.
Penny, A. J. and Dickens, R. J. 1986 Monthly Notices Roy. Astron. Soc. **220**, 845.
Vandenberg, D. A. and Bell, R. A. 1985 Astrophys. J. Suppl. **58**, 561.

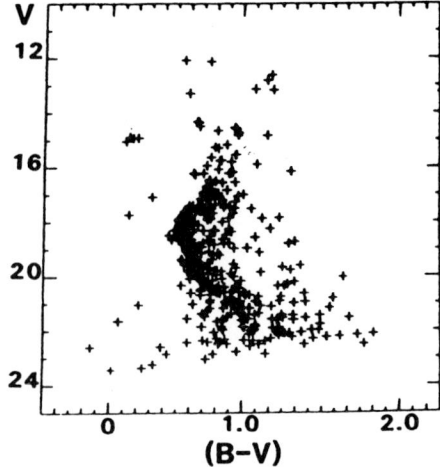

Fig. 1. ω Centauri color magnitude diagram.

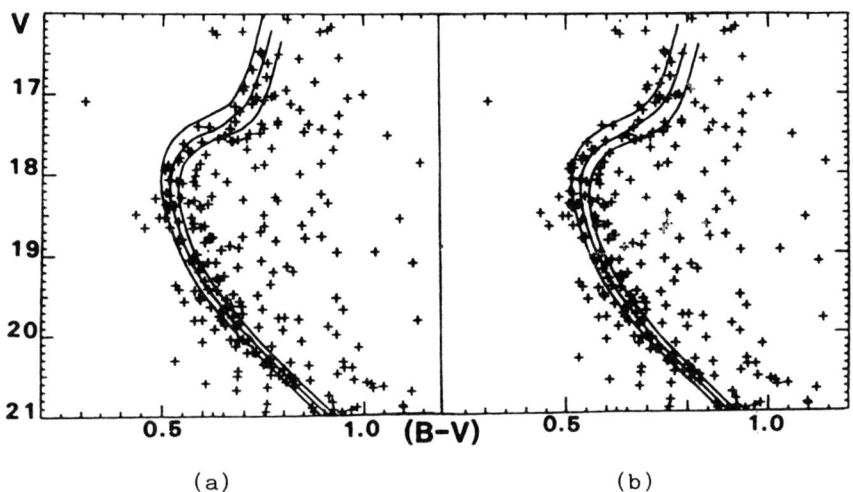

Fig. 2. Isochrones for [Fe/H] = -1.75, -1.5 and -1.23. In 2a the isochrones correspond to an age of 18 Gyr, the distance modulus being 13.9 and a reddening of E(B-V) = 0.14. In 2b they are for an age of 16 Gyr, a distance modulus of 14.1 and E(B-V) = 0.17.

THE AGES OF GLOBULAR CLUSTERS DERIVED FROM BVRI CCD PHOTOMETRY

Gonzalo Alcaino and William Liller

Isaac Newton Institute

Main sequence photometry has been obtained in the BV system for at least 36 galactic globular clusters (Peterson 1986). One of the most important objectives for these continued researches by several astronomical groups has been that of determining the ages of globular clusters. During the past five years, a general consensus has emerged that the spread in age among the known galactic globular clusters could be as small as 1 billion years; Burstein (1985) suggests that the average age of these objects lies between 14 and 17 Gys.

To obtain reliable ages for globular clusters, two aspects are basic: first, to secure accurate photometry to faint magnitudes; and second, to derive high-quality isochrones. The modern generation of electronic detectors such as the CCD have made improved photometry possible, especially at magnitudes near photographic plate limits. We believe as well that the measurements should be extended to longer wavelenghts, since there now exist BVRI synthetic isochrones provided by the work of VandenBerg and Bell (1985). Using several colors in matching observed and theoretical CMDs to derive cluster ages should be more accurate for several reasons: first, it reduces the importance of blanketing by absorption lines which are an important source of uncertainty in the computation of the isochrones and which are fewer in number in the red and infrared than in the blue. Secondly, the longer baseline in wavelength from the blue to the infrared should make isochrone fitting more accurate since details in the CMDs should stand out more clearly in, for example, (B-I) than in (B-V). Finally, by having several evaluations of the age of a given cluster, one can intercompare the results and arrive at a more reliable estimate of the accuracy of the final value.

With these considerations in mind, we have embarked on a program of CCD BVRI photometry of globular clusters near enough to the Earth for us to reach below the main sequence turn-off with medium-sized telescopes. We have at this moment completed the reductions of the following 5 globular clusters: NGC 104 (47 Tuc), NGC 2298, NGC 5139 (ω Cen), NGC 6121 (M4) and NGC 6362. All the BVRI CCD frames have been obtained with the 1.54 Danish telescope at ESO - La Silla. In order to minimize errors produced by photometric transfer and differential extinction we have established photoelectric standards in the same cluster fields with the 1 m telescope at La Silla. Using the ESO - Max Planck 2.2 m telescope at La Silla, we have as well obtained BVRI CCD frames of the following eight globular clusters: NGC 1851, NGC 1904, NGC 2808, NGC 3201, NGC 4372, NGC 4590, NGC 5946 and NGC 6139. The data are currently being reduced at La Silla and analyzed at the Isaac Newton Institute. The CMDs of the V mag versus the color indices, (B-V), (V-R), (R-I), (V-I) and (B-I) have been compared with the isochrones of Vanden-Berg and Bell (1985). Table I summarizes the basic results of the five clusters so far analyzed. We find that the ages derived for all of them are $17\pm1.5\times10^9$ y, again providing evidence that the globular cluster system is coeval, and that the epoch of the galactic contraction was short. These ages set a lower limit for the age of the universe, and thus an upper limit for the Hubble constant of $H_o < 58\pm5$ Kms^{-1} assuming $q_o = 0$.

REFERENCES

Alcaino, G. and Liller W. 1986a Astron. J. 91, 303.
Alcaino, G. and Liller W. 1986b Astron. Astrophys. 161, 61.
Alcaino, G. and Liller W. 1986c, submitted.
Burstein, D. 1985 Publ. Astron. Soc. Pacific 97, 89.
Harris, W. E. and Racine, R. 1979 Ann Rev. Astron. Astrophysics. 17, 241.
Peterson, C. J. 1986, private communication.
VandenBerg, D. A. and Bell, R. A. 1985 Astrophys. J. Suppl. 58, 561.

TABLE I

SUMMARY OF BASIC RESULTS

Cluster	V_{TO}	Color Turnoffs	ΔM(TO-HB)	$|Fe/H|$ Used for Isochrones	Age 10^9 y
NGC 104	17.60±0.1	B-V=0.56±0.02	3.5±0.2	-0.49	17.0±1.5
NGC 2298	19.50±0.1	B-V=0.60±0.02	3.4±0.2	-1.27	17.0±1.5
NGC 5139	18.30±0.15	B-V=0.55±0.03	3.8±0.2	-1.27	17.0±1.5
NGC 6121	16.80±0.1	B-V=0.81±0.02	3.40±0.2	-1.27	17.0±1.5
NGC 6362	18.75±0.1	B-V=0.50±0.02	3.40±0.2	-1.27	16.0±1.5

TURNOFFS AND AGES OF GLOBULAR CLUSTERS

F. Buonanno[1], C. E. Corsi[1] and F. Fusi Pecci[2]

Astronomical Observatory of Monte Mario, Rome   1
Department of Astronomy, Bologna   2

The way to arrive at (even relative) ages for globular clusters involves the determination of their (relative) distances. We would like to see a theory which would fit the absolute magnitudes of RR Lyrae stars as determined from observations (Sandage effect). We have examined a sample of 17 CM diagrams of galactic globular clusters, 11 of which were observed at ESO and reduced with the program, ROMAFOT and 6 of which were taken from the literature. In Fig. 1 the difference in bolometric magnitude between the turnoff point and the location of the zero-age horizontal branch (ZAHB), ΔV^{RR}_{TO} (bol) is plotted versus [Fe/H]. It turns out that $\Delta V^{RR}_{TO} \simeq \Delta V^{RR}_{TO}$ (bol) $+ 0.1 = 3.56 \pm 0.15$. We are faced with the problem of determining how the horizontal branch scales with metallicity in order to understand the constant value of 3.56 in this relation.

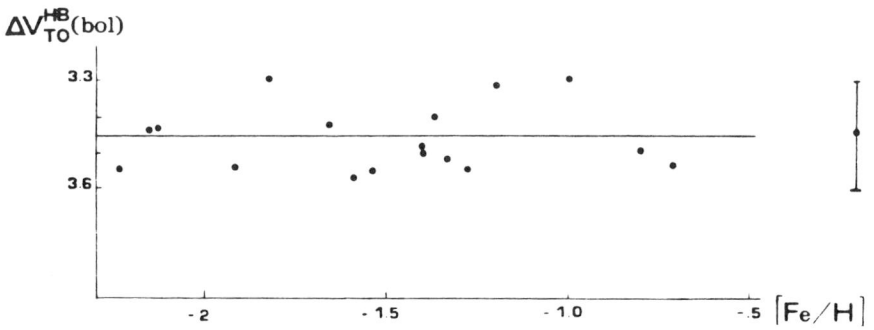

Fig. 1. The difference in bolometric magnitude between the turnoff point and the location of the ZAHB is plotted versus [Fe/H].

In studying the relative distances to the globular clusters we do the following: 1. We assume <u>for one single cluster</u> that the absolute magnitude is known and determine a distance modulus $(m - M)_V$ (possibly with a shift in the zero point). 2. We shift the lower main sequence of this cluster to the blue to take into account interstellar reddening. The main sequence of the reference cluster remains fixed in the M_v, $(B-V)_0$ plane. 3. We displace all the CM diagrams according to their reddening. 4. The CM diagrams are shifted in (B-V) color to adjust for differences in metallicity. 5. We determine relative distances by superimposing the lower main sequences. M^{RR}_v remains determined (with the quoted zero point error).

Sandage was the first to suggest this procedure in 1970. To carry out point 4 we used the following relation:

$$\Delta(B-V) = 0.38\Delta([Fe/H])^2 + 0.197\Delta[Fe/H]$$

interpolated using the models of **VandenBerg and Bell**. The fit to the distribution of the derived HB "absolute" magnitudes versus metallicity gives us a slope:

$$\Delta M^{HB}_v / \Delta[Fe/H] = 0.34.$$

This slope, together with the derived constancy of ΔM^{RR}_{TO}, strongly suggests a <u>constant age</u> for all the clusters in our sample.

The <u>internal errors</u> (calibrations, reddening, metallicity, level of the branches, estimate of the turnoff point, etc.) do not generally affect our conclusions since these are based on the properties of the sample as a whole. <u>External errors</u> (for example, an error in the models resulting in a metal-dependent color shift (cf. the first equation above) could alter our conclusions significantly.

THE DYNAMICS OF GLOBULAR CLUSTERS IN HIGH ECCENTRICITY ORBITS

R. K. Bhatia

Royal Observatory, Edinburgh

The aim of the present work is two-fold: 1) To study the effect of the tidal field of the Galactic bulge on a cluster in a high eccentricity plunging orbit with half-mass density nearly equal to the Roche density; 2) To estimate the tidal or limiting radius for such orbits.

Present evidence seems to support the idea that globular clusters move in high eccentricity orbits (e.g. Castellani and Melchiorri 1981; Caputo and Castellani 1984), although Innanen et al (1983) have argued for moderate eccentricity orbits. Tidal effects become very strong for clusters in high eccentricity orbits with small pericenter distances. For such an orbit, after pericenter passage, a cluster will lose appreciable mass and expand.

Representing the cluster by a 500-particle system, we have performed N-Body simulations using the code TIDAL kindly made available by Sverre Aarseth. The collision parameters are:

Ratio of cluster mass to mass of galactic bulge: 10^{-5}
Eccentricity of orbits studied: 0.6, 0.65, 0.68, 0.70, 0.8, 0.9.
Pericentre distance: Chosen such that the half-mass radius equals the tidal radius
          computed from King's formula (< 1 Kpc).

Fig. 1 displays the variation with eccentricity for a) $\Delta U/|U|$, the fractional change in the total energy of the system, b) $\Delta U_B/|U|$, the fractional change in energy of the bound system, c) $\Delta M/M$, the fractional change in mass, and d) the fractional change in half-mass density normalized to the density for e=0.9

Note the sharp increase near e=0.7; also the increase in $\Delta M/M$ falls below that for $\Delta U_B/|U|$, implying that the gain in energy is going more towards the expansion of the cluster rather than for expelling mass. Indeed, for e=0.6 the cluster disrupts completely.

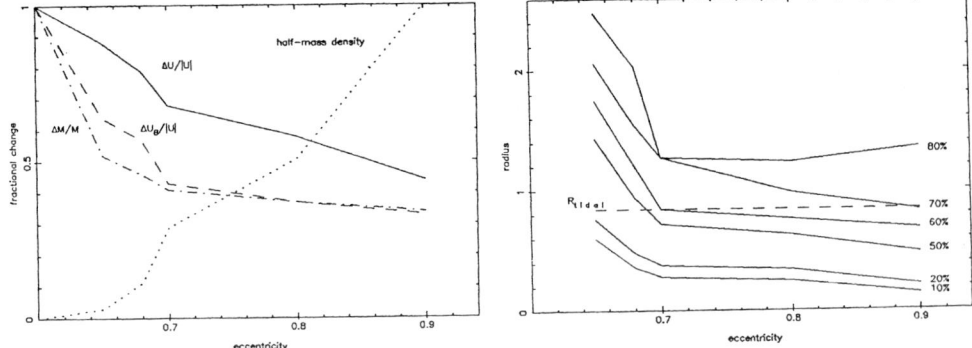

Fig.1. $\Delta U/|U|$ etc. vs. e   Fig.2 Radius containing 10% mass etc. vs. e

Fig. 2 displays the radius containing 10,20,50,70 and 80 per cent masses as a function of eccentricity. The dashed line indicates the theoretical tidal radius. The expansion increases with decreasing eccentricity: for e=0.68, the half-mass radius increases by a factor of 1.4 and the radius containing 80% mass by a factor of almost 3. Indeed, as the dotted line shows in Fig. 1, the half-mass density for e=0.68 is nearly 9 times less than for e=0.9. It is well known that halo globular clusters in the Galaxy have very low densities. Is it possible that some of them have had a close encounter with the Galaxy which has caused them to expand?

An idea of the tidal radius of the clusters in high eccentricity orbits can be obtained from Fig. 2. In all the cases, the final radius of the cluster is much larger than the theoretical tidal radius by a factor of up to 3. We therefore conclude that for halo clusters, the tidal radius computed from King's formula can be too small, depending on the eccentricity of the orbit and on the pericenter distance.

To summarise: 1) The final structure of a cluster after collision depends critically on the collision parameters and on the eccentricity of the orbit. If halo clusters move in orbits of high eccentricity, they can undergo considerable change in their structure after pericenter passage, leading to a considerable expansion of the cluster. 2) For such orbits, the tidal radius computed from King's formula is much smaller than that suggested by simulations.

REFERENCES

Caputo, F. and Castellani, V. 1984 Monthly Notices Roy. Astron. Soc. 207, 185.
Castellani, V. and Melchiorri, M. 1981 Astrophys. and Space Sci. 89, 289.
Innanen, K. A., Harris, W. E. and Webbink, R. F. 1983 Astron. J. 88, 338.

MASS DISTRIBUTIONS OF GALAXIES WITH GLOBULAR CLUSTER SYSTEMS

Kazutomo Takayanagi

Ryukoku University and Kyoto University

Density curves of polytrope index 5 were fitted to the surface density distributions of 26 globular cluster systems. Though we have shown only two of them in Fig. 1, the remaining systems resemble the above cases. No differences exist between elliptical and spiral galaxies.

If we can regard the globulars associated with a galaxy as tracers of mass distribution of the parent galaxy, this can be represented by that of the globular cluster system in spite of some discrepancy between the mass and the brightness distributions.

Assuming that the density distributions of the galaxies with globular cluster systems can be represented by those of the polytrope of index 5, we obtained masses of the galaxies by putting central velocity dispersions of these galaxies in the following formula;

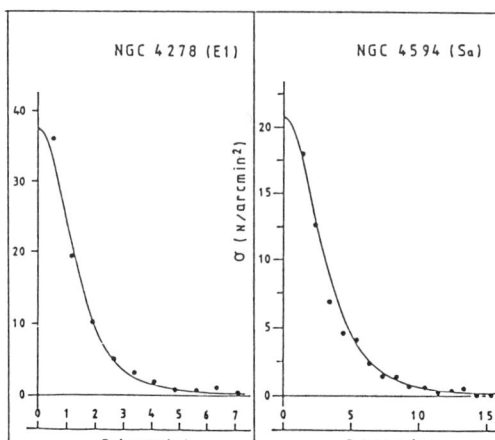

Fig.1. Surface density distributions of globular cluster systems with the best fit of the polytropes (solid lines).

$$m = \sqrt{3}\,\alpha^3 \rho_c = (2\sqrt{3}\,\alpha\,\overline{V_p^2})/G$$

where α is a scale factor which is obtained in the process of the fit mentioned above.

The results have been shown in Table 1. Masses of twelve ellipticals and one spiral (Sa) are order of $10^{12}\,m_\odot$. M87 is the most massive galaxy even in this study.

The masses obtained above and the various physical parameters which have been taken from catalogs and lists in other papers have been combined to examine the correlations among them though only two of them have been shown in Fig. 2, all the correlations obtained and their least square solutions are as follows;

Table 1
Masses of the galaxies with the globular cluster systems.

GALAXY (NGC) (1)	α (2)	$\overline{V^2}$ (Km/s) (3)	d (Mpc) (4)	MASS ($10^{10} m_\odot$) (5)
224	3.333	166	0.66	22
524	1.186	270	43	260
1052	0.800	204	24	56
1399	0.841	*250	22	81
2683	1.360	142	22	15
3226	0.597	207	20	36
3311	1.565	*250	55	380
3377	0.976	160	10	18
3379	0.800	218	13	35
3607	0.733	240	14	42
4278	1.155	243	14	68
4374	2.121	296	20	260
4406	3.259	256	20	300
4472	2.213	315	20	310
4486	3.983	335	20	610
4526	2.757	275	20	290
4565	1.302	136	20	34
4594	2.706	256	20	250
4621	2.684	225	20	190
4636	2.673	217	20	180
4649/4697	3.596	344	20	600
	1.788	186	20	87
5128	4.904	*100	5.0	17
5813	0.949	281	31	110
5846	1.163	250	29	150
Galaxy	3.142	*137	//	14

(1) $\log N_t = 0.91 \log m - 7.9$

(2) $\log m = 0.89 \log D_g^3 + 8.3$

(3) $\log m = -0.51 M_B + 1.2$

(4) $\log N_t = -0.37 M_B - 4.9$

(5) $\log N_t = 0.014 R_G + 2.4$

(6) $R_G = 2.3 D_g - 12$

(7) $\log N_t = 0.70 \log D_g^3 + 0.082$

(8) $m/L_B = 0.10 + 14$

(9) $M_B = -1.8 \log D_g^3 + 14$

where D_g means a geometrical mean of the major and the minor axis of the parent galaxy and R_G is the radius of the cluster system derived from the density curve.

We showed that the mass distributions of galaxies fit well the distributions of polytrope index 5. This fact suggests that the mass distributions of galaxies (halos) have been formed in the gas stage because the form of the polytropic density distribution is thought to be the result of collisions of the elementary particles of which it made.

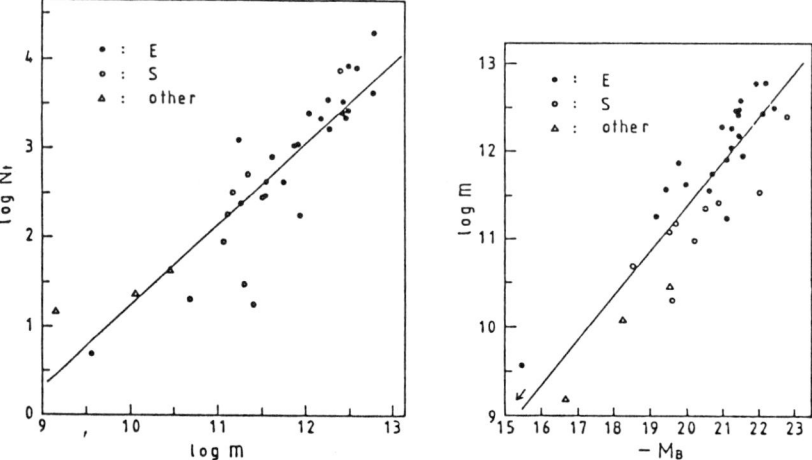

Fig. 2. Correlations of physical parameters included in our work.

THE DYNAMICS OF GLOBULAR CLUSTER SYSTEMS

Natarajan Ramamani

University of Edinburgh

ABSTRACT: This paper describes a project whose aim is to study the dynamics of a globular cluster system using an N-body code modified to include the gravitational field of an isothermal galaxy model. The galaxy and the globular cluster system have the same radii, are spherically symmetric and non-rotating. The evolution is to be followed up to a Hubble time.

1. INTRODUCTION

The dynamics of an isolated, self-gravitating N-body system has been reasonably well understood for single-mass models; but that of a subsystem like the globular cluster system has not received much attention. Studies of two- and multi-component systems has given some insight into the behaviour of subsystems (Aarseth 1985, Cohn 1985, and Inagaki 1985). An attempt to compare the globular cluster system of our galaxy with multi-component systems that have been studied so far shows that the relevant ratios of the individual masses and total masses of the two components, viz. the globular clusters and the stars constituting the parent galaxy fall well outside the range of ratios chosen by earlier studies. With (i) the mass of a globular cluster as $10^{(+5)}$ M$\odot$, (ii) the mass of the galaxy as $6 \times 10^{(+11)}$ M$\odot$ and (iii) the number of globular clusters as 200, the ratio of the stellar mass to globular cluster mass is $10^{(-5)}$ and the ratio of the mass of the globular cluster system to the mass of the parent galaxy is $3 \times 10^{(-5)}$.

2. AIM

We intend to follow the dynamical evolution of a self-gravitating N-body system under the additional gravitational field of a parent galaxy, after suitably modifying the simple N-body code to include its presence.

Initially it is planned to study (i) the density and velocity profiles, (ii) the rate of evolution using the radius of sphere containing a given fraction of total mass, (iii) velocity anisotropy, (iv) the distribution of orbital eccentricities, and (iv) the formation and disruption of binaries due to three-body encounters.

3. ASSUMPTIONS

1. The globular cluster system initially has (i) clusters uniformly distributed in radius, (ii) an isotropic velocity distribution and (iii) no rotation.

2. The parent galaxy (i) has a mass profile $M(r) = Kr$ out to some cutoff radius, (ii) is spherically symmetric and (iii) is non-rotating.

4. MODEL

The globular cluster system is modelled with 200 equal-mass particles and the galaxy with the value of K chosen to have a good fit with a standard mass model of our galaxy (Ostriker and Caldwell 1980). The cutoff radius is chosen to be 50 kpc for both.

5. RESULTS

This study has just begun and the preliminary tests show that the modified code conserves energy as well as the unmodified code does over a comparable period of time.

REFERENCES

Aarseth, S. J.   1985   in IAU Symposium No. 113, Dynamics of Star Clusters, J. Goodman and P. Hut, eds., Reidel, Dordrecht, p. 251.
Cohn, H.   1985   in IAU Symposium No. 113, Dynamics of Star Clusters, J. Goodman and P. Hut, eds., Reidel, Dordrecht, p. 161.
Inagaki, S.   1985   in IAU Symposium No. 113, Dynamics of Star Clusters, J. Goodman and P. Hut, eds., Reidel, Dordrecht, p. 189.
Ostriker, J. P. and Caldwell, J. A. R.   1980   in IAU Symposium No. 84, Large-Scale Characteristics of the Galaxy, W. B. Burton, ed., Reidel, Dordrecht, P. 441.

THIRD POSTER PAPER DISCUSSION

HANES: Welcome to the discussion of poster papers which fall under the somewhat fanciful title "Distant Globular Clusters". You will have seen in your inspection of these that in fact they represent quite a mixed lot. We have in total just over twenty papers to discuss and time is short, so let us begin. Our first paper is that of Zinnecker et al. There is a demonstrable consistency of colors between the nuclei and the extra-nuclear parts of some of the nucleated dwarfs in the Virgo cluster. On this basis, the authors suggest that Colin Norman's proposed mechanism of nucleus formation by mergers of clusters through dynamical friction may be credible. I have some small reservations about the results owing to the very difficult treatment for wavelength-dependent seeing in establishing nuclear colors. Comments and questions?

VAN DEN BERGH: At a given magnitude level, round dE galaxies have a higher probability of exhibiting nuclei than do flatter ones. This suggests that the nuclei might have been formed from low angular momentum gas.

ZINNECKER: I certainly agree that it is the rounder Virgo dwarf ellipticals that tend to have nuclei. However, the reason for this tendency could be either that the small ellipticity (indicative of little rotation) gives nuclei a chance to form (as suggested by Sidney van den Bergh) or that the existence of a nucleus <u>causes</u> the dwarf ellipticals to become secularly more spherical (as advocated by Colin Norman on the basis of the scattering of radial orbits off the central density cusp). It appears to be a chicken and egg situation. I have discussed this in a recent paper presented at the Rome Astrophysical Colloquium 1986.

DJORGOVSKI: If this merger picture is correct, then why do we see such nuclei in the diffuse dwarfs? They are the least luminous or least dense of all galaxies.

ZINNECKER: It seems that the low central mass density of the low surface brightness dwarf ellipticals allows star clusters to spiral to the center without being disrupted, while disruption will occur for galaxies with less shallow gravitational potentials.

SCHOMMER: George Lake and I looked at this dynamical timescale for the clusters in Fornax following a suggestion by Scott Tremaine some 10 years ago. We were worried that they all should be found in the nucleus, but when you treat the background density carefully, the dynamical friction time is ~1/2 Hubble time, which is consistent with the observed distribution.

HANES: Thank you. We move on now to a poster paper by Judy Cohen on the globular clusters of M 87. Her results have been discussed to a degree already particularly in John Huchra's review; let me just remind you that she finds small radii gradients in color both for the light and for clusters. More interestingly, she confirms Strom et al.'s finding that the median cluster color at given radius is everywhere bluer than the galaxy light, although she, in fact, started as a skeptic on this very point!

COHEN: The published version of this work will have luminosity functions and spatial distributions as well for three galaxies, NGCs 4406, 4472 and 4486.

LEE, H. M.: The color difference between globular clusters and underlying light may be consistent with the scenario of making the halos by destroying the globular clusters. The stars leaving the clusters are lower mass stars than those staying in the clusters.

TRIMBLE: What is the ratio of total Milky Way halo turnoff to that in the globular clusters?

VAN DEN BERGH: 100 - 1.

TRIMBLE: If so, then you cannot account for the field halo population, as a small factor of material shed from clusters. You need 99% of the total population, and so this cannot account for the field cluster color difference.

HANES: We now turn our attention to a paper by Lauer and Kormendy, one already published. They find the globular cluster distribution in M 87 to have a core an order of magnitude larger than that exhibited by the underlying light, and argue that this difference is innate. Comments?

LAUER: The core is large. We assume that the core is primordial only because we don't know how to make it now. We also believe that dynamic friction cannot make the observed nucleus by the central accretion of clusters; it fails by at least a factor of 10.

VAN DEN BERGH: Grillmair Pritchet and van den Bergh, find no radii variation of the luminosity function of globular clusters within the inner few arcmin. of M 87.

HANES: This would seem to confirm that there are no strongly mass-dependent disruptive processes as a function of radius within M 87.

HANES: Our next contribution by Held et al., reports U photometry of the globular clusters in M 87. These authors have detected the core alluded to in the previous paper, and perhaps I should recall here, that Rene Racine recognized the value of the U bandpass in sharpening the contrast of the clusters against the galaxy: a decade ago he detected the central surface density plateau on a 200-inch U plate.

Held et al. have nicely confirmed and quantified the effect, and will in due course present us with a fully calibrated U luminosity function.

We turn now to an exciting contribution by Thompson and Valdes: the discovery of globular clusters around the giant elliptical NGC 4874 in the Coma cluster. In his review talk this morning, Bill Harris also recorded success in this difficult endeavor, but there are some differences: Bill sees a central concentration, which Thompson and Valdes do not, and quotes a specific frequency of S ~ 30, while Thompson and Valdes find S ~ 2. Are these discrepancies worrying?

THOMPSON: The specific frequency determined by Frank Valdes and me for NGC 4874 is a <u>local</u> value and not a global value. In situations where the cluster population is more extended than the galaxy light, the local specific frequency increases as a function of radius. The global value mentioned by Bill Harris this morning is the most significant value scientifically but both values are "correct".

HARRIS, W.: I'd like to add first that Laird Thompson and I were aware of each other's results before the conference and the points of agreement were enough to make us each more willing to present them! My CCD material was taken at the prime focus of the CFHT with NGC 4874 centered on the chip, and so covers a wider field than Laird's material (but with lower resolution). I used the detected population in a rather wide annulus around 4874 and scaled it to an annulus of the same <u>linear</u> size around M 87, assuming Coma to be 5.5 times farther away than Virgo.

COHEN: Are you convinced that you have overcome the effects of clustering in the field of background galaxies?

HARRIS, W.: We're well aware of the problems involved in subtracting off background to find a residual excess of objects, and it's true that the density of faint background galaxies can easily vary by factors of 50% or so from one place to another. However, the detection of the cluster system in Coma is a factor of 3 or 4 above the background level at B > 25 and is not at all a marginal detection.

HUCHRA: Would either of you guys be brave enough to quote a distance modulus?

HARRIS, W.: No. The fit of the Coma luminosity function to that of M 87 is just a consistency fit and not a distance measurement.

HANES: Thank you. We next examine the presentation by Harris, Bothun and Hesser of the results of their search for globular clusters associated with edge-on spiral galaxies. Some systems have been detected, the most interesting being in NGC 7814, where the galaxy halo and the globular cluster system are both flattened perhaps by rotation. I invite questions.

SCHOMMER: I have a question concerning S vs. luminosity and environment. While Bill Harris this morning didn't speak about the late-type disk or irregular galaxies, we have in the Local Group a wide range of galaxy luminosities with well studied cluster systems. And the number of clusters increases steadily, so S remains reasonably constant.

HARRIS, H.: It is difficult to estimate the specific frequency S for spirals. When the projected cluster distribution differs from the spheroid light, a radius must be chosen at which to calculate S, or else assumptions must be made about the cluster distribution in the uncounted (central and outer) regions in order to estimate the total cluster population. In the late-type spirals NGC 5170 and NGC 5907, no central bulge is visible, and in all edge-on spirals decomposition of an exponential disk, thick disk, bulge and/or halo is difficult and model-dependent.

SCHOMMER: Do we know of any other group yet, which has been searched over a wide range of galaxy luminosities, or is Virgo the test case, at least for a cluster environment?

HARRIS, W.: I can't think of any other groups like the Local Group which have a similar or wider range of galaxy types and sizes, to measure cluster populations. It might be best to go right to Virgo to get a complete sample.

HANES: Hugh Harris has made the important point that the specific frequency S is at best ill-defined: do we normalize the cluster population against total galaxy mass, halo mass, or (perhaps equivalently) total or halo luminosity? For these disk systems in particular meaningful intercomparisons of S are difficult.

Globular clusters have also been detected in the S0 galaxy NGC 3115 and a luminosity function has been reported in a poster paper by Held and Capaccioli. It is exciting to see these results appear for so many new cluster samples. Would the authors care to comment?

HELD: We have not yet carried out tests for magnitude-dependent imcompleteness, and our luminosity function must be regarded as preliminary.

HANES: You do appear to reach the turnover in that function, which may provide a useful test of the universality of the globular cluster luminosity function.

That remark leads me naturally to our last "extragalactic" poster paper, by myself and Whittaker. We argue on the basis of Monte Carlo simulations that maximum likelihood statistical methods will permit globular cluster luminosity functions to be used as distance indicators to a precision of about 10% out to at least distance moduli of 32 mag from the ground. It is not even necessary to reach the turnover, provided the intrinsic shape of the luminosity function is indeed universal. Of course we would like as much reassurance as possible on that point!

STATLER Any claim that you can get H_o to 10% deserves to be challenged: your method depends on the assumption that the width of the distribution, σ, be universal. At the bottom of your poster you say that NGC 3379 suggests this may be true. Can you elaborate on this?

HANES: NGC 3379, a classical E0 galaxy, is the only one which has been studied deeply enough to permit the acid test. For it, $\sigma = 0.9 \pm 0.3$ formally, according to Pritchet and VandenBergh. This value is consistent with the value ($\sigma \simeq 1.2$) seen both in the Milky Way and in M 31. (The large error bar is because the sample is numerically small.)

LAUER: You say $\sigma = 0.9 \pm 0.3$ for N 3379, but for M 87 $\sigma = 1.8$. These are both ellipticals but they do not have universal σ.

HANES: I do not in fact claim $\sigma = 1.8$ for M 87; rather, I point out that within the uncertainties a range of σ values fit the deep vandenBergh et al. counts. My concern is that crowding, non-randomness in the background field, and the necessary incompleteness corrections at B > 24 may deform the observed function. I go into details in the submitted paper.

GRINDLAY: Would not a better use of your maximum liklihood technique be to establish the universality of the Gaussian form of the luminosity function and its value of sigma rather than to use it as a distance indicator?

HANES: That would be ideal but deriving σ really requires going well past the peak - it's not well constrained. It is very important to do this in a number of galaxies! For other galaxies, where we see the bright tail only, the only thing one can sensibly do is adopt σ and derive a distance.

KING: This morning the use of a Gaussian luminosity function made me uncomfortable; now you are really scaring me, because you rely so much more on the detailed accuracy of a Gaussian. Let's remember that Hubble's original luminosity function for galaxies was also Gaussian.

HANES: That's true, but the census of globular clusters in the Milky Way is quite different from the galaxy data Hubble worked with a half century ago. Moreover, we have deep complete functions in the M 31 halo. But the Gaussian merely represents an empirical fit, which works very well; I do not claim any special reason for its use other than practicality. It may eventually need modification.

Well, we must move on - back into our own Galaxy in fact, for a series of papers on color-magnitude diagrams. The first of these is by Samus, who presents photographic photometry for several globular clusters. One concern I have is that of calibration: two clusters are reported to have a 4 magnitude gap between the horizontal branch and the turnoff, and reference is made to a third, ω Cen, discussed by Cannon. But Cannon withdrew that conclusion, which was based on some simple error; Don VandenBergh has shown that a value of 3.5 mag seems to be nearly universal. Would the author care to comment?

SAMUS: I am not as strong as I used to be in my opinion that ages do differ, though I am not alone in regarding age as a possible second parameter. Photographic photometry can still be of some use, especially if we measure many more stars than do the observers using CCD's.

HANES: Certainly the panoramic detectors do provide very large samples, and we welcome these results.

Our next paper, by Buonanno et al., presents photometry for 10,000 stars in M 3. Interestingly, one sees here that the AGB, HB, and RGB stars all follow the mean luminosity distribution while the blue stragglers are *less* (rather than more) concentrated. This argues against their being mass-exchange binaries, which might settle towards the cluster center. I invite comments.

FUSI PECCI: Using plates it was impossible to get highly accurate photometry down to the turnoff in the central region of the cluster. This implies that we have no direct information on the presence, number, and spatial distribution of the blue stragglers there. Within the anulus we measured, BSs seem to be more frequent in the outer regions, contrary to the evidence found in NGC 5466 (a very loose cluster studied even in the very central region). A detailed comparison and discussion will be however possible only after the complete coverage of the central regions of M 3 we are currently obtaining with CCD cameras.

KING: I hope that you will use this material to make an explicit study of crowding corrections.

FUSI PECCI: Crowding has been of course a problem in the reductions and we made many tests and checks to reduce and evaluate its influence. We don't believe that crowding might have affected at any significant level the projected spatial distribution of blue stragglers we have derived.

HANES: In an attractive presentation Richer and Fahlman have described the study of three fields in M 5 at 8, 21, and 58 core radii. The similarity in CM diagrams argues against any metallicity gradient, and the luminosity functions are likewise similar to V = 23. Differences fainter than that may bespeak mass segregation.

RICHER: The evidence for mass segregation at the faint end must be still regarded as somewhat tentative because (1) the number of stars entering into the luminosity function for the outer field is quite small and (2) the incompleteness correction for the inner field is large (but well determined).

HANES: No further comments? Then I invite discussion of Piotto et al.'s study of 3,000 stars in M 30. Those authors find consistency with McClure et al.'s suggestions: this low metallicity cluster seems to have a steep luminosity function, although their data do not go quite deep enough to provide absolute confirmation.

ORTOLANI: Our points are limited in magnitude but reliable due to the large number of stars.

RICHER: Fahlman and I have a luminosity function for M 30 done with a CCD at CFHT and the slope of the fitted power low mass function is very flat, in fact the best fit is X = 0.0. This should not necessarily be taken as a counter example to the McClure et al. result that the more metal poor the cluster the steeper the mass function (M 30 is very metal poor.)

VANDENBERG: Regarding the consistency of observed LF's with the McClure et al. (1986) result that the slope of the mass function varies inversely with [Fe/H], it is important to note the following. Theoretical LFs show a turnup at the faint end $M_v \sim 7.5$, depending on metallicity) due to a flattening of the mass-luminosity relation - independent of the shape of the mass function. Therefore, if a turnup is observed, then it may be reasonable to assume that the present day mass function is similar to the IMF. Certainly, if a turnup is not observed, then there must have been significant evaporation of low mass stars. Therefore, to test whether the observed cluster LFs follow the McClure et al. relation, they must be extended to sufficiently faint magnitudes. Since the M 30 data do not reach $M_v = 7$, it is premature to conclude that it is an exception to the rule.

McCLURE: Is it not true Harvey that the incompleteness corrections in your last points are very large?

RICHER: Its a factor of 3 in the last point...but that can be dropped without changing the result at all.

McCLURE: A factor of 3 correction is very large!

VAN DEN BERGH: If memory serves we correctly the M 30 has the bluest integrated (B-V) color of any luminous globular cluster. A reddening as large as $E_{B-V} = 0.13$ would make its intrinsic $(B-V)_o$ color quite unusual.

HANES: I now draw your attention to an automated photometric study by Trimble and Irwin of NGC 6809. I can claim to have had a hand in this, since I took the AAT plates used in their work, which was accomplished using the APM in the other Cambridge. The derived main sequence is wide enough to admit the existence of binaries of the ω UMa type.

TRIMBLE: We set out to look for main sequence variables and found about eight candidates, several of which were found independently by M. Hazen. They have periods consistent with ω UMas. These should be followed up for light curves, but this is not likely to be done by any of us involved in the project.

HANES: I propose to discuss the next two papers together: each of them (one by Ortolani, the other by Noble et al.) reports a study of ω Cen. Ortolani presents tantalizing hints of the detection of white dwarfs in ω Cen, although I am puzzled that these occur in only one of his fields. The other study, based in part on 1-meter observations from South Africa, reports that the main sequence of ω Cen is wide below the turnoff; in this way it is like the giant branch, the width of which is well established.

ORTOLANI: Only one field (the innermost one) was fully reduced. The number of possible white dwarfs is consistent with theoretical predictions. Quasar contamination could be a problem.

NOBLE: I'd like to reiterate the intrinsic width of the main sequence, in this study shown to be present to below $V = 20$. Subsequent data to much fainter limits ($V = 24$) confirm this spread to those limits. These deeper data do not show white dwarfs, although they do reach the region where the very brightest would show up.

HANES: Thank you. Alcaino and Liller suggest, in our next paper, that age determinations for globular clusters are best made through use of V, (B-I) color magnitude diagrams. Isochrones are now available which incorporate the redder bandpasses, and their figures show that the shallower form of the turnoff makes fitting the data formally easier.

Da COSTA: I point out that CCD's have wide spectral response which often is not taken advantage of as much as it should be.

PENNY: I tried some (B-V) and (B-I) work and found little difference in resultant accuracy. I agree that (V-R) is a good regime.

Da COSTA: I find that observations in B and R give better precision than observing B, V. The errors in V and R are usually comparable, but for given change in T_{eff}, the change in (B-R) is almost twice as large as for (B-V).

HESSER: Don VandenBerg has emphasized the importance of matching isochrones in the subgiant region, yet in the 47 Tuc CM diagram shown the Alcaino/Liller data appear to cross the isochrones. Do you see that effect in any of your data?

Da COSTA: I cannot really comment, as I do not have any data of precision to equal that of Hesser and Harris' V, (B-V) data for 47 Tuc.

COHEN: You don't win using broad band photometry using CCD's beyond r. The atmospheric emission in the near infrared accessible to CCD's is so great that you cannot go deeper or get more accurate photometry than if you stay below 7500 Å.

HANES: Our last observational paper, by Buonanno et al., reports a study of cluster ages. These authors have dereddened and deblanketed cluster CMDs. Subsequent main sequence fitting of the clusters, one with another, then reveals the dependence of HB luminosity on metallicity, which in turn (via ΔV_{TO}^{HB}) yields cluster ages. They find ages of ~ 16 Gyr, with no significant spread.

BUONANNO: I want to remind you that, as there is no evidence of varying ΔV_{TO}^{HB}, constant ages for globular clusters imply the existence of the Sandage effect.

HANES: Finally, we turn our attention to several papers of a more theoretical nature. Bhatia has carried out N-body simulations of clusters in eccentric orbits around a point mass galaxy to aid our understanding of cluster disruption. Any comments?

BHATIA The expansion of a cluster depends on the eccentricity of the orbit. For $e = 0.6$ the cluster disrupts while for $e = 0.65$ the cluster expands. Perhaps this expansion can explain sparse clusters in the halo.

HANES: Next, I draw your attention to a contribution from Takayanagi, who has assumed that globular clusters trace the mass of a galaxy and that the distribution is best matched by polytropes of index 5 to derive masses for various galaxies. However, I am puzzled that he remarks upon the inadequacy of de Vaucouleurs-type ($r^{1/4}$) fits to the cluster distributions: my experience is that such fits are very successful.

Finally, then, I invite comments on a program described in a poster by Ramamani. N-body simulations of clusters within an isothermal galaxy will be studied. I gather that this is a report on work in progress?

RAMAMANI: Due to problems with energy conservation in the modified code, the results could not be presented. I hope to publish these in the near future.

HANES: Then we wish you success and look forward to seeing those results in due course.

And my thanks to all the participants - authors and attendees - for their evident interest in the poster papers we have been discussing this afternoon. We have seen some exciting results.

Chapter X

Poster Papers

Formation and Evolution of Globular Clusters

Judy Cohen fields the discussion after Roger Bell's talk

The poster papers received due attention and discussion

A MULTICOLOR CCD SURVEY OF SOUTHERN GLOBULAR CLUSTERS

Juan C. Forte[*]

Institute of Astronomy and Physics of Space, Buenos Aires
Faculty of Science, La Plata

Mariano Méndez[*]

Faculty of Science, La Plata

The existence of deviations of the observed photometric profiles with respect to the King models is a known fact (e.g., Newell and O'Neil, 1978; Aurière, 1983) and has been recently emphasized in the large survey by Kron et al., 1984. The origin of the light "excesses" (or defects) in the nuclear regions is, however, not very well understood. This work aims at clarifying this problem on the basis of multicolor ($BVRI_{KC}$) CCD observations.

Observations were carried out with the 0.9m telescope at CTIO, including a range of exposure times and also off-center images for a good estimate of the sky level. A set of H-alpha frames (taken with a narrow filter centered on $\lambda = 6563$ Å) was also secured. Data reduction and handling was performed with the "Quasi-Interactive Image Processing System" developed at La Plata Observatory.

The overall photometric centers of the clusters were determined using different techniques (profile folding on resolution degraded images, etc.) and each profile was fitted with a King curve. We found a very good agreement between our derived concentration radii and those given in the literature (e.g., $\pm 0\rlap{.}''5$).

A summary for some southern clusters is given in Table I. In brief, these clusters can be classified as: "Perfect" King clusters, clusters with light excesses, post-collapse objects (i.e. with slope -1 in the surface brightness vs. radius diagram; Djorgovski and King, 1984) and finally, clusters with defects of light. No significant gradients were found in the peripheral regions of any cluster and, when present, they were confined to the innermost regions, a situation similar to that found by Peterson, 1986, or Forte and Méndez, 1984.

An interesting result is that, after a preliminary remotion of seeing effects, the regions showing excesses of light, have very similar linear sizes. Adopting the scale of distances given by Webbink, 1984, the average half radius for 8 clusters results 0.14 ± 0.05 parsecs (for clusters spanning core radii from o.16 to 1.86 pc). Furthermore, the in-

---
[*] Visiting Astronomer, Cerro Tololo Interamerican Observatory, operated by AURA, under contract with the National Science Foundation.

tegrated absolute magnitudes of the excesses M_v are also similar, yielding $M_v = -3.1 \pm 0.4$. NGC 2808, a cluster observed with a different technique (Forte and Méndez, 1984) has also a comparable light excess with $M_v = -3.4$.

Unfortunately, integrated colors for these regions (subtracting the underlying King profiles) do not help in identifying their nature. While NGC 362 (and NGC 2808) have (B-V) colors redder than the average for the cluster, the remaining clusters show bluer colors in their nucleus. The color difference in B-V ranges from zero to -0.23. These values are not correlated with relaxation time, metallicity, absolute magnitude, or Mironov's index (taken from Zinn, 1980), which is a description of the horizontal-branch morphology. An alternative way to produce a nuclear "blueing" would be the presence of hot gas. However, a large range of "astrophysical" gas temperatures would originate an H-alpha contribution which, at our detectability levels, is not present in any cluster. The small linear sizes of the regions with light excesses demands a very careful treatment of seeing effects in order to reveal the existence of gas emission, which is under way.

We did not detect extended H-alpha emission in a number of clusters which have been reported as having emission in the literature. On the other hand, we found (by simulating concentric aperture photometry) that centering errors may originate spuriously large H-alpha excesses, particularly in small core radii clusters.

TABLE I
Southern Globular Clusters CCD Survey

Object	Type	H.W.	Mv(excess)	Δ(B-V)	Observations
NGC 362	Excess	0.130	-3.40	+0.26	
NGC 1261	Excess	0.216	-3.02	-0.11	
NGC 1851	Excess	0.170	-3.16	+0.08	
NGC 5824	King	--	--	--	
NGC 6266	Excess	0.148	-3.24	-0.23	
NGC 6388	Excess	0.160	-3.01	--	
NGC 6624	Post-collapse	0.086	-3.69	+0.00	Djorgovski and King
NGC 6723	Deffect	--	--	--	
NGC 6752	Post-collpase	0.144	-3.10	+0.01	Newly found
NGC 7099	Post-collpase	0.072	-2.46	-0.13	

H.W. : Half width of the light excess in pc.
Mv : Absolute visual magnitude of the excess.
Δ (B-V) : (B-V) difference between the excess and the cluster.

REFERENCES

Aurière, M. 1983 *Thesis, University of Pierre et Marie Curie*.
Djorgovski, S. G., King, I. R. 1984 *Astrophys. J. Letters* **277**, L49.
Forte, J. C. and Méndez. M. 1984 *Astron. J.* **89**, 648.
Kron, G. E., Hewitt, A.V. and Wasserman, L. H. 1984 *Publ. Astron. Soc. Pacific* **96**, 193.
Newell, E. B. and O'Neil, E. J. 1978 *Astrophys. J. Suppl.* **37**, 27.
Peterson, C. J. 1986 *Publ. Astron. Soc. Pacific* **98**, 192.
Webbink, R. F. 1985 1985 in *IAU Symposium No. 113, Dynamics of Star Clusters*, J. Goodman and P. Hut, eds., Reidel, Dordrecht, p. 541.
Zinn, R. 1980 *Astrophys. J.* **241**, 602.

THE STRUCTURE OF COLLAPSED CLUSTER CORES

Phyllis M. Lugger and Haldan Cohn

Indiana University

Jonathan E. Grindlay and Charles D. Bailyn

Harvard University

Paul Hertz

E. O. Hulburt Center for Space Research, NRL

ABSTRACT. In order to test the prediction that many Galactic globular clusters have undergone core collapse (Lightman 1982, Cohn and Hut 1984) and should therefore have central surface brightness cusps, we have obtained UBVR CCD frames of the cores of 72 clusters. We present and analyze U-band surface brightness profiles for three clusters: one "control cluster" with a normal flat core profile — NGC 6388 — and two with central power law cusps — NGC 6624 and M15 (NGC 7078).

1. INTRODUCTION

The predicted observational signature of a cluster that has undergone core collapse is a central surface brightness cusp — a surface brightness profile that continues to rise as a power law at small radii in contrast to a normal King type profile that becomes flat within the cluster core. During the past several years there has been mounting observational evidence for central surface brightness cusps in some globular clusters. Djorgovski and King (1986) have recently reported definite cusps in 21 clusters and possible cusps in 7 others out of 123 clusters surveyed, mostly using CCD data in the BVR bands.

2. OBSERVATIONS

We have concentrated on obtaining U-band frames of globular cluster cores, in addition to BVR frames, using RCA CCD chips on the CTIO 4 m, CFH 3.6 m, and KPNO #1 0.9 m telescopes. U-band surface brightness profiles are less dominated by individual red giants and are thus less "noisy" and more representative of the underlying mass distribution than are longer wavelength profiles (King 1985). To date we have obtained data for 72 clusters, with at least one complete set of UBVR frames for most clusters. We present U-band profiles for NGC 6388, which has been cited as a prototypical normal cluster by Djorgovski and King (1986), and for NGC 6624 and M15, which have both been reported to show central surface brightness cusps in several previous studies. We observed NGC 6388 and NGC 6624 from CTIO in May 1985 and observed M15 from KPNO in May 1984. We also observed the latter two from CFH in July 1986; analysis of these new data is underway.

3. RESULTS

We have determined cluster centers and surface brightness profiles from our CCD frames using software kindly provided by S. Djorgovski and I. King. The U-band images of NGC 6624 and M15 show central surface brightness "spikes" about 2-3" in radius which are centered within 0.5" of the autocorrelation cluster centers. Our profiles (Fig. 1b,c) indicate that these spikes represent the central few arcsec of power law cusps that extend out to about 10" in both clusters.

We first attempted to fit the entire profiles of the clusters with seeing-convolved King models, obtaining a reasonable fit for NGC 6388 (Fig. 1a). No acceptable King model fits to the entire profiles of NGC 6624 and M15 could be obtained. We next fit the inner profiles of the clusters with seeing-convolved power laws, varying the outer radius of the fit region from 5" to 40". While there is no particularly good fit of this model to the inner region of NGC 6388, it gives a good fit to the central regions of both NGC 6624 and M15 out to a radius of about 10", indicating that the the core radii of these clusters are unresolved at a level of about 0.75" (the seeing HWHM). The best fit power law slopes of -0.77 for NGC 6624 and -0.64 for M15 are significantly flatter than the value of -1 predicted for a post-collapse cluster of identical stars. This indicates the presence of nonluminous remnants — possibly massive white dwarfs — that are more massive than the stars that dominate the luminosity profile (Lee, this volume).

We find that King models provide excellent fits to the *outer* ($r \gtrsim 10"$) regions of NGC 6624 and M15 (Fig. 1b,c), as expected from computer simulations of clusters undergoing core collapse (Cohn 1980).

REFERENCES

Cohn, H. 1980 Astrophys. J. 242, 765.
Cohn, H. and Hut, P. 1984 Astrophys. J. Letters 277, L45.
Djorgovski, S. and King, I. R. 1986 Astrophys. J. Letters 305, L61.
King, I. R. 1985 in IAU Symposium No. 113, Dynamics of Star Clusters, J. Goodman and P. Hut, eds., Reidel, Dordrecht, p. 1.
Lightman, A. P. 1982 Astrophys. J. Letters 263, L19.

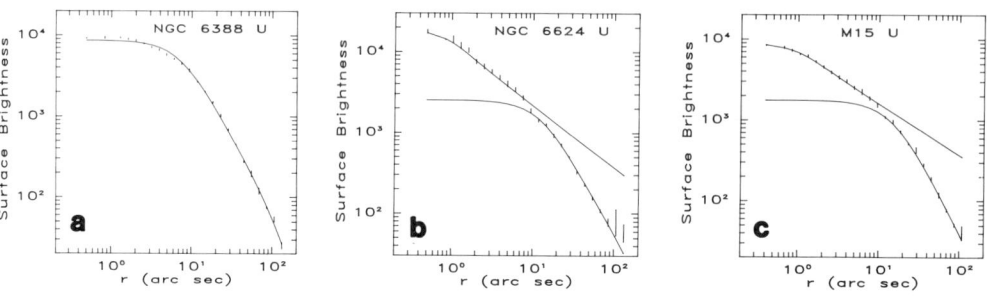

Fig. 1. King model fit to entire profile of NGC 6388; power law fits to inner regions and King model fits to outer regions of NGC 6624 and M15.

RADIAL VELOCITY STUDY OF NGC 6712

J. Grindlay, C. Bailyn, R Mathieu and D. Latham

Harvard/Smithsonian Center for Astrophysics

ABSTRACT: We report MMT Echelle radial velocity observations of 52 giants within 3 core radii of the center of NGC 6712. The mean radial velocity of these stars is -107.5 km/s, with a line of sight velocity dispersion of 4.0 km/s. We use these data, together with CCD photometry of the cluster, to derive a mass to light ratio for the center of the cluster of 0.7, an unusually low value.

1) VELOCITY DISPERSION

NGC 6712 is an unusual globular cluster in that it is the only cluster with a high luminosity X-ray source which is *not* centrally condensed. If tidal capture of neutron stars is the mechanism responsible for the X-ray binaries, as is generally accepted, one would expect that they would preferentially arise in dense clusters. NGC 6712 is therefore a particularly strong candidate for a cluster which is currently in a state of post-core-collapse reexpansion, since the X-ray source could then have been created when the cluster was denser at some time in the past (Grindlay, 1985). There may be dynamical differences between collapsing and reexpanding clusters (Bettweiser and Sugimoto 1984); with this in mind, we undertook a study of the radial-velocity dispersion of NGC 6712.

74 giants within three core radii of the center of NGC 6712 ($13.4 \leq m_V \leq 15.2$) were observed with the MMT Echelle spectrograph, to an accuracy of $\approx$ 1 km/s. Of the stars observed, 52 were found to have velocities between -116 km/s and -98 km/s and we consider these to be cluster members. Several of these stars were observed more than once; none showed any significant velocity variation. The other stars all had velocities more than 50 km/s away from these values. All of these stars were reobserved; no significant velocity changes were found, and we conclude that they are not members.

The 52 cluster members have a mean velocity of -107.5 km/s and a line of sight velocity dispersion of 4.0 km/s. The radial distribution of the velocity dispersion is consistent with an isothermal distribution. No sign of rotation is detected in the cluster.

2) MASS-TO-LIGHT RATIO

We have also obtained CCD photometry of the central regions of the cluster as part of a UBV photometric survey of X-ray globular clusters (Bailyn et al. 1986). The CCD frames allowed us to determine several other cluster parameters. We find that the central surface brightness in V is 18.48 magnitudes per square arcsecond, which is within 0.02 magnitudes of the value quoted by Webbink (1985).

From the radial-velocity distribution, central surface brightness, and core radius ($\approx 40''$ Cohn and Lugger, private communication) one can derive a value of M/L at the center of the cluster which is relatively model independent (see Richstone and Tremaine, 1986). Richstone and Tremaine's Equation 1 can be written in the form

$$\left(\frac{M}{L_V}\right)_0 = \eta\, 5.2 \times 10^{-32} \times \left(\frac{\sigma^2 d}{R_{hb} 10^{.4(D_V + V_\odot - l_0)}}\right)$$

where σ is the line of sight velocity dispersion in cm/s, d is the distance to the cluster in cm, R_{hb} is the half brightness radius of the cluster in arc seconds, D_V is the observed distance modulus to the cluster in V, $V_\odot = 4.8$ is the absolute V magnitude of the Sun, l_0 is the central surface brightness of the cluster in V magnitudes per square arc second, and η is a numerical factor very close to unity for a wide variety of dynamical models. Using the values given above of $\sigma = 4$km/s, $R_{hb} = 40''$, $l_0 = 18.48$, $D_V = 15.6$ and $d = 6.2$kpc (Webbink 1985), we derive a value of M/L of 0.7. The errors in this value come primarily from uncertainty in the velocity dispersion and the core radius, and are probably around 20%. This is one of the lowest mass-to-light ratios yet found for a globular cluster (see Pryor et al. 1987) which would indicate that there is relatively little dark matter in the center of NGC 6712.

ACKNOWLEDGEMENTS: We thank H. Cohn and P. Lugger for performing the surface density profile, and useful discussions. This work was partially supported by NSF grant NSF-AST-84-17846.

REFERENCES

Bailyn, C. D., Grindlay, J. E., Cohn, H. and Lugger, P. M. 1986, in preparation.
Bettweiser, E. and Sugimoto, D. 1984 Monthly Notices Roy. Astron. Soc. 208, 493.
Grindlay, J. E. 1985 in IAU Symposium No. 113, Dynamics of Star Clusters, J. Goodman and P. Hut, eds., Reidel, Dordrecht, p. 43.
Pryor, C., McClure, R. D. and Hesser, J. E. 1987 in IAU Symposium No. 126, Globular Cluster Systems in Galaxies, J. E. Grindlay and A. G. D. Philip, eds., Reidel, Dordrecht, p. 661.
Richstone, D. O. and Tremaine, S. 1986 Astron. J. 92, 72.
Webbink. R. F. 1985 in IAU Symposium No. 113, Dynamics of Star Clusters, J. Goodman and P. Hut, eds., Reidel, Dordrecht, p. 541.

A SURVEY OF GLOBULAR CLUSTER VELOCITY DISPERSIONS

Carlton Pryor[*]

University of Victoria and Vanderbilt University

Robert D. McClure[*], J. M. Fletcher[*] and James E. Hesser

Dominion Astrophysical Observatory
Hertzberg Institute of Astrophysics

We have used the Dominion Astrophysical Observatory radial velocity spectrometer on the 3.6m Canada-France-Hawaii telescope to obtain radial velocities, accurate to ~0.8 km/s, for ~20 stars in each of nine globular clusters. The stars are generally within three core radii of the cluster center. The cluster names and metallicities (the latter are averages of values in Zinn and West (1984), Pilachowski (1984), and Webbink (1985)) are given in Table I. This sample includes two clusters with cusps in their surface brightness profiles: NGC 6624 and 6681.

We fit multi-component King models with isotropic velocity dispersions to our velocity data and surface brightness profiles taken from the literature. The velocity data are fitted using Gunn and Griffin's (1979) maximum likelihood technique, which removes the contribution of the measurement uncertainty. The surface brightness profiles are fitted using least squares. Uncertainties in the cluster parameters are determined by fitting many simulated data sets. We do not include the large uncertainty in the cluster distances. The distances are calculated assuming that the RR Lyrae variables have $M_V = +0.6$. To be acceptable, a model must be a good fit to the surface brightness profile and the mass-to-light ratio (M/L) from the fit must agree with the M/L of the mass function used in the model.

Table I gives our best estimates of the absolute visual magnitude (M_V), the concentration parameter (c), the mean cluster velocity ($< v >$), the central projected velocity dispersion (PV_o), and the global and central M/L_V's. We define $(M/L_V)_o$ to be the ratio of the central mass and luminosity surface densities. These values are based on fits of multi-component models with mass function exponents, x, of 0.67, 1.35, and 2.00 $(dN \propto m^{-x} dlog(m))$. Often, all three exponents produce acceptable fits. If this results in a range of values for a parameter that is larger than the uncertainties in the individual fits, the range is given in the table.

Mass functions with exponents of 1.35 and 2.00 yield acceptable models for all of the clusters in our sample. An x of 0.67 is acceptable for all of the clusters

* Visiting Astronomer, Canada-France-Hawaii Telescope, operated by the National Research Council of Canada, the Centre National de la Recherche Scientifique of France, and the University of Hawaii.

except NGC 6171, 6218, and 6624. We have also tried models with $x=0.0$ and they are not consistent for the clusters in our sample. The theoretical M/L_V (3.2) is always at least 2σ larger than the value derived from the fit. This arises because the shallow mass function predicts a large number of white dwarfs with masses equalling or exceeding that of the giants (we assume that stars with initial masses less than 8 $M_\odot$ make white dwarfs). We have fitted models with an anisotropy radius of $10r_c$ to NGC 6624 and 6681 and, for these clusters at least, the acceptable x values do not depend strongly on the assumption of isotropy.

Table I.
Results From Fitting Isotropic Multi-Component King Models

NGC	[Fe/H]	M_V	c	# of stars	$<v>$ (km/s)	PV_o (km/s)	M/L_V ($M_\odot/L_\odot$)	$(M/L_V)_o$ ($M_\odot/L_\odot$)
288	−1.2	−6.71 ± .09	1.0	26	−45.9 ± 0.6	2.9 ± 0.5	2.6 ± 0.8	2.0 ± 0.6
4147	−1.8	−6.00 ± .07	1.7	9	183.2 ± 0.9	2.8 ± 0.9	2.8 ± 1.7	1.7 ± 0.9
5466	−1.9	−6.85 ± .08	1.0	20	107.7 ± 0.4	2.0 ± 0.4	1.7 ± 0.6	1.3 ± 0.5
6171	−0.9	−6.87 ± .15	1.7	18	−32.8 ± 0.7	2.9 ± 0.6	2.4 ± 0.9	1.0 ± 0.4
6218	−1.6	−7.08 ± .11	1.3	20	−45.0 ± 0.8	3.7 ± 0.7	1.9 ± 0.6	1.1 ± 0.4
6624	−0.7	−7.53 ± .14	2.3:	19	54.1 ± 1	5.4 ± 1	1.6-3.8	1.2 ± 0.5
6626	−1.5	−8.19 ± .13	1.8	19	15.9 ± 2	8.9 ± 1.5	1.5-5	1.6 ± 0.6
6681	−1.2	−7.23 ± .15	2.2:	18	218.8 ± 1	5.2 ± 1	1.5-5	2.1 ± 0.9
6809	−1.6	−7.41 ± .06	0.9	20	176.8 ± 0.9	4.2 ± 0.7	1.7 ± 0.6	1.4 ± 0.4

That none of our clusters are consistent with a flat mass function is surprising since McClure et al. (1986) found several clusters whose main sequence luminosity functions yielded $x \simeq 0.0$ ($x < 0.0$ when mass segregation corrections are applied, see Pryor, McClure, and Smith (1986)). NGC 6624 appears to have a steeper mass function than the correlation of x with metallicity found by McClure et al. would suggest. However, a main sequence luminosity function has not been measured.

Although the King models that we use underestimate the luminosity at the center of a cusp cluster, they also underestimate the mass there. In agreement with this, our central M/L's show little dependence on the concentration of the model used to derive them. The low central M/L of NGC 6624 suggests that giants make the dominant contribution to the density in its inner regions. This is consistent with the giants showing an r^{-1} profile (Djorgovski and King, 1984), since they would be the heaviest important component. On the other hand, NGC 6681 has the largest central M/L in our sample. Based on this sample of two, cusp clusters do not seem to have M/L's significantly different from those of the cluster population as a whole.

REFERENCES

Djorgovski, S. and King, I. R. 1984 Astrophys. J. Letters 277, L49.
Gunn, J. E. and Griffin, R. F. 1979 Astron. J. 84, 752.
McClure, R. D., VandenBerg, D. A., Smith, G. H., Fahlman, G. G., Richer H. B., Hesser, J. E., Harris, W. E., Stetson, P. B. and Bell, R. A. 1986 Astrophys. J. Letters 307, L47.
Pilachowski, C. A. 1984 Astrophys. J. 281, 614.
Pryor, C., McClure, R. D. and Smith, G. H. 1986 Astron. J. 92, in press.
Webbink, R. F. 1985 in IAU Symposium No. 113, Dynamics of Star Clusters, J. Goodman and P. Hut, eds., Reidel, Dordrecht, p. 541.
Zinn, R. and West, M. J. 1984 Astrophys. J. Suppl. 55, 45.

ANISOTROPY IN ω CENTAURI AND 47 TUCANAE

G. Meylan

University of California, Berkeley

ABSTRACT. The southern sky gives us the great opportunity to observe two among the brightest and nearest globular clusters of the Galaxy: ω Cen and 47 Tuc. For these giant clusters, we present the comparison between observations and King-Michie multi-mass dynamical models with anisotropy in the velocity dispersion. A more comprehensive description of this work is to be published (Meylan 1986a,b).

1. THE OBSERVATIONS

The present dynamical description uses both surface brightness and velocity dispersion profiles. Precise radial velocities have been obtained with the photoelectric spectrometer CORAVEL at the European Southern Observatory at Cerro La Silla, Chile, in collaboration with astronomers in Geneva, Marseilles, Copenhagen, and ESO (Mayor et al. 1983, 1986). The number of observations amounts in ω Cen to 540 measurements of 318 member stars, and in 47 Tuc to 371 measurements of 272 member stars, with typical uncertainties of 0.9 and 0.6 km/s, respectively.

2. THE MODEL

Models have been constructed in an approach nearly identical to Gunn and Griffin (1979). Heavy remnants (e.g. neutron stars), white dwarfs, and MS stars have been distributed into ten different subpopulations, each obeying the energy-angular momentum distribution function: $f_i(E, J) \propto \left[\exp(-A_i E) - 1\right]\exp(-\beta J^2)$. In the cluster center, thermal equilibrium is assumed in order to force A_i to be proportional to the mean mass of the stars in the subpopulation considered. A model is specified by a mass function exponent x, and by four parameters: the scale radius r_c, the scale velocity v_s, the central value of the potential W_o, and the anisotropy radius r_a. Beyond r_a, the velocity dispersion tensor is mostly radial.

3. THE RESULTS

Models have been calculated for a wide range of values of each parameter.

For ω Cen, the observations are well fitted only by models with strong anisotropy, with $r_a \simeq$ 2-3 r_c (Fig. 1). This is related to the large half-mass relaxation time $t_{rh} \simeq$ 20-30 10^9 yr ($t_r(0) \simeq 10^9$ yr). The mean value of the exponent x of the mass function is $\simeq 1.25$, close to Salpeter's 1.35. The heavy remnants represent from 0 to 9 % of the total mass, being anticorrelated with the white dwarfs. The mean total mass of the cluster is about $4\ 10^6 M_\odot$, giving a mean $M/L_V \simeq 3$.

In 47 Tuc, using the flat luminosity function of Harris and Hesser (1985 HH), the best models (Fig. 2) have an anisotropy radius $r_a \simeq 20\text{-}40\ r_c$. The high concentration involves $t_{rh} \simeq 3\ 10^9$ yr $(t_r(0) \simeq 10^7$ yr). Heavy remnants are not needed, in contrast to white dwarfs which represent always about 30% of the mass. The mean total mass is about $0.7\ 10^6 M_\odot$, giving a mean $M/L_V \simeq 1.8$.

REFERENCES

Mayor, M., Imbert, M., Anderson, J., Ardeberg, A., Baranne, A., Benz, W., Ishci, E., Lindgren, H., Martin, N., Maurice, E., Meylan, G., Nordström, B. and Prévot, L. 1983, 1986 <u>Astron. Astrophys. Suppl.</u> 54, 495 and in press.
Meylan, G. 1968a,b, in preparation.

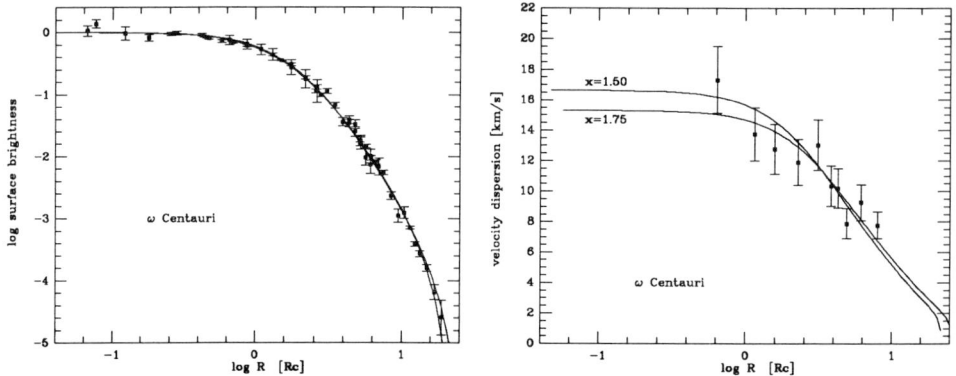

Fig. 1 ω Cen: observed and computed surface brightness and velocity dispersion profiles, for the models with $m_{hr} = 2.0$, $x = 1.50$ and 1.75, $W_\circ = 8.5$, and $r_a = 3.0$.

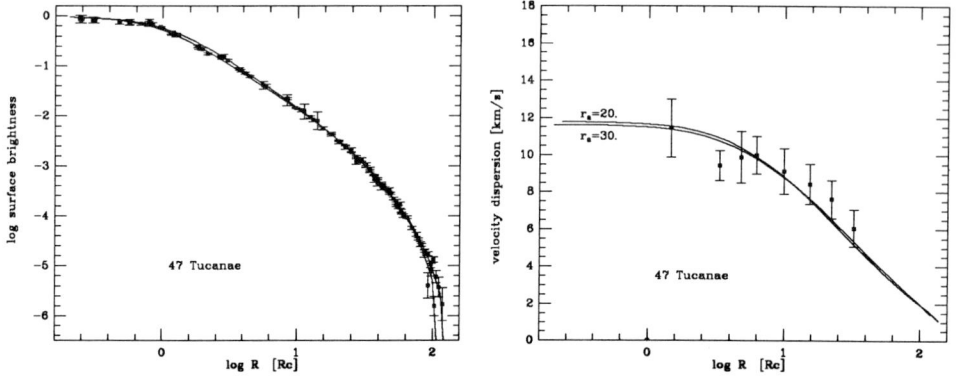

Fig. 2 47 Tuc: observed and computed surface bright. and velocity disp. profiles, for the models with $m_{hr} = 1.4$, $x = $ HH, $W_\circ = 10.$ and 9.5, and $r_a = 30.$ and $20.$

EVOLUTION OF GLOBULAR CLUSTERS INCLUDING A DEGENERATE COMPONENT

Hyung Mok Lee

Princeton University Observatory

Low mass X-ray sources observed in many globular clusters are usually interpreted as compact binaries with degenerate components (e.g., Hertz and Grindlay 1983). Degenerate stars can exist in globular clusters if the IMF contains a sufficiently large number of high mass stars. Since the main-sequence lifetime is a very steep function of stellar mass, most of degenerate stars can be regarded as primordial. If the typical mass of degenerate stars is higher than that of main-sequence stars, mass segregation makes the core crowded with degenerate stars. Tidally captured binaries between degenerates and main-sequence stars can abundantly form as the core density becomes very high.

We used the isotropic orbit-averaged Fokker-Planck code, originated from Cohn (1980), to follow the dynamical evolution of globular clusters containing primordial degenerate stars with individual masses higher than those of main-sequence stars. The total mass of the cluster is $10^5 M_\odot$ and the parameters for main-sequence stars are fixed at $m_{MS}=0.7 M_\odot$ and $R_*=0.57 R_\odot$. The mass ratio m_D/m_{MS} and the number ratio N_D/N_{tot} are regarded as free parameters. The interactions involving three-body binaries are treated as a collection of many small changes in energy and incorporated within the Fokker-Planck framework. The same technique as in Statler et. al. (1986) is used for the interactions involving tidally captured binaries.

Clusters having a sufficiently large population of degenerate stars may be approximated as single component clusters with point masses since the core is almost entirely dominated by degenerates due to the mass segregation process. Three-body binaries provide most of the energy to maintain the quasi-static post-collapse expansion. The structure of post-collapse clusters in this case becomes self-similar, and the physical parameters of such clusters follow simple power laws in time:

$$\rho_o \propto t^{-2} \; ; \; v_o \propto t^{-1/3} \; ; \; r_c \propto t^{2/3} \; ; \; r_h \propto t^{2/3},$$

where ρ_o is the central density, v_o is the central velocity dispersion,

r_c is the core-radius, and r_h is the half-mass radius. On the other hand, tidally captured binaries play a dominant role if there initially exist only a small number of degenerate stars. Numerical solutions similar to those of Statler et. al. (1986) are obtained in such cases.

Some clusters having very small core radii have flatter surface brightness profiles than that of an isothermal sphere. The logarithmic slopes of observed surface brightness profiles of "post-collapse" clusters lie between -0.75 and -1.25 (Djorgovski and King 1986; however, see Lugger et. al. 1987 for even flatter profiles), while the density distribution of post-collapse clusters (thus well relaxed) is expected to be close to that of isothermal sphere, which has logarithmic slope -1 for the surface density profile. Surface brightness profiles that are flatter than that of an isothermal sphere can be understood if these clusters have dynamically significant amounts of degenerate stars as envisaged in the present study. For example, the post-collapse cluster for the model with $m_D/m_{MS}=1.86$ and $N_D/N_{tot}=0.06$ has logarithmic slope -0.64, which is close to the surface brightness profile of M15 as measured by Lugger et. al. (1987). The post-collapse cluster for the model with $m_D/m_{MS}=1.25$ and $N_D/N_{tot}=0.04$ has a similar distribution for matter and light.

The number of tidally captured binaries in post-collapse clusters is typically several times 10^1. The inferred average number of compact binaries (cataclysmic binaries) from X-ray observations (~ 10 per galactic cluster according to Hertz and Wood 1985) is consistent with the average number of tidally captured binaries of our model if $\sim 20\%$ of the galactic globular clusters are in the post-collapse stage (Djorgovski and King 1986). More detailed results of this study will be published in The Astrophysical Journal.

I would like thank D. Merritt for careful reading of the manuscript.

REFERENCES

Cohn, H. 1980 Astrophys. J. 242, 765.
Djorgovski, S. and King, I. R. 1986 Astrophys. J. Letters 305, L61.
Hertz, P. and Grindlay, J. E. 1983 Astrophys. J. Letters 267, L83.
Hertz, P. and Wood, K. S. 1985 Astrophys. J. 290, 171.
Lugger, P. M., Cohn, H. N., Grindlay, J. E., Bailyn, C. and Hertz, P.
    1987 in IAU Symposium No. 126, Globular Cluster Systems in Galaxies
        J. E. Grindlay and A. G. D. Philip, eds., Reidel, Dordrecht, p. 657.
Statler, T. S., Ostriker, J. P. and Cohn, H. N. 1986, preprint.

EVOLUTION OF GLOBULAR CLUSTERS WITH TIDALLY-CAPTURED BINARIES THROUGH CORE COLLAPSE

Thomas S. Statler and Jeremiah P. Ostriker

Princteon University Observatory

Haldan N. Cohn

Indiana University

We present calculations of globular cluster evolution performed by a modified Fokker-Planck approach, in which binaries formed by tidal capture are followed explicitly, along with subsequent heating mechanisms. The cluster is simulated by a two component model, using the cross sections of Press and Teukolsky (1977) for tidal capture, those of Hut (1984) for the single-binary encounters and for distant binary-binary encounters, and those of Mikkola (1983) for the strong binary-binary encounters. The initial state of the cluster is a Plummer model with $N = 3 \times 10^5$ and scale radius $r_0 = 1.13\,\text{pc}$. All stars are identical, with mass $M_* = 0.7 M_\odot$ and $R_* = 0.57 R_\odot$. This gives an initial core radius $r_c = 0.8\,\text{pc}$, and one-dimensional dispersion $\sigma = 11.6\,\text{km s}^{-1}$. All binaries are assumed to be identical, with separation $a = 2.5 R_*$. There are no binaries in the cluster initially. Additional important effects, such as tidal truncation, tidal shocks, stellar evolution and mass loss, and stellar mergers, are not included.

The number of binaries expected to form prior to core collapse can be estimated by integrating the formation rate over the self-similar single-component solution (e.g., Cohn 1980). One finds $N_b = 90[N(0)/\ln \Lambda(0)](v_h/v_*)^{1.8}$, where v_h is the rms velocity at the half-mass radius and v_* is the escape velocity from the surface of the star. For the parameters typical of globular clusters, this gives $N_b \sim 10^3$. Fig. 1 shows the evolution of the core and half-mass radii in the full calculation with all heating processes included. The cluster bounces at a central density of about $1.5 \times 10^8 M_\odot \,\text{pc}^{-3}$. Bounce occurs when the core has shrunk by only a factor of 100 in radius, and several hundred stars remain in it. By comparison, the first hard binary formed by three-body processes is expected to appear when $N_{core} \sim 40$. We find that the three-body formation rate is always less than 10^{-4} of the tidal capture rate until late in the re-expansion.

A criterion for core bounce may be derived by equating the gravitational heating (due to ejection of stars) of the binaries to the losses to single stars in the core. One finds that, ignoring the time variation of $\ln \Lambda$, the core bounces when the ratio of central to half-mass dispersion reaches a value which is quite independent of cluster parameters. The numerical results indicate $(v_0/v_h)_{bounce} \approx 1.3$. During the re-expansion, gravitational heating always dominates over direct heating because the central potential becomes shallower and it becomes progressively easier to eject stars.

At late times, the evolution of the half-mass point is governed by the local conduction rate; this and conservation of energy lead to r_h scaling as $t^{2/3}$ and v_h as $t^{-1/3}$. Since the cluster evolves on a time-scale of many half-mass relaxation times, the region interior to r_h is very nearly isothermal, so that v_0 also scales as $t^{-1/3}$. Most of the energy from gravitational heating is deposited far out in the cluster, so that *heat is conducted inward* from the half-mass point into the core. The evolution of the central density and core radius is not easy to understand, and may be due to the (necessary!) non-self-similarity of the solution, and/or to cumulative numerical errors and the large time step employed. When a much smaller time step is used, this solution is found to become unstable to large-amplitude gravothermal oscillations (see Cohn, this volume).

In the idealized case of an isolated cluster made entirely out of main-sequence stars, these calculations show that binaries formed by tidal capture dominate the dynamics of core bounce and re-expansion. The applicability of this result to real systems is limited by our exclusion of processes known to be important, such as tidal effects in superelastic encounters (McMillan 1986) and stellar mergers and mass loss (Lee 1986). Nevertheless, it is clear that the objects formed by tidal capture, whether they be binaries or merged stars, influence the dynamics of the cluster long before three-body binaries can. In other words, for a cluster made predominantly of normal stars, it is the effects of intercations between stars of finite size, rather than point-mass effects, that dominate the dynamics as the cluster passes through core collapse.

REFERENCES

Cohn, H. N.   1980   Astrophys. J. 242, 765.
Hut, P.   1984   Astrophys. J. Suppl. 55, 301.
Lee, H. M.   1986   Ph.D. Thesis, Princeton University.
McMillan S.   1986   Astrophys. J. 306, 552.
Mikkola, S.   1983   Monthly Notices Roy. Astron. Soc. 203, 1107.
Press, W. H. and Teukolsky, S. A.   1977   Astrophys. J. 213, 183.

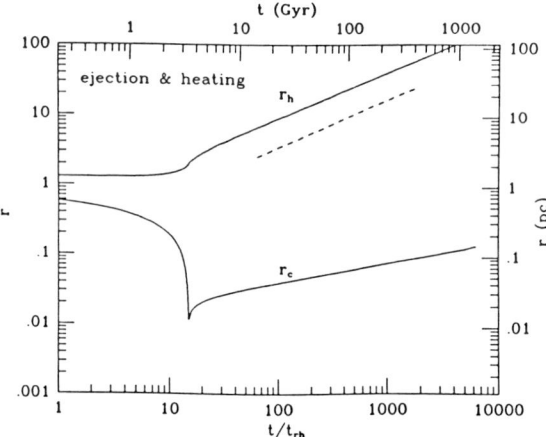

Fig. 1. Evolution of the core and half-mass radii. The dashed line indicates a logarithmic slope of 2/3.

BINARY INTERACTIONS IN STAR CLUSTERS

Stephen L. W. McMillan

Northwestern University

Over the past decade, a very considerable amount of effort in stellar dynamics has gone into the study of interactions between binary systems and other stars. The asymptotic analytic results obtained by Heggie (1975) for binary-single star encounters have been largely confirmed and extended by later numerical experiments (Hills 1975, Hut and Bahcall 1983). Binary-binary interactions have been studied by Mikkola (1983).

In numerical simulations of real star clusters, it is found that binaries can not always be regarded as interacting with only one other star or binary at a time. When the number of stars in the core of a stellar system is small, statistical fluctuations about mean-field quantities can be very large, so that all higher-order interactions tend to "turn on" almost simultaneously with the first appearance of triple effects. These interactions are very difficult to treat theoretically and are currently quite poorly understood. While N-body experiments offer an obvious way of studying binaries "in the field," their use still presents some difficulties. In an isolated, self-gravitating system containing a small number of bodies, the first hard binary to form usually comes to dominate the dynamics of the cluster and cause it to expand or even dissolve, severely limiting the number of binary interactions that can be studied. Accounts of the behavior of binaries in such circumstances tend to be largely anecdotal in nature.

I present here a quantitative account of over one thousand "close" encounters between binaries and other stars during the final stages of globular cluster core collapse. The hybrid integration is described in detail by McMillan (1986, "run 2"). Because the N-body system is only a small part of an extensive confining cluster, the large amount of energy liberated by hard binaries is relatively easily absorbed, allowing densities to remain high and conditions to remain suitable for the study of binary encounters over long periods of time. The run described here lasted for some 800 initial core relaxation times, corresponding to around 5500 internal time units. Every half time unit, the identity, binding energy (B), eccentricity, center of mass energy (E) and nearest neighbors of all significant bound pairs (i.e. those having binding energy greater than or comparable to the mean stellar kinetic energy, 1.5 kT) were stored for subsequent analysis. In addition, similar information on close triple systems involving the center of mass of any close pair (bound or unbound) was retained.

The sampling time interval is longer than most binary interactions, allowing them to be convenienty identified simply by checking for abrupt changes in a binary's orbital parameters or components. Once a significant encounter is found, the star(s) involved are easily determined from the stored data, and the event is classified according to the binding energy change (δB), the total number of stars involved (I), the local core conditions, and so on. A star is deemed an "interactor" if it is a component of, or a neighbor within five semi-major axes of, the binary before or after the interaction. For the small number of cases in which no obvious interactor is found, a value of I = 3 is assumed.

The overall statistics of the run, for encounters resulting in $|\delta B| > kT$, are illustrated in the table below, which lists the number of interactors found in various situations. In the first four columns, they are broken down by type of interaction (binary creation or destruction, flyby or exchange). In the next two, the binary's environment is taken into account, the local density ρ being divided into two ranges: "low" (less than five times the initial density) and "high" (near the peak density at core collapse, some ten to twenty times higher). During the course of the run, the system oscillated between the high- and low-density states approximately eight times as binary formation, heating and ejection governed the core dynamics. The final two columns subdivide the encounters by the average binding energy of the binary involved.

	Creation	Destruction	Flyby	Exchange	Low-ρ	High-ρ	$B \leq 8\,kT$	$B > 8\,kT$
I = 3	135	207	271	195	535	273	554	254
I = 4	60	104	81	84	166	163	295	34
I ≥ 5	36	64	58	56	105	109	194	20

Evidently a substantial fraction (~40%) of all binary encounters involve more than just three stars. For low-energy binaries or high ambient densities, the fraction is even higher (~50%). As expected, hard binaries tend to interact with just one star at a time, since a perturber has to approach within a few binary semi-major axes before a strong triple interaction can take place.

The value of I also affects the detailed statistical properties of the interactions. For relatively soft binaries (B < 20 kT), the distribution of $\Delta \equiv \delta B/B$ for I = 3 is significantly different from that for I $\geq$ 4, while the expected correlation between δB and δE, the binary's recoil energy, is almost entirely absent in the latter case, as the number of ways in which the recoil energy can be distributed increases. Even for triple encounters involving hard binaries, the Δ distribution is not consistent with the $(1+\Delta)^{-9/2}$ behavior found by Heggie, and δE falls well below the predicted value of $\delta B/3$ for $\delta B > 40$ kT. This may indicate that a considerable fraction of the "triples" are actually misclassified higher-order encounters. (Alternatively, these encounters might not have been completely over when sampled: proximity to a third star would tend to lower the calculated center of mass energy.)

The results of this preliminary investigation suggest that a sizeable fraction of all binary interactions in a star cluster actually involve four or more participants. As the binding energy of a binary increases, it becomes more likely that it will interact with just one other star at a time, but even hard binary encounters appear to be appreciably affected by the presence of additional nearby perturbers. As a result, such encounters may be somewhat more efficient at directly heating the cluster than was previously thought. At lower energies, extra participants in an interaction can substantially alter the distribution of the fractional binding energy change, and reduce almost to zero the correlation between binding energy change and binary recoil energy.

Note that these results refer to the specific problem of globular cluster core collapse, where velocities are somewhat higher (relative to the central potential) than for an isolated, self-bound system. It is not immediately clear how to scale the above conclusions to the latter situation. The routine accumulation of similar data in such integrations would be a welcome development.

REFERENCES

Heggie, D. C.   1975   Monthly Notices Roy. Astron. Soc. 173, 729.
Hills, J. G.   1975   Astrophys. J. 80, 809.
Hut, P. and Bahcall, J. N.   1983   Astrophys. J. 268, 319.
McMillan, S. L. W.   1986   Astrophys. J. 307, 261.
Mikkola, S.   1983   Monthly Notices Roy. Astron. Soc. 205, 733.

TIDAL EFFECTS ON STELLAR EVOLUTION IN CLOSE BINARIES FORMED IN GLOBULAR CLUSTERS

H. M. Antia, A. K. Kembhavi and A. Ray

Tata Institute of Fundamental Research

A mechanism of forming x-ray binaries by close collision of a neutron star and a normal star in a globular cluster core (GC) was proposed by Fabian, Pringle and Rees (1975). Press and Teukolsky (1977) (PT) made detailed computaions of tidal energy deposition in the non-radial modes of a main sequence (MS) star (approximated by a n = 3 ploytrope) and two-body tidal capture cross-section. Here, we correct numerical errors in PT for the n = 3 plytrope; extend the calculation to the n = 3/2 case which better approxiamtes a fully convective low mass star most abundant in GC; we discuss the effects of tidal energy dissipation on the evolution of the MS star and the binary orbit. The energy transferred to tides in the normal star during the encounter with a netron star (NS) is:

$$E_{tide} = (\frac{GM_*^2}{R_*}) (\frac{M_n}{M_*})^2 \sum_{\ell=2}^{\infty} (\frac{R_*}{R_{min}})^{2\ell+2} T_\ell(\eta)$$

Here M_* is the main sequence star's mass (=0.6$M_\odot$), R_* it's radius (=4.5 x 10^{10}cm), M_n the neutron star mass (=1.4 $M_\odot$), R_{min} the separation at periastron. The function $T_\ell(\eta)$ is defined in PT and we display its corrected values in Fig. 1 (here, $\eta = \{M_*/(M_* + M_n)\}^{1/2}(R_{min}/R_*)^{3/2}$). The n=3/2 polytrope gives a larger energy transfer compared to the n=3 case for similar collision parameters and also a larger cross-section. Fig. 2, which displays the ratio of tidal energy deposited in a spherical shell of unit thickness to the mean energy density, shows that a large part of the tidal energy is deposited near the surface for n=3/2 polytrope. In a fully convective star, the mechanical tidal energy is quickly dissipated due to the viscosity of turbulent eddies and gives a large thermal luminosity. We use the effective viscosity (cf Goldreich and Keeley, 1977) $\nu_t^{eff} = \nu_t/\beta^2$ reduced from the canonical value $\nu_t = (1/3) \ell_{mix} v_{conv}$ for turbulent eddies in the mixing length approximation whenever $\beta=\tau_{conv}/P_{osc}$ > 1 (here τ_{conv} = eddy turnover time, P_{osc} is tidal oscillations period). For typical MS stars in GCs, this is roughly 10^4 years. The resulting thermal luminosity at various times after first encounter is displayed in Fig. 3. Thermal luminosity transmitted from below the stellar surface is 700 times normal stellar luminosity after a few hundred years, causing the star to reach a new equilibrium at a large radius. An isolated low mass (0.6$M_\odot$) Pop II star (opacity of outer radiative layers due to H$^-$ atoms), transmitting 100$L_\odot$ expands to an equilibrium radius at R ≃ 10$R_\odot$ (cf Stein (1966) for L_*-R_* relation). But the normal star's Roche-lobe radius (R_c = 1.6R_*) after orbit circularization is well within this for the assumed binary mass ratio. Thus, the normal star quickly overflows to form a circumbinary envelope. The NS spirals in towards the core of the normal star under frictional drag of envelope. Spiral-in timescale

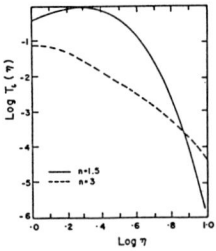 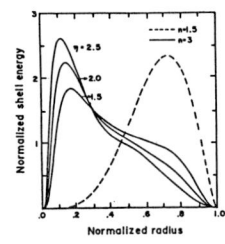 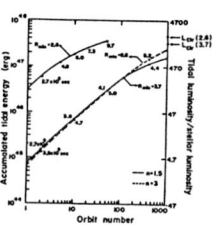

Fig. 1. Dimensionless function $T_\ell(n)$ for energy in quadrupole ($\ell=2$) tide
Fig. 2. Distribution of tidal energy vs stellar radius.
Fig. 3. Accumulated tidal energy vs orbit number. Time elapsed since 1st encounter marked on each curve. R_{min} initial periastron distance in R_*. Right ordinate indicates luminosity in units of normal L_* ($0.17 L_\odot$).

τ_D is related to orbital period P_{orb} (Paczynski, 1976) by: $\tau_D/P_{orb} \sim \langle\rho\rangle/\rho$; $\langle\rho\rangle$ = mean density of envelope inside NS orbit, ρ = density at the NS orbit. Further evolution of this common envelope binary depends on what fraction of the envelope mass is ejected, which in turn depends on whether the hydrodynamic ejection time-scale of a given mass of envelope is short or long compared to the thermal transport time across it (Taam et al (1978)). Four different end products of this binary are possible: (1) a detached binary where the MS star contracts to the original size inside Roche lobe before the NS orbit shrinks too much; (2) a contact x-ray binary where the common envelope is ejected fast compared to the Roche-lobe overflow of the MS star and where the orbital decay of the NS stops due to lessened friction; (3) Thorne-Zytkow Red Supergiant where the NS spirals in all the way to the center of the MS star without ejecting its envelope; and (4) a bare NS with a planetary nebula where NS ejects all matter in the MS star if common envelope density is too high. Thus, x-ray binaries form in 2-body tidal captures only part of the time but in such case the MS star starts filling the Roche-lobe soon after formation.

A.R. thanks VITA, Univ. of Virginia for hospitality and assistance.

REFERENCES

Fabian, A. C., Pringle, J. E. and Rees, M. J. 1975 Monthly Notices Roy. Astron. Soc. 172, 15.
Goldreich, P. and Keeley, D. A. 1977 Astrophys. J. 211, 934.
Paczynski, B. 1976 in IAU Symposium No. 73, Structure and Evolution of Close Binary Systems, P. Eggleton, S. Mitton and J. Whelan, eds., Reidel, Dordrecht, p. 75.
Press, W. H. and Teukolsky, S. A. 1977 Astrophys. J. 213, 183.
Stein, R. F. 1966 in Stellar Evolution, R. F. Stein and A. G. W. Cameron, eds., Plenum Press, New York, p. 23.
Taam, R. E., Bodenheimer, P. and Ostriker, J. P. 1978 Astrophys. J. 222, 269.

THE EFFECTS OF STELLAR EVOLUTION AND GALACTIC TIDES ON GLOBULAR CLUSTER EVOLUTION

D. F. Chernoff, M. D. Weinberg and S. L. Shapiro

Cornell University

1. DESCRIPTION OF MODEL

We investigate the evolution of globular clusters in the Galactic tidal field prior to core collapse. These multimass models incorporate mass loss by stellar evolution.

The relaxation is simulated by the 1-dimensional (energy) Fokker-Planck equation. We assume the mass included by an $R_g = 8\,\text{kpc}$ orbit produces the Galactic tide. The energy at the inner Lagrange point defines the tidal boundary. The number of stars in a given species is chosen according to a powerlaw IMF: the number of stars with mass between m and $m + dm$ is $\propto m^{-\beta}$. The 20 species used for these calculations have initial masses logarithmically spaced from 0.4 to $14.0\,M_\odot$. For purposes of modeling stellar evolution, we assume that mass is lost instantaneously from a star and the cluster at the end of its main-sequence lifetime leaving a $0.7\,M_\odot$ remnant. The main-sequence lifetime is taken to be $\tau_{ms} = 10\,\text{Gyr}\,(m/M_\odot)^{-3}$. We intend to investigate a range of orbital radii and cluster concentrations. In all cases presented here, the cluster mass is initially distributed as a $W_0 = 3.0$ ($\log c = 0.672$) King model with $M = 10^5\,M_\odot$. The tidal radius is initially $R_{tidal} = 60\,\text{pc}$. These parameters are summarized in Table I. For the IMF's considered, the stars with largest individual mass initially have the smallest density.

TABLE I

Initial conditions:
1). King Model with $W_0 = 3.0$ ($\log c = .672$), initial mass $10^5\,M_\odot$ and tidal radius 60 pc.
2). All models have 20 mass species, logarithmically spaced between 0.4 and 14 $M_\odot$, with equal velocity dispersions at t=0.
3). IMF of the form $n(m) \propto m^{-\beta}$.

Model	A.	B.	C.	D.
Mass spectrum	$\beta = 3.5$	$\beta = 3.5$	$\beta = 4.5$	$\beta = 3.5$
Galactic tide	yes	yes	yes	no
Stellar evolution (SE)	no	yes	yes	yes
Collapse time (Gyr)	6.2	47	62	137
Final mass ($M_\odot$)	9.0×10^4	1.1×10^4	2.2×10^4	8.5×10^4
SE mass loss ($M_\odot$)	0.0	1.4×10^4	5.2×10^3	1.4×10^4

2. PRELIMINARY RESULTS

In case A (no stellar mass loss) relaxation is the main physical effect. The different species segregate by mass and the heaviest stars dominate the subsequent collapse. Due to mass segration the lightest stars are lost in the greatest number to the Galactic tidal field.

In cases B-D (with stellar mass loss), heating due to work done in ejecting the gas from the cluster controls the early evolution. The central potentials of the cluster initially decrease. Much later, relaxation dominates the evolution. However, the range in mass per star for the evolved components is small once the core collapse begins and there is no significant mass segregation. The collapse is morphologically similar to a single component model.

3. SUMMARY AND PLANS FOR FUTURE WORK

The main findings from cases A-D are:
1). We find that the lowest mass stars are significantly depleted in number due to mass segration in the presence of a tidal field. The $0.4\,M_\odot$ stars decreased by 95% in number in case B.
2). Stellar evolution narrows the range of stellar masses and decreases the extent of mass segration.
3). The cluster may lose up to 90% of its initial mass for an IMF powerlaw index of $\beta = 3.5$. This suggests that a proto-cluster must have a much larger initial mass or a much higher concentration.
4). The heating due to stellar evolution dramatically increases the core collapse time, perhaps exceeding the Hubble time. A larger initial mass will probably exagerate this effect. Given the sensitivity of the collapse time to the IMF and the remnant distribution, more detailed stellar evolution scenarios will be investigated.

A parameter survey in M, β, cluster concentration and R_g is underway.

We thank Ira Wasserman for helpful discussion. This work was supported in part by National Science Foundation Grant No. AST 84-15162 to Cornell University. Computations supporting the research were performed on the Cornell Production Supercomputer Facility which is supported in part by the National Science Foundataion and the IBM Corporation.

THE SPATIAL DISTRIBUTION OF SPECTROSCOPIC BINARIES AND BLUE STRAGGLERS IN M 67

Robert D. Mathieu and David W. Latham

Harvard-Smithsonian Center for Astrophysics

Mathieu et al. (1986) have completed an extensive radial-velocity survey of over 100 late-type stars in M 67 with V < 12.8. The spatial distributions of the spectroscopic binaries and single stars (i.e. those stars without detected radial-velocity variation; many of these are undoubtedly binaries, albeit with lower secondary masses) are shown in Fig. 1. The distribution of the binaries is notably more centrally concentrated than the single stars. The two observed distributions derive from distinct parent distributions at the 98% confidence level. The projected half-mass radius of the binaries is 0.9 pc; the half-mass radius of the single stars is 2.4 pc. Indeed, 77% of the binaries lie within the single-star half-mass radius.

A reasonable explanation for the central concentration of an ensemble of stars in a stellar system is that 1) the stars are more massive than typical cluster members and 2) relaxation processes have produced a velocity distribution approaching energy equipartition and consequently mass segregation in the cluster. In order to test the viability of this explanation for the central concentration of the binaries, we have fit multi-mass isotropic equipartition King models to the entire cluster, including a 2 $M_\odot$ component to model the binaries. The value of 2 $M_\odot$ was chosen _a priori_ to be consistent with binaries that are somewhat less than twice the main-sequence-turnoff mass of 1.2 $M_\odot$. The cumulative distribution of this component is also shown in Fig. 1. The agreement of the model with the binary distribution is quite good. Thus the central concentration of the binaries is entirely consistent with them being more massive objects in a relaxed cluster. The relaxation time of M 67 derived from the average conditions inside the half-mass radius is 1×10^8 yr, short relative to the cluster age of 5×10^9 yr, so that indeed the cluster is expected to be well relaxed.

The central concentration of the blue stragglers is equally striking in Fig. 1. All of the blue stragglers lie within 3.2 pc of the cluster center; the half-mass radius is 1.1 pc. The blue straggler distribution is distinct from that of the single stars at better than the 99% confidence level. The central concentration of the blue stragglers is strong evidence that the blue stragglers do not comprise a relaxed population of objects with masses comparable to those of the single stars; the reasonable fit of the 2 $M_\odot$ component argues for the blue stragglers being substantially more massive than the main-sequence-turnoff stars. This argument is entirely independent of any assumptions concerning the internal physical state of the blue stragglers. An alternative explanation for the central concentration might be that the blue stragglers are not a more massive relaxed population but rather that they are preferentially formed in the cluster core. For this to be the case, the blue straggler lifetimes must be short, of order 10^8 yr or less.

The blue straggler distribution is notably similar to that of the binaries, in appealing concord with the hypothesis that some population of binaries are the progenitors of the blue stragglers. We note that the data presented here do not require the binarity of the blue stragglers themselves, an issue that remains controversial. The data require only that in the process of evolving from a binary system with normal cluster-member components to a blue straggler, whether through mass transfer, coalescence or some other mechanism, little mass is lost from the system. Indeed, even this requirement can be relaxed if the blue straggler lifetimes are short. On the other hand, the blue stragglers are more centrally concentrated than would be expected given masses derived from the mixed-stellar-interiors models computed by Saio and Wheeler (1980) for the blue stragglers of NGC 7789. However, without detailed models for the M 67 blue stragglers a definitive test of the extended lifetime scenario cannot be made.

REFERENCES

Mathieu, R. D., Latham, D. W., Griffin, R. F. and Gunn, J. E. 1986
    Astron. J., in press.
Saio, H. and Wheeler, J. C. 1980 Astrophys. J. 242, 1176.

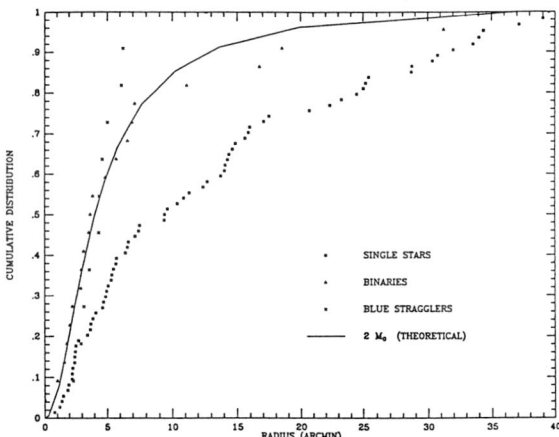

Fig. 1. Cumulative radial distributions of the 1.2 $M_\odot$ single stars, the spectroscopic binaries, the blue stragglers and a theoretical 2 $M_\odot$ component in equipartition.

EVIDENCE FOR MASS SEGREGATION IN NGC 5466

James M. Nemec

Palomar Observatory

Hugh C. Harris

Dominion Astrophysical Observatory

ABSTRACT: Forty-eight blue straggler stars have been discovered in NGC 5466, the only Galactic globular cluster known to contain an anomalous Cepheid of the sort found in dwarf galaxies. The stars were identified in color-magnitude diagrams constructed from photometry of deep photographic plates taken with the Canada-France-Hawaii 3.6 m telescope (calibrated with new UBV photoelectric photometry), and from point spread function photometry of CCD frames taken with the Palomar 5 m telescope. The stars typically have magnitudes $<V> \sim 19\overset{m}{.}1$ and colors $<B-V> \sim 0\overset{m}{.}2$. Forty-two of the 48 stars are situated inside of R=2.5 arcmin (see Fig.1), the projected radius containing half the cluster luminosity, and only six stars are found between 2.5 and 9 arcmin. A one-sided, two-sample Kolmogorov-Smirnov test (using the CCD data) establishes at the 98% significance level that the blue stragglers are more centrally concentrated than the subgiant stars of the same magnitude. By fitting multi-component King models to the projected radial distributions (Fig.2), the mean mass of the blue stragglers is shown to be ~ 1.5 to two times larger than the masses of the stars that contributed the light from which the core and tidal radii were derived (i.e. M(Blue Str.)=1.3±0.3 $M_\odot$). Because the central relaxation time for NGC 5466 is much less than the cluster age, the different radial distributions are attributed to mass segregation. A similar mass segregation is also observed in the globular cluster NGC 5053, where Nemec and Cohen (1986, in preparation) have recently identified $\sim$30 blue stragglers. The low stellar density and small escape velocity of NGC 5466 make a recent epoch of star formation (during which the blue stragglers might have formed as massive single stars) seem unlikely. Instead, the blue stragglers probably are either close binary systems that have transferred mass, or are coalesced stars. The very low frequency of stellar collisions expected in the center of NGC 5466 suggests that the blue stragglers are primordial binary systems. The simultaneous presence in NGC 5466 of the blue stragglers and the anomalous Cepheid V19, and their relative numbers, supports the hypothesis that there is an evolutionary connection between the two types of stars. By fitting theoretical isochrones to the photographic c-m diagram, NGC 5466 is estimated to have an age of 18±3 Gyr.

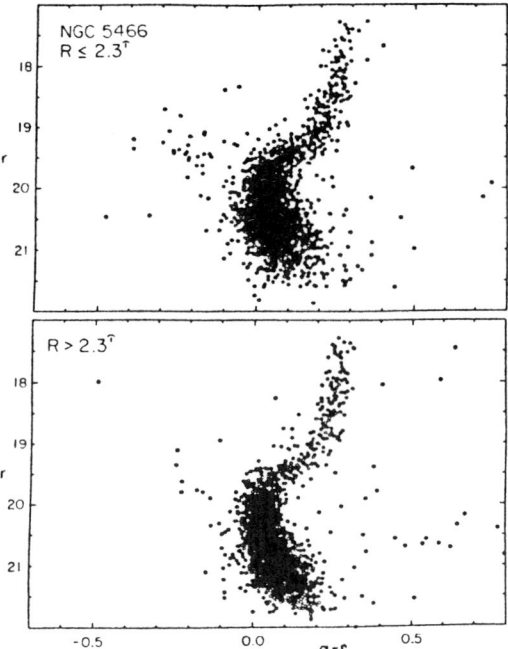

Fig.1 - C-M diagrams derived from photometry of the CCD frames taken with the 4-Shooter camera on the Palomar 5 m telescope. **Top panel**: 1963 stars interior to R=2.3 arcmin; **Bottom panel**: 1939 outer region stars. Note the pronounced central concentration of the blue stragglers.

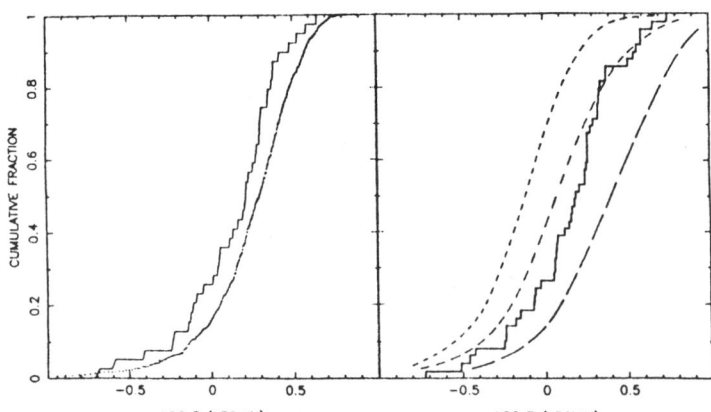

Fig.2 - **Left Panel**: Comparing the cumulatiave radial distributions of 38 blue stragglers (left) and 905 subgiants (right) in the same magnitude interval, measured on the CCD frames. The blue stragglers are significantly more centrally concentrated than are the subgiants. **Right Panel**: Comparing the radial distribution of all 49 blue stragglers with the distributions expected according to multi-component King models for stars of mass 0.8, 1.6 and 2.4 $M_\odot$. Interpolation shows that the dynamically derived mass of the blue stragglers is $\sim 1.3 \pm 0.3$ $M_\odot$.

ORIGIN AND RADIAL DISTRIBUTION OF FAINT BLUE HORIZONTAL-BRANCH STARS

Charles D. Bailyn and Jonathan E. Grindlay

Harvard/Smithsonian Center for Astrophysics

Haldan Cohn and Phyllis M. Lugger

Indiana University

ABSTRACT: We report the identification of 23 faint blue horizontal branch stars in Omega Centauri similar to those discussed by Buonanno et al. (1985) in M15. We find that these stars are significantly concentrated towards the center of the cluster with respect to other giants. We suggest that they may have formed from the collision of a main sequence star and a white dwarf.

1) THE DATA

Faint blue horizontal branch stars (FBHBs) form a distinct sequence in several globular clusters. These stars have unreddened $B - V \leq 0.0$ and extend down to 2.5 magnitudes in V below the horizontal branch (HB). It is not clear whether these stars are true HB stars which for some reason have smaller envelopes than other HB stars (Demarque and Eder 1985, Buonnano et al. 1985 hereafter BB, and references therein) or are produced by some other mechanism (Wesemael et al. 1982) in which case their position in the HR diagram as an apparent extension of the HB is fortuitous.

The large core radius of ωCen permits complete photometry of the HB much closer to the center of the cluster (in terms of r_c) than in BBs study of M15. Here we make use of our CCD photometry (UBV) of ωCen obtained at CTIO in May 1985. Analysis of three CCD frames centered at 2 and 5 r_c indicates that the ratio of FBHBs/BHBs increases sharply towards the center of the cluster. Specifically, we find 23 FBHBs and 92 BHBs in the B frame (centered at $\approx 2r_c$), while there are 28 BHBs and *no* FBHBs in the two frames centered further out (see Figure 1). A Monte Carlo simulation showed only 3 such configurations in 1000 trials.

2) FORMATION SCENARIO

We suggest that FBHBs might be formed by a collision between a white dwarf and a main-sequence star, an event which may result in an essentially unchanged white dwarf surrounded by an extended atmosphere of a few percent of the original mass of the main-sequence star (Shara and Regev 1986). Such an object would have a core of similar mass to other horizontal branch stars, but a much smaller envelope. Also, such objects would be formed preferentially in the dense cores of globular clusters where collisions would occur most frequently.

It has been suggested (e.g. Hills and Day 1976) that blue stragglers could form from similar collisions between two main sequence stars. The number of collisions that have occurred between stars in a globular cluster strongly depends on its past dynamical evolution since this determines the history of the core density of the cluster. Thus the presence of blue objects in the color-magnitude diagram of a cluster may contain information about its dynamical history.

ACKNOWLEDGEMENTS: We thank P. Stetson for making his superb program DAOPHOT available to us, and G. Da Costa for useful discussions. This work was supported by NSF grant NSF-AST-84-17846.

REFERENCES

Buonnano, R., Corsi, C. E. and Fusi Pecci, F. 1985 Astron. Astrophys. 145, 97.
Demarque, P. and Eder, J. -A. 1985 in Horizontal-Branch and UV-Bright Stars, A. G. D. Philip, ed., L. Davis Press, Schenectady, p. 91.
Hills, J. G. and Day, C. A. 1976 Astrophys. Letters 17, 87.
Shara, M. M. and Regev, O. 1986 Astrophys. J. 306, 543.
Wasmael, F., Winget, D. E., Cabot, W., van Horn, H. M. and Fontaine, G. 1982 Astrophys. J. 254, 221.

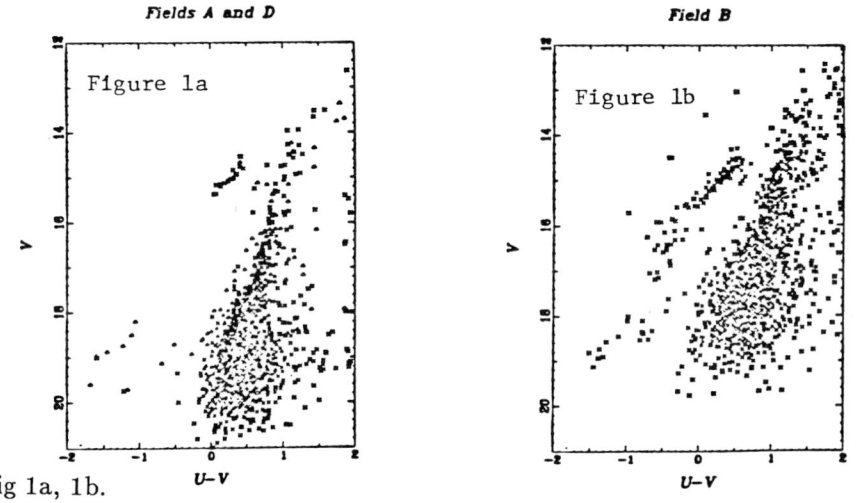

Fig 1a, 1b.
Color magnitude diagrams of ωCen. The B field is centered at $r = 2r_c$ while the A and D frames are centered near $r = 5r_c$. The absence of FBHBs in the outer fields can be clearly seen. The data are *not* photometric.

VARIABILITY OF ω CENTAURI BLUE STRAGGLERS: CLUES TO THEIR ORIGIN

G. S. Da Costa

Yale University Observatory

John Norris

Mt. Stromlo and Siding Spring Observatories

The four possible origins usually discussed for blue stragglers in stellar systems are (a) mass transfer in, or coalescence of, close binaries; (b) main sequence lifetime extension either through internal mixing or high non-thermal pressures; (c) complete mixing events in evolved stars at the core helium flash; and (d) ongoing star formation over long intervals. However, alternative (d) can be ruled out for the case of globular clusters because any residual gas in the cluster is very efficiently swept out as the cluster passes through the galactic disk, an event that occurs approximately every hundred million years. Hypotheses (a) and (b) both predict masses for the blue stragglers that exceed the turnoff mass but if alternative (c) is correct, then the blue stragglers should have masses less than or equal to the turnoff mass.

Can we determine the mass of a blue straggler? In general the answer is no, but the discovery of at least 3 dwarf cepheid variables in the ω Cen blue straggler sequence (see Jorgensen and Hansen 1984; Da Costa, Norris and Villumsen 1986) enables an estimate of the mass of these stars, relative to that of the RR Lyraes, to be made via the equation:

$$\log M(BS)/M(RR) = 2 \, \Delta\log Q - 2 \, \Delta\log P - 0.6 \, \Delta m(bol) - 6 \, \Delta\log T_{eff}$$

This calculation was first performed by Jorgensen and Hansen (1984) who derived a mean blue straggler mass of $1.2 \pm 0.2 \, M_\odot$ on assuming a mass of $0.6 \, M_\odot$ for the RR Lyraes. In making this calculation, they assumed on the basis of light curves derived from a small number of observations, that the dwarf cepheids were fundamental oscillators. The alternative assumption of first overtone pulsation however, yields a very different mass: $0.5 \pm 0.1 \, M_\odot$ (Da Costa, Norris and Villumsen 1986). Thus it is of some importance to improve the determinations of the light curves of these variables.

Figure 1 shows new light curves for the dwarf cepheid variables E39 which is a confirmed radial velocity member (Da Costa, Norris and Villumsen 1986), and NJL220 which has similar color and magnitude to E39. Based on the asymmetry of these light curves, we conclude that both are indeed fundamental oscillators. Consequently, they have masses exceeding the turnoff mass and their origin lies either with mass transfer in binaries or with main sequence lifetime extension in single stars. A search for radial velocity variations in the photometrically non-variable blue stragglers is required to distinguish between these ideas.

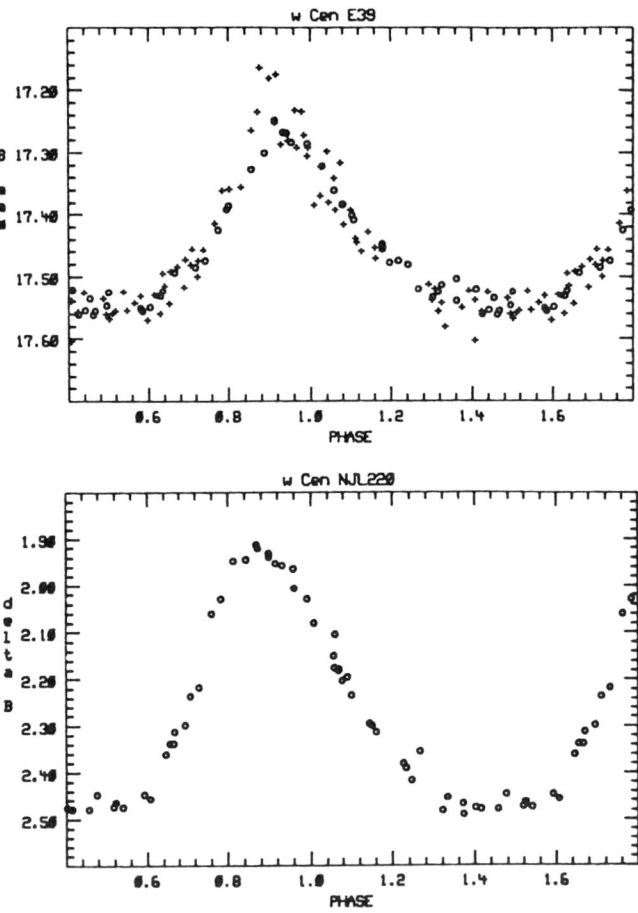

Fig. 1. The upper panel shows the light curve for the dwarf cepheid E39. The period is 82.70 minutes; open circles are CCD observations made over a 4 night interval, plus signs are photoelectric observations from the night following the CCD observations. The lower panel shows similar data for the star NJL220. The period is 67.95 minutes.

REFERENCES

Da Costa, G. S., Norris, J. and Villumsen, J. V. 1986 <u>Astrophys. J.</u>, in press.
Jorgensen, H. E. and Hansen, L. 1984 <u>Astron. Astrophys.</u> 113, 165.

A SEARCH FOR OPTICAL COUTERPARTS OF GLOBULAR CLUSTER X-RAY SOURCES

Michel Aurière

European Southern Obs.

Claude Chevalier
Sergio Ilovaisky

Obs. of Haute-Provence

Lydie Koch-Miramond

Center of Nuclear Studies

Jean-Pierre Cordoni

University of Montpellier

M 15:AC211/4U2131+11: THE OPTICALLY IDENTIFIED X-RAY SOURCE IN M 15

The identification of the optical counterpart of the M 15 (NGC 7078) X-ray source has been confirmed photometrically and spectroscopically (Aurière et al. 1984, 1985, 1986; Ilovaisky and Chevalier: 1985; Charles et al. 1985, 1986). It is designated M 15:AC211 from the Aurière and Cordoni catalogue (1981).

Rapid large amplitude brightness variations of M 15:AC211 in the U passband were discovered one year ago (Aurière et al. 1985; Ilovaisky and Chevalier 1985). These variations are very likely to be periodic but it turned out to be impossible to derive without ambiguity the period from series of data obtained at one site at once. We got recently new series of observations using both the 2m telescope of Pic du Midi Observatory and the 3.6m Canada-France-Hawaii telescope on the same nights. These new data are under reduction.

Averaged magnitude and colours for M 15:AC211 when bright are: U=14.6, U-B=-1.2, B-V=-0.1, all ± 0.2. Taking into account EXOSAT measurements (courtesy M. Redfern) we find a ratio L_x/L_{opt} of about 20 (X: 2 kev-11 kev; opt: 3000Å-7000Å). Both this low value and the large optical variations suggest that M 15:AC211/4U2131+11 is a binary system observed at high inclination like Her X1, 0921-63, 1822-37, 2129+47. This gives a natural explanation for the M 15 X-ray source being the only globular cluster source which has never been observed to burst: the main part of the X-ray emission could be shielded by the accretion disk; it is thus possible that the X-ray source is a very high luminosity non-bursting one. Obtaining the period and the exact shape of the brightness variations will enable one to know the nature of the companion (main sequence or post m.s. star?) and the nature of the variable UV emission (bright accretion disk as in 1822-37?; heated face of the companion as in 2129+47?).

SEARCH FOR OPTICAL COUNTERPARTS IN OTHER X-RAY GLOBULAR CLUSTERS

We have made recently a reinvestigation of the HRI Einstein satellite error boxes (Grindlay et al. 1984) for some of the southern X-ray globular clusters. We used a UV coated GEC CCD (Cullum et al. 1985) at the 2.2m telescope of La Silla (ESO). UBV images in good seeing conditions (FWHM of stellar profiles 0".8 in V, 1" in U) were obtained for 47 TUC, NGC 1851, NGC 6441, NGC 6624. They are now being reduced. A quick look examination showed that no object as prominent as M 15:AC211 was observed in the error boxes for the X-ray sources.

Two most interesting targets were NGC 1851 and NGC 6624 because a possible UV counterpart could have been observed with IUE (Grindlay 1983).

We present our provisional result for NGC 1851:
5 stars are resolved nearer than 3" from the HRI Einstein position for the X-ray source. None of them is either hot enough to explain the UV emission or variable. On the other hand, blue horizontal branch stars which could be responsible for the UV emission are found in the field.

Thus, the optical counterpart of the X-ray source in NGC 1851 does not appear to be detected. This suggests that, if this object is a neutron star binary, the companion is unlikely to be a heated post main-sequence star as could be the case for M 15 (Aurière et al. 1986).

REFERENCES

Aurière, M. and Cordoni, J. P. 1981 Astron. Astrophys. Suppl. **46**, 347.
Aurière, M., Le Fèvre, O and Terzan, A. 1984 Astron. Astrophys. **138**, 415.
Aurière, M., Cordoni, J. P. and Koch-Miramond, L. 1985 IAUC 4101.
Aurière, M., Maucherat, A., Cordoni, J. P., Fort, B. and Picat, J. P. 1986 Astron. Astrophys. **158**, 158.
Charles, P. A., Aurière, M., Ilovaisky, S. and Koch-Miramond, L. 1985 IAUC 4146.
Charles, P. A., Jones, D. C. and Naylor, T. Nature, submitted.
Cullum, M. Deireies, S., D'Odorico, S. and Reiss, R. 1985 Astron. Astrophys. **153**, L1.
Grindlay, J. E. 1983 Advances in Space Research 2, No. 9 p. 133.
Grindlay, J. E., Hertz, P. Steiner, J. E., Murray, S. S., and Lightman, A. P. 1984 Astrophys. J. Letters **282**, L13.
Ilovaisky, S. and Chevalier, C. 1985 IAUC 4146.

LOW LUMINOSITY GLOBULAR CLUSTER X-RAY SOURCES

Paul Hertz

Naval Research Laboratory

ABSTRACT: Two classes of globular cluster X-ray sources are known. Each consists of compact objects accreting material from a close binary companion. The brighter class has a neutron star primary, and the low luminosity class has a white dwarf primary. These sources formed by tidal capture of the compact object by a main sequence dwarf in the core of the globular cluster. Their presence and number has implications on the end points of stellar evolution in globular clusters and on the formation of binaries in cluster cores.

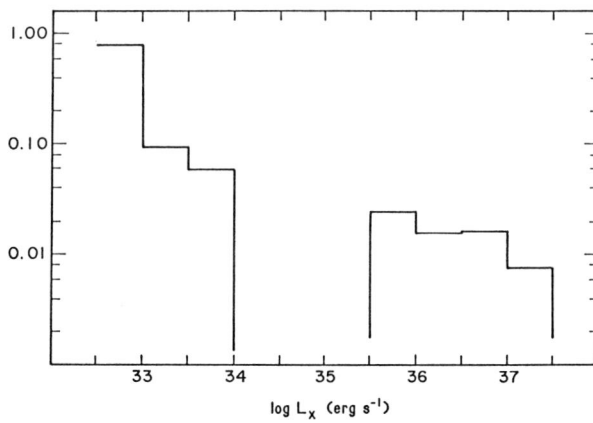

Fig. 1. Luminosity function of the brightest X-ray sources in globular clusters. From Hertz and Wood (1985).

Low luminosity globular cluster X-ray sources were discovered during a survey of galactic globular clusters with the *Einstein Observatory* (Hertz and Grindlay 1983b). These sources are 10^3 times less luminous than classical globular cluster X-ray sources. The luminosity function has a 1.5 decade gap between the two source classes. The low luminosity sources are ~100 times more numerous than the luminous sources (see Fig. 1). Every globular cluster has at least one source brighter than 10^{32} ergs s^{-1}.

Several considerations lead to the conclusion that the low luminosity sources are accreting white dwarfs in close binaries, similar to cataclysmic variables (Hertz and Grindlay 1983a): (i) the factor of ~1000 difference in X-ray luminosity is comparable to the difference in surface gravity between a neutron star and a white dwarf; (ii) the maximum luminosity observed agrees with the theoretical maximum predicted for accretion onto a low mass white dwarf; (iii) the low luminosity sources are located at larger radial offsets, indicating a lower mass than the luminous sources (Hertz 1985); (iv) the total number of sources observed is in agreement with the number expected from tidal capture of white dwarfs by main sequence dwarfs.

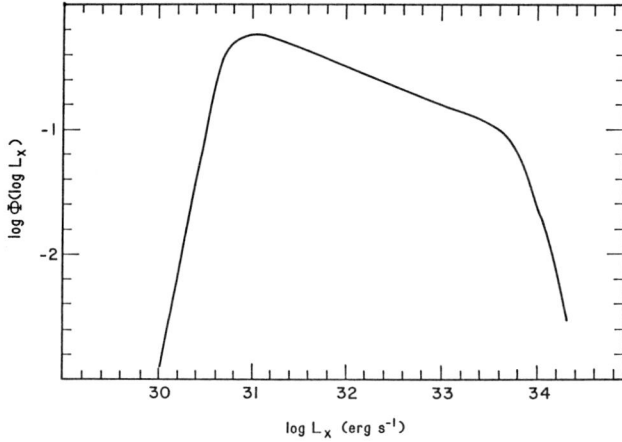

Fig. 2. Model luminosity function for low luminosity globular cluster X-ray sources. From Hertz and Wood (1985).

A simple model predicting the X-ray luminosity function for the low luminosity sources has been constructed (Hertz and Wood 1985); see Fig. 2. Comparison of the theoretical luminosity function with the data indicates remarkably good agreement and this agreement is insensitive to most input parameters. The model requires a typical cluster to have ~10 low luminosity sources (cf. Krolik 1984) and to have luminosities $L_x < 10^{34.5}$ ergs s^{-1}.

The low luminosity globular cluster X-ray sources are several orders of magnitude brighter in X-rays than galactic plane cataclysmic variables. This can be accounted for by the difference in the number of sources observed even if the X-ray luminosity functions are identical (Hertz 1985). The model predicts that X-ray surveys in the galactic plane will eventually discover X-ray luminous cataclysmic variables.

Optical identification of a globular cluster X-ray source will allow study of a tidally captured binary in the core or halo of a globular cluster. We have an ongoing program searching for the counterparts of low luminosity sources. The counterpart of the source in M15 has recently been discovered (Aurière et al. 1987).

REFERENCES

Aurière, M., Chevalier, C., Cordoni, J., Ilovaizky, S. and Koch-Miramond, L. 1987 in IAU Symposium No. 126, Globular Cluster Systems in Galaxies, J. E. Grindlay and A. G. D. Philip, eds., Reidel, Dordrecht, p. 683.
Hertz, P. 1985 in X-Ray Astronomy '84. M. Oda and R. Giacconi, eds., ISAS, Tokyo, p. 85.
Hertz, P. and Grindlay, J. E. 1983a Astrophys. J. Letters 267, L83.
Hertz, P. and Grindlay, J. E. 1983b Astrophys. J. 275, 105.
Hertz, P. and Wood, K. S. 1985 Astrophys. J. 290, 171.
Krolik, J. H. 1984 Astrophys. J. 282, 452.

EXOSAT OBSERVATIONS OF OMEGA CENTAURI

Lydie Koch-Miramond

Centre d'Etudes Nucléaires de Saclay

Michel Aurière

European Southern Observatory

1. THE OBSERVATIONS

We have undertaken an EXOSAT study of Omega Cen to tackle the two aspects of its X-ray emission revealed by the Einstein satellite (Hartwick et al. 1982) : i) diffuse emission which could be due to the interaction between the gas lost by the evolving cluster stars and the interstellar medium ; ii) dim point sources, the major part of them being white-dwarf binaries (Hertz and Grindlay 1983a,b) ; their nature is discussed by Verbunt et al (1984) and Hertz and Wood (1985). EXOSAT made two observations about 6 months apart : i) 1984, day 42 useful exposure time 39468s with Lexan 3000 filter and the Channel Multiplier Array (CMA) and exposure time 2744s with Polypropylene filter (PPL) ii) 1984, day 213, a useful exposure time of 48276s with the Lexan 3000 filter was obtained by Verbunt et al (1986). The two filters have a similar X-ray bandpass 0.062 keV and different UV leaks (Chiappetti and Giommi, 1985) : around 1900Å for the PPL filter, around 1500 Å for the Lexan filter.

2. THE UV EMISSION FROM OMEGA CEN

The emission at 1500Å and shorter seen by OAO2 and ANS is mainly due to blue horizontal-branch (BHB) stars but some "UV bright" stars are also present (Norris, 1974, Cacciari et al. 1984, Van Albada et al. 1981). The UV sensitivity of the CMA has enabled us to observe the two components of the UV emission of Omega Cen: diffuse emission due to faint unresolved BHB stars; point sources due to UV bright stars. We found that the radial variation of the UV diffuse emission agrees well with the King (1966) model proposed by DaCosta (1979; r_c = 2'63 ; r_t = 43'65) to give the best fit of the Omega Cen brightness in the B band.

Four UV stars are detected on days 42 and 213: ROA1, ROA2, ROA4 and ROA5701 (numbered according to the Woolley (1966)'s catalogue). The flux obtained for ROA2 and ROA5701 are in good agreement with the expected ones from ANS (Van Albada et al. 1981) and IUE (Cacciari et al. 1984).

3. X-RAY EMISSION OF OMEGA CEN

The extended hot gas emission discovered by Hartwick et al (1982) is not detected but our upper limit (f_x < 8 10^{-13} erg.cm^{-2}.s^{-1} in aperture 5 by

20 arcmin. assuming a thin thermal bremsstrahlung at $10^6 °K$) is not inconsistent with the Einstein result.

Three X-ray point sources are detected among 5 known from the Einstein catalogue. Only HGD on day 213 was rediscovered without using the previously known position, the count-rate being $(2.09 \pm .44)10^{-3} s^{-1}$.

3.1. Variability of HGD and HGA

The CMA count-rate ratios (day 213/day42) for HGD and HGA are respectively 3.1 and 4.5, with significance levels of 1.4 σ and 2.6 σ (Koch-Miramond and Aurière, 1986). It follows that HGD at least may be a large amplitude variable object.

3.2. Spectra and Luminosities

Whatever their assumed emitted spectrum (either a very soft black body or a hard thermal bremsstrahlung) we find that HGA and HGD can be as bright as 6 to 9 10^{33} erg.s^{-1} in the 0.2-4 keV range, i.e. brighter than any observed cataclismic variable (CV) in the galactic disc. (Cordova and Mason, 1983). Actually the Einstein IPC recalibrated data show no evidence of very soft spectra for these sources. On the contrary the best fits we have obtained for spectra of dim globular cluster sources exclude kT < 1.0 ; 0.3 ; 0.2 keV at the 90 percent confidence level, respectively for 47 TUC ; M22 (A) ; Omega CEN (HGA) sources.

4. CONCLUSIONS

HGA and HGD can have X-ray luminosities approching 10^{34} erg s^{-1} as exhibited by the central source in 47TUC. The two EXOSAT and Einstein observations require HGD (and may be HGA) to be a highly variable object. It could be a CV, since during outbursts, as observed for U Gem, Lx can increase by a factor 10 (Cordova and Mason, 1983).

REFERENCES

Cacciari, C., Caloi, V., Castellani, V. and Fusi Pecci, F. 1984 Astron. Astrophys. **139**, 285.
Chiappetti, L. and Giammi, P. 1985 Exosat Express **11**, 37.
Cordova, F. and Mason, K. 1983 in Accretion Driven Stellar X-Ray Sources, W. Lewin and van den Heuvel, eds., Cambridge Univ. Press, Cambridge.
Hartwick, F., Cowley, A. and Grindlay, J. 1982 Astrophys. J. Letters **254**, L11.
Hertz, P. and Grindlay, J. E. 1983a Astrophys. J. Letters **267**, L83.
Hertz, P. and Grindlay, J. E. 1983b Astrophys. J. **275**, 105.
Hertz, P. and Wood, K. 1985 Astrophys. J. **290**, 171.
Koch-Miramond, L. and Aurière, M. 1986, in preparation.
King, I. 1966 Astron. J. **71**, 276.
Norris, J. 1974 Astrophys. J. **194**, 109.
van Albada, T., de Boer, K. and Dickens, R. 1981 Monthly Notices Roy. Astron. Soc. **195**, 591.
Verbunt, F., van Paradijs, J. and Elson, R. 1984 Monthly Notices Roy. Astron. Soc. **210**, 899.
Verbunt, F., Shafer, R., Jansen, F., Arnaud, K. and van Paradijs, J. 1986 MPE Preprint 60.
Wooley, R. 1966 Roy. Obs. Annals **2**.

NEW METHODS FOR THE SEARCH FOR HOT GAS IN GLOBULAR CLUSTERS

Yu. N. Gnedin and T. M. Natsvlishvili

Pulkovo Observatory

ABSTRACT: Two new methods for the detection of hot gas in globular clusters are proposed: 1. measurement of the position change of a radio source due to refraction and 2. determination of the concentration of hot gas using the Faraday rotation of the polarization plane of the background radio source radiation.

Recently, an observational result was obtained, which can be interpreted as evidence for the presence of hot gas in globular clusters. A detection of diffuse X-ray emission from three globular clusters has been made: 47 Tuc, ω Cen and M 22 with the Einstein Observatory (Hartwick et al. 1982, Grindlay et al. 1984). Hartwick et al. (1982) supposed that these regions contained hot gas and that their luminosities were due to bremsstrahlung. They estimated the gas temperature and the emission measure (See Table III, in their paper.). The purpose of this report is to propose new methods of search for hot gas in globular clusters. A few of the methods may prove to be more sensitive than X-ray observations.

Distant (e.g. extragalactic) radio source emission from behind a globular cluster will undergo refraction if there is a sufficient amount of hot gas in the cluster. The effect will show up in the fact that the position of the background radio source will change with wavelength. We shall estimate the effect using the evaluations of hot gas densities from X-ray observations by Hartwick et al. (1982) and the formula in Wright and Nelson (1979)

$$\Delta\alpha = -(2\pi e^2 L/m\omega^2)\,\text{grad}_\perp n \quad (1)$$

where $\Delta\alpha$ is the angular displacement of the ray due to refraction, ω is the frequency of emission, e is the electron charge, n is the electron density and L is the characteristic length of the refraction region. The density gradient is calculated transversely to the line of sight.

Assuming the electron density varies with distance according to

$$n = n_o(r_c/r)^2 \quad (2)$$

where r_c is the radius of the globular cluster core, the angular displacement $\Delta\alpha$ is given by

$$(\Delta\alpha)" \approx 3 \times 10^{-8} n\lambda^2. \quad (3)$$

For example, the angular displacement of a radio source for $n = 2 \times 10^2$ and $\lambda = 20$ cm will reach $(\Delta\alpha) = (2 \times 10^{-3})"$, which is quite observable with the VLBI. In our opinion decameter interferometry is more suitable for this purpose. With a wavelength $\lambda = 20$ cm and densities $n = 3$ cm^{-3} and $n = 2 \times 10^2$ the angular displacement will be equal to 0.4" and 0.5" respectively.

Another effect which can be used for a determination of the hot gas density in a globular cluster is that of the Faraday rotation of the polarization plane of an extragalactic radio source projected onto the globular cluster field. The rotation angle is given by

$$\Delta\Psi = 2.62 \times 10^{-17} \lambda^2 nB \, L \quad (4)$$

where B is the magnetic field component along the path of view, L is the characteristic size of a hot gas region. The strength of the magnetic field may be estimated using the equipartition between the magnetic and thermal energy densities:

$$B_{\parallel} \sim B \, ; \, B^2/8\pi = nkT; \, B = \sqrt{8\pi nkT}. \quad (5)$$

Then from (4) and (5) it follows

$$\Delta\Psi \approx 1.5 \times 10^{-24} \lambda^2 n^{3/2} \sqrt{T} \, L. \quad (6)$$

With the hot gas parameters $T = 10^6$ K, $n = 0.6$ and $L = 7$ pc the measure of the Faraday rotation in the centimeter range ($\lambda = 2$ cm; 20 cm) reaches

$$\Delta\Psi = 0.05; \, \Delta\Psi = 5, \text{ respectively.}$$

This estimate shows that a noticeable Faraday depolarization of the background radio source radiation may take place if the magnetic field of the hot gas is random.

REFERENCES

Grindlay, J. E., Hortz, P., Steiner, J. E., Murray, S. S. and Lightman, A. P. 1984 Astrophys J. Letters 282, L13.
Hartwick, F. D. A., Cowley, A. P. and Grindlay, J. E. 1982 Astrophys. J. 254, L11.
Wright, C. S. and Nelson, G. J. 1979 Icarus 38, 123.

RADIAL VELOCITY PROFILES FOR ANISOTROPIC SPHERICALLY SYMMETRIC CLUSTERS: AN EXAMPLE

Herwig Dejonghe

The Institute for Advanced Study

A 1-parameter family of anisotropic models is presented. They all satisfy the Plummer law in the mass density, but have different velocity dispersions. Moreover, the stars are not confined to a particular subset of the total accessible phase space. This family is mathematically simple enough to be explored analytically in detail. The family is rich enough though to allow for a 3-parameter generalization which illustrates that even when both the mass density and the velocity dispersion profiles are required to be the same, a degeneracy in the possible distribution functions persists. The observational consequences of the degeneracy can be studied by calculating the observable radial velocity line profiles obtained with different distribution functions. It turns out that line profiles are relatively sensitive to changes in the distribution function. They therefore can be considered to be more natural observables when a determination of the distribution function is desired.

Be E the binding energy of a single star and L its total angular momentum. The fundamental integral equation that relates the distribution function $F(E,L)$ to the mass density reads

$$\rho(\psi,r) = 2\pi M \int_0^\psi dE \int_0^{2(\psi-E)} \frac{F(E,L)}{\sqrt{2(\psi-E) - L^2/r^2}} d(L^2/r^2). \qquad (1)$$

The double integration over the distribution function gives the mass density as a function of the potential ψ and the spherical radius r. It is immediately apparent that different $F(E,L)$ usually lead to different $\rho(\psi,r)$ but not necessarily to different $\rho(r) = \rho(\psi(r),r)$. Essentially, the problem comes to the necessity of knowing the full $\rho(\psi,r)$, while we have only the cut $\rho(\psi(r),r)$, provided by the observation of the mass density. We instead need infinitely many cuts, at least in principle. This means that the distribution function $F(E,L)$ is not uniquely determined by the mass density, and not even by the mass density together with the velocity dispersions.

It follows from (1) that anisotropic models can be constructed by expression of ρ as a function of both ψ and r. The simple form for the potential and mass density of the Plummer model (1911) suggests the following choice for the function $\rho(\psi,r)$

$$\rho_q(\psi,r) = \frac{3}{4\pi} \psi^{5-q}(1+r^2)^{-q/2}, \qquad (2)$$

with one parameter q, but leading for all q to the same mass density. The distribution function $F_q(E, L)$ that is consistent with $\rho_q(\psi, r)$ can be calculated analytically (see Dejonghe 1986a). It is expressed conveniently in terms of the hypergeometric function. Positiveness of $F_q(E, L)$ restricts the possible parameter values to $q \leq 2$. When $q < 2$, the distribution functions are nonnegative over the full region in phase space which is in principle accessible, and are zero only for $E = 0$. This is an interesting property in view of stability requirements. The limiting case $q = 2$ gives rise to a very simple distribution function (see Osipkov (1979)). It is also identical to one of the Merritt (1985) models (with anisotropy parameter $r_a = 1$).

All the spatial moments of the distribution function for general q, the marginal distributions and all the moments of the line of sight velocity distributions can be calculated analytically (Dejonghe 1986b). As an application, we can infer from the expression of the velocity dispersions that q has the same sign as Binney's anisotropy parameter β. We call a cluster with $q > 0$ a radial cluster and a cluster with $q < 0$ tangential. The well-known isotropic case has $q = 0$.

A line profile $\mathrm{lp}_{r_p}(v_p)$ at a particular projected radius r_p is defined as the distribution of the velocities v_p projected along the line of sight (and which are therefore the observed radial velocities). In order to calculate them, we need for every projected radius r_p and every projected velocity v_p to perform a triple integration: two of them integrate over three-dimensional velocity space, reducing its dimensionality to one, and the last integrates through the cluster along the line of sight. This is a somewhat tricky calculation, and it is worthwhile to try to avoid it by using the expansion

$$\mathrm{lp}_{r_p}(v_p)\, dv_p = (1 - v_p^2)^{5-\alpha}\left[c_0 + \sum_{i \geq 1} c_i(1 - v_p^2)^i\right] dv_p. \tag{3}$$

The first term follows from the asymptotical form of the tails of lp_{r_p}. The additional terms in c_i, $i \geq 1$ are correction terms: the coefficients c_i can be determined by the known moments of the line of sight velocity distribution.

At the centre, tangential clusters have a more peaked projected velocity distribution relative to radial clusters. At core radius though, the different types are virtually indistinguishable. At large radius, we see even a qualitative difference. Tangential clusters show bimodal lines, arising from the nearly circular orbits, with stars populating them in both senses in equal amount (no net streaming). Radial clusters persist in the old unimodal type of profile. Therefore, the conclusions to be drawn from the line profiles depend essentially on the region one is looking at.

As was already pointed out above, the indeterminacy in $F(E, L)$ comes essentially from the degeneracy $\rho(\psi, r) \equiv \rho(r)$. It is only natural to exploit this fact to construct distribution functions that all give rise to the same spatial mass density $\rho(r)$ and the same spatial velocity dispersions. Details about this construction are very technical and can be found in Dejonghe (1986b, preprint).

REFERENCES

Dejonghe, H.   1986a   Physics Reports 133, Nos. 3-4.
Dejonghe, H.   1986b   Monthly Notices Roy. Astron. Soc., in press.
Merritt, D.   1985   Astron. J. 90, 1027.
Osipkov, L. N.   1979   Sov. Astron. Letters 5, 77.
Plummer, H. C.   1911   Monthly Notices Roy. Astron. Soc. 71, 460.

LINEAR DENSITY WAVES IN GLOBULAR CLUSTERS

Yousef Sobouti

Shiraz University

ABSTRACT. It is often maintained that Antonov's equation, a linearization of the collisionless Liouville-Boltzmann equation, governs small perturbations of a stellar system. The variational integrals resulting from Antonov's equation are in six dimensional phase space. However, expanding the perturbations in the velocity coordinates and carrying out the integrals over the velocity components gives integrals in the three dimensional configuration space. Solutions in successive approximations lead to standing density waves. The first order equations involve a vector field $\xi(x)$ related to but not identical with the Lagrangian displacements of a volume element of the system. In this respect the problem is analogous to the linear oscillations of a fluid star. The analogy is exploited to provide a classification for the modes of oscillation and to obtain suitable data for variational calculations. The normal modes appear to be trispectral, in the sense that the associated vector field $\xi(x)$ is derived predominantly either from a scalar potential, a toroidal vector potential, or a poloidal vector potential. The eigenfrequencies of the radial ($\iota = 0$) and non-radial ($\iota = 1$) modes are calculated. The associated density waves are analyzed.

ON GRAVOTHERMAL OSCILLATIONS

Jeremy Goodman

Institute for Advanced Study

ABSTRACT: An isolated conducting gas sphere (an idealized model of a globular cluster) can have non-singular postcollapse similarity solutions for appropriate energy sources. Solutions have been found for a source mimicking three-body binaries; they are parametrized by N, the number of stars in the cluster, and are linearly stable if $N < 7000$, overstable if $7000 < N < 40,000$, and unstable if $N > 40,000$. The nonlinear development is always oscillatory and produces a typically lower central concentration than in the "equilibrium".

Previous work (e.g. Larson 1970, Hachisu et. al. 1978, Lynden-Bell and Eggleton 1980 [LBE], and Heggie 1984) has shown that models of conducting gas spheres are in good agreement with Fokker-Planck results for the pre- and post-collapse evolution of star clusters. Adding a local energy-generation rate per unit mass of the form (ρ = density, σ = velocity dispersion)

$$C\rho^a \sigma^b \tag{1}$$

to the structure equations laid out by LBE, Bettwieser and Sugimoto (1984 [BS]) found that post-collapse gas spheres undergo large variations in the central density (ρ_c), which they called "gravothermal oscillations" [GTOs].

Desiring to study GTOs further, I sought an "equilibrium" postcollapse solution that could be subjected to a linear stability analysis. Previously known self-similar postcollapse models (e.g. Hénon 1961, 1965; Inagaki and Lynden-Bell 1983) are all unstable, and with infinite growth rates, because of their infinite central densities. A fuller account of the work reported here will appear in Goodman (1987).

It is well known (Hénon 1965) that the half-mass radius, r_h, of an isolated postcollapse cluster expands as the two-thirds power of time, t. A non-singular postcollapse similarity solution must have a core radius, r_c, with the same scaling. It can be shown that self-similar solutions are possible if

$$6a + b = 5. \tag{2}$$

I assume that $a > 1/2$ so that most of the energy is generated in the core.

Energy-generation by three-body binaries satisfies (2) with $a = 2$ and $b = -7$. Solutions based on this particular energy source have been constructed. The solutions are indexed by a single dimensionless parameter, Γ, which is related to N by

$$\Gamma \approx \frac{2.4}{N^2}, \tag{3}$$

although the coefficient here conceals a logarithmic dependence on N and uncertainties in the binary efficiency. The structure of the solutions around the half-mass radius and beyond is almost independent of Γ; the solutions are nearly isothermal inside the half-mass radius. The age, t, of the self-similar solution is always approximately $11 t_{rh}$. But the smaller is Γ the more extreme are the central conditions:

$$r_c/r_h \approx 4.4/N^{2/3},$$
$$t_{rc}/t_{rh} \approx 40./N^{4/3}, \qquad (4)$$

where t_{rc} and t_{rh} are the central and half-mass relaxation times as defined by Spitzer and Hart (1971). Three-body binary creation becomes less efficient compared to two-body relaxation as N grows, whence a balance between energy production in the core and the overall cluster expansion can be achieved only by increasing ρ_c.

When described by the appropriate comoving variables, the solutions appear static, so that their stability can be analyzed in the same way as that of a true equilibrium; modes evolve however as t^s rather than as $\exp(st)$. The models are stable for $\log \Gamma > -7.3258$, or (by (3)), for $N < 7000$. But for $7000 < N < 40,000$ they are overstable, and for $N > 40,000$ they are unstable.

For very large N, the growth rate approaches $(\sim 1/(570 t_{rc}))$, and a separate calculation shows that an infinite but non-singular isothermal sphere without an energy source has the same growth rate. This and other considerations lead me to agree with BS that, asymptotically at least, the instability leading to gravothermal oscillations is the same as that discovered by Antonov (1962). The physical details in the overstable regime, however, are still obscure.

When non-linear, the instabilities become chaotic relaxation oscillations. By far the largest part of a cycle is spent near minimum density, since $1/t_{rc}$ and the evolution rate are then greatest. Consequently, if observed at a random time in its postcollapse evolution, a cluster with $N \gg 40,000$ would have a much lower central concentration than would the self-similar equilibrium for the same N. Comparison of these results with the King sequence suggest $\log(r_{tidal}/r_c) \sim 2 - 3$ independently of N if N is large, but this result is tentative as tidal limitation and other important physical complications have been omitted from the present models, and because a thorough exploration of parameter space has not been made.

Instability is also likely for other energy sources that are inefficient enough so that $r_h/r_c > 10^2$ at late times. This appears to be the case for tidal binaries (Statler et. al. 1986).

REFERENCES

Antonov, V. A. 1962 Vest. Lenigrad Univ 7, 135 (Trans. in IAU Symposium No. 113, Dynamics of Star Clusters, J. Goodman and P. Hut eds., Reidel, Dordrecht, p. 525.
Bettwieser, E. and Sugimoto, D. 1984 Monthly Notices Roy. Astron. Soc. 208, 439. [BS]
Goodman, J. 1987 Astrophys. J. 313, in press.
Hachisu, I., Nakada, Y., Nomoto, K. and Sugimoto, D. 1978 Prog. Theor. Phys 60, 393.
Heggie, D. C. 1984 Monthly Notices Roy. Astron. Soc. 206, 179.
Hénon, M. 1961 Annals d'Astrophys. 24, 369.
Hénon, M. 1965 Annals d'Astrophys. 28, 62.
Inagaki, S. and Lynden-Bell, D. 1983 Monthly Notices Roy. Astron. Soc. 205, 913.
Larson, R. B. 1970 Monthly Notices Roy. Astron. Soc. 147, 323.
Lynden-Bell, D. and Eggleton, P. P. 1980 Monthly Notices Roy. Astron. Soc. 191, 483, [LBE]
Spitzer, L. and Hart, M. H. 1971 Astrophys. J. 164, 399.
Statler, T. S., Ostriker, J. P. and Cohn, H. N. 1986, preprint.

COOLING AND FRAGMENTATION OF PROTO-CLOBULAR CLUSTER CLOUDS

Francesco Palla

Astrophysical Observatory Arcetri

Hans Zinnecker

Royal Observatory, Edinburgh

Recently, Fall and Rees (1985) have proposed a theory for the origin of globular clusters forming from the largely primordial gas in the protogalaxy. These authors have explained the typical masses of proto-globular cluster clouds ($\sim 10^6$ $M_\odot$) as gravitationally unstable condensations at temperature T $\sim 10^4$ in a hot protogalactic medium (T $\sim 10^6$ K) but they were not concerned with how these clouds would fragment into stellar masses (~ 1 $M_\odot$). In fact, their proto-globular cluster clouds are trapped at T $\sim 10^4$ K, and cannot cool to lower temperatures. However, substantial cooling must occur if these clouds are to form solar mass stars. It is known that under primordial conditions the only available cooling agent is molecular hydrogen, formed in the gas phase. Therefore, if sufficient molecular hydrogen is formed, it is possible to cool the gas well below T $\sim 10^4$ K. In the following we outline how non-equilibrium conditions lead to a larger H_2 abundance than derived by Fall and Rees, who assumed equilibrium conditions.

When the protocluster gas has cooled down from high temperatures ($\sim 10^6$ K) to temperatures of the order of 15000 K, hydrogen recombination starts to occur, but it will be out of equilibrium, because the cooling time due to Ly-α photons is shorter than the recombination time. At T = 10^4 K the relation between the recombination and Ly-α cooling time scales is 10(1-x) where x is the fractional degree of ionization. This shows that ionization equilibrium (where x would be $< 10^{-2}$ cannot be reached; therefore, a high ionization fraction (x < 0.9) is maintained at T $\sim 10^4$ K, and even below that temperature. Note that for high fractional ionization both time scales are much shorter than the free-fall time of the protocluster cloud ($t_{ff} \sim 10^{6.5}$ years at cloud densities of $\sim 10^2$ cm^{-3}). The high fractional ionization promotes the efficient formation of H^- and H^+_2 which both contribute to H_2 formation. [$H + H^+ \rightarrow H^+_2 + \gamma$; $H^+_2 + H \rightarrow H_2 + H^+$; $H + e \rightarrow H^- + \gamma$; $H^- + H \rightarrow H_2 + e$]

Fall and Rees did not consider the H^+_2 channel, but this is the driving reaction at the highest temperatures (8000 K $<$ T $<$ 10000 K) since H⁻ is rapidly destroyed by collisions with H^+, rather than reacting with H atoms to form H_2. We note that the chemistry in this situation is more complicated (see Palla and Zinnecker 1986), but the net result can be approximated as above. Self-shielding of H_2 against photodissociation in the external radiation field also turns out to be non-negligible, since the H_2 column densities across the protocluster clouds exceed the critical value (Federman et al. 1979). As a result of the more efficient H_2 formation process, the final molecular abundance can be as high as 10^{-2} - 10^{-3} rather than 10^{-4} - 10^{-6} predicted in the equilibrium scheme. (Mac Low and Shull 1985, and Shapiro and Kang 1986 have studies the formation of H_2 in the context of pregalactic shocks and galaxy formation, reaching similar conclusions.)

With such a high fractional H_2 abundance the gas can cool almost instantaneously (in a time very short compared to the free-fall time) to temperatures as low as 100 K. At this temperature the protocluster density will be 10^4 cm^{-3}, in pressure equilibrium with the surrounding hot and diluted halo gas. The H_2 molecules will survive at these densities, because collisional dissociation will not be operative at such low temperatures, despite the strong density dependence of the dissociation rate (Lepp and Shull 1983). The same applies to HD molecules, which are a much better coolant than H_2 at T < 150 K (Dalgarno and Wright 1972) and may allow the gas to cool down to a few tens of degrees Kelvin so that the Jeans mass drops below 1 $M_\odot$.

In principle, isothermal contraction (at T = 10^4 K) implies lowering the Jeans mass, and when the density is sufficiently high (~ 10^9 cm^{-3}) H_2 formation via three-body reactions (Palla et al. 1983) may occur. The associated cooling can lower the temperatures to T ~ 10^3 K, but in order to get a Jeans mass of order 1 $M_\odot$ would require very high densities (~ 10^{12} cm^{-3}) and fragmentation at these densities would lead to a cluster far too compact to resemble a typical globular cluster. This suggests that cooling and fragmentation should occur at a much earlier stage in the evolution of the proto-globular cluster cloud. We thank Dr. R. Stanga at ESO for making this effort possible. We would also like to thank Professor M. Rees for a timely preprint.

REFERENCES

Dalgarno, A. and Wright, E. L. 1972 Astrophys. J. Lett. 174, L49.
Fall, S. M. and Rees, M. J. 1985 Astrophys. J. 298, 18.
Federman, S. R., Glassgold, A. E. and Kwan, J. 1979 Astrophys. J. 227, 446.
Lepp. S, and Shull, J. M. 1983 Astrophys. J 270, 578.
Mac Low, M. M. and Shull 1986 Astrophys. J. 302, 585.
Palla, F., Salpeter, E. E. and Stahler, S. K. 1983 Astrophys. J. 271, 632.
Palla, F. and Zinnecker, H. 1986, in preparation.
Shapiro, P. R. and Kang, H. 1986 preprint (Univ. of Texas).

FORMATION OF GLOBULAR CLUSTERS AND THE FIRST STELLAR GENERATIONS

A. Di Fazio

Astronomical Observatory of Rome

Using a new theory of fragmentation via gravitational instability (Di Fazio 1986) a multi-fluid evolution model for a self-gravitating spherical gas cloud (Di Fazio 9186), new opacity functions suitable for media of low density and temperature (Capuzzo Dolcetta, Di Fazio and Palla 1985), the initial collapse, fragmentation and virialization of a <u>protogalactic</u> gas cloud is followed. The chosen initial conditions are: $M = 10^{11} M_\odot$, $R_o = 50$ kpc, $T_o = 10$ K, $Z = 0$ (only H + He), $M_{fragments}$ (t=0) = 0° (pure gas initially). The difference, with respect to Di Fazio (1986), consists in taking into account the formation and dissociation of H_2, together with its radiative processes. The calculations are performed in the approximation of LTE (besides carrying out an additional calculation with a NLTE-estimated correction factor in the abundance of H_2). The results are compared with Di Fazio (1986). Four generations of fragments are again obtained, with small differences in the mass ranges and slopes. Isothermal phases at lower temperatures (1800 - 2300 K) than before (Di Fazio 1986) are reached. The virialization time for the fragments does not vary, as expected. A mass function is obtained for each of the mentioned four generations. The second generation (peaking at $10^5 M_\odot$) represents (also the shape of the mass function agrees with the globular cluster one) the family of the proto-globular clusters. The third generation (peaking at ~ 100 $M_\odot$) is a valid candidate for the first stellar generation. The presence of H_2 allows the formation of lighter objects than in Di Fazio (1986): M can get as low as ~ 40 $M_\odot$, instead of 80 $M_\odot$. After the explosion of these objects (Δt ~ 2 Myr) a new, metal (C,O)-enriched generation of globular clusters is formed, which fragments into lighter "stars" (mass range [0.035, 100]$M_\odot$). The mass function shape resembles that of Pop II stars. (See Scalo 1985.)

It is interesting to note that the accounting for H_2 radiative processes contradicts Palla et al. (1983), which predicted very low mass stars (~ a few times 0.01 $M_\odot$) for the first stellar generation. We ascribe this difference to the following reasons: 1. the mentioned work does not conserve energy in its computation; 2. it <u>assumes</u> a continued collapse (so that the density is ever-increasing, helping the Jeans mass to decrease, even though the formation of stars would induce a stop in the collapses of the system, as the orbiting system of stars has a "stiffness" $\gamma = 5/3$ to contraction) without calculating the dynamics coupled to the thermodynamics; 3. as explained in Di Fazio (1986), it uses the total density, in place of the gas density alone as it would be correct, in the computation of the Jeans mass, so that the result is smaller. This is not acceptable, as in only a Jeans time (about 1.4 free-fall times in these cases) a predominant fraction of mass is in stars (so that $\rho/\rho_{stars} \ll 1$), and $\rho \sim \rho_{total}$ is no longer acceptable. The Jeans mass would rise again (even though the temperature behavior is isothermal), stopping the fragmentation. So, in conclusion we show that the introduction of H_2 cooling is important <u>only in getting a more precise estimate</u> of the temperatures, and thus of the fragment masses, but that H_2 cooling can <u>by no means</u> lower by orders of magnitude the characteristic masses obtained. The fundamental fact is that the Jeans mass (due both to the gas exhaustion and to the bounce of the system) rises back up in only 1.3 - 1.4 free-fall times.

REFERENCES

Capuzzo Dolcetta, R., Di Fazio, A and Palla, F. 1985 <u>Astron. Astrophys.</u> 145, 290.
Di Fazio, A. 1986 <u>Astron. Astrophys.</u> 159, 49.
Palla, F. Salpeter, E. E. and Stahler S. W. 1983 <u>Astrophys. J.</u> 271, 632
Scalo, J. M. 1985 in <u>Protostars and Planets, II</u>, D. C. Black, M. S. Matthews, eds., Univ. of Arizona Press, Tucson, p. 201.

FORMATION OF POPULATION III OBJECTS DUE TO COSMIC STRINGS

Tetsuya Hara and Shigeru Miyoshi

Kyoto Sangyo University

ABSTRACT. Behind the moving cosmic strings, wakes are formed from dark matter and baryon matter. The shock waves for baryon matter appear after the stage $z \simeq 790(\mu/10^{-6})$. These wakes will fragment into lumps of mass $M \simeq 10^8 (\mu/10^{-6})^3 (v_{st}/c)^3/(1+z)^{1.5} M_\odot$ where μ and $v_{st} \sim c$ are the line density in units of $G=c=1$ and the velocity of the strings.

1 INTRODUCTION

It has been discussed that the galaxies are formed by closed cosmic strings (Vilenkin 1985), while not yet so much attention has been paid to the formation of galaxies and other objects in the wakes behind moving cosmic strings, originally suggested by Silk and Vilenkin(1984).

Behind the moving string, the particles obtain the velocity, perpendicular to the direction of the string, as

$$v_t = 4\pi G \mu v_{st}/c^2 = 3.8 \times 10^5 (\mu/10^{-6}) \cdot (v_{st}/c) \text{ cm/sec} .$$

The stage at $v_t = v_s$ where v_s is the sound velocity is given by $1+z_s \simeq 790(\mu/10^{-6})$. Before this stage, the compression of baryon matter propagates with velocity v_s as sound wave into the surrounding space. After $1+z_s$, shock waves will appear behind the string. The temperature behind the strong adiabatic shock is $T \simeq 3 m_h \cdot \mu_m v_t^2/(16 k_B) \simeq 400(\mu/10^{-6})^2$ K , which is not high enough to ionize the baryon matter, so the cooling is mainly due to H_2 molecules which becomes efficient after $z \simeq 10^2$. The fragmented gas clouds will contract and stars of mass 10^{-1} $\sim 10^2 M_\odot$ will be formed (Palla et al. 1983).

2 WAKES OF DARK MATTER AND BARYON MATTER

Assuming that the wake is extending plane-like in the y-z plane infinitely, the equation of motion for dark matter (P=0) and baryon matter is given by

$$d^2x/dt^2 = -4\pi G \rho_b x/3 - 4\pi G \int (\rho - \rho_b) dx - 1/\rho_n \cdot dP/dx ,$$

where ρ, ρ_b and ρ_n are the total, background and baryon density, respectively. An example of the numerical calculations for $\Omega_d=0.9$ and $\Omega_b=0.1$ started from $1+z_i=10^2$ is given in Figure 1. Dark matter is calculated with 2000 particles for $x \geq 0$ region and baryon matter is

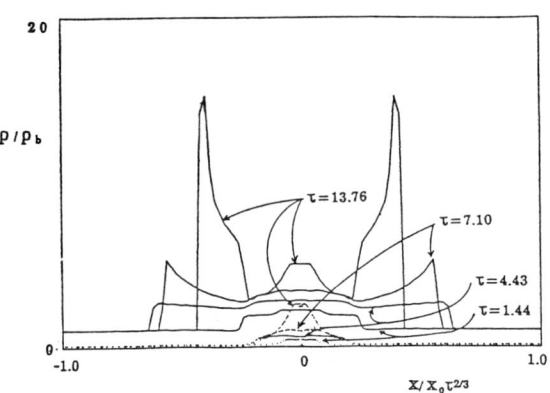

Fig. 1 The density of total matter including the dark and baryon matter. The dotted lines show the density of the baryon matter. The time is normalized as $\tau=t/t_i$ where t_i is the initial time. The abscissa is the distance from the symmetric plane, normalized by $x_0 \tau^{2/3}$ where $x_0 = 2vt/(3t_i)$. The ordinate is normalized by the background density ρ_b.

followed by 500 Lagrangean points.

For the dark matter, infinite density planes appear after the time $\tau \simeq 9$ and they are unstable for fragmentation with the time scale proportional to $\sqrt{\rho_b/\rho} t_i \tau$. For baryon matter, the density is much less than that of the dark matter at first, and hence the pressure is almost constant through the layer. After $\tau \geq 10$, the density increases by its self gravity and the gas pressure gradients appear. The gravitational instability may also occur in the central plane. The fragmented mass could be approximately estimated from the virial mass as

$$M \simeq v_t^2 \Delta x/G \simeq 10^8 (\mu/10^{-6})3(v_{st}/c)3/(1+z)1.5 \tau_f^{2/3}(\Delta x/(\tau_f^{2/3} x_0)) M_\odot ,$$

where τ_f is the fragmentation time. The criterions $k^2 c^2 \leq 4\pi G\rho$ and including the expanding effects give almost the same mass scale. Even after the fragmentation, dark clouds will oscillate through the central plane.

3 EFFECTS OF POPULATION III OBJECTS

Even though the width of the perturbed region $v_t t$ is small compared to the horizon ct, the energy supply from such Population III objects by shock waves or x-rays from the supernovae or disks around black holes will affect the surrounding space profoundly and other astronomical objects such as galaxies will be formed there (Ikeuchi 1981).

REFERENCES

Ikeuchi, S. 1981 Publ. Astron. Soc. Japan 33, 211.
Palla, F., Salpeter, E. and Stahler, S. 1983 Astrophys. J. 271, 632.
Silk, J. and Vilenkin, A. 1984 Phys. Rev. Letters 53, 1700.
Vilenkin, A. 1985 Phys. Reports 121, 263.

FOURTH POSTER DISCUSSION

KING: For convenience I have tried to group the papers of this section into a logical order by subject matter. The first group deals with surface photometry, in the papers A Multicolor CCD Survey of Southern Globular Clusters and The Structure of Collapsed Cluster Cores. I think that we should also include in our discussion the related material from George Djorgovski's invited paper this morning. All three of these studies are quite similar in their methods, and they even have a number of clusters in common. As far as I can see, their observations agree with each other rather well; where there are differences, they are differences of interpretation. Forte and Mendez put more emphasis on small deviations than I would, but I suspect that Djorgovski might agree more with them.

COHEN: Does your new set of U frames indicate the presence of cusps in those clusters where Djorgovski & King saw central cusps?

GRINDLAY: To the extent we have analyzed our new UBV CCD data beyond the first clusters reported in our paper, they agree with the Djorgovski and King results. We have good data on all of their candidate cusp clusters and are now deriving U-profiles and model fits for each.

LARSON: Djorgovski said this morning that globular clusters come in two "flavors", with and without cusps. Could there actually be a continuum of properties in this respect?

DJORGOVSKI: Yes, it is clearly possible that there are weak or invisible cusps in King-model-like clusters. That may be the explanation for high-concentration clusters, like NGC 5824, which should have collapsed by now, but look like "perfect" King models. In most cases one can clearly classify the profiles as either King model or PCC, but there are some intermediate or noisy cases, where we cannot tell.

KING: George and I work together, but we don't necessarily agree about everything. I do think that we are likely to have real intermediate cases. A problem, though, is that some of our data are rather noisy.

GRINDLAY: I agree (with Ivan) that cusps could be missed (even in U) in some clusters. The available resolution (both seeing and detector) is obviously important, and further cusps may become evident with studies such as the one we have initiated at the CFHT with both better seeing and the double density CCD chip. The factor-of-two pixel resolution is important.

KING: The next set of papers all deal with velocities, although they are rather different from each other. The paper Radial Velocity Study

of NGC 6712 deals with radial velocities in one cluster and notes that
the low velocity dispersion implies that there are not a lot of massive
remnants, whereas the cluster has an X-ray source. A Survey of
Globular Cluster Velocity Dispersions fits isotropic-velocity models to
a number of clusters and derives values of M/L, which do not appear to
correlate with anything. This does not directly contradict the
much-discussed result of McClure et al., but it does lead me to wonder
a little. Anisotropy in Omega Centauri and 47 Tucanae fits detailed
models to two clusters and determines values of the anisotropy. Here I
was pleased to see that Omega Centauri, which has about 30 times as
long a relaxation time, has much more anisotropy than 47 Tucanae.

WEBBINK: Regarding the lack of correlation of M/L with anything else,
as discussed in the poster paper by Pryor et al., this same effect
appears in Illingworth's survey. The cluster-to-cluster variations
appear to be real. I realize you do not have a very large sample of
clusters, but did you look for a correlation with a combination of
parameters?

PRYOR: I have checked the physically meaningful combinations that I
could think of and found no correlations. If you have any suggestions
of quantities to check, I would like to hear them.

KING: Did you include other people's velocities in your paper?

PRYOR: The clusters on our poster are only those for which we have
obtained data. However, including Illingworth's clusters does not
change the conclusion that there is no apparent correlation between M/L
and metallicity.

PENNY: The velocity anisotropies deduced by Meylan for ω Cen are of
the form that would reduce the discrepancies between the radial
velocities and proper motions in M 22 found by Kyle Cudworth.

CUDWORTH: 1) The M 22 proper motions are within a few core radii of
the cluster center, where anisotropy is not expected. We looked for
and did not find evidence for anisotropy in the proper motion data. 2)
Proper motions in 47 Tuc are currently being reduced and will go out to
a larger radius than the radial velocity data that has been published.

KING: Now that we have observed values of the anisotropy in ω Cen, 47
Tuc, M 3, and M 13, it ought to be possible to compare them with the
relaxation times and predict the anisotropies in other clusters.

KING: The next set of papers deals with dynamical evolution of
clusters. Evolution of Globular Clusters Including a Degenerate
Component and Evolution of Globular Clusters with Tidally-Captured
Binaries through Core Collapse deal with post-core-collapse evolution,
including the effect of binaries; The Formation and Evolution of
Three-Body Binaries in Evolved Globular Clusters and Tidal Effects on
Stellar Evolution in Close Binaries Formed in Globular Clusters treat

the formation and behavior of the binaries themselves; and The Effects of Stellar Evolution on Globular Cluster Dynamical Evolution follows evolution with a number of effects operating simultaneously.

KING: Would one of the authors please contrast the approaches of the first two papers?.

LEE, H. M.: The paper by Statler et al. showed that, using a single component model, tidally captured binaries can reverse the core collapse well before any dynamically important three-body binaries form. In my paper, the massive degenerate stars are included to get a wide range of solutions for post-collapse clusters. Especially, some clusters with very small cores (e.g., M 15) can be accounted for by this model.

KING: The paper by McMillan seems to increase the effect of binaries. Is that right?

MCMILLAN: The main point I wanted to make in this paper is that, if you want to include "three-body" binaries correctly in a simulation, it is inadequate only to consider three-body encounters. I find that as many as fifty percent of the encounters involving binaries (formation and subsequent interactions) involve a total of four or more stars and this has a substantial effect on the hardening and heating rates, even for quite hard pairs.

KING: The paper by Ray and Kembhavi improves the calculations of the tidal formation of binaries. In this connection, Bob Mathieu has an interesting empirical result on the rate at which orbits circularize.

MATHIEU: Two years ago Mayor and Mermilliod noted that all ~ 1 $M_\odot$ binaries in the Hyades with periods of less than 5.7^d had circular orbits, while those with longer periods all had eccentric orbits. They attributed this cutoff to tidal circularization processes - at 0.8 billion years binaries with periods less than 5.7 days have had time to tidally circularize. We have recently completed our radial velocity survey of M 67 and find a similar cutoff, but at 10.7 days. This longer period cutoff is in the expected sense given at tidal circularization integration. In fact, this result argues for a new clock with which to date clusters, one that is independent of stellar evolution theory and, when used in a relative sense, is also not sensitive to stellar interiors theory. Tsevi Mazeh and I have developed the clock mechanism from the circularization theory of Lecar et al. and applied it to the Hyades and M 67. Our preliminary results are ages of 0.2 - 0.4 billion years for the Hyades and 5 - 7.5 billion years for M 67, to be compared with stellar evolution ages of 0.8 billion and 5 billion years respectively. Ivan was not so very impressed with the accuracy, but I find it quite encouraging given the primitive state of the clock mechanism! More to the point at this session, the agreement is remarkable given the very simple state of tidal circularization theory; Lecar et al. expected at best order of magnitude accuracy from their theory. Our observations indicate that the theory may be in better shape than we had a right to expect, and add support to the application of similar theory to the problem of tidal-capture binaries. We are extending our binary survey to the young cluster M 35, where we predict a period cutoff of about 3 days.

KING: Would Dave Chernoff please explain the relation of his paper to his other papers at this symposium?

CHERNOFF: This paper explains the precollapse evolution of a globular cluster under the influence of 1) a tidal boundary 2) a mass spectrum and 3) stellar evolution. The results indicate that for a moderately bound cluster (W_o = 3) of 10^5 $M_\odot$ at the galactocentric location a substantial fraction of mass is lost from the cluster across the tidal boundary and collapse is significantly delayed.

CAPUZZO-DOLCETTA: With regard to Chernoff's paper on the influence of stellar evolution on globular cluster dynamics, I would point out that a lot of work (and results) on this topic were obtained several years ago by Angeletti, Giannone and me and published in the journals Astron. Astrophys. and Astrophys and Space Sci.

KING: The next set of papers deals with distributions that are segregated, presumably by stellar mass. The Spatial Distribution of Spectroscopic Binaries looks at binaries and blue stragglers in the "honorary globular cluster" M 67; Evidence for Mass Segregation in NGC 5466 looks at blue stragglers in NGC 5466 (and adds some preliminary information on NGC 5053); Origin and Radial Distribution of Faint Blue Horizontal-Branch Stars deals with faint BHB stars in M 15. There is a lot of agreement between these papers, about most of these stellar types being more centrally concentrated. The only thing we have seen that doesn't seem to fit is the blue-straggler result mentioned previously by Flavio Fusi Pecci. I would like to give him a few minutes to try to reconcile the discrepancy.

FUSI PECCI: From a detailed comparison of the HB morphologies of the galactic globular clusters, M 15, NGC 5466 and M 92 we have shown (Buonanno et al. 1985, Astron. Astrophys. 145, 97) that the BHB stars found in the blue horizontal branch of M 15 fainter than the HB gap are "bona fide" HB stars (by considering the population ratios over complete samples of the different branches and the different clusters). We have also proposed as a working hypothesis a naive interpretation of both the peculiar spatial distribution of blue HB stars we find in M 15 and in NGC 6752 (BHB stars are more frequent in the outer regions). The basic idea is that the central stellar density of the cluster (coupled with core rotation and the mass loss of each individual star) may have induced on each star subtle differential effects visible only during "special" evolutionary phases, such as the HB phase is. According to this suggestion one would expect (as a zeroth approximation): i, high stellar density: BHB stars more frequent in the outer regions; ii, low stellar density: BHB stars uniformly distributed or more frequent in central regions of the cluster.

GRINDLAY: I think the main point of disagreement between us is whether the radial dependence of faint blue horizontal-branch stars can even be discussed for very dense clusters, such as M 15 or NGC 6752. FBHB stars can simply not be recognized in the central few core radii of M 15 (although we are trying to do so with new high resolution CFHT observations) whereas they can be studied in lower density clusters such as, ω Cen. Dynamical mass segregation effects are expected to only be significant for the central few core radii.

NEMEC: A recent CM diagram for NGC 4833 (Nemec, Richer & Fahlman, 1986, in prep.), a relatively open, metal-poor globular cluster, shows a very faint blue horizontal branch. Does this, then, represent an exception to your trend that extended blue horizontal branches go with low concentration class?

FUSI PECCI: Is this connected with the existence and spatial distribution of blue stragglers? It is hard to answer as: i, many aspects may play a role: binary existence, frequency and origin; cluster central density; mass loss; metallicity; etc.; ii, the observational data are still insufficient to present any reliable scenario: NGC 5466, NGC 5053 - low stellar density - have BS with a spatial distribution peaked at the cluster center (Nemec et al., this Symposium), M 3 - high stellar density - displays a high BS frequency in the outer regions (Buonanno et al., this Symposium). My present opinion is that: i, the interpretation of HB morphology and HB star spatial distribution does not require mass segregation or the existence of a signigicant population of binaries, ii, BS presence and spatial distribution do probably require or imply both of them.

HESSER: To Grindlay: Have you examined the images of your blue stars in ω Cen, which were presumably located by an automatic star finding routine? Did their images look abnormal?

GRINDLAY: Yes, we have looked at the field blue horizontal-branch star images and they are real. We have also obtained spectra at CTIO on several of these stars and verified that they are indeed cluster FBHB stars.

KING: The next set of papers deals with particular kinds of stars, and to some extent with their origins. In discussing them, we should keep the preceding set in mind too. <u>Variability of Omega Centauri Blue Stragglers: Clues to their Origin</u> is an interesting attempt to connect blue stragglers with dwarf Cepheids, which are another kind of mystery star; and <u>A Search for Optical Counterparts of Globular Cluster X-ray Sources</u> is actually a pair of papers, one dealing with the identified optical counterpart of the X-ray source in M 15 and the other reporting on a less successful examination of some other clusters. And I think that along with these papers we should again take note of the paper

Candidate for the Recovered Nova of 1938 in the Globular Cluster M 14 on the M 14 nova.

WEBBINK: The position of the old nova candidate in M 14 in the HR diagram of that cluster would imply $M_v \sim +3$. Every other old nova this bright is extremely blue, making the color of this star (roughly that of a late K dwarf) quite anomalous.

MATHIEU: I have a question for Ron Webbink that I tried to ask him in Delhi, but I couldn't find him! Ron, what do you think blue stragglers are?

WEBBINK: I have no strong convictions regarding their true nature, though I don't think any of the alternatives are more compelling than the binary hypothesis. I would point out that the occurrence of blue stragglers more than two magnitudes above turnoff is not necessarily evidence against a binary origin, although it is frequently cited as such. A star which increases rapidly in mass - on a thermal timescale or faster - becomes very oversized and overluminous for its mass. Thus, if mass transfer or coalescence occurs fast enough, blue stragglers well above the two magnitude limit can be created, but the ratio of the number of blue stragglers more than two magnitudes above turnoff to the total number of blue stragglers should not exceed the ratio of thermal to nuclear timescales for those stars.

KING: Isn't that ratio quite small?

WEBBINK: It's about 1/10.

MCMILLAN: Another possible way of producing an anomalously blue star is the tidal capture process itself. After capture and circularization, the pulsational energy deposited in the component stars is radiated away. I have found that excess luminosities of over a hundred times that of the unheated star are quite possible. The star's response to this extra heat, which is deposited primarily near the surface, is a substantial increase (factor of two or three) in surface temperature, although an increase in radius ($\sim 50\%$-100%) also occurs. Thus, a much brighter, bluer object results.

KING: The next set of papers deals with the properties of X-ray sources in globular clusters, and also of the possibility of a related hot gas. Low Luminosity Globular Cluster X-ray Sources looks to me like a rather definitive discussion of low-luminosity X-ray sources in globular clusters. Exosat Observations of Omega Centauri reports four separate kinds of X-ray information for Omega Centauri, and New Methods of Search for a Hot Gas in Clusters discusses how hot gas might be looked for. I would like to start the discussion by asking Paul Hertz whether there is any correlation between low-luminosity X-ray sources and the presence of a central cusp.

HERTZ: A quick comparison shows no correlation. Low luminosity X-ray sources are correlated with tidal capture timescales. The presence of cusps may also be so correlated, but not at a level significant enough to convince anyone who knows anything about statistics.

KRON: M 22 is in the IRAS faint source catalog and is bright at 12.5μ.

COHEN: There is a previously undiscovered planetary nebula near the core of M 22 which is the IRAS point source (Neugebauer, Soifer, Gillett & Cohen, to be published).

KING: Next we have some more theoretical papers, which I separated out because they seem to be of a more abstract nature. Radial Velocity Profiles for Anisotropic Symmetric Clusters: An Example is an impressive but complicated discussion of a family of Plummer models. Personally I have always been peeved by Plummer models, because they don't look like real clusters at all. My first reaction to this paper was like that of the French general who witnessed the charge of the Light Brigade: "C'est magnifique, mais ce n'est pas la guerre!" But then I looked at it in a more responsible way. I think that the indeterminacies pointed out by de Jonghe must apply in a similar way to more realistic cluster models, and we should take serious note of the problem. Linear Density Waves in Globular Clusters deals with the possibility of density waves in globular clusters, on which I'm not capable of commenting. On Gravothermal Oscillations is far too long and difficult for me to have understood it in the brief time available, so I would like to ask Jeremy Goodman if he will make some brief remarks about it.

GOODMAN: I would like to emphasize that postcollapse clusters are expected to undergo large-amplitude oscillations in central density on a timescale of about one half-mass relaxation time. We do not yet know accurately how low the central density may become during the re-expansion phase (better calculations are needed), but it is possible that some apparently normal cores actually belong to postcollapse clusters. What was your question about M 67 here?

MATHIEU: M 67 is well fit by multi-mass King models, with no obvious irregularities in the core. But you must remember that the number of stars is small. If you look at Terlevich's or Aarseth's N-body simulations, while they show "oscillations" in the core, typically they involve only one or two binaries plus perhaps a few single stars. This would be difficult to distinguish. Indeed, several of our detected binaries are in the very center of the cluster and might be involved in such multiple-body interactions at the moment. (Jeremy then asked about the number of stars). The total mass of M 67 is of order 2000 $M_\odot$. The number of stars actually used in my dynamical analyses (set by the magnitude limit of V = 19) was about 800.

KING: Regarding the paper by Goodman, it appears that a cluster that has core oscillations will spend most of its time in a puffed-up state. Should we observe something unusual then, or should the cluster look like an ordinary equilibrium state?

GOODMAN: These oscillating clusters are certainly in dynamical equilibrium (they obey Jeans' theorem), but not in thermodynamic equilibrium; but neither are King models. They do have approximately Maxwellian velocity distributions.

KING: Finally, we have three contributions related to problems of formation of globular clusters. Fragmentation of Proto-Globular Clusters discusses molecular processes that the authors believe are relevant, and The Formation of Globular Clusters and the First Stellar Generation deals with successive stages of fragmentation. Formation of Population III Objects Due to Cosmic Strings tries to relate cosmic strings to cluster formation, and I can't make any intelligent comments on it at all. Do the authors of any of these papers have anything to add?

DI FAZIO: Just a short comment about my paper. It is interesting to note that the introduction of H_2 formation and H_2 cooling functions in the radiative calculations for the evolution of a proto-globular cluster and its fragmentation, does not alter dramatically the mass range of the mass spectrum of the first stellar generation formed. In fact, an isothermal phase is indeed reached at lower temperatures (due to H_2 cooling processes), but the ambient gas exhaustion due to the fragmentation process itself raises the Jeans mass back up, halting the fragmentation. All this results in a range $[40/1000]$ $M_\odot$, instead of $[80/3000]$ $M_\odot$ that was obtained (without H_2 cooling) in a previous calculation for the fragmentation of proto-globular clusters (DiFazio, 1986, Astron. Astrophys. **159**, 47).

ZINNECKER: I would just like to emphasize that there is a large uncertainty regarding the predicted physical conditions in the protogalaxy: Fall and Rees predict rather warm clouds, while we (Palla and Zinnecker) predict much cooler clouds due to strong molecular cooling. This is highlighted in our poster paper.

DJORGOVSKI: It is very interesting that the collapse time depends on the slope of the mass function. The slope according to McClure et al., correlates well with the metallicity. Ivan King and I checked whether there is any systematic difference in metallicity between the King model clusters and those with post-collapse cores. There isn't any.

Random Quotes

Irwin Shapiro opened the first IAU Symposium co-sponsored by the CfA

Core Collapse after the conference

RANDOM QUOTES

Let the speakers speak.

Once I am here I can talk about anything.

I am absolutely amazed you have finally come around.

I don't believe it, it's completely crazy.

Like the trial in <u>Alice in Wonderland</u>, I will present the conclusions first.

I will give this talk in the international language of astronomy - bad English.

Dutch is the official language of the Astronomical Mafia.

My apologies to the observers for speaking as a theoritician.

Sorry I don't know all the names, but I thought I recognized the ideas.

I don't think that the data are very good.

I don't think the theory is very good either.

This is not a confessional

There ain't much you can do about that.

I understand nothing about these three papers.

I will discuss a simple astronomical system, the universe as a whole.

I used to be very skeptical about the Washington system and wouldn't believe in the results even if George Washington did them. But I am becoming a convert.

Globular Clusters are being destroyed right now!

We will still do it faster than the cluster does it.

Peace and Harmony now reign over the globular cluster abundance scale.

This work was done by Buonanno and his collaborators Et and Al.

I don't know of any spectral synthesis program that discovered a Carbon star.

Caclcium is floating around.

Anybody who has a working hypothesis must work.

We had to destroy most of the data. There wasn't enough room on the disk to save it.

I know this code because it is mine.

When I came to the meeting I thought that for at least one day I could get away from solar neutrinos.

The important aspect of the Hubble Space Telescope is the stimulus it is providing.

This is <u>not</u> a ST photo of NGC 6397!

We attempted to go farther out in M 31. It was an incredible effort which we never finished.

Every chip has its own personality.

I know you would. You are going to work for me!

The velocity ellipsoid looks like a bagel.

$N \propto B(age)Lt$ (Bagel)

The Sun is ploughing through a sea of WIMPS.

A quasar is worth about 20 galaxies.

C'est magnifique, mais ce n'est pas la guerre.

Are you related to <u>that</u> man? I am related to Charley.

INDICES

Name Index   717

Object Index   723

Subject Index   728

   If a page number in the index listing is underlined it indicates
that the name is the name of an author of a paper (Name Index) or to an
object mentioned in the title of a paper (Object Index) or to a subject
listed in the title of a paper (Subject Index).

NAME INDEX

A

Aaronson, M.	163, 165, 169, 195 198, 267
Aarseth, S.	637, 709
Abell, G.	246
Adams, W.	8
Adur	422
Agassiz, L.	8
Aguilar, L.	138, 309
Alcaino, G.	118, 202, 453, 476, 561 598, 625, 633, 650-1
Aller, L.	26
Altner, B.	345-6, 501, 534, 671
Ames, A.	15, 21, 24
Andersen, J.	156
Angeletti, G.	706
Antia, H. M.	671
Antonov, V. A.	340, 693
Armandroff, T.	131, 169, 534
Arp, H.	623
Aurière, M.	66, 364, 683, 687

B

Baade, W.	29, 31, 33, 43, 134-5 587, 589, 598, 600
Bahcall, J. E.	178, 215, 290, 444 446-7, 455
Bailey, S. I.	4, 13, 20
Bailyn, C. D.	77, 345, 361, 365, 379 657, 659, 679
Baker, R. H.	26, 480
Balmer, J.	353
Baryshnikova, O. A.	620
Battistini, P.	547, 595
Baum, W.	252, 345
Bell, R. A.	79, 91-2, 106, 114 118-9, 486, 531, 533 563, 600, 630, 636
Berendzen, R.	30
Bettweiser, E.	379, 382
Bhatia, R. K.	322, 407, 567, 599 637, 651
Bidelman, W. P.	419
Bingham, E. A.	591
Binney, J. J.	692
Blazhko, S. N.	587
Blecha, A.	391, 543, 595
Blitz, L.	422
Boesgaard, A.	118
Bohlin, K.	4-5
Bohlin, R.	450
Bok, B. J.	20-1, 32
Bok, D.	540
Boltzmann, L.	693
Bond, H.	454
Bonoli, F.	541
Bothun, G. D.	613, 645
Bowen, I.	26
Bowie, F.	13
Bregman, J. E.	419
Brodie, J.	268, 426
Brosche, P.	525, 537

Buananno, R.	312, 445, 508, 510, 555 602, 621, 635, 648, 651, 707
Butler, D.	80
Buttress, J.	575, 599, 631
Buzzoni, A.	621

C

Cacciari, C.	450, 587, 600
Caloi, V.	508, 510
Cannon, A. J.	11, 14, 20, 151
Cannon, R. D.	195, 575, 603, 631, 648
Canterna, R.	81, 91
Capaccioli, M.	615, 625, 646
Capuzzo, R.	559, 597
Carlberg, R.	426
Carney, B. W.	48, 133, 146-8, 150 235-6, 429-30, 442, 517 536, 581, 589, 600
Cayrel, R.	60, 119, 131, 137, 235 252, 267, 281, 431, 442 596
Chandrasekhar, S.	272
Chernoff, D. F.	283, 296, 344, 673, 706
Chevalier, C.	683
Chrétien	543
Christian, C.	91, 186, 187, 195 202-4 464, 595
Clementini, G.	587
Clement, C. M.	18, 591
Cohen, J.	47, 60, 86-7, 131, 147 158, 172, 186, 235, 331 345, 407, 422, 430, 464 531, 534-5, 595, 598-9 605, 643-5 651, 703, 709
Cohn, H. N.	344-5, 365, 379, 386 390-1, 657, 660, 667-8 679
Colin, J.	521, 537
Conklin, E. K.	419
Conti, P. S.	419
Coolidge, J.	9
Copernicus, N.	29
Cordoni, J.-P.	66, 683
Corsi, C. E.	555, 621, 635
Cowley, A.	364
Crocker, D. A.	131, 507-8, 509, 535
Cudworth, K.	135, 422, 429-30, 523 537, 704
Curtis, H. D.	6-9, 477-8
Cushman, F.	13

D

Da Costa, G.	61, 66, 77, 114, 131 150, 162-4, 169, 217 235-6, 312, 365, 430 447, 454, 531, 535, 580 595, 598, 650-1, 681
Dapergolas, A.	573
Day	281, 392
De Carstens, M. E.	306
De Correbo, M. C.	306

De Jonghe, H. <u>691</u>, 709
De Rocha, A. 306
De Vaucouleurs, G. 31, 177, 651
Demarque, P. 32, 65, 106, 121, 131
163-4, 311, 505, 535
Dessaunet, V. H. 303
Di Fazio, A. 172, 185, 216, <u>699</u>, 710
Dickens, R. J. <u>591</u>, 631
Djorgovski, S. 66, <u>333</u>, 342, 345, 364
366, 378-9, 390, <u>527</u>, 537
643, 656, 658, 703
Dolcetta, R. C. 706
Duncan, D. 73
Dunlap, Mrs. 18
Dunlop, J. S. 603

E

Eddington, A. 5, 352, 359
Edlen, B. 26
Eggen, O. 437
Eggletom, P. 373
Einstein, A. 355-6, 364, 684-5, 687-9
Eliot, T. H. 28
Elson, R. 194-5, 202, 332, 403, 597
Emden 272

F

Faber, S. 267, 327
Fabry 414
Fahlman, G. G. 110, 537, <u>623</u>, 648-9
Falconer, Lady 18
Falconer, R. 18
Fall, M. 194-5, 202, 248-9, <u>323</u>
331-2, 403, 536, 585
597, 697-8, 710
Faraday, M. 421, 689-90
Faulkner, J. 422, 453
Federici, L. <u>541</u>
Feller 390
Ferrall, T. 81
Ferraro, F. 450
Fisher 262, 267, 327
Fletcher, J. M. <u>661</u>
Fokker 283-4, 368, 370, 373
376, 380-2, 384-5, 387
390-2, 394, 408, 665, 667
673, 695
Ford, H. 450
Forte, J. C. 476, <u>655</u>, 703
Francois, P. 596
Freeman, K. 140, 158, 365, 425
Frenk, C. S. 54
Frogel, J. 584
Fusi Pecci, F. 77, <u>173</u>, 185-6, 346, 186
450, 508, 510, <u>541</u>, <u>555</u>
<u>621</u>, <u>635</u>, 648, 706-7

G

Gaposchkin, C. P. 13-4, 16-7, 21, 25
32, 481
Gaposchkin, S. 481
Geffert, M. <u>525</u>, 537

Geisler, D. 86, 169, 533, <u>577</u>, <u>579</u>
599
Gerasimovic, B. 17
Giannone, P. 706
Gilllett, F. 147, 709
Gill, D. 3
Gingerich, O. 9, 16, 21, <u>23</u>, 31-3
478
Gliese, W. 469
Gnedin, Y. N. 215, 266, 296, 331, 364
421, 453, <u>689</u>
Goldberg, L. 26
Golev, V. K. <u>569</u>, 599
Goodman, J. 266, 296, 344, 381, 387
390-1, <u>695</u>, 709-10
Graham, J. 61, <u>151</u>, 204, 215, 583
596, 600
Gratton, R. <u>495</u>, 532-3
Green 370
Green, E. M. 122
Greenstein, J. 26
Greggio, L. <u>483</u>, 531, <u>555</u>
Griffin, R. 488
Griffiths, W. K. <u>575</u>, <u>603</u>, <u>631</u>
Grillmair 644
Grindlay, J. E. 66, 68, 77, 120, 185
281, 321, 344-5, <u>347</u>
359, 364-5, 380, 421
429, 466, 647, <u>657</u>, <u>659</u>
<u>679</u>, 703, 707, 712
Gunn, J. E. 72, 91, 245, 488, 605

H

Hale, G. E. 4-9, 24, 26, 477-8
Halley, E. 5, 43
Hanes, D. A. 31, 33, 185, 215, 249
252, 256, 267, 309, 595
<u>617</u>, 643-51
Hara, T. <u>701</u>
Harris, G. L. <u>205</u>, 535
Harris, H. C. 77, 146, 168-9, <u>205</u>
215-6, 236, 251, 266
535, 595, <u>613</u>, 645-6
651, <u>677</u>
Harris, W. E. 77, 81, <u>237</u>, 251-3
267, 281, 305, 309
426, <u>485</u>, 645-6
Hartmann 464
Hartwick, F. D. A. 62, 71, 364, 454
Hartzidimitriou, D. 204, <u>567</u>, <u>571</u>, <u>573</u>
596, 599
Hawkins 430
Haynes, S. 480
Hazen, M. L. 14, 249, 410, 476, 526
<u>593</u>, 600, 650
Heeschen, D. 18
Heggie, D. 670
Held, E. V. <u>609</u>, <u>615</u>, 644-6
Henon, N. 340, 379, 394-5
Herschel, J. 3-4, 9
Hertzsprung, E. 17, 73
Hertz, P. 345, <u>657</u>, <u>685</u>, 709
Hesser, B. 73
Hesser, G. 73

Hesser, J. E.   61-2, 73, 77-8, 169
                        205, 215, 251, 266, 454
                        485, 531, 535, 595, 613
                        645, 650-1, 661, 707
Hewitt   489
Hills   281, 392
Hoagland, H.   20
Hodgon, Miss   14
Hodge, P.   159-62, 167, 169, 171
                       270, 342, 407, 549, 595
Hoffleit, D.   21
Hogg, D. E.   419
Hogg, F.   12, 15-6, 18, 21
Hogg, H. Sawyer   2, 11-3, 15, 19, 21
                        25, 32, 491, 585
Holmes, Sherlock   411
Holweger, H.   493
Hoskin, M.   3, 16, 30
Hubble, E.   23, 25, 43, 202, 232
                        245, 255, 261, 267, 277
                        283, 324, 333, 340, 378
                        393, 398, 405, 469, 562
                                        634, 643
Huchra, J.   150, 215, 255, 266-8
                                       426, 644-5
Humphries, R.   430
Hut, P.   344

I

Illingworth, G.   158, 704
Ilovaisky, S.   683
Inagaki, S.   236, 344, 367, 378-9
                                        392, 407
Innanen, K.   296, 407, 423, 429-30
                                             469
Irwin, M. J.   627, 650

J

Jackson   267, 327
Janulis, R.   529, 537
Jeans, J.   271, 325, 698, 700, 710
Jones, R. V.   589

K

Kamper, K.   21
Kang   332
Kapteyn, J. C.   3, 7, 19, 24
Keable, C. J.   603
Kelvin, Lord   698
Kembavi, A. K.   671, 705
Kent, S.   345
Kepler, J.   425, 429
King, I.   48, 60, 70, 202, 251-2
                       281, 283-7, 292-3, 334
                       339-40, 343, 345-6, 352
                       364, 366-8, 390-1, 408
                       419, 421, 453, 464
                       487, 490, 531-3, 571
                       591-2, 598, 637-8, 647-8
                       655-8, 661-3, 673, 675
                       677, 696, 703-8, 710
Klemola, A. R.   526

Koch-Miramond, L.   364, 683, 687
Kolmogorov   141, 390, 492, 536, 567
                                             679
Kontizas, E.   573
Kontizas, M.   571, 573, 599
Koorneef, J.   450
Kormendy, J.   607, 644
Kraft, R.   146
Kron, R. G.   346, 489, 541, 709
Kruszewiski   615
Kulkarni, S.   364
Kurucz, R. L.   501-2, 517, 536, 589

L

Lagrange, J. K.   673
Laird, J. B.   136, 146, 442, 517, 536
Lake, G.   421, 643
Larson, R. B.   249, 311, 321-3, 703
Latham, D. W.   146-7, 442, 517, 536
                       589, 600, 659, 675
Lauer, T. R.   185, 607, 644, 647
Leavitt, H.   5
Lecar, M.   706
Leep, E. M.   493, 532
Lee, H. M.   378, 390, 644, 658, 665
                                             705
Lee, Y. -W.   505, 535
Lelievre, G.   610
Light, R. M.   223, 583-4
Liller, H.   249
Liller, W.   32-3, 77, 131, 268
                       359, 421, 561, 598
                       625, 633, 650-1
Lindsay   599
Liouville   693
Lowell, Pres.   8-9
Lugger, P. M.   333, 344-5, 379, 657
                                        660, 679
Lundmark, K.   31
Lupton, R. H.   488
Luyten, W.   17
Lyman   327
Lyman, T.   25
Lynden-Bell, D.   373, 430, 437
Lynga, G.   161, 167, 171, 595-600

M

Macchetto, D.   450
Malagnini, M. L.   541
Manduca, A.   563, 597
Markov, A.   14
Marshak, R.   26
Mateo, M.   161-2, 167, 169, 342
                                   403, 557, 597
Mathieu, R.   321, 422, 659, 675, 705
                                            708-9
Mayall, M. W.   13-4, 20-1
Mayor, M.   705
Mazeh, T.   705
McCarthy, M. F.   33, 171, 203, 476, 481
McCarthy, Senator J.   26, 28
McClure, R. D.   36, 476, 485, 487
                       531-2, 649, 661-2, 710

McMillan, S. L. W.   364, 391, <u>669</u>, 705, 708
Mendez, M.   345, <u>655</u>, 703
Mermilliod, J. C.   705
Merritt, D.   666
Metaxa, M.   <u>571</u>, 599
Meylan, G.   342, 344, <u>663</u>, 704
Michie   663
Mineur, H.   31
Mironov   656
Miyoshi, S.   <u>701</u>
Moffat, A. F. J.   <u>585</u>
Mould, J. R.   83, 162, 165, 267-8
Muller   493
Mussells, S.   24
Muzzio, J.   266, <u>297</u>, 309-10

N

Natsvlishvili, T. M.   <u>689</u>
Neil, R.   481
Nemec, J. M.   66, 77, 91, 106, 203, 235, 267-8, 281, 392, <u>591</u>, 598, <u>677</u>, 707
Neugebauer, G.   147, 709
Newton, I.   3, 634
Nieto, J.-L.   <u>609</u>, 615
Noble, R. G.   <u>575</u>, <u>631</u>, 650
Norman, C.   643
Norris, J.   60, 77, <u>93</u>, 101, 106, 118, 147, 312, 532, 534-6, <u>681</u>

O

Oemler, G.   158
Oke, J. B.   257, 267, <u>493</u>-4, 532
Okun'   453
Olszewski, E.   61, 152, <u>159</u>, 163, 168, 171-2, 235, 430
Oort, J.   26, 325
Oosterhoff   119, 218, 470, 591
Oren, A.   343
Orrall, F.   29
Ortolani, S.   <u>495</u>, 532, <u>625</u>, <u>629</u>, 649-50
Osmer, P.   188
Ostriker, J. P.   32, 47-8, 77, 146, 251, 271, 281-2, 296, 309, 321-2, 324, 331, 364, 386, 429, 442, <u>667</u>
Ozernoy, L.   332, 365, 454

P

Palla, F.   <u>697</u>, 710
Paltoglou   101
Papenhausen, P.   342, <u>565</u>, 598
Parasce, F.   450
Paraskevopolous, J. S.   13
Penner, H.   343
Penny, A. J.   575, <u>631</u>, 650, 704
Perot   414
Peterson, C. J.   346, <u>479</u>, <u>489</u>-90, 532, 534, 540

Peterson, R. C.   118, 135, 353, 426, 430, <u>523</u>-4, 537
Philip, A. G. D.   147, <u>511</u>, <u>513</u>, 536, 712
Pickering   33
Pickering, E. C.   7, 12-3, 24-5
Pickering, W. H.   13
Pilachowski, C. A.   <u>497</u>, 531-5, 537
Piotto, G.   <u>625</u>, 649
Planck, M.   283-4, 368, 370, 373, 376, 380-2, 384-5, 387, 390-2, 394, 408, 634, 665, 667, 673, 695
Plaskett, H. H.   25
Plaskett, J. S.   24
Plummer   395, 667, 691, 709
Potter, M.   <u>585</u>
Prevot, L.   <u>587</u>
Preston, G.   80
Pritchet   644, 647
Pryor, C.   331, 365, 532, <u>661</u>, 704
Pudritz, R.   249

Q

Quarta, M. L.   <u>495</u>, 532

R

Racine, R.   111, 215, 256-7, 476, 609, 644
Ramani, N.   204, <u>641</u>, 651
Rankin, J. E.   28, 32
Ratnatunga, K.   140, <u>519</u>, 536-7
Ray, A.   <u>671</u>-2, 705
Redfern, M.   683
Reed, B. C.   <u>489</u>, 532
Rees, M. J.   248, <u>323</u>, 331-2, 365, 453, 536, 697-8, 710
Renzini, A.   165, 171, 421, <u>443</u>, 453-4, <u>483</u>, 531, <u>555</u>, 598
Retterer   390
Richer, H. B.   32, 110, 158, 171, 202, 282, 322, 346, 537, 598, <u>623</u>, 648-9
Richey   543
Richstone   660
Richtler, T.   169, <u>553</u>, 596
Riecke, M.   195
Ritchey, G. W.   6
Robertson, H.P.   26, 433
Roberts, M. S.   253, 410-<u>1</u>, 421-2
Roche, E.   276, 343, 353-4, 637, 672
Rood, R. T.   118, 131, 419, <u>507</u>, <u>509</u>, 535
Roosevelt, F. D.   29
Rose, J. A.   <u>499</u>, 533
Rosino, L.   <u>625</u>
Rucinski, S.   17
Rupin, M. P.   136, <u>517</u>
Russell, H. N.   4-5, 7-9, 17, 73, 477-8, 480

S

Saha, M.   26, 146, 531

Salpeter, E.	124, 395, 629, 664	Swope, H.	15, 21
Samus, N.	<u>511</u>, 535, <u>619</u>, 648	T	
Sandage, A.	68, 78, 91, 119-20, 133		
	147, 188, 190, 202, 262		
	363, 437, 590, <u>621</u>, 635	Takayanagi, K.	<u>639</u>, 651
Santin, P.	<u>541</u>	Tammann, G. A.	262
Sargent, W.	118	Terlevich	709
Schalen, C.	19	Thomas, H. -C.	126-7
Schilt, J.	17-8	Thompson, L. A.	<u>611</u>, 645
Schlesinger, F.	7	Thorne, K.	72, 91, 245, 365, 605
Schneider, D. P.	<u>455</u>		672
Schommer, R. A.	91, 118, 146, 163, 166	Thuan, T.	72, 91, 245, 605
	168-9, 171, 186, 188	Tremaine, S.	215, 426, 643, 660
	200, 251, 342, 407, 421-2	Trimble, V.	31, 118, 453, 532, <u>627</u>
	533, 537, <u>565</u>, <u>577</u>, 595-6		644, 650
	598, 643, 646	Tripicco, M. J.	<u>499</u>, 533
Schuster	298	Trumpler, R. J.	18, 20, 24
Scopes, Professor	27	Tucholke, H. -J.	<u>525</u>-6, 537
Seares, F. H.	4, 8, 479-80	Tully, B.	262, 267, 327
Searle, L.	185, 257, <u>563</u>, 597		
Seggewiss, W.	158, 169, <u>553</u>, 596-7	U	
Seidel	163-4		
Seitzer, P.	223, 425, 430, <u>581</u>, 583	Updegraff, M.	479
	600		
Shapiro, S. L.	<u>283</u>, 332, <u>673</u>	V	
Shapley, A.	33		
Shapley, H.	3-5, 7-9, 11-21, 23-25	Valdes, F.	<u>611</u>, 645
	27-33, 37, 73, 134, 151	Van de Kamp, P.	19
	159, 233, 237, 271, 477-81	Van den Bergh, S.	114, 185, 236, 322
Shapley, W.	2, 16, 32-33		407, 421, 426, <u>467</u>
Shapley, M. B.	17		643-4, 649
Shara, M. M.	<u>585</u>, 600	Van Maanen, A.	6, 16
Sharples, R.	205-6, 208, 210-1	Van Rhijn, P. J.	19
	212, 215, <u>545</u>, 595	Van Wyck, U.	481
Shawl, S. J.	<u>491</u>, 532	Vandenberg, D. A.	68, 91, <u>107</u>, 118-20
Shklovsky	215		134, 266, 311, 454
Shortley, G.	26		468, <u>485</u>, 531, 619
Silchenko, O. K.	620		630, 636, 647-50
Slater, J.	26	Velikovsky, I.	28-9
Slipher, V. M.	6	Vergne, M. M.	300
Smirnov	141, 492, 536, 567, 679	Villumsen, J. V.	77
Smith, G. H.	485, <u>503</u>, 532, 534	Vlasov	271, 283, 285
Smith, H. A.	81, 171, 531, 536, <u>563</u>	Vuosalo, C.	344
	597, 599	Vyssotsky, A.	16
Sneden, C.	<u>497</u>, 532		
Sobouti, Y.	<u>693</u>	W	
Soifer, B. T.	147, 709		
Soneira, R.	178	Walker, A. D.	14, 433
Spassova, N. M.	<u>569</u>, 599	Wallace, H.	33
Spite, F.	106, 118, 137, 435, 596	Wallerstein, G.	77, 92, 146, 148
Spite, M.	106, 137, 435, 596		309, 442, <u>493</u>-4, 531-2
Spitzer, L.	340	Ward, D.	190
Stalin, J.	23	Wasserman, I.	489, 674
Staneva, A. V.	<u>569</u>, 599	Watanabe, J.	551, 595
Stanga, R.	698	Webbink, R. F.	36, <u>49</u>, 60, 236, 332
Statler, T. S.	146, 386, 391, 647, 667		426, 490, 537, 704, 708
	705	Wehlau, A.	585
Steigman, G.	118	Weinberg, M. D.	<u>673</u>
Stetson, P. B.	222, <u>485</u>, 531, 583, 680	Welther, B. L.	14-5, <u>477</u>
Straizys, V.	<u>529</u>, 537	Wesselink, A.	134-5, 587, 589, 600
Strömgren, B.	513, 553	West	615
Strom, S.	644	Westerlund, B.	161, 167, 171
Struve, O.	31	Whipple, F.	32
Sugimoto	379	White, R. E.	54, 429, <u>491</u>, 532
Sweigart, A. V.	<u>483</u>, 531, <u>555</u>	Whittaker, D. G.	<u>617</u>, 646

Wielen, R.   <u>393</u>, 407-8, 597
Williams, E. T. R.   13, 16, 21
Wirtanen, C. A.   185
Wirth, A.   178
Wlerick, G.   610
Wooley, R.   469
Wren, C.   5
Wright, F.   159, 160, 162, 167, 171

Y

Yamagata, T.   <u>551</u>, 595
Young, A.   11
Young, C.   11-2

Z

Zdanavicius, K.   <u>515</u>, 537
Zinnecker, H.   32, 47, 148, <u>195</u>, 204
                 267, 296, 331, 442, 536
                 596, 600, <u>603</u>, 643, <u>697</u>
                                      710
Zinn, R.   <u>37</u>, 47-8, 60, 77, 86
                  91, 119, 158, 202, 222
                 251, 282, 331, 365, <u>505</u>
                             535-6, 584
Zworkin   131
Zytkow, A. N.   365, 672

OBJECT INDEX

A

AREAS
  Bok (in LMC) .... <u>561</u>
  Carina 220
  Galactic Bulge 529, 637
  Galactic Equator 529
  Galactic Pole 142
    See NGP, SGP
  Orion 312
    Trapezium 312, 315
  SA 127 519-20
  Sagittarius 4
  Sagittarius-Scorpius 14-5
  South Galactic Pole 140, 468
  North Galactic Pole 429

C

CLUSTERS (Galactic)
  Globular
    121SC03 = ESO 121-03 598
    AM-1 = E 1 .... 39, 68-9, 167
    AM-4 68-9
    E 3 .... 66, 68-9, 167
    Eridanus 39, 69
    IC 4499 .... <u>591</u>-2
    M 02   See NGC 7089
    M 03   See NGC 5272
    M 04   See NGC 6121
    M 05   See NGC 5904
    M 09   See NGC 6333
    M 10   See NGC 6254
    M 12   See NGC 6218
    M 13   See NGC 6205
    M 14   See NGC 6405
    M 15   See NGC 7078
    M 22   See NGC 6656
    M 28   See NGC 6626
    M 30   See NGC 7099
    M 53   See NGC 5024
    M 55   See NGC 6809
    M 56   See NGC 6779
    M 62   See NGC 6266
    M 68   See NGC 4596
    M 70   See NGC 6681
    M 71   See NGC 6838
    M 79   See NGC 1904
    M 80   See NGC 6093
    M 92   See NGC 6341
    M107   See NGC 6171
    NGC 0104 = 47 Tuc   41, 43, 62, 66
                          70, 77, 79, 81-7
                          89, 92, 97, 99, 101
                          106, 111, 115-6, 119
                          125, 127, 131, 198
                          223, 225, 355, 446
                          454, 464, 471, 485-6
                          488, 497, 525-6, 533
                          561, 583, 629-30
                          <u>633</u>-4, 650-1, 653
                          663, 664, 684, 688-9
                          703

NGC 0288   41, 65, 81, 87
                   507-8, 619, 662
NGC 0362   41, 87, 92, 103
                   125, 127, 425, 429
                   525-6, 656
NGC 1261   656
NGC 1851   65, 223, 356, 361
                   425, 429, 505-6, 562
                   634, 656, 684
NGC 1904 = M 79   65, <u>501</u>-2, 562
                   634
NGC 1960   599
NGC 2298   561-2, 633-4
NGC 2419   39, 65, 203
NGC 2808   65, 103, 142, 513
                   562, 634, 656
NGC 3201   86, 103, 142, 562
                   <u>629</u>-30, 634
NGC 4147   51, 525-6 662
NGC 4372   562. 634
NGC 4590 = M 68   113-6, 118-9
                   488, 562, 634
NGC 4833   65, 87, 142
                   513, 706
NGC 5024 = M 53   425, 429, 525
NGC 5033   66, 77, 598, 619
                   620, 677, 706
NGC 5139 = W Cen   65-6, 77, 80, 94
                   95, 99, 100, 102-3
                   106, 127, 221, 355
                   356, 364, 419, 422
                   464, <u>503</u>, 534, 561-2
                   626, <u>629</u>-30, <u>631</u>-4
                   648, 650, <u>663</u>-4
                   679-0, <u>681</u>, <u>687</u>-9
                   703, 706-7
NGC 5272 = M 03   77, 80, 87, 218
                   312, 425, 429, 447
                   450, 471, 507, 509-10
                   <u>511</u>, 523, 525, 591-2
                   620, <u>621</u>-2, 648, 706
NGC 5466   51, 65-7, 72, 77
                   281, 392, 469, 525
                   599, 648, 662, <u>677</u>-8
                   705-6
NGC 5824   355, 656, 703
NGC 5904 = M 05   70, 88, 103, 110-1
                   346, 488, <u>497</u>-8
                   505-8, 525, <u>623</u>, 648
NGC 6093 = M 80   178, 585
NGC 6121 = M 04   65, 70, 88, 98
                   103, 488, 493-4
                   505-6, <u>513</u>-5, 537
                   561-2, 526, 633-4
NGC 6139   562, 634
NGC 6171 = M107   68, 86, 110, 662
NGC 6205 = M 13   20, 52, 64, 70, 80
                   88, 99, 101, 103
                   142, 455-6, 458-9
                   460-1, 469, 488
                   509-10, <u>511</u>, <u>513</u>-4
                   523, 525, 586
NGC 6207 = Gamale 52
NGC 6218 = M 12   70, 485-6, 525-6
                   619, 662

NGC 6229	72, 345
NGC 6254 = M 10	525, 619
NGC 6266 = M 62	345, 656
NGC 6273	492
NGC 6284	65
NGC 6293	65
NGC 6333 = M 09	65
NGC 6341 = M 92	70, 80, 85, 91
	96-7, 103, 116
	120, 142, 222, 228
	345, 469, 485-6, 507
	509-10, 511, 515, 523
	525, 583, 590, 705
NGC 6352	86
NGC 6362	70, 88, 485-6
	561-2, 633-4
NGC 6388	357, 471, 593
	656-8
NGC 6397	71, 84, 142, 388
	456, 462-4, 531
NGC 6402 = M 14	585-6, 600, 707
NGC 6441	356, 361, 684
NGC 6553	88
NGC 6659	471, 593
NGC 6624	347-8, 350, 356-8
	360-1, 464, 656-8
	661-2
	684
NGC 6626 = M 28	364, 662
NGC 6652	471, 593
NGC 6656 = M 22	64, 80, 94, 99
	103, 142, 221, 355
	419, 422, 429, 493-4
	515, 524, 688-9, 703
	708
NGC 6681 = M 70	661-2
NGC 6712	80, 89, 345, 356-7
	361, 365, 529, 537
	659-60
	703
NGC 6723	80, 505-6, 656
NGC 6752	64-6, 70, 99, 100
	103, 127, 142, 464
	488, 507, 509-10
	535, 656, 705-6
NGC 6779 = M 56	525, 529, 537
NGC 6809 = M 55	103, 425, 513-4
	627-8, 650, 662
NGC 6838 = M 71	38, 43, 51, 66
	80-1, 83, 86-7, 89
	493, 497-8, 515
	529-32, 537, 657
NGC 6934	103, 525
NGC 7006	72, 110, 345
NGC 7078 = M 15	65, 70, 80-1, 89
	106, 112-3, 115, 137
	218, 223-5, 345, 347-8
	351-3, 356-61, 388
	414, 485-6, 488, 507
	525, 590, 592, 626
	657-8, 679, 683-4
	705-6
NGC 7089 = M 02	223, 346, 525, 619
NGC 7099 = M 30	625-6, 649, 656
Pal 03	38-9
Pal 04	38-9, 48, 69

Pal 05	64, 68, 102
	110-1
Pal 12	110-1
Pal 13	69, 110-1
Pal 14	39, 599
Pal 15	430, 537
Open	
h and X Per	567
Hyades	267, 704
NGC 0752	531
NGC 2158	189
NGC 2168 = M 35	705
NGC 2682 = M 67	80, 87, 189, 675
	704-5, 708
NGC 6705 - M 11	531
NGC 7789	531, 676
CLUSTERS (Non-Galactic)	
Fornax 02	221
M 31	
Bo 158	471
Mayall II	342
M 33	
C 27	198
C 39	197
H 21	198
U 83	198
M 87	
IV 94	267
Magellanic Clouds	
121SC03	160-1, 167, 171
	565-6
E 2	159-61, 167
Hodge 11	153
Lindsay 01	164
Lindsay 11	153, 575, 599
Lindsay 54	574
LW 047	160, 167-8
LW 177	167
LW 195	167-8
LW 207	160, 167-8
LW 255	572
NGC 0121	153-4, 164, 166
	555
NGC 0152	555
NGC 0330	553, 596
NGC 0458	565-6, 598
NGC 1466	160-1, 165, 555
NGC 1756	555
NGC 1786	166
NGC 1831	555
NGC 1835	155, 166
NGC 1841	160, 165, 555
NGC 1850	577
NGC 1866	577-8
NGC 1868	160, 555
NGC 1987	555-6
NGC 2070	596
NGC 2107	555
NGC 2108	555
NGC 2134	555
NGC 2136	568
NGC 2137	568
NGC 2164	555, 577
NGC 2173	555
NGC 2209	198, 555-6

NGC 2210	153, 155, 166
NGC 2213	<u>579</u>-80, 600
NGC 2249	555
NGC 2257	153-5, 160-1, 166-7
NGC 5128 clusters	535
Reticulum System	160-1

G

GALAXIES (Single)
Cen A See NGC 5128	
Dwarf	
And I	468
And II	468
And III	468
Carina	165, 217-9, 228-9
	231-2, 235
Draco	64, 66, 77, 116
	119-20, 217-9, 222
	225-32, 235, <u>581</u>
	584, 592
Fornax	24, 32, 217, 219
	221, 223, 225-9
	231-2, 235, 239-40
	242, 312, 443
	<u>583</u>-4
Leo	217
Leo I	219, 229
Leo II	218-9, 229
Sculptor	24, 32, 66, 77
	217,-9, 221, 223-6
	228-32
Ursa Minor	66, 217-9, 221-2
	225-32, 235
Virgo Cluster	261-2

I Zw 18	314
IC 1613	550
IC 3475, Virgo Cluster	603
M 31 = Andromeda	6, 23-4, 26-7, 29
	31, <u>173</u>-4, 176-82
	185-88, 195-8, 205
	207, 217, 225-6, 237
	244, 256, 277, 306
	342, 345, 443, 449-50
	453, 468, 470-2, 547
	549, 550, 551, 569-70
	595-6
	599, 614, 647
M 32	277, 306, 378
M 33	179, 186, <u>187</u>
	188-200, 202-4
	210, 251, 313
	443, 549, 595-6
M 49, Virgo Cluster	242, 252, 264
	305-6, 309-10
M 81	595
M 82	314
M 87 = NGC 4486	210, 215, 239, 241-8
	252, <u>255</u>-60, 262-4
	267, 301-2, 304, 306
	309-10, 321, 328-9
	469-70, 546, 603-4
	<u>605</u>-6, <u>607</u>, <u>609</u>
	639

	644-5, 647
M 101	6, 314
Magellanic Clouds	16, 20, 61, <u>151</u>-3
	155, <u>159</u>, 165, 167-8
	172, 202, 210, 233, 236
	322,329, 331-2, 342
	443, 448, 472, 481
	484, 531, <u>553</u>-4, <u>555</u>-6
	559, <u>563</u>, <u>565</u>, 599
30 Dor	159, 313, 596
Constellation III	159
Large	126, 155, 158, 160-2
	171, 179, 188, 192
	194-200, 202-3, 219
	313, 315, 402-3, 407
	443, 549, 553, 555-6
	<u>557</u>, <u>561</u>, <u>567</u>, <u>571</u>
	<u>577</u>, <u>579</u>, 595-8, 600
Bar	159, 167, 169
Bok Region	<u>561</u>
H II region	313
Small	5, 160, 162, 164
	166, 169, 171, 198-200
	219, 430, 443, 481
	525, 553, 555, 559
	563-4, 569, <u>573</u>, <u>575</u>
Stream	50, 162
Milky Way	3-6, 8, 24, 29, <u>37</u>
	43, 49, 126, 138
	140-2, 153, 164
	177-80, 182, 185-8
	194, 198-9, 203-5
	207, 209, 230, 237
	239, 244, 251, 253
	284, 292, 313, 316
	321, 326, 329, 331
	340, 342, 365, 403
	423, 443-4, 448, 450
	468, 531, 536, 551
	556-7, 614, 638, 641-2
	644, 647-8
NGC 0055	595
NGC 0147	225-6, 236, 468
NGC 0185	236
NGC 0205	225-6, 236, 468
NGC 0300	598
NGC 0330	596
NGC 0891	614
NGC 1399	242, 246, 248
NGC 1705	314
NGC 2070	596
NGC 2403	<u>541</u>
NGC 2683	613-4
NGC 3109	<u>543</u>
NGC 3115	<u>615</u>-6, 646
NGC 3379	647
NGC 3717	613-4
NGC 4278	639
NGC 4406	606, 644
NGC 4472	606, 644
NGC 4565	614
NGC 4594	614, 639
NGC 4874	246-8, 252, 468
	<u>611</u>-2, 645

NGC 4889	246-7, 252
NGC 5128 = Cen A	205-12, 215-6, 244
	535, 545-6, 595
NGC 5170	613-4, 646
NGC 5866	613-4
NGC 5907	613-4, 646
NGC 6166	472
NGC 6822	550
NGC 7814	242, 251, 613-4
Virgo Cluster	
VCC 1185	604
VCC 1348	604
VCC 1539	604
GALAXIES (Clusters of)	
Coma	237, 239, 246-7
	252, 309, 468, 611-2
	645
Fornax	246
Hydra I	246
Local Group	245, 313, 646
Virgo	148, 239-40, 242
	244-7, 252, 255-7
	259-62, 300, 305
	328, 342, 603, 605
	643, 645-6

S

SOLAR SYSTEM	
Earth	63
Jupiter	322
Saturn	3
Sun	4, 37, 322, 448
	519-20, 529, 660
SOURCE	
X-Ray	
4U1916-05	360
4U2127+49	351
Cyg, X-3	351
HCA	688
Her, X1	683
HGO	688
M 15 AC 211	351, 353, 356, 683
M 15 4U2127+12	351
NGC 6624 4U1820-30	350-1
STARS (in Galactic Field)	
1823-37	683
2129+47	683
And, S	587
And, SW	587-8, 600
Ari, X	589-90
Cap, YZ	587-8
Cas, A	468
Cas, Mu	137
Cen, Z	6
Cep, Del	4
Cet, XZ	236
Cordoni and Aurière	
210	502
214	502
Cyg, Eps	497
Del, Alp	502
Dra, CM	138

Dra, SW	587-90
For, SS	587-8
Gem, U	688
HD 025329	114
HD 02935 = Eri, 49	501
HD 064090	114
HD 103095	114, 136
HD 134439	114
HD 134440	114
HD 161817	512
HD 201891	114
HR 5340 = Alp Boo = Arcturus	495-6
	498, 533
Ind, Nu	136
Leo, AQ	592
Phe, RV	587-8
Planetary Nebula	
49+88 1	137
61+14 1	137
108-76 1	137
M 15 K 648	137
Ser, VY	589
Sgr, V 440	587-8
Thorne-Zytkov Objects	365
STARS (in Clusters)	
M 03 = NGC 5272	
IV 18	512
182	511-2
M 05 = NGC 5904	
III 18	498
IV 34	498
M 13 = NGC 6205	
II 67	99
II 76	99
16	511-2
M 14 = NGC 6402	
Nova	585-6, 600
M 15 = NGC 7078	
I 51	353
M 22 = NGC 6656	
III 3	493-4
IV 102	493-4
M 71 = NGC 6838	
56	498
95	498
M 79 = NGC 1904	
Star A	502, 511
M 92 = NGC 6341	
II 26	511-2
NGC 0104 = 47 Tuc	
V9	81
NGC 5466	
V19	219, 677
NGC 5139 = W Cen	
II 67	101
E 39	66, 681
NJL 220	681
ROA 001	687
ROA 002	687
ROA 004	687
ROA 043	101-2
ROA 053	101-2
ROA 100	101-2

ROA 150	101-2
ROA 253	219
ROA 5701	687
NGC 6752	
CL 025	100
CL 166	100

SUBJECT INDEX

A

ABSORPTION	16, 18, 20
	24, 352, 481
	492
ABUNDANCES	
anomalies	93-104, 422, 536
chemical inhomogeneity	93
CMD's	61, 71, 221, 600
CNO	67, 72, 94-9
	127, 136, 493, 507
dispersions	93-104, 225
	503, 581
galactic radius	9, 67, 141, 536
giants	79, 93-100
	497, 515
globular clusters	79, 493
helium	69, 112, 484
	508, 509, 535
primordial	69, 118, 121, 433
	435
iron	493, 495, 498, 532
kinematics	37-60, 519
oxygen	468, 493, 496, 533
pollution	317, 496
primordial	100, 103, 137
	148, 325, 328
	431, 432-41, 536
scale	81, 87, 141, 467
	495, 497
TiO	83, 86
vs. age	42, 69, 72, 108
	134, 155, 167, 590
ACADEMY, AMERICAN	26
ACADEMY, NATL. (NAS)	7, 477
AGB stars	94, 165, 221
	445, 484, 503
	584, 648
AGE	
variations	153
vs. abundances	42, 69, 108-10
	119, 172, 468
	559, 564, 578, 597
vs. galact. radius.	69, 110-11
ANISOTROPIC MODELS	488, 662-3, 691
	704
AUTOCORRELATION CTRS.	336

B

BAADE-WESSELINK METHOD	135, 587, 589
	600
BALMER LINES	353, 511
BIG BANG	432
BINARIES,	
3-body formation	665, 667, 695
	705
capture	274, 347-9, 354
	357, 380, 385
	665, 667, 705, 708
common envelope	354, 671-2
eclipsing	4, 480
hard	274, 339, 349
	379, 524, 585
	669, 675
heating effects	382, 670, 705
main sequence	68, 349, 679
mass transfer	66, 219, 350
	581, 648, 681, 708
primordial	321, 380, 677
BLACK HOLES, MASSIVE	275, 380, 394
	399, 403, 702
BLAZHKO EFFECT	587
BLUE STRAGGLERS	66, 77, 192, 228-9
	281, 392, 471
	573, 581, 599
	621, 648, 675
	676-8, 681, 682
	707-8
mass	675-7, 681

C

CALCIUM TRIPLET	534, 535
CARBON STARS	101, 158
CATACLYSMIC VARIABLES	348, 355, 358
	666, 685-8
CCD's	see INSTRUMENTS
	456, 485
CENTRAL RELAXATION	286, 382, 413
CEPHEIDS	5, 23, 31, 66
	146 ,188, 435, 597
anomalous	219, 677
dwarf	681
type II	220
CH BANDS	504
CH STRENGTHS	94
CLUMP GIANTS	162, 531, 555
CLUSTERS	
binary interactions	669
CLUSTERS, GLOBULAR	133, 217
abundances	79
ages	107, 561, 635
binaries	274, 339
bulge	67, 176, 185
	189, 293, 339
	360, 468, 527
	534, 551, 637
central density	102, 383, 487
	490, 532, 665, 709
central surface	
brightness	490, 660
chemical inhomogeneity	93
CMD's see COLOR MAGNITUDE DIAGRAMS	
collapse	273, 281, 286, 293
color excesses	529
color gradients	343, 346, 356, 358
	361, 648, 655-6
concentration	
parameter	341, 367, 664, 696
cores	657
core radii	183, 271, 340
	384, 397, 489, 532
	660, 667, 679, 692
density waves	693

destruction	281, 317, 332		667, 691-2, 704
	393-4, 396-9, 401-2	velocity distribution	488, 523
	618, 637, 643		642
diffuse gas	411-20, 687-90	velocity, space	<u>525</u>
disk	37, 60, 188	CLUSTERS, OPEN	296, 312, 393
	271, 316, 331		403, 472
	517, 551, 590, 614	CLUSTER SYSTEM	205-213, <u>237</u>, 238-249
disruption	47, 275-6, 281		251-3, 255-64, <u>271</u>
	283-4, 324, 340		541-546, 549-574
	359-60, 369, 597	abundances	38, 245, 262
	651		305, 436-8, 521
distances, astrometric	<u>523</u>	ages	<u>107</u>, 165, 194, 203
distances, indicators	<u>617</u>		454, 468, 633
dynamics	<u>637</u>	colors	209, 245, 251, 470
evolution	271, 283-93, 339-40	dynamics	<u>641</u>
	347-349, 354, 359-61	luminosity functions	85, 195, 207
	364-6, <u>367</u>, 368-77		243-4, 247-8, 251
	<u>379</u>, 380-89, 391-2, 393-406		405, 444, 456
	443, 657-664, <u>665</u>, 666		541, 544, 617, 646
	<u>667</u>, 668, 672, <u>673</u>	orbits	50, 57, 261, 276
	674, 677-88, 695-6		284, 332, 425, 430
ellipticities	491, 569-70		526, 637-8, 651
evaporation	273, 340, 367, 599	radial distribution	245, 251, 304
expansion	273, 356, 365, 386		468, 470, 543
	407, 469, 659, 667		546, 639, 642, 651
	709	rotation	38, 54-5, 155
faint stars, photom.	<u>619</u>		203, 211, 321
flattening	102, 472, 491		<u>521</u>, 522, 537
	532, 599		545, 645
formation	187, 202, 213	swapping	277, <u>297</u>, 298-306
	237, 246-7, 264		309, 470
	286, <u>311</u>, 312-29, 536	velocity dispersion	49, 54-5, 211
	553, 558, 596, 618		263
	697-8, 701-2, 710	velocity ellipsoid	54-5, 60
gas	<u>689</u>	X-ray binaries	<u>347</u>
interstellar matter in	<u>411</u>	z distribution	38, 60, 188, 204
kinematics	<u>49</u>		216, 316, 340
luminosity function	<u>121</u>		545-6
mass spectrum	373, 376, 387, 395	COLOR MAGNITUDE DIAGRAMS	<u>61</u>, 62-77, 443
mass, total	102, 158, 323, 331		467, 503-10
	397		515, 621, 623, 648
metal-rich	<u>495</u>, <u>499</u>	abundances	600, 623, 625, 631
models, photometric	<u>559</u>	ages	61, 68, 227, 444
obscured	471, <u>527</u>, 528, 537		531, 623, 625, 627
orientation	<u>491</u>		629, 631, 634, 650
origin	<u>323</u>	gaps	127, 131, 505-8
primordial composition	<u>431</u>	morphology	61, 64, 127, 411
proper motion	<u>525</u>		627, 679
proto-globular cluster clouds	<u>697</u>, <u>699</u>	delta-V	84, 108, 155, 619
rotation	388, 659		623
summary paper	<u>467</u>	(B-V)o,g	84, 449
surface photometry	298, <u>333</u>, 334-43	Magellanic	
	346, 355, 655-8	globular clusters	162-70
	661, 666, 703	photographic	67, 152, 454, 625
survey	<u>489</u>	COMPUTER	
system	<u>37</u>	Cray SDSC	89
evolution	<u>271</u>	Ridge 32C	89
kinematics	<u>49</u>	VAX	63
tidal heating	<u>283</u>	CN STRENGTHS	67, 81, 94, 106
tidal radii	60, 271		493, 499, 504
tidal shocks	275, 283-95, 289, 324		505, 532, 580
turn-offs	<u>635</u>	CNO ABUNDANCES	see ABUNDANCES
velocity dispersion	357, 365		493, 507
	659-60, <u>661</u>, 665		

CO BANDS	95, 534
COLLISIONS, STARS	274, 350
COMA CLUSTER	239, 246, 468, 611
concentric aperture photometry	334, 489
concentration parameter	367, 489
luminosity function	
globular clusters	611-12, 645
radial distribution	612, 645
CONVECTION	94-5, 108, 127
	587, 671
COOLING FLOWS	
clusters	249, 328
CORE COLLAPSE	273, 333, 356
	367, 379, 391
	456, 464, 657-8
	666-70, 673-4, 706
CORE OSCILLATIONS	339, 380, 382-84
	390, 667, 710
COSMIC RAYS	442
COSMIC STRINGS	701, 710
CRITICAL DENSITY	370
CROWDING CORRECTIONS	531, 565, 625
	629, 648
CUSPS	
cluster cores	338, 340, 358, 388
	469, 655-8, 661-2
	703, 708-9

D

DAOPHOT	See PROGRAMS
DARK MATTER	230, 249, 297
	312, 323, 331
	453, 660, 701-2
DENSITY WAVES	693, 709
DISK	
abundances	137, 437, 534
globular clusters	See CLUSTERS, GLOBULAR
mass functions	
globular clusters	71, 467, <u>487</u>
stability	315
stars, kinematics	519, 537
DISRUPTION OF GLOBULAR CLUSTERS	See CLUSTERS, GLOBULAR
and the halo	47, 146, 405
	536, 644
DISTANCE SCALE	113, 120, 152
	164, 202, 207
	262, 450, 453
	481, 498, 524
	598, 617, 635, 646
DUST	
toward globular clusters	419, 422
	527, 529
DWARF GALAXIES	<u>217</u>, 218-33, 306, 312
	322, 342, 378
	430, 468, 550
	600, 603-4, 643
abundances	584
CMD's	227, 581, 583-4
Draco	<u>581</u>
formation	236
Fornax	<u>583</u>
globular clusters	235, 583
DWARF/GIANT RATIO	499, 533, 534
DYNAMICAL FRICTION	242, 276, 470
	603, 607, 643

E

EDDINGTON LIMIT	352, 354
EINSTEIN OBSERVATORY	352, 684
EQUIPARTITION	375

F

FIGURES	
abundances	
ages	72, 113-5
depletion	96-7
dispersion	95, 98, 100
	102-3
field	135, 139-41
	436, 518
giants	96, 100, 112
globular clusters	113, 518
	522
helium	109, 124-5
luminosity funct.	70
primordial	434, 440
binaries	672, 676
blue stragglers	67, 676, 678
Cepheids	682
cluster	
disruption	293, 396-9
	401-2
cluster systems	
abundances	199, 438
ages	193-4, 558
catalogs	528
colors	550
luminosity funct.	197, 207
	303, 612
	616, 640
radial profile	341, 418, 604
	608, 610, 612
	616, 639
rotation	44, 522
velocity	546
clusters, globular	
abundances	113, 518, 522
ages	113
colors	209, 211, 359
evolution	292-3, 368-9
	371-6, 383-6
	396-9, 658, 668
flattening	103
images	15, 335-7
	458-63, 568
orbits	53, 638
proper motion	53
star luminosity funct.	486
	620, 626
	630
structure	336-8, 341, 572
	664, 668

surface photom. 337-8, 502
 658, 664
color magnitude diagrams 62, 110-1
 504, 566, 586
 620, 622, 625
 630, 632, 678
 680
 cluster systems 550, 560
 field stars 62, 576, 628
 galaxies 584
 halo globular clusters 69, 110-1
 115, 576
 morphology 65
 number of papers 61
 turnoffs 64, 71, 113
 622
Coma cluster
 globular clusters 247, 612
core collapse 375-6, 383-6
 658
dark matter 702
disk heating 291-2
galaxies
 dwarf 220
 dwarf, CMD's 116, 224, 226
 584
 masses 260
giant molecular clouds 287-8
halo stars, kinematics 139-40
 520
horizontal branch
 faint blue stars 67, 512, 514
 680
 morphology 65, 506
 zero age 108-9, 112
 508, 635
HCO, historical photos 12-3, 15
isochrones 72, 108, 114-5
 566, 625, 630
 632
luminosity functions
 globular clusters. 197, 207
 244, 247
 stars 70, 124-5, 486
 620, 626, 630
M 31 globular clusters
 abundances 180
 distribution 183
 flattening 342
 searches 174, 183
M 33 globular clusters
 ages 193-4
 colors 193
 distribution 191
M 87 globular clusters
 abundances 263
 dynamics 259-60
 radial profile 241, 259-60
 608, 610
Magellanic Clouds 160-1
 globular clusters
 abundances 199, 564
 age 163, 166, 168
 194, 558, 564

 574
 CMD's 168
 mass functions
 stars 488
 Milky Way
 globular clusters 15, 335-7
NGC 5128
 globular clusters 208-9, 211-2
 RR Lyrae stars, data 44, 220
 592
S (specific freq.) 240
Shapley, photos 12, 27-8
spectra, giants 95, 100, 102
 353, 502, 510
 512
tidal capture, binaries 672
tidal
 radius 638
 shocks 288, 292
Virgo cluster 249, 604
X-ray sources
 globular cluster 352-3, 685-6
FK4 COORD. SYS. 52
FOKKER PLANCK 284, 368, 373
 380, 392, 394
 408, 665, 667, 695

G

G BAND 81, 94, 563
GALACTIC CENTER 365, 469, 492
GALAXIES
 cD 242, 252, 472
 clusters of 599
 dwarf spherical 24, 32, 52, 61
 120, 217-35, 296, 421
 elliptical 176, 205, 311, 321
 421, 468, <u>603</u>
 605, 639
 formation <u>311</u>, 312-29
 binary 325
 globular cluster systems <u>237</u>
 dissolution of <u>393</u>
 in different galaxies <u>613</u>
 lenticular 615
 masses <u>423</u>, 424-25, 639-42, 651
 Milky Way
 mass distribution 140, 526, <u>639</u>
 nuclei 603, 643
 spirals 176, 296, 313
 477, 613
GIANT BRANCH
 width 100, 222, 236
 312, 504, 534
GIANT MOLECULAR CLOUDS 276, 283, 317
 360, 394, 398
 407, 597
GRAVITATIONAL RADIATION 354
GRAVITY, SURFACE 498, 508, 509
 514, 577, 587
GRAVOTHERMAL INSTAB. 283, 367
 369-70, 381
 <u>695</u>, 696

H

H II regions	314, 324, 544
	596, 687-90, 697
HALO	
abundances	135-7, 439, 496
	517, 536
ages	134-5
clusters	326, 329, 638
	651
kinematics	49
formation	107, 237, 248
	281, 311, 324
	405, 536, 604
	644
gas	311, 326, 414
	419, 421, 640
	698-700
giants	519
gradients	39, 45, 140, 406
	520
kinematics	133, 138-40, 263
	407, 426, 470
	520, 525
stars (field)	133
populations	37, 39, 133-35
	241, 470
HORIZONTAL BRANCH	
absolute mag.	524, 584, 636
abundances	509, 593, 636
	651
bimodal distrib.	505, 507, 508, 509
	535
	705
blue stars	118, 122, 353
	505, 509, 511
	513, 535, 684
	705
faint	77, 147, 349, 357
	505, 679-80, 706-7
distance scale	113
lifetimes	446, 448
morphology	41, 65, 92, 116
	449, 469, 535, 619
	656, 679, 707
stars	
origin	679
radial distribution	349, 413
	471, 679, 680
strong He	509
vs. ms delta V	108, 119, 435
zero age	109, 113, 505, 635
	635
HELIUM	
abundances	See ABUNDANCES
flash	94, 104
ignition	531
HIERARCHICAL TRIPLES	348, 364, 669
HUBBLE SPACE TELESCOPE	See INSTRUMENTS
HUBBLE CONSTANT	43, 262, 267
	634, 647

I

IC 4499	591
IMPULSE APPROXIMATION	287, 400
INDICES	
spectral, abund.	81, 85, 180
	186, 496, 511
	513, 515, 535
	553, 579
INITIAL MASS FUNCTION	122, 194, 313
	317, 340, 365
	437, 457, 649
	673, 699, 710
INSTITUTIONS	
AAAS	21
AAS	17, 21, 30
AAT	205, 216, 245, 627
AAVSO	17
Allegheny Observatory	7
American Academy of Arts and Sciences	19, 21, 26
American Philosophical Society	26
Amherst College	18
Anglo-Australian Obs.	545
Applied Research Corp.	501
Associated Universities Inc.	419
Astronomical Council, Moscow	511, 619
Astronomical Inst., Munster	525
Astronomical Inst., Rome	559
Astronomical Obs., Asiago	629
Astronomical Obs., Bologna	587
Astronomical Obs., Marseille	587
Astronomical Obs., Monte Mario	635
Astronomical Obs., Rome	495, 555
	621, 699
Astronomical Obs., Trieste	541
Astronomisches Rechen Inst.	393
Astrophys. Obs., Arcetri	697
Astrophys. Obs., Asiago	495, 625
AURA	205, 497, 655
Australian Nat. University	93
Besançon Observatory	521
Bulgarian Academy of Sciences	569
California Inst. of Tech.	493, 591, 605
	613, 677
Cambridge University, UK	3, 25
Carnegie Inst. of Washington	151
Carnegie Southern Obs.	21
Center for Astrophysics	23, 255, 333
	347, 477, 517
	527, 589, 593
	659, 675, 679
Center for Nuclear Studies, France	683
	687
Center for Radiophys. and Space Research	283
Center for Scientific Research, France	661
CFHT Corp	73, 187, 189, 247
	355, 464, 485, 605
	609, 613, 657, 661
	677, 683, 703
CITA	306, 423

Columbia University	17	Inst. of Space Physics	
CONICET	306	Frascati	495
Copenhagen Observatory	663	Inst. of Theo. Phys., Moscow	453
Cornell University	283, 293, 673-4	Johns Hopkins University	323
CTIO	73, 205-6, 209-11 216, 225, 344, 355-6 485, 507-8, 510, 513 528, 557, 577, 579 583, 585, 600, 613 655, 663, 679	KPNO	62, 73, 173, 188-9 355, 485, 497, 507-8 510, 541, 549, 581 583, 657
David Dunlap Obs.	11, 17-8, 686, 591	Kyoto Sangyo University	701
Dept. of Astronomy,		Kyoto University	639
Bologna	173, 431 541 555, 621	La Plata Obs.	655
		Lab for Astrophys., Athens	571
Dominion Astrophysical Obs	18, 61, 205 467, 485, 490 503, 607, 613 661, 677	Laws Obs, Univ. of Missouri	4
		Leeds University	603
		Lick Observatory	5-7, 18, 51 120, 344, 524
		Loiano Obs	541
ESO	63, 553, 555 561-2, 596, 609 615, 629, 633-4 663, 683-4, 687	Maidanak Mountain Obs.	529
		Marseille Obs.	663
		McDonald Obs.	18, 497
		McMaster University	237, 485
		Michigan State University	563
Faculty of Science		Michigan Summer School	25
La Plata	655	MMT Obs.	524, 589
Geneva Observatory	543, 663	Mount Holyoke College	11-3, 18
Groningen State Prison	3	Mt. Stromlo and Siding	
Harvard College	3, 7-9, 12, 16 18, 23-5, 28, 30 32, 255, 333-4, 347 477, 517, 526-7, 589 593, 659, 675, 679	Spring Obs.	93, 681
		Mt. Wilson Obs.	4, 6-7, 16, 26 24-5, 29, 51
		Mt. Wilson & Las Campanas Obs.	563, 621
		Munster Astron. Inst.	526
Observatory	7, 11, 13-5, 21 25-6, 151, 524	NASA	128, 361, 483 488, 494, 505 555
Radcliffe Grad. Sch.	16	National Acad. of Sciences	7-8
South African Station	24	Nat. Cent. for Atmos. Res.	494
Southern Station	526	National Optical Astron. Obs.	497, 577 579, 583, 611
Summer School	25		
University	657	Nat. Res. Council, Canada	490, 661
Herzberg Inst. of Astrophys.	61, 205 485, 503, 661	NRAO	18, 411, 419
		National Univ., La Plata	297
House Un-American Comm.	26, 28, 32	Naval Res. Lab.	657, 685
Hurlburt Center for Space		Northwestern Univ.	669
Research	657	NSERC, Canada	249, 426
IAU	25-7, 29-31 73, 162, 169 364, 367, 380-1 387, 391, 421	NSF	45, 58, 89, 128 143, 169, 200, 205 293, 361, 419, 494 497, 505, 524, 584 655, 660, 674, 683
IBM Corp	674	Observatorium Hoher List	525
Indiana University	379, 657, 667, 679	Observatory of Athens	573
Institute for Adv. Study	455, 519, 691 695	Observatory of Haute Provence	431
Institute for Astron, Hawaii	58, 499 611	Observatory of Paris	431, 610
		Observatory of Pic du Midi	609, 683
Inst. for Astron. Geophys.		Obs. of Univ. of Bonn	525, 553
Sao Paulo	495	Osmania University	567, 637
Institute of Astron,		Padua Observatory	609
Buenos Aires	655	Palomar Observatory	60, 190, 217 456, 493, 591 605, 677
Institute of Astron,			
Cambridge	323, 627		
Institute of Astron,		Princeton University	4, 17, 32, 607
Padua	609, 615	Observatory	665, 667
Inst. of Isaac Newton	561-2, 625, 633-4	Pulkovo Observatory	15, 17, 52, 689
Institute of Physics,		Queen's Univ., Kingston	617
Vilnius	515, 529		

Radcliffe College	14, 16, 25	USSR Academy of Sciences	619
RAS	25	US Naval Observatory	205, 613
Royal Greenwich Obs.	591	Van Vleck Observatory	511, 513
Royal Obs., Edinburgh	567, 575, 603	Vanderbilt University	661
	631, 637, 697	Vatican Observatory	481
Rozhen Nat. Obs, Bulgaria	569	Whipple Observatory	311
Rutgers University	565, 577	Yale	
Rutherford Appleton Lab.	591, 631	Astronomy Dept.	311
Ryukoku University	639	Univ. Observatory	17, 32, 37
Science Clubs of America	21		121, 217, 505
Secretaria de Estado de Ciencia	306		583, 681, 683
Shiraz University	693	Yerkes Observatory	523-4
Sigma Xi	21, 26	York University	423
Smithsonian Institution	523-4		
Sofia University	569	INSTRUMENTS	
Space Telescope Sci. Inst	323, 431, 450	0.3 m telescope	
	585, 587	Harvard Southern Station	526
Steward Observatory	159, 491	0.6 m telescope	
St. Mary's University	489	Harvard College Obs.	14
Tata Institute	671	Harvard Southern Station	526
Tokyo Astronomical Obs.	551	0.9 m telescope	
Tombaugh Observatory	491	CTIO	655, 657
UNESCO	23	1.0 m telescope	
Union College	511, 513	ESO	562
University of Arizona	159, 491, 524	Maidanak	529
University of Athens	571, 573	South African	631, 650
University of Bologna	483, 547	1.2 m telescope	
University of British Col.	623	Schmidt, UKSTU	569, 571, 573-4
University of California Berkeley	32, 487, 490	1.5 m telescope	217
	663	CTIO	531, 577, 579-80
Irvine	627	ESO	553, 543, 561
University of Cambridge, UK	58		633-4
University of Chicago	32, 523	Loiano	541
University of Edinburgh	567, 571, 573	Mt. Wilson	4-5
	603, 641	1.8 m telescope	
University of Hawaii	190, 499, 611	Perkins	345
	661	2.0 m telescope	
University of Illinois	49	Pic du Midi	683
University of Kansas	491	Rozhen Obs.	569
University of Kyoto	367	2.2 m telescope	
University of Leeds	575, 631	ESO	562, 614, 634
University of Maryland	79, 627	KPNO	507, 615
University of Michigan	32	La Silla	507
University of Missouri	4, 479, 489-90	2.4 m telescope	
University of Montpellier	683	Hubble Space Tel.	48, 63, 66, 69
University of Montreal	585		72, 155, 202
University of Munster	525		334, 340, 342, 361
University of North Carolina	133, 143		431-3, $\underline{443}$, 445
	497, 517, 581		447, 449, 453
	589		$\underline{455}$, 456-7, 460-2, 464
University of Padua	609, 615, 625		469, 472, 586, 599
University of Rome	559	2.5 m telescope	
University of Texas	497	du Pont Telescope	563, 603
University of Tokyo	551	Newton	603
University of Toronto	11, 18, 21	3.6 m telescope	
	585, 591	AAT	206, 209, 211, 520
University of Victoria	107, 485, 661		524, 545, 585
University of Virginia	507, 509, 672		615, 627, 631
VITA	672	CFHT	189, 239, 247
University of Washington	493, 549, 557		252, 359, 485
University of Waterloo	205		607, 609, 611
University of Western Ontario	585		613, 615, 645
			649, 657, 661
			677, 683

Radial Vel. Spectrometer	661		559, 622, 656
ESO	553, 629	ISOCHRONES	64, 107, 114-6
IDS	553		172, 444, 483
4.0 m telescope	63, 545		555, 565, 619
CTIO	485, 507, 557		623, 634, 650
	565, 580, 585	abundances	108, 557, 580
	600, 625, 657	ISOTHERMAL	271, 326, 357
Echelle Spectrograph	600		696, 699
KPNO	188-9, 342, 485		
	497, 541, 549	**J**	
	585, 600, 625		
	657	JEANS MASS	317, 325, 437
Echelle Spectrograph	497		698, 700, 710
IIDS	189, 507	JETS	215, 266, 609
4.5 m telescope			
MMT	258, 268, 353, 524, 589	**K**	
Digital Speedometer	589		
Echelle Spectrograph	430, 659	K LINE SPECTRA	80, 85, 511
5.0 m telescope			536, 563
Palomar	80, 456, 458, 464	KINEMATICS	49-55, 133-4
	605, 677-8	vs. Fe/H	38, 43, 138
Four-shooter	456, 605, 678	KING MODELS	339, 366, 367
Spectrograph	493		391, 489, 565
6.0 m telescope			627, 655-8, 673
USSR	511, <u>619</u>		675, 677, 696
ANS satellite	501, 687		
APM, Cambridge	650	**L**	
Bruce Telescope	217		
CASPIEC spectrograph, ESO	596	LMC GLOBULAR CLUSTERS	See MAGELLANIC
CCD	42, 62-4, 67-72, 75		CLOUDS
	91, 110, 122, 152, 168-9	LUMINOSITY FUNCTION	<u>485</u>
	191, 202, 204, 210-1, 215, 223	age	123-25, 444
	228, 244, 247, 252, 256,-7	clusters	<u>121</u>, 185, 607, 610
	306, 334, 337-8, 357, 364		612, 615-7, 620
	407, 445, 447, 449-50, 456-7		626-7, 629, 644
	485, <u>493</u>, 495, 497, 513, 528		647, 649
	532, 545, 549, 555-7, 561, 562	Fe/H	70, 123, <u>485</u>, 531
	565, 579, 581, 583, 585, 603		662, 664
	607, 611, 613, 615, 623, 627	integrated	56
	629, 631, 633, 634, 655	stars	61, 65, 69-70
	657-9, 677-9, 684, 703		121-8, 446, 454
Camera	648-9		<u>485</u>, 487, 620, 625
Photometry	660		
Spectroscope	210-1, 650	**M**	
CORAVEL	587, 600, 660		
Einstein Observatory	355, 501, 684-5	M 03	511, <u>621</u>
	689	M 04	<u>513</u>
EXOSAT	350, 422, <u>687</u>	M 05	497, <u>625</u>
Herschel telescope	3	M 13	511, <u>513</u>
HIPPARCOS	526	M 14	<u>585</u>
IRAS	527-8, 537, 709	M 30	<u>625</u>
IUE	<u>501</u>	M 31	
McMullen Camera	543	catalogs	<u>547</u>
Microdensitometer		clusters	<u>173</u>
COSMOS	206, 208	blue	<u>549</u>, 550, 595
Geneva Obs.	543	globular	29, 31, 173-81
MDM-6, Rozhen	569		187, 210, 342
PDS	122, 526, 542		449-50, 464, 547-50
OAO2	687		<u>551</u>, 552, 596
SIT	215	age	181, 549
INTEGRATED LIGHT OBS.	85, 122, 152	completeness	173, 185, 470
	180, 447, 450		547-8, 551
	484, 499, 533	ellipticity	<u>569</u>, 599
		kinematics	179

luminosity funct.	178, 185	
	196, 647	
metallicity	178, 551	
radial distrib.	176, 183, 595	
	599	
surface photom.	342, 569	

M 33
clusters	
globular	187
	187-200, 595
	605-10
ages	188, 192, 203
kinematics	189, 197-200
luminosity dist.	196, 202
metallicity	198, 200
spatial dist.	190, 550, 551-2

M 55   513
M 67   675
M 71   497
M 79   501
M 87
halo	260, 639
nucleus	205
globular clusters	210, 239, 255
	256-64, 302, 468
	605, 607, 609
	644
abundance	258, 261, 329
colors	256, 606
luminosity funct.	607, 609, 644
	647
radial distrib.	241, 257, 310
	605, 607, 609

M 92   511

MAGELLANIC CLOUDS   20, 61, 151-69
     187, 402, 430
     481, 483, 553-69
Bok region	561
density profiles, clusters	571
distance modulus	555, 561, 575
	579, 598
field	553, 572, 576, 597
globular clusters	151
abundance	557, 558, 563, 577
	579, 596
ages	153-55, 165-69, 468
	531, 555, 557, 578
	563, 567, 573
	578, 597
CMD's	555, 560, 561
	575, 579, 598, 599
dynamics	342, 565, 572
intermediate age	159, 210, 403
	425, 448, 554-5
formation	553
giant branch	555
Large Cloud	151-69, 188, 198
	210, 402, 553-68
	571-4, 577-80, 597
binary clusters?	567
luminosity funct.	570, 597
masses	158, 559
old	152-5, 166, 332, 393
profiles	342, 571, 599

Small Cloud	151-6, 573, 575, 599
clusters	573
Lindsay I	575
spatial	551, 569
stream	162

MAIN SEQUENCE   91, 354, 445
     468, 481, 485
     499, 627, 629, 671
fitting	435, 635, 651
luminosity	61, 67, 171

MASS FUNCTION
galaxy	70-1, 699
globular clusters	376, 531

MASS LOSS
clusters	413, 554, 674
stars	352, 412, 507
	327, 668, 674, 676

MASS SEGREGATION   128, 343, 456
     532, 565, 598
     623, 648-9, 665
     674, 677, 707

MCCARTHY, SENATOR   28
METALLICITY DISTRIBUTION   517
MILKY WAY   3, 4, 326, 423, 442
nucleus	365, 423
rotation	423, 426
size	5, 237, 424-5
	429, 469, 530

MIXING, ABUNDANCES   99, 101, 106
     120, 169, 494

MOLECULAR BANDS   504
MOVING GROUPS   324
MULTI-MASS MODELS   373, 658
M/L, CLUSTERS   56, 133, 213
     230, 357, 365
     425, 659-62, 664
     704

N

NGC 104 = 47 Tuc	663
NGC 2213	579
NGC 2403	541
NGC 3109	543

NGC 3115
globular clusters	615
luminosity funct.	616
radial distrib.	616

NGC 3201   629
NGC 4874
globular clusters	611

NGC 5128
globular clusters	205, 206-13, 535
	545, 595
colors	209
luminosity function	207-9, 570
metal gradient	210

NGC 5139 = W Cen   503, 629, 631
     663, 681, 687
NGC 5466   679
NGC 6712   659
NEBULAE
planetary	147, 165, 412
	709

spirals	5, 24	VRI	534
NEUTRON STARS	349, 351, 354	Washington System	72, 75, 82, 91
	364, 381, 468		577, 579-80, 599
	487, 663, 672	PLACES	
	685	Asiago, Italy	495, 625, 629
NOVAE	6, 412, 414	Athens, Greece	571, 573
	421, 585-6, 600	Berkeley, Cal.	339, 487, 663
	708	Bologna, Italy	173, 431, 541
N-BODY APPROACHES	299, 386, 392		547, 555, 587
	394, 407, 637		621
	641, 651, 669	Bonn, W. Germany	525, 553
		Boston	28, 73
O		British Columbia, Canada	623
		Brookline, Mass.	19
OOSTERHOFF GROUPS	119, 146, 218	Bulgaria	569
	470, 591	Buenos Aires	655
OXYGEN, ABUNDANCES	116, 119	California, Southern	25
		Cambridge, Mass.	12, 19, 25
P		Cambridge, UK	9, 323, 627
		Cape of Good Hope	4
		Cerro Silla, Chile	663
PHOTOMETRY		Chalfont, UK	9
BRI	603	Commander Hotel, Cambridge	19
BV	191, 196-7, 199, 202	Connecticut	11
	565, 581, 629, 633	Dayton, Tennessee	27
BVR	188	Dublin, New Hampshire	18
BVRI	561, 512, 633	Edinburgh, UK	206, 567, 571
	634		573, 575, 603
BVRK	598		631, 637, 641
CCD	485, 561, 583, 623, 629		697
	631, 633, 655, 659	England	21
DDO	82	France	21
Four-Color	72, 84, 511, 513	Frascati, Italy	495
	514, 535-6	Garden St, Cambridge, Mass.	25
Four-Color and H Beta	70-1	Harvard Yard	25
Four-Color, Simulated	553	Hawaii	499
H Beta	200	India	21, 30
Infrared	71, 85, 87-9, 92	Irvine, Cal.	627
	179, 195, 199, 210	Jamaica	13
	228, 503, 528, 600	Kingston, Canada	617
	627	Kyoto, Japan	367, 639
photographic	545	La Palma	603
Q39	86-9, 91-2	La Plata, Argentina	297, 655
Searle-Zinn	83	La Silla, Chile	561-2, 629, 633-4
Strömgren	See Four-Color		684
Thuann-Gunn	72, 75, 83-4	Las Campanas, Chile	21
	91, 245, 605	Leeds, UK	575, 603, 631
TiO	83, 87	Leiden, Holland	17
U	609, 644	Lincolnshire, UK	3
UBR	256	London, UK	25
UBV	82, 84, 91-2, 94, 104	Lowell, Mass.	11
	110-6, 118, 179, 210	Marseille, France	587
	498, 538, 545, 555-6	Massachusetts	18, 33
	559, 585-6 609, 623	Mauna Kea	67, 611
	644, 649, 651, 656	Merrimac River, Mass.	11
	660, 677, 679, 684	Mexico	21
UBVR	557, 657	Minnesota	430
UBVRI	71, 82, 192-3, 210	Mississippi	32
	364, 503, 650	Monte Mario, Italy	635
UVJI	599	Montreal, Canada	585
VBRI	583	Moscow, USSR	453, 511
VI	225	Munster, W. Germany	525
Vilnius	515, 529, 537	Netherlands	17

New Delhi, India	708
New York, NY	9, 29
Padua, Italy	609, 615, 625
Paris, France	27, 31
Pic du Midi	609
Pittsburgh	32
Poland	29
Princeton, N. J.	11, 607
Rome, Italy	25, 29, 31, 495, 555, 559, 699
Ryukoku, Japan	639
Sao Paulo, Brazil	495
Schenectady, NY	167
South Africa	3, 217, 650
South Hadley, Mass.	11-2
Stonehenge, UK	3
Texas	18
Tokyo, Japan	551
Toronto, Canada	18, 585, 591
Toulouse, France	609
Trieste, Italy	541-2
Uzbekistan, USSR	529
Vatican	21
Victoria, British Columbia	18, 661
Vilnius, Lithuania	515, 529
Washington, DC	7-9, 28
Western Ontario, Canada	585
PLUMMER MODEL	395, 691, 709
POPULATION III	701
POST-AGB STARS	165, 171, 443, 528
POST-CORE COLLAPSE	274, 283, 293, 339, 346, 356, 364, 387, 407, 469, 659, 695-6, 709
PRECESSION PERIODS	351
PROFILES, CLUSTERS	335, 367, 390, 655-8, 666, 691
PROGRAMS	
ADDSTAR	565
DAOPHOT	62, 71, 128, 334-5, 343, 345, 357, 485, 565, 583, 585, 680
PSF determination	334
star stripping	335, 345
FOCAS	611
IHAP	609
INVENTORY	615, 625
MIDAS	63, 609
MOAN	543
RICHFIELD	26
ROMAPHOT	635
TIDAL	637
PROPER MOTIONS	
globular clusters	50, 52, 57, 60, 63, 422, 525, 704
stars	133, 352, 469, 497, 523-4
PROTO-GALAXY FORMATION	326 536, 697-8

Q

QSO's	60, 650

R

RADIAL VELOCITIES	33, 49, 135, 171, 206, 467, 523, 589, <u>659</u>, 660, 691, 705
RADIATIVE COOLING	327, 331, 698-9
RADIO OBSERVATIONS	414, 689-90
RED GIANTS	33, 153, 412, 481, <u>483</u>, 531, 648
mass loss	411-12
REDDENING	3, 185, 209, 355, 430, 529, 537, 562, 598, 629, 649
RELAXATION TIME	378, 598, 663, 675
ROCHE LOBE	354, 671-2
ROTATIONAL FLATTENING	45
RR LYRAE STARS	31, 44, 134, 153, 218, 228, 429, 505, <u>587</u>, <u>589</u>, <u>591</u>, 681
Fe/H	40, 42, 80, 84, 103, 116, 470, 588, 591, 593
absolute mag.	523, 537, 587-90, 600, 661
delta S	44, 80, 221
kinematics	57, 164
periods	106, 146, 470, 591-94, 600
R-PROCESS	137, 439

S

S (specific freq.)	239, 252, 267, 296, 468, 544, 612, 613, 645-6
SANDAGE RELATION	590, 651
SECOND PARAMETER	41, 65, 322, 450, 469, 507, 510
vs. age	41, 47, 469, 648
SHAPLEY	3-5, 7-9, 11-21, 23-33, 73, 151, 237, 271, <u>477</u>, <u>479</u>, 480, <u>481</u>
SHAPLEY, DEBATE	<u>3</u>, 7-8, 478
SHAPLEY, ERA	<u>11</u>
SHAPLEY, IMPACT	<u>23</u>
SMC FIELD	See MAGELLANIC CLOUDS
SMC GLOBULAR CLUSTERS	See MAGELLANIC CLOUDS
SPECTROSCOPY	
high disp.	79-91, 497, 532, 659
low disp.	80-1, 553
STAR FORMATION	313, 316-7, 324, 442, 557, 596

STARS	See	AGB
		BINARIES
		BLUE STRAGGLERS
		CARBON
		CATACLYSIC VARIABLES
		CEPHEIDS
		CLUMP GIANTS
	FIGURES (Halo, HB, RR Lyrae)	
		HORIZONTAL BRANCH
		NEUTRON
		RR LYRAE
	TABLES (halo, RR Lyrae)	
		WHITE DWARFS
binaries		5, 66, 146, 321
carbon		101, 195, 210
		236, 575
coalesced		6, 668, 676-7, 680
field, Fe/H		48, 138
Nova in M 14		__585__
pulsating		5
standards		63, 77, 501
		532, 615, 619
variables, short period		__593__
velocity ellipsoid		139
W Ursa Majoris		17, 627, 650
STELLAR		
core rotation		103, 508, 510
evolution		62-63, 443-50
		484, 531, 560
		673-4, 705-6
STRINGS		
cosmic		__701__
SUPERNOVAE		137, 317, 321
		324, 327, 437
		439-40, 702
SYNTHETIC SPECTRA		80, 95, 204
		495
S-PROCESS		100, 104, 137
		439

T

TABLES	
abundances	
field	142
globular	88, 142, 494
	496, 498, 515
primordial	433
big bang, sequence	432-3
black holes, massive	404
cluster	
ages	634
colors	88, 542, 656
diffuse gas	415-7
disruption	404
evolution	673
flattening	570
giants	498
orbits	526
star luminosity funct.	622
structure	662, 673
swapping	301
cluster systems	
abundances	181, 262

catalogs	238, 613-4
rotation	54, 143
velocity dispersion	56
velocity ellipsoid	143, 526
abundances	88, 142, 494
	496, 498, 515
	593, 634, 662
color mag. diags.	542, 634, 656
diffuse matter	415-7
distance moduli	257, 562
galaxies	
dwarf spherical	
M/L	231
Velocity Dispersion	231
Field *, abund.	142
Field *, vel. ellips.	143
	520
ellipticals	248
late type	541, 613-4
masses	640
M 31 globular clusters	
abundance	179, 181, 262
candidates	175
catalogs	174-5, 552, 570
flattening	570
kinematics	179
M 33 globular clusters	
catalogs	201
velocities	201
M 87 globular clusters	
abundance	262
dynamics	261
Magellanic Clouds	
globular clusters	
abundance	154, 578
age	154, 562, 566
	578
dynamics	566
RR Lyraes	154
globular clusters	238, 248
	640
halo stars, kinematics	520
luminosity function	622
M/L values	231, 662
Stars	
RR Lyrae, data	588
S (specific freq.)	238, 248
Virgo cluster	257
TEST PARTICLES	263, 299
THERMAL INSTABILITY	327
THOMAS PEAK	126, 131
THORNE-ZYTKOW OBJECT	365, 672
TIDAL	
capture binaries	274, 348-61, 364
	381, 659, 665
	667, __671__, 672, 686
	696, 705
capture, dormant	354, 360, 672
dissolution time	396, 402
heating	__283__, 284-93
radii	50, 57, 60, 271
	343, 395, 424
	429, 430, 492

shocks (cf.clust)	638 275, 333 340, 377, 388
stripping	297, 301, 425 470, 531, 673-4
TiO BANDS	83, 533
TRIAXIALITY	276
TRIPLE SYSTEMS	347, 669-70

U

UNESCO	23

V

VIRGO CLUSTER, M 87	237, 244, 643

W

WHITE DWARFS	
cluster	66, 134, 346 354, 364, 381 411, 446, 455 487, 629, 650 658, 662-3, 665 679, 685-6
WIMPS	448, 453, 469
WIND	
cluster	413-4
hot gas	352
stellar	327, 554

X

X-RAY	
binaries, (globular clust.)	<u>347</u>
burst sources	283, 289, 347, 360 683
halos	260, 702
sources	
central	66, 347, 380 471, 501, 585 659, 671, 683 704, 707-8
diffuse	414, 419, 421 429, 687-90, 708
globular clusters	<u>683</u>, <u>685</u>
low luminosity	501, 585, 665 685, 708-9

Addresses of Participants

The editors - J. Grindlay (R) and A. G. Davis Philip - after finishing their editing task in Cambridge (photo by Sandra Grindlay).

Dr. Harold D. Ables
Flagstaff Station
U.S. Naval Obs.
P.O. Box 1149
Flagstaff, AZ
86002

Dr. Bharat Adur
Nehru Center
Dr. Annie Besant Road
Worli, Bombay 400 018
INDIA

Dr. Gonzalo Alcaino
Instituto Isaac Newton
Casilla 8-9, Correo 9
Santiago
CHILE

Mr. Bruce Altner
Code 684
Lab for Astronomy and Solar Physics
Goddard Space Flight Center
Greenbelt, MD
20771

Dr. Johannes Andersen
Mail Stop 19
Center for Astrophysics
60 Garden Street
Cambridge, MA
02138

Dr. Taft Armandroff
Dept. of Astronomy
Yale University
P.O. Box 6666
New Haven, CT
06511

Dr. Michel Aurière
European Southern Observatory
Karl-Schwarzschild Strasse 2
D-8046 Garching bei Muenchen
WEST GERMANY

Prof. John N. Bahcall
School of Natural Science
Inst. for Advanced Study
Olden Lane, Bldg. E
Princeton, NJ
08540

Mr. Charles Bailyn
Dept. of Astronomy
Harvard University
60 Garden Street, MS-10
Cambridge, MA
02138

Dr. Pierluigi Battistini
Dipartimento di Astronomia
Universita' di Bologna
Via Zamboni 33
40126 Bologna
ITALY

Dr. William A. Baum
Lowell Observatory
Mars Hill Road
Flagstaff, AZ
86001

Prof. Roger A. Bell
Astronomy Program
Univ. of Maryland
College Park, MD
20742

Dr. Rajiv Bhatia
Royal Observatory
Blackford Hill
Edinburgh EH9 3HJ
U.K. (SCOTLAND)

Dr. Andre Blecha
Observatoire de Geneve
Ch. des Maillettes 51
CH-1290 Sauverny
SWITZERLAND

Dr. Howard E. Bond
Space Telescope Science Institute
3700 San Martin Drive
Homewood Campus
Baltimore, MD
21218

Jean Brodie
Astronomy Dept. and Space Sci. Lab
University of California
Berkeley, CA
94720

Dr. Roberto Buonanno
Osservatorio Astronomico su Monte Mario
Vile del Parco Mellini 84
00136 Rome
ITALY

John Buttress
c/o Dr. W.K. Griffiths
Dept. of Physics
University of Leeds
Leeds LS2 9JT
U.K. (ENGLAND)

Dr. Carla Cacciari
Space Telescope Science Institute
3700 San Martin Drive, J.H.U.
Homewood Campus
Baltimore, MD
21218

Dr. Roberto Capuzzo Dolcetta
Osservatorio Astronomico di Roma
viale del Parco
Mellini 84
I-00136 Roma
ITALY

Prof. Bruce Carney
Dept. of Phys. and Astron.
Phillips Hall 039A
Univ. of North Carolina
Chapel Hill, NC
27514

Dr. Roger Cayrel
Observatoire de Paris
61, Avenue de L'Observatoire
75014 Paris,
FRANCE

Dr. David Chernoff
222 CRSR
Cornell University
Ithaca, NY
14853

Dr. Carol A. Christian
Canada-France-Hawaii
  Telescope Crp.
P.O. Box 1597
Kamuela, HI
96743

Gisella Clementini
Space Tel. Sci. Inst.
3700 San Martin Dr.
Homewood Campus
Baltimore, MD
21818

Prof. Judith Cohen
105-24
California Institute of Technology
Pasadena, CA
91125

Prof. Haldan N. Cohn
Astronomy Dept.
Indiana University
Swain West 319
Bloomington, IN
47405

Dr. Jacques Colin
Observatoire de Besancon
41 bis, Avenue de l'Observatoire
25044 Besancon CEDEX
FRANCE

Dr. Deborah A. Crocker
Department of Astronomy
Univ. of Virginia
Box 3818 University Station
Charlottesville, VA
22903

Mr. Ken Croswell
Mail Code 10
Center for Astrophysics
60 Garden St.
Cambridge, MA
02138

Dr. Kyle Cudworth
Yerkes Observatory
Box 258
Williams Bay, WI
53191-0258

Prof. Gary Da Costa
Dept. of Astronomy
Yale University
P.O. Box 6666
New Haven, CT
06511

Dr. Robert J. Davis
Mail Stop 20
Center for Astrophysics
60 Garden Street
Cambridge, MA
02138

Dr. Herwig Dejonghe
The Institute for Advanced Study
School of Natural Sciences, Room E301
Princeton, NJ
08540

Prof. Pierre Demarque
Yale University Observatory
260 Whitney Avenue
P.O. Box 6666
New Haven, CT
06511

Dr. Alberto Di Fazio
Osservatorio Astronomico di Roma
viale del Parco
Mellini 84
I-00136 Roma
ITALY

Dr. S. G. Djorgovski
Center for Astrophysics
60 Garden Street, MS-20
Cambridge, MA
02138

Dr. Andrea Dupree
Center for Astrophysics
60 Garden Street, MS-15
Cambridge, MA
02138

Mr. Charles A. Field
Univ. of Massachusetts
P.O. Box 208
Amherst, MA
01002

Ms. Eileen Friel
Lick Observatory
Santa Cruz, CA
95064

Dr. Flavio Fusi Pecci
Dipartimento di Astronomia
Universita di Bologna
Casella Postale 596
I-40100 Bologna
ITALY

Dr. Douglas P. Geisler
CTIO
Casilla 603
La Serena
CHILE

Prof. Owen Gingerich
Center for Astrophysics
60 Garden Street, MS-9
Cambridge, MA
02138

Prof. Yurii N. Gnedin
Main Astronomical Observatory
196140 Leningrad
M-140 Pulkovo
U.S.S.R.

Dr. Jeremy Goodman
The Institute for Advanced Study
Building E
Princeton, NJ
08540

Dr. J. A. Graham
Dept. of Terrest. Magnetism
Carnegie Inst. of Washington
5241 Broad Branch Rd., N.W.
Washington, DC
20015

Dr. Raffaele Gratton
Osservatorio Astronomico di Roma
Viale del Parco Mellini, 84
00132 Roma,
ITALY

Dr. Laura Greggio
Dipartimento di Astronomia
Universita di Bologna
Casella Postale 596
I-40100 Bologna
ITALY

Dr. W. K. Griffiths
Dept. of Physics
University of Leeds
Leeds LS2 9JT
U.K. (ENGLAND)

Prof. Jonathan E. Grindlay
Center for Astrophysics
60 Garden Street, MS-6
Cambridge, MA
02138

Dr. David A. Hanes
Physics Department
Queen's University
Stirling Hall
Kingston, Ontario K7L 3N6
CANADA

Dr. Tetsuya Hara
Dept. of Physics
Kyoto Sangyo Univ.
Kamigamo-Motoyama
Kita-ku
Kyoto 603
JAPAN

Dr. Gretchen Harris
University of Waterloo
Waterloo, Ontario
CANADA

Dr. Hugh Harris
U.S. Naval Observatory
P.O. Box 1149
Flagstaff, AZ
86001

Prof. William Harris
Dept. of Physics
McMaster Univ.
1280 Main Street West
Hamilton Ontario L8S 4M1
CANADA

Dr. Lee Hartmann
Center for Astrophysics
60 Garden Street, MS-15
Cambridge, MA
02138

Dr. Despina Hatzidimitriou
Royal Observatory
Blackford Hill
Edinburg EH9 3HJ
U.K. (SCOTLAND)

Prof. Kim Innanen
Physics Dept.
York University
4700 Keele Street
North York, Ont. M3J 1P3
CANADA

Prof. Kenneth Janes
Boston University
Dept. of Astronomy
725 Commonwealth Ave.
Boston, MA
02215

Mr. Kenneth Jones
P. O. Box 45
West Tisbury, MA
02575

Mr. Rodney V. Jones
c/o Dr. Bruce W. Carney
Dept. of Physics and Astronomy
Phillips Hall 039A
Univ. of North Carolina
Chapel Hill, NC
27514

Dr. Gopal C. Kilambi
Center of Advanced Study
  in Astronomy
Osmania University
Hyderabad-500 007 (A.P.),
INDIA

Prof. Ivan King
Astronomy Dept.
Univ. of California
Berkeley, CA
94720

Mrs. Jacqueline Kloss
174 Brattle Street
Cambridge, MA
02138

Dr. Lydie Koch-Miramond
Service d'Astrophysique
CEN Saclay
F-91191 GIF/Yvette Cedex
FRANCE

Dr. Martha L. Hazen
Mail Stop 41
Center for Astrophysics
60 Garden Street
Cambridge, MA
02138

Dr. Enrico V. Held
Istituto di Astronomia
Vicolo dell'Osservatorio, 5
35122 Padova
ITALY

Dr. Paul Hertz
Code 4121.5
Space Science Division
Naval Research Laboratory
Washington, DC
20375-5000

Dr. James E. Hesser
Dominion Astrophys. Obs.
5071 W. Saanich Road
Victoria, B.C. V8X 4M6
CANADA

Dr. Paul W. Hodge
Astronomy Dept., FM-20
University of Washington
FM-20
Seattle, WA
98195

Prof. John Huchra
Center for Astrophysics
60 Garden Street, MS-19
Cambridge, MA
02138

Dr. Shogo Inagaki
Dept. of Astronomy
Faculty of Science
University of Kyoto
Kyoto 606
JAPAN

Dr. Gerald E. Kron
2929 Poni Moi Road
Honolulu, Hawaii
96815

Dr. Robert L. Kurucz
Center for Astrophysics
60 Garden St., MS-16
Cambridge, MA
02138

Dr. John B. Laird
Dept. of Physics and Astronomy
Phillips Hall 039A
Univ. of North Carolina
Chapel Hill, NC
27514

Prof. Richard Larson
Dept. of Astronomy
Yale University
P.O. Box 6666
New Haven, CT
06511

Dr. David W. Latham
Mail Stop 20
Center for Astrophysics
60 Garden Street
Cambridge, MA
02138

Dr. Tod R. Lauer
Peyton Hall
Princeton University Observatory
Princeton, NJ
08544

Dr. Hyung Mok Lee
Princeton University Observatory
Peyton Hall
Princeton, NJ
08544

Mr. Young-Wook Lee
Dept. of Astronomy
Yale University
P.O. Box 6666
New Haven, CT
06511

Mr. Robert M. Light
Dept. of Astronomy
P. O. Box 6666
New Haven, CT
06511

Dr. William Liller
Casilla 437
Vina del Mar
CHILE

Prof. A. Edward Lilley
Mail Stop 42
Center for Astrophysics
60 Garden Street
Cambridge, MA
02138

Mr. Piang-Xing Liu
c/o Prof. Kenneth Janes
Boston University
Dept. of Astronomy
725 Commonwealth Ave.
Boston, MA

Prof. L. Lo Cascio
Istituto di Fisica
Universita' di Palermo
via Archirafi, 36
90123 Palermo
ITALY

Dr. Phyllis Lugger
Dept. of Astronomy
Indiana University
Swain West 319
Bloomington, IN
47405

Prof. Gosta Lyngå
Lund Observatory
Box 43
S-221 00 Lund
SWEDEN

Mrs. Florence Menzel
1010 Memorial Drive
Cambridge, MA
02138

Dr. Georges Meylan
Berkeley Astronomy Dept.
523 Campbell Hall
University of California
Berkeley, CA
94720

Dr. Romas Mitalas
Astronomy Dept.
University of Western Ontario
London, Ont.N6A 3K7
CANADA

Prof. George S. Mumford
P. O. Box 267
Dover, MA
02030

Mr. Stephen Murray
Lick Observatory
University of California
Santa Cruz, CA
95064

Dr. Juan C. Muzzio
Observatorio Astronomico
Paseo Del Bosque
1900 La Plata
ARGENTINA

Dr. James M. Nemec
Dept. of Geoph. and Astron.
University of B.C.
Vancouver, B.C.   V6T 1W5
CANADA

Richard G. Noble
c/oDr. W.K. Griffiths
Dept. of Physics
University of Leeds
Leeds LS2 9JT
U.K. (ENGLAND)

Dr. Maria Lucia Malagnini
Dipartimento di Astronomia
Universita' Degli Studi di Trieste
Via Tiepolo, 11
I-34131 Trieste
ITALY

Dr. Laurence A. Marschall
Mail Code 20
Center for Astrophysics
60 Garden St.
Cambridge, MA
02138

Dr. Mario Mateo
Dept. of Astronomy, FM-20
University of Washington
Seattle, WA
98195

Dr. Robert Mathieu
Center for Astrophysics
60 Garden Street, MS-19
Cambridge, MA
02138

Dr. Martin F. McCarthy
Vatican Observatory
Vatican City State
I-00120
EUROPE

Dr. Robert D. McClure
Dominion Astrophysical Observatory
5071 W. Saanich Road
Victoria, BC V8X 4M6
CANADA

Dr. Steve McMillan
Dept. of Physics and Astronomy
Northwestern University
2145 Sheridan Road
Evanston, IL
60201

Dr. Mariano Mendez
Observatorio Astronomico de La Plata
Paseo del Bosque
1900 La Plata
ARGENTINA

Dr. John Norris
Mt. Stromlo and Siding Spring Obs.
Australian National University
Private Bag, Woden P.O.
Canberra ACT 2606
AUSTRALIA

Dr. Edward Olszewski
Steward Observatory
Univ. of Arizona
Tucson, AZ
85721

Dr. Sergio Ortolani
Osservatorio Astronomico Asiago
36012 Asiago (Vicenza)
ITALY

Prof. Jeremiah P. Ostriker
Princeton Univ. Obs.
Peyton Hall
Princeton, NJ
08544

Prof. Costas Papaliolios
Department of Physics
Harvard University
Cambridge, MA
02138

Mr. Paul Papenhausen
Dept. of Physics and Astronomy
Serin Physics Lab
Rutgers University
Piscataway, NJ
08854

Dr. Alan Penny
Space Telescope Science Institute
3700 San Martin Drive
Homewood Campus
Baltimore, MD
21218

Dr. Charles J. Peterson
Dept. of Physics and Astronomy
University of Missouri
Columbia, MO
65211

Dr. Ruth Peterson
Visiting Scientist, Whipple Obs.
c/o Steward Obs., Univ. of Ariz.
Tucson, AZ
85719

Dr. A. G. Davis Philip
1125 Oxford Place
Schenectady, NY
12308

Dr. Catherine Pilachowski
KPNO
P.O. Box 26732
Tucson, AZ
85726

Mr. Marc Pound
c/o Prof. Kenneth Janes
Boston University
Dept. of Astronomy
725 Commonwealth Ave.
Boston, MA
02215

Dr. Carlton Pryor
Dept. of Physics
Univ. of Victoria
P. O. Box 1700
Victoria, B.C. V8W 2Y2
CANADA

Dr. Rene Racine
Departement de Physique
Université de Montréal
C.P. 6128, Succ. "A"
Montréal, P.Q. H3C 3J7
CANADA

Dr. Natarajan Ramamani
R. No. 6320, Dept. of Mathematics
James Clerk Maxwell Building
The King's buildings
Mayfield Road, Edinburgh EH9 3JZ
U.K. (SCOTLAND)

Dr. Kavan U. Ratnatunga
The Institute for Advanced Study
Room E206
School of Natural Sciences
Princeton, NJ
08540

Dr. Alak Ray
Tata Institute of Fundamental Research
Homi Bhabha Road
Bombay 400 005
INDIA

Prof. B. Cameron Reed
Saint Mary's University
Dept. of Physics
Halifax, Nova Scotia B3H 3C3
CANADA

Prof. Martin Rees
Institute of Astronomy
Madingley Road
Cambridge CB3 0HA
U.K. (ENGLAND)

Prof. Alvio Renzini
Dipartimento di Astronomia
Universita di Bologna
Casella Postale 596
I-40100 Bologna
ITALY

Dr. Harvey Richer
Dept. of Geophysics and Astronomy
University of British Columbia
2075 Westbrook Mall
Vancouver, BC, V6T 1W5
CANADA

Dr. Morton S. Roberts
NRAO
Edgemont Rd.
Charlottesville, VA
22901

Dr. Robert T. Rood
Box 3818 University Station,
Charlottesville, VA
22903

Prof. James A. Rose
Dept. of Phys. and Astron.
Phillips Hall 039A
Univ. of North Carolina
Chapel Hill, NC
27514

Dr. Lucio Rossi
Consiglio Nazionale delle Ricerche
Istituto di Astrofisica Spaziale
C.P. 67
00044 Frascati
ITALY

Prof. Nicolai N. Samus
Astronomical Council
Academy of Sciences of U.S.S.R.
48 Pyatnitskaya
Moscow 109017
U.S.S.R.

Dr. Helen Sawyer-Hogg
98 Richmond Street
Richmond Hill, ON
L4C 3Y4
CANADA

Dr. Rudolph E. Schild
Mail Stop 19
Center for Astrophysics
60 Garden St.
Cambridge, MA
02138

Prof. Robert Schommer
Dept. of Phys. and Astron.
Rutgers University
P.O. Box 849, Serin Phys. Lab
Piscataway, NJ
08854

Prof. Dr. Wilhelm Seggewiss
Observatium Hoher List
D-5568 Daun
WEST GERMANY

Dr. Patrick Seitzer
KPNO
Box 26732
Tucson, AZ
85726

Mr. Willis H. Shapley
3040 "P" Street, N.W.
Washington, DC
20007

Dr. Ray Sharples
Anglo-Australian Observatory
Epping Laboratory
P. O. Box 296
Epping, NSW 2121
AUSTRALIA

Dr. Graeme Smith
Dominion Astrophysical Observatory
5071 West Saanich Road
Victoria, BC V8X 4M6
CANADA

Dr. Horace A. Smith
Dept. of Physics and Astronomy
Physics-Astronomy Building
Michigan State University
East Lansing, MI
48824-1116

Prof. Yousef Sobouti
Physics Dept. and Biruni Obs.
Shiraz University
Shiraz
IRAN

Dr. N. M. Spassova
Dept. of Astronomy
Lenin Street 72
Sofia 1784
BULGARIA

Dr. Thomas S. Statler
Princeton University Observatory
Peyton Hall
Princeton, NJ
08544

Dr. Theodore P. Stecher
Code 680
NASA-Goddard Space Flight Ctr.
Greenbelt, MD
20771

Prof. Vytas L. Straizys
Astrophysical Department
Institute of Physics
Pozelos Str. 54
Vilnius 232600
Lithuania
U.S.S.R.

Dr. Allen V. Sweigart
Code 681
NASA Goddard Space Flight Center
Greenbelt, MD
20771

Dr. Kazutomo Takayanagi
Department of Astronomy
Kyoto Univ.
Dept. of Nat. Sci.
Ryukoku University
Kyoto
JAPAN

Dr. Laird A. Thompson
Institute for Astronomy
University of Hawaii
2680 Woodlawn Drive
Honolulu, HI
96822

Dr. Virginia Trimble
Dept. of Physics and Astronomy
University of Maryland
College Park, MD
20742

Dr. Hans-Joachim Tucholke
Astronomisches Institut
Universitaet Muenster
Domagkstrasse 75
D-4400 Muenster
GERMANY (WEST)

Dr. Don VandenBergh
Physics Dept.
University of Victoria
P.O. Box 1700
Victoria, BC
V8W 2Y2
CANADA

Dr. Sidney van den Bergh
Dominion Astrophy. Observatory
5071 W. Saanich Road
Victoria, BC
V8X 4M6
CANADA

Dr. Merle Walker
Lick Observatory
University of California
Santa Cruz, CA
95064

Prof. George Wallerstein
Astron. Dept. FM-20
Univ. of Washington
Seattle, WA
98195

Dr. Jun-ichi Watanabe, M.S.
Dept. of Astronomy
Faculty of Science
University of Tokyo
Bunkyo-ku, Tokyo 113
JAPAN

Prof. R. Webbink
Dept. of Astronomy
Univ. of Illinois
1011 W. Springfield Ave.
Urbana, IL
61801

Dr. Amelia Wehlau
Dept. of Astronomy
University of Western Ontario
London, Ontario, N6A 3K7
CANADA

Ms. Barbara L. Welther
Mail Stop 9
Center for Astrophysics
60 Garden Street
Cambridge, MA
02138

Dr. Raymond E. White
Steward Observatory
University of Arizona
Tucson, AZ
85721

Prof. Charles A. Whitney
Mail Stop 16
Center for Astrophysics
60 Garden Street
Cambridge, MA
02138

Prof. Roland Wielen
Astronomisches Rechen-Institut
Moenchhofstr. 12-14
D-6900 Heidelberg 1
W. GERMANY

Mr. John G. Wolbach
Mail Stop 50
Center for Astrophysics
60 Garden Street
Cambridge, MA
02138

Dr. K.V. Zdanavicius
Astronomijos Observatorijos
Vilnius 31
Cuirlionio 29
Lithuania
U.S.S.R.

Prof. Robert J. Zinn
Dept. of Astronomy
Yale University
P.O. Box 6666
260 Whitney Avenue
New Haven, CT
06514

Dr. Hans Zinnecker
Royal Observatory
Blackford Hill
Edinburgh EH9 3HJ
U.K. (SCOTLAND)